Anatomy and Physiology

SHORT VERSION

Laboratory Textbook

Fifth Edition

Harold J. Benson
Pasadena City College
Stanley E. Gunstream
Pasadena City College
Arthur Talaro
Pasadena City College
Kathleen P. Talaro
Pasadena City College

WCB Wm. C. Brown Publishers

Book Team

Editor *Colin H. Wheatley*
Developmental Editor *Jane DeShaw*
Production Editor *Kay J. Brimeyer*
Designer *Jeff Storm*
Photo Editor *Robin Storm*
Art Processor *Jodi K. Wagner*

WCB **Wm. C. Brown Publishers**

President *G. Franklin Lewis*
Vice President, Publisher *George Wm. Bergquist*
Vice President, Operations and Production *Beverly Kolz*
National Sales Manager *Virginia S. Moffat*
Group Sales Manager *Vincent R. Di Blasi*
Vice President, Editor in Chief *Edward G. Jaffe*
Marketing Manager *John W. Calhoun*
Advertising Manager *Amy Schmitz*
Managing Editor, Production *Colleen A. Yonda*
Manager of Visuals and Design *Faye M. Schilling*
Production Editorial Manager *Julie A. Kennedy*
Production Editorial Manager *Ann Fuerste*
Publishing Services Manager *Karen J. Slaght*

WCB Group

President and Chief Executive Officer *Mark C. Falb*
Chairman of the Board *Wm. C. Brown*

Anatomy and Physiology Laboratory Textbooks

Authors	Titles
Benson, Gunstream, Talaro, Talaro	Complete Version—Cat, 5th edition Intermediate Version—Cat, 2nd edition Intermediate Version—Fetal Pig, 1st edition Short Version, 5th edition
Gunstream, Benson, Talaro, Talaro	Essentials Version
Benson, Talaro	Human Anatomy, 3rd edition

Some of the laboratory experiments included in this text may be hazardous if materials are handled improperly or if procedures are conducted incorrectly. Safety precautions are necessary when you are working with chemicals, glass test tubes, hot water baths, sharp instruments, and the like, or for any procedures that generally require caution. Your school may have set regulations regarding safety procedures that your instructor will explain to you. Should you have any problems with materials or procedures, please ask your instructor for help.

Cover illustration: Carlyn Iverson

Copyeditor: Sara K. Fisher

Library of Congress Catalog Card Number: 91-71681

ISBN 0-697-08689-5

Printed in the United States of America by Wm. C. Brown Publishers, 2460 Kerper Boulevard, Dubuque, IA 52001

10 9 8 7 6 5 4 3

Contents

Contents

Preface

This fifth edition of the *Anatomy and Physiology Laboratory Textbook, Short Version,* differs from the fourth edition in the following ways: (1) the Histology Atlas has been changed and expanded, (2) three new experiments pertaining to the use of Intelitool computerized hardware have been added, (3) most eponyms have been replaced with more acceptable terminology, (4) many new illustrations have been added, and (5) additional color has been incorporated where it is needed.

Two significant changes in the Histology Atlas will make this new version more functional. First of all, we have been able to increase its coverage by adding new photomicrographs pertaining to the skin, receptors, eye, ear, tooth development, salivary glands, digestive tract, spleen, urinary bladder, gallbladder, lymphoid tissues, and reproductive organs. Secondly, we have incorporated on each page of the Atlas a brief text to call the student's attention to significant structures. In the previous edition only photomicrographs with labels were provided on each page. These changes necessitated a complete reorganization of the entire Atlas.

As is the case with many other institutions, Pasadena City College has come around to using the Intelitool Data Acquisition Systems in many of its laboratory sections. The use of computerized physiological equipment hasn't obsoleted the use of chart recorders, but it does provide a very convenient and often simpler way to do certain experiments. Exercises 23, 46, and 59 provide instructions for using the three Intelitool systems designated as the Physiogrip, Cardiocomp, and Spirocomp. Software for these three systems is available for most types of computers. A brief description of the Intelitool system is provided in Exercise 16.

Since most of the newly published anatomy and physiology lecture textbooks have been deemphasizing eponyms in favor of more descriptive terminology, we, too, have decided to conform to this practice. Instead of terms such as Eustachian tube, Stensen's duct, and Volkmann's canal, we have adopted the more acceptable terms such as auditory tube, parotid duct, and perforating canal. For cross-referencing we have included the eponyms in parentheses after the newer terms that are set in bold type.

In addition to the above changes there are many minor textual changes which have been made to improve clarity. Some minor changes have also been made on the Laboratory Report sheets.

The remainder of the book remains much the same as the previous edition with a materials list provided at the beginning of each experiment and Laboratory Reports in the back of the book. Appendixes of supplemental information follow the Laboratory Reports. A Reading References list and Index complete the book.

An Instructor's Handbook is available to users of this book that provides setup information and time allotments for each experiment and answers to the questions on the Laboratory Reports. A set of eighty Kodachrome 2″ × 2″ histology slides is also available, at no cost, to institutions that adopt the manual. A legend for using the slides accompanies the set.

Our appreciation is extended to the following who have made extensive suggestions for improvement: John P. Messick of Missouri Southern College, Herbert House of Elon College, Clarence Wolfe of Northern Virginia Community College, and Karen A. Carlberg of Eastern Washington University. For help in developing the Intelitool experiments we are indebted to Wendy Johnston on our staff at Pasadena City College and Randall R. Steinle at Intelitool. Dan Trubovitz, formerly of Pasadena City College, was very helpful with his critique of the Laboratory Reports and Instructor's Handbook. In addition to the above, we wish to thank all the other unnamed individuals that contacted us about desired changes. Although we have been unable to make all the changes requested, most have been incorporated into this edition.

Some of the laboratory experiments included in this text may be hazardous if materials are handled improperly or if procedures are conducted incorrectly. Safety precautions are necessary when you are working with chemicals, glass test tubes, hot water baths, sharp instruments, and the like, or for any procedures that generally require caution. Your school may have set regulations regarding safety procedures that your instructor will explain to you. Should you have any problems with materials or procedures, please ask your instructor for help.

Introduction

These laboratory exercises have been developed to provide you with a basic understanding of anatomical and physiological principles that underlie medicine, nursing, dentistry, and other related health professions. Laboratory procedures that reflect actual clinical practices are included wherever feasible. In each exercise you will find essential terminology that will become part of your working vocabulary. Mastery of all concepts, vocabulary, and techniques will provide you with a core of knowledge crucial to success in your chosen profession.

During the first week of this course your instructor will provide you with a schedule of laboratory exercises in the order of their performance. There is an implied expectation that you will have familiarized yourself with the content of each experiment prior to that week's session, thus ensuring that you will be properly prepared so as to minimize disorganization and mistakes.

The *Laboratory Reports* coinciding with each exercise are located at the back of the book. They are perforated for easy removal; be sure to remove each sheet as necessary. This will facilitate data collection, completion of answers, and grading. Your instructor may give further procedural details on the handling of these reports.

The exercises in this laboratory guide consist essentially of four kinds of activities: (1) illustration labeling, (2) anatomical dissections, (3) physiological experiments, and (4) microscopic studies. The following suggestions should be helpful in performing these assignments.

Labeling The activity of labeling illustrations is essentially a determination of your understanding of the written text. Since all labeled structures are explicitly described in the manual, all that is necessary is to read the manual very carefully. Incorrectly labeled illustrations usually indicate a lack of comprehension.

Once the illustrations are labeled they can be useful to you in two other ways. First, the illustrations may be used for reference purposes in dissections or examinations of anatomical specimens. This is particularly true in the skeletal and nervous systems. Secondly, the illustrations can be used for review purposes. If the number legend of the labels is covered over as you mentally attempt to name the structures on the illustration, you can easily determine your level of understanding. Periodic reviews of this type during the semester will be very helpful.

In general, the labeling of illustrations will usually be performed *prior* to coming to the laboratory. In this way the laboratory time will be used primarily for dissections, experimentation, or microscopic examinations.

Dissections Rats and frogs will be used in many laboratory periods for anatomical and physiological studies. Occasionally, the brains, hearts, lungs, and other organs of sheep or cattle will also be used. If cadavers are available for study, dissection activity will probably be minimal: their high cost and scarcity usually limit widespread use.

When using live animals in experimental procedures it is imperative that they be handled with great care. Consideration must be exercised to minimize pain in all experiments on vertebrate animals. Inconsiderate or haphazard treatment of any animal will not be tolerated.

Physiological Experiments Before performing any physiological experiments be sure that you understand the overall procedure. Reading the experiment prior to entering the laboratory will help a great deal.

Handle all instruments carefully. Most pieces of equipment are expensive, may be easily damaged, and are often irreplaceable. The best insurance against breakage or damage is to thoroughly understand how the equipment is expected to function.

Maintain astuteness in observation and record keeping. Record data immediately; postponement detracts from precision. Insightful data interpretation will also be expected.

Cellular and Histological Studies Since most physiological phenomena occur at the cellular level, a great deal of your time will be devoted to studying various cells and tissues. The Histology Atlas, which is located in the middle of the book on pages 183 through 218, will be constantly referred to. In many exercises specific assignments will be made to examine prepared slides of tissues of various organ systems. The photomicrographs in the Histology Atlas will be used to help you to identify the various significant cellular structures. If drawings are required, they should be executed with care.

Safeguards in Hematological Studies Exercises 38 through 42 pertain to blood slide staining, hemoglobin measurements, blood typing, etc. The blood you will be working with will be from your own finger, or your laboratory partner's finger. Only a drop will be used at a time. When performing these experiments it is very important that you observe certain sanitary precautions to prevent: (1) self infection from environmental contaminants, and (2) transmission of blood infections such as hepatitis, AIDS, etc., from person to person. If the following safeguards are rigidly observed, there is very little chance of injury to yourself or others:

- Sponge down your table top at the beginning of the period with an appropriate disinfectant that is available in the laboratory.
- Wash your hands with soap and water before and after doing any blood tests.
- If you are perforating another person's finger for a blood sample, avoid contact with the blood, or use rubber gloves.
- Use disposable lancets only one time. Dispose of the lancet into a receptacle that contains disinfectant. Do not toss used lancets into the waste basket!
- Before perforating the finger for a blood sample, disinfect the skin with alcohol.
- At the end of the period wash up all equipment with soap and water. All pipettes must be cleaned with solutions drawn from the pipette cleaning kits.
- Scrub down the table top at the end of the period and wash your hands again with soap and water. A final rinse of the hands with disinfectant will provide additional protection.

If your career preparation is for nursing, medical and dental assisting, dental hygiene, or some other related medical endeavor, you must develop sanitary habits that become reflexly second nature. Learning the above sanitary safeguards, and adhering to them under all conditions, is a good place to start.

Laboratory Efficiency Success in any science laboratory situation requires a few additional disciplines:

1. Always follow the instructor's verbal comments at the beginning of each laboratory session. It is at this time that difficulties will be pointed out, group assignments will be made, and procedural changes will be announced. Take careful notes on substitutions or changes in methods or materials.
2. Don't be late to class. Since so much takes place during the first ten minutes, tardiness can cause confusion. If you are tardy don't expect the instructor to be very helpful.
3. Keep your work area tidy at all times. Books, bags, purses, and extraneous supplies should be located away from the work area. Tidiness should also extend to assembly of all apparatus.
4. Abstain from eating, drinking, or smoking within the confines of the laboratory.
5. Report immediately to the instructor any injuries that occur.
6. Be serious-minded and methodical. Horseplay, silliness, or flippancy will not be tolerated during experimental procedures.
7. Work independently, but cooperatively, in team experiments. Attend to your assigned responsibility, but be willing to lend a hand where necessary. Participation and development of laboratory techniques are an integral part of the course.

Laboratory Reports When seeking answers to the questions and problems on the Laboratory Reports, work independently. The effort you expend to complete these reports is as essential as doing the experiment. The easier route of letting someone else solve the problems for you will handicap you at examination time. You are taking this course to learn anatomy and physiology. No one else can learn it for you.

1 Anatomical Terminology

Anatomical description would be extremely difficult without specific terminology. A consensus prevails among many students that anatomists synthesize multisyllabic words in a determined conspiracy to harass the beginner's already overburdened mind. Naturally, nothing could be further from the truth.

Scientific terminology is created out of necessity. It functions as a precise tool which allows people to say a great deal with a minimum of words. Conciseness in scientific discussion not only saves time, but it usually promotes clarity of understanding as well.

Most of the exercises in this laboratory manual employ the terms defined in this exercise. They are used liberally to help you to locate structures that are to be identified on the illustrations. If you do not know the exact meanings of these words, obviously you will be unable to complete the required assignments. First of all, read over all of the material carefully; then read the specific assignments for this exercise.

Relative Positions

Descriptive positioning of one structure with respect to another is accomplished with the following pairs of words. Their Latin derivations are provided to help you understand their meanings.

Superior and Inferior These two words are used to denote vertical levels of position. The Latin word *super* means *above;* thus, a structure that is located above another one is said to be superior. Example: The nose is *superior* to the mouth.

The Latin word *inferus* means *below* or *low;* thus, an inferior structure is one that is below or under some other structure. Example: The mouth is *inferior* to the nose.

Anterior and Posterior Fore and aft positioning of structures are described with these two terms. The word anterior is derived from the Latin *ante,* meaning *before.* A structure that is anterior to another one is in front of it. Example: Bicuspids are *anterior* to molars.

Anterior surfaces are the most forward surfaces of the body. The front portions of the face, chest, and abdomen are anterior surfaces.

Posterior is derived from the Latin *posterus,* which means *following.* The term is the opposite of anterior. Example: The molars are *posterior* to the bicuspids.

Cephalad and Caudad When describing the location of structures of four-legged animals, these terms are often used in place of anterior and posterior. Since the word cephalad pertains to the head (Greek: *kephal,* head), it may be used in place of anterior. The word caudad (Latin: *cauda,* tail) may be used in place of posterior.

Dorsal and Ventral These terms, as used in comparative anatomy of animals, assume all animals, including man, to be walking on all fours. The dorsal surfaces are thought of as *upper* surfaces, and the ventral surfaces as *underneath* surfaces.

The word *dorsal* (Latin: *dorsum,* back) not only applies to the back of the trunk of the body, but may also be used in speaking of the back of the head and the back of the hand.

Standing in a normal posture, man's dorsal surfaces become posterior. A four-legged animal's back, on the other hand, occupies a superior position.

The word *ventral* (Latin: *venter,* belly) generally pertains to the abdominal and chest surfaces. However, the underneath surfaces of the head and feet of four-legged animals are also often referred to as ventral surfaces. Likewise, the palm of the hand may also be referred to as being ventral.

Proximal and Distal These terms are used to describe parts of a limb with respect to the point of reference such as the attachment of the appendage to the trunk of the body. *Proximal* (Latin: *proximus,* nearest) refers to that part of the limb nearest to the point of attachment. Example: The upper arm is the *proximal* portion of the arm.

Distal (Latin: *distare,* to stand apart) means just the opposite of proximal. Anatomically, the distal portion of a limb or other part of the body is that portion of the structure that is most remote from the point of reference (attachment). Example: The hand is *distal* to the arm.

Medial and Lateral These two terms are used to describe surface relationships with respect to the median line of the body. The *median line* is an imaginary line on a plane which divides the body into right and left halves.

The term *medial* (Latin: *medius,* middle) is applied to those surfaces of structures that are closest to the median line. The medial surface of the arm, for example, is the surface next to the body because it is closest to the median line.

As applied to the appendages, the term *lateral* is the opposite of medial. The Latin derivation of this word is *lateralis,* which pertains to *side.* The lateral surface of the arm is the outer surface, or that surface furthest away from the median line. The sides of the head are said to be lateral surfaces.

Body Sections

To observe the structure and relative positions of internal organs it is necessary to view them in sections that have been cut through the body. Considering the body as a whole, there are only three planes to identify. Figure 1.1 shows these three sections.

Sagittal Sections A section parallel to the long axis of the body (longitudinal section) which divides the body into right and left sides is a *sagittal section.* If such a section divides the body into equal halves, as in figure 1.1, it is said to be a *midsagittal section.*

Frontal Sections A longitudinal section which divides the body into front and back portions is a *frontal* or *coronal section.* The other longitudinal section seen in figure 1.1 is of this type.

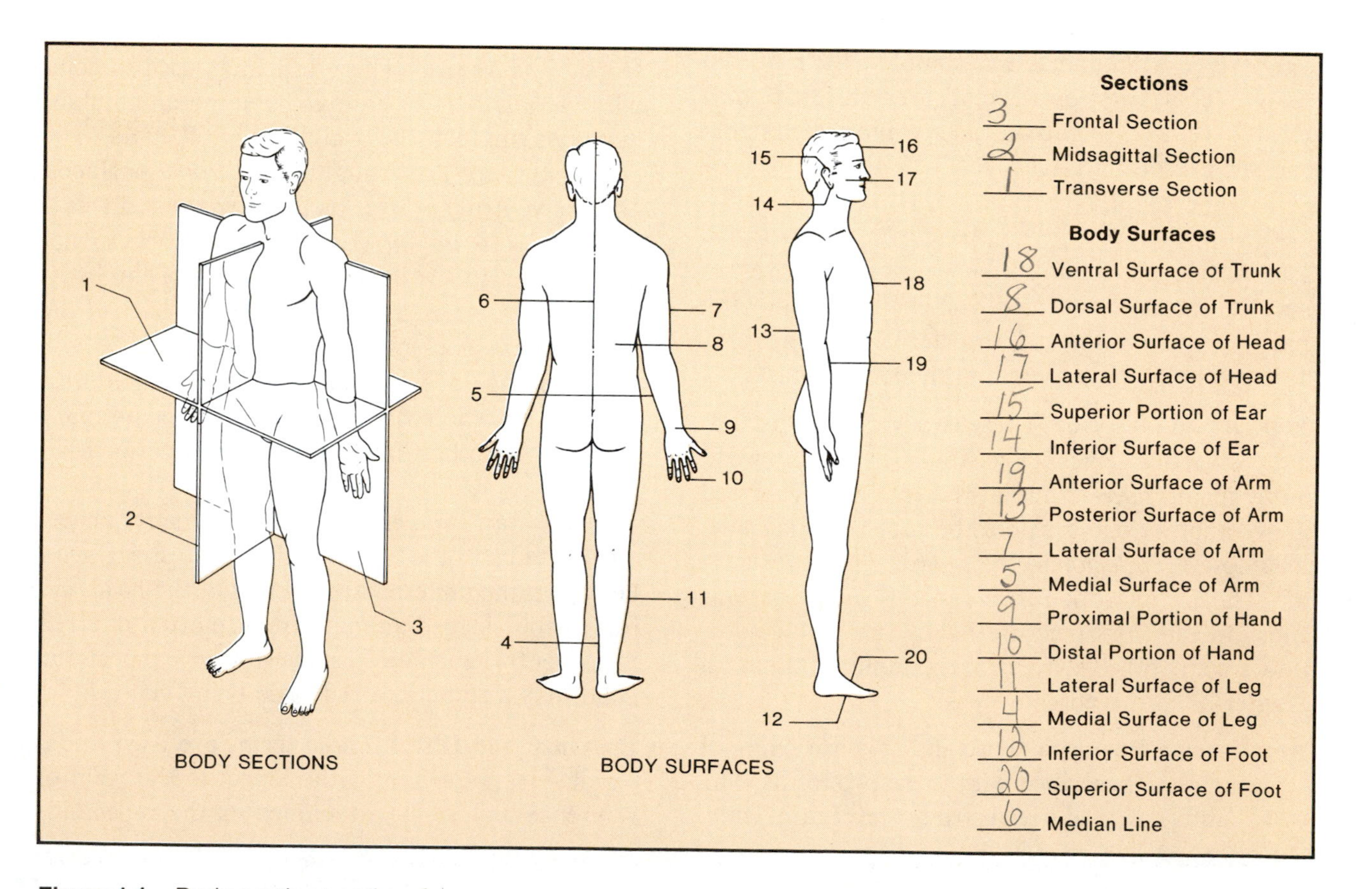

Figure 1.1 Body sections and surfaces.

Transverse Sections Any section which cuts through the body in a direction which is perpendicular to the long axis is a *transverse* or *cross section.* This is the third section which is shown in figure 1.1. It is parallel to the ground.

Although these sections have been described here only in relationship to the body as a whole, they can be used on individual organs such as the arm, finger, or tooth.

Assignment:

To test your understanding of the above descriptive terminology, identify the labels in figure 1.1 by placing the correct numbers in front of the terms to the right of the illustrations. Also, record these numbers on the Laboratory Report.

Regional Terminology

Various terms such as flank, groin, brachium, and hypochondriac have been applied to specific regions of the body to facilitate localization. Figures 1.2 and 1.3 pertain to some of the more predominantly used terminology.

Trunk

The anterior surface of the trunk may be subdivided into two pectoral, two groin, and the abdominal regions. The upper chest region may be designated as **pectoral** or **mammary** regions. The anterior trunk region not covered by the ribs is the **abdominal** region. The depressed area where the thigh of the leg meets the abdomen is the **groin.**

The posterior surface or dorsum of the trunk can be differentiated into the costal, lumbar, and buttocks regions. The **costal** (Latin: *costa,* rib) portion is that part of the dorsum which lies over the rib cage. The lower back region between the ribs and hips is the **lumbar** or **loin** region. The **buttocks** are the rounded eminences of the rump formed by the gluteal muscles; this is also called the **gluteal** region.

The side of the trunk which adjoins the lumbar region is called the **flank.** The armpit region which is between the trunk and arm is the **axilla.**

Arm

To differentiate the parts of the arm the term **brachium** is used for the upper arm and **antebrachium** for the forearm (between the elbow and wrist). The

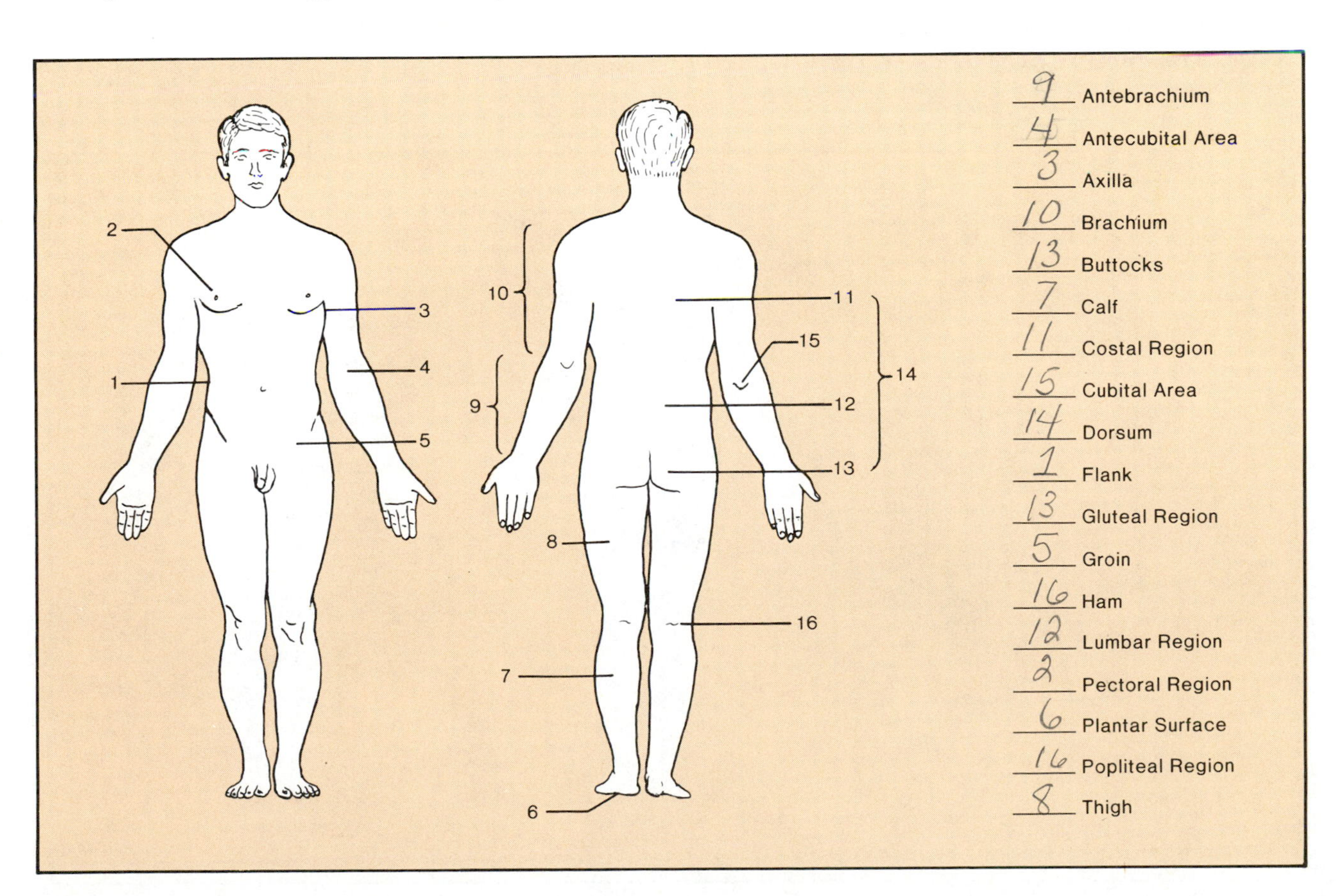

Figure 1.2 Regional terminology.

elbow area which is on the posterior surface of the arm is the **cubital** area. That area on the opposite side of the elbow is the **antecubital** area. It is also correct to refer to the entire anterior surface of the antebrachium as being antecubital.

Leg and Foot

While the upper portion of the leg is designated as the **thigh,** the lower fleshy posterior portion is called the **calf.** Between the thigh and calf on the posterior surface, opposite to the knee, is a depression called the **ham** or **popliteal** region. The sole of the foot is the **plantar** surface.

Abdominal Divisions

The abdominal surface may be divided into quadrants or into nine distinct areas. To divide the abdomen into nine regions one must establish four imaginary planes: two that are horizontal and two that are vertical. These planes and areas are shown in figure 1.3. The **transpyloric plane** is the upper horizontal plane which would pass through the lower portion of the stomach (pyloric portion). The **transtubercular plane** is the other horizontal plane, which touches the top surfaces of the hipbones (iliac crests). The two vertical planes, or **right** and **left lateral planes,** are approximately halfway between the midsagittal line and the crests of the hips.

The above planes describe the following nine areas. The **umbilical** region lies in the center, includes the navel, and is bordered by the two horizontal and two vertical planes. Immediately above the umbilical area is the **epigastric,** which covers much of the stomach. Below the umbilical zone is the **hypogastric** or *pubic area.* On each side of the epigastric are a right and left **hypochondriac.** Beneath the hypochondriac areas are the right and left **lumbar** areas. On each side of the hypogastric area are the right and left **iliac** areas.

Assignment:

Label figures 1.2 and 1.3 and transfer these numbers to the Laboratory Report.

Laboratory Report

After transferring all the labels from figures 1.1 through 1.3 to the proper columns on Laboratory Report 1,2, answer the questions that pertain to this exercise.

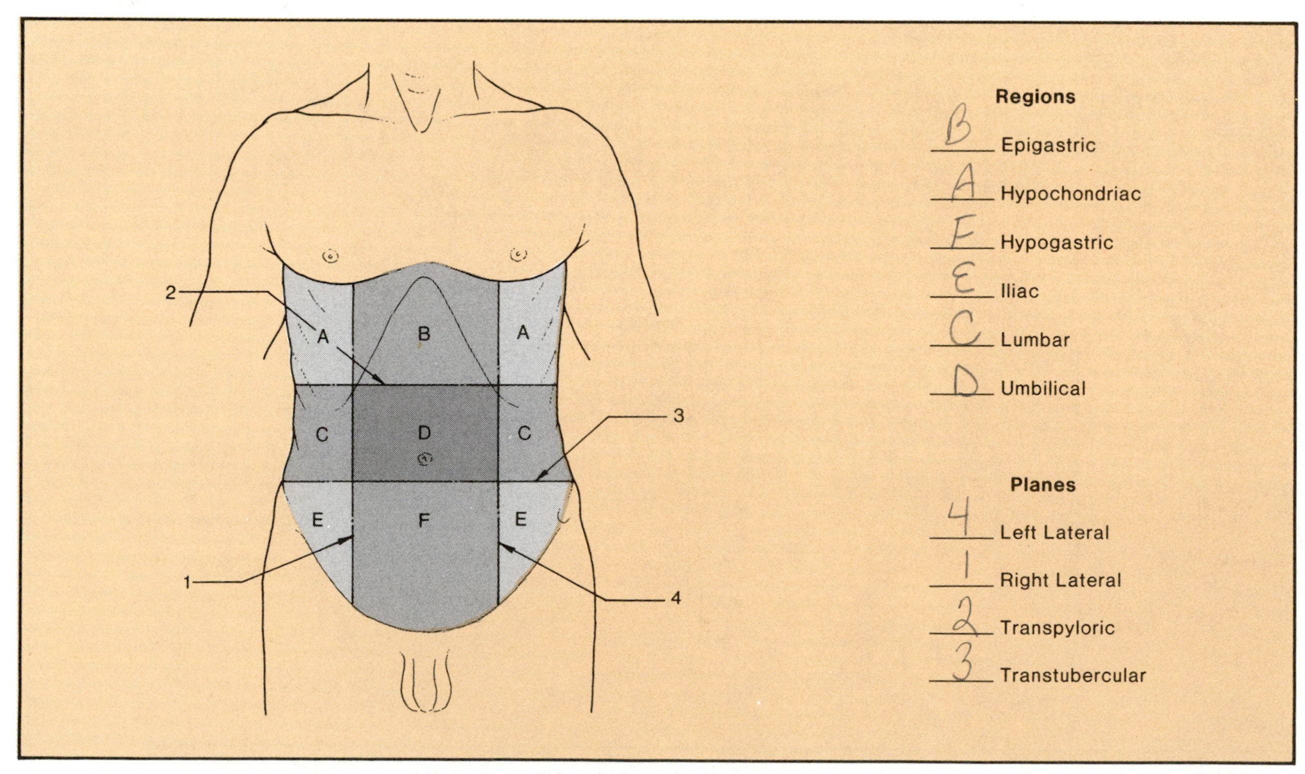

Figure 1.3 Abdominal regions.

Body Cavities and Membranes 2

All the internal organs, or *viscera,* are contained in body cavities that are completely or partially lined with smooth membranes. The relationships of these cavities to each other, the organs they contain, and the membranes that line them will be studied in this exercise.

Body Cavities

Figure 2.1 illustrates the seven principal cavities of the body. The two major cavities are the dorsal and ventral cavities. The **dorsal cavity,** which is nearest to the dorsal surface, includes the cranial and spinal cavities. The **cranial cavity** is the hollow portion of the skull that contains the brain. The **spinal cavity** is a long tubular canal within the vertebrae which contains the spinal cord. The **ventral cavity** is the large cavity which encompasses the chest and abdominal regions.

The superior and inferior portions of the ventral cavity are separated by a dome-shaped thin muscle, the **diaphragm.** The **thoracic cavity,** which is that part of the ventral cavity superior to the diaphragm, is separated into right and left compartments by a membranous partition or septum called the **mediastinum.** The lungs are contained in these right and left compartments. The heart, trachea, esophagus, and thymus gland are enclosed within the mediastinum. Figure 2.1 reveals the relationship of the lungs to the structures within the mediastinum. Note that within the thoracic cavity there exist right and left **pleural cavities** which contain the lungs, and a **pericardial cavity** which contains the heart.

The **abdominopelvic cavity** is that portion of the ventral cavity which is inferior to the diaphragm. It consists of two portions: the abdominal and pelvic

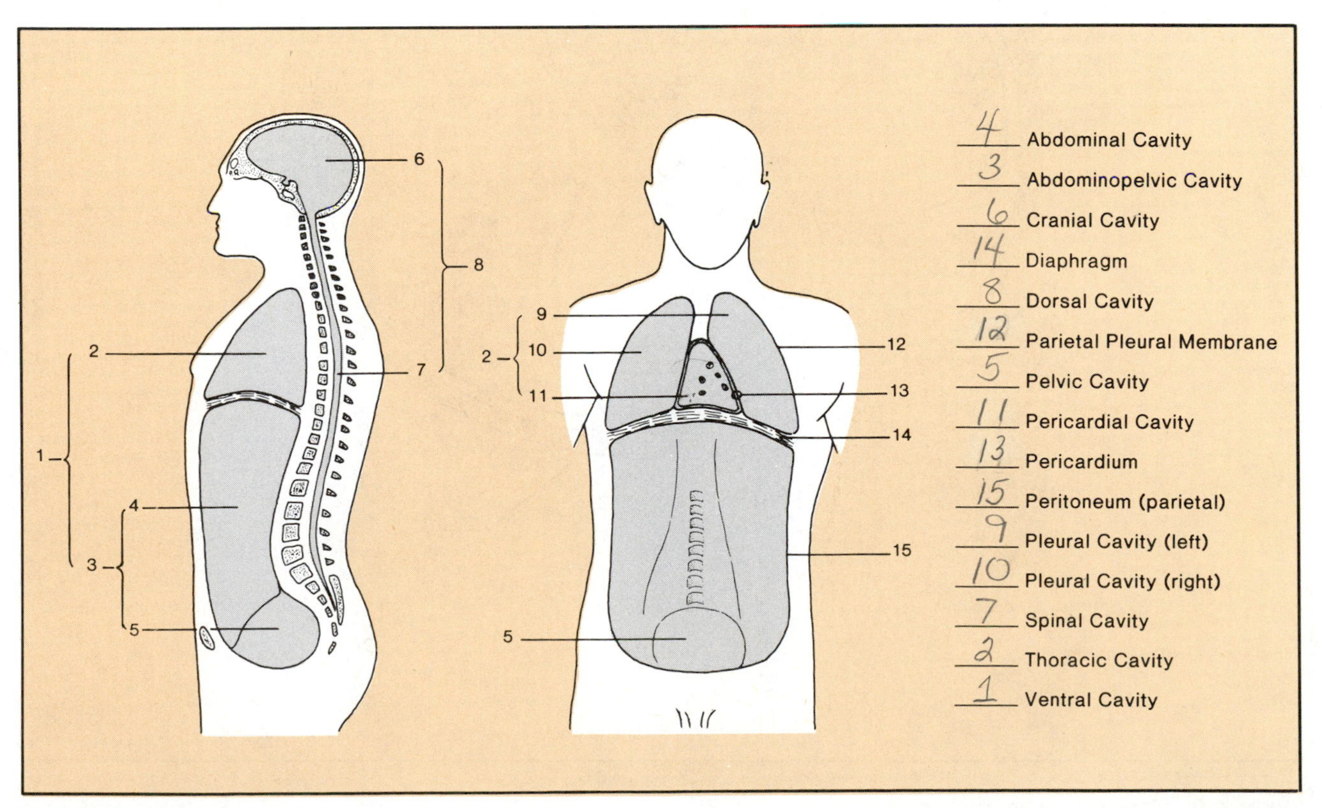

Figure 2.1 Body cavities.

cavities. The **abdominal cavity** contains the stomach, liver, gallbladder, pancreas, spleen, kidneys, and intestines. The **pelvic cavity** is the most inferior portion of the abdominopelvic cavity and contains the urinary bladder, sigmoid colon, rectum, uterus, and ovaries.

Body Cavity Membranes

The body cavities are lined with *serous membranes* which provide a smooth surface for the enclosed internal organs. Although these membranes are quite thin they are strong and elastic. Their surfaces are moistened by a self-secreted *serous fluid* which facilitates ease of movement of the viscera against the cavity walls.

Thoracic Cavity Membranes

The membranes that line the walls of the right and left thoracic compartments are called **parietal pleurae** (*pleura,* singular). The lungs, in turn, are covered with **visceral** (pulmonary) **pleurae.** Note in figure 2.2 that these pleurae are continuous with each other. The potential cavity between the parietal and visceral pleurae is the **pleural cavity.** Inflammation of the pleural membranes results in a condition called *pleurisy.*

Within the broadest portion of the mediastinum lies the heart. It, like the lungs, is covered by a thin serous membrane, the **visceral pericardium,** or **epicardium.** Surrounding the heart is a double-layered fibroserous sac, the **parietal pericardium.** The inner layer of this sac is a serous membrane that is continuous with the epicardium of the heart. Its outer layer is fibrous, which lends considerable strength to the structure. A small amount of serous fluid produced by the two serous membranes lubricates the surface of the heart to minimize friction as it moves within the parietal pericardium. The potential space between the visceral and parietal pericardia is called the **pericardial cavity.**

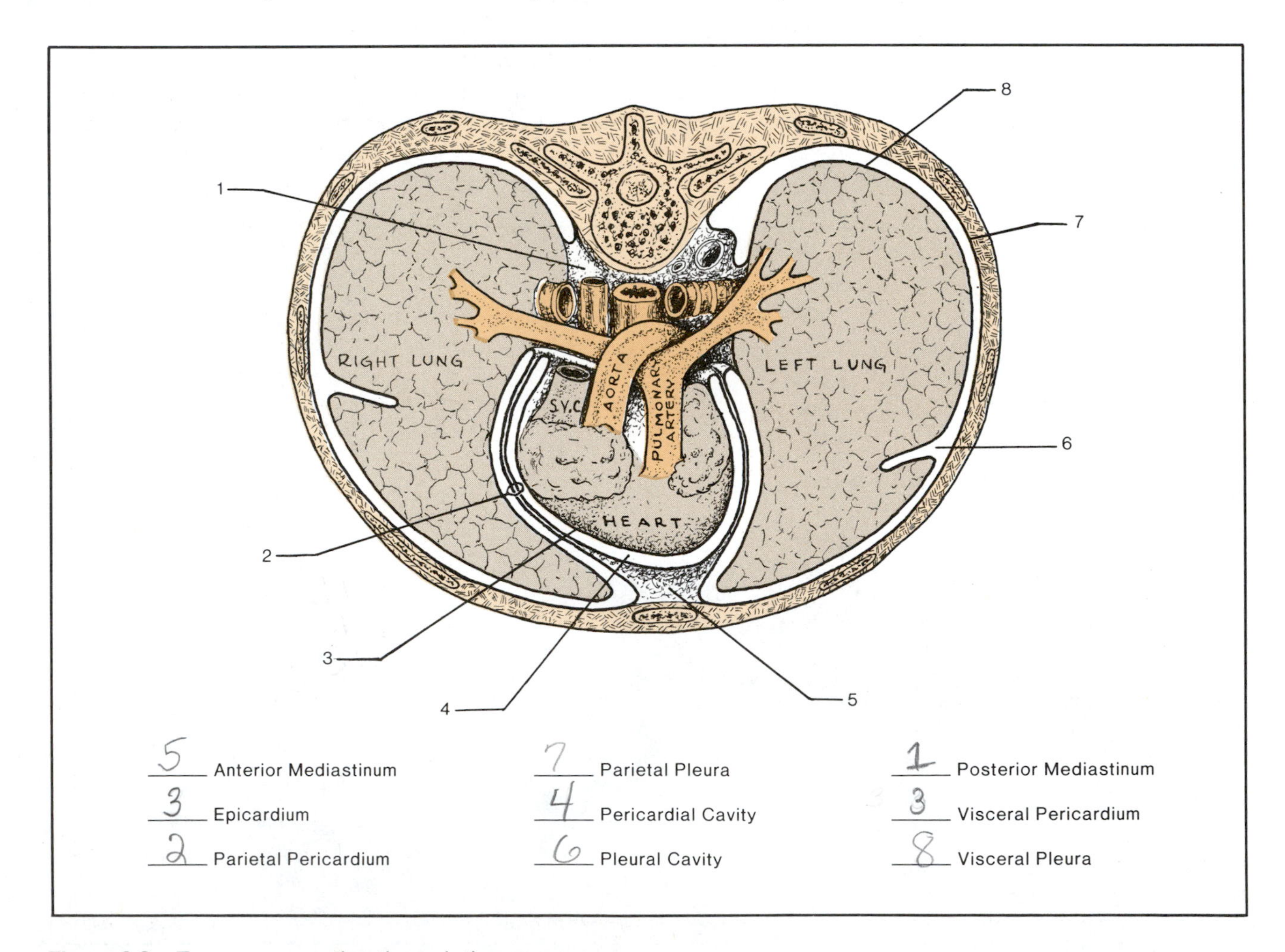

Figure 2.2 Transverse section through thorax.

Abdominal Cavity Membranes

The serous membrane of the abdominal cavity is the **peritoneum.** It does not extend deep down into the pelvic cavity, however; instead its most inferior boundary extends across the abdominal cavity at a level which is just superior to the pelvic cavity. The top portion of the urinary bladder is covered with peritoneum. In addition to lining the abdominal cavity, the peritoneum has double-layered folds called **mesenteries,** which extend from the dorsal body wall to the viscera, holding these organs in place. These mesenteries contain blood vessels and nerves that supply the viscera enclosed by the peritoneum. That part of the peritoneum attached to the body wall is the **parietal peritoneum.** The peritoneum that covers the visceral surfaces is **visceral peritoneum.** The potential cavity between the parietal and visceral peritoneums is called the **peritoneal cavity.**

Extending downward from the inferior surface of the stomach is a large mesenteric fold called the **greater omentum.** This double membrane structure passes downward from the stomach in front of the intestines, sometimes to the pelvis, and back up to the transverse colon where it is attached. Because it is folded upon itself it is essentially a double mesentery consisting of four layers. Protuberance of the abdomen in obese individuals is due to fat accumulation in the greater omentum. A smaller mesenteric fold, the **lesser omentum,** extends between the liver and the superior surface of the stomach and a short portion of the duodenum. Illustration B of figure 2.3 shows the relationship of these two omenta to the abdominal organs.

Assignment:

Label figures 2.1, 2.2, and 2.3.

Laboratory Report

Complete the second half of Laboratory Report 1,2.

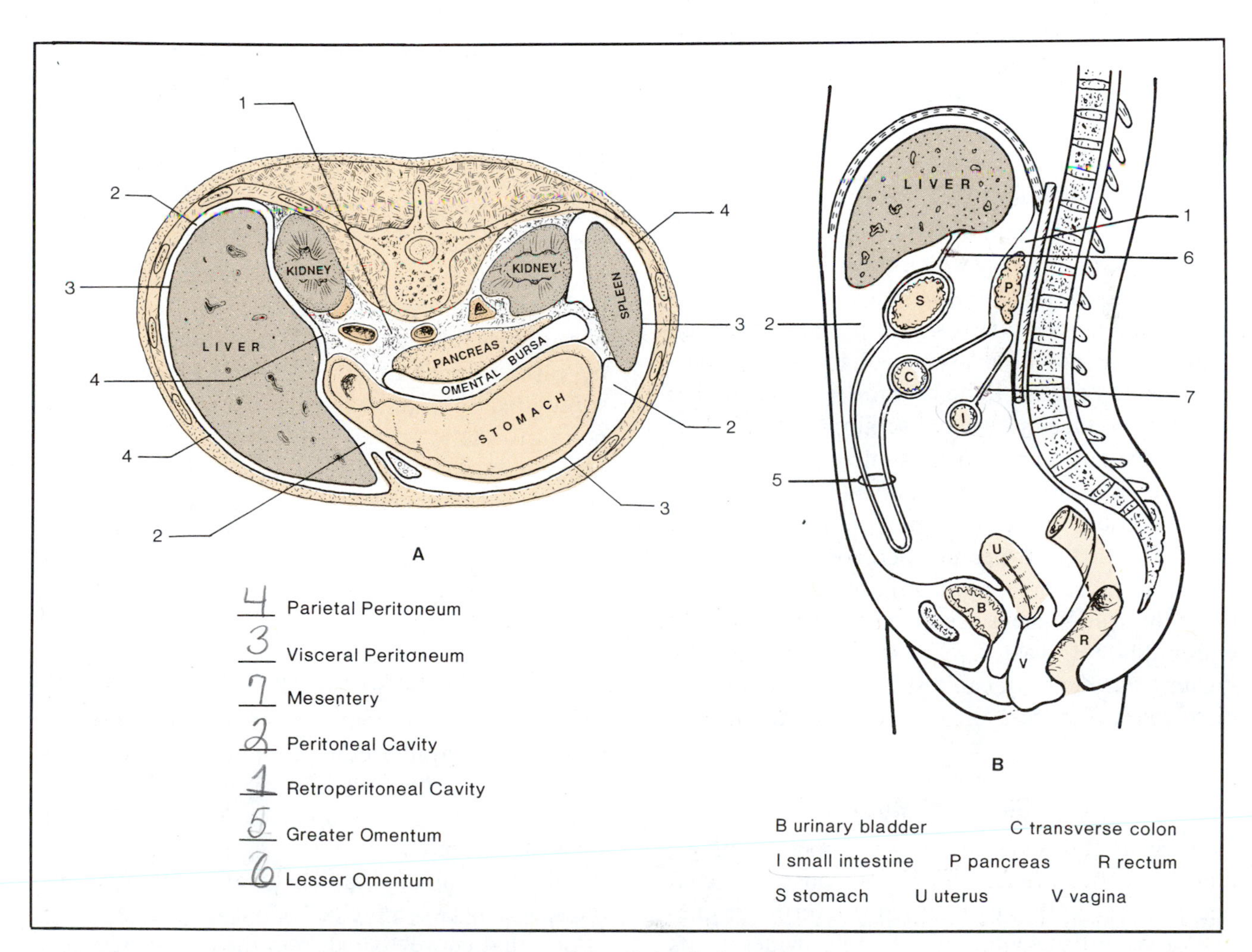

Figure 2.3 Transverse and longitudinal sections of the abdominal cavity.

3 Organ Systems: Rat Dissection

During this laboratory period we will dissect a freshly killed rat to perform a cursory study of the majority of the organ systems. Since rats and humans have considerable anatomical and physiological similarities, much will be learned here about human anatomy.

Before beginning the dissection, however, it will be necessary to review the eleven systems of the body. A brief description of each system follows. Keep in mind that an **organ** is defined as a structure composed of two or more tissues which performs one or more physiological functions. A **system,** on the other hand, is a group of organs that are directly related to each other functionally. Answer the questions on the Laboratory Report prior to doing the dissection.

The Integumentary System

The Latin word *integumentum* means covering. The surface of the body which consists of skin, hair, and nails comprises the integumentary system.

The integumentary system's principal function is to prevent bodily invasion by harmful microorganisms. In addition to being a mechanical barrier, the skin produces sweat and sebum (oil) that contains some antimicrobial substances.

The skin also aids in temperature regulation and excretion. The evaporation of sweat cools the body. The fact that perspiration contains many of the same excretory products found in urine indicates that the kidneys are aided by the skin in the elimination of wastes. Severe skin damage by burning, incidentally, can result in loss of fluids and dissolved substances; result: electrolytic imbalances.

The Skeletal System

The organs of this system consist of bone, cartilage, and connective tissue. The various bones are connected to each other by ligaments to form a framework that protects vital organs and provides points of attachment for muscles.

In addition to the above functions, the red marrow in bones gives rise to the various types of blood cells.

The Muscular System

Attached to the skeletal framework of the body are muscles that make up nearly half of the weight of the body. The ability of the elongated multinucleated cells of muscles to shorten when stimulated by nerve impulses, enables them to move parts of the body in walking, eating, breathing, and other activities.

The Nervous System

The nervous system consists of the brain, spinal cord, nerves, and receptors. While the nervous system has many functions, such as muscular control, regulation of circulation and breathing, etc., its basic function is to facilitate adaptability: i.e., the ability to adjust to external and internal environmental changes.

To be able to adapt to environmental changes there are, first of all, a multitude of different kinds of **receptors** throughout the body that are activated by various kinds of stimuli. A receptor may be stimulated by changes in pressure, chemicals, temperature, sound waves, light, etc. Once activated, nerve impulses pass along **conduction pathways** (nerves and spinal cord) to **interpretation centers** in the brain where recognition and evaluation of stimuli occur. Correct responses, which may be muscular, are then achieved by the nervous system via outgoing conduction pathways.

The Circulatory System

The circulatory system consists of the heart, arteries, veins, capillaries, blood, and spleen. **Arteries** are thick-walled vessels that carry blood from the heart to the microscopic **capillaries** of all organs. **Tissue fluid,** containing nutrients and oxygen, leaves the blood through the capillary walls and passes into the spaces between the cells. **Veins** are large blood vessels that convey blood from the capillaries back to the heart.

Thus, we see that this system provides transportation of various materials from one part of the body to another. In addition to the transportation of nutrients, oxygen, and carbon dioxide, the circulatory system transports hormones from glands, metabolic wastes from cells, and excess heat from muscles to the skin.

In addition to transporting all these substances, the blood is the body's primary defense against microbial invasion. The presence of phagocytic (cell-eating) white blood cells, antibodies, and special enzymes in blood prevents invading organisms from destroying the body.

The **spleen** is an oval structure on the left side of the abdominal cavity that acts to some extent as a blood reservoir. It also plays an important role in the removal of fragile red blood cells.

The Lymphatic System

The lymphatic system is a network of lymphatic vessels that returns tissue fluid from the intercellular spaces of tissues to the blood. This system is also responsible for the absorption of fats from the intestines. Although carbohydrates and proteins are absorbed directly into the blood through the intestinal wall, fats must pass first into the lymphatic system and then into the blood.

Once tissue fluid enters the lymphatic vessels it is called **lymph.** As this fluid moves through the lymphatic vessels it passes through nodules of lymphoid tissue called **lymph nodes.** Stationary phagocytic cells in these nodes remove bacteria and other foreign material, purifying the lymph before it is returned to the blood. This lymphoid tissue, which is found throughout the body in organs such as the liver, spleen, thymus, adenoids, appendix, Peyer's patches, and bone marrow is collectively referred to as the **reticuloendothelial system.**

The Respiratory System

The respiratory system consists of two portions: (1) the air passageways and (2) the respiratory portion. The actual exchange of gases between the blood and the air occurs in the respiratory portion. The lungs contain many tiny sacs called **alveoli,** which greatly increase the surface area for the transfer of oxygen and carbon dioxide in breathing. The passageways consist of the **nasal cavity, nasopharynx, larynx, trachea,** and **bronchi.**

The Digestive System

The digestive system includes the **mouth, salivary glands, esophagus, stomach, small intestine, large intestine, rectum, pancreas,** and **liver.** Its function is to convert ingested food to molecules that are small enough to pass through the intestinal wall into the blood and lymphatic system. The conversion of large food molecules to absorbable molecules is achieved by **enzymes** that are produced by various organs within this system. In addition to ingestion, digestion, and absorption of food, this system functions in the elimination of undigested materials.

The Urinary System

Cellular metabolism produces waste materials such as carbon dioxide, excess water, nitrogenous products, and excess metabolites. Although the skin, lungs, and large intestine assist in the removal of some of these wastes from the body, the majority of these products are removed from the blood by the **kidneys.**

To assist the kidneys in excretion of these products are the **ureters,** which drain the kidneys, the **urinary bladder,** which stores urine, and the **urethra,** which drains the bladder.

The Endocrine System

This system consists of a number of widely dispersed **endocrine glands** that dispense their secretions directly into the blood. These secretions, that are absorbed directly into capillaries within the glands, are called **hormones.**

Hormones perform such functions as integrating various physiological activities (metabolism and growth), directing the differentiation and maturation of the ovaries and testes, and regulating specific enzymatic reactions.

The endocrine system includes the **pituitary, thyroid, parathyroid, thymus, adrenal, pancreas, pineal glands, ovaries,** and **testes.** It even includes certain tissue of the placenta and digestive tract lining.

The Reproductive System

Continuity of the species is the function of the reproductive system. Spermatozoa are produced in the testes of the male, and ova are produced by the ovaries of the female. Male reproductive organs include the **testes, penis, scrotum, accessory glands,** and various ducts. The female reproductive organs include the **ovaries, vagina, uterus, uterine** (Fallopian) **tubes,** and **accessory glands.**

Laboratory Report

Complete the Laboratory Report for this exercise.

It is due at the beginning of the laboratory period in which the rat dissection is scheduled.

Rat Dissection

You will work in pairs to perform this part of the exercise. Your principal objective in the dissection is to *expose the organs for study, not to simply cut up the animal.* Most cutting will be performed with scissors. Whereas the scalpel blade will be used only occasionally, the flat blunt end of the handle will be used frequently for separating tissues.

Materials:

freshly killed rat
dissecting pan (with wax bottom)
dissecting kit
dissecting pins

Skinning the Ventral Surface

1. Pin the four feet to the bottom of the dissecting pan as illustrated in figure 3.1. Before making any incision examine the oral cavity. Note the large **incisors** in the front of the mouth that are used for biting off food particles. Force the mouth open sufficiently to examine the flattened **molars** at the back of the mouth. These teeth are used for grinding food into small particles. Note that the **tongue** is attached at its posterior end. Lightly scrape the surface of the tongue with a scalpel to determine its texture. The roof of the mouth consists of an anterior **hard palate** and a posterior **soft palate.** The throat is the **pharynx,** which is a component of both the digestive and respiratory systems.
2. Lift the skin along the midventral line with your forceps and make a small incision with scissors as shown in figure 3.2. Cut the skin upward to the lower jaw, turn the pan around, and complete this incision to the anus, cutting around both sides of the genital openings. The completed incision should appear as in figure 3.3.
3. With the handle of the scalpel, separate the skin from the musculature as shown in figure 3.4. The fibrous connective tissue which lies

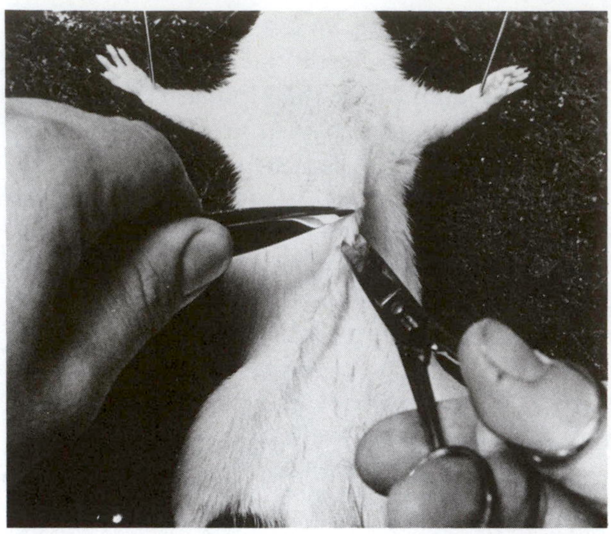

Figure 3.1 Incision is started on the median line with a pair of scissors.

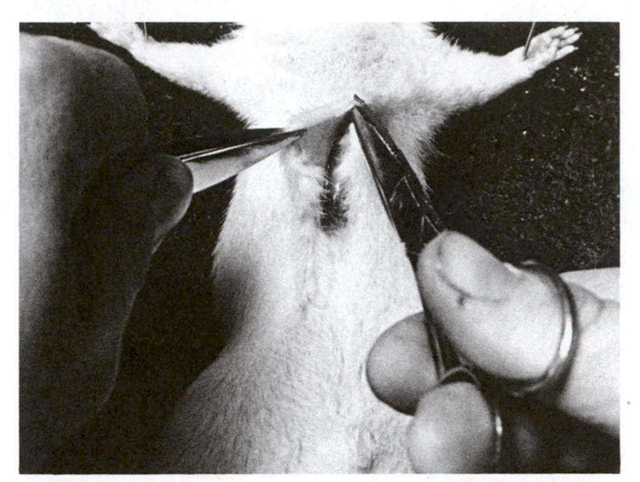

Figure 3.2 First cut is extended up to the lower jaw.

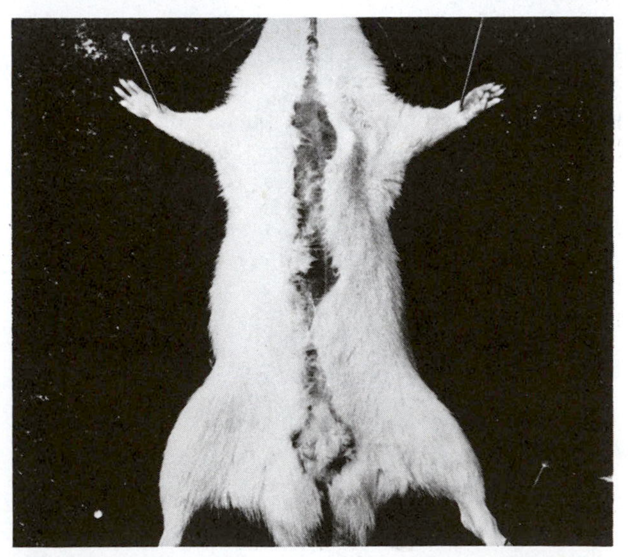

Figure 3.3 Completed incision from the lower jaw to the anus.

between the skin and musculature is the **superficial fascia** (Latin: *fascia,* band).

4. Skin the legs down to the "knees" or "elbows" and pin the stretched-out skin to the wax. Examine the surfaces of the **muscles** and note that **tendons,** which consist of tough fibrous connective tissue, attach the muscles to the skeleton. Covering the surface of each muscle is another thin gray feltlike layer, the **deep fascia.** Fibers of the deep fascia are continuous with fibers of the superficial fascia, so that considerable force with the scalpel handle is necessary to separate the two membranes.
5. At this stage your specimen should appear as in figure 3.5. If your specimen is a female, the mammary glands will probably remain attached to the skin.

Opening the Abdominal Wall

1. As shown in figure 3.5, make an incision through the abdominal wall with a pair of scissors. To make the cut it is necessary to hold

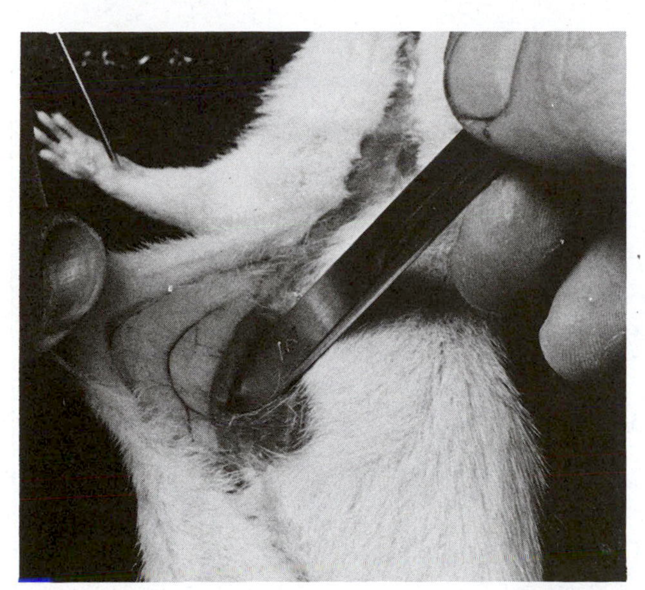

Figure 3.4 Skin is separated from musculature with scalpel handle.

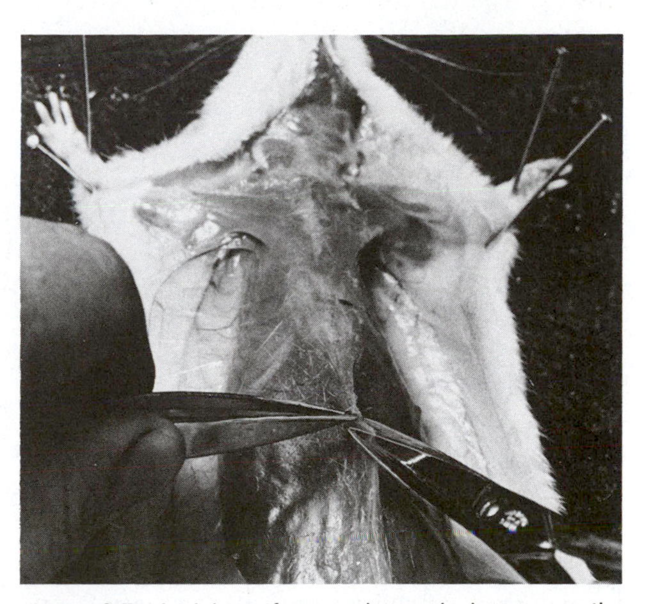

Figure 3.5 Incision of musculature is begun on the median line.

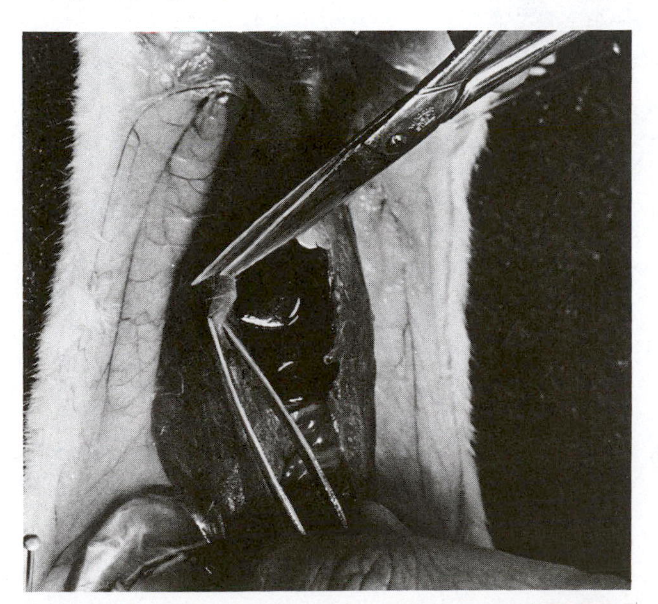

Figure 3.6 Lateral cuts at base of rib cage are made in both directions.

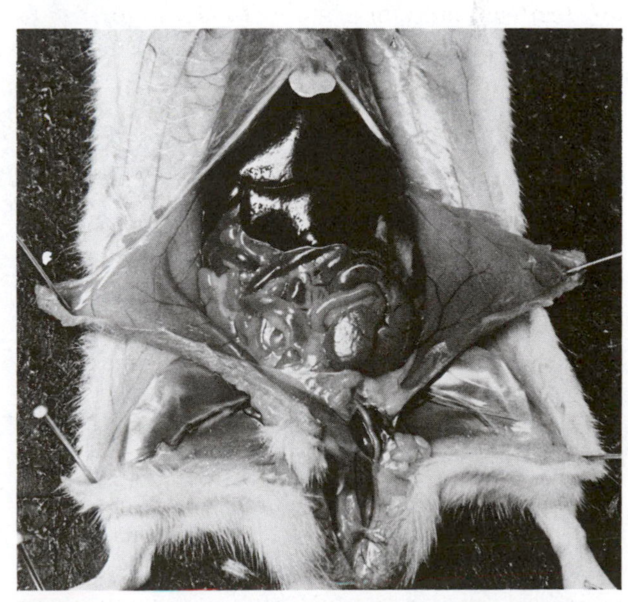

Figure 3.7 Flaps of abdominal wall are pinned back to expose viscera.

the muscle tissue with a pair of forceps. **Caution:** Avoid damaging the underlying viscera as you cut.

2. Cut upward along the midline to the rib cage and downward along the midline to the genitalia.
3. To completely expose the abdominal organs make two lateral cuts near the base of the rib cage—one to the left and the other to the right. See figure 3.6. The cuts should extend all the way to the pinned-back skin.
4. Fold out the flaps of the body wall and pin them to the wax as shown in figure 3.7. The abdominal organs are now well exposed.
5. Using figure 3.8 as a reference, identify all the labeled viscera without moving the organs out of place. Note in particular the position and structure of the **diaphragm.**

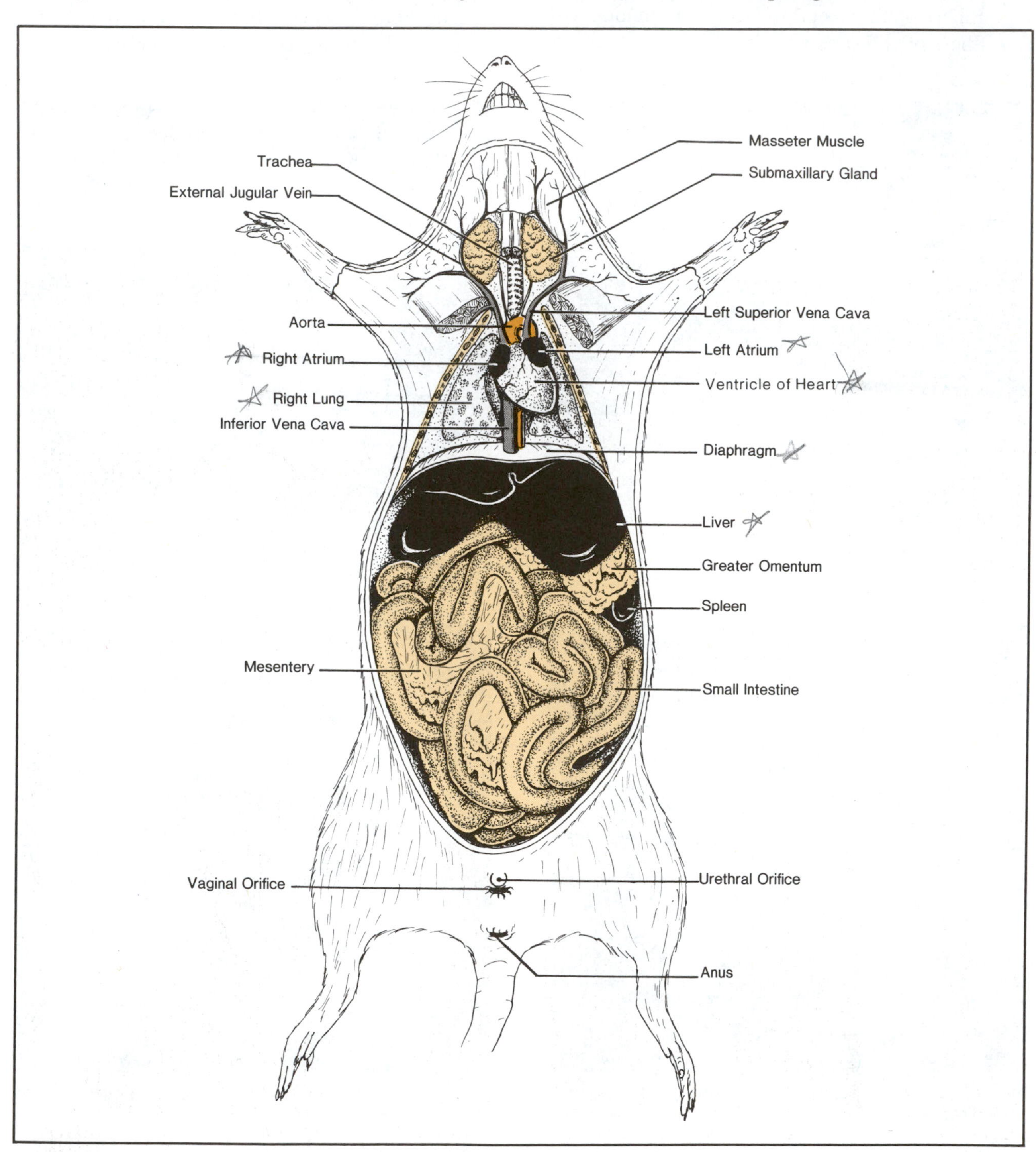

Figure 3.8 Viscera of a female rat.

Examination of Thoracic Cavity

1. Using your scissors, cut along the left side of the rib cage as shown in figure 3.9. Cut through all of the ribs and connective tissue. Then, cut along the right side of the rib cage in a similar manner.
2. Grasp the xiphoid cartilage of the sternum with forceps as shown in figure 3.10 and cut the diaphragm away from the rib cage with your scissors. Now you can lift up the rib cage and look into the thoracic cavity.
3. With your scissors, complete the removal of the rib cage by cutting off any remaining attachment tissue.
4. Now examine the structures that are exposed in the thoracic cavity. Refer to figure 3.8 and identify all the structures that are labeled.
5. Note the pale-colored **thymus gland,** which is located just above the heart. Remove this gland.
6. Carefully remove the thin **pericardial membrane** that encloses the **heart.**
7. Remove the heart by cutting through the major blood vessels attached to it. Gently sponge away pools of blood with *Kimwipes* or other soft tissues.
8. Locate the **trachea** in the throat region. Can you see the **larynx** (voice box) which is located at the anterior end of the trachea? Trace the trachea posteriorly to where it divides into two **bronchi** that enter the **lungs.** Squeeze the lungs with your fingers, noting how elastic they are. Remove the lungs.

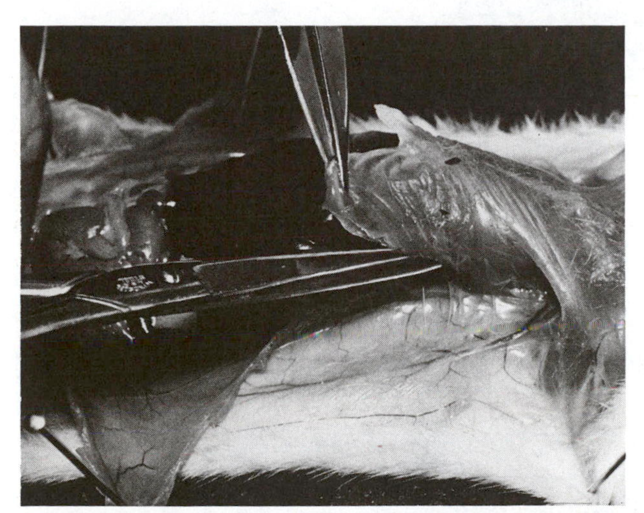

Figure 3.9 Rib cage is severed on each side with scissors.

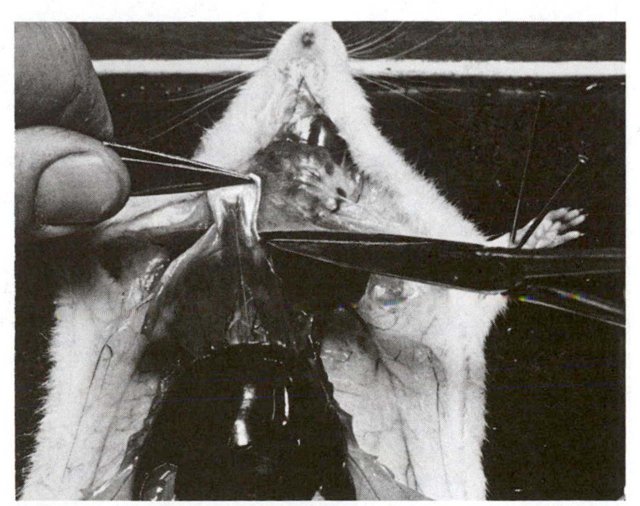

Figure 3.10 Diaphragm is cut free from edge of rib cage.

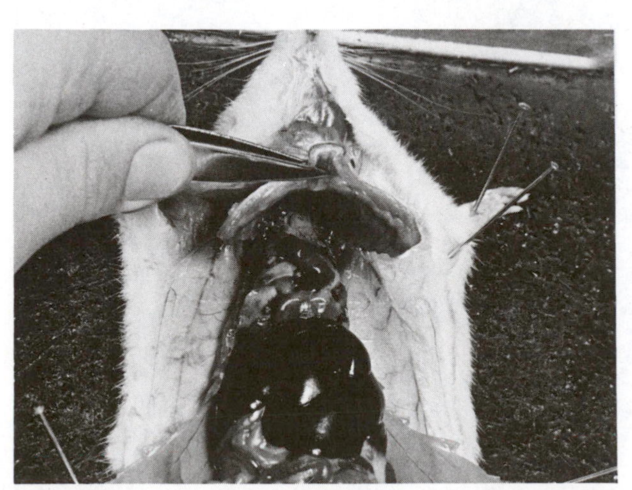

Figure 3.11 Thoracic organs are exposed as rib cage is lifted off.

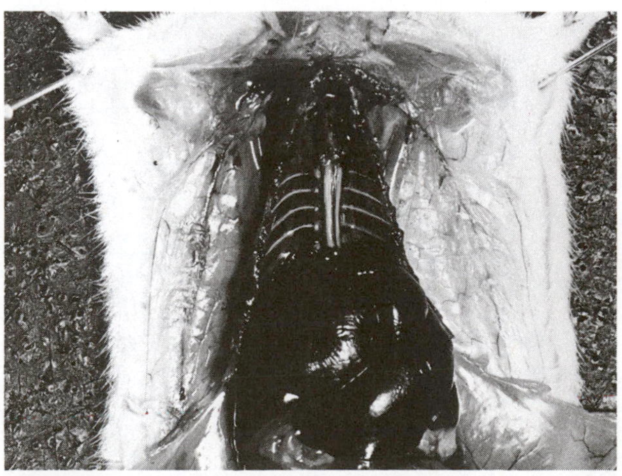

Figure 3.12 Specimen with heart, lungs, and thymus gland removed.

9. Probe under the trachea to locate the soft tubular **esophagus** that runs from the oral cavity to the stomach. Excise a section of the trachea to reveal the esophagus as illustrated in figure 3.13.

Deeper Examination of Abdominal Organs

1. Lift up the lobes of the reddish brown liver and examine them. Note that rats lack a **gallbladder.** *Carefully excise the liver* and wash out the abdominal cavity. The stomach and intestines are now clearly visible.
2. Lift out a portion of the intestines and identify the membranous **mesentery** which holds the intestines in place. It contains blood vessels and nerves that supply the digestive tract. If your specimen is a mature healthy animal the mesenteries will contain considerable fat.

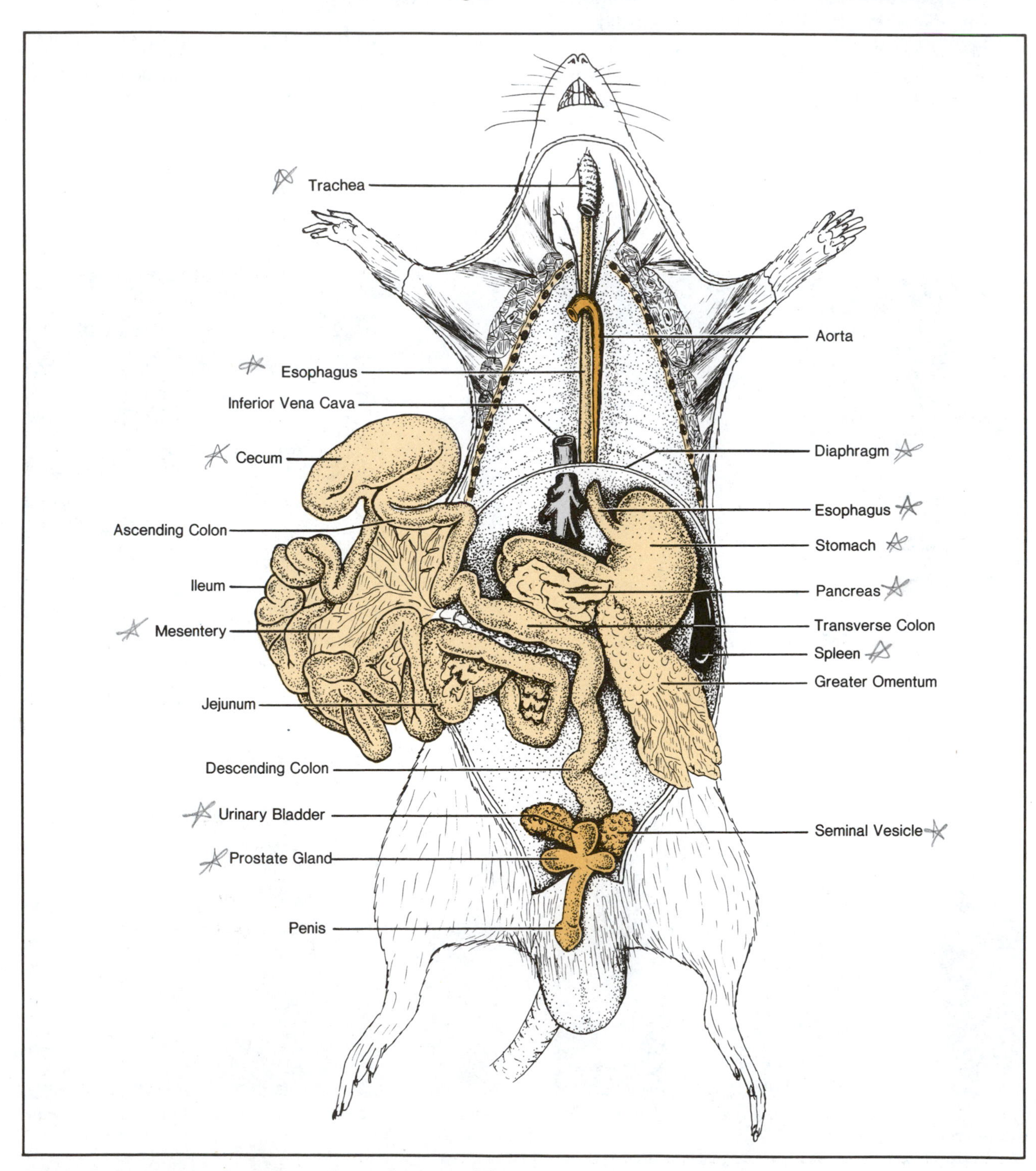

Figure 3.13 Viscera of a male rat (heart, lungs, and thymus removed).

3. Now lift the intestines out of the abdominal cavity, cutting the mesenteries, as necessary, for a better view of the organs. Note the great length of the small intestine. Its name refers to its diameter, not its length. The first portion of the small intestine, which is connected to the stomach, is called the **duodenum.** At its distal end the small intestine is connected to a large saclike structure, the **cecum.** The *appendix* in man is a *vestigial* portion of the cecum. The cecum communicates with the **large intestine.** This latter structure consists of the **ascending, transverse, descending,** and **sigmoid** divisions. The last of these portions empties into the **rectum.**
4. Try to locate the **pancreas** which is embedded in the mesentery alongside the duodenum. It is often difficult to see. Pancreatic enzymes enter the duodenum via the **pancreatic duct.** See if you can locate this minute tube.
5. Locate the **spleen** which is situated on the left side of the abdomen near the stomach. It is reddish brown and held in place with mesentery. Do you recall the functions of this organ?
6. Remove the digestive tract by cutting through the esophagus next to the stomach and through the sigmoid colon. You can now see the descending **aorta** and the **inferior vena cava.** The aorta carries blood posteriorly to the body tissues. The **inferior vena cava** is a vein that returns blood from the posterior regions to the heart.
7. Peel away the peritoneum and fat from the posterior wall of the abdominal cavity. *Removal of the fat will require special care to avoid damaging important structures.* This will make the kidneys, blood vessels, and reproductive structures more visible. Locate the two **kidneys** and **urinary bladder.** Trace the two **ureters** which extend from the kidneys to the bladder. Examine the anterior surfaces of the kidneys and locate the **adrenal glands,** which

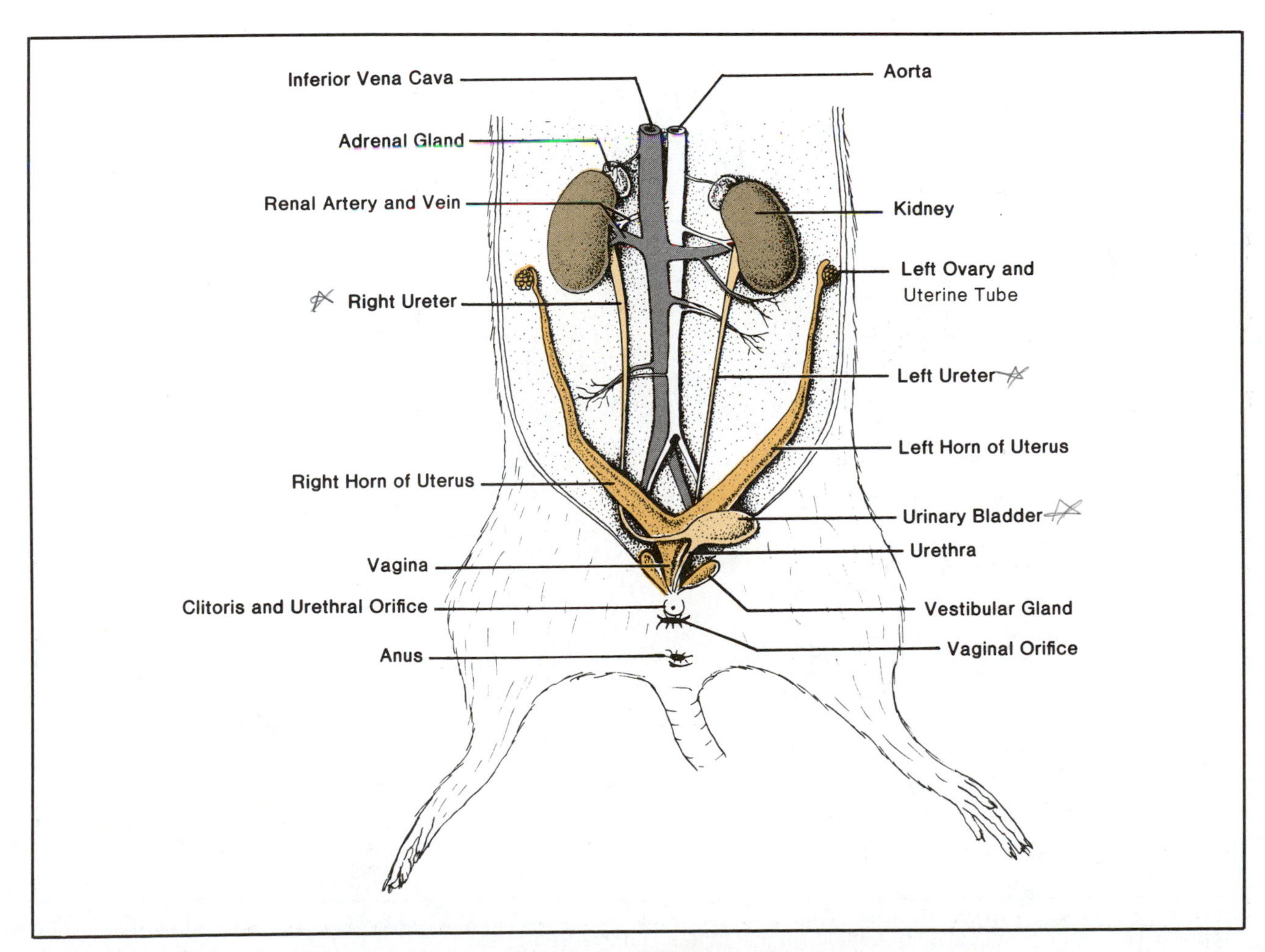

Figure 3.14 Abdominal cavity of a female rat (intestines and liver removed).

are important components of the endocrine gland system.

8. **Female.** If your specimen is a female compare it with figure 3.14. Locate the two **ovaries,** which lie lateral to the kidneys. From each ovary a **uterine tube** leads posteriorly to join the **uterus.** Note that the uterus is a Y-shaped structure joined to the **vagina.** If your specimen appears to be pregnant, open up the uterus and examine the developing embryos. Note how they are attached to the uterine wall.

9. **Male.** If your specimen is a male compare it with figure 3.15. The **urethra** is located in the **penis.**

   Apply pressure to one of the **testes** through the wall of the **scrotum** to see if it can be forced up into the **inguinal canal.**

   Carefully dissect out the testis, **epididymis,** and **vas deferens** from one side of the scrotum and, if possible, trace the vas deferens over the urinary bladder to where it penetrates the **prostate gland** to join the urethra.

In this cursory dissection you have become acquainted with the respiratory, circulatory, digestive, urinary, and reproductive systems. Portions of the endocrine system have also been observed. Five systems (integumentary, skeletal, muscular, lymphatic, and nervous) have been omitted at this time. These will be studied later. If you have done a careful and thoughtful rat dissection, you should have a good general understanding of the basic structural organization of the human body. Much that we see in rat anatomy has its human counterpart.

Clean-up Dispose of the specimen as directed by your instructor. Scrub your instruments with soap and water, rinse, and dry them.

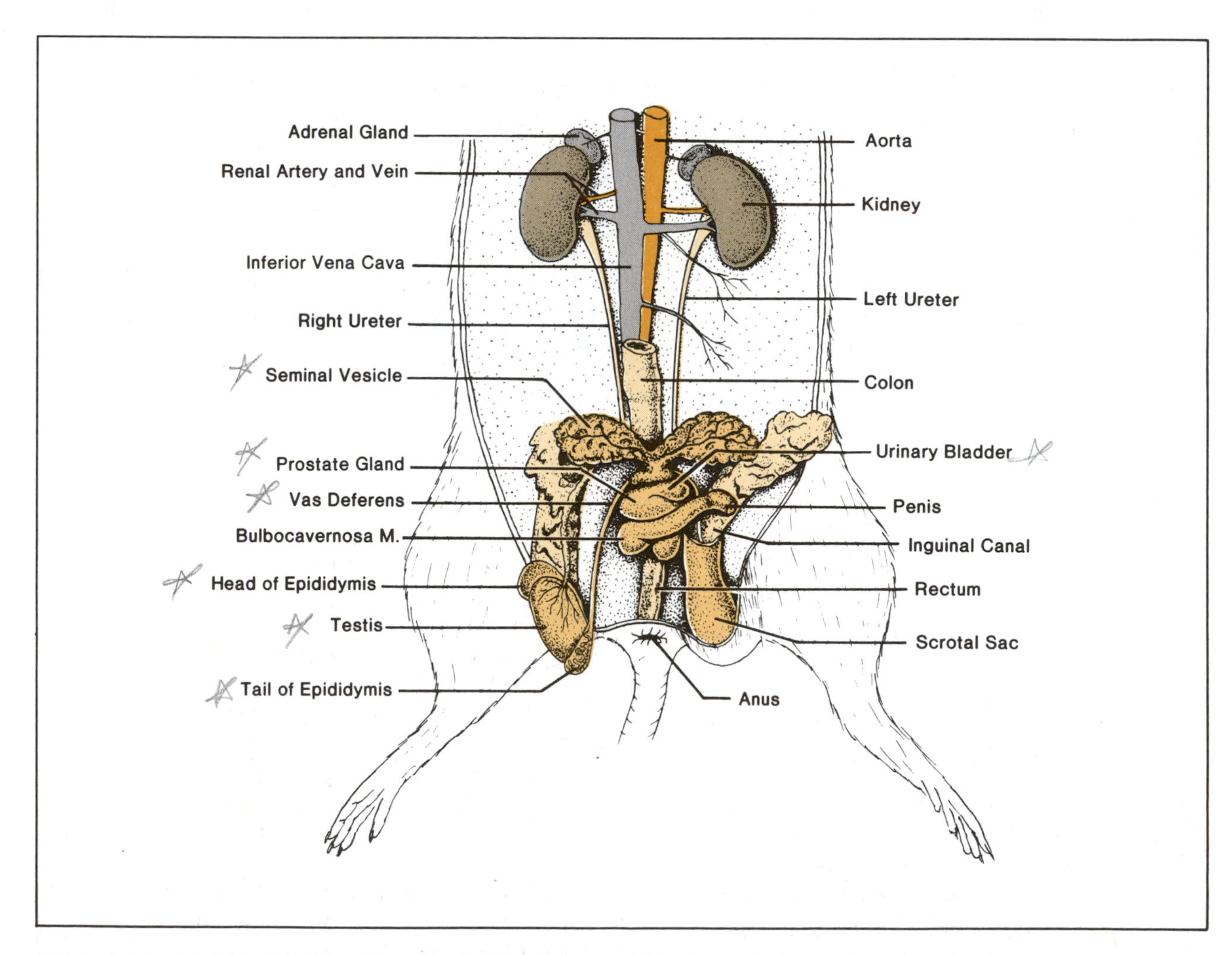

Figure 3.15 Abdominal cavity of a male rat (intestines and liver removed).

Microscopy 4

Since the next few exercises pertain to microscopic studies of various types of cells, it is essential that some orientation be provided concerning acceptable procedures in microscopy. It is verly likely that you have had some previous experience using a microscope, so we will provide here only a brief summary of what you should know.

Microscopy can be fascinating or frustrating. Success in this endeavor depends to a considerable extent on how well you understand the mechanics and limitations of your microscope. It is the purpose of this exercise to outline procedures that should minimize difficulties and enable you to get the most out of your subsequent attempts at cytological microscopy.

Before entering the laboratory to do any of the studies on cells and tissues, it would be desirable if you answered all the questions on the Laboratory Report after reading over the entire exercise. Your instructor may require that the Laboratory Report be handed in prior to doing any laboratory work.

Care of the Instrument

Microscopes represent considerable expense and can be damaged rather easily if certain precautions are not observed. The following suggestions cover most hazards.

Transport When carrying your microscope from one part of the room to another, use both hands when holding the instrument, as illustrated in figure 4.1. If it is carried with only one hand and allowed to dangle at your side, there is always the danger of collision with furniture or some other object. And, incidentally, under no circumstances should one attempt to carry two microscopes at one time.

Clutter Keep your work station uncluttered while doing microscopic work. Keep unnecessary books, lunches, and other unneeded objects away from work area. A clear work area promotes efficiency and results in fewer accidents.

Electric Cord Microscopes have been known to tumble off of tabletops when students have entangled a foot in a dangling electric cord. Don't let the light cord on your microscope dangle in such a way as to hazard entanglement.

Lens Care At the beginning of each laboratory period check the lenses to make sure they are clean. At the end of each lab session be sure to wipe any immersion oil off the oil immersion lens if it has been used.

Dust Protection In most laboratories dustcovers are used to protect the instruments during storage. If one is available, place it over the microscope at the end of the period.

Components

Before we discuss the procedures for using a microscope, let's identify the principal parts of the microscope as illustrated in figure 4.4.

Framework All microscopes have a basic frame structure which includes the **arm** and **base.** To this framework all other parts are attached. On many of the older microscopes the base is not rigidly attached to the arm as is the case in figure 4.4; instead, a pivot point is present which enables one to tilt the arm backward to adjust the eyepoint level.

Stage The horizontal platform that supports the microscope slide is called the *stage.* Note that it has a clamping device, the **mechanical stage,** which is used for holding and moving the slide around on the stage. Note, also, the location of the **mechanical stage control** in figure 4.4.

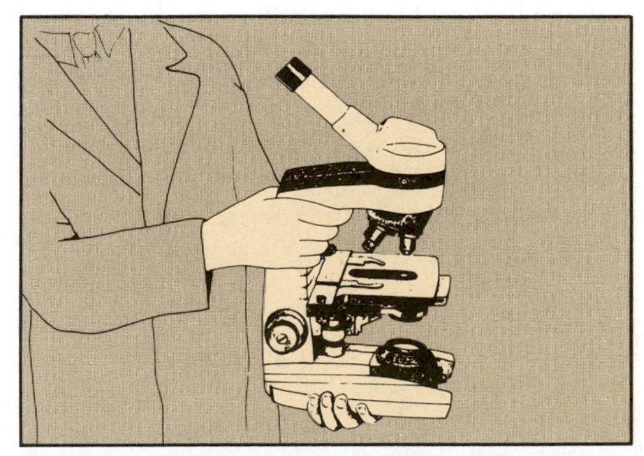

Figure 4.1 The microscope should be held firmly with both hands while carrying it.

Light Source In the base of most microscopes is positioned some kind of light source. Ideally, the lamp should have a voltage control to vary the intensity of light. The microscope in figure 4.4 has the type of lamp that is controlled by a detached transformer unit (not shown). The light source on the microscope base in figure 4.2 has an electronic voltage control that is built into the lamp housing.

Many microscopes utilize a **neutral density filter,** in addition to a voltage control, to further reduce light intensity.

Lens Systems Three lens systems are present in all laboratory microscopes: the oculars, objectives, and the condenser.

The **ocular,** or eyepiece, which is at the top of the instrument, consists of two or more internal lenses and usually has a magnification of 10×. Although the microscope in figure 4.4 has two oculars (binocular), a microscope often has only one.

Three or more **objectives** are usually present. Note that they are attached to a rotatable nosepiece which makes it possible to move them into position over a slide. Objectives on most laboratory microscopes have magnifications of 10×, 45×, and 100×, designated as **low power, high-dry,** and **oil immersion,** respectively.

The third lens system is the **condenser,** which is located under the stage. It collects and directs the light from the lamp to the slide being studied. The **condenser adjustment knob** shown in figure 4.4 enables one to move the condenser up and down. A **diaphragm lever** is also provided to control the amount of light exiting from the condenser under the stage.

Focusing Knobs The concentrically arranged **coarse adjustment** and **fine adjustment knobs** on the side of the microscope are used for bringing objects into focus when studying an object on a slide.

Ocular Adjustments On binocular microscopes one must be able to change the distance between the oculars, and to make diopter changes for eye differences. On most microscopes the interocular distance is changed by simply pulling apart or pushing together the oculars. To make diopter adjustments, one focuses first with the right eye only. Without touching the focusing knobs, diopter adjustments are then made on the left eye by turning the knurled **diopter adjustment ring** on the left ocular until a sharp image is seen. One should now see sharp images with both eyes.

Resolution

The resolution limit, or **resolving power,** of a microscope lens is a function of its numerical aperture, the wavelength of light, and the design of the condenser. The maximum resolution of the best microscopes is around .2 μm. This means that two small objects that are 0.2 μm apart will be seen as separate entities; objects closer than that will be seen as a single object.

The maximum amount of resolution from a lens system is possible if: (1) a blue filter is used over the light source, (2) the condenser is kept at its highest position, and (3) the diaphragm is not stopped down too much. When using the oil immersion lens, oil between the lens and object on the slide improves resolution.

Lens Care

Unless all lenses are kept scrupulously clean, maximum resolution cannot occur. When cleaning lenses observe the following suggestions:

Cleaning Tissues Only lint-free, optically safe tissues should be used to clean lenses. Tissues free of

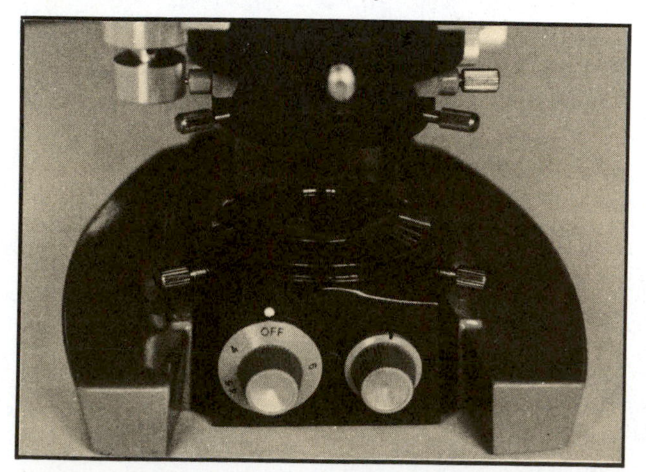

Figure 4.2 The left knob controls voltage; the other knob controls the neutral density filter.

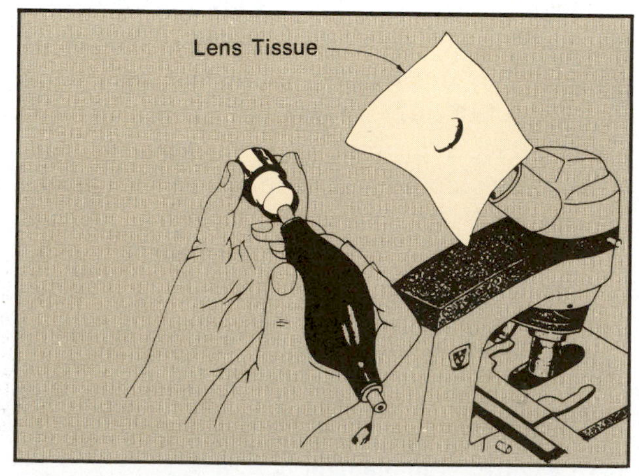

Figure 4.3 After cleaning the lenses, a blast of air from an air syringe removes residual lint.

abrasive grit fall in this category. Booklets of lens tissue are most widely used for this purpose. Although several types of boxed tissues are also safe, *use only the type of tissue that is recommended by your instructor.*

Solvents Various liquids can be used for cleaning microscope lenses. Green soap with warm water works very well. Xylene is universally acceptable. Alcohol and acetone are also recommended, but often with some reservations. Acetone is a powerful solvent that could possibly dissolve the lens mounting cement in some objective lenses if it were used too liberally. When it is used it should be used sparingly. Your instructor will inform you as to what solvents can be used on the lenses of your microscope.

Oculars The best way to determine if your eyepiece is clean is to rotate it between the thumb and forefinger as you look through the microscope. A rotating pattern will be evidence of dirt.

If cleaning the top lens of the ocular with lens tissue fails to remove the debris, one should try cleaning the lower lens with lens tissue and blowing off any excess lint with an air syringe. *Whenever*

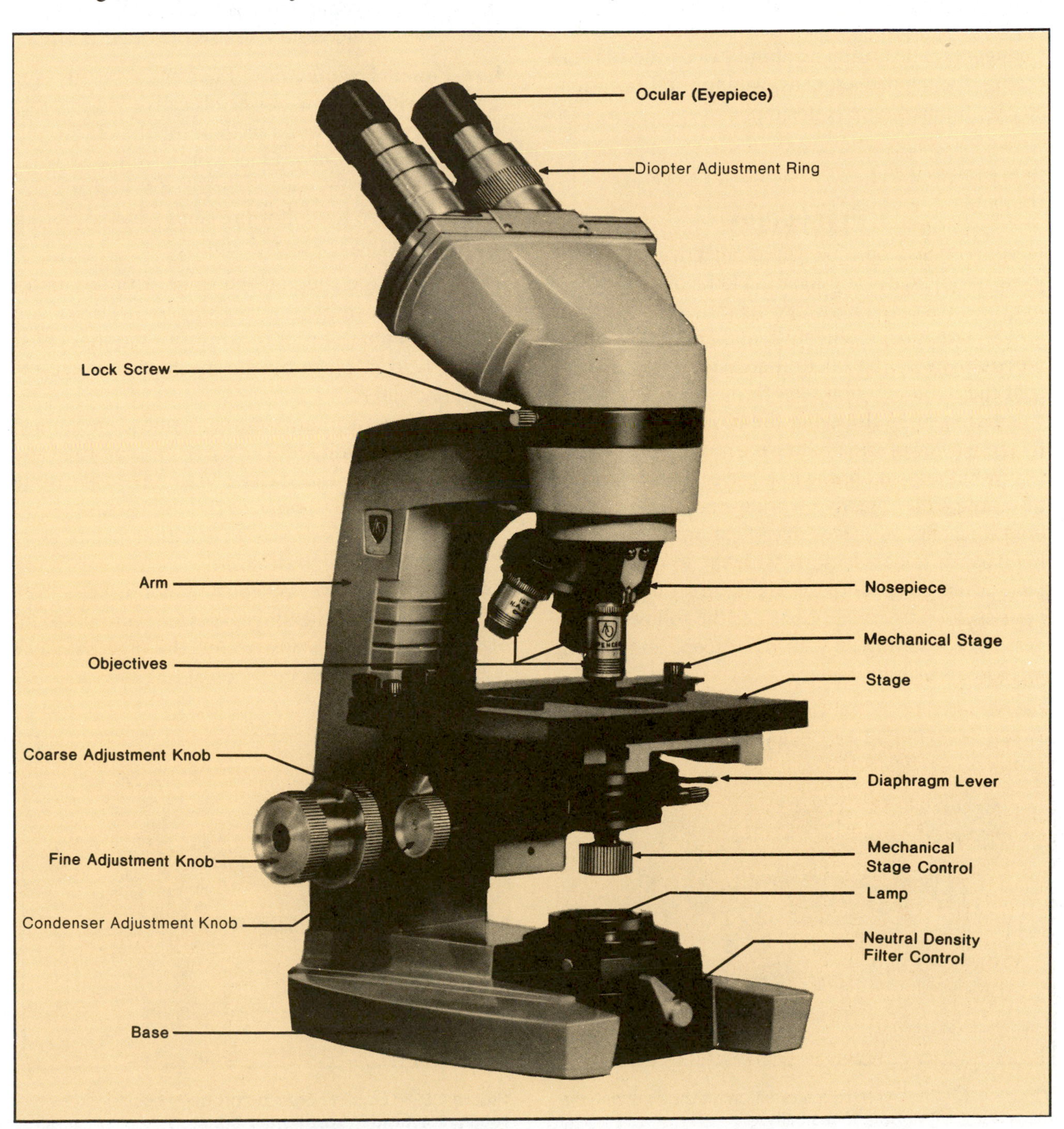

Figure 4.4 The compound microscope.

the ocular is removed from the microscope, it is imperative that a piece of lens tissue be placed over the open end of the microscope as illustrated in figure 4.3.

Objectives Objective lenses often become soiled by materials from slides or fingers. A piece of lens tissue moistened with green soap and water or one of the other solvents mentioned will usually remove whatever is on the lens. Sometimes a cotton swab with a solvent will work better than lens tissue. At any time that the image on a slide is blurred or cloudy, assume at once that the objective you are using is soiled.

Condenser Dust often accumulates on the top surface of the condenser; thus, wiping it off occasionally with lens tissue is desirable.

Procedures

If your microscope has three objectives you have three magnification options: (1) low-power or 100× magnification, (2) high-dry magnification which is 450× with a 45× objective, and (3) 1000× magnification with the oil immersion objective. Note that the magnification seen through an objective is calculated by multiplying the power of the ocular by the power of the objective.

Whether you use the low-power objective or the oil immersion objective will depend on how much magnification is necessary. Generally speaking, however, it is best to start with the low-power objective and progress to the higher magnifications as your study progresses. Consider the following suggestions for setting up your microscope and making microscopic observations.

Viewing Setup If your microscope has a rotatable head, such as the one being used by the two students in figure 4.5, there are two ways that you can use the instrument. Note that the student on the left has the arm of the microscope near him, and the other student has the arm away from her. With this type of microscope, the student on the right has the advantage in that the stage is easier to observe. Note, also, that when focusing the instrument she is able to rest her arm on the table. The manufacturers of this type of microscope intended that the instrument be used in the way demonstrated by the young lady. If the head is not rotatable, it will be necessary to use the other position.

Low-Power Examination The main reason for starting with the low-power objective is to enable you to explore the slide to look for the object you are planning to study. Once you have found what you are looking for, you can proceed to higher magnifications. Use the following steps when exploring a slide with the low-power objective:

1. Position the slide on the stage with the material to be studied on the *upper* surface of the slide. Figure 4.6 illustrates how the slide must be held in place by the mechanical stage retainer lever.
2. Turn on the light source, using a *minimum* amount of voltage. If necessary, reposition the slide so that the stained material on the slide is in the *exact center* of the light source.
3. Check the condenser to see that it has been raised to its highest point.
4. If the low-power objective is not directly over the center of the stage, rotate it into position. Be sure that as you rotate the objective into position it clicks into its locked position.

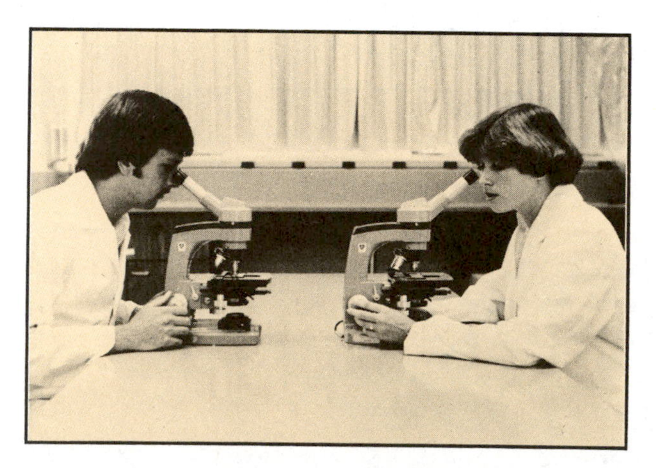

Figure 4.5 The microscope position on the right has the advantage of stage accessibility.

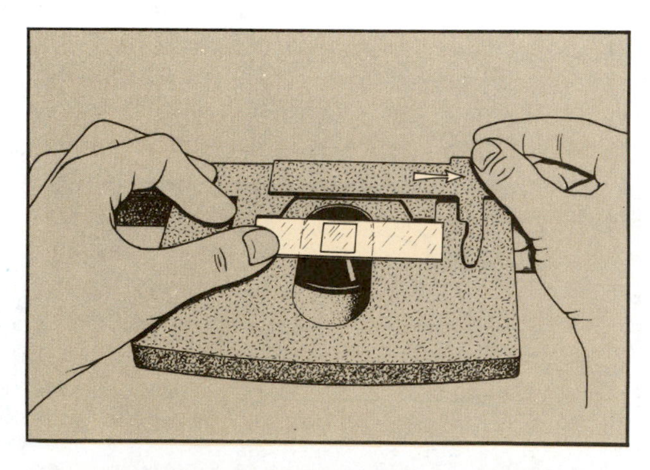

Figure 4.6 The slide must be properly positioned as the retainer lever is moved to the right.

5. Turn the coarse adjustment knob to lower the objective *until it stops*. A built-in stop will prevent the objective from touching the slide.
6. While looking down through the ocular or oculars, bring the object into focus by turning the fine adjustment focusing knob. Don't readjust the coarse adjustment knob. If you are using a binocular microscope it will also be necessary to adjust the interocular distance and diopter adjustment to match your eyes.
7. Manipulate the diaphragm lever to reduce or increase the light intensity to produce the clearest, sharpest image. Note that as you close down the diaphragm to reduce the light intensity, the contrast improves and the depth of field increases.
8. Once an image is visible, move the slide about to search out what you are looking for. The slide is moved by turning the knobs that move the mechanical stage.
9. Check the cleanliness of the ocular, using the procedure outlined above.
10. Once you have identified the structures to be studied and wish to increase the magnification, you may proceed to either high-dry or oil immersion magnification. However, before changing objectives, *be sure to center the object you wish to observe.*

High-Dry Examination To proceed from low-power to high-dry magnification, all that is necessary is to rotate the high-dry objective into position and open up the diaphragm somewhat. It may be necessary to make a minor adjustment with the fine adjustment knob to sharpen up the image, but *the coarse adjustment knob should not be touched.*

If a microscope is of good quality, only minor focusing adjustments are needed when changing from low power to high-dry because all the objectives will be **parfocalized.** Nonparfocalized microscopes do require considerable focusing adjustments when changing objectives.

High-dry objectives should only be used on slides that have cover glasses; without them, images are often unclear. When increasing the lighting, be sure to open up the diaphragm first instead of increasing the voltage on your lamp; reason: lamp life is greatly extended when used at low voltage. If the field is not bright enough after opening the diaphragm, feel free to increase the voltage. A final point: keep the condenser at its highest point.

Oil Immersion Techniques The oil immersion lens derives its name from the fact that a special mineral oil is interposed between the lens and the microscope slide. The oil is used because it has the same refractive index as glass, which prevents the loss of light due to the bending of light rays as they pass through air. The use of oil in this way enhances the resolving power of the microscope. Figure 4.7 reveals this phenomenon.

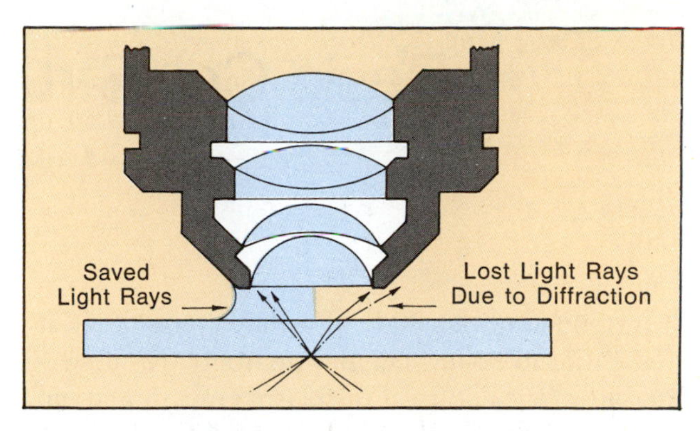

Figure 4.7 Immersion oil, having the same refractive index as glass, prevents light loss due to diffraction.

With parfocalized objectives one can go to oil immersion from either low power or high-dry. Once the microscope has been brought into focus at one magnification, the oil immersion lens can be rotated into position without fear of striking the slide.

Before rotating the oil immersion lens into position, however, a drop of immersion oil is placed on the slide. If the oil appears cloudy it should be discarded.

When using the oil immersion lens it is best to open the diaphragm as much as possible. Stopping down the diaphragm tends to limit the resolving power of the optics. In addition, the condenser must be at its highest point. If different colored filters are available for the lamp housing, it is best to use blue or greenish filters to enhance the resolving power.

Using this lens takes a little practice. The manipulation of lighting is critical. Before returning the microscope to the cabinet, all oil must be removed from the objective and stage. Slides with oil should be wiped clean with lens tissue, also.

Laboratory Report

Before entering the laboratory for your first encounter with the microscope, answer the questions on the Laboratory Report. Preparation on your part prior to going to the laboratory will greatly facilitate your understanding. Your instructor may wish to collect this report at the beginning of the period on the first day that the microscope is to be used in class.

5 Basic Cell Structure

As a prelude to the study of cellular physiology, it is essential to review cellular anatomy. An understanding of the nature of the plasma membrane and the roles of organelles such as mitochondria, ribosomes, and endoplasmic reticulum is of fundamental importance.

In the first portion of this exercise, a summary of the present knowledge of cellular structure and function is presented. This is followed by laboratory studies of epithelial cells and living microorganisms.

Since much of the discussion in this exercise pertains to cellular structure as seen with an electron microscope, and since our laboratory observations will be made with a light microscope, there will be some cellular structures that cannot be seen in great detail in the laboratory.

To get the most out of this exercise *it is recommended that the questions on the Laboratory Report be answered prior to entering the laboratory.*

The Basic Design

The study of cellular structure (**cytology**) has advanced at a rapid rate over the past decades with the use of electron microscopy and special techniques in the study of cellular physiology. The old notion of the cell as a "sac of protoplasm" has been supplanted by a much more dynamic model. A cell is now viewed as a highly complex miniature computer with an integrated, compartmentalized ultrastructure capable of communicating with its environment, altering its shape, processing information, and synthesizing a great variety of substances. It is largely through our advancing knowledge of the cellular organelles that this new viewpoint has emerged.

Figure 5.1 is a view of a pancreatic cell as it might appear if magnified by an electron microscope. This cell has been chosen for our study because it lacks the degree of specialization that is seen in many other cells, and so it can be considered illustrative of generalized cell structure. It contains the majority of organelles present in body cells and is also representative of an actively secretory cell.

Although it is not possible to see the fine structure of many organelles with our ordinary light microscope, we are including descriptions of their appearance and current theories of function.

As the various cell structures are discussed in the following text, identify them in figure 5.1.

Plasma Membrane

The outer surface of every cell consists of an extremely thin, delicate *plasma membrane* or *cell membrane.* Chemical analyses of plasma membranes indicate that phospholipids, glycolipids, and proteins are the principal constituents. Lipids account for one-half the mass of plasma membranes, proteins the other half.

Examination with an electron microscope reveals that all cell membranes have a similar basic trilaminar structure, the so-called *unit membrane.* The current interpretation of the molecular organization of this membrane proposes that the lipid molecules form a double fluid layer in which protein and glycoprotein molecules are embedded. Figure 5.1 depicts an enlarged view of this fluid mosaic model (Singer-Nicholson 1972). Note that some of the proteins protrude externally, others protrude internally, and some extend through both sides of the membrane. This asymmetrical architecture appears to account for the external and internal differences in receptor sites that affect the permeability characteristics of the membrane.

Cell membranes are *selectively permeable* in allowing certain molecules to pass through easily and preventing other molecules from gaining entrance to the cell. Although cell membranes play both active and passive roles in molecular movements, experimental evidence seems to indicate that the movement of all molecules, including water, are assisted by the membrane. Exercise 7 pertains to the forces involved in permeability.

Nucleus

Near the basal portion of the cell is a large spherical body, the *nucleus*. It is shown in figure 5.1 with a portion of its outer surface cut away. The substance of this body *(nucleoplasm)* is surrounded by a double-layered **nuclear envelope** which, unlike other membranes, is perforated by pores of significant size. These openings provide a viable passageway between the cytoplasm and nucleoplasm.

A cell that has been stained with certain dyes will exhibit darkly stained regions in the nucleus called *chromatin granules*. These granules are visible even with a light microscope. They represent highly condensed DNA molecules that comprise part of the chromosomes. Current theories suggest that increased condensed chromosomal material indicates a metabolically less active cell.

The most conspicuous substructure of the nucleus is the **nucleolus.** Although nucleoli of different cells vary considerably in number and structure, they all consist primarily of RNA, and function chiefly in the production of ribonucleoprotein for the ribosomes.

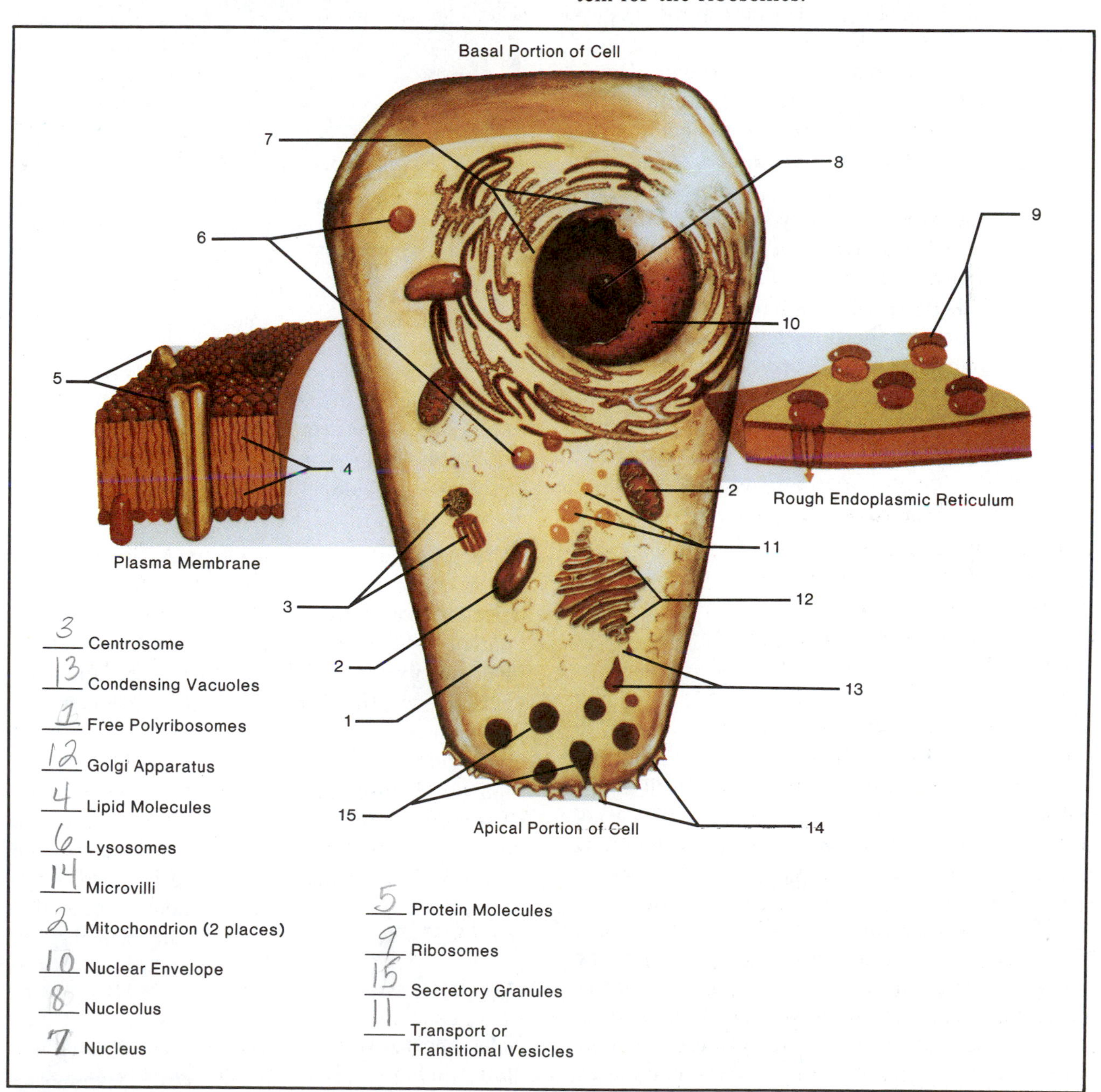

Figure 5.1 The microstructure of a cell.

Cytoplasmic Matrix

Between the plasma membrane and nucleus is an area called the *cytoplasmic matrix,* or *cytoplasm.* This region is a heterogenous aggregation of many components involved in cellular metabolism. The cytoplasm is structurally and functionally compartmentalized by the numerous organelles, many of which are hollow and enclosed by unit membranes. A description of each of these cytoplasmic organelles follows.

Endoplasmic Reticulum The most extensive structure within the cytoplasm is a complex system of tubules, vesicles, and sacs called the *endoplasmic reticulum (ER).* It is a double-layered unit membrane that is somewhat similar to the nuclear and plasma membranes; in some instances it is continuous with these membranes. It appears to function, in part, as a microcirculatory system for the cell, providing a passageway for intracellular transport of molecules.

Detailed studies of cells have shown that ER may have surfaces that are smooth or rough. ER that is designated as **rough endoplasmic reticulum (RER)** is a fluid-filled canalicular system studded with ribosomes. It is the type illustrated in the magnified view on the right side of figure 5.1. RER is believed to be an important site of protein synthesis and storage. It is more developed in cells that are primarily secretory in function. **Smooth endoplasmic reticulum (SER)** also consists of multilayered unit membrane compartments, but lacks ribosomes. It is thought that SER is active in various types of biosynthesis.

Ribosomes As stated above, ribosomes are small bodies attached to the surface of the RER. They are also scattered throughout the cytoplasm in rosettes and chains called **free polyribosomes.** Since they are only 170 Angstrom units (17 nanometers) in diameter they can be resolved only by electron microscopy.

The components for each ribosome originate in the nucleolus of the nucleus. They pass from the nucleolus through the pores of the nuclear membrane into the cytoplasm where they unite to form the completed ribosomal particle. Each ribosome consists of about 60% RNA and 40% protein. Their proposed structure is shown in figure 5.1.

Ribosomes are sites of protein synthesis and are particularly numerous in actively synthetic cells.

Figure 5.2 A computer-generated visualization of microtubules, vesicles, and mitochondria in a nerve cell. Vesicles and mitochondria are being propelled by the microtubules.

There is evidence to indicate that free polyribosomes are involved in synthesis of proteins for endogenous use; attached (RER) ribosomes are implicated in the synthesis of protein to be transported extracellularly (i.e., secretion).

Mitochondria The bean-shaped bodies with double-layered walls in figure 5.1 are *mitochondria.* These unit membrane structures are approximately 1 × 2–3 micrometers in size. Note that the inner membranes of these bodies have infoldings, called *cristae,* that serve to increase the inner surface area.

Mitochondria contain DNA and are able to replicate themselves. They also contain their own ribosomes. They are sometimes referred to as the "power plants" of the cell, since it is here that the energy-yielding reactions of oxidative respiration occur. Simply stated, these reactions oxidize glucose ($C_6H_{12}O_6$) to yield energy in the form of ATP:

$$C_6H_{12}O_6 + 6O_2 \rightarrow 6CO_2 + 6H_2O + ATP + Heat$$

Golgi Apparatus This unit membrane organelle is quite similar in basic structure to SER. In figure 5.1 it appears as a layered diamond-shaped stack of vesicles; in other cells it may appear differently. Although not shown in figure 5.1, it is connected to a portion of the RER.

Its primary role relates to the packaging, movement, and completed synthesis of products to be released by secretory cells. Proteins synthesized by the RER are pinched off into ovoid sacs, or **transitional vesicles** (label 11), which then coalesce with the cisternae of the Golgi apparatus proper. Here, further synthesis and completed processing of the secretory product take place. Completely processed materials accumulate at the apical end of the organelle and form **condensing vacuoles.** These membrane-bound bodies move to the apex of the cell where, as **secretory granules,** they are ultimately exocytosed. Figure 5.1 illustrates this process and how the granules fuse with the plasma membrane. The fate of the exocytosed product depends upon the cell type.

Lysosomes Oval sacs, such as the one in the upper left part of the cell in figure 5.1, contain digestive enzymes and are called *lysosomes.* These sacs consist of a single unit membrane and are believed to form by the pinching off of sacs from the Golgi apparatus. The contents of these organelles vary from cell to cell, but typically they include enzymes that hydrolyze proteins and nucleic acids.

The function of these organelles is not known for certain. It appears, however, that they act as disposal units of the cytoplasm, for they often contain fragments of mitochondria, ingested food particles, dead microorganisms, worn-out red blood cells, and any other debris that may have been taken into the cytoplasm. When the lysosome has performed its function, it is expelled from the cell through the plasma membrane.

In dying cells, the membrane of the lysosome disintegrates to release the enzymes into the cytoplasm. The hydrolytic action of the enzymes on the cell hastens the death of the cell. It is for this reason that these organelles have been called "suicide bags."

Centrosome At the far left of figure 5.1 are a pair of microtubule bundles that are arranged at right angles to each other. Each bundle, called a *centriole,* consists of nine triplets of microtubules arranged in a circle. The two centrioles are collectively referred to as the *centrosome.* Because of their small size, centrosomes appear as very small dots when observed with a light microscope. During mitosis and meiosis in animal cells centrioles play a role in the formation of spindle fibers that aid in the separation of chromatids. They also function in the formation of cilia and flagella in certain types of cells.

Cytoskeletal Elements

It now appears that a living cell has an extensive permeating network of fine fibers that contribute to cellular properties such as contractility, support, shape, and translocation of materials. Microtubules and microfilaments fall into this category.

Microtubules Molecules of protein *(tubulin),* arranged in submicroscopic cylindrical hollow bundles, the *microtubules,* are found dispersed through the cytoplasm of most cells. They converge especially around the centrioles to form aster fibers and the spindles that are seen during cell division (Exercise 6). They also play a significant role in the transport of vesicles from the Golgi apparatus to other parts of the cell.

Nerve cells that have long processes, called axons, utilize microtubules to transport vesicles and mitochondria. Movement of these organelles is from the cell body to the axon terminal, as well as from the terminal back to the cell body.

Figure 5.2 is a computer-generated image of how these microtubules might look where they enter an axon of a nerve cell. Note that various sized vesicles are propelled along the microtubules, and that these vesicles move along the microtubules in both directions, *simultaneously.* The energy for vesicle transport is supplied by mitochondria. A portion of one is seen in the lower right quadrant; another is shown in the upper left quadrant.

Microfilaments All cells show some degree of contractility. Cytoskeletal elements responsible for this phenomenon are the *microfilaments.* These filaments are composed of protein molecules arranged in solid parallel bundles. They are most highly developed in muscle cells that are adapted for contraction. In other types of cells, they are present as interwoven networks of the cytoplasm and are relatively inconspicuous even in electron photomicrographs. Besides aiding in contractility, they also appear to be involved in cell motility and support.

Cilia and Flagella

Hairlike appendages of cells that function in providing some type of movement are either *cilia* or *flagella.* While cilia are usually less than 20 micrometers long, flagella may be thousands of micrometers in length. Cells that line the respiratory tract and uterine tubes have cilia (see illustration D in figures HA–2 and HA–31 of the Histology Atlas). The only human cells with flagella are the spermatozoa. Both cilia and flagella originate from centrioles, which explains why they both have microtubular internal structure similar to the centrioles.

Microvilli

Certain cells, such as the one in figure 5.1, have tiny protuberances known as *microvilli* on the apical or free surfaces. They may appear as a fine **brush border** (illustration C, figure HA–2) or as **stereocilia** (illustration B, figure HA–34). Both of these may be mistaken for cilia with a light microscope.

Each microvillus is an extension of cytoplasm enclosed by the plasma membrane. Microvilli are very common on cells lining the intestine and kidney tubules where they function to increase surface area of the cells for absorption of water and nutrients. Microvilli are not motile.

Laboratory Assignment

After labeling figure 5.1 and answering the questions on the Laboratory Report pertaining to cell structure and function, do the following two cellular studies in the laboratory. The epithelial cell will be removed from the inner surface of the cheek and examined under high-dry magnification. The study of microorganisms (primarily protozoa) is performed here to observe the activity of living cells. Ciliary and flagellar action, amoeboid movement, secretion and excretion of cellular products, and cell division can be observed in viable cultures containing a mixture of protozoans.

Materials:

microscope slides
cover glasses
toothpicks
IKI solution
mixed cultures of protozoa
medicine dropper
microscope

Epithelial Cell Prepare a stained wet mount slide of some cells from the inside surface of your cheek as follows:

1. Wash a microscope slide and cover glass with soap and water.
2. Gently scrape some cells loose from the inside surface of your cheek with a clean toothpick. It is not necessary to draw blood.
3. Mix the cells in a drop of IKI solution on the slide.
4. Cover with a cover glass.
5. After locating a cell under low power of your microscope, make a careful examination of the cell under high-dry magnification. Identify the *nucleus, cytoplasm,* and *cell membrane.*
6. Draw a few cells on the Laboratory Report, labeling the above structures.

Microorganisms Prepare a wet mount of some protozoans by placing a drop of the culture on a slide and covering it with a cover glass. Be sure to insert the medicine dropper all the way to the bottom of the jar for your study sample since most protozoans will be found there.

Examine the slide first under low power, then under high-dry. Study individual cells carefully, looking for nuclei, cilia, flagella, vacuoles of ingested food, excretory vacuoles, etc. If you wish to identify the organisms being observed, consult Appendix E for representative types. Since algae are often present in protozoan cultures, there are also some illustrations of algae in the appendix.

Laboratory Report

Answer the questions on Laboratory Report 5,6 that pertain to this exercise.

Mitosis 6

Growth of body structures and the repair of specific tissues occur by cell division. While many cells in the body are able to divide, many do not beyond a certain age. Cell division is characterized by the equal division of all cellular components to form two daughter cells that are identical in genetic composition to the original cell. Cleavage of the cytoplasm is referred to as *cytokinesis;* nuclear division is *karyokinesis,* or *mitosis.* Cells that are not undergoing mitosis are said to be in a "resting stage," or *interphase.*

During this laboratory period we will study the various phases of mitosis on a microscope slide of hematoxylin-stained embryonic cells of *Ascaris,* a roundworm. The advantage of using *Ascaris* over other organisms is that it has so few chromosomes.

To facilitate clarity in understanding this process it has been necessary to combine photomicrographs and diagrams in figures 6.1 and 6.2. However, before we get into the mitotic stages, let's see what takes place in the interphase.

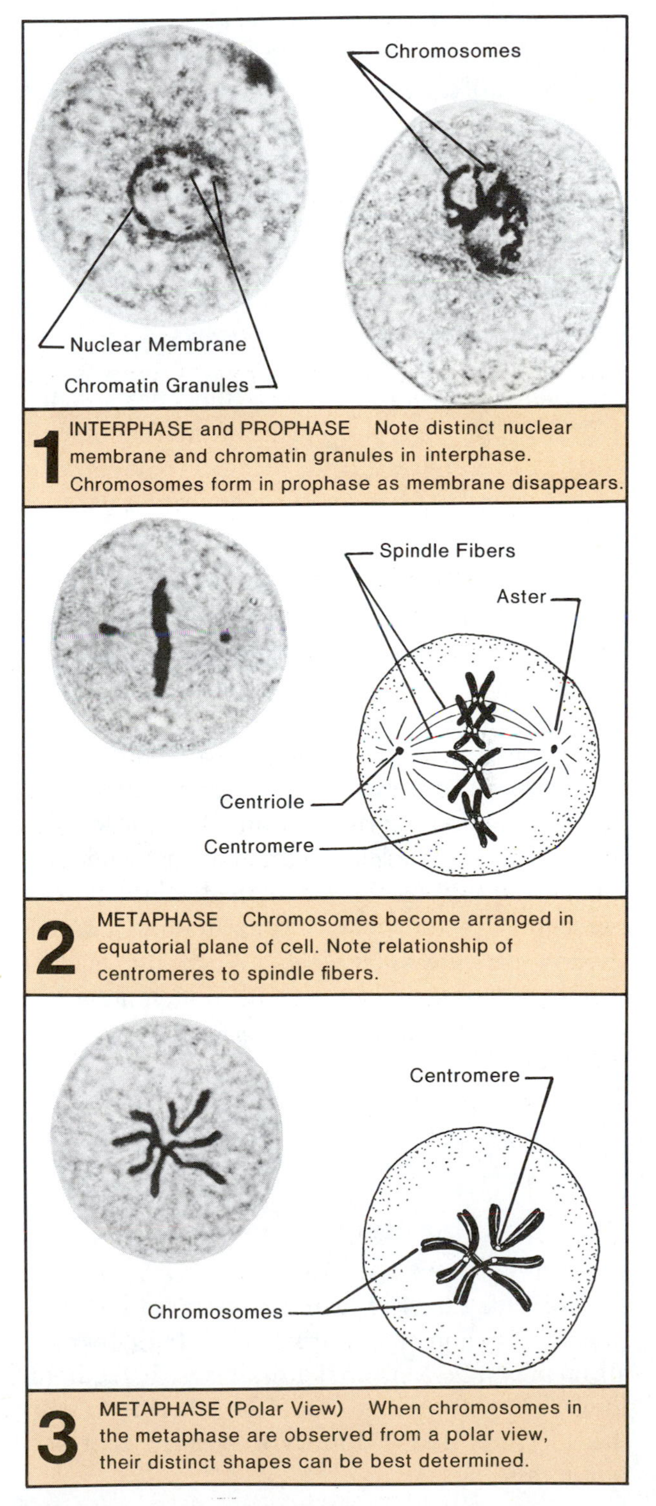

Figure 6.1 Early stages of mitosis.

The Interphase

Although a cell in the interphase stage doesn't appear to be active, it is carrying on the physiological activities that are characteristic of its particular specialization. The cell may be in one of two different phases: the G_1 phase or the S phase (G_1 for growth, S for synthesis). After a previous division, cells in the *S phase* begin to synthesize DNA, with each double-stranded DNA molecule replicating itself to produce another identical molecule. Most cells in the S phase will accomplish this replication within twenty hours and then divide. If division does not occur within this time frame, the cell is said to be in the *G phase* and will not divide again. Note in illustration 1, figure 6.1, that cells in the interphase stage exhibit a more or less translucent nucleus with an intact nuclear membrane.

Stages of Mitosis

The process of mitotic cell division is a continuous one once the cell has started to divide. However, for discussion purposes, the process has been divided into a series of four recognizable stages: prophase, metaphase, anaphase, and telophase. A description of each phase follows.

Prophase In this first stage of mitosis the nuclear membrane disappears, **chromosomes** become visible, and each **centriole** of the centrosome moves to opposite poles of the cell to become an **aster.** The dark-staining chromosomes are made up of DNA and protein. Each aster consists of a centriole and astral rays (microtubules). By the end of the prophase some of the astral rays from each aster extend across the entire cell to become the **spindle fibers.**

From the early prophase to the late prophase the chromosomes undergo thickening and shortening. During this time each chromosome has a double nature, consisting of a pair of **chromatids.** This dual structure is due to the replication of the DNA molecules during the interphase.

Metaphase During this stage the chromosomes migrate to the equatorial plane of the cell. Note in illustration 2, figure 6.1, that each chromosome has a **centromere** which is the point of contact between each pair of chromatids and a spindle fiber.

Anaphase As indicated in illustration 1 of figure 6.2, the individual chromatids of each chromosome are separated from each other and move apart along a spindle fiber to opposite poles of the cell.

Telophase This stage begins with the appearance of a **cleavage furrow.** It is much like the prophase in reverse in that the chromosomes become less distinct and a new nuclear membrane begins to form around each set of chromosomes. In the late telophase the spindle fibers disappear and the centrioles duplicate themselves. When the nuclear membranes and newly formed cell membrane are complete the two new structures are referred to as **daughter cells.**

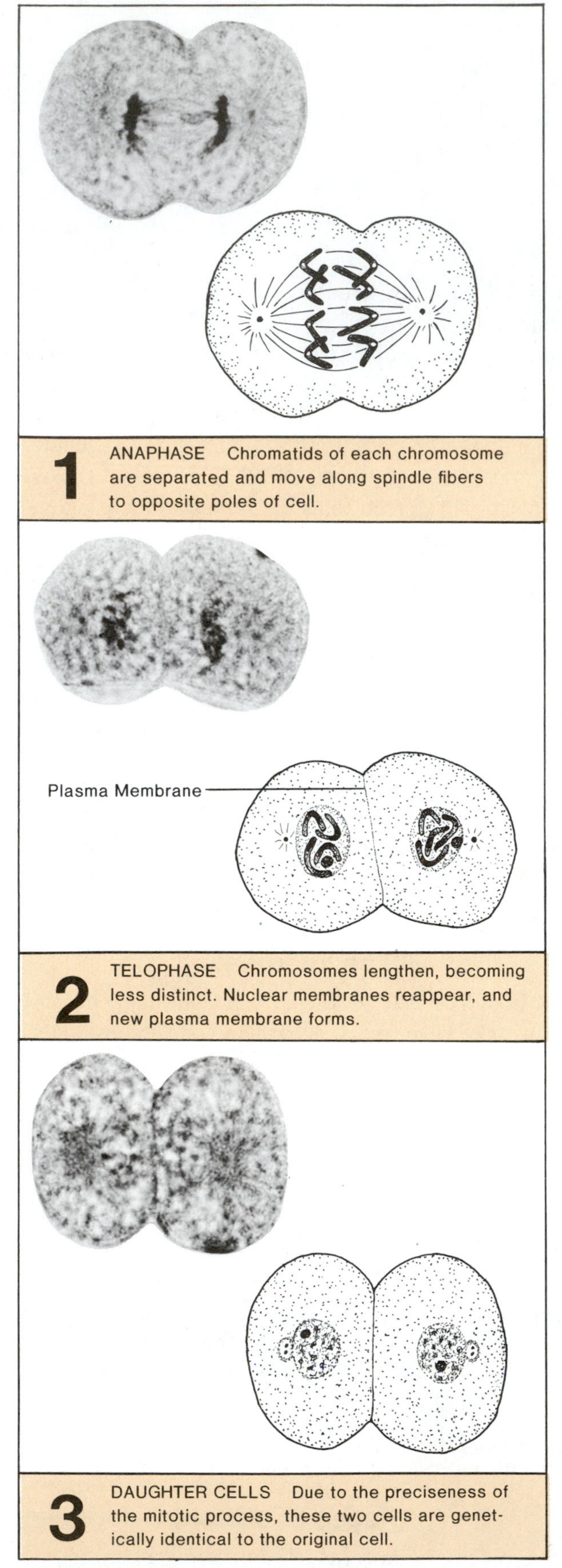

Figure 6.2 Later stages of mitosis.

Laboratory Assignment

Examine a prepared slide of *Ascaris* mitosis (Turtox slide E6.24). It will be necessary to use both the high-dry and oil immersion objectives. Identify all of the stages and make drawings, if required. Identifiable structures should be labeled. Answer the questions on Laboratory Report 5,6 that pertain to mitosis.

Osmosis and Cell Membrane Integrity 7

Every cell of the body is bathed in a watery fluid that contains a mixture of molecules that are essential to its survival. This fluid may be the plasma of blood or the tissue fluid in the interstitial spaces. In either case, these molecules, whether water, nutrients, gases, or ions, pass in and out of the cell through the plasma membrane. Some molecules, usually of small size, are able to diffuse passively through the cell membrane from areas of high concentration to low concentration. Organic molecules, such as glucose and amino acids, and certain ions move through the plasma membrane either with or against a concentration gradient by active transport, which requires energy expenditure by the cell. Movement of the molecules occurs if the membrane is viable and undisturbed.

The integrity of plasma membranes throughout the body is due largely to the fact that body fluids are isotonic. In this exercise a study will be made of diffusion and osmosis, and the effects of hypertonic and hypotonic solutions on the plasma membrane of red blood cells.

Demonstration Setups:

Brownian Movement (India Ink Preparation)
Osmosis (Thistle Tube Setup)

Molecular Movement

All molecules, whether in a gas or liquid state, are in constant motion. As molecules bump into each other, directions are changed, causing random dispersal. Although direct observation of molecular movement is impossible due to the invisibility of molecules to the naked eye, one can observe this activity indirectly by both microscopic and macroscopic techniques. Two such observations will be made here.

Brownian Movement

In 1827 a Scottish botanist, Robert Brown, observed that extremely small particles in suspension in the protoplasm of plant cells were in constant vibration. He erroneously assumed, at first, that this movement was a characteristic peculiar to living cells. Subsequent studies revealed, however, that this vibratory movement was caused by invisible water molecules bombarding the small visible particles. This type of movement became known as *Brownian movement*.

Demonstration

The simplest way to observe this phenomenon is to examine a wet mount slide of india ink under the oil immersion objective of a microscope. India ink is a colloidal suspension of carbon particles in water, alcohol, and acetone.

Examine the demonstration setup that has been provided in the laboratory. This type of movement is of particular interest to microbiologists, since it must be differentiated from true motility in bacteria. Nonmotile bacteria are small enough to be displaced in this manner by water molecules.

Diffusion

If a crystal of some soluble compound is placed on the bottom of a container of water, molecules will disperse from the crystal to all parts of the container. The molecules are said to have spread throughout the water by *diffusion*. Movement of the molecules away from the original site of high concentration to low concentration is due to the fact that the reduced number of obstructing molecules in the lower concentration area favors movement of molecules into that area. The rate of diffusion is variable and depends on ambient temperature and molecular size. Given sufficient time, diffusion eventually achieves even dispersal of all molecules throughout the container.

To observe this type of molecular activity, one can use crystals of colored compounds such as potassium permanganate (purple) and methylene blue. Figure 7.1 illustrates how to set up such a demonstration. Methylene blue has a molecular weight of 320. Potassium permanganate has a molecular weight of 158. A Petri plate of 1.5% agar-agar will be used. This agar medium is 98.5% water and allows freedom of movement for molecules. Prepare such a plate as follows and record your observations on the Laboratory Report.

Materials:

crystals of potassium permanganate and methylene blue
Petri plate with about 12 ml of 1.5% agar-agar

1. Select one crystal of each of the two different chemicals and place them on the surface of the agar medium about 5 cm apart. Try to select crystals of similar size.
2. Every 15 minutes, for one hour, examine the plate to measure the extent of diffusion.
3. Record the measurements on the Laboratory Report.

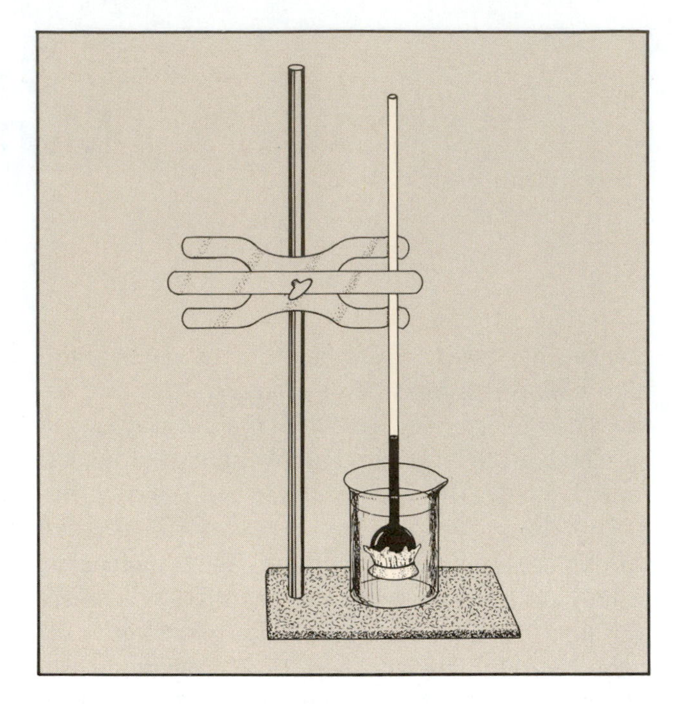

Figure 7.2 Osmosis setup.

Osmotic Effects

Water molecules, because of their small size, move freely through cell membranes into and out of the cell. Water acts as a vehicle which enables ions and other molecules to move through the membrane. This movement of water molecules through a semipermeable membrane is called *osmosis.*

Although water molecules are always moving both ways through the membrane, the predominant direction of flow is determined by the concentrations of solutes on each side of the membrane. If a funnel containing a sugar solution is immersed in a beaker of water and separated from the water with a semipermeable membrane, as in figure 7.2, more water molecules will enter the funnel through the membrane than will leave the sugar solution. As was observed in the diffusion experiment, molecules tend to move from high concentration to low concentration areas. In this case, the water molecules are of greater concentration in the beaker than in the sugar solution. The inward flow of water molecules produces an upward movement of the sugar solution. The force that would be required to restrain this upward flow is called the *effective osmotic pressure.*

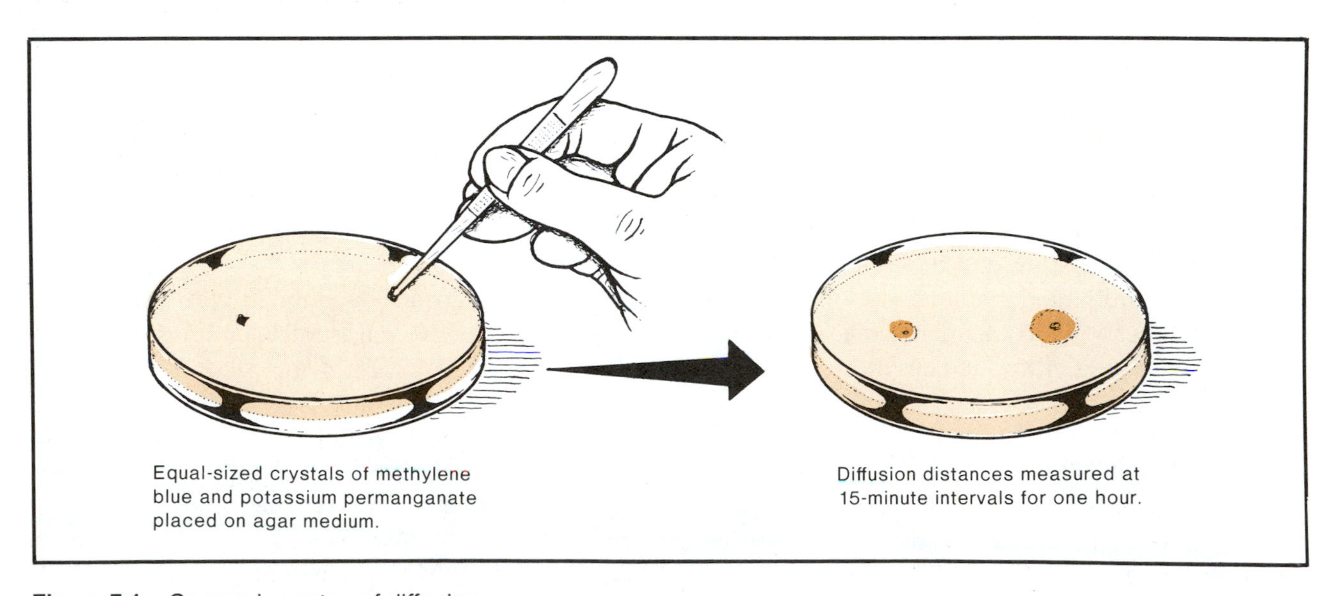

Figure 7.1 Comparing rates of diffusion.

Solutions that contain the same concentration of solutes as cells are said to be **isotonic solutions.** Blood cells immersed in such solutions gain and lose water molecules at the same rate, establishing an *osmotic equilibrium.* See figure 7.3.

If a solution contains a higher concentration of solutes than is present in cells, water leaves the cells faster than it enters, causing the cells to shrink and develop cupped (crenated) edges. This shrinkage is called *plasmolysis,* or *crenation.* Such solutions of high solute concentration are called **hypertonic solutions** and are said to have a higher osmotic potential than isotonic solutions.

A solution that has a lower solute concentration than is present in cells is said to be a **hypotonic solution.** In such solutions water flows rapidly into the cells, causing them to swell *(plasmoptysis)* and burst *(lyse)* as the plasma membrane disintegrates. Lysis of red blood cells is called *hemolysis.*

To observe the effects of the various types of solutions on red blood cells, we will follow the procedures outlined in figure 7.4. Blood cells will be added to various concentrations of solutions. The effects of the solutions on the cells will be determined macroscopically and microscopically. Proceed as follows:

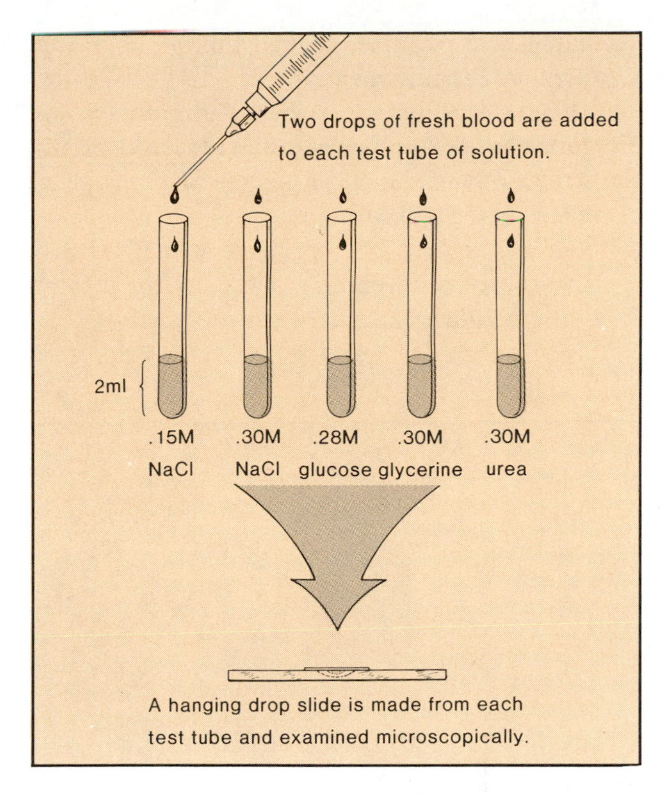

Figure 7.4 Routine for making cell suspensions.

Materials:

5 serological test tubes and test tube rack
small beaker of distilled water (50 ml size)
2 depression microscope slides, cover glasses
wax pencil, Vaseline, and toothpicks
1 serological pipette (5 ml size)
cannister for used pipettes
mechanical pipetting device
syringes, needles, Pasteur pipettes
fresh blood
solutions of .15M NaCl, .30M NaCl, .28M glucose, .30M glycerine, and .30M urea

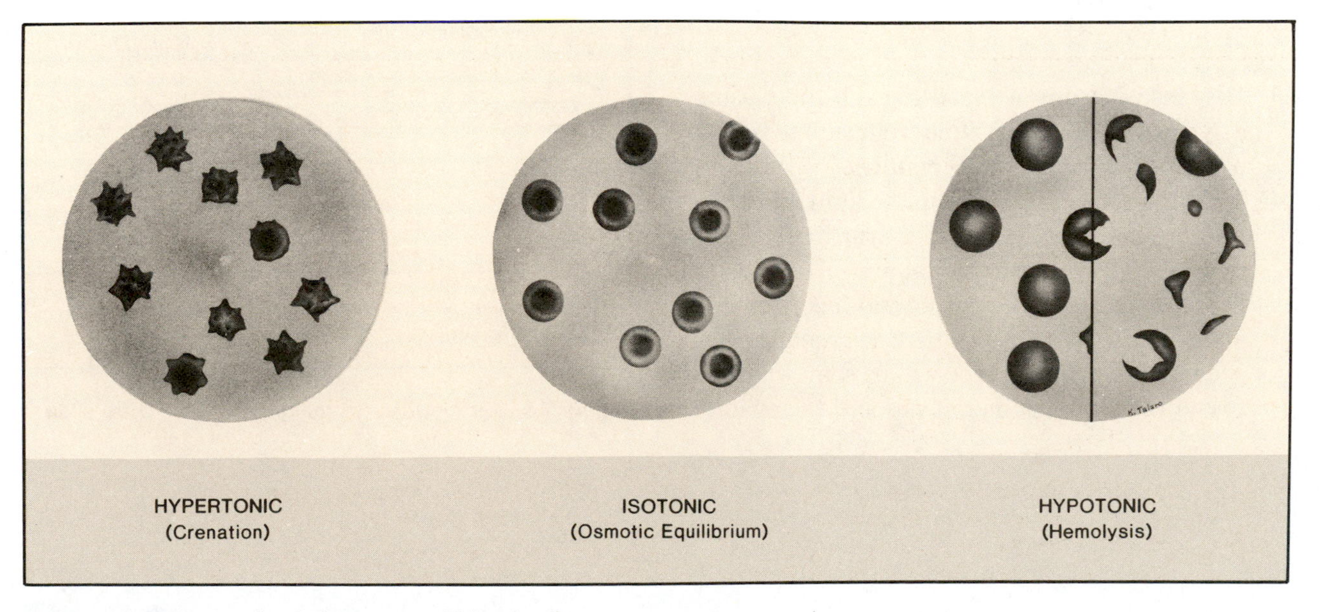

Figure 7.3 Effects of solutions on red blood cells.

1. Label five clean serological tubes **1** to **5** and arrange them sequentially in a test tube rack.
2. With a 5 ml pipette, deliver 2 ml of each solution to the appropriate tube (**1:** .15M NaCl, **2:** .30M NaCl, **3:** .28M glucose, **4:** .30M glycerine, and **5:** .30M urea).

   See figure 7.5 for delivery method. To insure purity of solutions, rinse out the pipette with distilled water between each delivery.

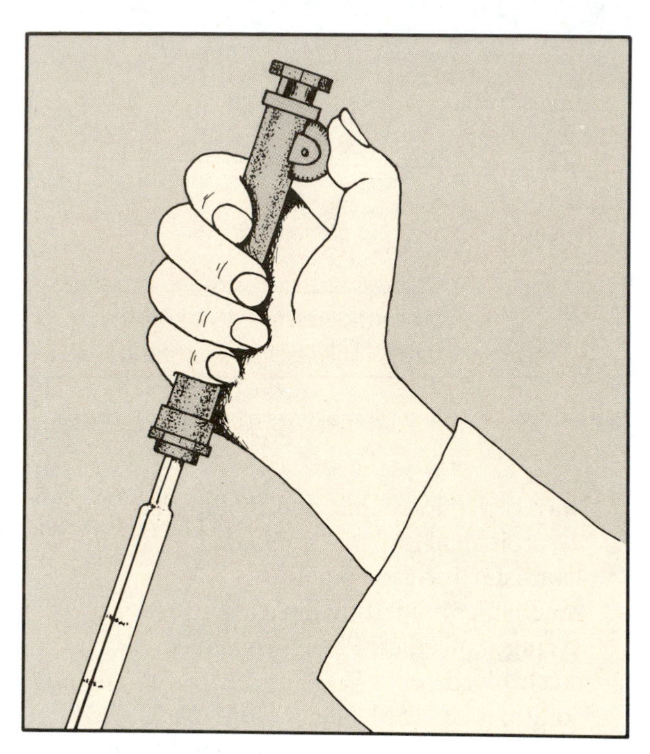

Figure 7.5 Mechanical pipetting device works best when controlled with thumb.

3. Dispense 2 drops of blood to each tube, using a syringe. Shake each tube from side to side to mix, then let stand for **5 minutes.**
4. Hold the rack of tubes up to the light and compare them. If the solution is transparent, **hemolysis** has occurred. If you are unable to see through the tube, **no hemolysis** has occurred and the cells should be intact. If crenation has occurred, the appearance will be somewhat between the clarity of hemolysis and the opacity of osmotic equilibrium. Record your results on the Laboratory Report.
5. Make a hanging drop slide from each tube as follows:
   a. Place a tiny speck of Vaseline near each corner of a cover glass. See figure 7.6.
   b. With a Pasteur pipette transfer a drop of the cell suspension to center of cover glass.
   c. Place depression slide on cover glass and quickly invert.
6. Examine each slide under high-dry and record your results on the Laboratory Report.

Laboratory Report

Complete the Laboratory Report for this exercise.

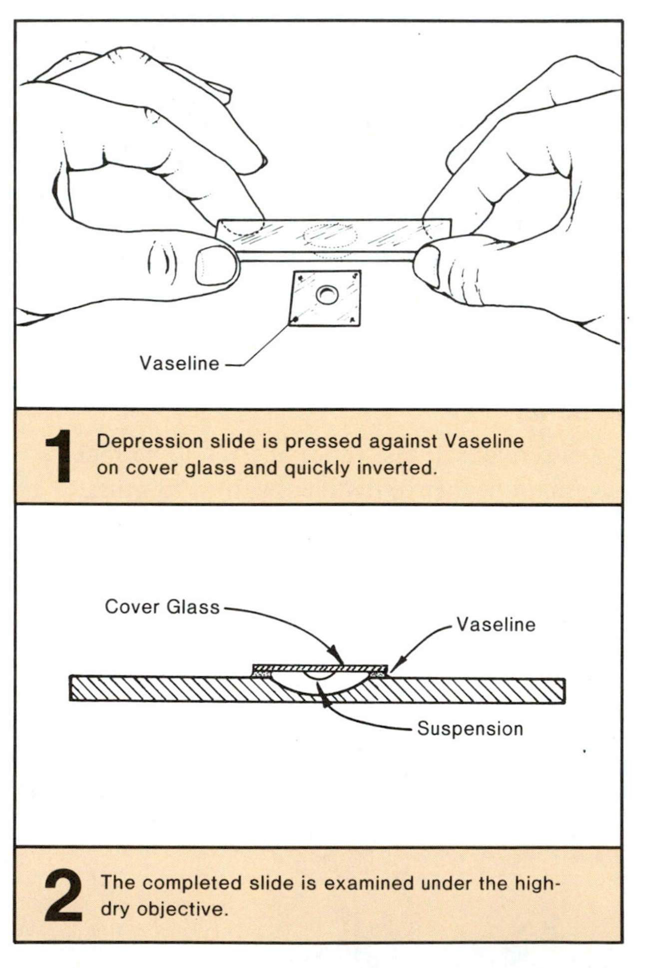

Figure 7.6 Procedure for making hanging drop slides.

Epithelial Tissues 8

Although all cells of the body share common structures such as nuclei, centrosomes, and Golgi apparatus, they differ considerably in size, shape, and structure according to their specialized functions. An aggregate of cells that are similar in structure and function is called a *tissue.* The science which relates to the study of tissues is called *histology.*

In this exercise we will study the different types of epithelial tissues. Prepared microscope slides from portions of different organs will be available for study. Learning how to identify specific epithelial tissues on slides that have several types of tissue will be part of the challenge of this exercise.

Epithelial tissues are aggregations of cells that perform specific protective, absorptive, secretory, transport, and excretory functions. They often serve as coverings for internal and external surfaces and they rest upon a bed of connective tissue. Characteristics common to all epithelial tissues are as follows:

- The individual cells are closely attached to each other at their margins to form tight sheets of cells lacking in extracellular matrix and vascularization.
- The cell groupings are oriented in such a way that they have an apical (free) surface and a basal (bound) region. The basal portion is closely anchored to underlying connective tissue. This thin adhesive margin between the epithelial cells and connective tissue is called the **basement lamina.** Although this structure was formerly referred to as the "basement membrane," it is not a true membrane. Unlike true membranes the basement lamina is acellular. In reality, it is a colloidal complex of protein, polysaccharide, and reticular fibers.

Differentiation of the various epithelia is illustrated in the separation outline, figure 8.1. The basic criteria for assigning categories are cell shape, surface specializations, and layer complexity. The three divisions *(simple, pseudostratified,* and *stratified)* are based on layer complexity. A discussion of each follows.

Simple Epithelia

Epithelial tissues that fall in this category are composed of a single cell layer that extends from the basement lamina to the free surface. Figure 8.2 illustrates three basic kinds of simple epithelia.

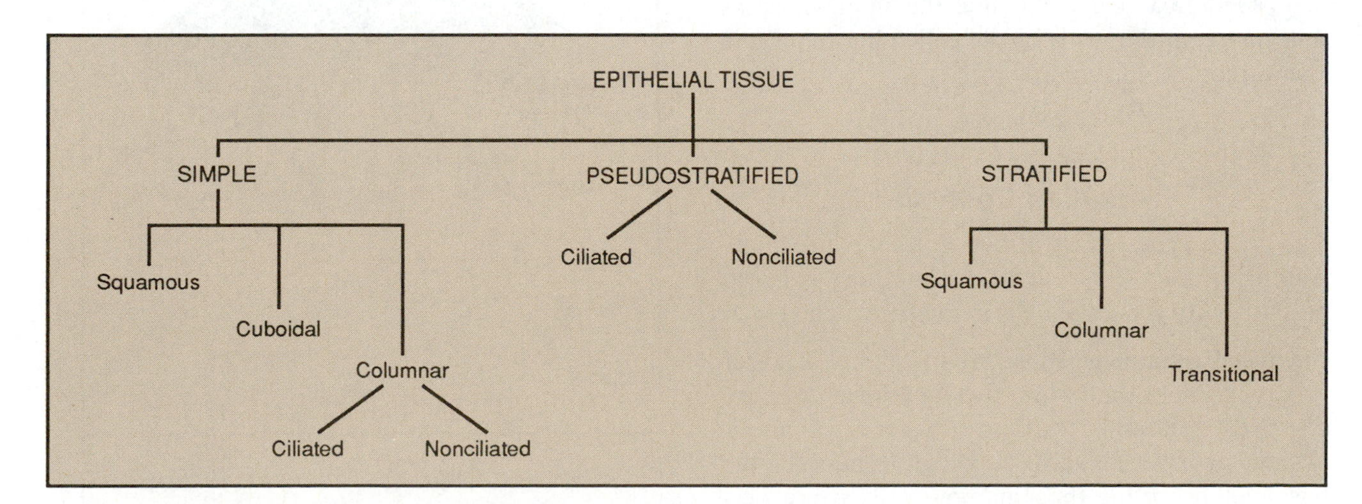

Figure 8.1 A morphologic classification of epithelial types.

Simple Squamous Epithelia These cells are very thin, flat, and irregular in outline. An example of this type is seen in illustration 1, figure 8.2. They form a pavementlike sheet in various organs that perform filtering or exchange functions. Capillary walls, alveolar walls in the lungs, the peritoneum, the pleurae, and blood vessel linings consist of simple squamous epithelia.

Cuboidal Epithelia Cells of this type are stout and blocklike in cross section, and hexagonal from a surface view (illustration 2, figure 8.2). Several glands (thyroid, salivary, pancreas), the ovary, and the lens of the eye (its capsule) contain cuboidal tissue.

Simple Columnar Epithelia Illustration 3, figure 8.2, reveals that cells of this type are elongated between their apical and basal surfaces; on their free surfaces they are polygonal. They may be functionally specialized for protection, secretion, or absorption. Specialized secretory cells in columnar epithelia, called **goblet cells,** produce the protective glycoprotein, **mucus.**

The linings of the stomach, intestines, and kidney collecting tubules consist of *plain* (nonciliated) columnar epithelial tissue. *Ciliated* columnar cells are seen lining the respiratory tract, uterine tubes, and portions of the uterus. Goblet cells may be present in both plain and ciliated columnar epithelia.

Another modification of the free surfaces of columnar cells is the presence of **microvilli.** These structures are seen as a **brush border** when observed with a light microscope under oil immersion. Columnar cells that line the small intestine and make up the walls of the collecting tubules in the kidney exhibit microvilli.

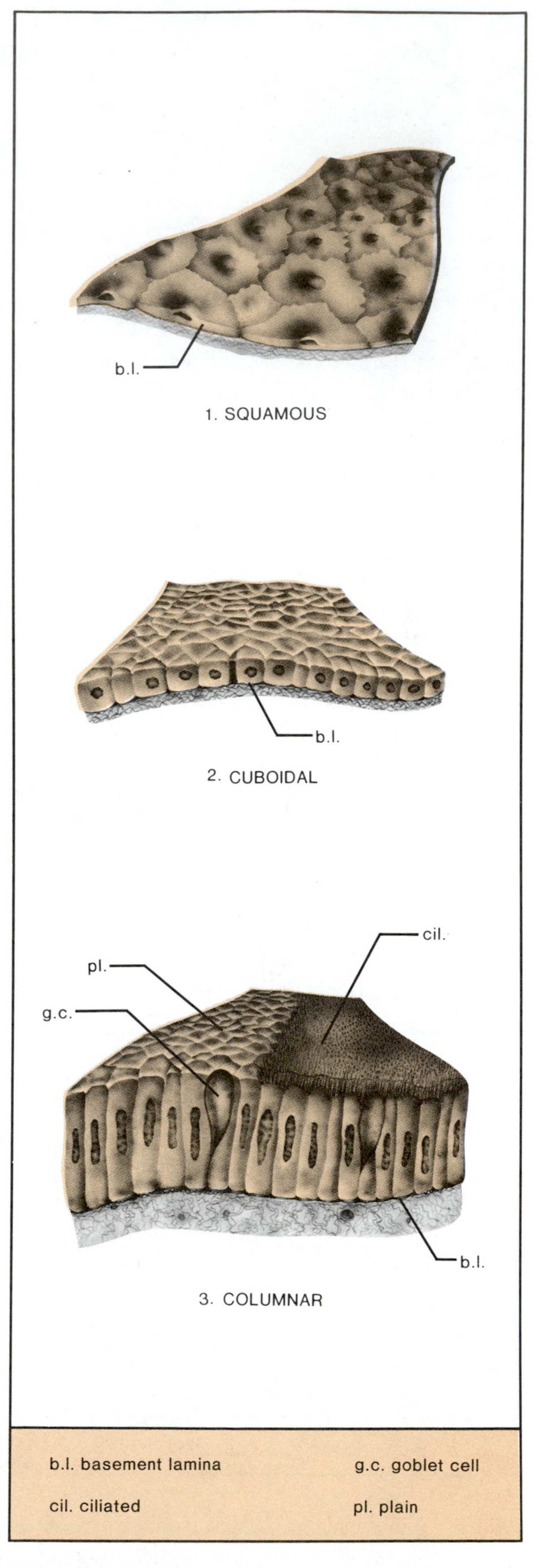

Figure 8.2 Simple epithelia.

Stratified Epithelia

The three types of stratified epithelia are illustrated in figure 8.3. Categorization here is based on shape differences of cells in the superficial layer.

Stratified Squamous Illustration 1, figure 8.3, is of this type. Note that while the superficial cells are distinctly squamous, the deepest layer is columnar; in some cases this layer is cuboidal. In between the basal cell layer and the squamous cells are successive layers of irregular and polyhedral cells.

Protection is the chief function of this type of tissue. Exposed inner and outer surfaces of the body, such as the skin, oral cavity, esophagus, vagina, and cornea, consist of stratified squamous epithelia.

Stratified Columnar This type of epithelium is shown in illustration 2, figure 8.3. Note that although the superficial cells are distinctly columnar, the deeper cells are irregular or polyhedral. Note also that the superficial columnar cells are variable in height.

Protection and secretion are the chief functions of this type of tissue. Distribution of stratified columnar is limited to some glands, the conjunctiva, the pharynx, a portion of the urethra, and the anus.

Transitional A unique characteristic of this type of stratified epithelium is that the surface layer consists of large, round, dome-shaped cells which may be binucleate. The deeper strata are cuboidal, columnar, and polyhedral.

Note in illustration 3, figure 8.3, that the deeper cells are not as closely packed as other stratified epithelia. Distinct spaces can be seen between the cells. This looseness of cells imparts a certain degree of elasticity to the tissue. Organs such as the urinary bladder, ureters, and kidneys (the calyces) contain transitional epithelium that enables distension due to urine accumulation.

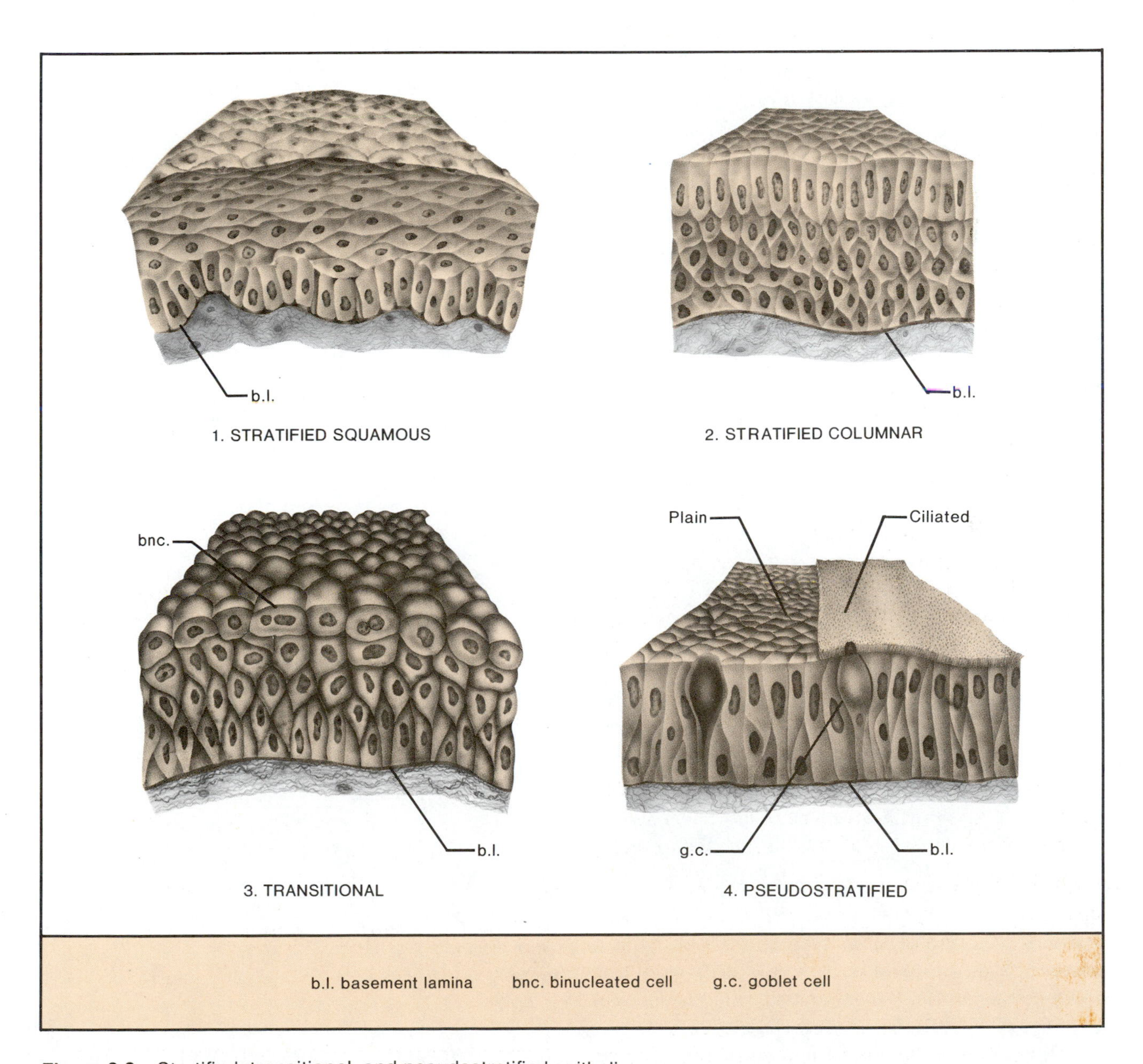

Figure 8.3 Stratified, transitional, and pseudostratified epithelia.

Pseudostratified Epithelia

This epithelium presents a superficial stratified appearance because of the staggered nuclei as well as two different cell orientations. It is only one layer thick, however. Close examination will reveal that every cell is in contact with the basal lamina, but only the columnar types extend to the free surface. The smaller cells wedged between them have no free surface.

Two types are shown in illustration 4 of figure 8.3: plain and ciliated. The *plain,* or nonciliated, is found in the male urethra and parotid gland. The *ciliated* type lines the trachea, bronchi, auditory tube, and part of the middle ear. Since both types frequently produce mucus, they possess goblet cells.

Laboratory Assignment

Do a systematic study of each kind of epithelial tissue by examining prepared slides that should show the kinds of tissue you are looking for. Remembering that epithelial cells have a free surface and a basement lamina, always look for the tissue near the edge of a structure.

Note in the materials list below that *preferred* and *optional* lists of slides are given. The preferred list is of slides that are limited to the type of tissue being studied. If these are available, use them. If they are not available, use the optional slides, which are somewhat more difficult to use. The numbers listed in parentheses after each tissue are Turtox code numbers of slides that should reveal structures illustrated in the Histology Atlas.

Materials:

preferred slides: squamous (H1.1), stratified squamous (H1.14), cuboidal (H1.21), simple columnar (H1.31), pseudostratified ciliated (H1.32), and transitional (H1.41)

optional slides: skin (H11.12 or H11.14), trachea (H6.41), stomach (H5.415 or H5.425), ileum (H5.531), kidney (H9.11 or H9.15), and thyroid gland (H14.11, H14.12, or H14.13)

Squamous Epithelium Look for the flattened cells that are representative of this kind of tissue. Use figure HA–1 in the Histology Atlas for reference. The exfoliated cells are the same ones you studied in Exercise 5. Explore each slide first with the low-power objective before using the high-dry or oil immersion objectives. Make drawings if required.

Columnar Epithelium Consult figure HA–2 in the Histology Atlas for representatives of this group of epithelial cells. To see the **brush border** or **cilia** on these cells it will be necessary to study the cells with the high-dry or oil immersion objectives. Can you identify the **basement lamina** on each tissue?

Cuboidal Epithelium Illustrations A and B, figure HA–3, reveal the appearance of cuboidal tissue. If Turtox slide H1.21 is unavailable, use a slide of the thyroid gland for this type of tissue. Turtox slide H1.21 is usually made from the uterine lining of a pregnant guinea pig.

Transitional Tissue Illustrations C and D of figure HA–3 and illustration C, figure HA–30, provide good examples of this type of tissue. Note that the cells seem to be loosely arranged. Look for **binucleate cells.**

Ciliated Pseudostratified Columnar Epithelium The best place to look for this type of tissue is in a cross section of the trachea. Illustration D, figure HA–2, is the epithelium of the trachea.

Laboratory Report

Answer the questions on Laboratory Report 8,9 that pertain to the epithelial tissues.

9 Connective Tissues

Connective tissues include those tissues which perform binding, support, transport, and nutritive functions for organs and organ systems. Characteristically, all connective tissues have considerable amounts of nonliving extracellular substance that holds and surrounds various specialized cells.

The extracellular material, or **matrix,** is composed of fibers, fluid, organic ground substance, and/or inorganic components intimately associated with the cells. This matrix is a product of the cells in the tissue. The relative proportion of cells to matrix will vary from one tissue to another.

Several systems of classification of connective tissues have been proposed. Figure 9.1 reveals the system that we will use here. It is based on the nature of the extracellular material; in some cases there is a certain amount of overlapping of categories.

Connective Tissue Proper

Histologically, *connective tissue proper* is composed primarily of protein fibers, special cells, and a ground substance that varies among the several types. The fibers may differ in protein composition and density. From the standpoint of composition, fibers consist of either collagenous or elastin protein. There are three basic types of fibers: collagenous, reticular, and elastic. **Collagenous,** or *white fibers,* are relatively long, thick bundles seen in most ordinary connective tissues in varying amounts. **Reticular fibers** constitute minute networks of very fine threads. Although both collagenous and reticular fibers are composed of collagen, the reticular fibers stain more readily with silver dyes *(argyrophilic).* **Elastic fibers,** also called *yellow fibers,* are often found in the connective tissue stroma of organs that must yield to shape changes. These are the only fibers to contain elastin protein. All three types of fibers are seen in areolar tissue (figure 9.2).

The **ground substance** usually consists of complex peptidoglycans which form an amorphous solution or gel around the cells and fibers. Chondroitin sulfate and hyaluronic acid are frequent components.

Cells or formed elements of connective tissue proper are of many types; examples are fibroblasts, adipose cells, mast cells, macrophages, and other blood cells, both fixed and wandering.

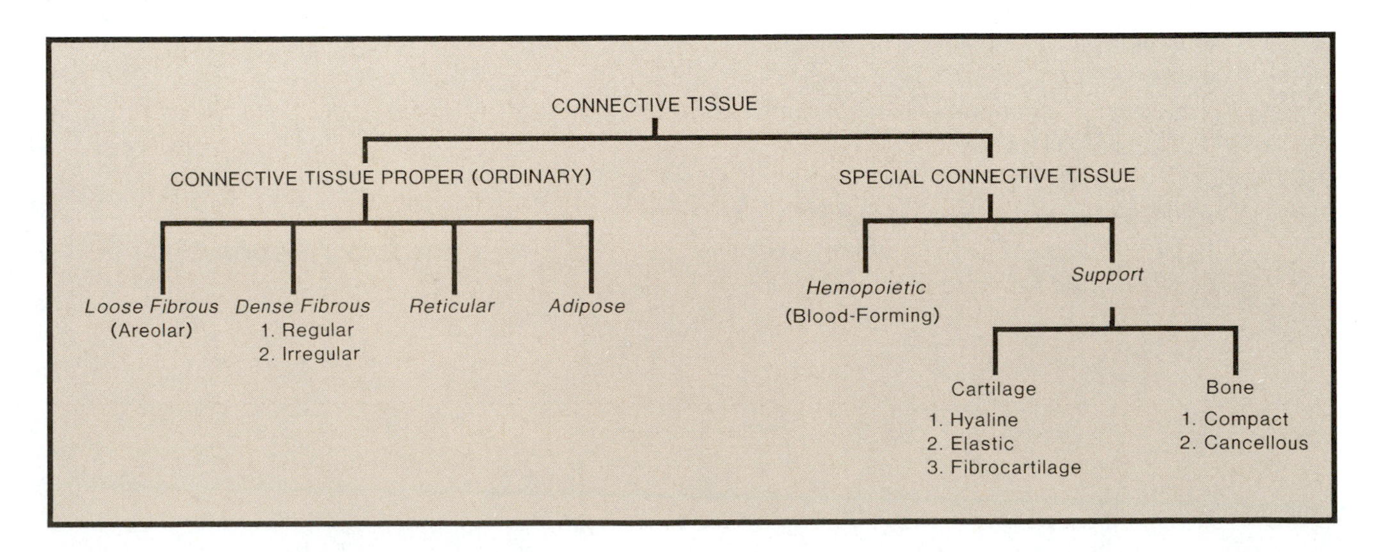

Figure 9.1 Types of connective tissue.

Loose Fibrous *(Areolar)* Connective Tissue

This is a widespread tissue that is interwoven into the stroma of many organs. The cells and extracellular substances of this tissue are very loosely organized. Because of the preponderance of spaces in this tissue, it is designated as *areolar* (a small space). It is highly flexible and capable of distension when excess extracellular fluid is present. The left-hand illustration in figure 9.2 depicts its generalized structure. Note the presence of **mast cells, macrophages,** and **fibroblasts.**

Three important functions are served by loose connective tissue: (1) it provides flexible support and a continuous network within organs, (2) it furnishes nutrition to cells in adjacent areas due to its capillary network, and (3) it provides an arena for activities of the immune system. It may be found beneath epithelia, around and within muscles and nerves, and as part of the serous membranes. The deep and superficial fasciae encountered in the rat dissection are of this type.

Adipose Tissue

Fat or adipose cells can be found in small groupings throughout the body; however, as they constitute a storage depot for fat, they frequently accumulate in large areas to make up the bulk of body fat. These latter areas are made up, essentially, of *adipose tissue.* There are various types of adipose tissue, but ordinary adipose is that most commonly found in humans. See figure 9.2.

Fat cells are very large and characterized by a spherical or polygonal shape; as much as 95% of their mass may be stored fat. As the process of fat deposition ensues, the cell cytoplasm becomes reduced and thin, and the vacuole, filled with lipids, appears as a large open space with a thin periphery of cytoplasm. The nucleus is displaced to one side, producing a signet ring appearance. A fine network of reticular fibers exists between the cells.

Fat tissue serves as a protective cushion and insulation for the body, in addition to being a potential source of energy and heat generator. Its rich vascular supply points to its relatively high rate of metabolism and turnover.

Reticular Tissue

This tissue (figure 9.3) is generally regarded as a network of reticular fiber elements within certain organs. As the fibers show a unique pattern and staining reaction, they can be identified as the supporting framework of many vascular organs such

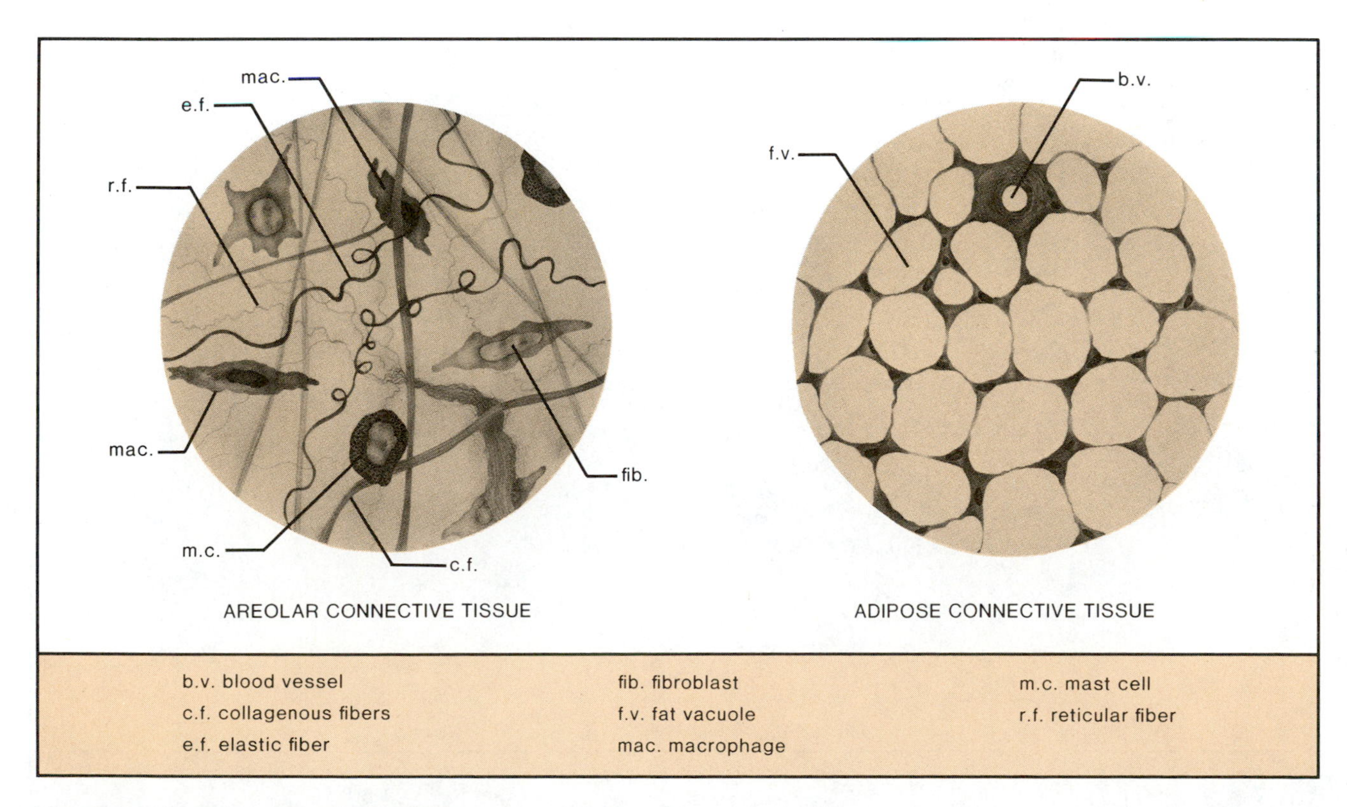

Figure 9.2 Areolar and adipose tissues.

as the liver, lymphatic structures, hemopoietic tissue, and basement laminas. The left-hand illustration in figure 9.3 reveals the appearance of the reticulum of a lymph node.

Dense Fibrous *(White)* Connective Tissue

Tissue in this category differs from the loose variety in having a predominance of fibrous elements and a sparseness of cells, ground substance, and capillaries. According to the arrangement of fibers this tissue can be separated into two groups: regular dense tissue and irregular dense tissue. The right-hand illustration in figure 9.3 reveals both types.

Dense regular connective tissue is the type found in tendons, ligaments, fasciae, and aponeuroses. It consists of thick groupings of longitudinally organized collagenous fibers, and some elastic fibers. These tough strands have enormous tensile strength and are capable of withstanding strong pulling forces without stretching. Fibroblasts are the principal cells, although they are quite scarce.

Dense irregular connective tissue has the same kind of fibers that are woven into flat sheets to form capsules around certain organs. The sheaths that encircle nerves and tendons are of this type. A large portion of the dermis also consists of the dense irregular type.

Supporting Connective Tissues

As indicated in figure 9.1, the supporting connective tissues fall in the category of special connective tissues which include hemopoietic, cartilage, and bone tissues. Since our only concern with hemopoietic tissue will be with the formed elements in blood (Exercise 38), we will devote the remainder of this exercise to cartilage and bone tissues.

Cartilage and bone are adapted for the bearing of weight. Structural characteristics shared by these tissues include (1) a solid, flexible, yet strong extracellular matrix, (2) cells borne in matrix cavities called **lacunae,** and (3) an external covering capable of generating new tissue.

Cartilage

In general, cartilage consists of a stiff, plastic matrix that has lubricating as well as weight-bearing capability. As a result, it is found in areas requiring support and movement (skeleton and joints).

Basically, all three types of cartilage in man consist of cartilage cells, or **chondrocytes,** embedded within a matrix that contains ground substance and fibers. The proportion devoted to matrix

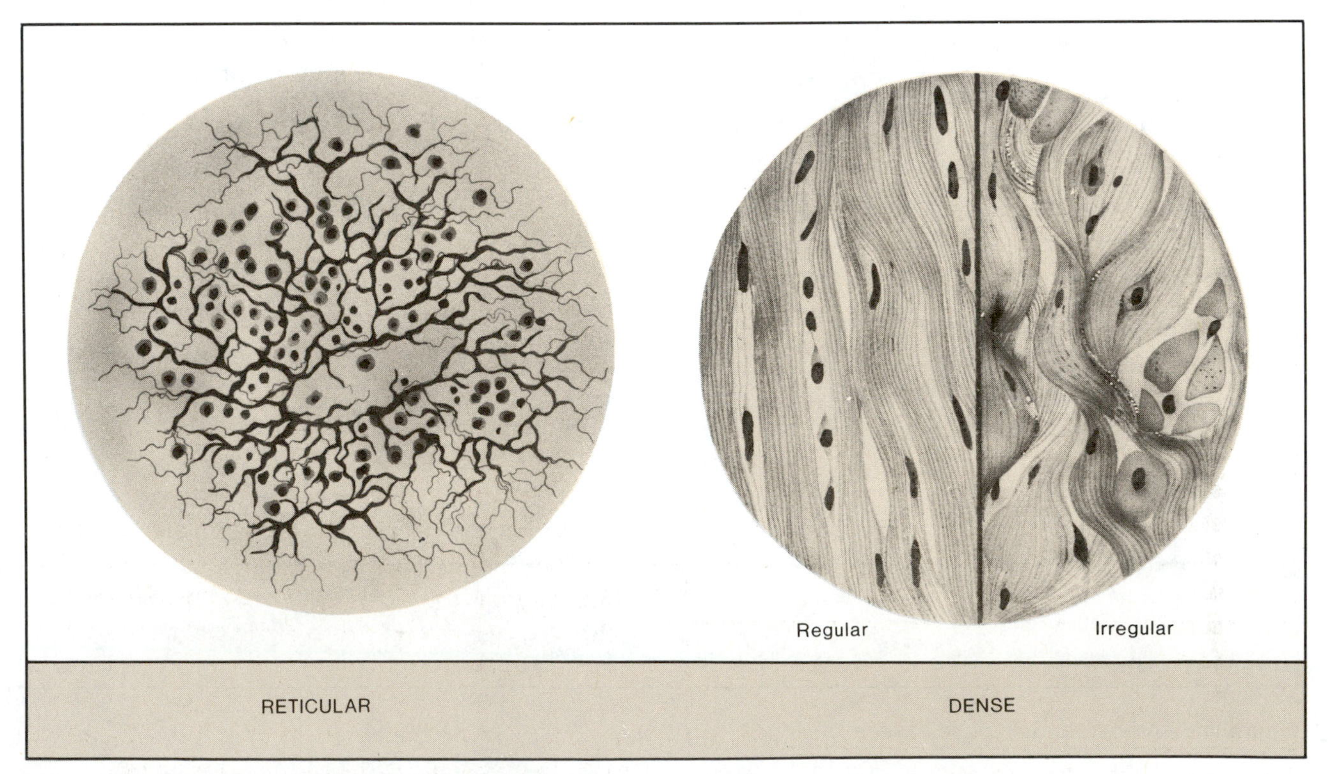

Figure 9.3 Reticular and dense connective tissues.

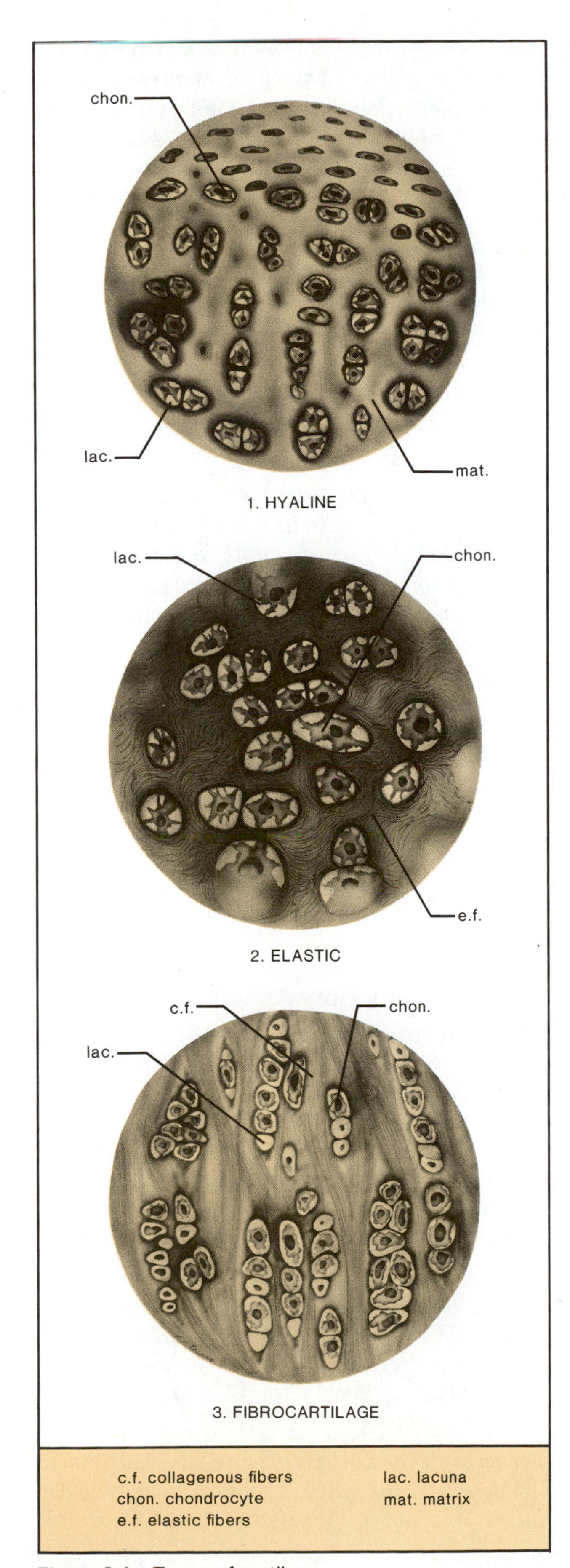

Figure 9.4 Types of cartilage.

is much greater than that for chondrocytes. Unlike other connective tissues, cartilage is devoid of a vascular supply and receives all nutrients through diffusion. The three classes of cartilage are shown in figure 9.4.

Hyaline cartilage (illustration 1) is a pearly white, glasslike tissue that makes up a very large part of the fetal skeleton which is gradually replaced by bone, except for areas in the joints, ear, larynx, trachea, and ribs. The matrix is a firm, homogenous gel made up of chondroitin sulfate and collagen. Cells occur singly or in "nests" of several cells, the sides of which may be flattened. Often the areas directly around lacunae appear denser in sections.

Elastic cartilage (illustration 2) is similar in basic structure to hyaline cartilage, but differs in that it contains significant amounts of elastic fibers in its matrix. The result is a tissue that has highly developed flexibility and elasticity. The external ear (pinna), epiglottis, and the auditory tube are reinforced with this type of cartilage.

Fibrocartilage (illustration 3) differs from the other two types in that its chondrocytes are arranged in groupings between bundles of collagenous fibers. It serves a useful cushioning function in strategic joint ligaments and tendons. It is the major component, for instance, of the intervertebral disks of the vertebral column and of the symphysis pubis.

Bone Tissue

The tissue that makes up the bones of the skeleton meets all the criteria of connective tissue, yet it has many striking and unique features. It is hard, unyielding, very strong, and light. It is found in those parts of the anatomy that require weight-bearing, protection, and storage capacity.

In a sense, bone can be visualized as a living organic cement. It consists of specialized cells, blood vessels, and nerves that are reinforced within a hard ground substance made up of organic secretions impregnated with mineral salts. It arises during embryonic and fetal development from cartilage and/or fibrous connective tissue precursors. The initial organic matrix consists of collagen fibers that serve as a framework for the gradual deposition of calcium and phosphate salts by special bone cells, the *osteoblasts*. Far from being an inactive tissue, it is continuously being modified and reconstructed by both metabolic and external influences.

In macroscopic sections of bones, two frameworks are apparent: the solid, dense **compact bone,** which makes up the outermost layer, and the more porous **spongy,** or *cancellous,* **bone,** which is located internally. Each of these variations has a distinctive histological character.

Compact Bone

Upon close inspection of a cut section of the sternum viewed from a three-dimensional perspective (figure 9.5 in illustration A), it is apparent that compact bone (label 2) is permeated by a microscopic framework of tunnels, channels, and interconnecting networks that are surrounded by a hard matrix. Within this hollow network exist the living substances of bone that facilitate its nourishment and maintenance.

The functional and structural unit of compact bone is a cylindrical component called the **osteon** or Haversian system (label 9, illustration B). In the center of each osteon is a hollow space, the **central** (Haversian) **canal,** which contains one or two blood capillaries. Surrounding each central canal are several concentric rings of matrix called the **lamellae. Osteocytes,** which are responsible for secreting the lamellae, can be seen within small hollow cavities, the **lacunae.** Note that these cavities, which are oriented between the lamellae, have many minute hollow tunnels, called **canaliculi,** that radiate outward, imparting a spiderlike appearance. The canaliculi contain protoplasmic processes of the osteocytes.

If one follows the various levels of structure shown in figure 9.5, it should become evident that the network comprises a continuous communication system from the central canal to the lacunae and between adjacent lacunae via the canaliculi. Thus, the entrapped osteocytes can receive nourishment and exchange materials within the hard space of the matrix. Intimate contact between the protoplasmic processes of adjacent osteocytes through the canaliculi makes all of this possible.

Groups of osteons lie in vertical array with adjacent lamellae separated at lines of demarcation called **cement lines.** Continuity between the central canals of adjacent osteons is achieved by **perforating** (Volkmann's) **canals** that penetrate the bone obliquely or at right angles.

Spongy Bone

This type of bone lies adjacent to compact bone and is continuous with it; there is no distinct line of demarcation between the two regions. Histologically, it presents a lesser degree of organization than compact bone. Illustration A reveals that its outstanding feature is a series of branching, overlapping plates of matrix called **trabeculae.** These are oriented so as to produce large, interconnecting cavelike spaces. These spaces function well in storage and as pockets to hold the blood-forming cells of the bone marrow; they also function in weight reduction. Note that the trabeculae are also randomly punctuated by the osteocyte-holding spaces, or lacunae, and that blood vessels meander through the large spaces between the trabeculae, bringing nourishment to nearby osteocytes.

The Periosteum

Contiguous with the outer layer of compact bone, and tightly adherent to it, is a thick, tough membrane called the *periosteum.* It is composed of an outer layer of fibrous connective tissue and an inner *osteogenic layer* which serves as a source of new bone-forming cells and provides an access for blood vessels. The periosteum is anchored tightly to compact bone by bundles of collagenous fibers that perforate and become firmly embedded within the outer lamellae. These minute attachments are called **Sharpey's fibers.**

Assignment:

Label figure 9.5.

Laboratory Assignment

Do a systematic study of each type of connective tissue by examining prepared slides that are available. Note that Histology Atlas references are indicated for each type of tissue.

Materials:

prepared slides: areolar (H2.13), adipose (H2.51), white fibrous (H2.115), yellow (elastic) fibrous (H2.125), reticular (H2.31), hyaline cartilage (H2.61), fibrocartilage (H2.63), elastic cartilage (H2.62), bone, x.s. (H2.735), and developing bone (H2.79)

Connective Tissue Proper (figure HA–4) Examine slides of areolar, adipose, fibrous, and reticular connective tissues, identifying all the structures shown in figure HA–4. Mast cells, which are seen in areolar tissue, secrete heparin and histamine. Macrophages are amoeboid cells that ingest bacteria, dead cells, and other materials.

Cartilage (figure HA–5) Study the three different types of cartilage, noting their distinct differentiating characteristics.

Bone (figure HA–6) When studying a slide of developing membrane bone try to differentiate the osteoblasts from osteoclasts. Note that an osteoclast is a large multinucleate cell with a clear area between it and the bony matrix.

Laboratory Report

Complete Laboratory Report 8,9 by answering the questions that pertain to this exercise.

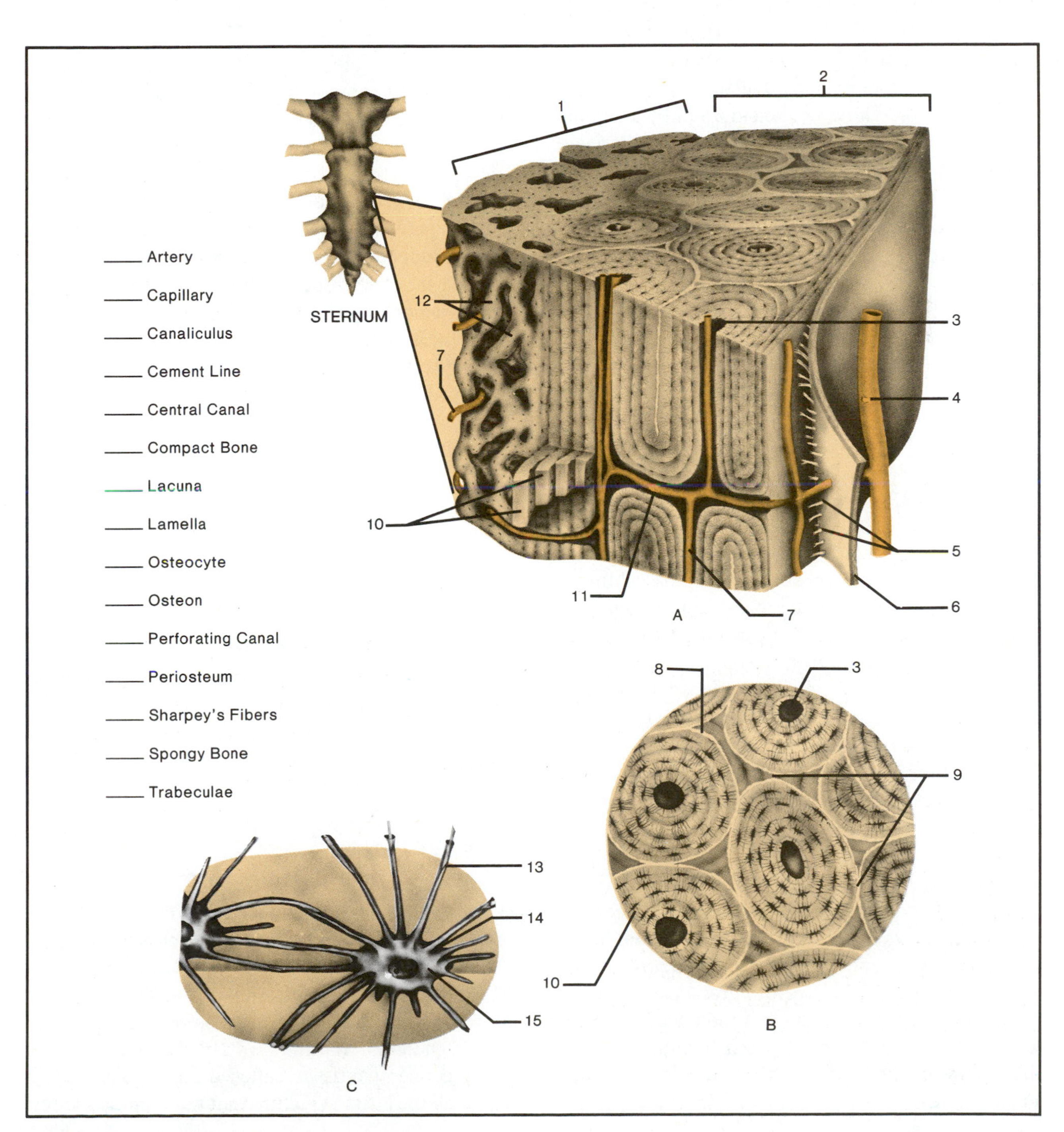

Figure 9.5 Bone tissue.

10 The Integument

Since it is constructed of epithelial and connective tissues, a study of the skin at this time provides one with an opportunity to review experiences of the last two laboratory periods. During this laboratory period prepared slides of the skin will be available for study. Prior to examining the slides, however, figure 10.1 should be labeled from the descriptive text that follows. Note that the skin consists of an outer multilayered **epidermis** and a deeper **dermis.**

The Epidermis

The enlarged section of skin on the right side of figure 10.1 is the *epidermis.* Note that it consists of four distinct layers: an outer **stratum corneum,** a thin translucent **stratum lucidum,** a darkly stained **stratum granulosum,** and a multilayered **stratum spinosum** (mucosum).

All four layers of the epidermis originate from the deepest layer of cells of the stratum spinosum. This deep layer, which lies adjacent to the dermis, is called the **stratum germinativum.** The columnar cells of this deep layer are constantly dividing to produce new cells that move outward to undergo metamorphosis at different levels. The brown skin pigment, *melanin,* which is produced by stellate *melanocytes* of the stratum germinativum, is responsible for skin color. Skin color differences are due to the amount of melanin present.

The stratum corneum of the epidermis consists of many layers of the scaly remains of dead epithelial cells. This protein residue of dead cells is primarily *keratin,* a water-repellent material. As the cells of the stratum spinosum are pushed outward, they move away from the nourishment of the capillaries, die, and undergo *keratinization.* The *eleidin* granules of the stratum granulosum are believed to be an intermediate product of keratinization. The translucent stratum lucidum consists of closely packed cells with traces of flattened nuclei.

The Dermis

This layer is often referred to as the "true skin." It varies in thickness of less than a millimeter to over six millimeters. It is highly vascular and provides most of the nourishment for the epidermis. It consists of two strata, the papillary and reticular layers.

The outer portion of the dermis, which lies next to the epidermis, is the **papillary layer.** It derives its name from numerous projections, or **papillae,** which extend into the upper layers of the epidermis. In most regions of the body these papillae form no pattern; however, on the fingertips, palms, and soles of the feet they form regularly arranged patterns of parallel ridges that improve frictional characteristics in these areas.

The deeper portion, or **reticular layer,** contains more collagenous fibers than the papillary layer. These fibers greatly enhance the strength of the skin. The surface texture of suede leather is, essentially, the reticular layer of animal hides.

Subcutaneous Tissue

Beneath the dermis lies the subcutaneous tissue, or **hypodermis;** it is also referred to as the *superficial fascia.* It consists of loose connective tissue, nerves, and blood vessels. One of its prime functions is to provide attachment for the skin to underlying structures.

Hair Structure

Hair *(pili)* consists of keratinized cells that are compactly cemented together. Each shaft of hair (label 1) is surrounded by a tube of epithelial cells, the **hair follicle.** The terminal end of the hair shaft, or **root,** is enlarged to form an onion-shaped region called the **bulb.** Within the bulb is an involution of loose connective tissue called the **hair papilla.** It is through the latter structure that nourishment enters

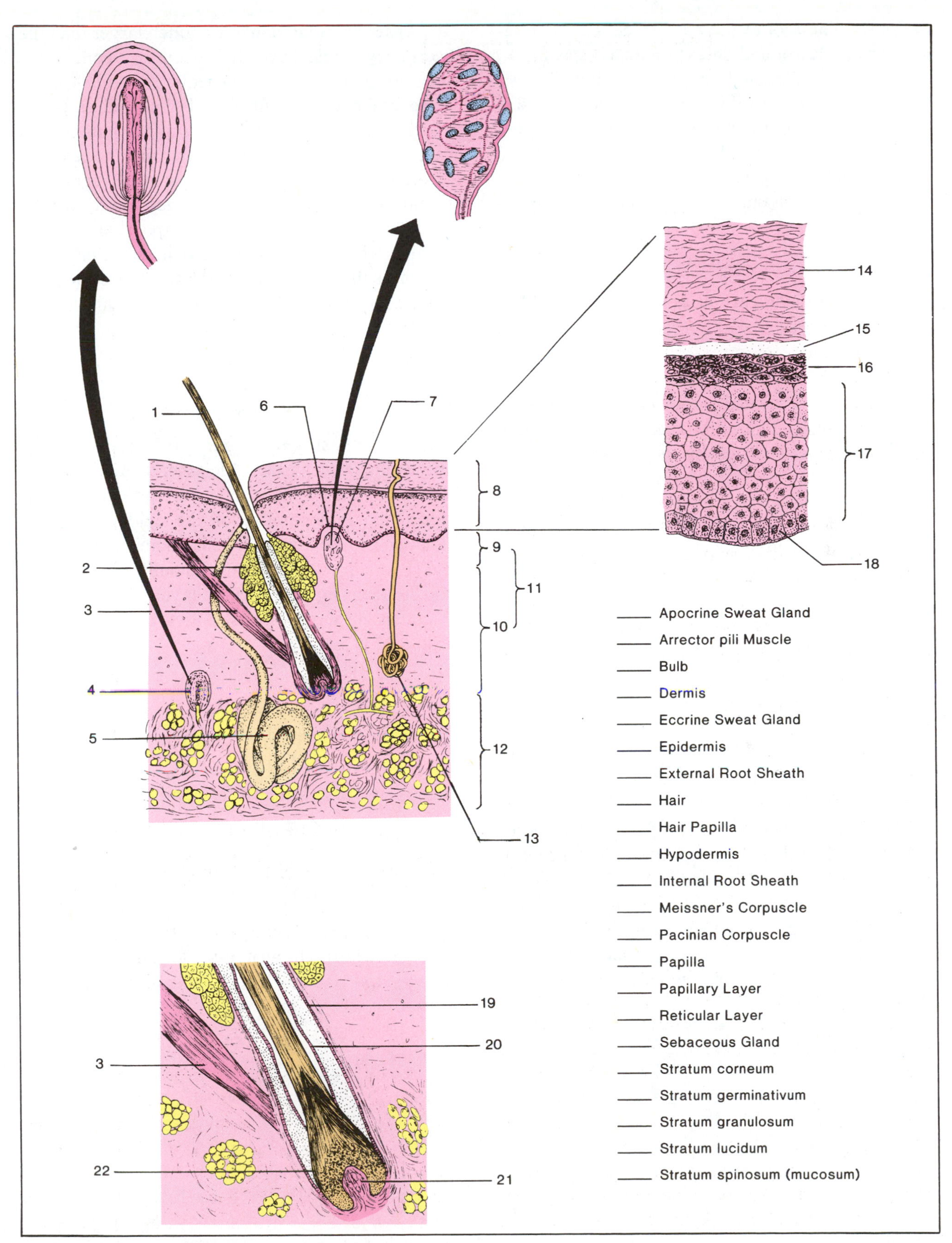

___ Apocrine Sweat Gland
___ Arrector pili Muscle
___ Bulb
___ Dermis
___ Eccrine Sweat Gland
___ Epidermis
___ External Root Sheath
___ Hair
___ Hair Papilla
___ Hypodermis
___ Internal Root Sheath
___ Meissner's Corpuscle
___ Pacinian Corpuscle
___ Papilla
___ Papillary Layer
___ Reticular Layer
___ Sebaceous Gland
___ Stratum corneum
___ Stratum germinativum
___ Stratum granulosum
___ Stratum lucidum
___ Stratum spinosum (mucosum)

Figure 10.1 Skin structure.

the shaft. The root of the hair is encased in an **internal root sheath** and an **external root sheath.**

Extending diagonally from the wall of the hair follicle to the epidermis is a band of smooth muscle fibers, the **arrector pili muscle.** Contraction of these muscle fibers causes the hair to move to a more perpendicular position, causing elevations on the skin surface commonly referred to as "goose pimples."

Glands

Two kinds of glands are present in the skin: sebaceous and sweat.

Sebaceous Glands

These glands are located within the epithelial tissue that surrounds each hair follicle. An oily secretion, called *sebum,* is secreted by these glands into the hair follicles and out onto the skin surface. Secretion is facilitated to some extent by the force of the arrector pili muscles during contraction. Sebum keeps hair pliable and helps to waterproof the skin.

Sweat Glands

Sweat glands are of two types: eccrine and apocrine. The small sweat glands that empty directly out through the surface of the skin are **eccrine sweat glands.** These glands are simple tubular structures that have their coiled basal portions located deep in the dermis. Except for the lips, glans penis, and clitoris, they are widely distributed throughout the body.

Although the composition of all eccrine secretions is similar, there are two different controlling stimuli. Almost everyone is aware that the sweat glands of some parts of the body, such as the palms and axillae, are affected by emotional factors. Glands in other regions, however, such as the forehead, neck, and back, are regulated primarily by thermal stimuli.

Apocrine sweat glands are much larger than the eccrine type and have their secretory coiled portions located in the hypodermis. Instead of emptying out onto the surface of the epidermis, all apocrine glands empty directly into a hair follicle canal. These glands are found in the axillae, scrotum of the male, female perigenital region, external ear canal, and nasal passages. While eccrine sweat is watery, the secretion of apocrine glands is a thick white, gray, or yellowish secretion. Malodorous substances in apocrine sweat are the principal contributors to body odors. Psychic factors rather than temperature changes primarily affect apocrine secretions.

Receptors

The receptors shown in figure 10.1 are Meissner's and Pacinian corpuscles. **Meissner's corpuscles** are located in the papillary layer of the dermis, projecting up into papillae of the epidermis. They function as receptors of touch. **Pacinian corpuscles** are spherical receptors with onionlike laminations and lie deep in the reticular layer of the dermis. Pacinian corpuscles are sensitive to variations in sustained pressure.

Laboratory Assignment

After labeling figure 10.1 and answering the questions on the Laboratory Report, proceed as follows:

Materials:

prepared slide of the skin (H11.11, H11.12, or H11.14)

Examine prepared slides of sections through the skin and identify the structures seen in figures 10.1, HA–9, HA–10, and HA–11. Make drawings, if required.

11 The Skeletal Plan

This laboratory period will be devoted to the study of the skeleton as a whole, and the detailed structure of a typical bone. Prior to performing the bone dissection, label figures 11.1 and 11.2.

Materials:

fresh beef bones, sawed longitudinally
articulated human skeleton

Long Bone Structure

Figure 11.1 shows a long bone, the *femur,* which has been sectioned to reveal its internal structure. Linearly, it consists of an elongated shaft, the **diaphysis,** and two enlarged ends, the **epiphyses.** Where the epiphyses meet the diaphysis are growth zones called **metaphyses.** During the growing years a plate of hyaline cartilage, the **epiphyseal disk,** exists in this area. As new cartilage forms on the epiphyseal side it is destroyed and then replaced by bone on the diaphyseal side. At maturity this area becomes completely ossified and linear growth ceases.

Note that the central portion of the diaphysis is a hollow chamber, the **medullary cavity.** Lining this cavity is a thin membrane called the **endosteum,** that actually extends into the Haversian canals. Filling the medullary cavity and a portion of the cancellous bone is a fatty material, the **yellow marrow.** The cancellous bone of the epiphyses of this bone (and the humerus) contain **red marrow** in adults. The epiphyses of other long bones in the adult contain only yellow marrow. Most red marrow in adults is found in the ribs, sternum, and vertebrae.

Except for the areas of articulation, the entire bone is covered with a **periosteum.** The articular surfaces of each epiphysis are covered with smooth **articular cartilage** (hyaline type).

Assignment:

Figure 11.1: Identify the labels in this illustration.

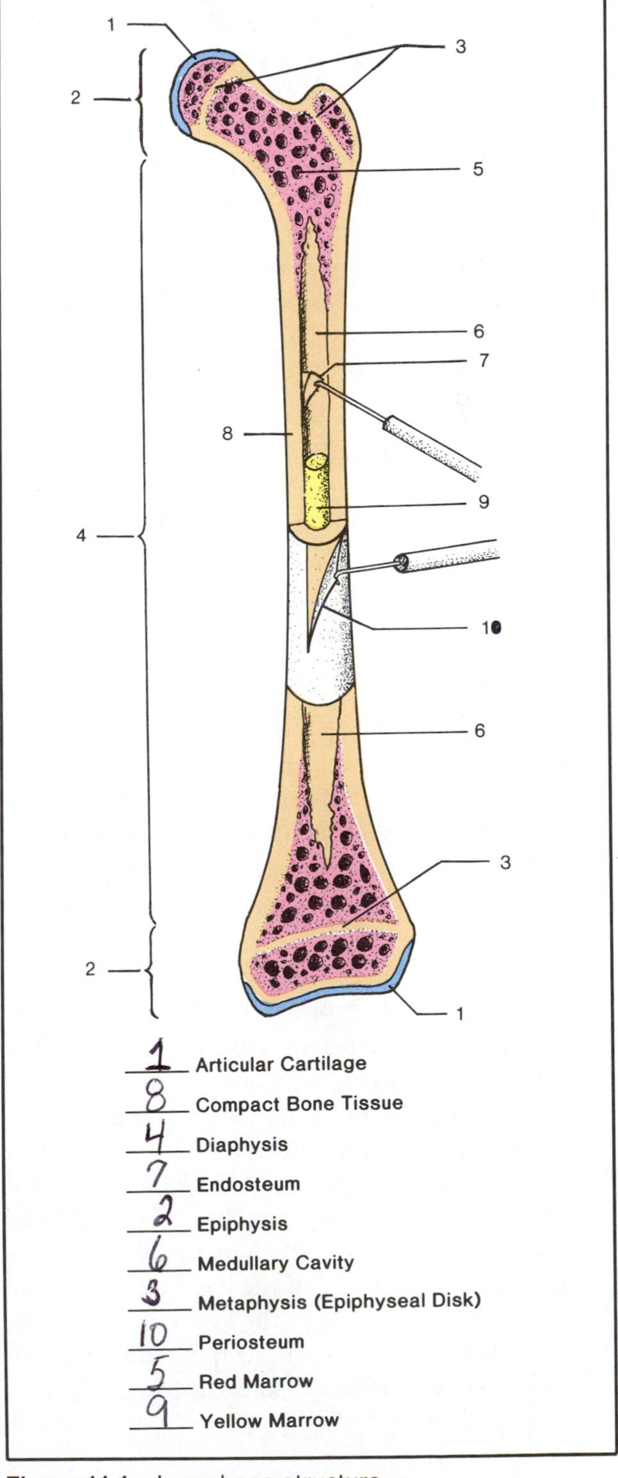

Figure 11.1 Long bone structure.

Beef Bone Study: Examine a freshly cut section of bone. Identify all structures shown in figure 11.1. Probe into the periosteum near a torn ligament or tendon; note the continuity of fibers between the periosteum and these structures. Probe into the marrow and note its texture.

Parts of the Skeleton

The adult skeleton consists of 206 named bones and many smaller unnamed ones. They vary considerably in configuration: some are long, others are short; flat, irregular, and round (sesamoid) shapes are also present.

The bones of the skeleton fall into two main groups: those that make up the axial skeleton and those forming the appendicular skeleton. Figure 11.2 identifies only the major portions of the skeleton; more extensive description will follow in subsequent exercises.

The Axial Skeleton

The parts of the axial skeleton are the **skull, hyoid bone, vertebral column** (spine), and the **rib cage.** The hyoid bone is a horseshoe-shaped bone that is situated under the lower jaw. The rib cage consists of twelve pairs of **ribs** and a **sternum,** or breastbone.

The Appendicular Skeleton

This portion of the skeleton includes the upper and lower extremities. The upper extremities consist of the shoulder girdles and arms. Each **shoulder girdle** consists of a **scapula** (shoulder blade) and **clavicle** (collarbone). Each arm consists of an upper portion, the **humerus,** two forearm bones, the **radius** and **ulna,** and the **hand.** The radius is lateral to the ulna.

The lower extremities consist of the pelvic girdle and legs. The **pelvic girdle** is formed by two bones, the **ossa coxae,** which are attached to the base of the vertebral column (sacrum) and to each other on their anterior surfaces. The joint where the ossa coxae are united on the median line is the **symphysis pubis.** Each leg consists of a femur, tibia, fibula, patella, and foot. The **femur** is the long bone of the upper part (thigh) of the leg. The **tibia** (shinbone) is the largest bone of the lower portion of the leg. The **fibula** (calf bone) parallels the tibia, lateral to it. The **patella,** or kneecap, is a roundish *(sesamoid)* bone.

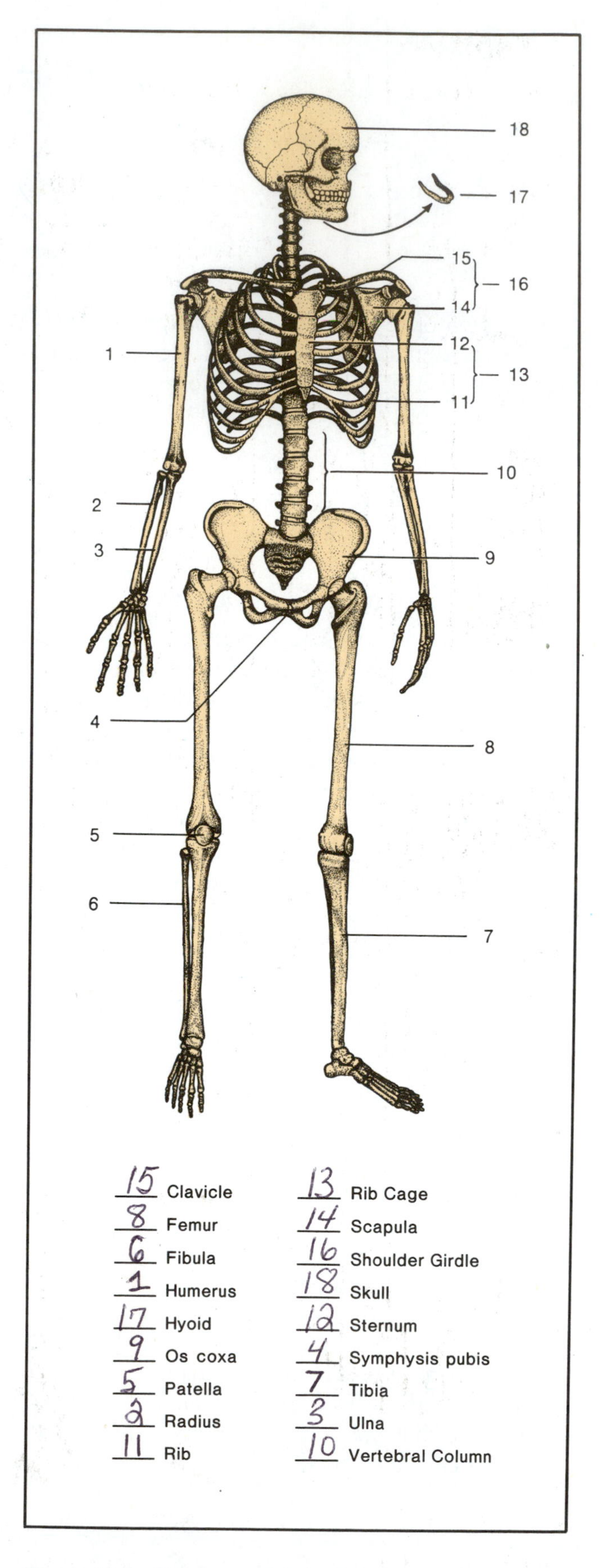

Figure 11.2 The human skeleton.

Assignment:

Label the parts of the skeleton in figure 11.2.

Bone Fractures

Various terms are used to describe different kinds of bone fractures. Fractures that do not penetrate the skin or mucous membranes are said to be **closed,** or **simple** fractures. On the other hand, those that do break through are said to be **open,** or **compound** fractures.

Fractures may also be complete or incomplete. **Incomplete** fractures are the type in which the bone is split, splintered, or only partially broken. Illustrations A, B, and C in figure 11.3 are of this type. When a bone partially breaks through on one side as a result of bending, it is often referred to as a **greenstick** fracture. Linear splitting of a long bone may be referred to as a **fissured** fracture.

Complete fractures are those in which the bone is broken clear through. If the break is at right angles to the long axis, it is considered to be a **transverse** fracture. Breaks that are at an angle to the long axis are termed **oblique** fractures. If a fracture results from torsional forces, it may be referred to as a **spiral** fracture.

If a piece of bone is broken out of the shaft it is a **segmental** fracture. Complete fractures, in which two or more fragments are seen, are designated as **comminuted** fractures. **Displaced** fractures are defined as those in which pieces of bone are forced out of alignment (illustration G). When one part of a bone is jammed into another part of the same bone, it may be referred to as a **compacted** fracture (illustration J). **Compression** fractures (not shown) are often seen in the vertebral column in which vertebrae are crushed by vertical forces.

Assignment:

Identify the types of fractures illustrated in figure 11.3.

Complete the Laboratory Report for this exercise.

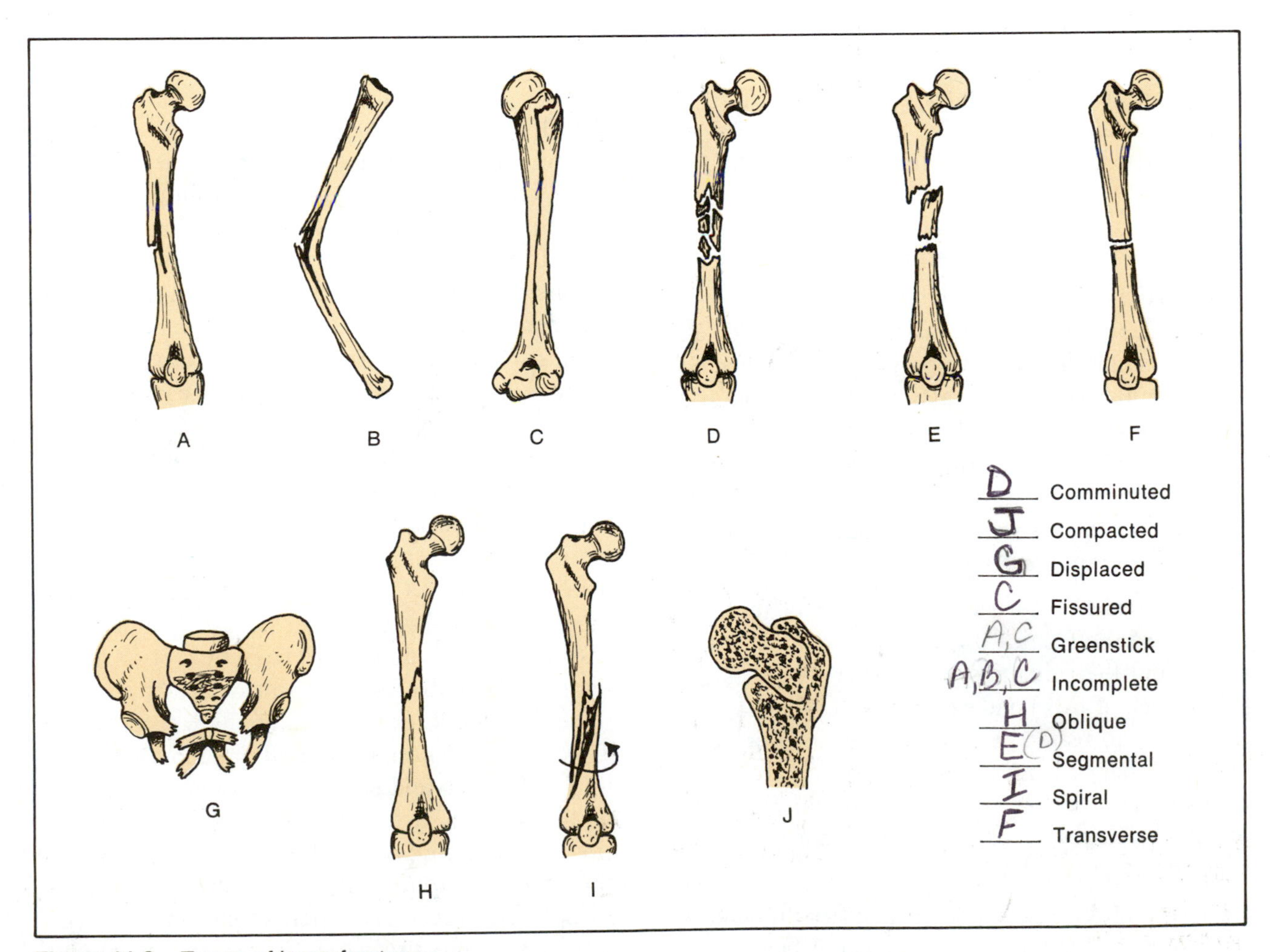

Figure 11.3 Types of bone fractures.

12 The Skull

For this study of the skull, specimens will be available in the laboratory. As you read through the discussion of the various bones identify them first in the illustrations and then on the specimens. Compare the specimens with the illustrations to note the degree of variance.

When handling laboratory skulls be very careful to avoid damaging them. Some parts are very delicate and easily broken when probed. **Pencils should never be used as probes or pointers;** instead, use a metal probe, or better yet, a pipe cleaner. If a metal probe is used, **touch the bones very gently** to avoid bone perforation where bone is thin.

Materials:

- whole and disarticulated skulls
- fetal skulls
- metal probe or pipe cleaner

The Cranium

The portion of the skull that encases the brain is called the *cranium.* It consists of the following bones: a single frontal, two parietals, one sphenoid, one occipital, two temporals, and an ethmoid. All these bones are joined together at their margins by irregular interlocking joints called *sutures.* The lateral and inferior aspects of the cranium are illustrated in figures 12.1 and 12.2. A sagittal section of the cranium is seen in figure 12.8.

Frontal The anterior superior portion of the skull consists of the frontal bone. It forms the eyebrow ridges and the ridge above the nose. The most inferior edge of this bone extends well into the orbit of the eye to form the **orbital plates** of this bone. On the superior ridges of the eye orbits are a pair of openings, the **supraorbital foramina.** Foramina (*foramen,* singular) are openings in bones for passage of nerves and blood vessels.

Parietals Directly posterior to the frontal bone on the sides of the skull are the parietal bones. The lateral view of the skull actually shows only the left parietal bone. The right parietal is on the other side of the skull. The right and left parietals meet on the midline of the skull to form the **sagittal suture.** Between the frontal and each parietal bone is another suture, the **coronal suture.** Two semicircular bony ridges that extend from the forehead (frontal bone) and over the parietal bone are the **superior temporal line** and **inferior temporal line.** These ridges form the points of attachment for the longest muscle fibers of the *temporalis* muscle. Reference to figure 24.1 shows the position of this muscle (label 2). It is the upper extremity of this muscle that falls on the superior temporal line.

Temporals On each side of the skull, inferior to the parietal bones, are the temporals. These bones are colored yellow in figure 12.1. Each temporal is joined to its adjacent parietal by the **squamosal suture.** A depression, the **mandibular** *(glenoid)* **fossa,** on this bone provides a recess into which the lower jaw articulates. Pull the jaw away from the skull to note the shape of this fossa. The rounded eminence of the mandible that fits into this fossa is the **mandibular condyle.** Just posterior to the mandibular fossa is the ear canal, or **external acoustic meatus** (*acoustic:* hearing; *meatus:* canal or passage).

The temporal bone has three significant processes: the zygomatic, styloid, and mastoid. The **zygomatic process** is a long slender process that extends forward, anterior to the external acoustic meatus, to form a bridge to the cheekbone of the face. The **styloid process** is a slender spinelike process that extends downward from the bottom of the temporal bone to form a point of attachment for some muscles of the tongue and pharyngeal region. This process is often broken off on laboratory specimens. The **mastoid process** is a rounded eminence on the inferior surface of the temporal just posterior to the styloid process. It provides anchorage for the *sternocleidomastoideus* muscle of the neck. Middle ear infections which spread into the cancellous bone of this process are referred to as *mastoiditis.*

Sphenoid The pink-colored bone seen in the lateral view of the skull, figure 12.1, is the sphenoid bone. Note in the inferior view, figure 12.2, that this bone extends from one side of the skull to the other. Identify the two greater wings, the orbital surfaces and the pterygoid processes of the sphenoid. Those parts that are on the sides of the skull in the "temple" region are the greater wings of the sphenoid. On the ventral surface (figure 12.2) are the pteryoid processes of the sphenoid to which the pterygoid muscles are attached. The orbital surfaces of the sphenoid (figure 12.4) make up the posterior walls of each eye orbit.

Ethmoid On the medial surface of each orbit of the eye is seen the ethmoid bone (label 6, figure 12.1). This bone forms a part of the roof of the nasal cavity and closes the anterior portion of the cranium.

Examine the upper portion of the nasal cavity of your laboratory skull. Note that the inferior portion of the ethmoid has a downward extending **perpendicular plate** on the median line. Refer to figure 12.8 (label 3). This portion articulates anteriorly with the nasal and frontal bones. Posteriorly, it articulates with the sphenoid and vomer. On each side of the perpendicular plate are irregular curved plates, the **superior** and **middle nasal conchae.** They provide bony reinforcement for the fleshy upper nasal conchae of the nasal cavity.

Occipital The posterior inferior portion of the skull consists primarily of the occipital bone. It is joined to the parietal bones by the **lambdoidal suture.** Examine the inferior surface of your laboratory skull and compare it with figure 12.2. Note the large **foramen magnum** which surrounds the brain stem in real life. On each side of this opening is seen a pair of **occipital condyles** (label 18). These two condyles rest on fossae of the *atlas,* the first vertebra of the spinal column. All *condyles* on bones are smooth, knucklelike processes that fit into *fossae* (depressions) of other bones.

Assignment:

Label all bones of the cranium in figure 12.1. Facial bones will be labeled later.

Label all cranial bones in figure 12.2. Bones of hard palate and all foramina will be labeled later.

Floor of the Cranium

The inside surface of the skull reveals other structural details of the ethmoid, sphenoid, and temporal bones of significance. Remove the top of your laboratory skull and compare the floor of the cra-

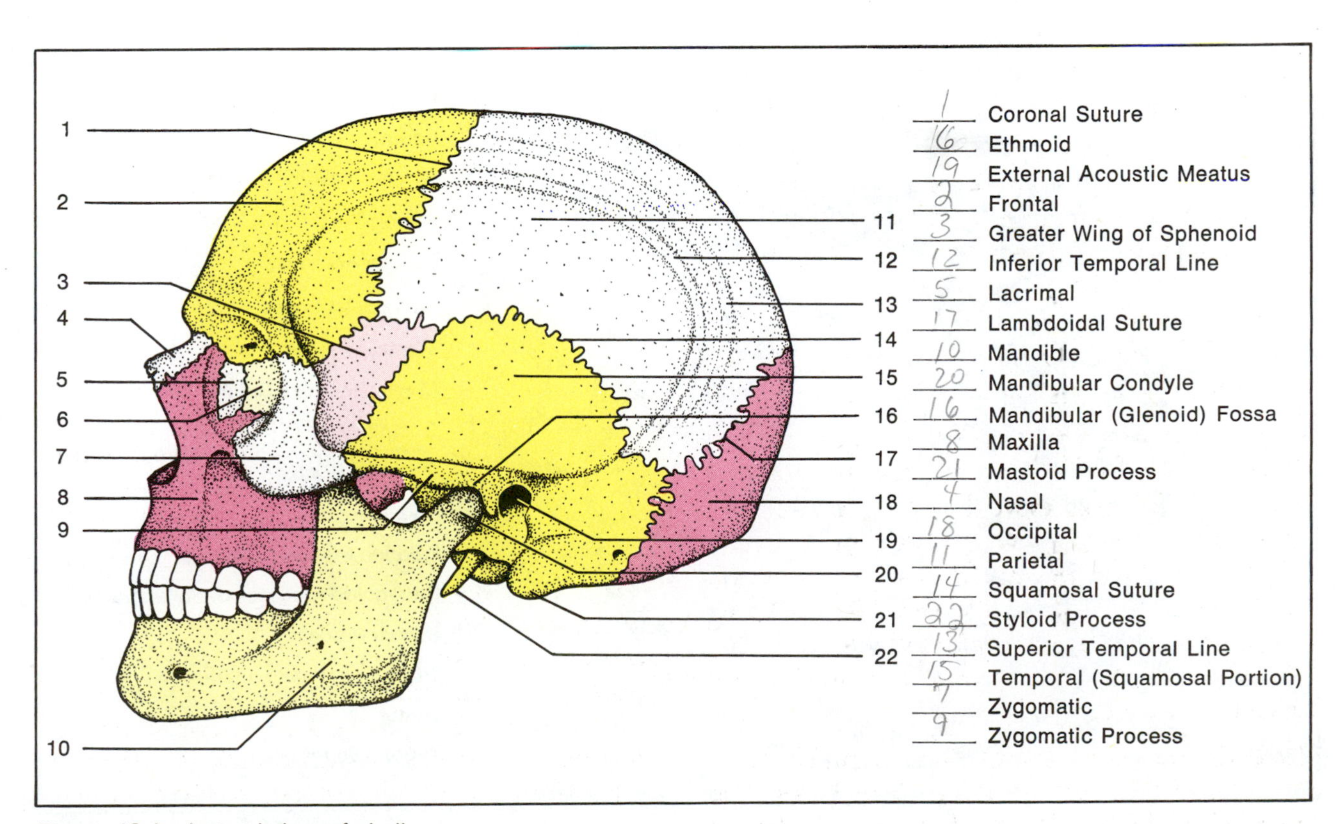

Figure 12.1 Lateral view of skull.

nium with figure 12.3 to identify the following structures.

Cranial Fossae As you look down on the entire floor of the cranium note that it is divided into three large depressions called *cranial fossae.* The one formed by the orbital plates of the frontal is called the **anterior cranial fossa.** The large depression made up mostly of the occipital bone is the **posterior cranial fossa.** It is the deepest fossa. In between these two fossae is the **middle cranial fossa** which is at an intermediate level. The latter involves the sphenoid and temporal bones.

Ethmoid The ethmoid bone in the anterior cranial fossa is seen as a pale yellow structure between the orbital plates of the frontal bone. Note that it consists of a perforated horizontal portion, the **cribriform plate,** and an upward projecting process, the **crista galli** (cock's comb). The holes in the cribriform plate allow branches of the olfactory nerve to pass from the brain into the nasal cavity. The crista galli serves as an attachment for the *falx cerebri* (label 1, figure 31.1).

Sphenoid Observe that on the median line of the sphenoid there is a deep depression called the **hypophyseal fossa.** This depression contains the pituitary gland *(hypophysis)* in real life. Posterior to this fossa is an elevated ridge called the **dorsum sella.** The two spinelike processes anterior and lateral to the hypophyseal fossa that project backward are the **anterior clinoid processes.** The outer spiny processes of the dorsum sella are the **posterior clinoid processes.** The hypophyseal fossa, dorsum sella, and clinoid processes, collectively, make up the **sella turcica,** or "Turkish saddle."

Temporals The significant parts of the temporal bone to identify in figure 12.3 are the petrous and squamous portions, the carotid canal, and the internal acoustic meatus. The **squamous portion** of the temporal is that thin portion that forms a part of the side of the skull. The **petrous portion** (label 8) is probably the hardest portion of the skull. It contains the hearing mechanism of the ear.

On the medial sloping surface of the petrous portion is seen an opening to the **internal acoustic meatus.** This canal contains the facial and cranial nerves. The latter pass from the inner ear region to the brain.

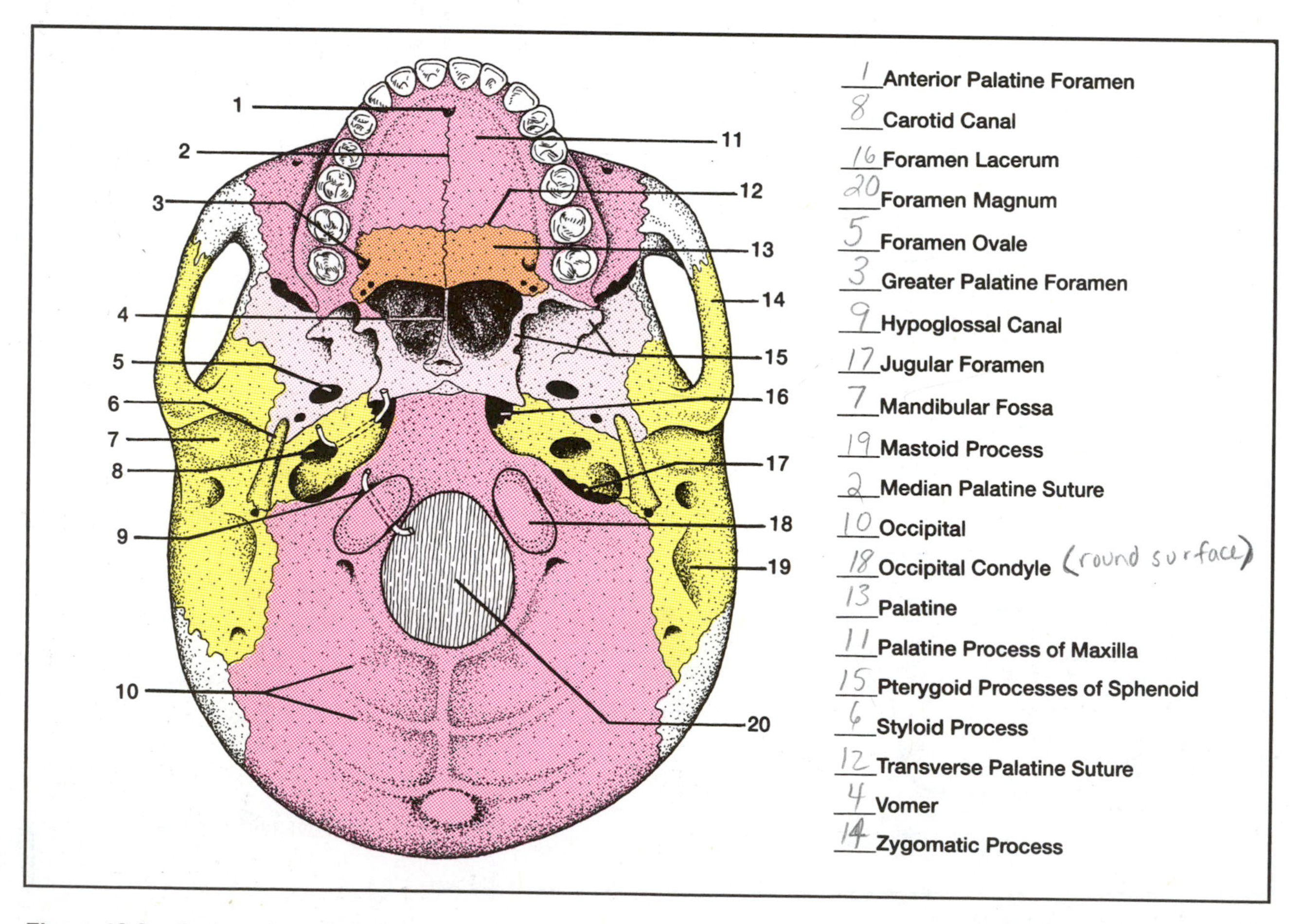

Figure 12.2 Bottom view of skull.

The **carotid canal** is a passageway running through the petrous portion of the temporal bone which allows the internal carotid artery of the neck to pass through the skull into the brain. The brain-side opening to this canal can be seen on the anterior margin of the petrous portion of the temporal. Insert a pipe cleaner or straightened-out paper clip into this foramen and note where it comes out on the ventral side of the skull.

Other Foramina In addition to the above foramina the following major foramina on the floor of the cranium should be identified: foramen lacerum, foramen ovale, optic foramen, jugular foramen and hypoglossal canal.

The **foramen lacerum** is a large jagged-edged foramen located on each side of the hypophyseal fossa of the sphenoid bone. Lateral to this foramen in the sphenoid bone is an oval opening, the **foramen ovale.** This foramen provides a passageway for the mandibular nerve. Just anterior to the anterior clinoid processes are a pair of **optic foramina** (label 11, figure 12.3) which provide passageways (*optic canals*) for the optic nerves to the eyes. Between the medial margin of the petrous portion of the temporal and the occipital bone is seen an irregular **jugular foramen.** Blood from the brain drains into the internal jugular vein through this opening. The inner openings to the **hypoglossal canals** are seen on each side of the foramen magnum. Insert a probe (paper clip) into one of these canals and note that it passes through the base of the occipital condyle. This canal provides an exit for the 11th and 12th cranial nerves (spinal accessory and hypoglossal). Be able to identify any of these foramina that pass through to the inferior surface of the skull.

Assignment:

Label figure 12.3.

Label the foramina in figure 12.2.

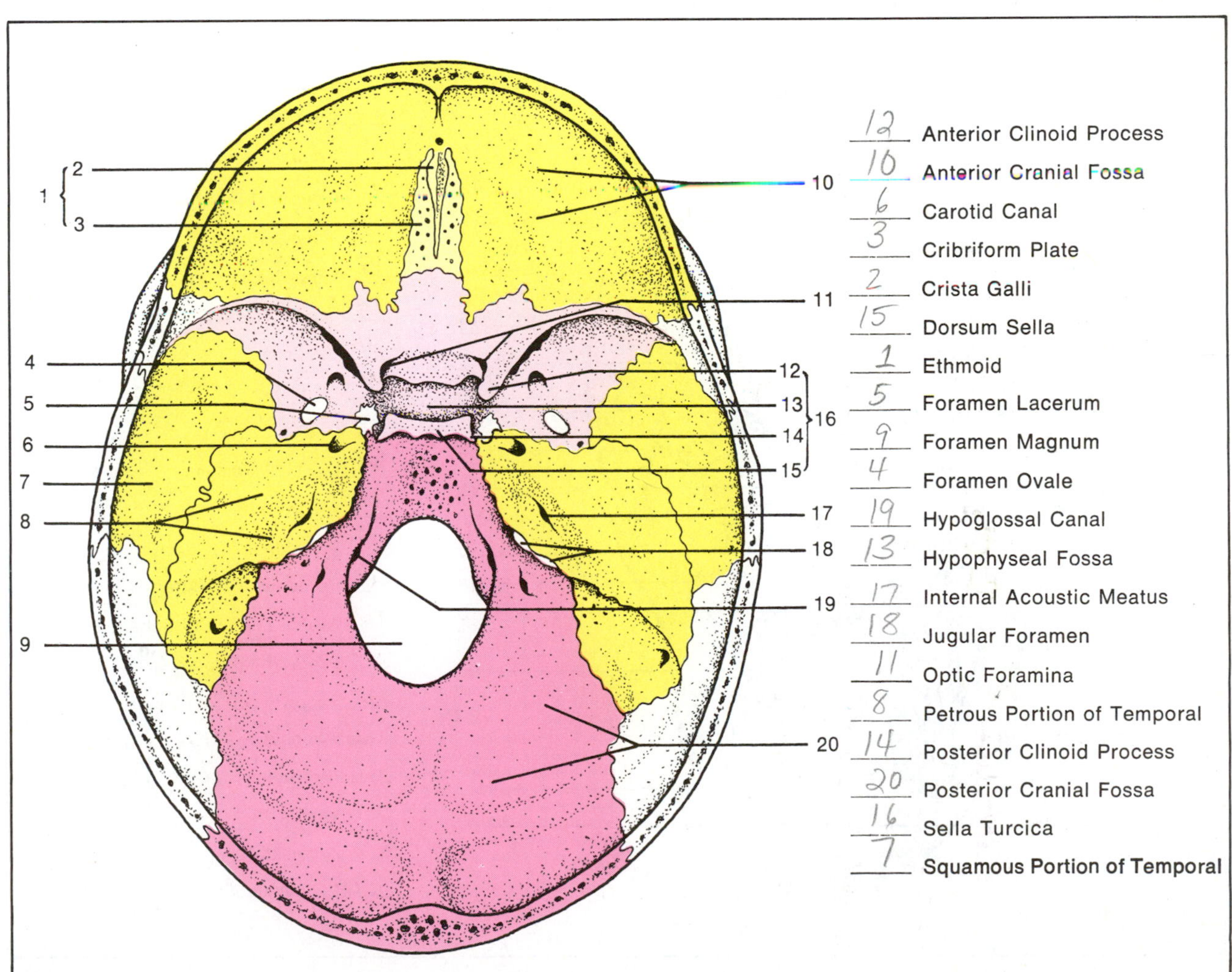

Figure 12.3 Floor of cranium.

The Face

The bones that make up the anterior portion of the skull constitute the facial bones. Except for two bones, the vomer and mandible, all of them are paired. Figure 12.4 reveals the majority of the facial bones.

Maxillae The upper jaw consists of two maxillary bones (maxillae) that are joined by a suture on the median line. Remove the mandible from your laboratory skull and examine the hard palate. Compare it with figure 12.2. Note that the anterior portion of the hard palate consists of two **palatine processes of the maxillae.** A **median palatine suture** joins the two bones on the median line.

The maxillae of an adult support sixteen permanent teeth. Each tooth is contained in a socket, or *alveolus.* That portion of the maxillae that contains the teeth is called the **alveolar process.**

Three significant foramina are seen on the maxillae: two infraorbital and one anterior palatine. The **infraorbital foramina** are situated on the front of the face under each eye orbit. Nerves and blood vessels emerge from each of these foramina to supply the nose. The **anterior palatine foramen** is seen in the anterior region of the hard palate just posterior to the central incisors.

Palatines In addition to the palatine processes of the maxilla, the hard palate also consists of two palatine bones. These bones form the posterior third of the palate. Locate them on your laboratory specimen. Note that each palatine bone has a large **greater palatine foramen** and two smaller **lesser palatine foramina.**

Assignment:

Label the parts of the hard palate in figure 12.2.

Zygomatics On each side of the face are two zygomatic *(malar)* bones. They form the prominence of each cheek and the inferior, lateral surface of each eye orbit. Each zygomatic has a small foramen, the **zygomaticofacial foramen.**

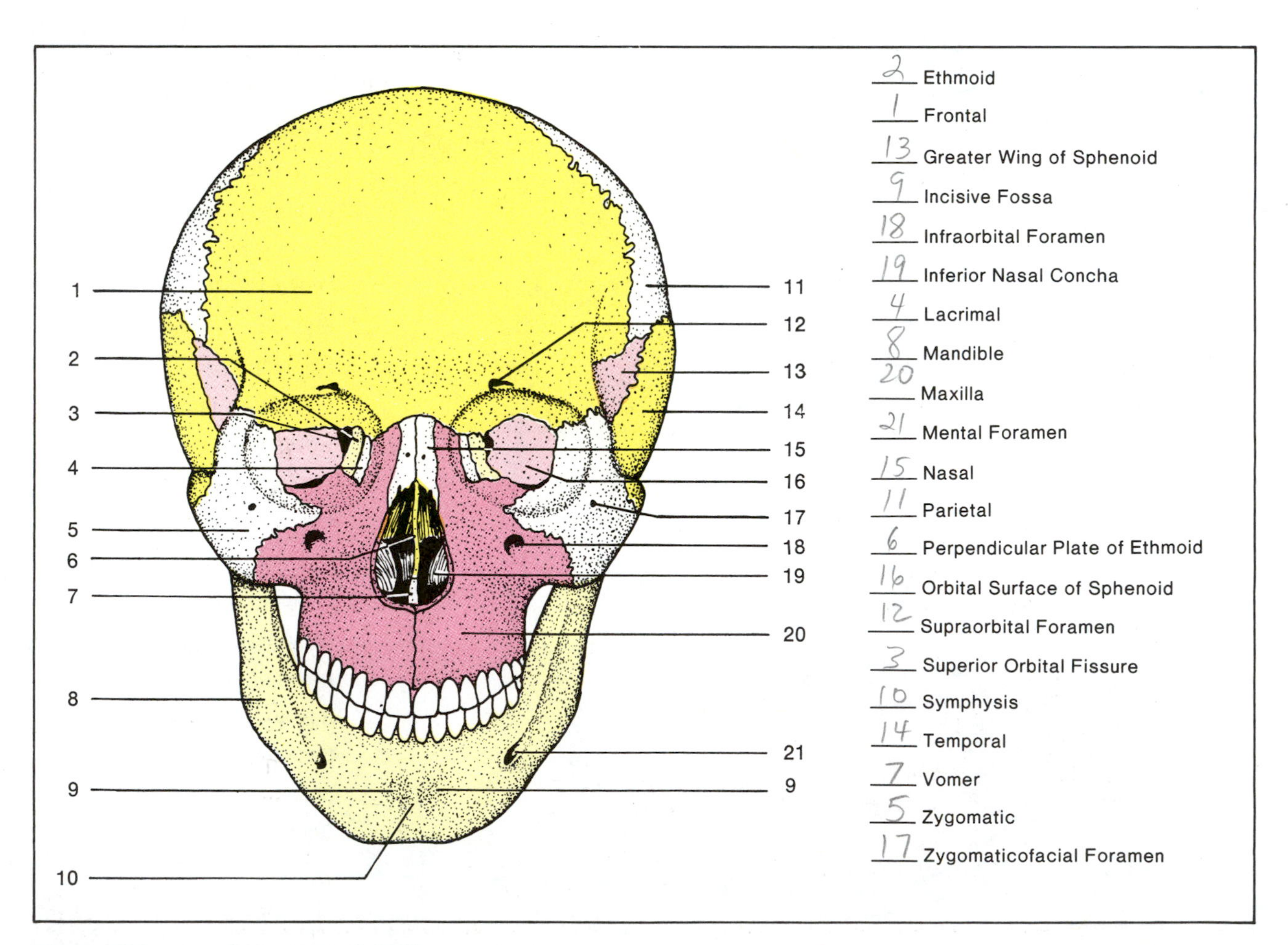

Figure 12.4 Anterior aspect of skull.

Lacrimals Between the ethmoid and upper portion of the maxillary bones are a pair of lacrimal (*lacrima:* tear) bones—one in each orbit. Each of these small bones has a groove which allows the tear ducts from the lacrimal glands of the eye to pass down into the nasal cavity.

Nasals The bridge of the nose is formed by a pair of thin, rectangular nasal bones.

Vomer This thin bone is located in the nasal cavity on the median line. Its posterior upper edge articulates with the back portion of the perpendicular plate of the ethmoid and the rostrum of the sphenoid. The lower border of the vomer is joined to the maxillae and palatines. The *septal cartilage* of the nose extends between the anterior margin of the vomer and the perpendicular plate of the ethmoid. Locate this bone on figures 12.2, 12.4, and 12.6 as well as on your laboratory specimen.

Inferior Nasal Conchae The inferior nasal conchae are curved bones attached to the walls of the nasal fossa. They are situated beneath the superior and middle nasal conchae which are part of the ethmoid bone.

Mandible The only bone of the skull that is not fused as an integral part of the skull is the lower jaw, or *mandible.* Figure 12.5 reveals the anatomical details of this bone.

It consists of a horizontal portion, the **body,** and two vertical portions, the **rami.** Embryologically, the mandible forms from two centers of ossification, one on each side of the face. As the bone develops toward the median line, the two halves finally meet and fuse to form a solid ridge. This point of fusion on the midline is called the **symphysis** (label 10, figure 12.4). On each side of the symphysis are two depressions, the **incisive fossae.**

The superior portion of each ramus has a condyle, a coronoid process, and notch. The **mandibular condyle** occupies the posterior superior terminus of the ramus. The process on the superior anterior portion of the ramus is the **coronoid process.** This tuberosity provides attachment for the *temporalis* muscle. Between the mandibular condyle and the coronoid process is the **mandibular notch.** At the posterior inferior corners of the mandible, where the body and rami meet, are two protuberances, the **angles.** The angles provide attachment for the *masseter* and *internal pterygoid* muscles.

A ridge of bone, the **oblique line,** extends at an angle from the ramus down the lateral surface of the body to a point near the mental foramen. This bony elevation is strong and prominent in its upper part, but gradually flattens out and disappears, as a rule, just below the first molar. On the internal

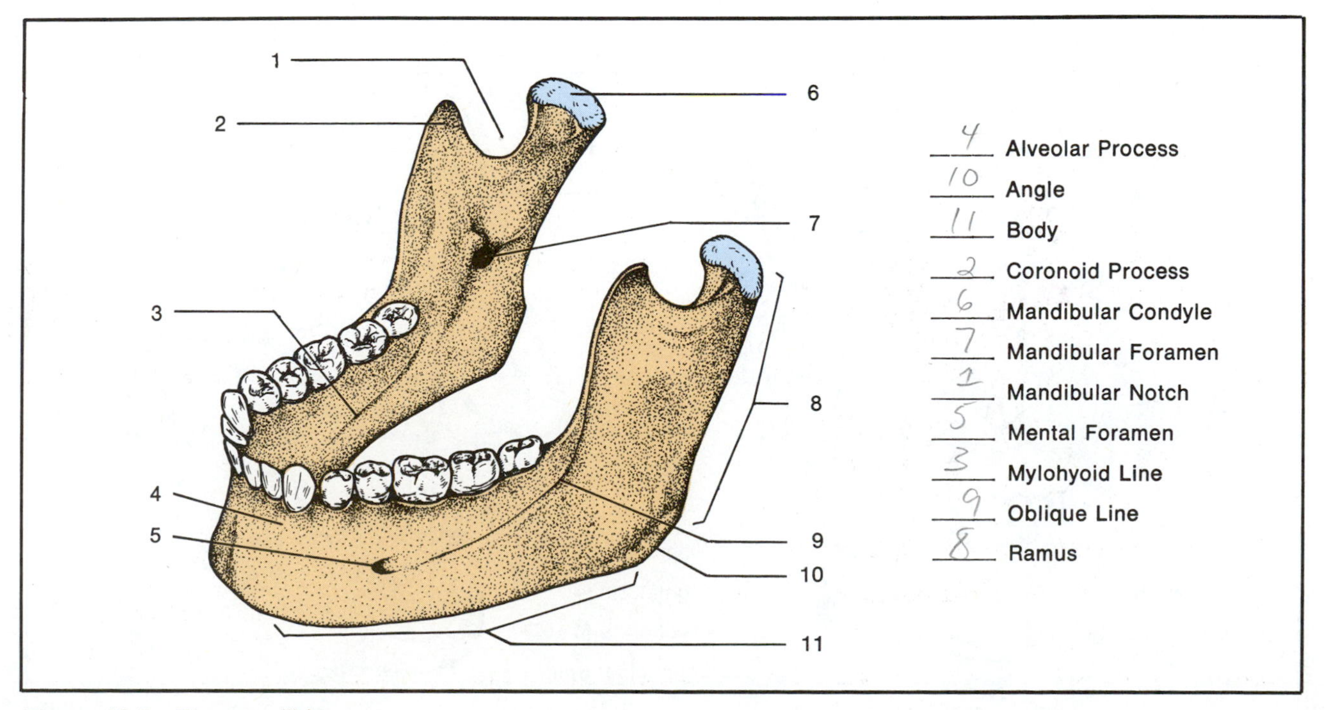

Figure 12.5 The mandible.

(medial) surface of the mandible is another diagonal line, the **mylohyoid line.** It extends from the ramus down to the body. To this crest is attached a muscle, the *mylohyoid,* which forms the floor of the oral cavity. The bony portion of the body that exists above this line makes up a portion of the sides of the oral cavity proper.

Each tooth lies in a socket of bone called an *alveolus.* As in the case of the maxilla, the portion of this bone that contains the teeth is called the **alveolar process.** The alveolar process consists of two compact tissue bony plates, the *external* and *internal alveolar plates.* These two plates of bone are joined by partitions, or *septae,* which lie between the teeth and make up the transverse walls of the alveoli.

On the medial surfaces of the rami are two foramina, the **mandibular foramina.** On the external surface of the body are two prominent openings, the **mental foramina** (*mental:* chin).

Assignment:

Label figures 12.4 and 12.5.

Label the bones of the face in figure 12.1.

The Paranasal Sinuses

Some of the bones of the skull contain cavities, the *paranasal sinuses,* which reduce the weight of the skull without appreciably weakening it. All of the sinuses have passageways leading into the nasal cavity and are lined with a mucous membrane similar to the type that lines the nasal cavities. The paranasal sinuses are named after the bones in which they are situated. Figures 12.6 and 12.8 show the locations of these cavities. Above the eyes in the forehead are the **frontal sinuses.** The largest sinuses are the **maxillary sinuses,** which are situated in the maxillary bones. These sinuses are also called the *antrums of Highmore.* The **sphenoidal sinus** is the most posterior sinus seen in figure 12.6. It is also shown in figure 12.8 (label 8). Between the frontal and sphenoidal sinuses are a group of small spaces called the **ethmoid air cells.**

Assignment:

Label figures 12.6 and 12.8.

Answer the questions on the Laboratory Report that pertain to the bones of the face.

The Fetal Skull

The human skull at birth is incompletely ossified. Figure 12.7 reveals its structure. These unossified membranous areas, called *fontanels,* facilitate compression of the skull at childbirth. During labor the bones of the skull are able to lap over each other as the infant passes down the birth canal without causing injury to the brain.

There are six fontanels joined by five areas where future sutures of the skull form. The largest fontanel is the **anterior fontanel,** a somewhat diamond-shaped membrane that lies on the median line at the junction of the frontal and parietal bones.

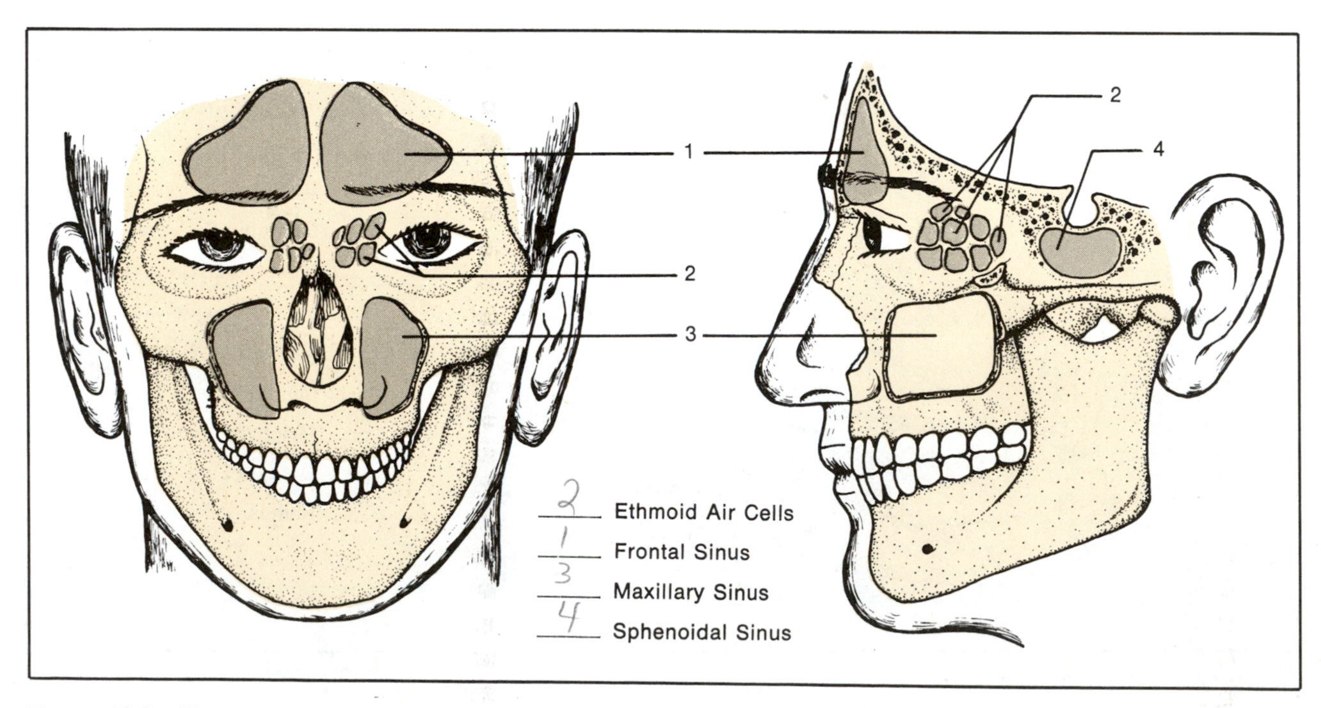

Figure 12.6 The paranasal sinuses.

The **posterior fontanel** is somewhat smaller and lies on the median line at the junction of the parietal and occipital bones. Between these two fontanels on the median line is a membranous area where the future sagittal suture of the skull will form.

On each side of the skull, where the frontal, parietal, sphenoid, and temporal bones come together behind the eye orbit, is an **anterolateral fontanel.** Between the anterior and anterolateral fontanels can be seen a membranous line that is the area where the future coronal suture will develop. The most posterior fontanel on the side of the skull is the **posterolateral fontanel** that lies at the junction of the parietal, temporal and occipital bones. Between the anterolateral and posterolateral fontanels is a membranous line that will develop into the **future squamosal suture.** Ossification of these fontanels and membranous future sutures is usually completed in the two-year-old child.

Assignment:

Label figure 12.7.

Examine a fetal skull, identifying all of the structures in figure 12.7.

Complete the Laboratory Report for this exercise.

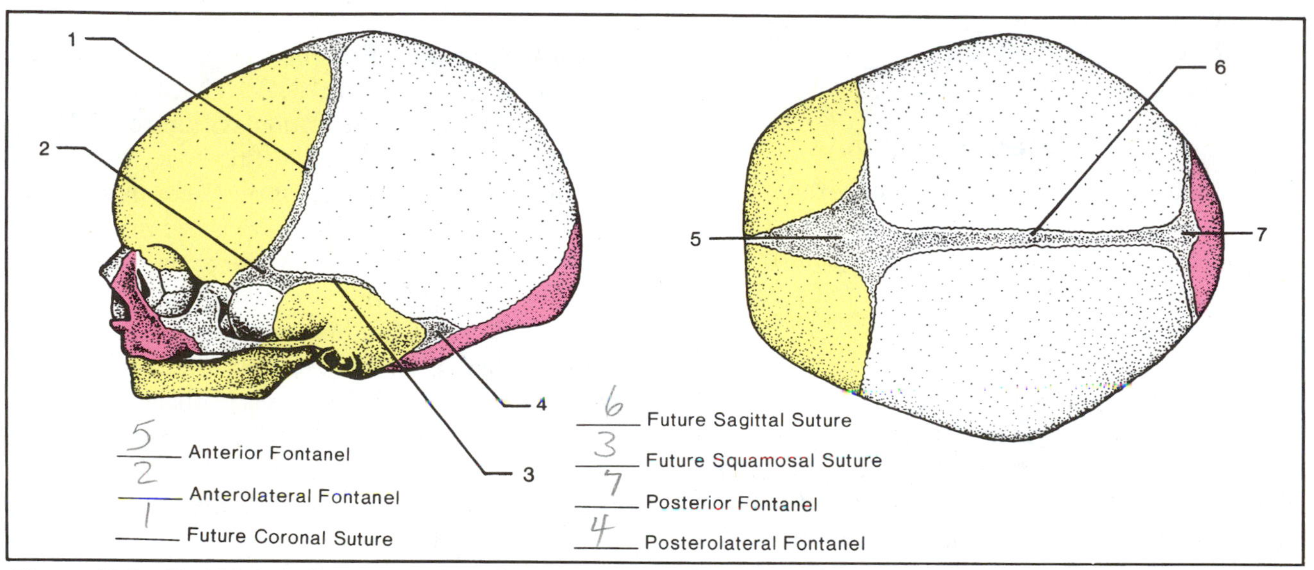

Figure 12.7 The fetal skull.

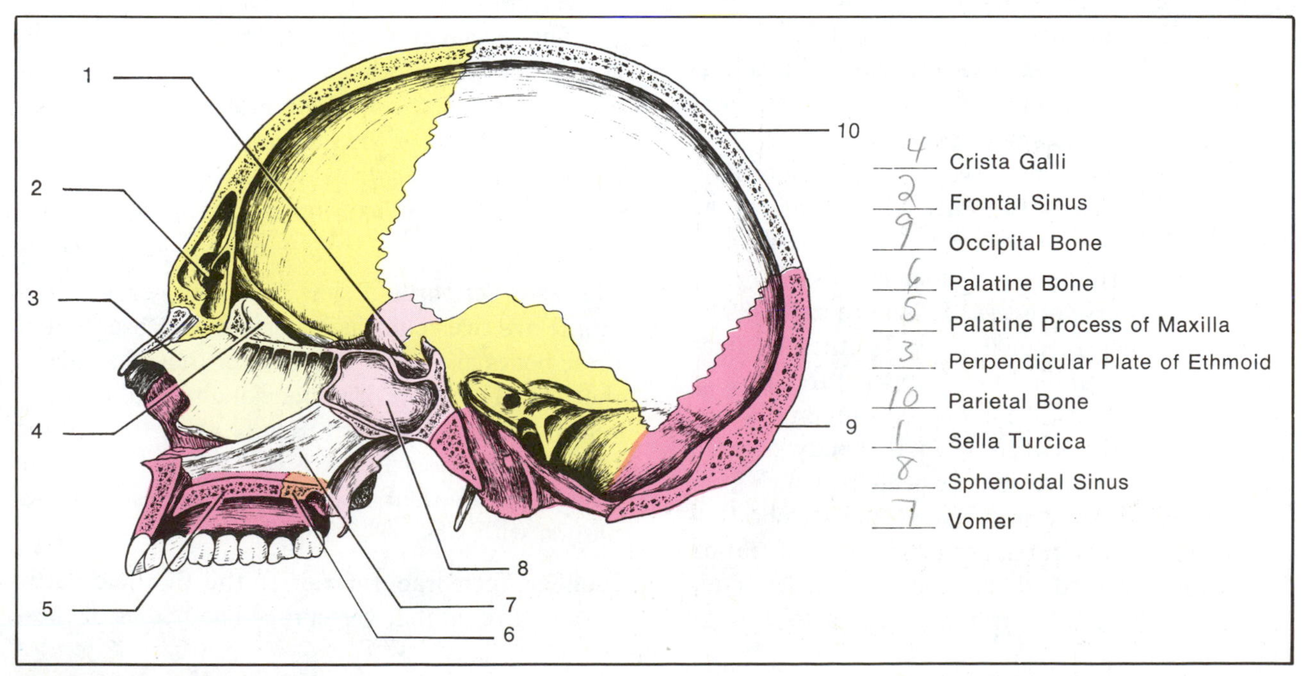

Figure 12.8 Sagittal section of skull.

13 The Vertebral Column and Thorax

The skeletal structure of the trunk of the body will be studied in this exercise. The vertebral column, ribs, sternum, and hyoid bone make up this part of the body.

Materials:

skeleton, articulated
skeleton, disarticulated
vertebral column, mounted

The Vertebral Column

The vertebral column consists of thirty-three bones, twenty-four of which are individual movable vertebrae. Figure 13.1 illustrates its structure. Note that the individual vertebrae are numbered from the top.

The Vertebrae

Although the vertebrae in different regions of the vertebral column vary considerably in size and configuration, they do have certain features in common. Each one has a structural mass, the **body,** which is the principal load-bearing contact area between adjacent vertebrae. The space between the surfaces of adjacent vertebral bodies is filled with a fibrocartilaginous **intervertebral disk.** The collective action of these twenty-four disks imparts a vital cushion effect to the spinal column.

In the center of each vertebra is an opening, the **vertebral** or **spinal foramen,** which contains the spinal cord. Projecting out from the posterior surface of each vertebra is a **spinous process.** On each side is a **transverse process.** These processes of the spinal column are joined to each other by ligaments to form a unified flexible structure. Various muscles of the body are anchored to them.

Extending backward from the body of each vertebra are two processes, the **pedicles,** which form a portion of the bony arch around the vertebral foramen. These structures are labeled in illustration C. The opening formed between the pedicles of adjacent vertebrae allows spinal nerves to emerge from the spinal cord. These openings are the **intervertebral foramina** (label 26) of the vertebral column.

The posterolateral portions of each vertebra consist of two broad plates, the **laminae** (label 7). The two pedicles and two laminae constitute the **neural arch.**

Cervical Vertebrae The upper seven bones are the cervical vertebrae of the neck. The first of these seven is the **atlas.** Illustration A, figure 13.1, reveals the superior surface of this bone. Note that the spinal foramen is much larger here than on the lumbar vertebrae. Its larger size is necessary to accommodate a short portion of the brain stem which extends down into this space. Note that on each side of the spinal foramen is a depression, the **superior articular surface** (facet), which articulates with the skull. Which processes of the skull fit into these depressions?

Within each transverse process is a small **transverse foramen.** These foramina are seen only in the cervical vertebrae. Collectively, they form a passageway on each side of the spinal column for the vertebral artery and vertebral vein.

The second cervical vertebra is called the **axis.** It differs from all other vertebrae in having a vertical protrusion, the **odontoid process** (dens), which provides a pivot for the rotation of the atlas. When the head is turned from side to side, movement occurs between the axis and atlas around this process.

Thoracic Vertebrae Below the seven cervical vertebrae are twelve thoracic vertebrae. Observe that these bones are larger and thicker than the ones in the neck. The superior surface of a typical thoracic vertebra is seen in illustration C. A distinguishing feature of these vertebrae is that all twelve of them have facets on their transverse processes for articulation with ribs.

Lumbar Vertebrae Inferior to the thoracic vertebrae lie five lumbar vertebrae. The bodies of these

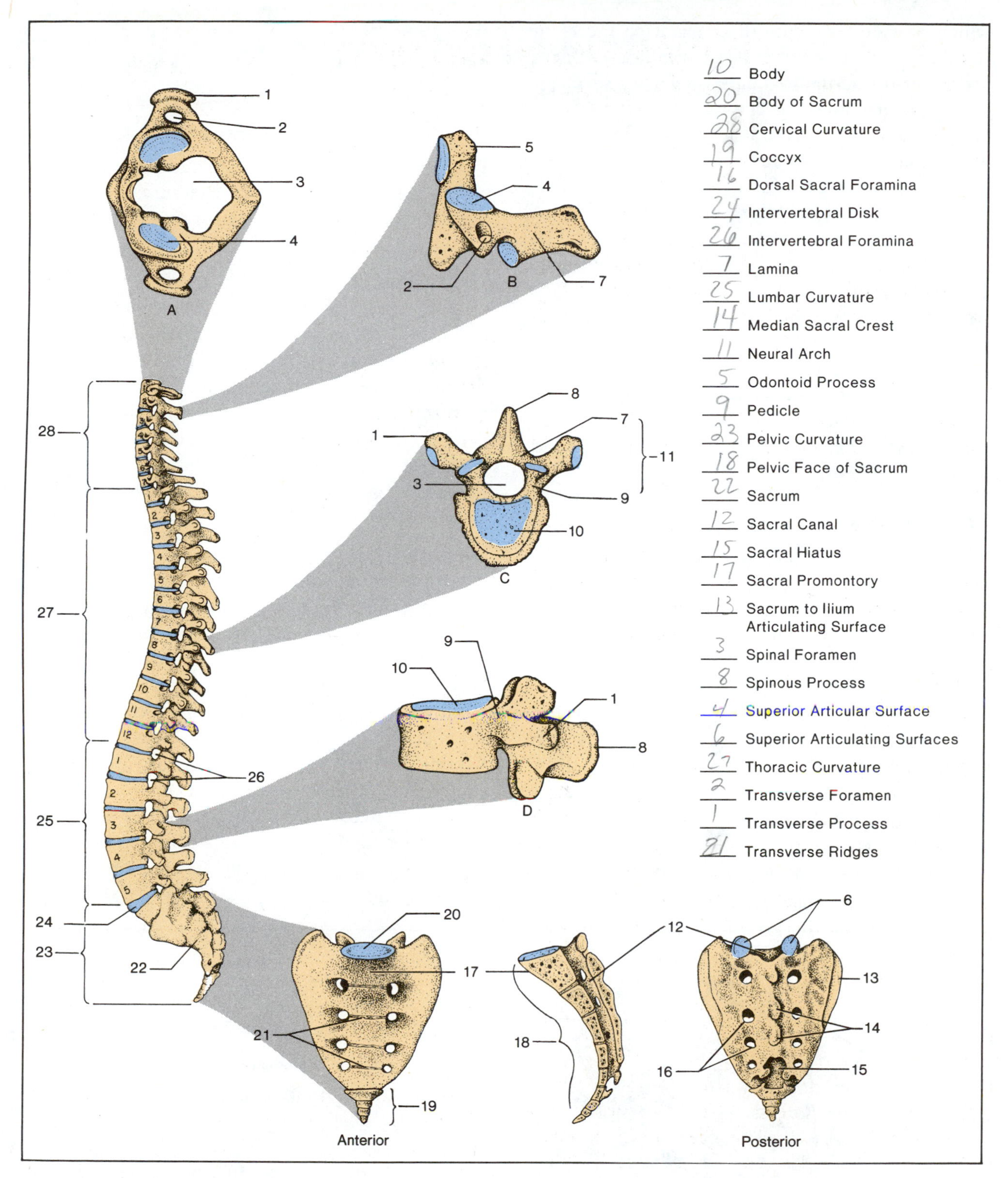

Figure 13.1 The vertebral column.

bones are much thicker than those of the other vertebrae due to the greater stress that occurs in this region of the vertebral column. Illustration D is of a typical lumbar vertebra.

The Sacrum

Inferior to the fifth lumbar vertebra lies the sacrum. It consists of five fused vertebrae. Note that there are two oval **superior articulating surfaces** (facets) which provide contact with articulating facets on the fifth lumbar vertebra. On its lateral surfaces are a pair of **sacrum to ilium articulating surfaces.** The intervertebral disk between the fifth lumbar vertebra and the sacrum contacts the flat surface of the **body of the sacrum.**

Note how the anterior aspect, or **pelvic face,** of the sacrum curves backward and that the body of the first sacral vertebra forms a protrusion called the **sacral promontory.** Observe, also, that four **transverse ridges** can be seen on the pelvic face which reveal where the five vertebrae are fused together.

Identify the **median sacral crest** (label 14) and the **dorsal sacral foramina** on the posterior surface. The neural arches of the fused sacral vertebrae form the **sacral canal,** which exits at the lower end as the **sacral hiatus.**

The Coccyx

The "tailbone" of the vertebral column is the coccyx. It consists of four or five rudimentary vertebrae. It is triangular in shape and is attached to the sacrum by ligaments.

Spinal Curvatures

Four curvatures of the vertebral column, together with the intervertebral disks, impart considerable springiness along its vertical axis. Three of them are identified by the type of vertebrae in each region: the **cervical, thoracic,** and **lumbar curves.** The fourth curvature, which is formed by the sacrum and coccyx, is the **pelvic curve.**

Assignment:

Label figure 13.1.

The Thorax

The sternum, ribs, costal cartilages, and thoracic vertebrae form a cone-shaped enclosure, the *thorax.* Its components are illustrated in figures 13.2 and 13.3.

The Sternum The sternum, or breastbone, consists of three separate bones: the upper **manubrium,** the middle **body** *(gladiolus),* and the lower **xiphoid** *(ensiform)* **process.** A sternal angle is formed where the inferior border of the manubrium articulates with the body. On both sides of the sternum are notches (facets) where the sternal ends of the costal cartilages are attached. Note that the second rib fits into a pair of *demifacets* (*demi,* half) at the sternal angle.

The Ribs There are twelve pairs of ribs. The first seven pairs attach directly to the sternum by costal cartilages and are called **vertebrosternal** or **true ribs.** The remaining five pairs are called **false ribs.** The upper three pairs of false ribs, the **vertebrochondral ribs,** have cartilaginous attachments on their anterior ends, but do not attach directly to the sternum. The lowest false ribs, the **vertebral** or **floating ribs,** are unattached anteriorly.

Figure 13.3 illustrates the structure of a central rib, a lateral view of a thoracic vertebra, and articulation details. Although considerable variability exists in size and configuration of the var-

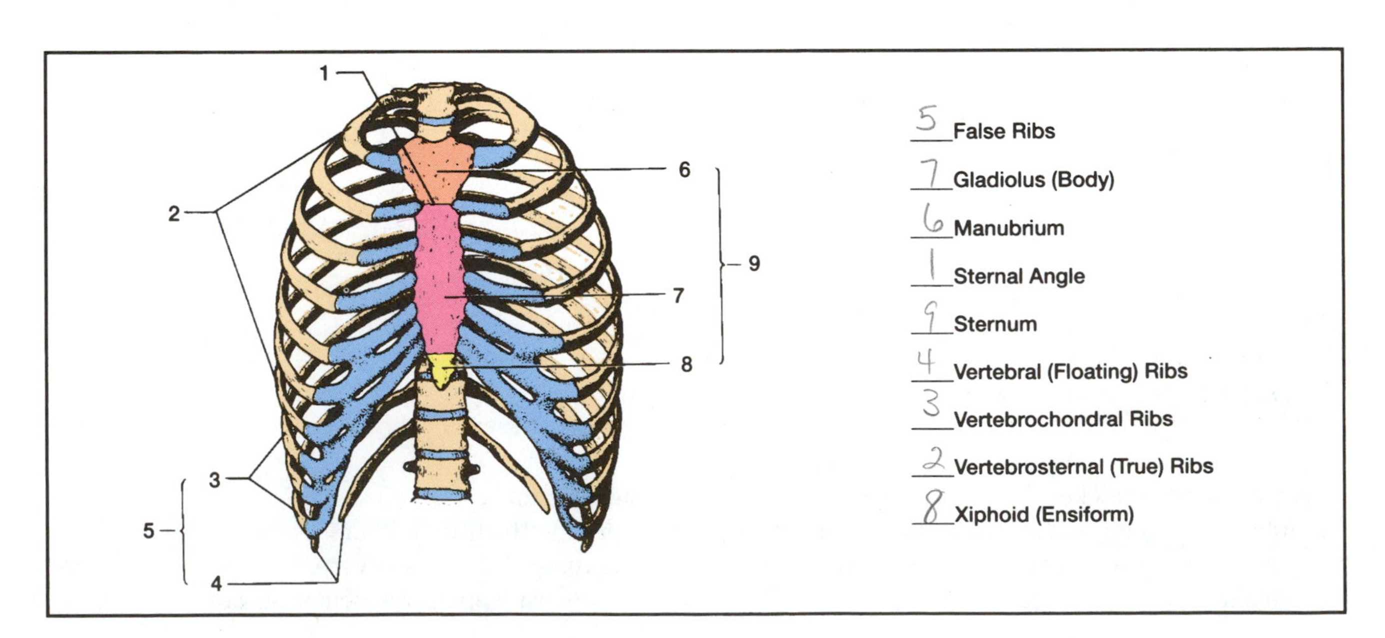

Figure 13.2 The thorax.

ious ribs, the central rib reveals structures common to most ribs.

The principal parts of each rib are a head, neck, tubercle, and body. The **head** (label 15) is the enlarged end of the rib that articulates with the vertebral column. The **tubercle** (label 17) consists of two portions: an **articular portion** and a **nonarticular portion.** The **neck** is a flattened portion, about 2.5 cm long, between the head and tubercle. The **body** is the flattened curved remainder of the rib.

Note that the head of this rib has two **articular facets** on its medial surface that enable contact with two adjacent vertebrae. Between these two facets is a roughened **interarticular crest** to which liga-

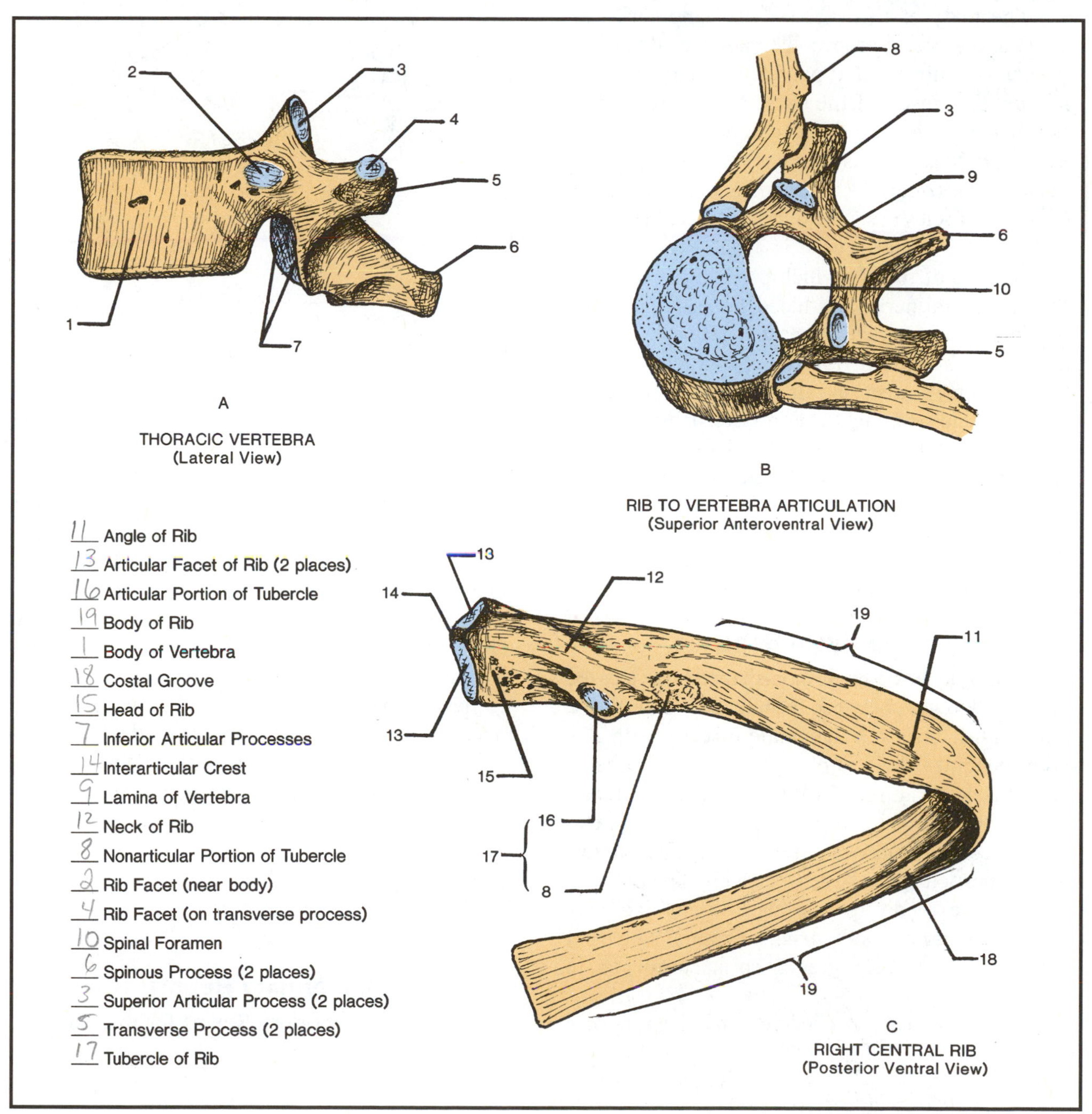

Figure 13.3 Rib anatomy and its articulation.

mentous tissue is anchored. The 1st, 2nd, 10th, 11th, and 12th ribs have only a single articular facet on their heads.

Articulation of each rib occurs at two points on each vertebra, as shown in illustration B. Note that the articular portion of the tubercle contacts a facet on the transverse process.

The body of this rib has two landmarks: an angle and a costal groove. The **angle** is a ridge on the external surface that provides anchorage for the *iliocostalis* muscle of the back. Note that the portion between the angle and tubercle is rough and irregular; it is on this surface that another back muscle, the *longissimus dorsi,* is attached. Both of these muscles are shown in figure 26.2 (labels 8 and 9). The **costal groove** is a depression on the ventral side of the rib which provides a recess for the intercostal nerve and blood vessels.

In addition to revealing the location of the two rib facets, the lateral view of the thoracic vertebra (illustration A) also reveals the **superior articular process** which contacts the inferior articular process of the vertebra above it. On its underside are seen the two **inferior articular processes.**

Assignment:

Label figures 13.2 and 13.3.

The Hyoid Bone

The hyoid bone is a horseshoe-shaped bone located in the neck region between the mandible and larynx. Although it does not articulate directly with any other bone, it is held in place by various ligaments and muscles. Figure 13.4 illustrates its structure.

The bone consists of five segments: a body, two greater cornua, and two lesser cornua. The massive central portion of the bone is the **body.** The two long arms that extend out from each side of the body are the **greater cornua** (*cornu,* singular). Note that the distal end of each greater cornu terminates in a tubercle.

The **lesser cornua** are the conical eminences that are located on the sides of the body superior to the bases of the greater cornua. These structures are separate bone entities that are connected to the body, and occasionally to the greater cornua, by fibrous connective tissue. These joints are usually diarthrotic (freely movable), but occasionally become fused (ankylosed) in later life.

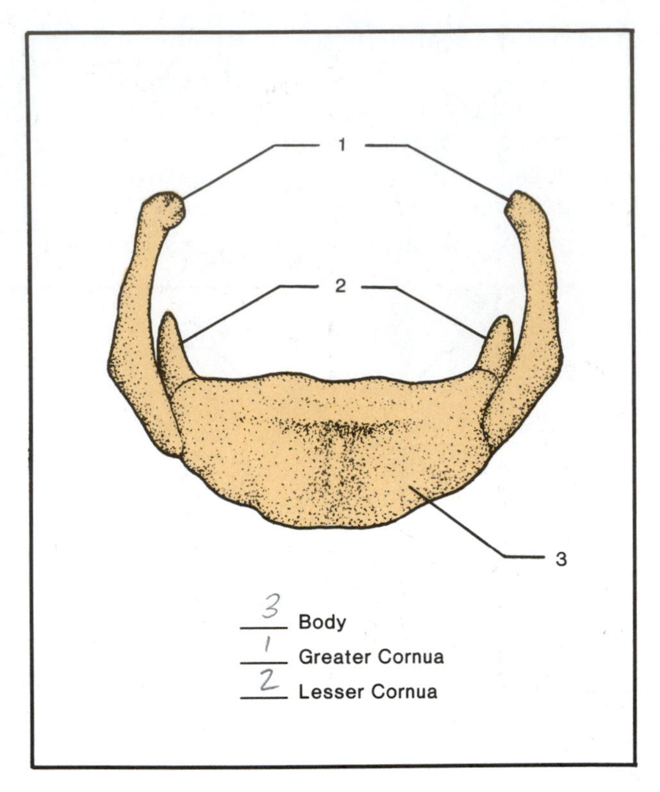

Figure 13.4 The hyoid bone.

The following muscles have points of attachment on the hyoid bone: *mylohyoideus, sternohyoideus, omohyoideus, thyrohyoideus, digastricus, hyoglossus, genioglossus,* and *constrictor pharyngis medius.*

Assignment:

Label figure 13.4.

Laboratory Report

Complete the Laboratory Report for this exercise.

The Appendicular Skeleton 14

In this exercise a study will be made of the individual parts of the upper and lower extremities. For this exercise the following materials should be available:

Materials:

skeleton, articulated
skeleton, disarticulated
male pelvis and female pelvis

The Upper Extremities

The upper extremities consist of the shoulder girdle and arm. As you locate the following structures in figures 14.1 and 14.2 compare the illustrations with the bones on a complete skeleton.

Shoulder Girdle Figure 14.2 illustrates the right arm and its attachment to the trunk. The **clavicle** is a slender S-shaped bone which articulates with the manubrium of the sternum on its medial end and the acromion process of the scapula on its lateral end.

The **scapula** of the shoulder girdle is a triangular bone which has a socket, the **glenoid cavity,** into which the head of the humerus fits. The scapula is not attached directly to the axial skeleton; rather, it is loosely held in place by muscles providing more mobility to the shoulder.

Figure 14.1 illustrates the anatomical details of the scapula. Note that on its posterior surface there is an elongated diagonal ridge called the **scapular spine** (label 4), above which is a depression, the **supraspinatus fossa;** inferior to the spine is another depression, the **infraspinatus fossa.** On the anterior surface of the bone is a large depression called the **subscapular fossa** (label 13). These fossae provide anchorage for many shoulder muscles.

Two prominent processes, the acromion and coracoid, are best seen on the lateral aspect. The **acromion process** (label 10) lies posterior and superior to the glenoid cavity. The **coracoid process** lies superior and anterior to the same cavity.

The scapula is bounded by three margins. The **vertebral** (medial) **margin** is the convex margin on the left side of the posterior aspect. This border extends from the **superior angle** (label 2) down to the **inferior angle** at the bottom. The **axillary** (lateral) **margin** makes up the border opposite to the vertebral border; it extends from the glenoid cavity down to the inferior angle. The third border is the **superior margin** that extends from the superior angle to a deep depression called the **scapular notch.**

Assignment:

Label figure 14.1.

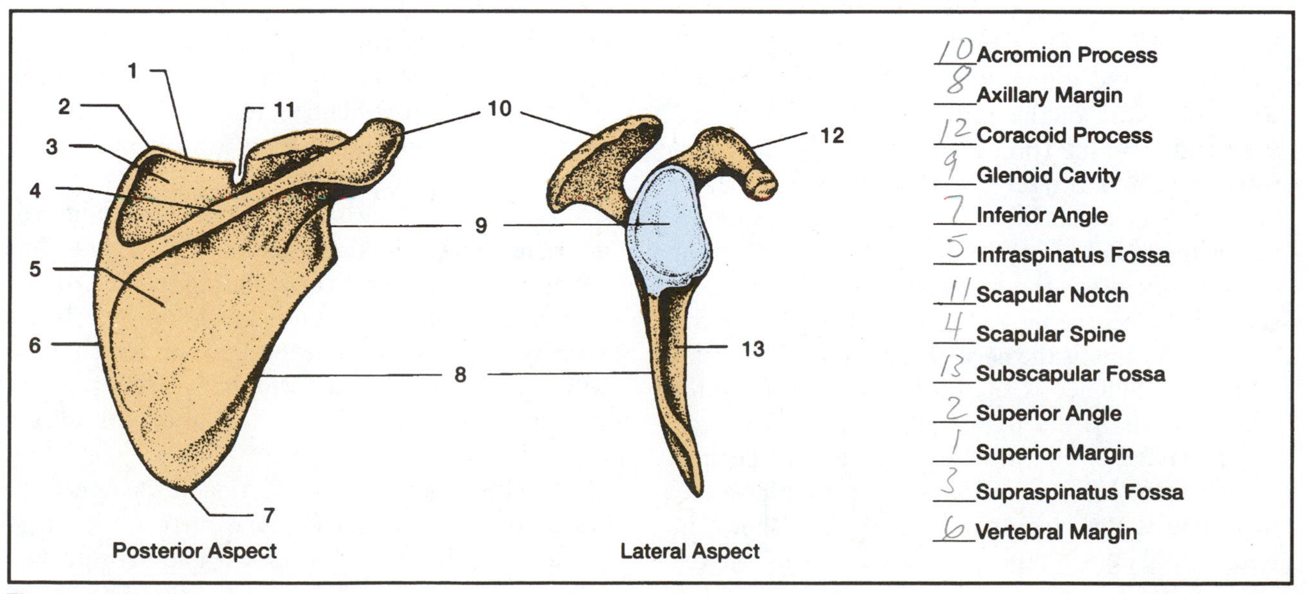

Figure 14.1 The scapula.

Upper Arm The skeletal structure of the upper arm consists of a single bone, the **humerus.** It consists of a shaft with two enlarged extremities. The smooth rounded upper end which fits into the glenoid cavity of the scapula is the **head.** Just below the head is a narrowing section called the anatomic neck. Inferior to the head and anatomic neck are two eminences, the greater and lesser tubercles. The **greater tubercle** is the larger process which is lateral to the **lesser tubercle.** Below these two tubercles is the **surgical neck,** so-named because of the frequency of bone fractures in this area.

The surface of the shaft of the humerus has a roughened raised area near its midregion which is the **deltoid tuberosity.** In this same general area is also an opening, the **nutrient foramen.**

The distal terminus of the humerus has two condyles, the capitulum and trochlea, which contact the bones of the forearm. The **capitulum** is the lateral condyle that articulates with the radius. The **trochlea** is the medial condyle that articulates with the ulna. Superior and lateral to the capitulum is an eminence, the **lateral epicondyle.** On the opposite side (the medial surface) is a larger tuberosity, the **medial epicondyle.** Above the trochlea on the anterior surface is a depression, the **coronoid fossa.** The posterior surface of this end of the humerus has a depression, the **olecranon fossa** (not shown in fig. 14.2).

Forearm The radius and ulna constitute the skeletal structure of the forearm. Figure 14.2 shows the relationship of these two bones to each other and the hand.

The **radius** is the lateral bone of the forearm. The proximal end of this bone has a disk-shaped **head** which articulates with the capitulum of the humerus. The disklike nature of the head makes it possible for the radius to rotate at the upper end when the palm of the hand is changed from one position to another (pronation-supination, see illustrations H and I, figure 18.1). A few centimeters below the head on the medial surface is an eminence, the **radial tuberosity.** This process is the point of attachment for the *biceps brachii,* a muscle of the arm. The region between the head and the radial tuberosity is the **neck** of the radius. The distal lateral prominence of the radius which articulates with the wrist is the **styloid process.**

The **ulna,** or elbow bone, is the largest bone in the forearm. The proximal posterior prominence of this bone is the **olecranon process** (not shown in figure 14.2). Within this process is a depression, the **semilunar notch,** which articulates with the trochlea of the humerus. The eminence just below the semilunar notch on the anterior surface is the **coronoid process** (label 24). Where the head of the radius contacts the ulna is a depression, the **radial notch.** The lower end of the ulna is small and terminates in two eminences: a large portion, the **head,** and a small **styloid process.** The head articulates with a fibrocartilaginous disk which separates it from the wrist. The styloid process is a point of attachment for a ligament of the wrist joint.

Hand Each hand consists of a carpus, a metacarpus, and phalanges. The **carpus,** or wrist, consists of eight small bones arranged in two rows of four bones each. The **metacarpus,** or palm, consists of five metacarpal bones. They are numbered from one to five, the thumb side being one. The **phalanges** are the skeletal elements of the fingers. They are distal to the metacarpal bones. There are three phalanges in each finger and two in the thumb.

Assignment:

Label figure 14.2.

The Lower Extremities

Pelvic Girdle The two hipbones (*ossa coxae*), articulate in front to form a bony arch called the *pelvic girdle.* The back of this arch is formed by the union of the hipbones with the sacrum and coccyx. The enclosure formed within this bony ring is the **pelvis** (Latin: *pelvis,* basin). Figure 14.4 illustrates one half of the pelvis and the right leg.

Figure 14.3 illustrates the lateral and medial views of the right os coxa. The large circular depression into which the head of the femur fits is the **acetabulum.** Note that each os coxa consists of three fused bones (ilium, ischium, and pubis) that are joined together in the center of the acetabulum. Although these ossification lines are readily discernible in young children, they are generally obliterated in the adult.

Two articulating surfaces (labels 15 and 16) are visible on the medial aspect of the os coxa. The upper one is the **sacrum articulating surface,** which interfaces with the sacrum; the lower one is the **symphysis pubis articulating surface** which joins with the other os coxa to form the symphysis pubis. The line of juncture between the ilium and sacrum is the sacroiliac joint.

The **ilium** is the yellow portion of the bone. On the greater portion of its medial surface is a large concavity called the **iliac fossa.** The **arcuate line** (label 11) is a ridge of demarcation below the iliac fossa.

The upper margin of the ilium, which is called the **iliac crest,** extends from the **anterior superior spine** (label 9) to the **posterior superior spine** (label 2). Just below the latter process is the **posterior inferior spine.** An anterior inferior spine is seen on the opposite margin below the anterior superior spine.

The reddish colored portion of the os coxa is the **ischium.** Note that it has two distinct processes: a small eminence, the **ischial spine,** and a larger one, the **tuberosity** of the ischium, which makes up the lower bulk of the bone. It is on this tuberosity that one sits on. Just below the ischial spine is a depression, the **lesser sciatic notch.** The larger depression superior to it is the **greater sciatic notch.**

The orange colored bone is the **pubis.** The

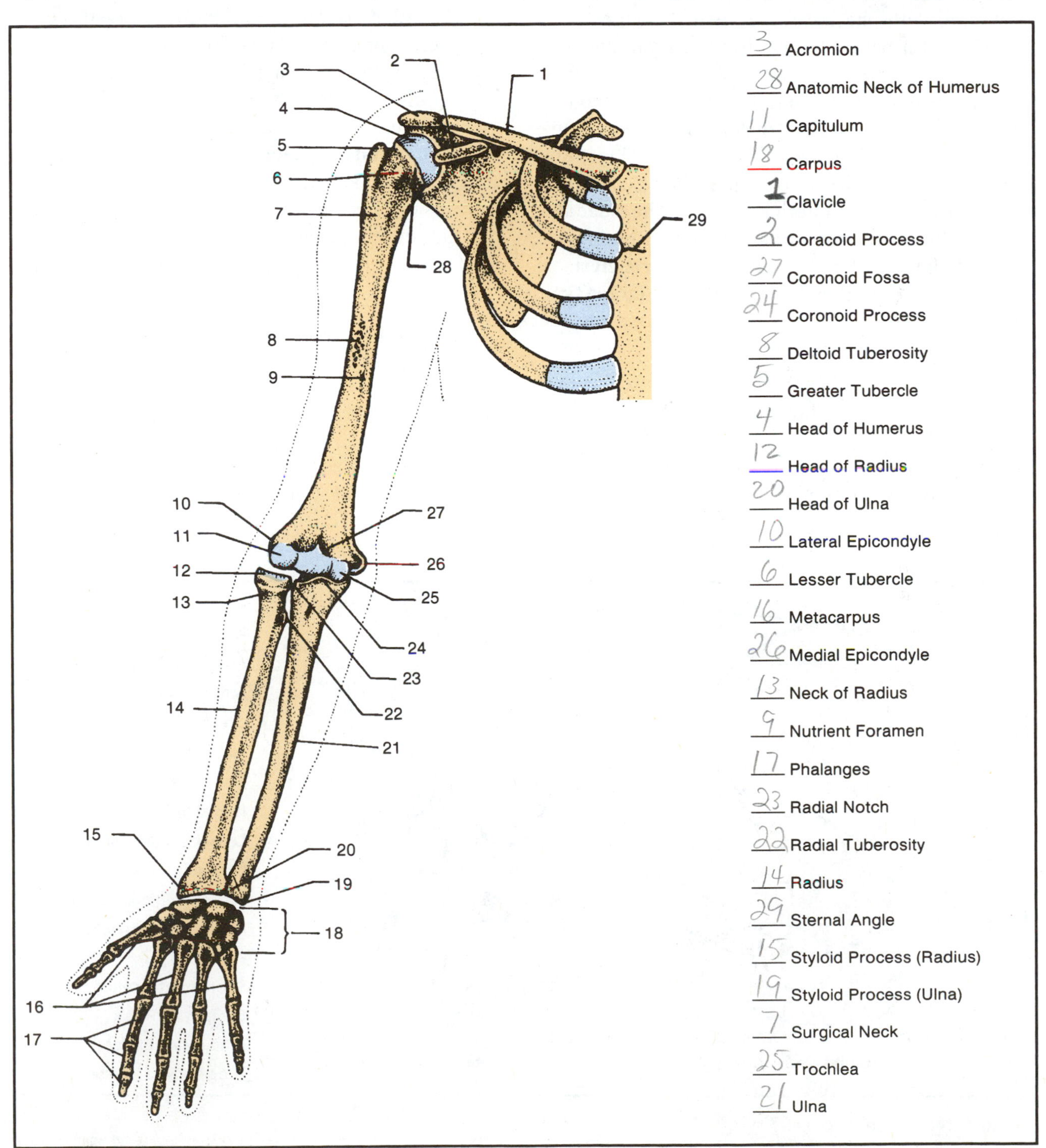

Figure 14.2 The arm and shoulder girdle.

opening surrounded by the pubis and ischium is the **obturator foramen.**

Assignment:

Label figure 14.3.

Compare a male pelvis with a female pelvis and answer questions on Laboratory Report 14,15 that pertain to their anatomical differences.

Upper Leg Skeletal support of the thigh is achieved with one bone, the **femur.** Its upper end consists of a hemispherical **head,** a **neck,** and two eminences, the greater and lesser trochanters. The **greater trochanter** is the large process on the lateral surface. The **lesser trochanter** is located further down on the medial surface. A ridge, the **intertrochanteric line,** lies obliquely between the two trochanters on the anterior surface. The ridge between these two trochanters on the posterior surface of the femur is the **intertrochanteric crest.**

The lower extremity of a femur is larger than the upper end and is divided into two condyles: the **lateral** and **medial condyles.**

Lower Leg The tibia and fibula constitute the skeletal structure of the lower leg. The **tibia,** or shinbone, is the stronger bone of this part of the leg. Its upper portion is expanded to form two condyles and one tuberosity. The condyles (label 7 and 19, figure 14.4) are named according to their location: **medial** and **lateral condyles.** The **tibial tuberosity** is located just below these condyles on the anterior surface of the tibia. The distal extremity of the tibia is smaller than the upper portion. A strong process, the **medial malleolus,** of the distal end forms the inner prominence of the ankle. The tibia is somewhat triangular in cross section, with a sharp ridge on its anterior surface, the **anterior crest.**

The **fibula** is lateral to the tibia and parallel to it. The upper extremity, or **head,** articulates with the tibia, but it does not form a part of the knee joint. Below the head is the **neck** of the fibula. The lower extremity of the fibula terminates in a pointed process, the **lateral malleolus,** which lies just under the skin forming the outer ankle bone. Locate this process on your own leg. Like the tibia, the fibula has an **anterior crest** extending down its anterior surface.

Foot Each foot consists of a tarsus, metatarsus, and phalanges. The **tarsus,** or ankle, consists of seven tarsal bones. The *calcaneus,* or heelbone, is the largest tarsal bone. The tibia of the leg articulates with the *talus,* a tarsal bone on top of the foot.

The **metatarsus,** or instep, consists of five elongated **metatarsal** bones. They are numbered one

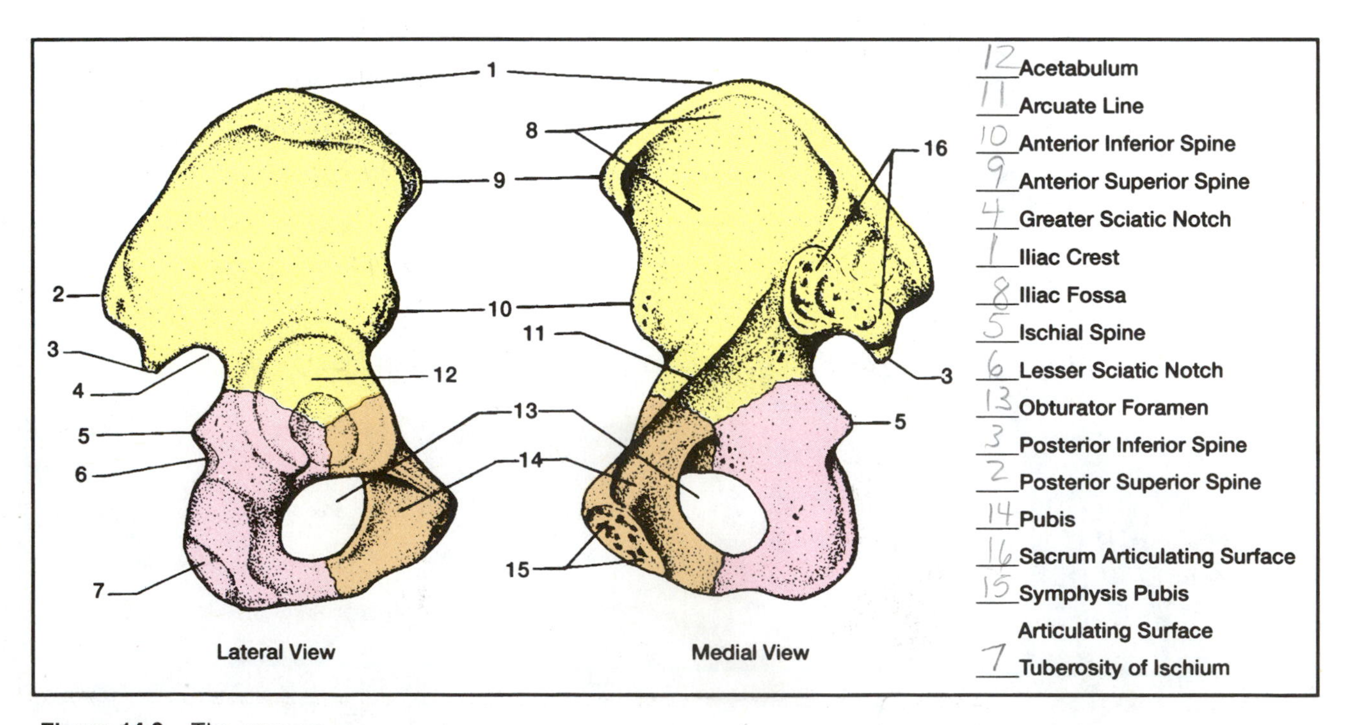

Figure 14.3 The os coxa.

through five, number one being on the medial side of the foot.

The **phalanges** are the bones of the toes. There are two phalanges in the great toe and three in each of the other toes.

Assignment:

Label figure 14.4.

Answer the remaining questions on the first part of the combined Laboratory Report 14,15 that pertain to this exercise.

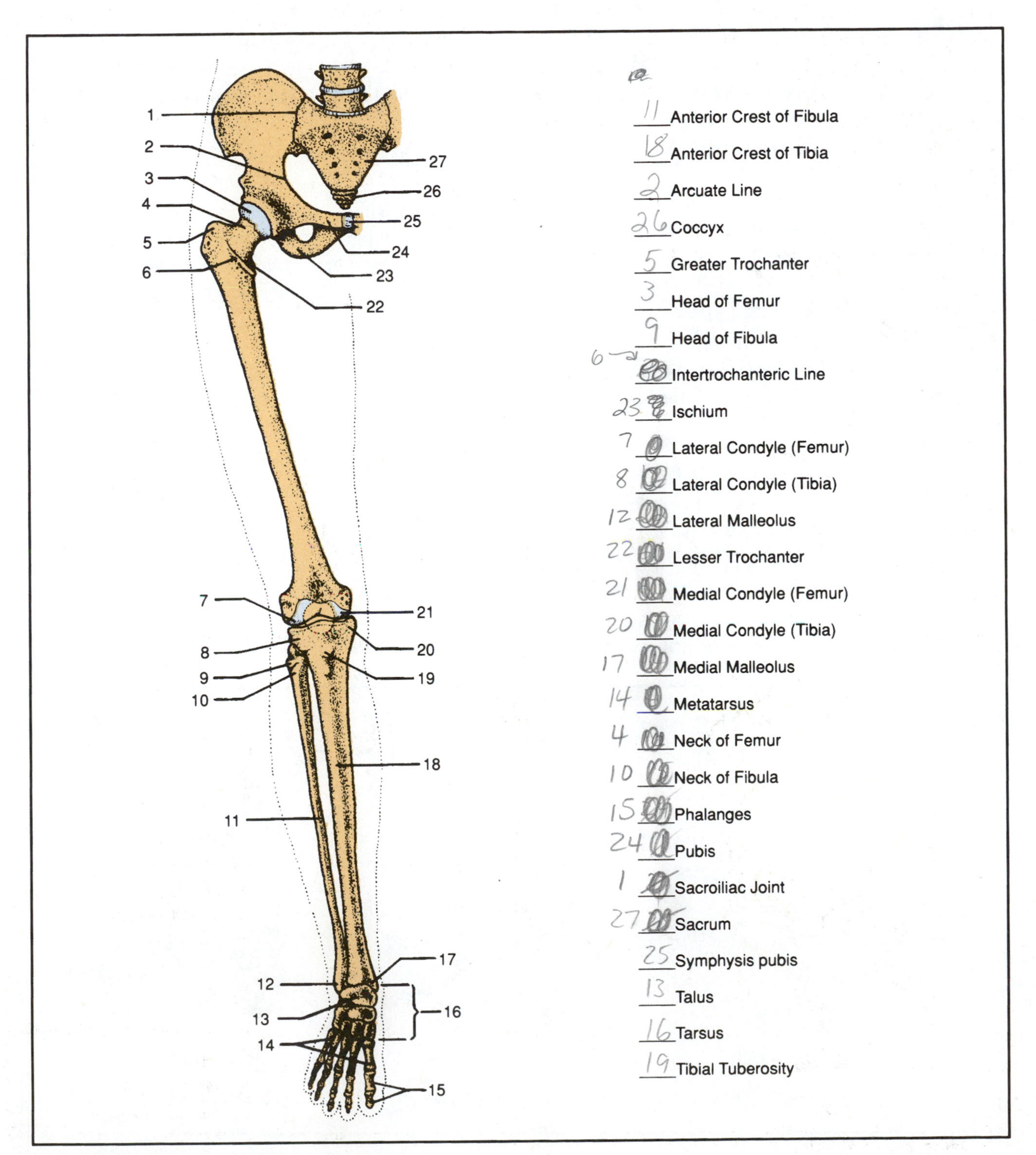

Figure 14.4 The leg and pelvic girdle.

15 Articulations

During this laboratory period we will study the basic characteristics of three types of joints, and the specifics of the shoulder, hip, and knee joints.

Materials:

fresh knee joint of cow or lamb, sawed through longitudinally

The Basic Types

All joints in the body fall into one of the three categories illustrated in figure 15.1. Note that classification is based on the degree of movement.

Immovable Joints

The lack of movement in these *synarthrotic* joints is due to the bonding effect of fibrous connective tissue or cartilage. Two types exist:

Sutures These irregular joints that exist between the flat bones of the cranium are held together by **fibrous connective tissue.** This tissue is continuous with the **periosteum** (label 2) on the external surface of the bone and the **dura mater** on the inner surface.

Synchondroses This type of immovable joint utilizes cartilage instead of fibrous tissue as the bonding tissue. The metaphyses of long bones in children are joints of this type. As was noted in Exercise 11, these growth areas in children consist of hyaline cartilage during the growing years.

Slightly Movable Joints

Slightly movable, or *amphiarthrotic,* joints are either symphyses or syndesmoses.

Symphyses Slightly movable joints that have a pad of fibrocartilage between the bones fall into this category. The intervertebral joints and symphysis pubis are representative.

Note in the intervertebral joint of illustration B that the bony surfaces are covered with **articular cartilage** (hyaline type). A pad of **fibrocartilage** (label 6) provides a cushion. These joints are held together with a fibroelastic **capsule.**

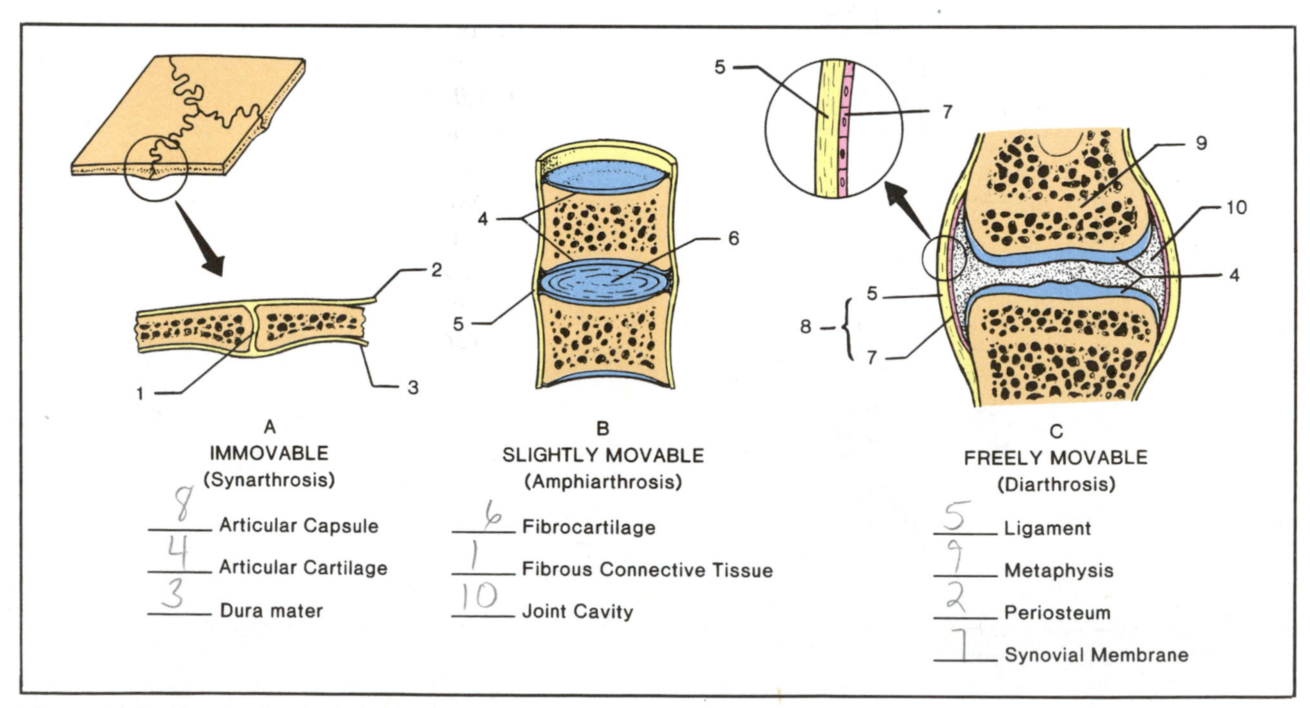

Figure 15.1 Types of articulations.

Syndesmoses Slightly movable joints that lack fibrocartilage and are held together by an interosseus ligament are of this type. A good example is the attachment of the fibula to the tibia, as seen in figure 15.4.

Freely Movable Joints

Articulations that move easily are designated as *diarthrotic,* or *synovial,* joints. Note in illustration C (figure 15.1) that the bone ends are covered with smooth **articular cartilage.** Holding the joint together is a fibrous **articular capsule** that consists of an outer layer of **ligaments** and an inner lining of **synovial membrane.** The latter membrane produces a viscous fluid, *synovium,* which lubricates the joint. These joints also contain sacs, or **bursae,** of synovial tissue. Fibrocartilaginous pads may also be present. Six different kinds of synovial joints are present in the body.

Gliding Joints The articular surfaces in these joints are nearly flat or slightly convex. Bones that exhibit gliding action are seen in between carpal bones of the wrist, and tarsals of the ankle.

Hinge Joints Joints such as the elbow, ankle, and knee that move in one direction only are of this type. Movement between the occipital condyles of the skull and the atlas is also hingelike.

Condyloid Joints Joints consisting of an oval-shaped head, or condyle, that moves in an elliptical cavity can move in two directions. The wrist is a typical condyloid joint.

Saddle Joints Like the condyloid joints, these joints allow movement in two directions. The bones' ends differ, however, in that each bone end is convex in one direction and concave in the other direction. The joint between the thumb metacarpal and adjacent carpal bone is of this type.

Pivot Joints A joint in which movement is rotational around an axis is a pivot type. Rotation of the atlas around the odontoid process of the axis is a good example.

Ball and Socket Joints These joints have angular movement in all directions. The shoulder and hip joints are representative.

Assignment:

Label figure 15.1.

The Shoulder Joint

The shoulder joint is the most freely movable articulation of the body. Its structure is revealed in figure 15.2. Note that the articulating surfaces of the humerus and glenoid cavity are covered with **articular cartilage.**

Two bursae are seen in this joint: the subdeltoid and subacromial. The **subdeltoid bursa** is the upper one that lies just under the **deltoid muscle** (label 1). It minimizes friction between the deltoid and humerus. The **subacromial bursa** is seen below the epiphysis of the humerus.

Other structures seen in this joint are the **supraspinatus muscle,** which is attached to the epiphysis of the humerus, the **clavicle,** and the **acromion process** (label 3).

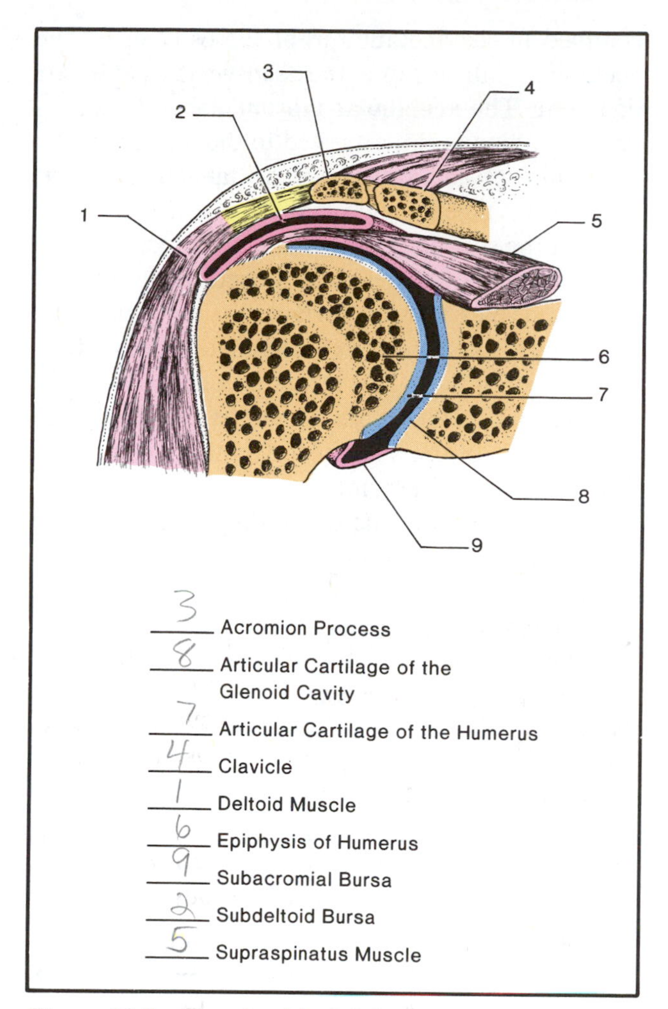

Figure 15.2 The shoulder joint.

Assignment:

Label figure 15.2.

The Hip Joint

The hip joint, another ball and socket joint, is shown in figure 15.3. The rounded head of the femur is confined in the acetabulum of the os coxa by the acetabular labrum and the transverse acetabular ligament. The **acetabular labrum** (label 8) is a fibrocartilaginous rim attached to the margin of the acetabulum. The **transverse acetabular ligament** (label 11) is an extension of the acetabular labrum. Note in the sectional view that a structure, the **ligamentum teres femoris,** is attached to the middle of the curved condylar surface of the femur. The other end of this ligament is attached to the surface of the acetabulum. This structure adds nothing to the strength of the joint; instead, it contributes to nourishment of the head of the femur and supplies synovial fluid to the joint.

The entire joint, including the preceding three structures, is enclosed in an **articular capsule.** This capsule consists of longitudinal and circular fibers surrounded by three external accessory ligaments. The three accessory ligaments are the iliofemoral, pubocapsular, and ischiocapsular. The **iliofemoral ligament** (label 7) is a broad band on the anterior surface of the joint. This ligament is attached to the **anterior inferior iliac spine** at its upper margin and the **intertrochanteric line** on its lower margin. Adjacent and medial to the iliofemoral ligament lies the **pubocapsular ligament.** The posterior surface of the capsule is reinforced by the **ischiocapsular ligament.** Lining the inside of the capsule is the **synovial membrane** which provides the lubricant for the joint.

Movements of flexion, extension, abduction, adduction, rotation, and circumduction of the thigh are readily achieved through this joint. This is made possible by the unique angle of the neck of the femur and the relationship of the condyle to the acetabulum. Excessive backward movement of the body at the joint is controlled to a great extent by the iliofemoral ligament, taking some of the strain from certain muscles.

Assignment:

Label figure 15.3.

The Knee Joint

Two views and a sagittal section of the knee are shown in figure 15.4. Although the action of this joint has been described earlier as being essentially hingelike, it is by no means a simple hinge. The curved surfaces of the condyles of the femur allow rolling and gliding movements within the joint. There is also some rotary movement due to the nature of hip and foot alignment.

The knee joint is probably the most highly stressed joint in the body. To absorb some of this stress are two *semilunar cartilages,* or *menisci,* in each joint. As revealed in the sagittal section, these fibrocartilaginous pads are thick at the periphery and thin in the center of the joint, providing a deep recess for the condyles. The **lateral meniscus** lies between the lateral condyle of the femur and the tibia; the **medial meniscus** is between the medial condyle and the tibia. These menisci are best seen in the anterior and posterior views of figure 15.4. Anteriorly and peripherally, they are connected by a **transverse ligament.**

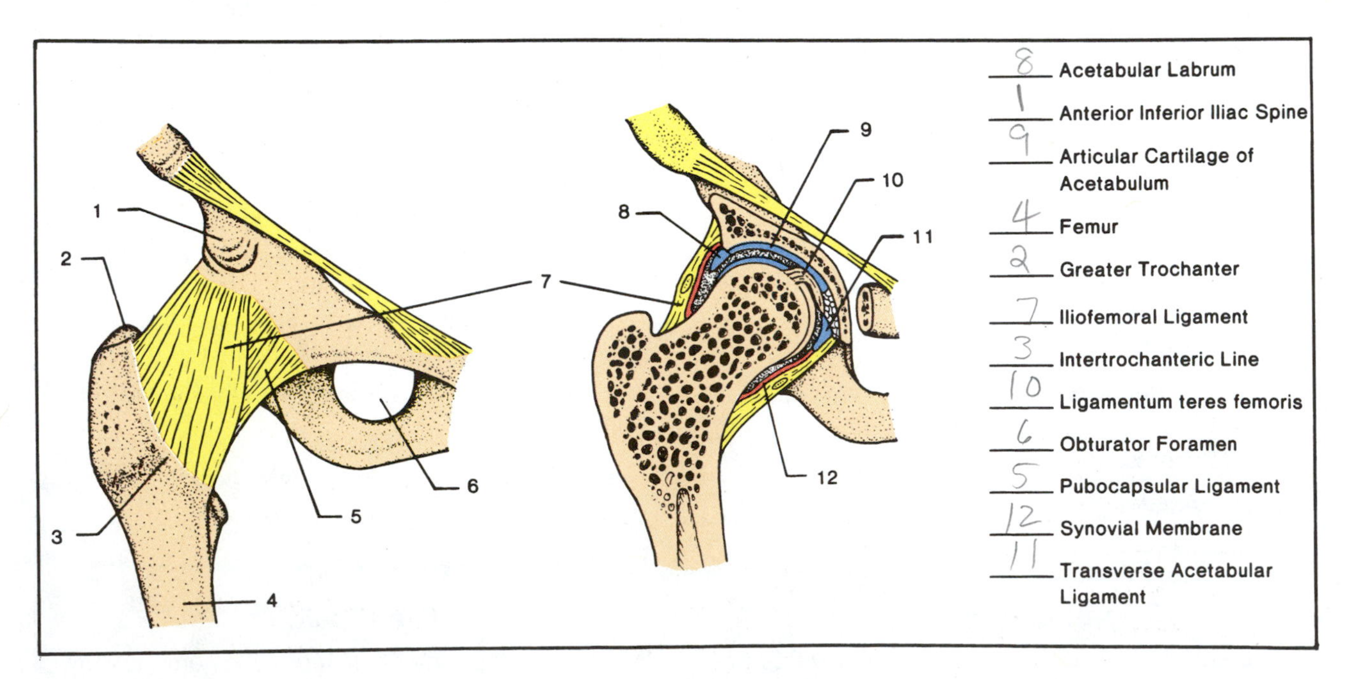

Figure 15.3 The hip joint.

The entire joint is held together by several layers of ligaments. Innermost are the two cruciate and two collateral ligaments. The **posterior** and **anterior cruciate ligaments** form an X on the median line of the posterior surface, with the posterior cruciate ligament being outermost. The anterior cruciate ligament is the outermost one seen on the anterior surface when the leg is flexed. The **fibular collateral ligament** is on the lateral surface, extending from the lateral epicondyle of the femur to the head of the fibula. The **tibial collateral ligament** extends from the medial epicondyle of the femur to the upper medial surface of the tibia. In addition to these four ligaments are the *oblique* and *arcuate popliteal ligaments* on the posterior surface. These are not shown in figure 15.4.

Encompassing the entire joint is the *fibrous capsule.* It is a complicated structure of special ligaments united with strong expansions of the muscle tendons that pass over the joint.

Observe in the sagittal section that the kneecap is held in place by an upper **quadriceps tendon** and a lower **patellar ligament.** The inner surfaces of these latter structures are lined with **synovial membrane.** The space between the synovial membrane and femur is the **suprapatellar bursa.**

It is significant that although this joint is capable of sustaining considerable stress, it lacks bony reinforcement to prevent dislocations in almost any direction. It relies almost entirely on soft tissues to hold the bones in place. It is because of this fact that knee injuries are so commonplace today among the general populace, as well as professional athletes. It is a vulnerable joint particularly to lateral and rotational forces. An understanding of its anatomy should alert one to its limitations.

Assignment:

Label figure 15.4.

Laboratory Assignment

Animal Joint Study Examine the knee joint of a cow or lamb that has been sawed through longitudinally. Identify as many of the structures as possible that are shown in figure 15.4.

Laboratory Report Complete the combined Laboratory Report 14,15.

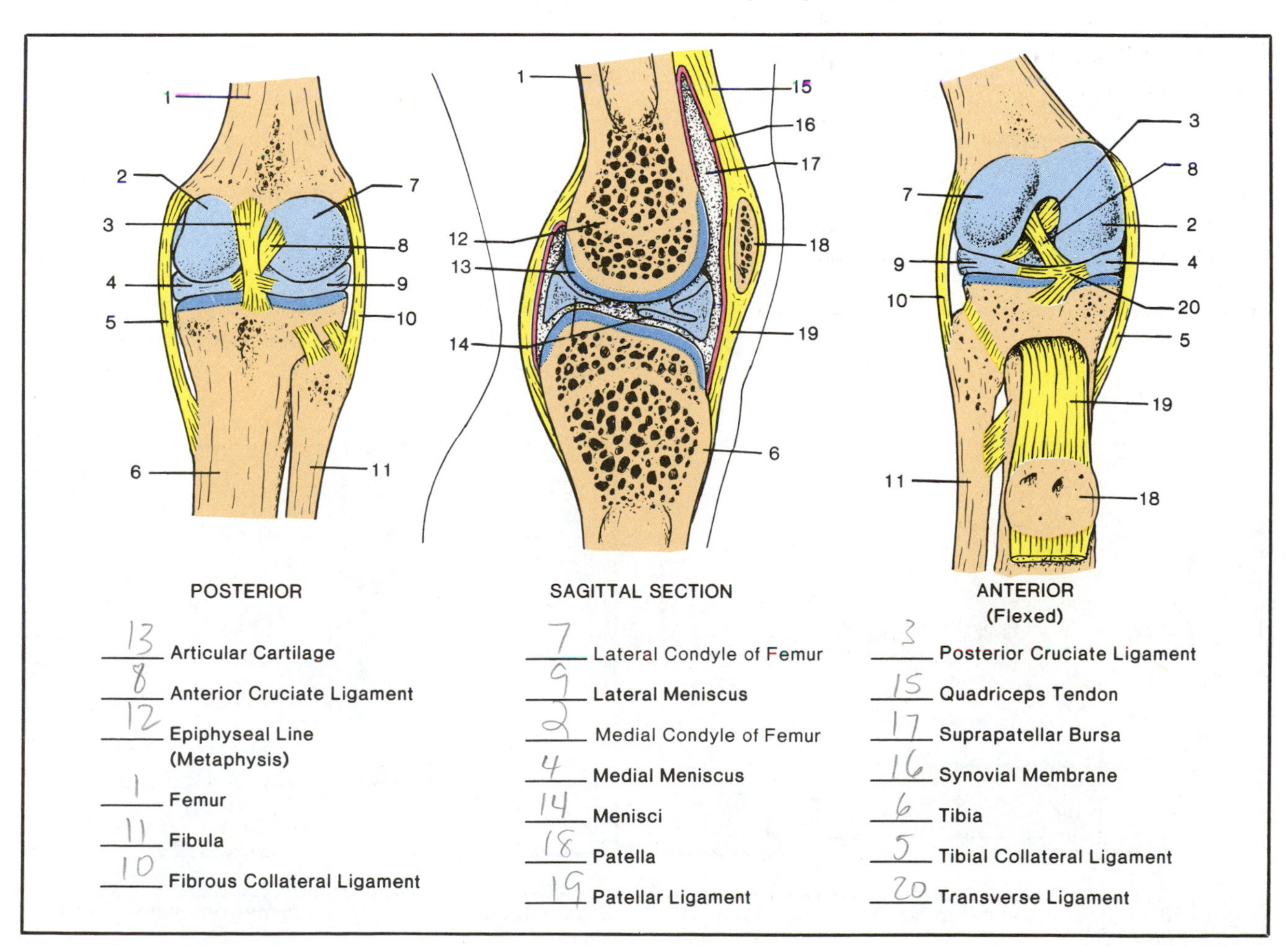

Figure 15.4 The knee joint.

16 Electronic Instrumentation

Various types of electronic equipment will be used in subsequent experiments that pertain to muscle contraction, heart action, respiration, etc. It is the purpose of this exercise to present a basic understanding of how this equipment is set up and how it works.

Figure 16.1 illustrates the arrangement of equipment that might be used in a typical instrumentation setup. Note that the biological phenomenon being studied, whether it produces an electrical or mechanical signal, must be picked up by either an electrode, or an input transducer. If the signal is electrical, an **electrode** will be used. Nonelectrical signals, however, must be converted to electrical signals before they can be measured by the setup. For the conversion of nonelectrical signals, such as temperature or pressure, an **input transducer** must be used.

Since biological phenomena often produce weak signals, the next unit in the system to receive the signal is the **amplifier,** where amplification takes place. Finally, the amplified signals are converted to some form of visual or audio display with an **output transducer,** such as a chart recorder, oscilloscope, computer monitor, or some other such type of instrument.

The considerable variety of components that one might encounter in a college laboratory precludes a complete description here of all types of equipment. Our main concern in this exercise will be to make a statement concerning the principal types that you are likely to encounter in this laboratory manual.

Electrodes

Two basic types of electrodes will be used in our experiments: pickup and stimulating electrodes. Pickup electrodes are the type discussed above that are used to receive a signal. EEG (electroencephalograph) and GSR (galvanic skin response) electrodes are of this type (see figure 16.2). Both of these electrodes work best when an electrode jelly, such as Biogel, is used on the skin to improve conductivity.

Electrodes that induce electrical stimulation into tissues to elicit a response are referred to as *stimulating electrodes.* This type will be used in some of our muscle experiments. An electronic stimulator must be used with this type of electrode.

Input Transducers

An *input transducer,* by definition, is any device that converts a nonelectrical form of energy to an electrical signal. The principal input transducers that we will use will be for measurements of force, pulse, and sound.

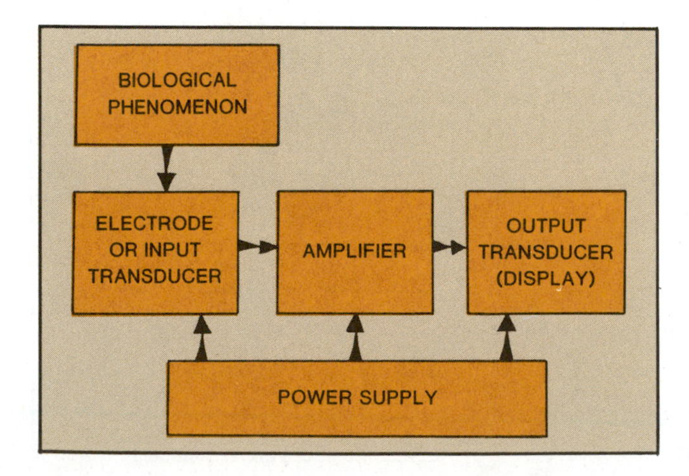

Figure 16.1 Equipment setup for monitoring biological phenomena.

Figure 16.2 EEG and GSR electrodes with electrode jelly.

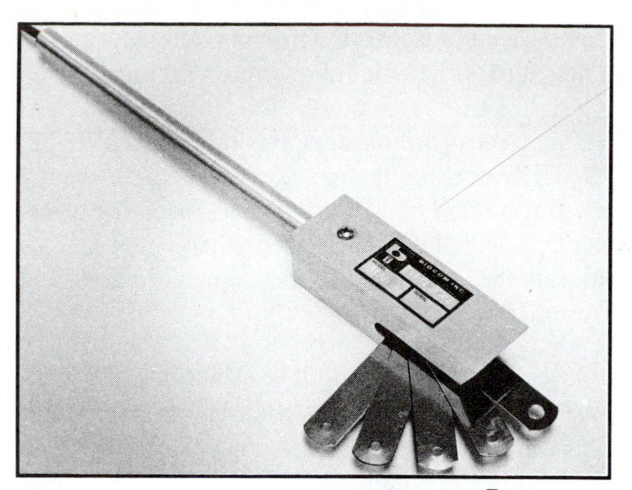

Figure 16.3 Bending of the steel leaves on this type of force transducer changes the resistivity in a stress-sensitive resistor in the transducer.

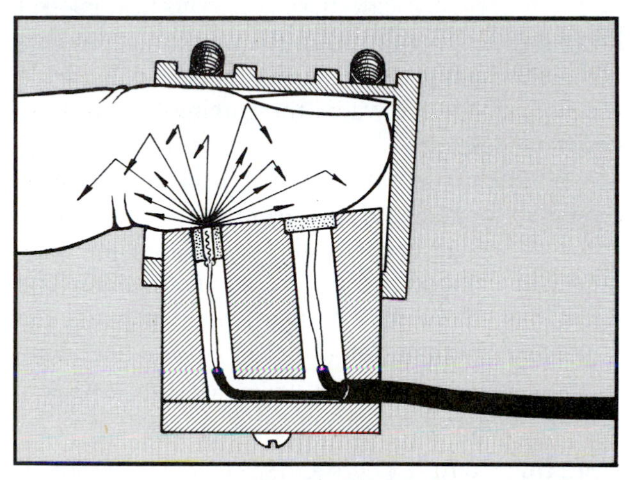

Figure 16.4 This pulse transducer utilizes a photosensitive resistor to detect differences in light intensities as blood surges through the tissues.

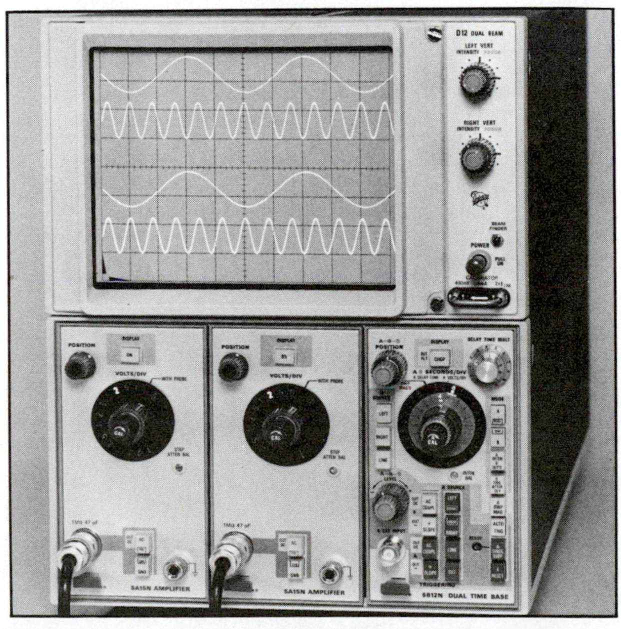

Figure 16.5 A dual-beam oscilloscope, manufactured by Tektronix, works well for two-channel experiments.

Force Transducers

Transducers of this type are able to convert the mechanical force exerted by a muscle into an electrical signal. Figure 16.3 illustrates the type that will be used in our muscle experiments. Since the force exerted by a muscle may vary from a weak pull to a very strong one, the transducer has a number of flexible leaves that can be used individually, or together. Weak forces may require only a single leaf; strong forces can be matched with two or more leaves. Figure 22.1 illustrates a setup in which this type of transducer is used.

The element in transducers of this type that makes this possible is a resistor that is affected by the stress applied to the leaves. As deformation occurs in the spring leaves, due to loading, proportional bending also occurs in the resistor. The induced stretching, or compression of the resistor changes its resistivity. Resistivity changes, in turn, result in changing electrical signals that pass to the amplifier, accomplishing the conversion of mechanical activity to electrical activity.

One complication with this type of transducer is that the resistor is part of a Wheatstone bridge that must be electronically balanced before each experiment is performed. This process of balancing the transducer resistor is outlined in Appendix C and will be utilized in those experiments where it is required.

Pulse Transducers

The pulse transducer illustrated in figure 16.4 utilizes a photoresistor and small light source to detect the presence or absence of the pulse in the finger. As blood surges through the finger (the pulse), the amplitude of light is altered, resulting in changes of reflected light that strike the photoresistor. The variations in light intensity striking the photoresistor generate a variable signal, which is fed to the amplifier and output transducer for interpretation.

Microphones

Microphones that pick up sound waves and convert them to electronic signals are used for studying heart and breathing sounds. An audio monitor (loud speaker) is often used in such setups.

Output Transducers

Chart recorders, oscilloscopes, audio monitors, and computer monitors are the principal kinds of output transducers used in the physiology laboratory today. In the past, chart recorders have been the instrument of choice; however, the computer with its monitor and a good printer is rapidly replacing many of the chart recorder applications. Two popular types of recorders will be discussed here as well as the cathode ray tube which is the basis for the oscilloscope and computer monitor.

The Cathode Ray Tube

The cathode ray tube (CRT) is an output transducer that visually displays input voltage signals on a fluorescent screen. Television screens, oscilloscopes (figure 16.5), and computer monitors are, basically, cathode ray tubes. Figure 16.6 illustrates the mechanics of a CRT. Note that the electrical potential difference, or **voltage,** of an input signal is expressed vertically, and the time base, or **sweep,** is shown horizontally. A stream of electrons, generated by an electron gun (cathode) at the back of the tube, is fed first between two vertical deflection plates and then between two horizontal plates. The deflection of the electron beam by the Y and X plates determines the trace that is shown on the screen by the stream of electrons that cause fluorescence on the screen of the tube.

The Unigraph

The Gilson Unigraph, figure 16.7, is a compact chart recorder that can be used for monitoring blood pressure, pulse rate, heart action, and many other physiological activities. Because of its small size it has been used in many of the experimental setups in this manual. Some brief comments about each of its controls follow.

Mode Selection Control This control knob (label 6 in figure 16.7) has the following six settings: EEG, ECG, CC-Cal, DC-CAL, DC, and Trans. For making electrocardiograms, the knob is set at ECG. The EEG setting is for electroencephalography (brain waves). The Trans setting is used for transducers. The Cal setting is used when one wishes to calibrate the instrument for certain recordings, such as electrocardiograms.

Styluses Instead of using ink, the stylus of a Unigraph is heated and marks the moving paper with a blue line as the heat acts on a heat sensitive chemical in the paper.

Two styluses are present: **a recording stylus** (label 4) and an **event marker** (label 3). Since stylus temperature is critical, there is a **stylus heat control knob** (label 8) to regulate the temperature. This control is usually set at the *two o'clock position.* If the styluses are too warm the tracing becomes excessively wide.

While the recording stylus is used to record the amplified input signal, the event marker is used to record when an event begins, or when some new pertinent influence affects the recorded event. The event may be recorded manually by depressing the white **event push button** (label 12) at the beginning of the event, or it may be recorded automatically if an **event synchronization cable** is used.

Centering Control (Zero Offset Control) Note on the left side of the control panel in figure 16.7 that

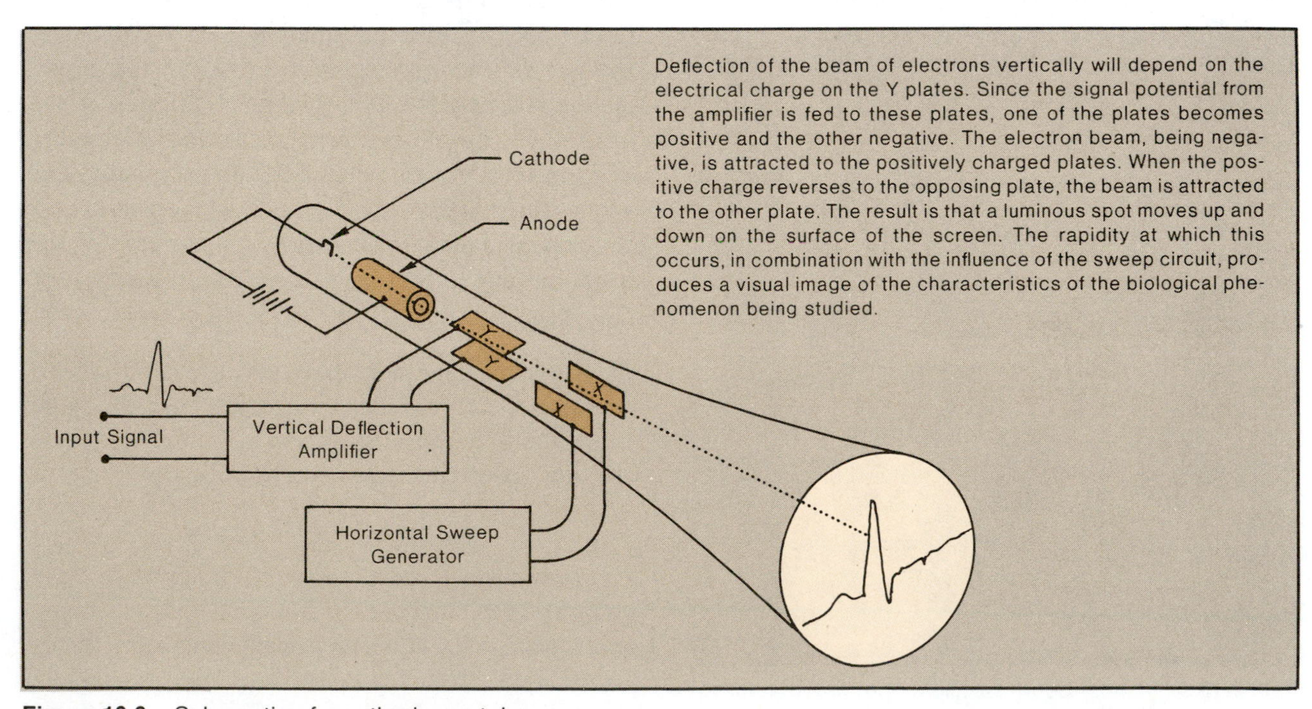

Figure 16.6 Schematic of a cathode ray tube.

there is a knob labeled CENTERING. This control is used for moving the baseline of the recording stylus to the most desirable position. It should be kept in mind that the most desirable location of the baseline is not always the center of the paper; it may be near one of the edges.

Chart Speed There are two speeds at which the paper can move on the Unigraph: 2.5 mm and 25 mm per second. The speed is controlled by the **chart speed control lever** (label 1). When pointed toward the end of the instrument, it is set at the faster speed. To change the speed to slow the lever is rotated 180° in the opposite direction. For the chart to move at all, however, the chart control switch has to be in the Chart On position.

Toggle Switches The Unigraph has four toggle switches, two of which are extensively used in its operation. The **main power switch** (label 7) is a small one with ON and OFF labels near it. It is located near the stylus heat control.

The large gray plastic switch (label 9) is the **chart control switch.** It has three settings: STBY, Chart On, and Stylus On. In the STBY (standby) position, the chart does not move and the stylus is deactivated. In the Stylus On position, the stylus heats up, but the chart does not move. At the Chart On position the chart moves and the stylus is activated.

The other two small toggle switches at the other end of the unit come into play only occasionally. They should be set at the NORM position for most experiments.

Sensitivity Controls Sensitivity of the Unigraph is controlled by the **gain control** (label 5) and the **sensitivity knob,** which has the label SENS near it. The gain control is calibrated in MV/CM (millivolts per centimeter) and it has a range of 2, 1, .5, .2, and .1 millivolts. The least sensitive position is 2; the greatest sensitivity is .1. When the sensitivity knob is turned clockwise the gain at any setting is increased toward the next level.

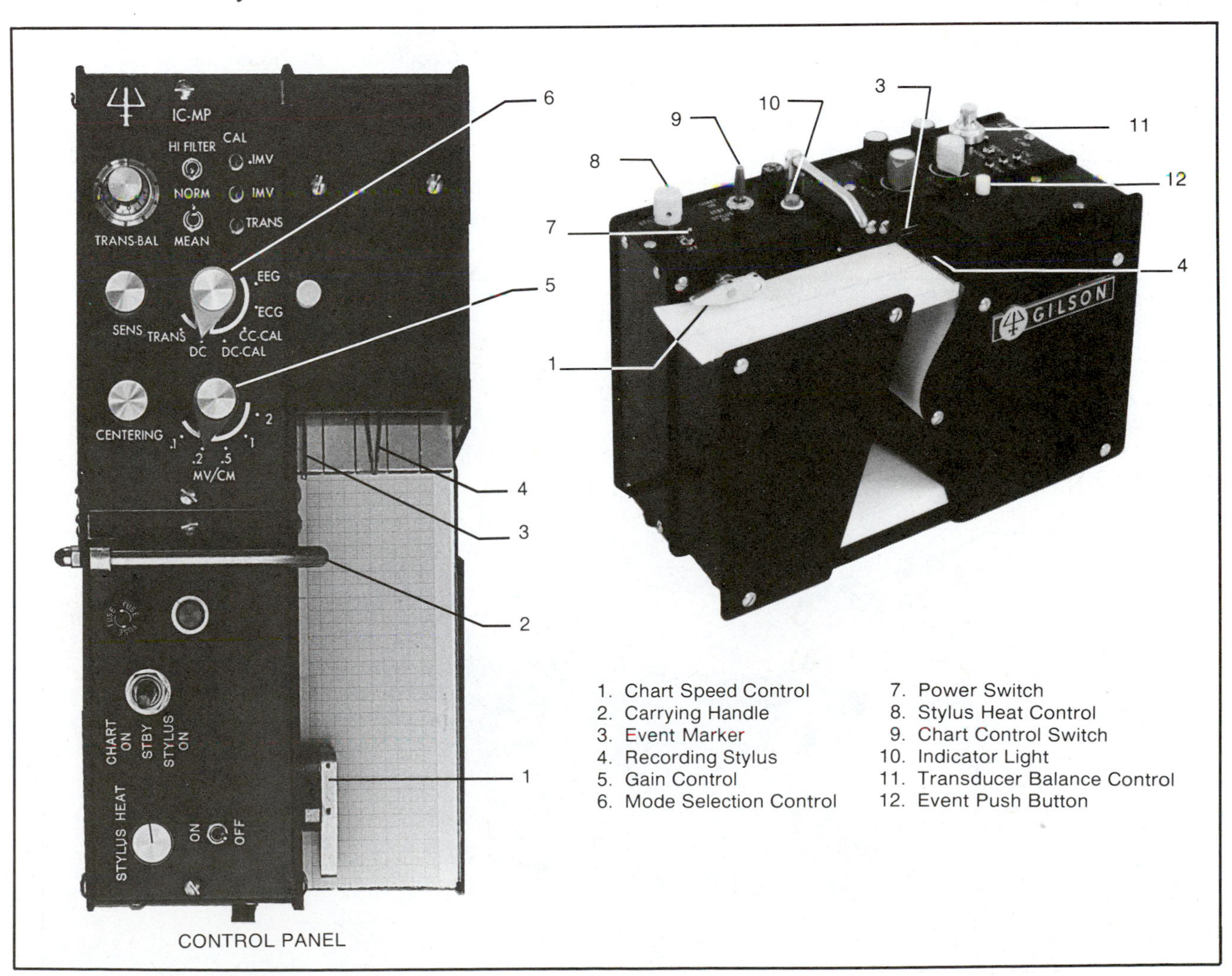

Figure 16.7 The Gilson Unigraph.

Unigraph Operational Procedures

When setting up the Unigraph in an experiment, the following points should be kept in mind.

Input The cable end of electrodes or transducer must be inserted into an input receptacle in the end of the Unigraph. The cable end must be compatible with the Unigraph. For some experiments it will be necessary to use an adapter unit between the cable end and the Unigraph.

Synchronization Cable If an electronic stimulator is to be used, it is desirable to have an event synchronization cable between the stimulator and the Unigraph.

Power Source The power cord on this instrument is three pronged. This means that it should only be used in a 110V grounded outlet. Before plugging it in, however, the power switch should be OFF and the chart control switch at STBY.

Stylus Heat Once the unit is plugged in, the power should be turned ON and the stylus heat control set at the two o'clock position.

After allowing a few minutes for the stylus to heat up, the trace should be tested by putting the chart control switch (c.c.) in the Chart On mode. If the line on the chart is too light, the darkness can be increased by turning the stylus heat control knob clockwise. This testing should be done with the chart speed set at the slow rate.

Balancing If a force transducer is going to be used it will be necessary to balance it, using the procedure outlined in Appendix C.

Calibration When monitoring EEG, ECG, or EMG it will be necessary to follow calibration instructions that are provided in Appendix C.

Mode Control Before running the experiment be sure to select the proper mode.

Sensitivity Control This control will usually be set first at a low level and then increased to the level that works best.

End of Experiment When finished with the experiment *always* return all controls to their least sensitive settings, or in the OFF position. The chart control switch should be left at STBY.

The Narco Physiograph®

When it is desirable to monitor two or more physiological activities simultaneously, one must use a two channel recorder, such as the Gilson Duograph, or a polygraph. While these units are often considerably larger than the Unigraph, they are basically the same in operation. Since the Narco Physiograph® will be used in several experiments in this manual, its features are described here.

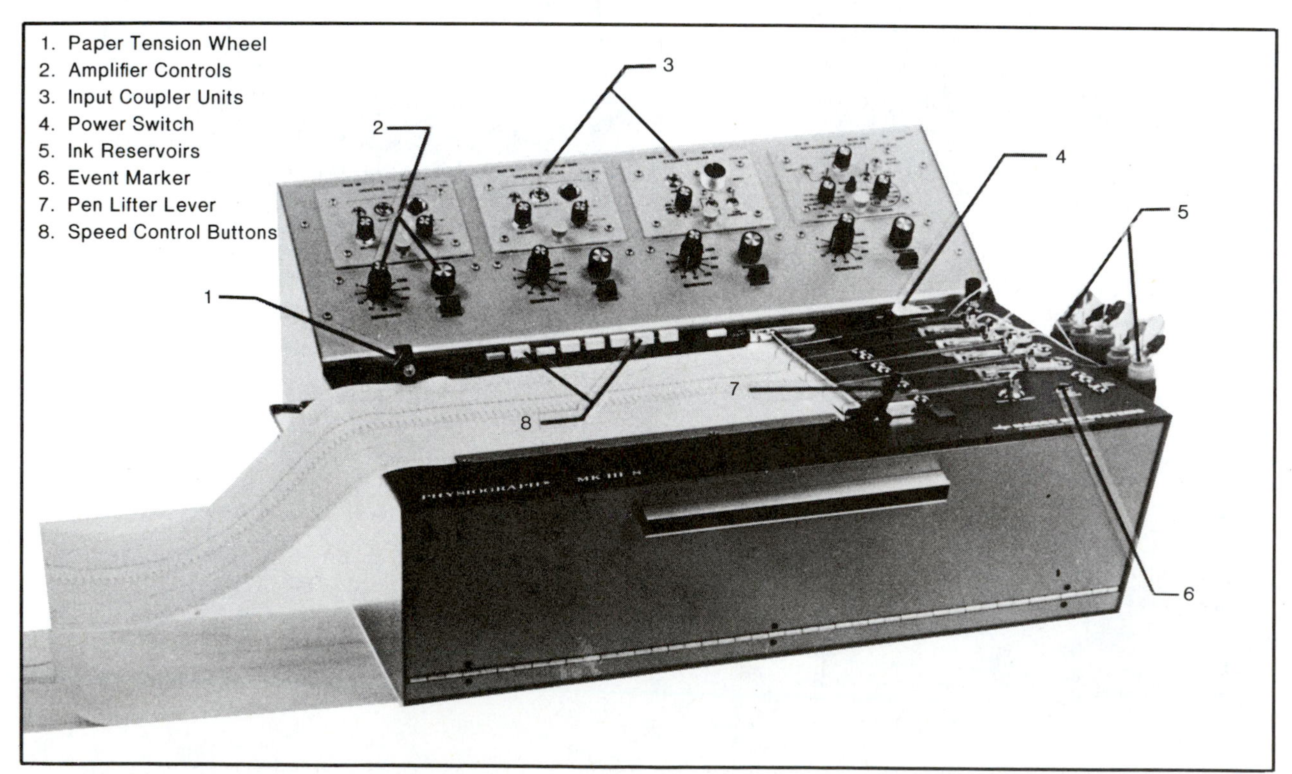

Figure 16.8 The Physiograph® Mark IIIS.

Figure 16.8 is a photograph of a four channel Narco Mark IIIS Physiograph®. It is manufactured by Narco Bio-Systems, Houston, Texas. Before utilizing this instrument in the laboratory, familiarize yourself with the following components.

Input Couplers Note that on the upper portion of the sloping control panel are four units called *input couplers* (label 3). It is into these units that the transducers or electrodes feed the input signals that are to be monitored. Couplers are, essentially, preamplifier units that modify the signals before they enter the amplifier units (label 2).

Couplers of various types are available from the manufacturer. At the present time Narco Bio-Systems produces thirteen different kinds of couplers. The Universal Coupler is probably the most widely used type on these units. It is equivalent to the Gilson Unigraph in that it can be used for monitoring ECG, EEG, EMG, and other DC and AC potentials. Other couplers that are available are the Strain Gage Coupler, GSR Coupler, Transducer Coupler, Temperature Coupler, etc.

Amplifier Controls Note in figure 16.8 that each input coupler has its individual cluster of amplifier controls located below it. There are two knobs and one push button on each of these clusters.

The control on the left, with nine settings, adjusts the sensitivity of the channel. It has an outer portion that enables one to select the desired millivolt setting: 1, 2, 5, 10, 20, 50, 100, 200, 500, or 1000. At setting 1, it takes only one millivolt to produce one centimeter pen movement; when set at 1000, it takes 1000 millivolts to produce the same amount of pen travel. Thus, we see that the 1 millivolt setting is the most sensitive setting available here. To adjust the sensitivity between any of these settings, one rotates the inner portion of this control. When the inner knob is completely clockwise, the sensitivity is identical to the reading on the dial. Counterclockwise rotation of the inner knob decreases the sensitivity.

The knob on the right with POSITION under it, is used for repositioning the pen to locate the baseline of your recording in the most desirable spot. The location of the baseline will be determined by the degree of pen travel. For example, if the pen deflection is to be as much as 6 centimeters, it would be desirable to set the baseline substantially below the center of the arc of the pen. The total range of deflection of the pen is approximately 8 centimeters.

The push button below the positioning knob is used for starting the recording process. When depressed, this button is in the ON mode which activates the pen. To stop the recording process one merely presses the button again, causing it to come up to a higher level which is the OFF position.

Ink Pens Note that the Mark IIIS has five ink pens: one for each of the four couplers and one to record one second time intervals at the bottom of the chart. Each pen is supplied by ink from five separate **ink reservoirs** (label 5). These pens can be raised from the paper with the **pen lifter lever** (label 7). This lever actuates a metal bar that lifts all the pens simultaneously off the paper. *The pens should always be kept in the raised position when recording is not taking place.*

Ink reaches each pen through a small tube that leads from its ink reservoir. To get the ink to flow to the pen it is necessary to first raise the reservoir 2–3 centimeters, and then squeeze the rubber bulb at the top of the reservoir as the index finger is held over the small hole at the tip of the rubber bulb. Once the tube is full of ink the fingertip is released before the bulb is released. When it becomes necessary to increase or decrease the flow of ink during recording, one simply raises or lowers the reservoir slightly.

To prevent clogging of pens it is necessary to remove all ink from the pens at the end of each

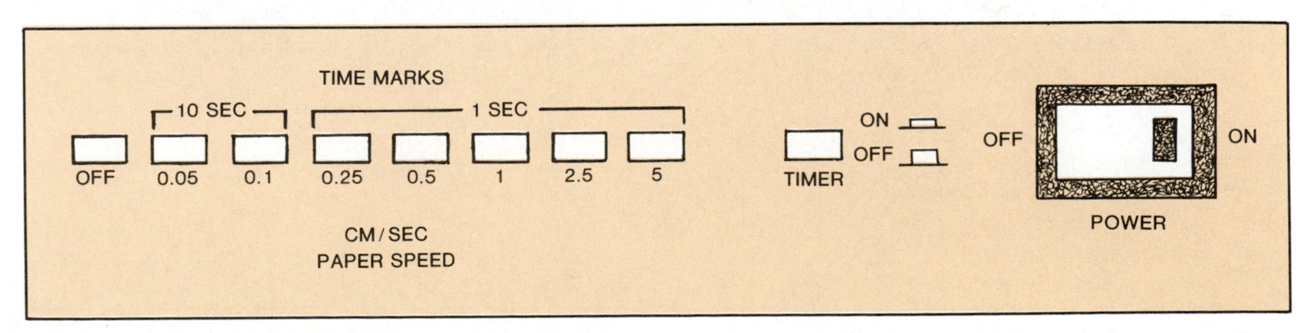

Figure 16.9 Push-button control panel of the Mark IIIS.

laboratory period. With the pens in the raised position, the ink is drawn back into the reservoir by squeezing the reservoir bulb, placing the index finger over the hole on the bulb, and then releasing pressure on the bulb. The vacuum, thus created, will draw all the ink back into the reservoir.

Chart Speed The Mark IIIS Physiograph® has seven push buttons (figure 16.9) that allow the following chart speeds: 0.05, 0.1, 0.25, 0.50, 1.0, 2.5, and 5.0 cm/sec. It also has an OFF push button on the left and a TIMER push button on the right end of the series.

To set the desired speed, one depresses the TIME button first (the up position is OFF), then the speed selection button is depressed, and finally, the OFF button is depressed (the up position for this button is ON). To change the speed, all one has to do is push a different speed button. To stop the paper movement, the OFF button is pressed. As soon as the paper stops moving, the pens should be raised with the pen lifter lever. At the end of the period it is important that both the OFF and TIME buttons be in the OFF position.

Physiograph® Operational Procedures

Although each experiment performed on the Physiograph® will differ in certain respects, there are some procedures that should always be followed. It is these general procedures that are outlined here:

Dust Cover The dust cover should be removed from the unit, neatly folded, and placed somewhere where it will not be in the way or damaged. Because laboratory dust can affect some electronic components over a period of time, always keep the unit covered during storage.

Recording Paper If there is no recording paper in the paper compartment, place a stack of paper in the compartment which is accessed through the hinged front panel of the instrument. Feed the paper up through the opening provided in the top right margin of the paper compartment and then slide the paper under the paper guides on each side.

Draw the paper across the top of the recording surface and slide it under the **paper tension wheel** (label 1) after lifting the wheel slightly by pulling the black lever straight up. After the paper is in place, release the lever. The paper should now be in firm contact with the paper drive mechanism and will move when the proper controls are activated.

Power Cord The power cord is a loose component that should be inserted first into the right side of the Physiograph, and then into a grounded electric outlet.

Power Switch Turn on the white power switch (label 4). Note that a small lighted indicator in the switch reveals that the unit has been turned on. This switch should be left in the ON position during the entire laboratory period and turned OFF at the end of the period.

Ink Pens Fill the ink pens, observing the procedure outlined previously.

Paper Speed Set the speed of the paper by first setting the TIME push button in the ON mode and depressing the speed push button called for in the experiment. Do not press the OFF button until you are ready to start recording.

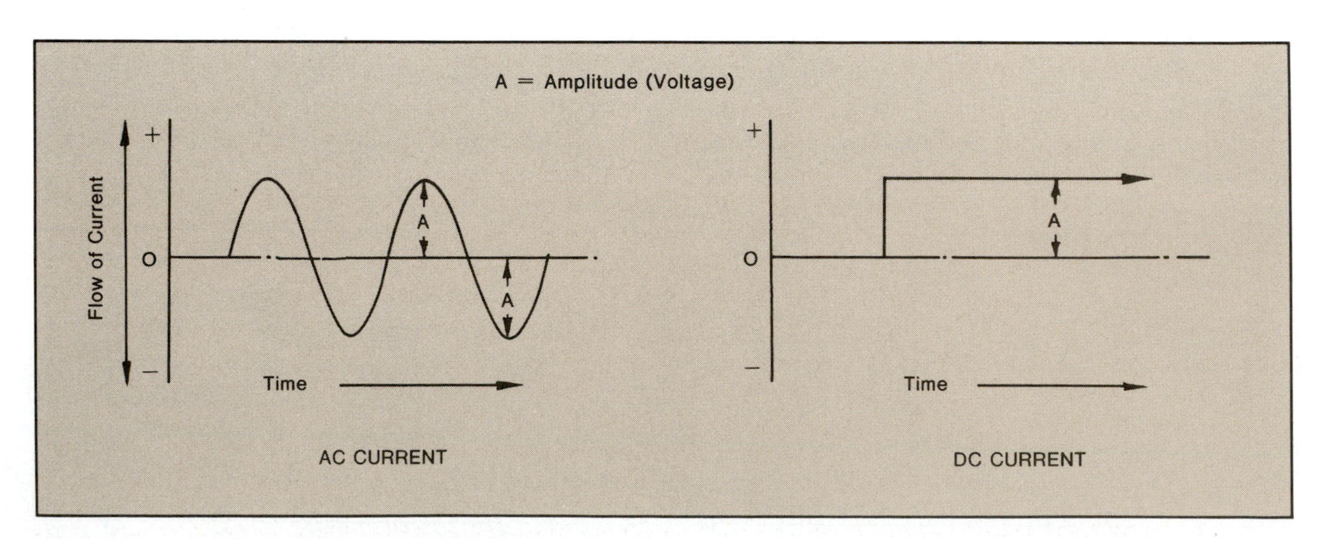

Figure 16.10 Comparison of AC and DC current.

Coupler Attachment Insert the cable from the experimental setup into the receptacle of the proper input coupler.

Amplifier Setting Set the Sensitivity control at the sensitivity level recommended in the experiment. In most experiments a low sensitivity will be used first.

Transducer Balancing If a transducer is to be used it will have to be balanced.

Calibration In certain experiments, such as recording an EEG, it will be necessary to calibrate the instrument.

Monitoring Once all the hookups have been completed, lower the pens to the paper and press the OFF button which will start the paper moving, and then press the RECORD buttons on each amplifier cluster that is involved in the experiment. The events occurring in the experiment will now record.

The Electronic Stimulator

Stimulation of nerve and muscle cells in the laboratory is most readily accomplished with an electronic stimulator. Although we will focus our attention here on the Narco and Grass stimulators, it is the general principles of stimulator operation that we will be primarily concerned with here.

The electronic stimulator accomplishes several things that are essential to controlled experiments.

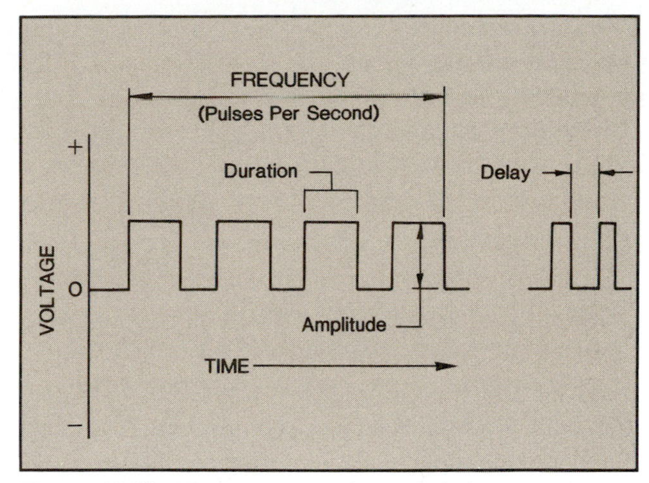

Figure 16.12 Square wave characteristics.

First of all, it converts alternating current (AC) to direct current (DC). As stated previously, DC moves continuously in one direction only; AC, on the other hand, reverses its direction repeatedly and continuously. The advantage of DC stimulation is that there is an abrupt rise of voltage from zero to maximum and then a sudden return to zero when a pulse of electricity is applied to a nerve or muscle. This type of voltage control produces a **square wave** instead of a sine wave. The differences are seen in figure 16.10.

In addition to converting AC to DC, the stimulator, through a variety of controls and switches, can vary the voltage, duration, and frequency of pulses that are fed to a specimen. With the increase of **voltage,** the **amplitude,** or height, of the wave form is increased. The length of time from

Figure 16.11 Control panel of Grass electronic stimulator.

the start to end of a single pulse of electrical stimulus is the **duration,** and the number of pulses delivered per second is called the **frequency.** Some stimulators, such as the Grass SD9 can also vary the time between twin pulses; this time interval is designated as the **delay.** Figure 16.12 illustrates these various characteristics.

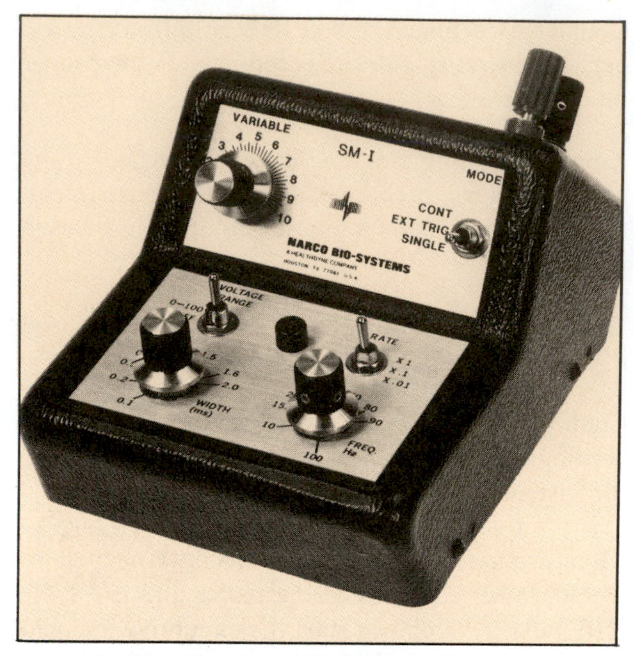

Figure 16.13 The Narco Bio-Systems SM-1 stimulator.

Control Functions

Figure 16.11 and 16.13 are of the Grass SD9 and the Narco SM-1 electonic stimulators. Although the control panels of the two instruments differ in configuration and terminology they both accomplish essentially the same functions.

Power Switch

The power switch for the SD9 is located at the bottom of the front panel near the middle. Immediately above the switch is a red indicator light that lights up when the power is turned on.

The power switch for the SM-1 is located on the back panel. Like the SD9, the SM-1 has a red indicator light on the horizontal front panel.

Voltage Settings

The intensity of any stimulus is varied by increasing or decreasing the voltage. Note that voltage on the SD9 is determined by two settings: a round knob in the upper right-hand corner and a decade (multiplier) switch below it. The voltage range here is from 0.1 to 100 volts.

Voltage adjustments on the SM-1 are made with the VOLTAGE RANGE switch and the VARIABLE control knob. The voltage range switch has three settings: 0–10, 0–100, and OFF. When the voltage range switch is set at the 0–100 setting, each digit on the variable dial is multiplied by 10; thus, the range on this scale is from 0 to 100 volts. When the voltage range switch is set on 0–10, the voltage output is between 0 and 10 volts. In the OFF mode no electrical stimulus can occur. The lowest voltage output for this stimulator is 50 millivolts.

Duration Settings

The duration of each electrical pulse is measured in milliseconds (msec). On the SD9 this stimulus characteristic is controlled by the DURATION knob that has a four-position decade switch below it. The combination of these two controls yields a range of 0.02 to 200 milliseconds (msec). The duration setting should not exceed 50% of the interval between pulses.

On the SM-1 duration is regulated by the WIDTH control, which is the left-hand knob on the horizontal panel. Its range is from 0.1 to 2.0 msec.

Mode Settings

Since stimulators can produce single stimuli, continuous pulses, and even twin pulses, there has to be a way of selecting the mode preferred for a particular experiment.

The two mode selection switches on the SD9 are located under the STIMULUS label. The right-hand MODE switch has three positions: repeat, single, and off. In the REPEAT position repetitive impulses are produced according to the setting on the Frequency control. To produce single pulses, the switch lever is depressed manually to the SINGLE position and released. One pulse is administered for each downward press of the lever; a spring returns the switch lever to the OFF position.

The other three-position switch under the stimulus label on the SD9 is used for twin pulses and hooking up with another stimulator. The MOD position is used when another stimulator is used in tandem with this one to modulate its output. When it is desirable to have the stimulator produce two pulses close together, the TWIN PULSES setting is used. When twin pulses or a second stimulator are not used, the setting should be on REGULAR. For most of our experiments this switch should be set at the regular setting.

The MODE switch on the SM-1 has three positions: continuous, single, and external triggering. When in the CONTinuous position the stimulator will produce repetitive pulses according to the setting on the Frequency control. When the switch lever is pressed to SINGLE, a single pulse will be delivered. The EXT TRIG position is used when an external triggering device is used. Such a device is plugged into a receptacle in the back of the stimulator.

Frequency Settings

The frequency control on a stimulator is set only when repetitive pulses are administered. In other words, the mode must be set at repeat on the SD9, or continuous on the SM-1.

On the SD9 the FREQUENCY control with its decade switch has a range of 0.20 to 200 pulses per second (PPS).

On the SM-1 the frequency range is from 0.1 Hz to 100 Hz. A RATE switch, which is calibrated X1, X.1, and X.01 must be used in conjunction with the round knob to get the proper setting. Incidentally, the Hz (Hertz) unit means essentially the same thing as PPS. One Hz represents one cycle per second.

Delay Settings

Stimulators that can produce twin pulses have a DELAY control to regulate the time interval between the pairs of pulses. The delay on the SD9 can be varied from 0.02 to 200 milliseconds, utilizing the control knob and four-place decade switch. On this stimulator the delay control can be used in either the single or repetitive mode. When the repeat mode is used, the delay time should not exceed 50% of the period between each set of pulses to avoid overlap.

Since the Narco SM-1 stimulator does not produce twin pulses, there is no need for a delay control.

Output Switches on SD9

In addition to the above controls the Grass SD9 has two switches positioned under the OUTPUT label that can modify the wave form. Note that the three-position slide switch on the right is for selecting MONOphasic, BIphasic, or DC. Figure 16.14, middle illustration, shows the differences between these three types of current. The proper setting for most experiments will be in the MONO position. However, when tissues are stimulated for a long period of time with the monophasic wave form, hydrolysis occurs, causing gas bubbles to form around the electrodes. For these longer duration experiments the biphasic wave form is preferred.

The DC position is seldom used due to the formation of electrolytes and gas bubbles in the preparation. Activation of the DC position overrides all other controls except voltage and mode.

The POLARITY switch to the left of the other switch is used for reversing the direction of electron flow. In the NORMAL position the red output binding post is positive (+) with respect to the

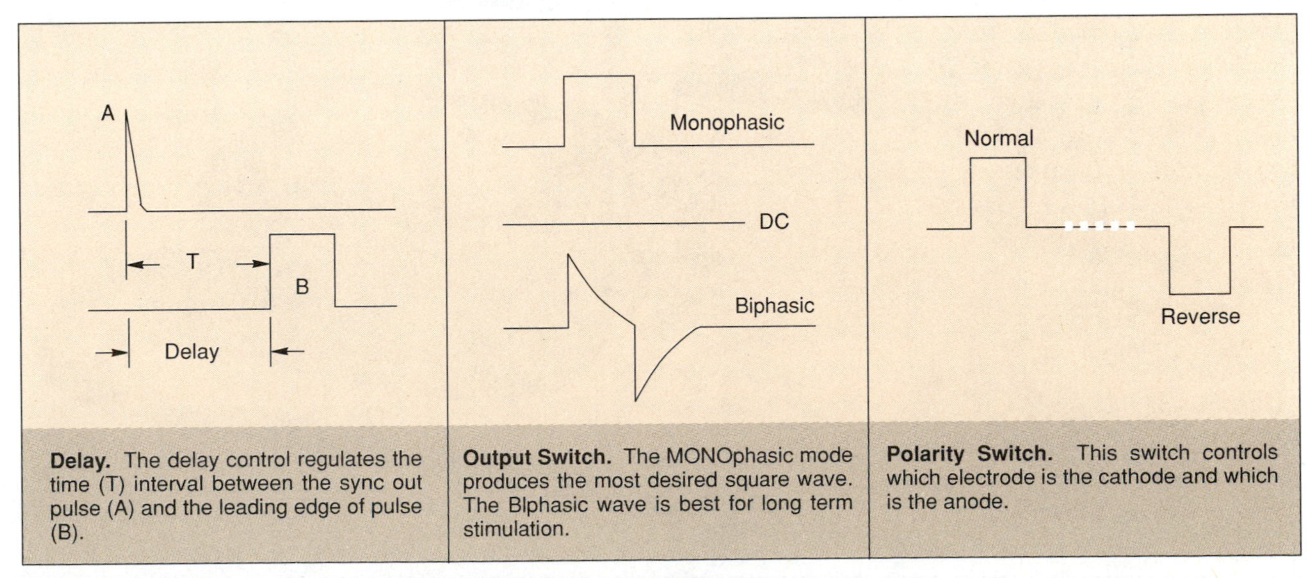

Delay. The delay control regulates the time (T) interval between the sync out pulse (A) and the leading edge of pulse (B).

Output Switch. The MONOphasic mode produces the most desired square wave. The BIphasic wave is best for long term stimulation.

Polarity Switch. This switch controls which electrode is the cathode and which is the anode.

Figure 16.14 Wave form modifications (Grass SD9 stimulator).

black binding post. In the REVERSE position the red binding post becomes negative with respect to the black binding post. See the right-hand illustration in figure 16.14.

Using Computerized Hardware

Three experiments in this laboratory manual utilize transducers with a computer to monitor physiological activities. The hardware and software for performing these experiments are manufactured by Intelitool, Inc., of Batavia, Illinois. In Exercise 23 the **Physiogrip**® is used to monitor muscle contraction; the **Cardiocomp**® is used in Exercise 46 to monitor electrocardiograms; and the **Spirocomp**® is used in Exercise 59 to do spirometry experiments. A fourth system called the Flexicomp® is available, but it is not included in this manual. Since all of these systems have certain facts in common, we will discuss here the general procedures in setting up the equipment and performing the experiments.

An Intelitool data acquisition/analysis system is an excellent way to perform certain experiments if a compatible computer and printer are available. Since the equipment is designed to be used only on the human body (with safety), no animals are needed or sacrificed. Although student computer knowledge is helpful, it is not required: the software instructions are such that, with minimum instructions, the experiments are easy to perform. As one performs each experiment, data is accumulated, automatically, and calculations are instantly made by the software. With a printer that has graphics capability, the results of the experiment can be printed out to produce a permanent record. There is one limitation to these systems, however: they are all single channel systems. If one wishes to monitor two physiological activities, simultaneously, such as ECG and respiration, one must use a Duograph or polygraph.

Figure 16.15 illustrates the equipment that is used in Exercise 58 to perform breathing experiments (spirometry). A wet spirometer with hose attached, mouthpieces, computer, printer, and software are all the equipment that is needed. The Spirocomp transducer is an interface box that is bolted to the scale arm that has a pulley on it. As the subject breathes into the spirometry hose, volumes are seen on the monitor of the computer, as the air chambered bell moves. Mathematical computations are automatically graphed by the software to determine pertinent information such as tidal volume, expiratory reserve volume, and vital capacity.

Equipment Setups

Assembling the various components for each experiment consists primarily of connecting the Intelitool transducer unit to the computer with the proper cable. Only one system, the Physiogrip, uses

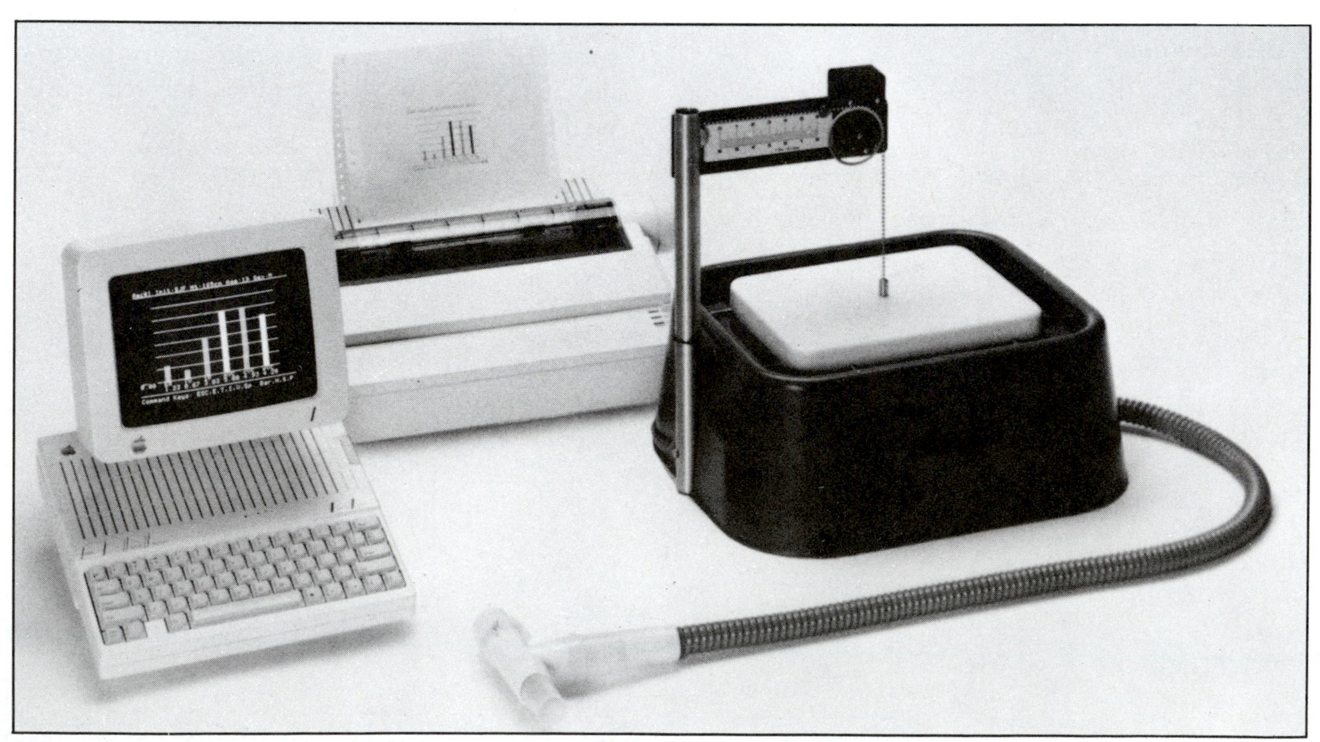

Figure 16.15 Intelitool setup needed for spirometry experiments.

an electronic stimulator, which is coupled to the subject's arm via a fitting on the front of the Physiogrip (see figure 23.1). All other systems have a single cable that leads to the computer.

Experimental Procedure

Since the software programs for the different systems have much in common it is possible to outline the general procedure.

Start Up Once the equipment is all hooked up, the program disk is inserted into the computer and the computer is turned on. If the computer is an Apple IIe, IIc, or IIGS, the caps lock key must be in the down position. To get started one simply presses any key and the program takes over.

Main Menu The main menus of the different systems vary in the number of selections, but all of them have the following items in common:

MAIN MENU

( ) Experiment Mode
( ) Review/Analyze
( ) Save Current Data
( ) Load Old Data File
( ) Disk Commands

No Data in Memory

In place of Experiment Mode, the item may be Run Spirocomp or Electrocardiogram, depending on which experiment is being performed. The experiment is always begun by selecting Experiment Mode or its equivalent first. The Review/Analyze and Save Current Data selections can only be made if experimental data is in the memory. If one attempts to select these items, the No Data in Memory flashes to tell the operator that data must be in the system before selection can be made. To select any item under the Main Menu, one must use certain command keys.

Command Keys Menu selections are made by moving the cursor up or down until it is in front of the item desired. If the computer is an Apple IIe, IIc, or IIGS, the cursor is moved up with the **I key** and down with the **M key.** On other computers the cursor is moved up and down with the **arrow keys** (up arrow raises cursor, down arrow lowers it). Pressing the **Return key** makes the selection. Other command keys are as follows:

Y/N - yes, no (for answering questions).
P - print (activates printer)
Esc - the escape key terminates the current operation and returns control to higher level menu.

Calibration As is true of transducers used with chart recorders, it is also necessary to calibrate some Intelitool hardware; exception: Cardiocomp. Instructions for calibration of the Physiogrip and Spirocomp are provided in Excercises 23 and 59. The procedures are relatively simple since the software provides most of the guidance needed.

Review/Analyze Once the experiment has been performed, or a file has been loaded from a disk, the operator can select Review/Analyze. In this mode the data put into the system are reviewed. The number of selections that are possible in this mode vary with the system that is being used.

Disk Commands This selection from the Main Menu will bring up a number of selections pertaining to Initializing, Locking and Unlocking Files, Deleting Files, and Renaming.

Laboratory Report

Answer all the questions on the Laboratory Report for this exercise.

17 Muscle Structure

The movement of limbs and other structures by skeletal muscles is a function of the contractility of muscle fibers and the manner of attachment of the muscles to the skeleton. In this exercise our concern will be with the detailed structure of whole muscles and their mode of attachment to the skeleton.

Figure 17.1 reveals a portion of the arm and shoulder girdle with only two muscles shown. Several other muscles of the upper arm have been omitted that would obscure the points of attachment of these muscles. The longer muscle which is attached to the scapula and humerus at its upper end is the **triceps brachii.** The shorter muscle on the anterior aspect of the humerus is the **brachialis.**

Muscle Attachments

Each muscle of the body is said to have an origin and insertion. The **origin** of the muscle is the immovable end. The **insertion** is the other end, which moves during contraction. Contraction of the muscle results in shortening of the distance between the origin and insertion, causing movement at the insertion end.

Muscles may be attached to bone in three different ways: (1) directly to the periosteum, (2) by means of a tendon, or (3) with an aponeurosis. The upper end of the brachialis (the origin) is attached by the first method, i.e., directly to the periosteum. The insertion end of this muscle, however, does have a short tendon for attachment. A **tendon** is a band or cord of white fibrous tissue that provides a durable connection to the skeleton. The tendon of the triceps insertion is much longer and larger due to the fact that this muscle exerts a greater force in moving the forearm. An example of the aponeurosis type of attachment is seen in figure 28.1, page 126. Label 5 in this figure is the aponeurosis of the external oblique muscle. An **aponeurosis** is a broad flat sheet of glistening pearly-white fibrous connective tissue that attaches a muscle to the skeleton or another muscle.

Although most muscles attach directly to the skeleton there are many that attach to other muscles or soft structures (skin) such as the lips and eyelids. In Exercises 25 through 29 the precise origins and insertions of various muscles will be identified.

Microscopic Structure

Illustrations B and C in figure 17.1 reveal the microscopic structure of a portion of the brachialis muscle. Note that the muscle is made up of bundles, or **fasciculi,** of skeletal muscle cells. A single fasciculus is shown, enlarged, in illustration C. Protruding upward from the upper cut surface of the fasciculus is a portion of an individual muscle cell, or **muscle fiber.** Note that between the individual muscle fibers of each fascicle is a thin layer of connective tissue, the **endomysium.** This connective tissue enables each muscle fiber to react independently of the others when stimulated by nerve impulses. Surrounding each fascicle is a tough sheath of fibrous connective tissue, the **perimysium.**

Surrounding the fasciculi of the muscle is another layer of coarser connective tissue that is called the **epimysium.** Exterior to the epimysium is the **deep fascia,** which covers the entire muscle. Since this muscle lies close to the skin, the deep fascia of this muscle would have fibers that intermesh with the superficial fascia that lies between the muscle and the skin. The connective tissue fibers of the deep fascia are also continuous with the tissue in the tendons, ligaments, and periosteum. Illustration A reveals the continuity of the connective tissue of the endomysium, perimysium, and epimysium. A fasciculus of muscle fibers is shown bracketed in this illustration.

Laboratory Assignment

Label figure 17.1.

Examine some muscles in a freshly killed or embalmed animal, identifying as many of the aforementioned structures as possible.

Complete the first portion of combined Laboratory Report 17,18.

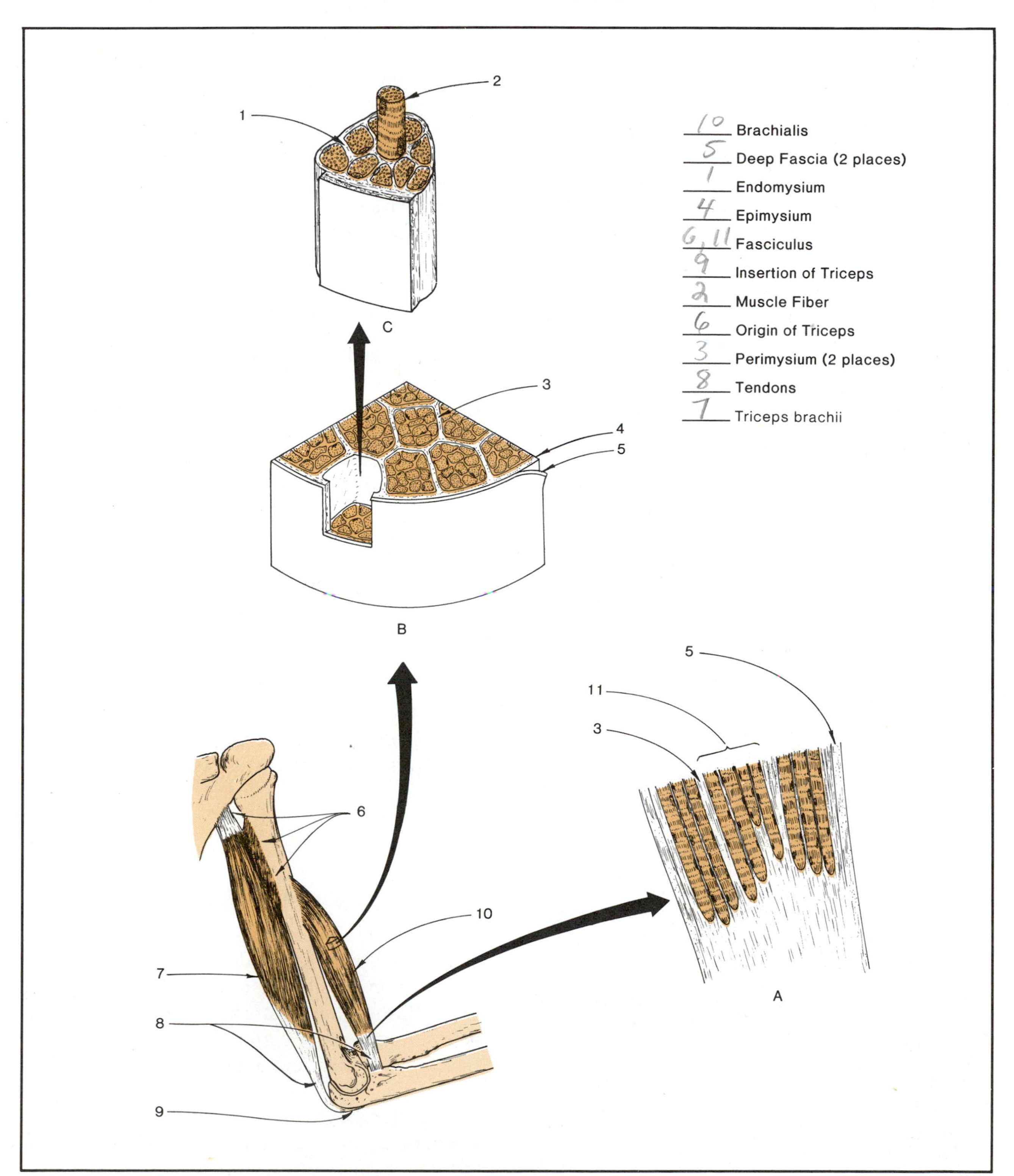

Figure 17.1 Muscle anatomy and attachment.

18 Body Movements

The action of muscles through diarthrotic joints results in a variety of types of movement. The nature of movement is dependent on the construction of the individual joint and the position of the muscle. Muscles working through hinge joints result in movements that are primarily in one plane. Ball and socket joints, on the other hand, will have many axes through which movement can occur; thus, movement through these joints is in many planes. The different types of movement are as follows.

Flexion When the angle between two parts of a limb is decreased the limb is said to be *flexed.* Flexion of the arm takes place at the elbow when the antebrachium is moved toward the brachium. The term *flexion* may also be applied to movement of the head against the chest and the thigh against the abdomen.

When the foot is flexed upward, the flexion is designated as being **dorsiflexion.** Movement of the sole of the foot downward, or flexion of the toes, is called **plantar flexion.**

Extension Increasing the angle between two portions of a limb or two parts of the body is *extension.* Straightening the arm from a flexed position is an example of extension. Extension of the foot at the ankle is essentially the same as plantar flexion. When extension goes beyond the normal posture, as in leaning backward, the term **hyperextension** is used.

Abduction and Adduction The movement of a limb away from the median line of the body is called *abduction.* When the limb moves toward the median line, *adduction* occurs. These terms may also be applied to parts of a limb, such as the fingers and toes, by using the longitudinal axes of the limbs as points of reference.

Rotation and Circumduction The movement of a bone or limb around its longitudinal axis without lateral displacement is *rotation.* Rotation of the arm occurs through the shoulder joint. The head can be rotated through movement in the cervical vertebrae. If the rotational movement of a limb through a freely movable joint describes a circle at its terminus, the movement is called *circumduction.* The arm, leg, and fingers can be circumducted.

Supination and Pronation Rotation of the antebrachium at the elbow affects the position of the palm. When the palm is raised upward from a downward facing position, *supination* occurs. When the rotation is reversed so that the palm is returned to a downward position, *pronation* takes place.

Inversion and Eversion These two terms apply to foot movements. When the sole of the foot is turned inward or toward the median line, *inversion* occurs. When the sole is turned outward, *eversion* takes place.

Sphincter Action Circular muscles such as those around the lips *(orbicularis oris)* and the eye *(orbicularis oculi)* are called sphincter muscles. When their fibers shorten they close the opening.

Muscle Grouping

For reasons of simplicity, muscles are often studied individually, as in figure 18.1. It should be kept in mind, however, that seldom, if ever, do they act singly. Rather, muscles are arranged in groups with specific functions to perform, i.e., flexion and extension, abduction and adduction, supination and pronation.

The flexors are the prime movers, or **agonists.** The opposing muscles, or **antagonists,** contribute to smooth movements by their power to maintain tone and give way to movement by the flexor group. Variance in the tension of the flexor muscles results in a reverse reaction in the extensor muscles.

Muscles that assist the agonists to reduce undesired action or unnecessary movement are called **synergists.** Other groups of muscles that hold structures in position for action are called **fixation muscles.**

Laboratory Report

Identify the types of movement in figure 18.1 and complete the last portion of Laboratory Report 17,18.

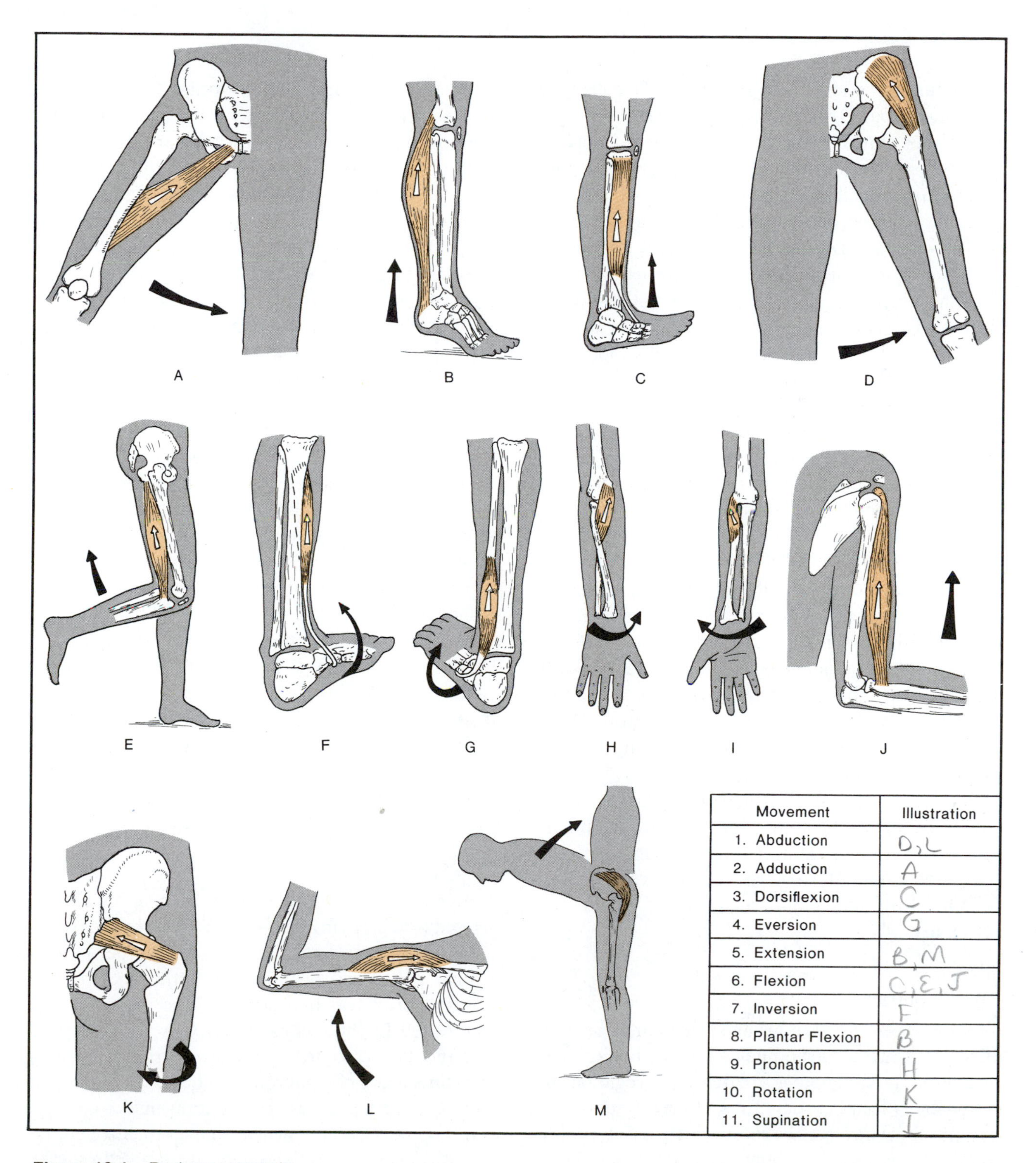

Movement	Illustration
1. Abduction	D, L
2. Adduction	A
3. Dorsiflexion	C
4. Eversion	G
5. Extension	B, M
6. Flexion	C, E, J
7. Inversion	F
8. Plantar Flexion	B
9. Pronation	H
10. Rotation	K
11. Supination	I

Figure 18.1 Body movements.

19 Nerve and Muscle Tissues

The characteristics of nerve and muscle cells that sets them apart from other cells is that they are capable of transmitting electrochemical impulses along their membranes. This property is called **excitability.** Excitability is a function of the cell membrane's electrical potential. Although all cells in the body possess membrane potentials, only nerve and muscle cells utilize this characteristic to generate action potentials (impulses) along their membranes. The end result is that nerve cells provide for neural coordination and muscle cells shorten to provide for movement.

During this laboratory period you will have an opportunity to study different kinds of neurons and three kinds of muscle tissue. Prior to examining the tissues under the microscope, label figure 19.1 and answer questions on the Laboratory Report.

Neurons

Since neurons perform different roles in different parts of the nervous system they exist in a variety of sizes and configurations. In spite of this diversity all neurons have much in common. Figures 19.1 and 19.2 illustrate the principal types.

For an overview of the basic structure of all neurons, the large peripheral neurons in figure 19.1 provide an excellent study. The cross section through the spinal cord reveals the relationship of these neurons to the spinal cord and to each other.

Three kinds of neurons are shown here: sensory, motor, and interneurons. These three neurons, together with the receptor and effector, comprise a **reflex arc.**

The Perikaryon

The cell body or **perikaryon** of a neuron usually contains a rather large nucleus, Nissl bodies, and neurofibrils. **Nissl bodies** are dense aggregations of rough endoplasmic reticulum scattered through the cytoplasm. They stain readily with basic aniline dyes such as toluidine blue or cresyl violet. They represent sites of protein synthesis. Some Nissl bodies are shown in the perikaryon of the motor neuron in figure 19.1. **Neurofibrils** are slender protein filaments that extend through the cytoplasm parallel to the long axis. They resemble roadways that circumvent Nissl bodies. It is significant that these minute fibers converge into the axon to form the core of the nerve fiber.

The most remarkable features of neurons are their cytoplasmic processes: the axons and dendrites. Traditional terminology once categorized these processes purely from a morphological sense; however, numerous exceptions require more accurate and universal definitions which emphasize their functional attributes. A **dendrite** or **dendritic zone** functions in receiving signals from receptors or other neurons and plays an important part in integration of information. An **axon** is a single elongated extension of the cytoplasm that has the specialized function of transmitting impulses away from the dendritic zone. The cell body can be located at any position along the conduction pathway. In the motor neuron of figure 19.1 the dendrites are the short branching processes of the perikaryon; the axon is the long single process. Although a neuron will usually have several dendrites, there will be only one axon. Some neurons, such as the *amacrine cells* of the retina, have no axon at all.

Fiber Construction

The neural fibers of all peripheral neurons are enclosed by a covering of **neurolemmocytes** (Schwann cells). This covering extends from the perikaryon to the peripheral termination of the fiber. In the larger peripheral neurons, such as those seen in figure 19.1, the neurolemmocytes are wrapped around the axon in a unique manner to form a **myelin sheath** of lipoprotein. The outer portion of the neurolemmocytes that contain nuclei and cytoplasm make up the **neurolemma,** or neurolemmal sheath. Those axons that have a thick myelin sheath

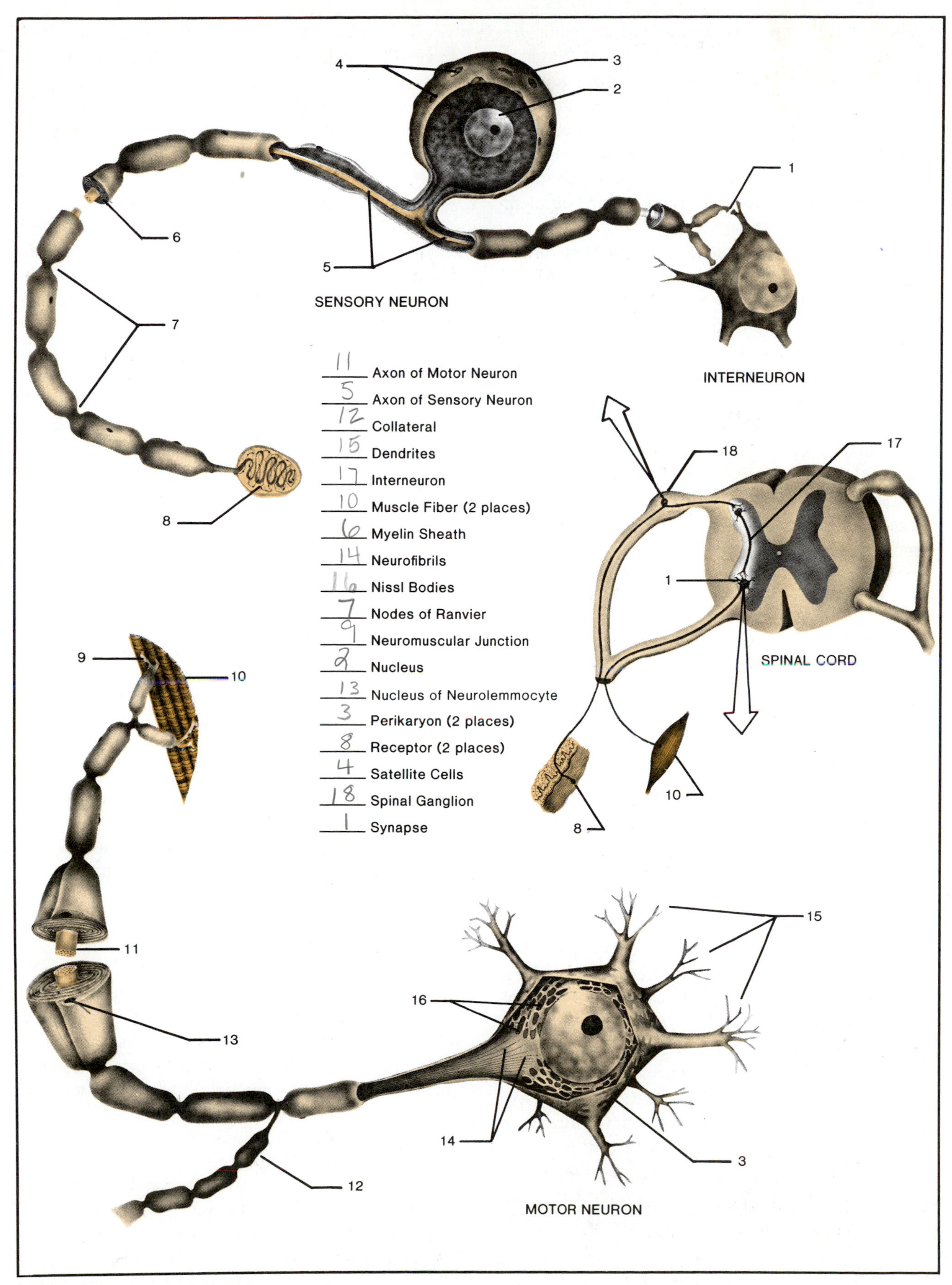

Figure 19.1 Sensory and motor neurons.

with a neurolemma are designated as being *myelinated*. The smaller peripheral neurons that have fibers enclosed in neurolemmocytes, but lack the myelin sheath, are said to have *unmyelinated* fibers. One of the most important functions of neurolemmocytes is to facilitate the regeneration of damaged fibers. When a fiber is damaged, regeneration occurs along the pathway formed by the neurolemmocytes.

The myelin sheath and neurolemma are formed during embryological development by the folding of the neurolemmocytes around the axons. The axon of the motor neuron in figure 19.1 reveals an enlarged cross-section of this lipoprotein sheath and neurolemma. The myelin sheath and neurolemma are interrupted at regular intervals by **nodes of Ranvier,** which are points of discontinuity between successive neurolemmocytes; each internode consists of a single neurolemmocyte. Each node of Ranvier represents a place where the exposed axon lacks the myelin and neurolemma. The nodes of Ranvier play a significant role in facilitating the speed of nerve impulse transmission.

Types of Neurons

The variability in size, shape, and arborization of neurons is considerable. Neurons may be multipolar, bipolar, unipolar, or pseudounipolar. Some have perikarya as small as 4 micrometers; others may be as large as 150 micrometers. Some have perikarya that are encapsulated by satellite cells; others are not encapsulated. Many have myelinated fibers, others do not. A brief survey of some of the types of neurons follows:

Motor Neurons As indicated in figure 19.1, these *multipolar* neurons have their perikarya located in the gray matter of the CNS. They are also referred to as *efferent neurons* since they carry nerve impulses away from the CNS. The axons of these neurons are myelinated and terminate in skeletal muscle fibers. They function in the maintenance of voluntary control over skeletal muscles.

The perikarya of these neurons are somewhat angular or star-shaped rather than ovoid. The cytoplasm in a properly stained perikaryon reveals a nucleus, many Nissl bodies, and neurofibrils. Many short branching dendrites are present that make synaptic connections with axons of interneurons or sensory neurons. Illustrations C and D of figure HA–12 in the Histology Atlas reveal the appearance of these cells when observed under oil immersion optics.

The myelinated axons of these neurons often have **collaterals** emanating at right angles from a node of Ranvier. Collaterals are branches that provide innervation to additional muscle fibers, other neurons, and even the same neuron in some instances.

Sensory Neurons These neurons are also called *ganglion cells* because their perikarya are located in **spinal ganglia** (label 18, figure 19.1) outside the CNS. Since they carry nerve impulses from the receptor toward the CNS, they are also referred to as *afferent neurons*.

As illustrated in figure 19.1, the perikaryon of a sensory neuron is ovoid in shape and is covered with a capsule of **satellite cells.** These cells are continuous with neurolemmocytes and have the same relationship to these neurons that certain neuroglia (oligodendrocytes) have to cells of the CNS. Satellite cells are not neuroglia, however, since they have a different embryological origin. Illustration E, figure HA–12 of the Histology Atlas, reveals what satellite cells look like in a sectioned sensory ganglion.

Although sensory neurons appear to be unipolar, histologists classify them as pseudounipolar on the basis of their embryological development. Note that the entire fiber is myelinated from the **receptor** in the skin to the synaptic connection in the spinal cord. The entire myelinated fiber is designated as an **axon** even though part of it functions like a dendrite in carrying nerve impulses toward the cell body. A **synapse** (label 1, figure 19.1) is the point where the terminus of an axon of one neuron adjoins the dendritic zone of another neuron.

Interneurons (*Association and Internuncial*) These multipolar unmyelinated neurons (figure 19.1) exist in large numbers in the gray matter of the CNS. They are essentially the *integrator cells* between sensory and motor neurons. Most reflexes involve one or more interneurons between the sensory and motor neurons. However, some automatic responses, such as the knee jerk reflex, are monosynaptic in that they lack interneurons.

Assignment:

Label figure 19.1.

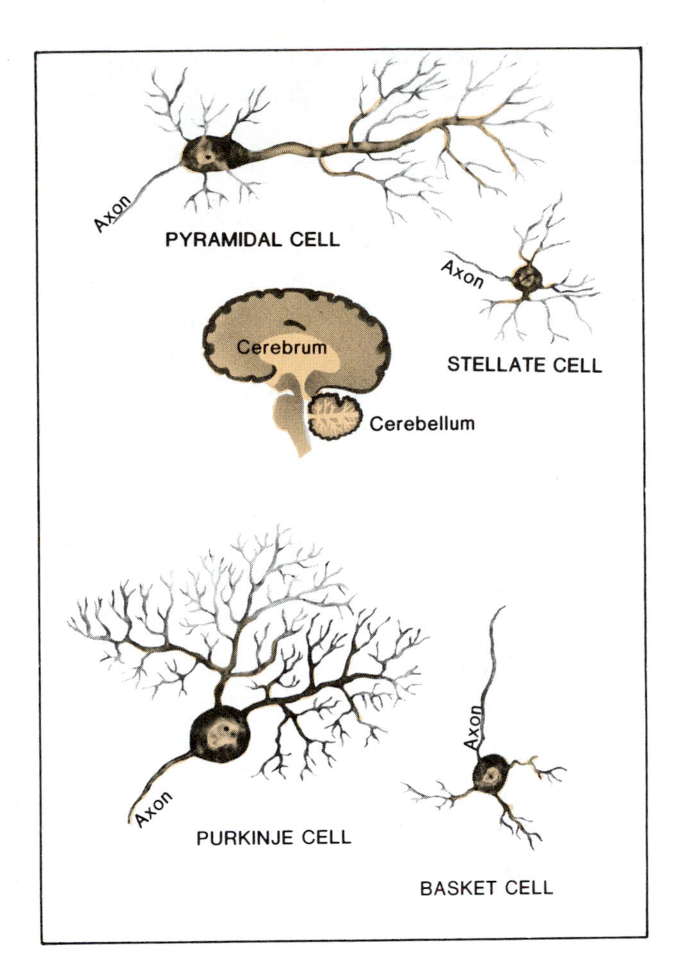

Figure 19.2 Neurons of the brain.

Pyramidal Cells These multipolar neurons are the principal cells of the cortex of the cerebrum. The upper neuron in figure 19.2 is of this type. Figure HA–10 of the Histology Atlas reveals actual photomicrographs of these cells.

The perikarya of these neurons are somewhat triangular to impart a pyramid-like shape. A unique characteristic of these cells is that they have two separate sets of dendrites: (1) a long dendritic trunk-like structure with many branches that ascends vertically in the cortex, and (2) several basal branching dendrites that extend outward from the surfaces of the perikaryon. The axon of each of these neurons can be differentiated from the dendrites by its smoother surface.

Purkinje Cells The large multipolar neuron in the lower left-hand corner of figure 19.2 is a Purkinje cell. The perikarya of these neurons are located deep within the molecular layer of the cerebellum (see figure HA–11, Histology Atlas). The large branching dendritic processes of these cells fill up most of the space in the molecular layer. The axons of these neurons extend inward into the granule cell layer of the brain.

Stellate Cells (*Golgi Type II*) These small neurons are found in both the cerebrum and cerebellum, where they act as linkage (association) cells for the pyramidal cells of the cerebrum and the Purkinje cells of the cerebellum. Both multipolar and bipolar types exist. See figures 19.2, HA–10, and HA–11.

Basket Cells These multipolar neurons are a variety of large stellate cells that are located deep in the molecular layer of the cerebellum in close association with the Purkinje cells (figure 19.2 and HA–11). They have short, thick, branching dendrites and a long axon. The axons of these neurons usually have five or six collaterals that make synaptic connections with dendrites of the Purkinje cells. Like the stellate cells, these cells function as association neurons.

Histological Studies of Nerve Tissues

After answering the questions on the Laboratory Report that pertain to cells of the nervous system, examine prepared slides of the nervous system using oil immersion optics whenever necessary. The following procedure will involve most of the cells discussed in this exercise.

Materials:

prepared slides of: x.s. spinal cord (H10.341 or H10.342), spinal cord smear (H4.11), spinal ganglion (H4.25), cerebral cortex (H10.11 or H10.12), cerebellum (H10.23)

Spinal Cord Examine the cross section of a rat spinal cord under low power and identify the structures labeled in figure HA–12. Study several of the large *motor neurons* in the gray matter with the high-dry objective.

Also, study a smear of ox spinal cord tissue and compare the cells with the photomicrograph in illustration D, figure HA–12.

Ganglion Cells Study a slide of a section through a spinal ganglion and look for *satellite cells* that surround the sensory neurons. Refer to illustration E, figure HA–12.

Cerebral Cortex Examine a slide of the cerebral cortex. Look for *pyramidal* and *stellate cells.* Refer to figure HA–13. Try to differentiate dendrites from axons on the pyramidal cells.

Note the presence of spiny processes called "gemmules" that are seen on the branches of the dendrites. These processes greatly increase the surface area of the dendrites, allowing large neurons to receive as many as 100,000 separate axon terminals or synapses.

Cerebellum Examine a slide of the cerebellum and look for *Purkinje cells, stellate cells,* and *basket cells.* Refer to figure HA–14.

Muscle Fibers

As stated above, excitability in muscle cells manifests itself in **contractility.** This property is made possible by the presence of contractile protein fibers within the cells. The shortening of muscle cells, due to this property, is responsible for the various movements of appendages and organs of the body. Because of their elongated structure, muscle cells are often referred to as **muscle fibers.**

On the basis of cytoplasmic differences, there are two categories of muscle fibers: striated and smooth. **Striated muscle** is characterized by regularly spaced transverse bands called *striae.* **Smooth muscle** lacks these transverse bands and has other distinctions to set them apart. Striated muscle cells are further subdivided into skeletal and cardiac muscle. **Skeletal muscle** cells are multinucleated, or *syncytial.* Attached primarily to the skeleton, they are responsible for voluntary body movements. **Cardiac muscle** cells, which make up the muscle of the heart wall, are not syncytial and have less prominent striations than skeletal muscle.

Smooth Muscle

Smooth muscle fibers are long, spindle-shaped cells as illustrated in figure 19.3. Each cell has a single elongated nucleus that usually has two or more nucleoli. When studied with electron microscopy, the cytoplasm reveals the presence of fine linear filaments of the same protein composition that are seen in striated muscle.

Smooth muscle cells are found in the walls of blood vessels, walls of organs of the digestive tract, the urinary bladder, and other internal organs. They are innervated by the autonomic nervous system and, thus, are under *involuntary nervous control.*

Skeletal Muscle

Muscle fibers of this type are long, cylindrical, and multinucleated (figure 19.4). Large numbers of these fibers are grouped together into bundles called **fasciculi.** The individual cells within each fascicle are separated from each other by a thin layer of connective tissue called **endomysium.** This separateness of muscle fibers enables each cell to respond independently to nerve stimuli.

Examination of the cytoplasm, or **sarcoplasm,** with a compound microscope reveals a series of distinct dark and light **striae.** Surrounding each cell

Figure 19.3 Smooth muscle fibers, teased.

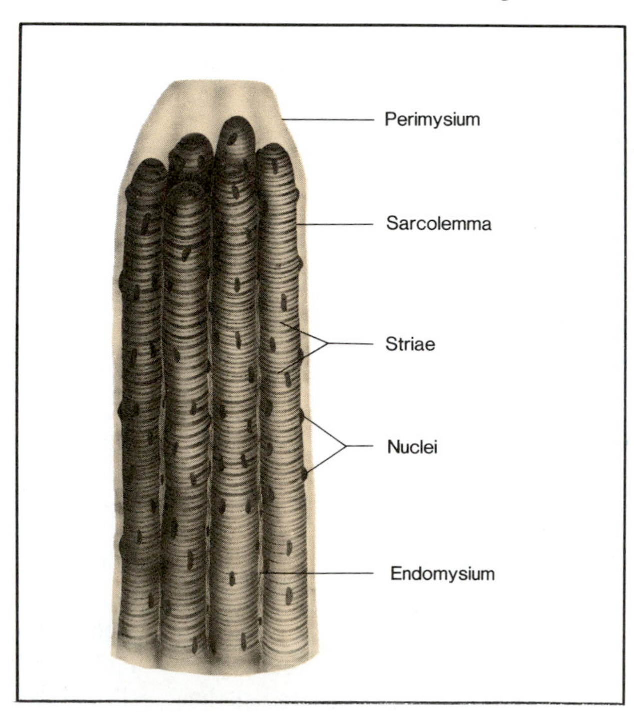

Figure 19.4 A fascicle of skeletal muscle fibers.

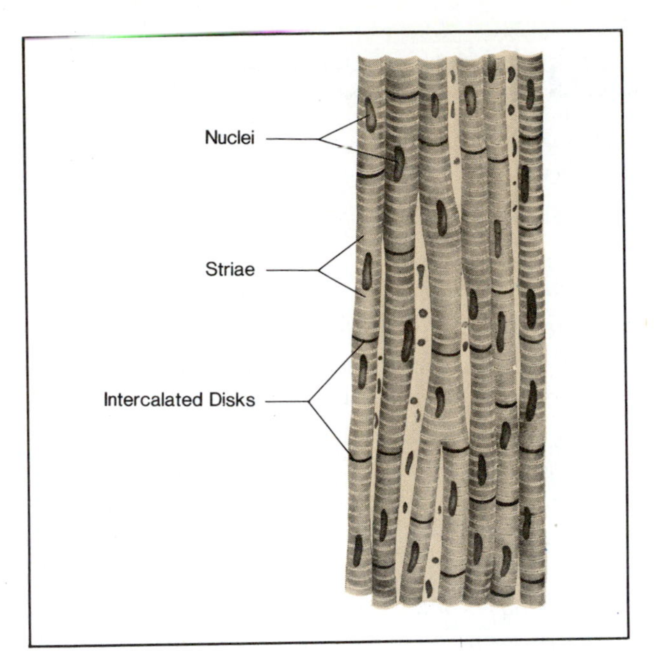

Figure 19.5 Cardiac muscle fibers.

is a plasma membrane called the **sarcolemma.** Note in figure 19.4 that the nuclei of each cell lie in close association with the sarcolemma.

Although striated muscle is referred to as skeletal and voluntary, there are certain muscles of this type that are not attached to bone or subject to voluntary control. Examples: (1) striated muscles of the tongue and external anal sphincter have no skeletal attachment; and (2) striated muscles of the pharynx and upper esophagus have no skeletal attachment and are not voluntarily controlled. In spite of these differences, these aberrational examples are morphologically indistinguishable from other skeletal muscle.

Cardiac Muscle

Muscle cells of this type are found in the wall of the heart and in the walls of the venae cavae where they join the right atrium of the heart. A unique characteristic of cardiac muscle is its ability to contract rhythmically and continuously as a result of intrinsic cellular activity. Morphologically, cardiac muscle is readily distingushed from skeletal and smooth fibers; yet it shares some characteristics of each type.

Cardiac muscle differs from skeletal muscle in the following ways:

- Instead of being anatomically syncytial, cardiac muscle is made up of separate cellular units that are separated from each other by **intercalated disks** (cell membranes). These disks stain somewhat darker than transverse striae.
- Some fibers of cardiac cells are bifurcated to form a branched three-dimensional network instead of forming only long straight cylinders.
- The elongated nuclei of each cellular unit lie deep within the cells instead of near the surface as in skeletal muscle.
- Cardiac tissue is not normally subject to voluntary control.

Histological Studies of Muscle Tissues

After answering the questions on the Laboratory Report that pertain to the characteristics of the various types of nerve and muscle tissue, examine prepared slides to identify the various kinds of cells. Consult the appropriate illustrations in the Histology Atlas for assistance in locating and identifying the various types of tissue.

Materials:

prepared slides of: striated muscle, teased
striated muscle, x.s. and l.s. (H3.13)
smooth muscle, teased (H3.21)
muscle to tendon (H7.13)
smooth muscle, sectioned (H3.22)
heart muscle, teased (H3.31)
heart muscle, l.s. (H3.32 or H3.33)

Skeletal Muscle Scan a slide of striated muscle tissue with low power to find a muscle fiber with the most pronounced striae. Increase the magnification to 1000X (oil immersion) and see if you can differentiate the A and I bands. Refer to illustration A, figure HA–7. Note the position of the nuclei in these fibers. Also examine a cross-section of this tissue to identify the endomysium. If a slide of muscle to tendon (H7.13) is available examine it under high-dry magnification and refer to illustration C, figure HA–7.

Cardiac Muscle Scan a slide of this tissue with low power, looking for pronounced striae and intercalated disks. Increase the magnification to 1000× to observe the nature of the striae and disks. Can you find a place on the slide where bifurcation of the fibers is seen? Consult illustration A, figure HA–8, of the Histology Atlas.

Smooth Muscle Study slides of this type of muscle tissue in the same manner as above. Finding individual fibers, separated from others, is not always as easy as it might seem. Compare your observations to illustrations B and C in figure HA–8.

Laboratory Report

Complete the Laboratory Report for this exercise.

20 The Neuromuscular Junction

It was observed in Exercise 19 that motor neurons emerge from the spinal cord through the anterior roots of spinal nerves to innervate skeletal muscle fibers throughout the body. Each large myelinated nerve fiber branches many times through collaterals to stimulate from 3 to 2,000 skeletal muscle fibers. At the end of each branch is a **neuromuscular junction** where the nerve fiber contacts the muscle fiber. This juncture is usually made at approximately the midpoint of the fiber so that the action potential generated in the fiber will reach the opposite ends of the fiber at about the same time. Normally there is only one junction per muscle fiber.

Figure 20.1 reveals the structure of a neuromuscular junction as determined by electron microscopy. Chemical studies that have taken place over the past several decades have uncovered a series of events that occur here. Our present understanding of these events differs to some extent with our views of only a few years ago, and it is certain that future research will reveal a more complete picture as present questions are answered.

After you have labeled the structures in this illustration and answered the questions on the Laboratory Report, you will have an opportunity to study microscope slides of the neuromuscular junction. Since the configuration revealed in figure 20.1 is the result of electron microscopy, don't expect to see this degree of detail on your slides.

Note in figure 20.1 that the end of the neuron consists of a cluster of knoblike structures that lie within invaginations of the **sarcolemma** (label 7) of the muscle fiber. These invaginations are **synaptic gutters;** the knoblike projections are called **axon terminals.**

Between the axon terminal and lining of the synaptic gutter is a gap, the **synaptic cleft,** which is approximately 20 μm wide and is filled with a gelatinous ground substance. The **presynaptic membrane** (label 5) and the **postsynaptic membrane** (label 4) form the surfaces on each side of this cleft. Note, also, that the surface area of the postsynaptic membrane is considerably increased by the presence of further infoldings to form **subneural clefts** (junctional folds).

The transmission of an impulse from the axon terminal across the synaptic cleft involves a chemical *neurotransmitter* rather than electrical transmission. The basic sequence of events are (1) depolarization of the neurolemma (an action potential), with sodium ions flowing into the neuron and potassium ions out of the neuron; (2) inward flow of calcium ions; (3) secretion of a neurotransmitter *(acetylcholine)* into the synaptic cleft by the presynaptic membrane; (4) hookup of acetylcholine molecules to receptors on the postsynaptic membrane; (5) development of an action potential on the sarcolemma of the muscle fiber; and (6) contraction of the muscle fiber.

Since the early 1950s it was believed by most neuroscientists that the acetylcholine (ACh) that bridges the synaptic cleft was exocytosed by the **synaptic vesicles** (label 3). It was thought that the action potential in the neuron caused synaptic vesicle membranes to fuse with the presynaptic membrane, releasing the ACh content of the vesicles. This theory seemed logical since the vesicles do contain ACh and they do fuse with the presynaptic membrane. More recent studies by Yves Dunant, Maurice Israel, et al., however, contradict this simple picture.

The present concept is that ACh passes directly from the cytoplasm of the axon terminal through the nerve cell membrane. The releasing mechanism appears to be a compound, most likely a protein, that is embedded in the presynaptic membrane. The protein probably acts as a valve, enabling ACh to pass through when the action potential reaches the area.

Once the action potential is initiated in the sarcolemma of the muscle fiber the following events occur within the fiber: (1) the action potential moves inward to the **T-tubule system** (label 2); (2) calcium ions, which permit contraction of the

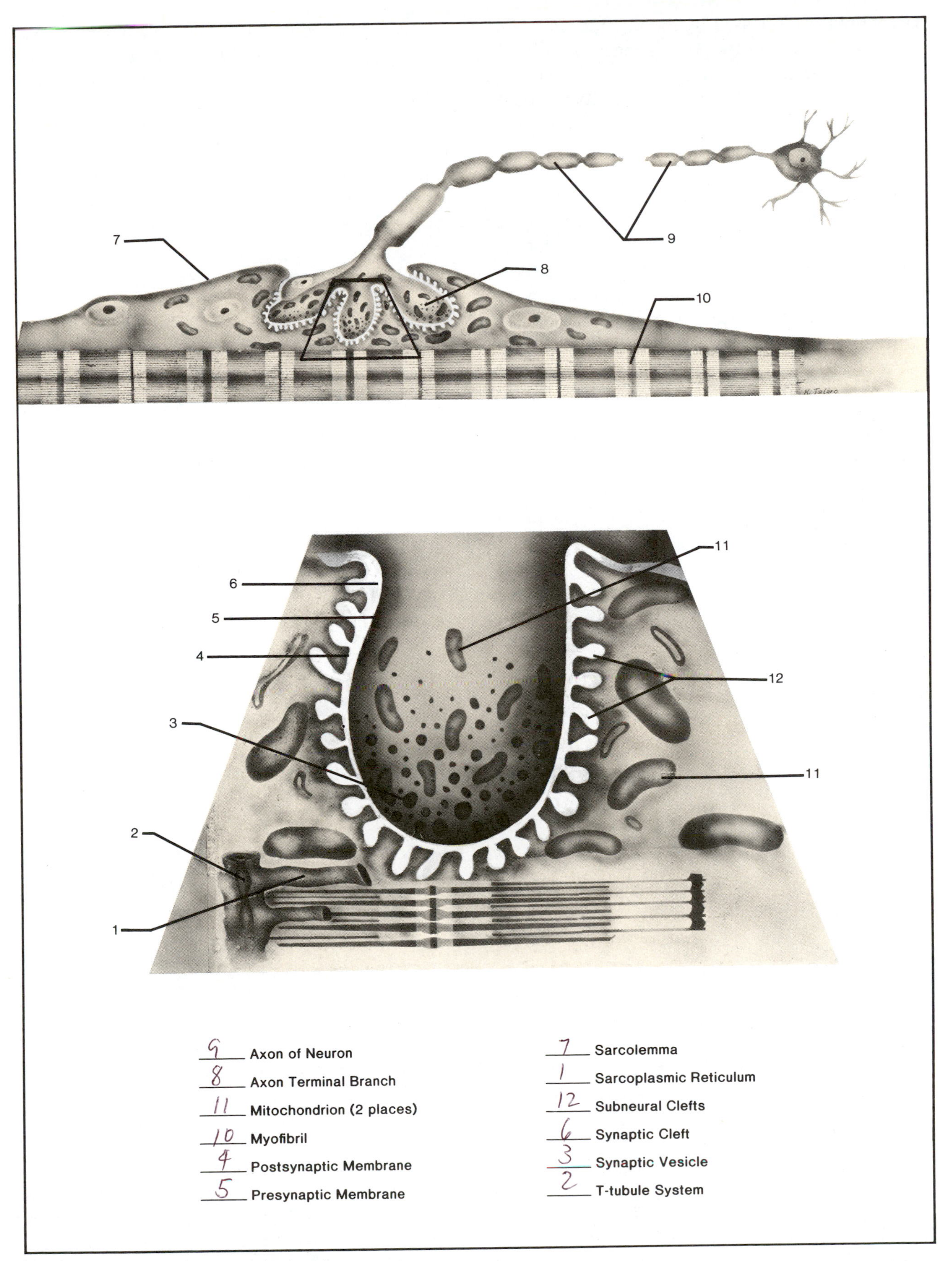

Figure 20.1 The neuromuscular junction.

myofibrils, are released by the **sarcoplasmic reticulum** (label 1); and (3) ACh is inactivated by the enzyme *acetylcholinesterase* within 1/500 of a second. The energy for muscle fiber contraction is supplied by ATP from mitochondria in the area.

Now back to the synaptic vesicles . . . what appears to be their role in this chain of events? According to Dunant and Israel, it appears that these structures do two things: (1) they release stored ACh when the cytoplasmic pool of this neurotransmitter has been exhausted due to sustained nervous activity, and (2) they accumulate calcium ions to help terminate the release of ACh. These two conclusions are based on the fact that the synaptic vesicles begin to leak ACh and take on calcium ions *only* when neurons are under prolonged stimulation.

The action of acetylcholinesterase on ACh is to break it down into acetate and choline. This enzyme is found on the other surfaces of the presynaptic and postsynaptic membranes as well as in the ground substance of the synaptic cleft.

Once the choline and acetate diffuse back into the axon terminal they are resynthesized back into ACh. The enzyme that catalyzes the synthesis of ACh is *choline acetyltransferase*. This enzyme is produced in the perikaryon of the neuron, and travels down microtubules of the axon to the terminal where it is stored in the cytoplasm.

These events, which depict how the nerve impulse bridges the gap from the axon terminal to a skeletal muscle fiber, are identical for nerve to nerve transmission, and for nerve to secretory cell as well.

Reference:

Dunant, Y.,. and M. Israel. "The Release of Acetylcholine." *Scientific American* 252 (4)(April 1985).

Laboratory Assignment

After labeling figure 20.1 examine a microscope slide of motor nerve endings (H4.31) under low-power and high-dry magnification. Consult illustration D, figure HA–7, for reference. Complete the first portion of Laboratory Report 20,21.

The Physicochemical Nature of Muscle Contraction 21

The microscopic changes that occur during skeletal muscle contraction will be observed here in a controlled experimental setup. The physicochemical nature of muscle contraction will be reviewed first, for clarification of this complex phenomenon.

Skeletal muscle contraction involves many components within the muscle cell. The entire process is initiated by an action potential that reaches the neuromuscular junction. As we learned in Exercise 20, acetylcholine initiates depolarization of the muscle fiber. The depolarization proceeds along the surface (sarcolemma) of the muscle fiber and deep into the sarcoplasm by way of the T-system tubules of the sarcoplasmic reticulum.

Figure 21.1 illustrates the ultrastructure of a single fiber. Note that the sarcoplasm of the fiber consists of many long tubular **myofibrils,** the **sarcoplasmic reticulum,** and **mitochondria. Nuclei** are seen in abundance beneath the **sarcolemma.** The large number of mitochondria provides all the ATP that is needed by the myofibrils for contraction.

Note, also, that a myofibril is constructed of two kinds of **myofilaments:** actin and myosin. **Myosin** filaments are the thickest filaments and are the principal constituents of the dark **A band.** They are slightly thicker in the middle and taper toward both ends. The thinner **actin** filaments extend about 1 μm in either direction from the **Z line** and make up the lighter **I band.**

The shortening of myofibrils to achieve muscle fiber contraction is due to the interaction of actin and myosin in the presence of calcium, magnesium, and ATP. Two other protein molecules plus troponin and tropomyosin are also involved.

As the action potential moves along the sarcolemma and T-tubule system, large quantities of calcium ions are released by the sarcoplasmic reticulum. As soon as these ions are released, attractive forces develop between the actin and myosin causing the ends of the actin fibers to be pulled toward each other, as shown in the bottom of figure 21.1

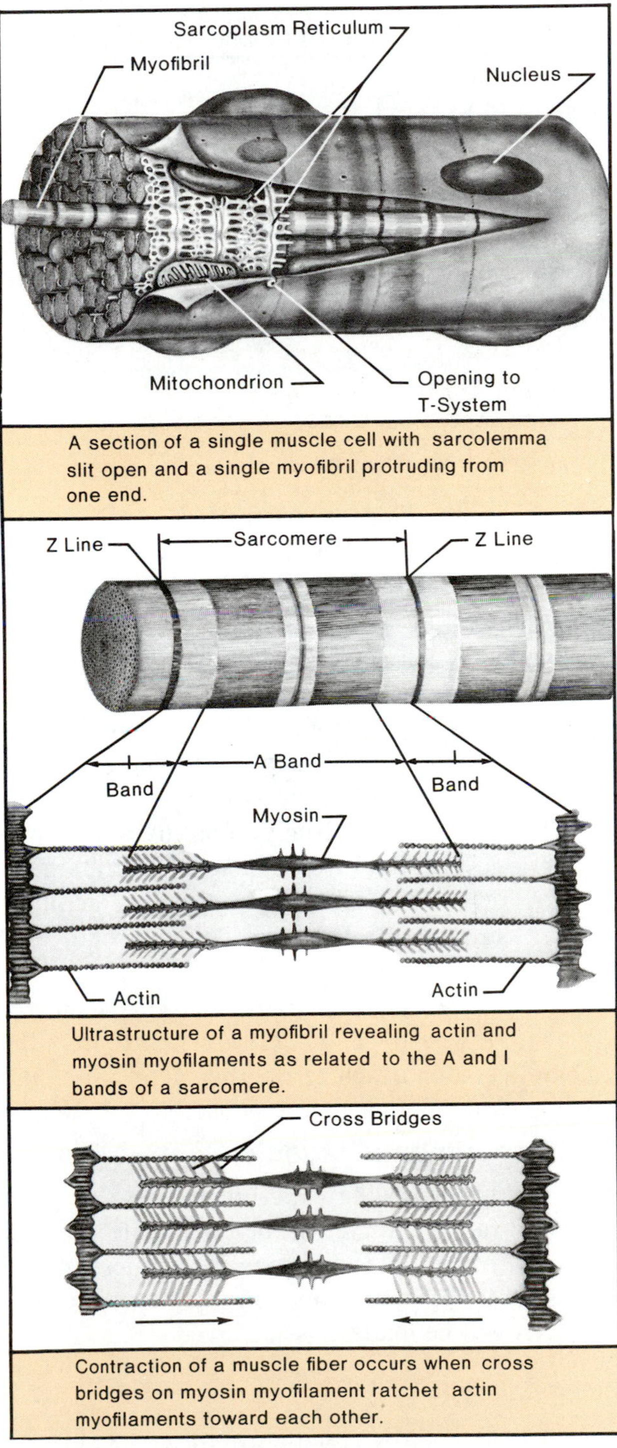

Figure 21.1 Skeletal muscle fiber ultrastructure.

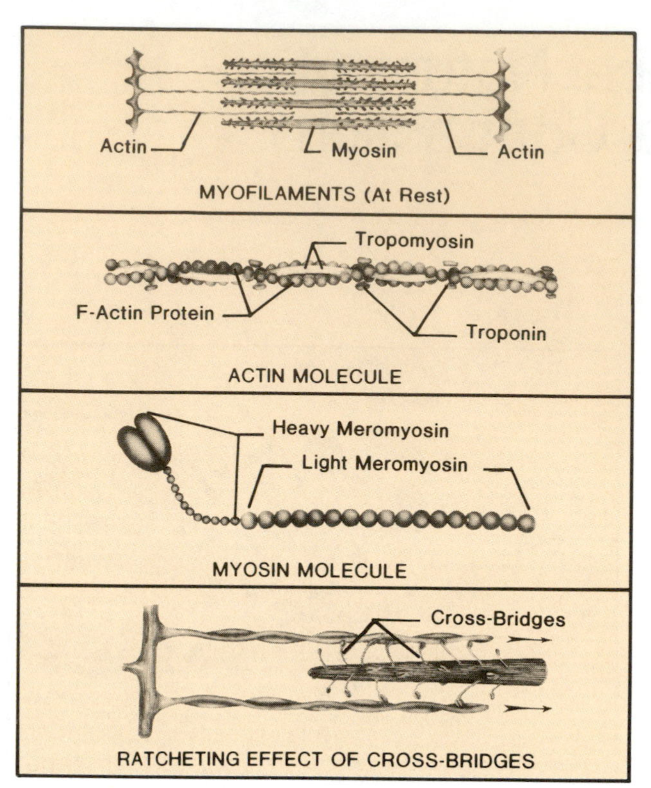

Figure 21.2 Myofilament structure.

The most plausible explanation that we have for the movement of actin myofilaments is proposed by the **ratchet theory.** The sequence of events is as follows:

1. Calcium ions combine with **troponin** (figure 21.2), a protein molecule on the actin.
2. Calcium-bound troponin causes **tropomyosin** on the actin myofilament to sink deep into the fiber and expose active attractive sites.
3. The "heads" of heavy meromyosin (cross-bridges) on the myosin molecule form a bond with the active sites and pull on the actin filament with a ratchetlike power stroke. ATP furnishes the energy.
4. After the first power stroke, ATP combines with the head, enabling it to separate from the active site, attach to another active site, and perform another power stroke.

Experimental Muscle Contraction

Perform the following experiment in which glycerinated rabbit muscle will be induced to contract with the aid of ATP and ions of magnesium and potassium. Measurements and microscopic examinations will be made.

Materials:

test tube of glycerinated rabbit psoas muscle (muscle is tied to a small stick)
dropping bottle of 0.25% ATP in triple distilled water
dropping bottles of 0.05M KCl, and 0.001M $MgCl_2$
3 microscope slides and cover glasses, scissors, forceps, dissecting needle, Petri dish, plastic ruler, microscopes

1. Remove the stick of muscle tissue from the test tube and pour some of the glycerol from the tube into a Petri dish.
2. With scissors, cut the muscle bundle into segments about 2 cm long, allowing the pieces to fall into the Petri dish of glycerol.
3. With forceps and dissecting needle, tease one of the segments into very thin groups of muscle fibers; single fibers, if obtained, will demonstrate the greatest amount of contraction. Strands of muscle fibers exceeding 0.2 mm in cross-sectional diameter are too thick and should not be used.
4. Place one of the strands on a clean microscope slide and cover with a cover glass. Examine under a high-dry or oil immersion objective. Sketch appearance of striae on the Laboratory Report.
5. Transfer three or more of the thinnest strands to a minimal amount of glycerol on a second microscope slide. Orient the strands straight and parallel to each other.
6. Measure the lengths of the fibers in millimeters by placing a plastic ruler underneath the slide. Use a dissecting microscope, if available. Record lengths on the Laboratory Report.
7. With the pipette from the dropping bottle, cover the fibers with a solution of ATP and ions of potassium and magnesium. Observe the reaction.
8. After 30 seconds or more, remeasure the fibers and calculate the percentage of contraction. Has width of the fibers changed? Record the results on the Laboratory Report.
9. Remove one of the contracted strands to another slide. Examine it under a compound microscope and compare the fibers with those seen in step 4. Record the differences on the Laboratory Report.
10. Place some fresh fibers on another clean microscope slide and cover them with a solution of ATP and distilled water (no potassium or magnesium ions). Does contraction occur?
11. Record all data on the last portion of combined Laboratory Report 20,21.

Muscle Contraction Experiments: Using Chart Recorders

22

The following phenomena of skeletal muscle contraction will be studied here: the **simple twitch, recruitment,** and **wave summation.** All of these physiological activities will be observed on frog muscle tissue of a freshly killed frog. Although the frog will be dead, the tissue will still be viable and respond to electrical stimulation. Frog tissue is used here in place of mammalian tissue because of its tolerance to temperature change and adverse handling. The results are similar to what would be seen in more carefully controlled mammalian experiments.

The study of skeletal muscle physiology requires the use of considerable equipment and careful handling of tissue. First of all, a frog must be carefully prepared to keep the tissues viable. Secondly, several pieces of equipment must be properly arranged and hooked up. Finally, stimuli must be administered to the skeletal muscle in a manner which will elicit the desired responses.

Due to the complexity of the overall procedure, and the need to complete all tests within a limited laboratory period, it will be necessary to perform the experiment as a team effort. The class will be divided into four or five teams of five or six students. When the teams are determined, the members of each group can decide among themselves which roles they prefer to perform on the team. The responsibilities of the various members on each team are outlined in table 22.1 on page 100.

Note that figures 22.1 and 22.2 illustrate two different setups for this experiment. Although one setup utilizes Gilson and Grass equipment, the other is essentially a Narco Bio-Systems setup. The type of equipment that is available in your laboratory will determine which setup is to be used. It is also possible to interchange some of the compatible components so that your particular setup is not exactly like either illustration. Your instructor will indicate the actual arrangement that is to be used.

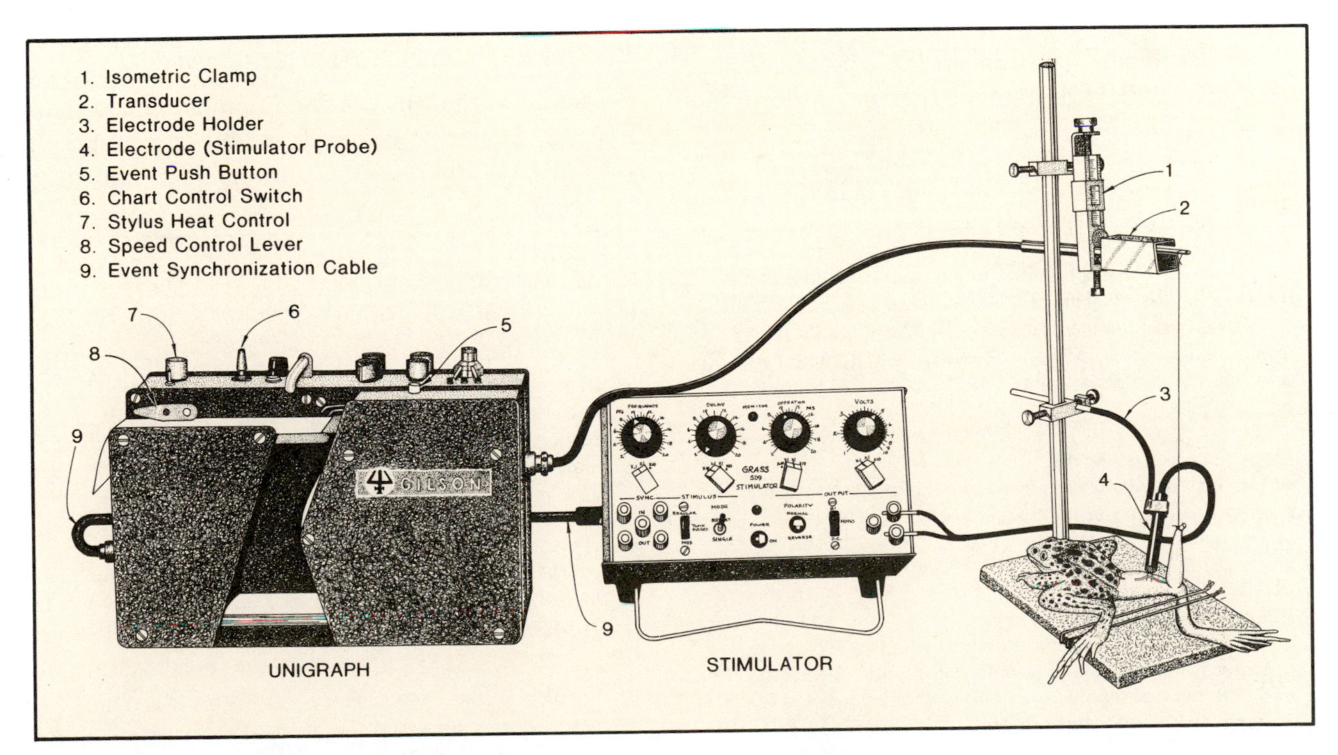

Figure 22.1 Equipment setup with Gilson recorder.

Materials:

For dissection:
- small frog (one per team)
- decapitation scissors
- scalpel, small scissors, dissecting needle, and forceps
- squeeze bottle of frog Ringer's solution
- spool of cotton thread

For Gilson setup:
- Unigraph (Duograph or polygraph)
- electronic stimulator
- event synchronization cable
- electrode and electrode holder
- T/M Biocom #1030 force transducer
- ring stand
- 2 double clamps
- isometric clamp (Harvard Apparatus, Inc.)
- galvanized iron wire and pliers

For Narco setup:
- Physiograph® IIIS with transducer coupler
- SM-1 electronic stimulator
- stimulus switch and myograph
- transducer stand and frog board
- myograph tension adjuster
- event marker cable
- sleeve electrode and pin electrodes
- dissecting pins

Preparations

While one team is setting up the equipment, another team will decapitate the frog and perform the dissection.

Frog Dissection

To hold the frog for decapitation, grip it with the left hand so that the forefinger is under the neck and the thumb is over the neck. Place the frog under a water tap, allowing cool water to flow onto its head and body. This tends to calm the animal and helps to wash away the blood.

Insert one blade of a sharp heavy-duty scissors into the mouth, well into the corner; the other scissors' blade should be poised over a line joining the tympanic membranes (eardrums). With a swift clean action, snip off the head. Immediately after this, force a probe down into the spinal canal to destroy the spinal cord. The frog should become limp and show no signs of reflexes.

Strip the skin off the right or left hind leg following the instructions in figures 22.3 through 22.5. *If the Gilson setup is used, use the right hind leg. If the Narco setup is used, use the left leg.*

With a sharp dissecting needle, tear through the fascia and muscle tissue of the thigh to expose the sciatic nerve and insert a short length (10″ long) of cotton thread under it. This thread will be used for manipulation of the nerve. Figures 22.6 through 22.8 outline the desired procedure. *Be careful in the way you handle the nerve.* The less it is traumatized or comes in contact with metal, the better. *Keep the exposed tissues moist with frog Ringer's solution.*

Equipment Hookup *(Gilson Arrangement)*

Set up the various pieces of equipment as shown in figure 22.1. The power cords of both the Unigraph

Table 22.1 Team member responsibilities.

EQUIPMENT ENGINEER
Hooks up the various components (recorder, stimulator, transducer, etc.). Balances the transducer and calibrates recorder. Operates the stimulator. Responsible for correct use of equipment to prevent damage. Dismantles equipment and returns all components to proper places of storage at end of period.

ENGINEER'S ASSISTANT
Works with engineer in setting up equipment to ensure procedures are correct. Makes necessary tension adjustments before and during experiment. Applies electrode to nerve and muscle for stimulation.

RECORDER
Labels events produced on tracings. Keeps a log of the sequence of events as experiment proceeds. Sees that each member of team is provided with chart records for attachment to Laboratory Report.

SURGEON
Decapitates and piths frog. Removes skin on leg and exposes sciatic nerve. Responsible for maintaining viability of tissue. Attaches string to nerve and muscle end.

HEAD NURSE
Assists surgeon as needed during dissection. Keeps exposed tissues moist with Ringer's solution. Helps to keep work area clean.

COORDINATOR
Oversees entire operation. Communicates among team members. Reports to instructor any problems that seem to be developing that seem insurmountable. Double-checks the setup and cleanup procedures. Keeps things moving.

and stimulator must be plugged into a grounded 110-volt outlet. Note that an event synchronization cable connects the stimulator to the Unigraph. Attach the Harvard isometric clamp to the ring stand with a double clamp, keeping the adjustment screw of the isometric clamp upward. Attach the shaft of the transducer to the lower part of the isometric clamp and plug the transducer cord into the Unigraph. Put another double clamp on the ring stand shaft at the lower level and clamp the shaft of the electrode holder (label 3, figure 22.1) in place. Plug the jacks of the electrode into the stimulator at the position shown in the illustration, but do not fix the electrode to the clamp at this time since the electrode will have to be maneuvered about in the first tests.

With the completion of all the necessary connections, *balance the transducer and calibrate the Unigraph* according to the instructions in Appendix C.

Equipment Hookup *(Narco Arrangement)*

Set up the various pieces of equipment as shown in figure 22.2. In this setup a stimulus switch is used between the stimulator and the frog. This switch makes it possible to switch from direct muscle stimulation to nerve stimulation without disturbing the setup.

Note that an event marker cable connects the stimulator with the Physiograph®. The receptacle for this cable is on the stimulator back; it is labeled MARKER OUTPUT. The other end of the cable is inserted into the middle receptacle of the transducer coupler.

After affixing the myograph tension adjuster to the transducer stand, attach the myograph to the support rod, and tighten the thumbscrew to hold it in place. Insert the free end of the myograph cable into the transducer coupler.

Plug the power cords for the Physiograph® and SM-1 stimulator into grounded 110-volt outlets. Insert the jacks for the pin electrodes into the right side of the stimulus switch and the jacks for the sleeve electrode into the left side of the stimulus switch. Also, attach the cable from the stimulus switch to the output posts of the SM-1 stimulator.

With the completion of all the hookups, *balance the channel and calibrate the myograph according to instructions in Appendix C.* You are now ready to begin muscle and nerve stimulations.

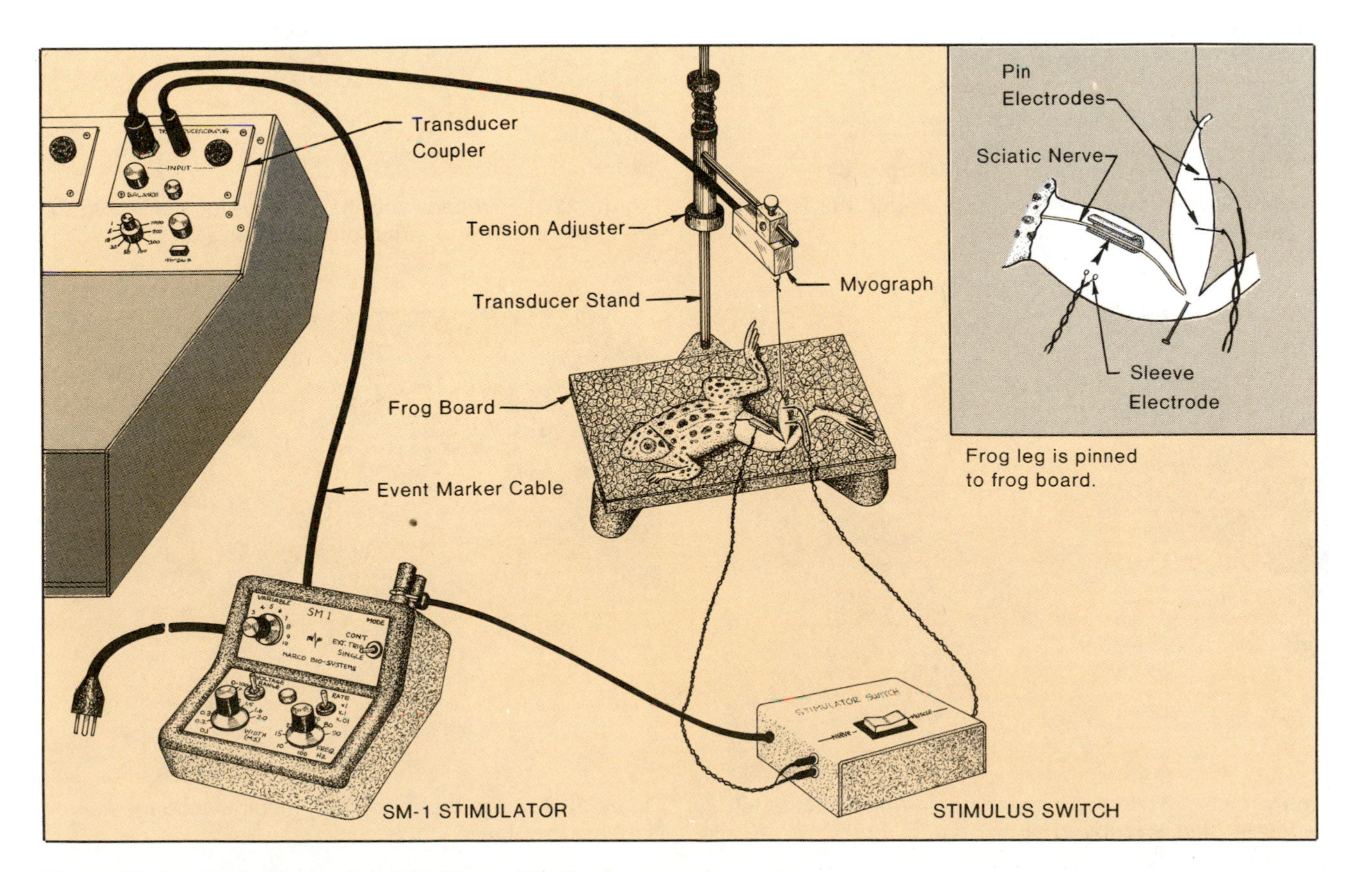

Figure 22.2 Equipment setup with Narco Bio-Systems equipment.

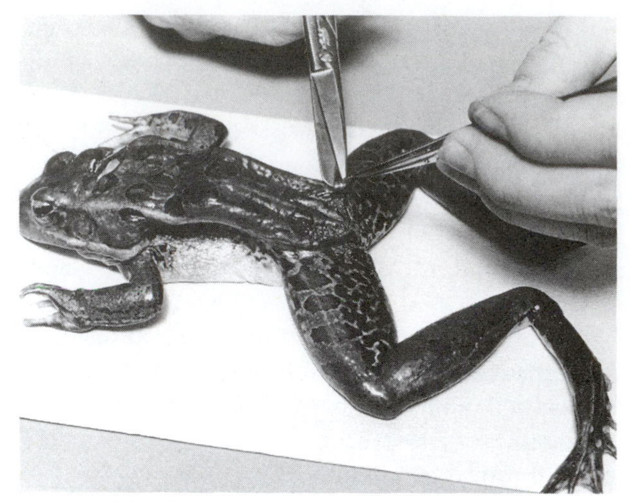

Figure 22.3 The first incision through the skin is made at the base of the thigh.

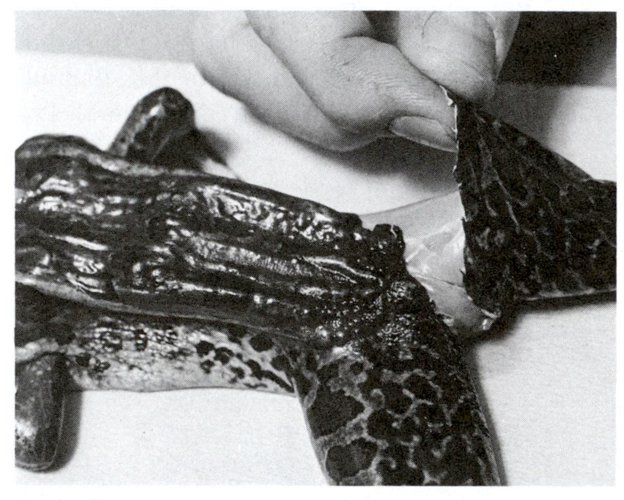

Figure 22.4 After the skin is cut around the leg, it is pulled away from the muscles.

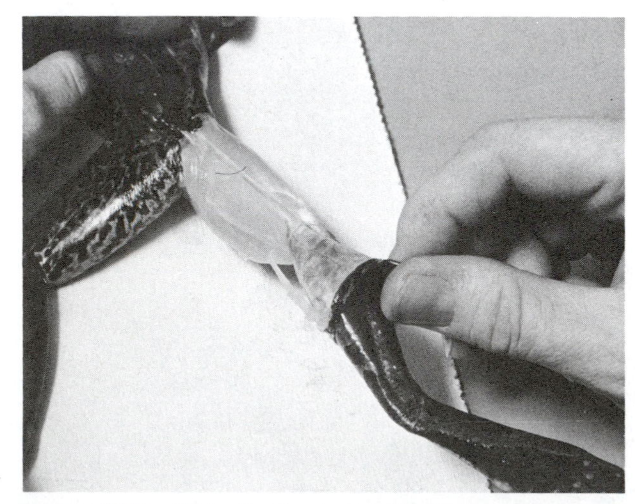

Figure 22.5 While the body is held firmly with the left hand, the skin is stripped off the entire leg.

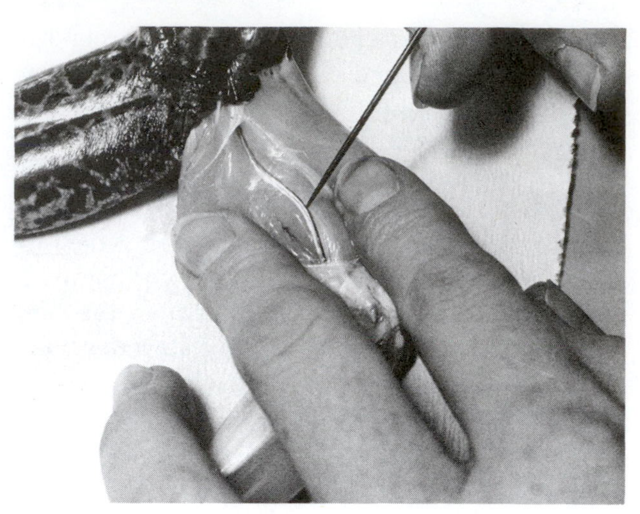

Figure 22.6 The muscles of the thigh are separated with a dissecting needle to expose the sciatic nerve.

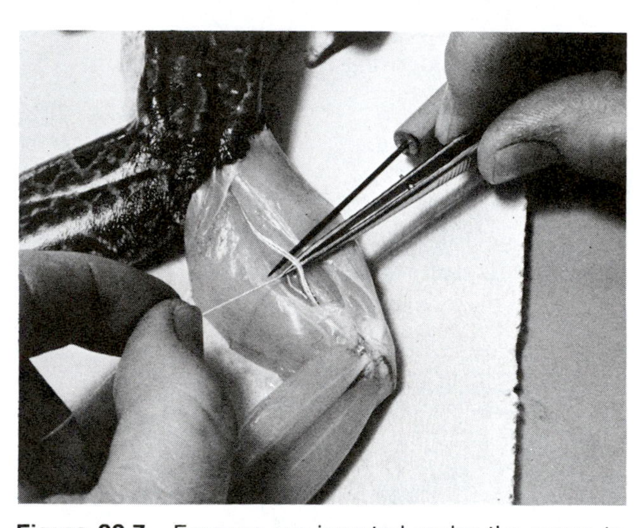

Figure 22.7 Forceps are inserted under the nerve to grasp the end of the string on the other side.

Figure 22.8 The thread used for manipulating the nerve is pulled through.

The Threshold Stimulus

Before severing the Achilles tendon on the muscle and attaching the muscle with string to the transducer to create the complete setup shown in figures 22.1 or 22.2, we are going to stimulate the intact muscle and nerve to determine the minimum voltage that will elicit a simple twitch. A stimulus that barely induces a muscle twitch is called the **threshold stimulus.** A stimulus of less than threshold strength is designated as being **subliminal.** Proceed as follows to compare the threshold differences between muscle and nerve stimulation. *Be sure to keep the muscle and nerve moist with frog Ringer's solution.*

Gilson Procedure

1. Set the Duration control at 15 milliseconds and the Voltage control at 0.1 volt. The Stimulus switches should be set on REGULAR and OFF. The Polarity switch should be on NORMAL and the Output switch on MONO.
2. Turn on the Power switch and depress the Mode switch to SINGLE, while holding the electrode against the belly of the gastrocnemius muscle as illustrated in figure 22.9. You are now administering a single pulse of 0.1 volt of 15 msec duration to the muscle. This is the minimum voltage possible with the Grass stimulator. If the muscle twitches, reduce the Duration to the lowest value that will produce a twitch. In this manner we can record 0.1 volt as the **threshold stimulus** on the Laboratory Report.

   If the muscle does not twitch at 0.1 volt, 15 msec, increase the voltage at increments of 0.1 volt until a twitch is seen. Record this voltage as the **threshold stimulus** on the Laboratory Report.
3. Now, place the foot in a flexed position and proceed to stimulate the muscle by gradually increasing the voltage until the foot extends completely, due to muscle contraction. Record this voltage on the Laboratory Report.
4. Cut the nerve at point A indicated in figure 22.10, and position the stimulator probe under the nerve as indicated in this illustration. The nerve is cut to prevent extraneous reflexes.
5. Place the foot in a flexed position.
6. Return the Voltage control to 0.1 volt and stimulate by pressing the Mode lever to SINGLE. If no twitch is seen, proceed as previously to stimulate the nerve at increasing increments of 0.1 volt until a twitch is visible. Record this voltage as the threshold stimulus by nerve stimulation.
7. Now, determine the minimum voltage that causes extension of the foot by stimulating the nerve. Record this voltage on the Laboratory Report. If extension occurs with a stimulus of 0.1 volt, try reducing the Duration in the same manner that you used for direct muscle stimulation.

Narco Procedure

1. Set the Width control at 2 msec, and the Voltage at 0.1 volt (Range switch at 0–10, Variable control at its lowest value).

Figure 22.9 The belly of the gastrocnemius muscle is stimulated with hand-held electrode (Gilson setup).

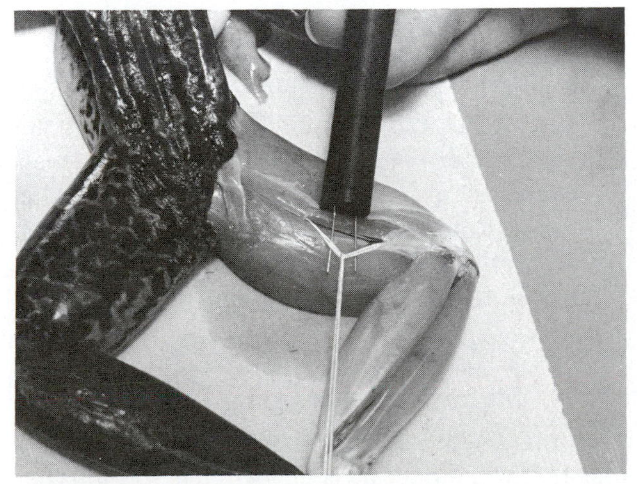

Figure 22.10 Nerve is cut at point A in both setups. Note that probe is under the nerve for Gilson setup.

2. Insert the pin electrodes into the muscle about one centimeter apart.
3. Press the MUSCLE side of the button on the Stimulus switch to insure that the muscle will get stimuli from the stimulator.
4. Now, deliver a single stimulus to the muscle by depressing the Mode switch on the stimulator to SINGLE. If the muscle twitches, reduce the Width to the lowest value that will produce a twitch. In this manner we can record 0.1 volt as the **threshold stimulus** on the Laboratory Report.

   If the muscle does not twitch at 0.1 volt, 2 msec, increase the voltage at increments of 0.1 volt increments until a twitch is seen. Record this voltage as the **threshold stimulus** on the Laboratory Report.
5. Now, place the foot in a flexed position and proceed to stimulate the muscle by gradually increasing the voltage until the foot extends completely, due to muscle contraction. Record this voltage on the Laboratory Report.
6. Carefully position the rubber tubing sleeve around the nerve and insert the sleeve electrode within the tubing next to the nerve. Use the string to manipulate the nerve, and *be sure to bathe the nerve in frog Ringer's solution.*
7. Press the NERVE side of the button on the Stimulus switch to make it possible to stimulate the nerve.
8. Cut the nerve at point A indicated in figure 22.10 to prevent extraneous reflexes.
9. Place the foot in a flexed position.
10. Return the voltage to 0.1 volt and stimulate by pressing the Mode lever to SINGLE. If no twitch is seen, proceed as previously to stimulate the nerve at increasing increments of 0.1 volt until a twitch is visible. Record this voltage as the threshold stimulus by nerve stimulation.
11. Now, determine the minimum voltage that causes extension of the foot by stimulating the nerve. Refer to illustration B, figure SR–1 in Appendix D for Physiograph® sample record. Record this voltage on the Laboratory Report.

Myogram of Muscle Twitch

Now that we have determined the threshold stimulus that will induce a visible twitch on the muscle, we are ready to produce a *myogram,* or tracing, of the twitch using a transducer and recorder. Figure 22.11 illustrates what a myogram of a human

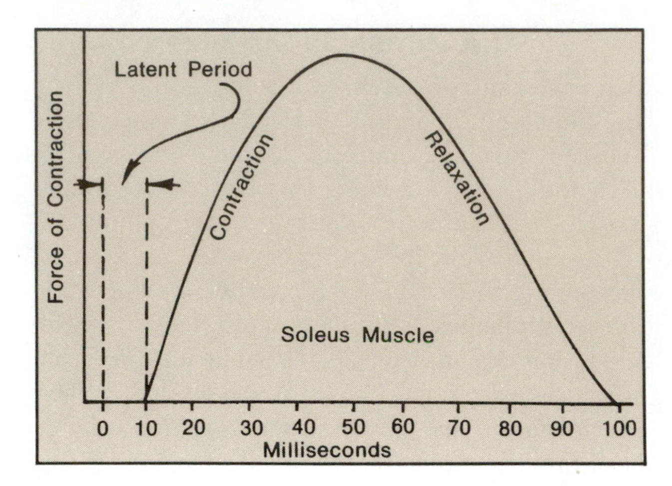

Figure 22.11 A simple twitch of the human soleus muscle.

muscle, the *soleus,* looks like. Although the frog muscle twitch will have the same general configuration, it will have a different amplitude and spread.

Note in figure 22.11 that the myogram of a muscle twitch has three distinct phases: a latent period, the contraction phase, and the relaxation phase. The **latent period** is a very short time lapse between the time of stimulation and the start of contraction. In most muscles of the body, it lasts about 10 milliseconds. In the **contraction phase** the muscle shortens due to the chemical changes that occur within the cells. Different muscles have different durations in this phase. After the contraction phase has reached its maximum, the muscle returns to its former relaxed state during the **relaxation phase.**

Proceed as follows to attach the frog's leg to the transducer, and to record a myogram.

Thread Hookup

Free the connective tissue that holds the gastrocnemius muscle to adjacent muscles with a dissecting needle. Insert a pair of forceps under the muscle near the Achilles tendon (figure 22.12) to grasp one end of an 18″ length of cotton thread. Pull the thread through with the forceps and tie it to the tendon a short distance away from where the tendon will be severed. Now, with a pair of scissors, cut off the tendon distal to where it is attached to the string (figure 22.13).

If the Gilson setup is being used, tie the free end of the thread to two leaves (the stationary one and the one next to it) of the transducer. If the Narco myograph is being used, tie a loop in the thread and slip it onto the little hook.

To prevent the leg from moving during contraction, pin the leg down to the cork board with a pin (Narco setup, figure 22.2). For the Gilson setup, it will be necessary to secure the leg to the base with wire. *When handling these transducers, do so gently. Undue stress to the leaves or hook can cause damage.*

Adjust the tension on the thread with the fine adjustment knob on the tension adjuster just enough to take the slack off the thread.

Electrode Positioning

For the Narco setup, check the sleeve electrode to make sure the electrode is in the proper position.

For the Gilson setup, position the electrode prongs so that the nerve rests on the prongs. Maneuver the electrode holder in such a way that the electrode is held firmly in place. Note that the stem of the holder consists of a soft metal that can be bent into any position.

Although the pin electrodes are shown in figure 22.2, they will not be used in the remainder of the experiment.

Recorder Settings

Unigraph Turn on the power switch. Place the chart control (c.c.) lever at STBY, the stylus heat control knob at the two o'clock position, the speed selector lever at the slow position (opposite of position shown in figure 22.1), the gain control on 2 MV/CM, and the mode control on TRANS (transducer).

Now, place the c.c. lever at Chart On and note the position of the line produced on the chart. The line should be about one centimeter from the nearest margin of the paper. If it is not at this position, relocate the stylus with the centering knob. Now, stop the chart by placing the c.c. lever at STBY. Proceed to Stimulator Settings below.

Physiograph® Turn on the power switch. Make sure that the myograph is calibrated so that 100 grams of tension is equivalent to 4 centimeters of pen deflection (Appendix C).

After the initial attachment of the muscle to the myograph and *with minimal tension on the myograph,* set the baseline to the centerline with the position control knob.

Place the record button in the ON position and gradually increase the tension by moving the myograph upward with the myograph tension adjuster. Tension should be increased until the pen has moved upward 0.5 centimeters (you have applied 12.5 grams of tension).

By using the Balance control, return the pen to centerline.

Now, by using the position control, move the pen to any desired baseline; normally, this would be 2 centimeters below the centerline.

Be sure to periodically bathe the muscle and nerve preparation with frog Ringer's solution.

Stimulator Settings

Grass Stimulator Set the Duration control at 15 msec and the Voltage at the level that produced flexion of the leg by nerve stimulation in the pre-

Figure 22.12 Thread is drawn through the space behind the Achilles tendon by inserting forceps through to grasp the thread.

Figure 22.13 After the thread is securely tied to the Achilles tendon, the tendon is severed between the thread and joint.

vious experiment. Check the switches to make sure that the Mode switch is set at OFF, pulses are REGULAR, Polarity is on NORMAL, and the right-hand slide switch is on MONO.

SM-1 Stimulator Set the Width control at 1 msec and the Voltage at the level that produced flexion of the leg by nerve stimulation in the previous experiment.

Recording

You are now ready to produce a simple muscle twitch myogram on the chart of your recorder and determine the durations of each of the three phases in the contraction cycle. Proceed as follows:

Unigraph

1. Place the c.c. lever at Chart On. The paper should be moving at the slow rate (2.5 mm/sec).
2. Depress the Mode switch to SINGLE and observe the tracing on the chart. The stylus travel should be approximately 1.5 centimeters. Adjust the sensitivity control to produce the desired stylus displacement.
3. If the stylus travel remains insufficient by adjusting the sensitivity knob, increase the sensitivity by changing the Gain control to 1 MV/CM and readjusting the sensitivity knob again.
4. Change the speed of the paper to 25 mm per second by repositioning the speed control lever 180° to the left.
5. Administer 25 to 30 stimuli to provide enough chart material so that each member of your team will have at least five inches of chart for attachment to the Laboratory Report sheet.
   Note: If your setup does not have an event sync cable it will be necessary to depress the event marker button on the Unigraph simultaneously with depression of the Mode lever on the stimulator.
6. Calculate the duration of each of the periods, knowing that one millimeter on the chart is equivalent to 0.04 seconds (40 milliseconds).

Physiograph®

1. Start the paper moving at 0.1 cm/sec, and lower the pens onto the recording paper. It is not necessary that the timer be turned on.
2. Press the Mode switch to SINGLE and observe the tracing on the chart. The pen travel should be approximately 1.5 centimeters. Adjust the sensitivity control to produce the desired pen travel.
3. Change the chart speed to 2.5 cm per second.
4. Administer 25 to 30 stimuli to provide enough chart material so that each member of your team will have enough chart to attach to the Laboratory Report sheet. Refer to illustration A, figure SR–1, Appendix D, for Physiograph® sample recording.
5. Calculate the duration of each of the periods on the contraction cycle and report your results on the Laboratory Report.

Summation

The ability of whole muscles to exert different degrees of pull is achieved primarily by *summation.* Two types of summation are known to occur: multiple motor unit and wave summation. **Multiple motor unit summation** is an increased force caused by the recruitment of additional motor units. It is also known as *recruitment* or *spatial summation.* **Wave summation** is due to an increase in the frequency of nerve impulses, and is also referred to as *temporal summation.* An attempt will be made in this portion of the experiment to demonstrate both types.

Multiple Motor Unit Summation

Once the threshold stimulus has been determined for producing a twitch, it will be observed that increasing the strength of the stimulus (by voltage increase) will cause increasingly greater degrees of contraction. This phenomenon is schematically illustrated in figure 22.14. As the diagram reveals, the force of contraction is a function of the number of motor units that are stimulated.

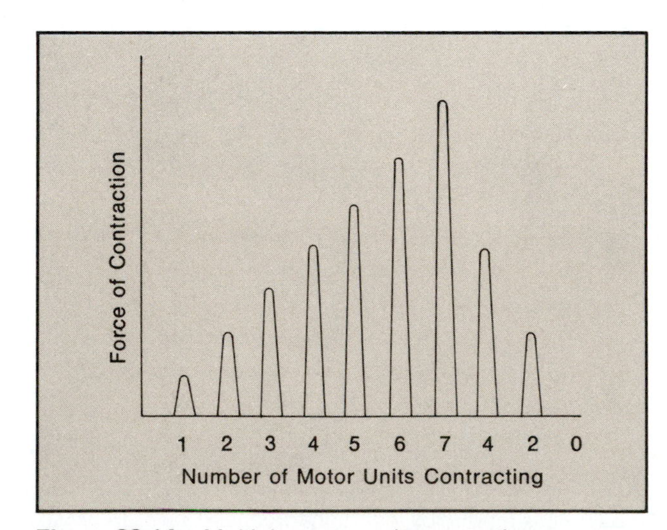

Figure 22.14 Multiple motor unit summation.

A **motor unit** consists of a group of muscle fibers that is innervated by a single motor neuron (motoneuron). Small muscles that react quickly and precisely may have as few as two or three muscle fibers per motor unit. Large muscles that lack a fine degree of control, such as the gastrocnemius, may have as many as 1,000 fibers per motor unit. These large muscles are well adapted to a posture-sustaining function.

The muscle fibers of adjacent motor units are arranged in an overlapping fashion so that each unit lends support to its neighbors. When some nerve fibers to a muscle are destroyed, as in poliomyelitis, **macromotor units** are formed by extensive arborization of the ends of surviving motoneurons to provide neural connections to all muscle fibers. These large motor units may be four or five times as large as normal units.

As voltage increases are administered, a degree of contraction is reached that cannot be exceeded by further voltage increase. This maximum contraction, called the **maximal response,** is due to the fact that all motor units are activated and no further contraction by individual stimuli can be produced. The lowest voltage that produces a maximal response is called the **maximal stimulus.** Proceed as follows to demonstrate recruitment, maximal stimulus, and maximal response.

1. **Unigraph:** For this unit set the speed control at the slow speed (2.5 mm/sec) and the c.c. switch at STYLUS ON. All other settings should be left as they were for the previous experiment. **Physiograph®:** Set the paper speed at 0.25 cm/sec, and lower the pens onto the paper, but do not start the paper at this time.
2. **Stimulator:** Except for the voltage, leave all other settings as they were in the previous experiment. Return the voltage controls to 0.1 volt.
3. **Electrodes:** Check the electrodes on the frog's nerve to see that they are in good contact with the nerve. Remember to keep the tissues moist with frog Ringer's solution.
4. With the chart still, administer a stimulus to the nerve by depressing the Mode lever to SINGLE. Look for movement of the stylus on the chart. Advance the chart approximately 5 mm, and stop it again.
5. Raise the voltage by 0.2 volt and produce another single stimulus. Continue this process of advancing the chart 5 mm and increasing the voltage by 0.2 volt until a maximal response has been achieved. *Be sure to record all voltages on the chart.*
6. Repeat this overall process five or more times to provide enough chart material for all members of your team.

Wave Summation

If motor unit summation were the only way in which maximum contraction could be achieved by muscles they would have to be much larger to accomplish the work they are able to do. Another phenomenon, called wave summation, plays a significant role in increasing the amount of muscle contraction.

Wave summation in skeletal muscle occurs when the muscle receives a series of stimuli in very rapid succession, as illustrated in figure 22.15. If the rate of stimulation is kept very slow, only single twitches will occur; however, as seen in figure 22.15 the single twitches produced at 35/sec will produce some summation. Note that when the frequency is stepped up to 70 pulses per second, considerable summation occurs, and 200 pulses per second produces sustained or *complete tetanization.*

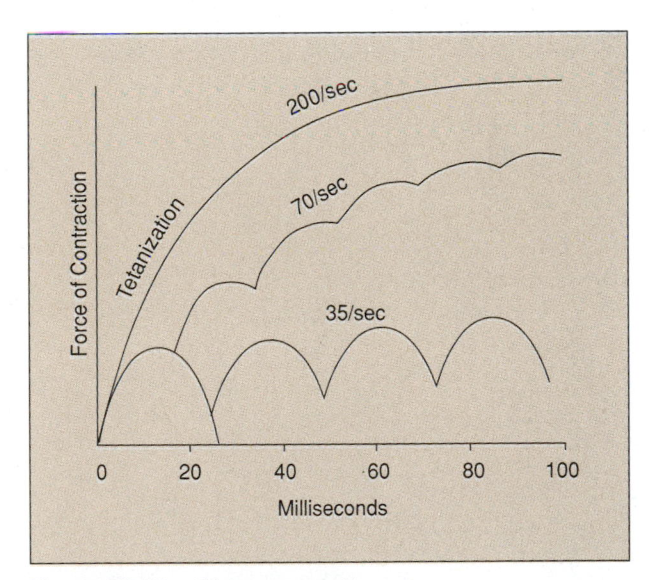

Figure 22.15 Wave summation.

In this portion of the experiment, we will use the Frequency control to regulate the pps or Hz.

Unigraph Setup Use the following procedure if you are using the Unigraph-Grass setup.

1. Set the Voltage at the previously determined maximal stimulus, and the Frequency at 1 pps. Place stimulator Mode switch at OFF.

2. Put the c.c. switch at Chart On with chart speed at slow setting (2.5 mm/sec).
3. Move Mode switch on stimulator to REPEAT. Note that monitor lamp on the stimulator is blinking.
4. Record the tracing for 10 seconds, return the stimulator Mode switch to OFF, and the c.c. switch at STBY. Let the muscle rest for **2 minutes.**
5. Increase the Frequency to 2 pps, start the paper moving, and record for another **10 seconds.** Stop the chart (STBY), and rest the muscle for another 2 minutes.
6. Now, double the frequency to 4 pps for another 10 seconds. Rest again for **2 minutes.**
7. Continue to double the frequency, recording for 10 seconds, and resting for 2 minutes, until the muscle goes into complete tetany.
8. After resting the muscle for a few minutes, repeat the experiment enough times to provide a record for each member of your team.

Physiograph® Setup Use the following procedure if you are using the Narco Physiograph® setup.

1. Set the Voltage at the previously determined maximal stimulus, and the Frequency at 1 Hz.
2. Lower the pens and start the chart paper moving at a speed of 0.25 cm per second.
3. Place the stimulator Mode switch on CONT. Note that the monitor lamp on the stimulator is blinking.
4. Record the tracing for **10 seconds,** and return the stimulator Mode switch to OFF. Stop the chart movement and raise the pens. Let the muscle rest for **2 minutes.**
5. Increase the Frequency to 2 Hz, set the stimulator on CONT, lower the pens, and start the paper moving again. Record for another 10 seconds, return the stimulator Mode switch to OFF, stop the paper and raise the pens. Allow the muscle to rest another 2 minutes.
6. Now, double the frequency to 4 Hz, turn on the stimulator to CONT again, lower the pens, and start the paper moving. After recording for another 10 seconds, shut down the system again and rest the muscle for another 2 minutes.
7. Continue to double the frequency, recording for 10 seconds, and resting the muscle each time for 2 minutes until the muscle goes into complete tetany.
8. Refer to illustration C, figure SR–1, Appendix D for sample Physiograph® records.
9. After resting the muscle for a few minutes, repeat the experiment enough times to provide a record for each member of your team.

Laboratory Report

Complete the Laboratory Report and attach the charts from the above experiment to it.

Muscle Contraction Experiments: Using Computerized Hardware 23

The same phenomena of muscular contraction studied in Exercise 22 will be studied here utilizing an Intelitool Physiogrip™ which is controlled by computer software. The setup is illustrated in figure 23.1. Note that electrical stimuli will be applied to nerves in the right arm that control flexion of the third and fourth fingers. With either the third or fourth finger on the trigger of the Physiogrip, flexor response is fed into the computer for data accumulation and evaluation. With this setup the following phenomena of muscle contraction will be studied:

- threshold stimulus
- motor unit summation (recruitment)
- wave summation
- tetany
- single muscle twitch
- fatigue

Although this experiment may be performed individually, it is much easier to perform if students work in pairs. By working in pairs, one individual can be the subject while the other student reads the instructions and operates the stimulator and computer.

Note in the materials list on the next page that among the items used in this experiment are copies of the *Physiogrip User* and *Physiogrip Lab Manuals* that are supplied with the equipment. Although you may not need them, the manuals will be available for reference.

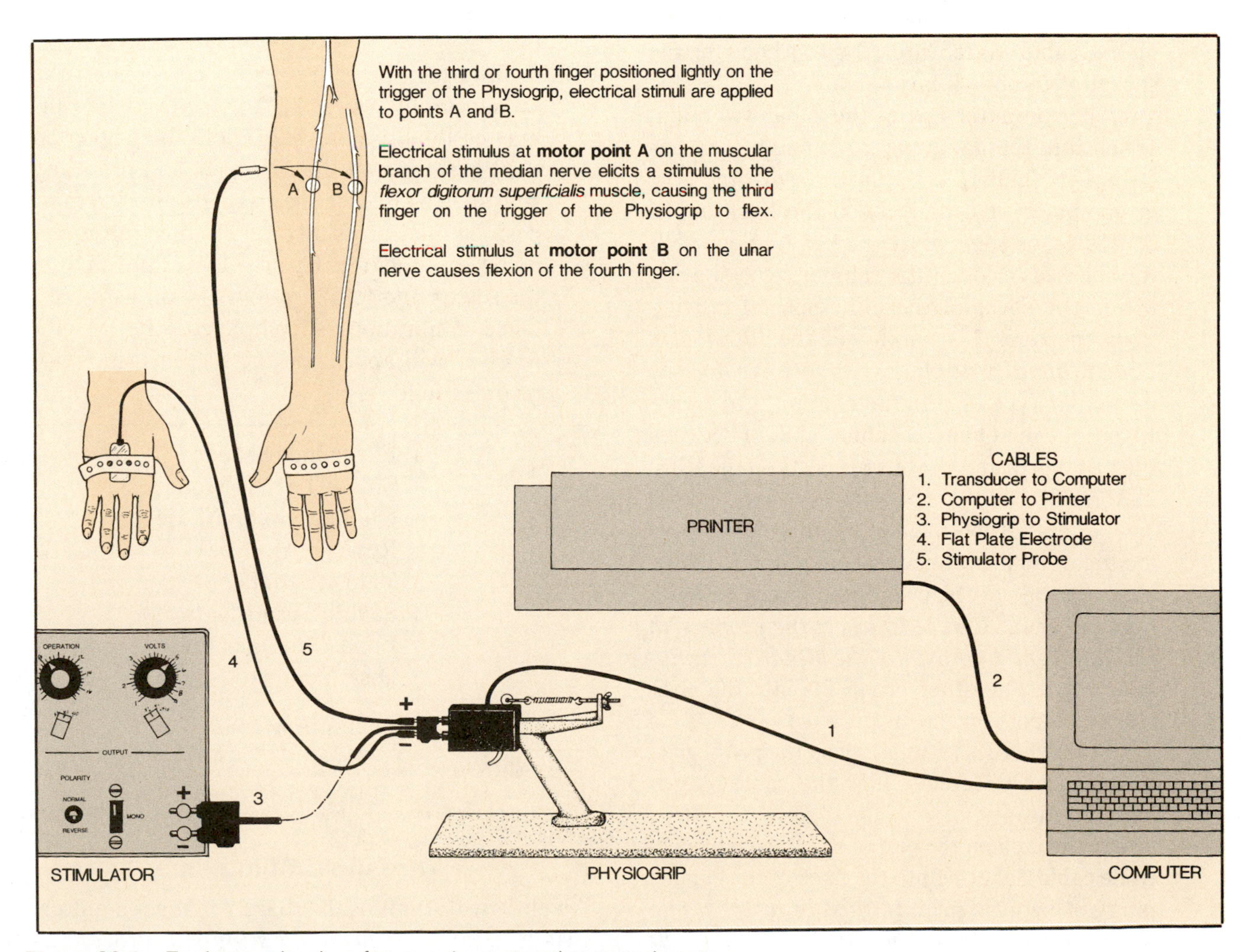

Figure 23.1 Equipment hookup for muscle contraction experiments.

Materials:

Physiogrip transducer on wooden base
Physiogrip program disk
Physiogrip Lab Manual
Physiogrip User Manual
blank disk for saving data
electronic stimulator, computer, and printer
flat plate electrode cable and rubber strap
stimulator probe cable
dual banana cable, gameport cable
electrode gel

Equipment Hookup

Arrange the various components illustrated in figure 23.1 in a manner that will make it convenient to use all pieces of equipment. Hook up the electronic components with the appropriate cables, observing the following precautions:

1. Plug in the computer, printer, and stimulator to *grounded* 110 volt electrical outlets.
2. *With the computer turned off,* connect the **game port cable** (label 1) between the black box of the pistol grip and the computer. Be sure that the connector pins on the transducer end of the cable match up with the holes in the socket of the black box.

   The computer end of the cable will be inserted into the game port. Although the location of the game port will vary with the type of computer, it usually is on the back of the computer; on some of the earlier Apple models it is located inside of the cabinet, necessitating lifting the cover of the unit. Special instructions on pages 3, 4, and 5 of the *Physiogrip User Manual* will clarify its location for the particular type of computer used.
3. Install the **dual banana cable** (label 3) between the pistol grip and the output posts on the stimulator.

   Note that both ends of this cable have identical dual-pronged banana plugs. Note, also, that one of the prongs on each plug has a bump near it. *Be sure to insert the prong with the adjacent bump into the negative stimulator post.* The other prong fits into the positive (red) post on the stimulator.

   Since there is no polarity on the pistol grip, it makes no difference how the cable plug is inserted into it.
4. Insert the jack on the end of the **flat-plate electrode cable** (label 4) into the back of the banana plug at the pistol grip. *It must be inserted into the negative side of the plug which has the bump near it.* Note that this plate electrode, which is the ground, is strapped to the back surface of the right hand. Before strapping it to the hand place an ample amount of electrode gel to the plate so that good electrical contact occurs.
5. Insert the jack of the **stimulator probe cable** (label 5) into the other hole in the back of the banana plug at the pistol grip.

Getting Started

Now that all components are hooked up, you are ready to begin the experimentation. Proceed as follows:

1. Insert the program diskette into drive A and the data diskette into drive B.
2. Turn on the computer. If you are using an Apple IIe, IIc, or IIGS, be sure to depress the caps lock key, since all command keys are functional only in the caps locked mode.
3. Press any key and the Physiograph program will take over. A group of questions will be asked which you must answer concerning the configuration. "Yes" responses are confirmed by pressing the Y key; "No" responses with the N key.
4. Calibration procedures will occur next on the screen. Only two steps are involved in calibrating the Physiogrip: (a) hold the trigger on the Physiogrip all the way back and press any key, and (b) release the trigger to its foremost position, and press any key. If the transducer is out of adjustment you will be informed on the screen and told how to make adjustments.
5. Once calibration is completed, the MAIN MENU will appear on the screen, which appears as follows:

( ) Experiment Mode
( ) Single Twitch Mode
( ) Change Sweep Mode
( ) Review/Analyze
( ) Velocity Plot
( ) Save Current Data
( ) Load Old Data File
( ) Disk Commands

(No Data in Memory)

Note: Selections of Review/Analyze, Velocity Plot, and Save Current Data in the Main Menu are possible *only* if there are experimental data in the memory. If no data are in the memory "No Data in Memory" will flash on the screen.

Threshold Stimulus

An electrical stimulus that barely induces a muscle twitch is called a *threshold stimulus.* Proceed as follows to locate the motor point and determine the threshold stimulus of the subject:

1. Set the stimulator at 20 volts, 1 pulse per second, and 1 millisecond duration.
2. With the subject completely relaxed, apply the probe to the skin of the right arm in the area of the motor point. Use considerable pressure for good contact. If no tingling is felt, increase the voltage until a sensation is detectable.
3. Move the probe around the area until the middle finger flexes. It may be necessary to increase the voltage to 70 or 80 volts before the muscle reacts.
4. Select "Change Sweep Mode" from the Main Menu and select **15 seconds per sweep.**
5. Select "Run Experiment Mode" from Experiment Mode in Main Menu.
6. Set the stimulator on **Single Mode,** and place the middle finger on the trigger of the Physiogrip.
7. Reduce the voltage to a point where there is no response on the screen; then increase the voltage to a level where you see a slight response. Record this voltage as the threshold limit.
8. If further help is needed consult pages 7–14 of the *Physiogrip Lab Manual* (PLM).

Spatial Summation (Recruitment)

With your finger still on the trigger, demonstrate *spatial summation* with the following procedure:

1. Increase the voltage by 10 volts above the threshold stimulus and administer a single stimulus. Note the increased contraction as more motor units respond.
2. Repeat stimulating the motor point by successively increasing the voltage by small increments until all motor units are recruited (**maximal contraction**). Note the stair stepping in the strength of responses. Do not increase the voltage once maximum recruitment has occurred. **Do not exceed 100 volts.**
3. Return to the Main Menu (use ESC key) and select Review/Analyze.
4. Measure the heights of the contraction caused by the threshold stimulus and maximal contract. (Reference: PLM, p. 14).

Wave Summation and Tetany

Wave summation occurs when the frequency of stimulation is increased to the extent that the motor units do not have time to relax. *Tetany* occurs when the contractions fuse to produce a steady state of contraction. Proceed as follows to demonstrate these two phenomena:

1. Select **3 seconds per sweep** from "Change Sweep Speed" of Main Menu.
2. Select "Run Experiment Mode" from Experiment Mode menu.
3. Position the middle finger on the trigger of the Physiogrip, and the probe on the motor point.
4. Set the stimulator as follows: continuous (multiple) mode, 1 msec duration, 1 pps frequency, and sufficient voltage to produce a response that is 25% of the screen.
5. Gradually increase the frequency of stimulus up to a tetanic contraction. This should be accomplished within two screens (6 seconds). Do this several times, sustaining tetanic contractions for 1–2 seconds.
6. Go back to the Main Menu and select "Review/Analyze."
7. Determine and record the frequency of stimulation as the point where a tetanic contraction was reached. (Reference: PLM, p. 17).

Single Muscle Twitch

In this demonstration of a single muscle twitch you will evaluate the durations of each phase of the phenomenon.

1. Select "Run Single Twitch Mode" from the Single Twitch Mode of the Main Menu.
2. Set the stimulator as follows: continuous (multiple) mode, lowest possible frequency, voltage of maximal stimulus (which was determined above), and duration of 1 msec.
3. Place the middle finger on the trigger and the probe on the motor point of the arm.
4. Increase the voltage until the response is 75% of the screen.
5. Press the T key and, instead of the screen registering the response, data will be recorded into the memory of the computer. Make as many frames of data as you wish up to 15 frames.
6. Select "Review Velocity Plot" from Velocity Plot of the Main Menu and study the double plot.
7. Return to the Main Menu and select "Review/Analyze." By moving the vertical line from left to right with the right arrow cursor key, determine the millisecond durations of the **latent period, contraction period, relaxation period,** and **entire twitch.**

Laboratory Report

Since no Laboratory Report is available for recording this experiment, write up this report in a manner required by your instructor.

24 Electromyography

Since the early experiments of the 1920s, when muscle contraction proved to be accompanied by bioelectricity, electromyography (hereafter EMG) has become firmly established in physiology. Electrical activity associated with muscle contraction arises from nerve and muscle action potentials of the motor unit. Depolarizations initiate contraction while repolarizations accompany relaxation; together, this activity produces voltage changes that are detectable at the body surface with skin electrodes, or intramuscularly with needle electrodes. To medicine, electromyography is valuable in the assessment of muscle dysfunction. For the student, EMG study can shed light on muscle physiology.

The type of recordings obtained in EMG depends upon the selection of electrodes. Recordings with needle electrodes inserted into muscle yield precise information concerning individual motor unit action potentials. This procedure permits the diagnostician to distinguish between myopathic and neurogenic disorders. Because of the skill required, not to mention the hazard of infection and legal implications, student applications are limited to surface electrodes. Although only a general impression of whole muscle groups is feasible with surface electrodes, single muscles may sometimes be detected.

Individual action potentials cannot be readily distinguished with skin electrodes since the region beneath the electrode often consists of several motor units. Figure 24.1 illustrates the nature of a typical EMG. The high-frequency, irregular, overlapping spikes of an EMG of this type are called an **interference pattern.** The voltage spread of these potentials may be as great as 50 millivolts.

With sufficient amplification, an interference pattern may be detectable even in muscles that are completely relaxed. This spontaneous activity is associated with the maintenance of normal muscle tone. As such a muscle is activated, however, the

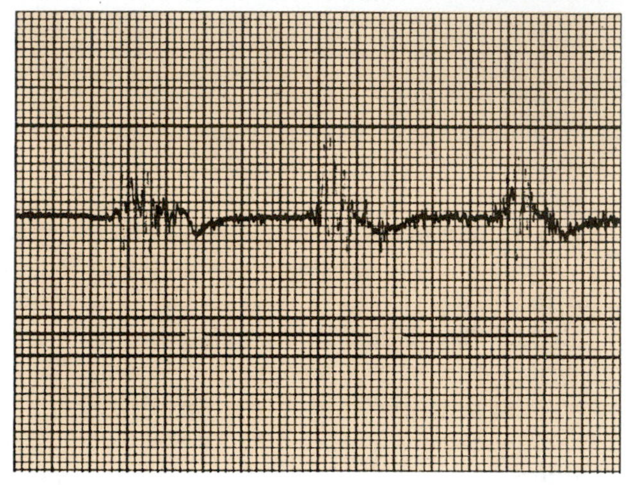

Figure 24.1 Electromyogram.

interference pattern becomes exaggerated.

To display an EMG, one might use an oscilloscope, audio monitor, or chart recorder. Although the Unigraph is the recorder of choice in this experiment because of its portability, a polygraph with an EMG module is excellent. A Unigraph in which the IC-MP module has been replaced by an IC-EMG module can also be used.

The class will be divided up into a number of teams of three or four students each. While one individual prepares the Unigraph, another will attach the electrodes to the subject, and assist in recording information on the chart.

Materials:

- Unigraph, Duograph, or polygraph
- handgrip dynamometer (Stoelting #19117)
- 3-lead patient cable for Unigraph
- 3 skin electrodes (Gilson self-adhering #E1081K)
- adhesive pads for electrodes
- electrode gel (EKG or other)
- Scotchbrite pads (grade #7447)
- alcohol or alcohol swabs

Preparations

Equipment Hookup

While the subject is being readied for EMG recording, connect the 3-lead cable to the Unigraph. Plug the power cord into an outlet. Turn on the power switch, the stylus heat control, and check out the intensity of the stylus line on the paper at slow speed.

If the Unigraph is not provided with a special EMG module, the EEG setting should be used. This mode works very well at lower gain levels for EMG monitoring. To be able to quantify the EMG in terms of millivolts, it will be necessary to calibrate the instrument according to instructions in Appendix C.

If a Gilson polygraph is used, the following modules can be used for EMG monitoring: IC-MP, IC-UM, and IC-EMG. If all modules are present, use the IC-EMG module.

Subject Preparation

Figure 24.2 illustrates the placement of three electrodes on the arm. The two recording red-wired electrodes are placed within an inch of each other on the belly of the major flexor muscle of the forearm (*flexor digitorum superficialis*). The third electrode is a ground and should be placed some distance away over a bony area.

Prior to placing the electrodes on the arm, scrub the skin areas gently with a Scotchbrite pad and disinfect the skin with 50%–70% alcohol. The removal of some of the dead cells from the skin surface greatly facilitates conductivity.

Consult figures 24.3 through 24.6 to see how the self-adhesive pads and electrode gel are placed on the electrodes prior to attaching them to the skin. There are several other kinds of electrodes that can be used. Figure 24.7 reveals how one might substitute three ECG plate electrodes on the arm. As in the case of the other electrodes, it is necessary to scrub first with Scotchbrite and then apply electrode gel before strapping on the electrodes.

When the electrodes are firmly attached to the skin, connect the wires to the appropriate wire of the 3-lead cable. These cables usually have alligator clips and are color coded to accept the correct electrodes. Be sure that the ground electrode is attached to the black ground clip of the 3-lead cable.

Monitoring

Three phenomena of muscle activity will be studied with this setup: spontaneous activity, recruitment, and fatigue. If a Duograph, dual-beam oscilloscope, or polygraph is available, it will also be possible to monitor EMGs of agonist and antagonist muscles, simultaneously.

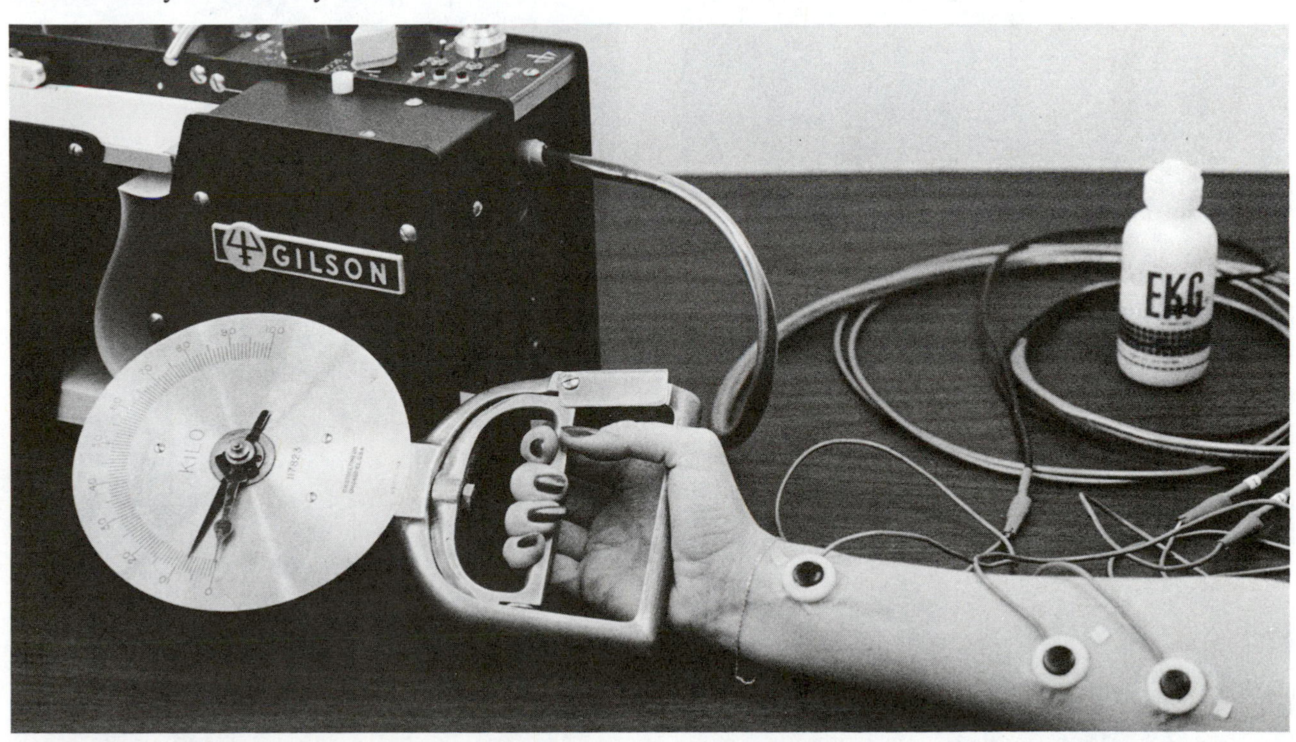

Figure 24.2 EMG monitoring setup.

Spontaneous Activity

With the arm completely relaxed and fully supported on the tabletop, turn on the Unigraph (high speed) and observe if any evidence of motor unit activity is discernible at the lowest gain. Increase the sensitivity by changing the gain settings and turning the sensitivity control. Determine the magnitude and frequency of peaks.

Recruitment

Instruct the subject to clench his or her fingers to form a fist and note the burst of activity to form a typical EMG interference pattern on the chart. To demonstrate recruitment, have the subject grip a hand dynamometer, as in figure 24.2. Record, first, a few bursts at 5 kilograms. Then, double the force to 10 kilograms. Continue to increase the force by 5- or 10-kilogram increments to maximum, recording each force magnitude on the chart with a pen or pencil. Do you see an increase in amplitude on the chart as evidence of recruitment of motor units? Produce enough tracings for all members of your team.

Fatigue

With the Unigraph turned off, direct the subject to perform some handgripping work until the hand muscles are thoroughly fatigued. When the subject cannot continue, take the dynamometer away and allow the subject's arm to recover on the table-top. Start the recorder and monitor any electrical activity. How does the EMG of fatigue compare with that of spontaneous activity?

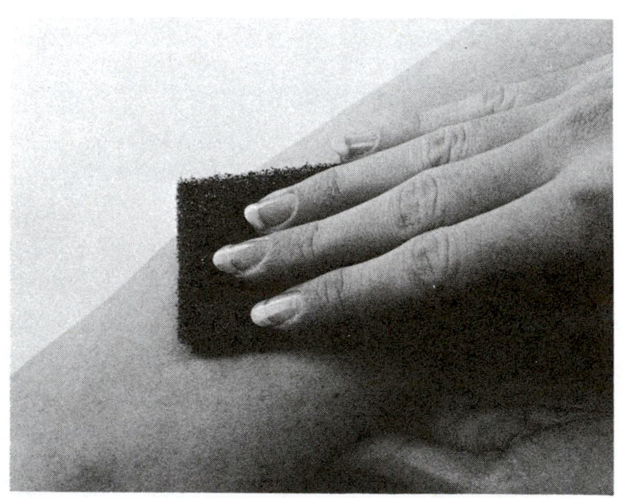

Figure 24.3 Skin surface is rubbed gently with a Scotchbrite pad to improve skin conductivity.

Figure 24.4 An adhesive disk is applied to the clean, dry undersurface of the skin electrode.

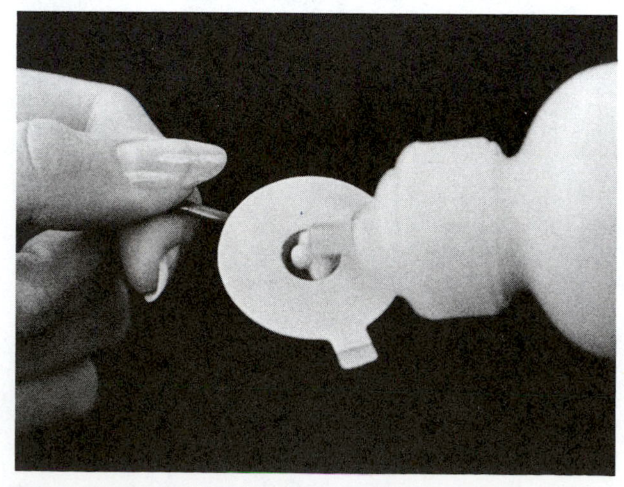

Figure 24.5 Electrode gel is applied to the depression of the skin electrode.

Figure 24.6 The top covering of the adhesive disk is removed and the electrode is placed on the skin.

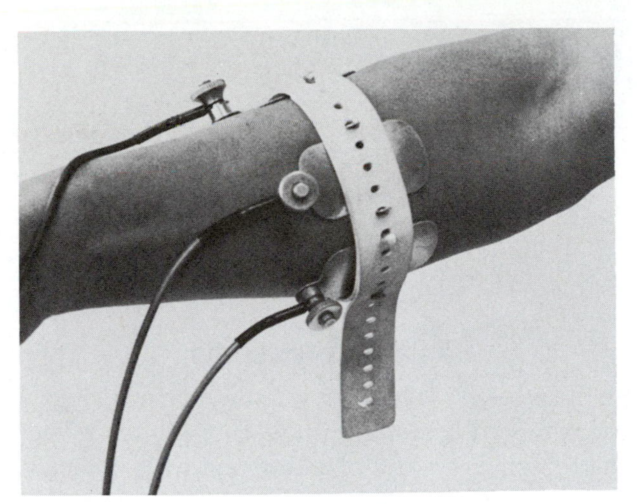

Figure 24.7 An alternate method of electrode application.

Agonist vs. Antagonist Muscles

Opposite movements are achieved by muscle groups called agonists and antagonists. Flexion of the forearm against the upper arm is achieved by contraction of the *biceps brachii.* Extension occurs when the *triceps brachii* contracts. In this case, the biceps is an agonist and the triceps is the antagonist. The action of these two muscles must be coordinated so that when one contracts the other relaxes. Activity in flexors cannot be accompanied by simultaneous activity in extensors, or rigidity will occur instead of flexion or extension. Coordination may be demonstrated by recording the activity of both muscles simultaneously with a Duograph, polygraph, or dual-beam oscilloscope.

To demonstrate coordination, hook up three electrodes to each muscle of one arm through separate channels of whatever type of equipment that is available. Have the subject execute the following maneuvers:

1. Flexion of arm against resistance. Assistant restrains subject's fist while flexion is attempted.
2. Extension against resistance.
3. Isometric tension (contraction of both agonist and antagonist, simultaneously).

Laboratory Report

Complete the last portion of combined Laboratory Report 22,24.

25 Head and Neck Muscles

The principal muscles of the head and neck will be studied in this exercise. Depending on the availability of materials, muscle manikins or cadavers may be used for laboratory studies. The head muscles are grouped here according to their functions.

Scalp Movements

Removal of the scalp reveals an underlying muscle known as the **epicranial.** It consists of the **frontalis** in the forehead region, an **occipitalis** in the occiput region, and an aponeurosis extending between these two over the top of the skull.

The occipitalis has its *origin* on the mastoid process and occipital bone; its *insertion* is on the aponeurosis. The frontalis takes its *origin* on the aponeurosis and its *insertion* on the soft tissue of the eyebrows.

Action: Contraction of the frontalis causes horizontal wrinkling of the forehead and elevation of the eyebrows. Action of the occipitalis causes the scalp to be pulled backward.

Sphincter Action

Circular muscles that close openings are called *sphincters.* The eyes and mouth are surrounded by muscles of this type.

Orbicularis oris This circular muscle lies just under the skin of the lips. It takes its *origin* in various facial muscles, the maxilla, mandible, and septum of the nose. Its *insertion* is on the lips.

Action: It closes the lips in various ways: by compression over the teeth, or by pouting and pursing them; utilized in kissing.

Orbicularis oculi Each eye is surrounded by one of these sphincters. Each muscle *arises* from the nasal portion of the frontal bone, the frontal process of the maxilla, and the medial palpebral ligament. It *inserts* within the tissues of the eyelids (palpebrae).

Action: Blinking and squinting.

Facial Expressions

Emotions such as pleasure, sadness, fear, and anger are expressed by certain muscles of the face. The following five muscles play different roles in facial expressions.

Zygomaticus This muscle extends diagonally from the zygomatic bone to the corner of the mouth. Its *origin* is on the zygomatic; its *insertion* is on the edge of the orbicularis oris.

Action: Contraction of these muscles draws the angles of the mouth upward and backward as in smiling and laughing.

Quadratus labii superioris This thin muscle, consisting of three heads, lies between the eye and upper lip. Its *origin* is on the upper part of the maxilla and part of the zygomatic bone. Its *insertion* is mainly on the superior margin of the orbicularis oris, and partially on the alar region of the nose.

Action: Furrowing of the upper lip occurs when the entire muscle contracts; result: expression of disdain or contempt. Expression of sadness occurs if only the infraorbital head contracts.

Quadratus labii inferioris This small muscle extends from the lower lip to the mandible. Its *origin* is on the mandible, and its *insertion* is on the lower margin of the orbicularis oris.

Action: It pulls the lower lip down in irony.

Triangularis This muscle is lateral to the quadratus labii inferioris. Its *origin* is on the mandible. Its *insertion* is on the orbicularis oris.

Action: Since it is an antagonist of the zygomaticus, it depresses the corner of the mouth.

Platysma This broad sheetlike muscle extends from the mandible over the side of the neck. Only a portion of it is seen in figure 25.1. Figure 25.2 (label 1) reveals its entirety. Since its *insertion* is on the mandible and muscles around the mouth, it profoundly affects facial expressions. Contraction causes the lower lip to move backward and downward, expressing horror.

Masticatory Movements

The chewing of food (mastication) involves five pairs of muscles. Four pairs move the mandible and one pair helps to hold the food in place. All of these muscles are revealed in illustrations A and B, figure 25.1.

Masseter Of the four muscles that move the mandible, the masseter is most powerful. It extends from the angle of the mandible to the zygomatic. Only the lower portion of this muscle is seen in illustration A; its entirety is seen on the large illustration. Its *origin* is on the zygomatic arch, and it *inserts* on the mandible.

Action: It raises the mandible.

Temporalis The large fan-shaped muscle on the side of the skull is the temporalis. It *arises* on portions of the frontal, parietal, and temporal bones and is *inserted* on the coronoid process of the mandible.

Action: It acts synergistically with the masseter to raise the mandible. It can also cause retraction of the mandible when only the posterior fibers of the muscle are activated.

Pterygoideus internus *(Internal pterygoid)* The position of this muscle has led some to refer to it as the "internal masseter." It lies on the medial surface of the ramus, taking its *origin* on the maxilla, sphenoid, and palatine bones. Its *insertion* is on the medial surface of the mandible between the mylohyoid line and the angle. See illustration B, figure 25.1.

Action: It works synergistically with the masseter and temporalis to raise the mandible.

Pterygoideus externus *(External pterygoid)* This muscle is the uppermost one in illustration B. It *arises* on the sphenoid bone and *inserts* on the neck of the mandibular condyle and the articular disk of the joint.

Action: Acting together, the two external pterygoids can cause the mandible to move forward (protrusion), and downward. Individually, they can move the mandible, laterally, to the right or left.

Buccinator The horizontal muscle that partially obscures the teeth in illustration B is the buccinator. It is situated in the cheeks (*buccae*) on each side of the mouth. It *arises* on the maxilla, mandible, and a ligament, the **pterygomandibular raphé** (label 17). This ligament joins the buccinator to the **constrictor pharyngis superioris.** The buccinator *inserts* on the orbicularis oris.

Action: It compresses the cheek to hold food between the teeth during chewing.

Assignment:

Locate all these muscles on a muscle manikin or cadaver.

Label figure 25.1.

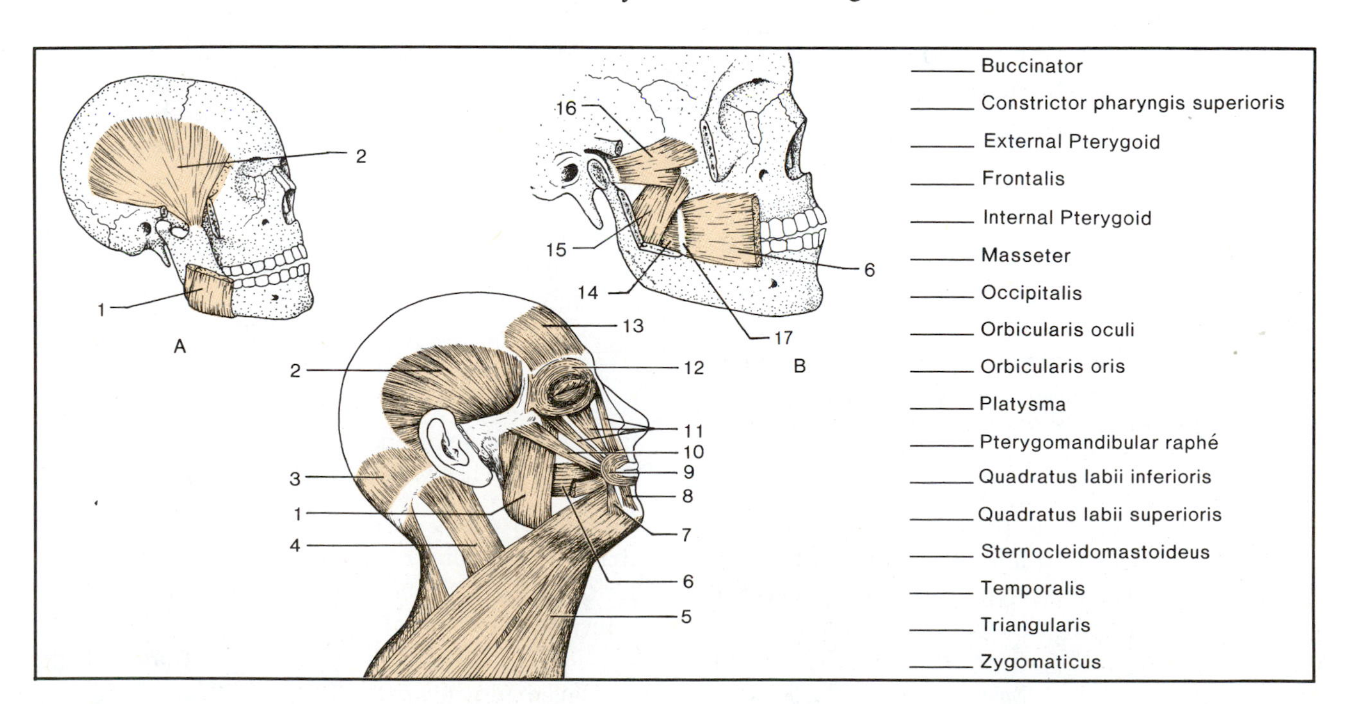

Figure 25.1 Head muscles.

Neck Muscles

Figure 25.2 illustrates a majority of the major muscles of the neck region. Illustrations A and B are of surface muscles. The other two illustrations reveal deeper muscles that are exposed when the surface muscles are removed.

Surface Muscles

The principal surface muscles of the neck are the platysma on the anterolateral surfaces, and the trapezius on the back of the neck.

Platysma It was noted on page 116 that this muscle takes its *insertion* on the mandible and the muscles around the mouth. Its *origin* is primarily the fascia that covers the pectoralis and deltoideus muscles of the shoulder region.

Action: Primarily to draw the lower lip backward and downward; assists to some extent in opening the jaws.

Trapezius This large triangular muscle occupies the upper shoulder region of the back. Only its upper portion functions as a neck muscle. It *arises* on the occipital bone, the ligamentum nuchae, and the spinous processes of the seventh cervical and all thoracic vertebrae. The **ligamentum nuchae** is a ligament that extends from the occipital bone to the seventh cervical vertebra, uniting the spinous processes of all the cervical vertebrae. The *insertion* of this muscle is on the spine and acromion of the scapula and the outer third of the clavicle.

Action: Pulls the scapula toward the median line (adduction); raises the scapula, as in shrugging the shoulder, and draws the head backward (hyperextension), if the shoulders are fixed.

Deeper Neck Muscles, Posterior

Illustration C reveals three prominent muscles on the posterior surface of the neck. Another prominent muscle in this region is the levator scapulae, which has been omitted here for clarity. It will be described in the next exercise.

Splenius capitis This muscle lies just beneath the trapezius and medial to the levator scapulae. Only the left splenius is shown in illustration C. Its *origin* is on the ligamentum nuchae and the spinous processes of the seventh cervical and upper three thoracic vertebrae. Its *insertion* is on the mastoid process.

Action: Hyperextension of the head if both muscles act together; however, if only one muscle contracts, the head is inclined and rotated toward the contracting muscle.

Semispinalis capitis This muscle is the larger of two muscles shown on the right side of the neck. Its *origin* consists of the transverse processes of the upper six thoracic vertebrae and the articular processes of the lower four cervical vertebrae. The *insertion* is on the occipital bone.

Action: Same as splenius capitis.

Longissimus capitis This muscle is lateral and slightly anterior to the semispinalis capitis. The *origin* of this muscle is on the transverse processes of the first three thoracic vertebrae and the articular processes of the lower four cervical vertebrae. The *insertion* is on the mastoid process.

Action: Same as the above two muscles.

Deeper Neck Muscles, Anterior

Illustration D, figure 25.2, depicts a majority of the anterior neck muscles revealed when the platysma is removed.

Sternocleidomastoideus This large neck muscle derives its name from the skeletal components that provide its anchorage. Its *origin* is located on the manubrium of the sternum and the sternal (medial) end of the clavicle. The *insertion* is on the mastoid process.

Action: Simultaneous contraction of both sternocleidomastoids causes the head to be flexed forward and downward on the chest. Independently, each muscle draws the head down toward the shoulder on the same side as the muscle.

Sternohyoideus This long muscle extends from the hyoid bone to the sternum and clavicle. The left sternohyoid is exposed in illustration D by the removal of the lower portion of the left sternocleidomastoid. The sternohyoid *arises* on a portion of the manubrium and clavicle. It *inserts* on the lower border of the hyoid bone.

Action: Draws the hyoid bone downward.

Sternothyroideus This muscle is somewhat shorter than the sternohyoideus. It *originates* on the posterior surface of the manubrium and *inserts* on the inferior edge of the thyroid cartilage of the larynx.

Action: Draws the thyroid cartilage of the larynx downward.

Omohyoideus The long curving muscle that is seen on the left side of the neck which extends from the hyoid bone into the shoulder region is the omohyoid. It *arises* from the upper surface of the scapula and *inserts* on the hyoid bone, lateral to the sternohyoid.

Action: Draws the hyoid bone downward.

Digastricus Two of these narrow V-shaped muscles are visible under the mandible in illustration D. Each muscle consists of anterior and posterior bellies with the central portion attached to the hyoid bone by a fibrous loop.

The posterior belly *arises* from the medial side of the mastoid process of the temporal bone. The anterior belly *arises* on the inner surface of the body of the mandible near the symphysis. The point of *insertion* is the fibrous loop on the hyoid bone.

Action: Raises the hyoid bone and assists in lowering the mandible. Acting independently, the anterior belly can pull the hyoid bone forward; the posterior belly can pull it backward.

Mylohyoideus The two mylohyoids extend from the medial surfaces of the mandible to the median line of the head to form the floor of the mouth. They lie just superior to the anterior bellies of the digastric muscles. Each mylohyoid *arises* along the mylohyoid line of the mandible and *inserts* on a median fibrous raphé that extends from the symphysis menti of the mandible to the hyoid bone.

Action: Raises the hyoid bone and tongue.

Assignment:

Locate all these muscles on a muscle manikin or cadaver.

Label figure 25.2.

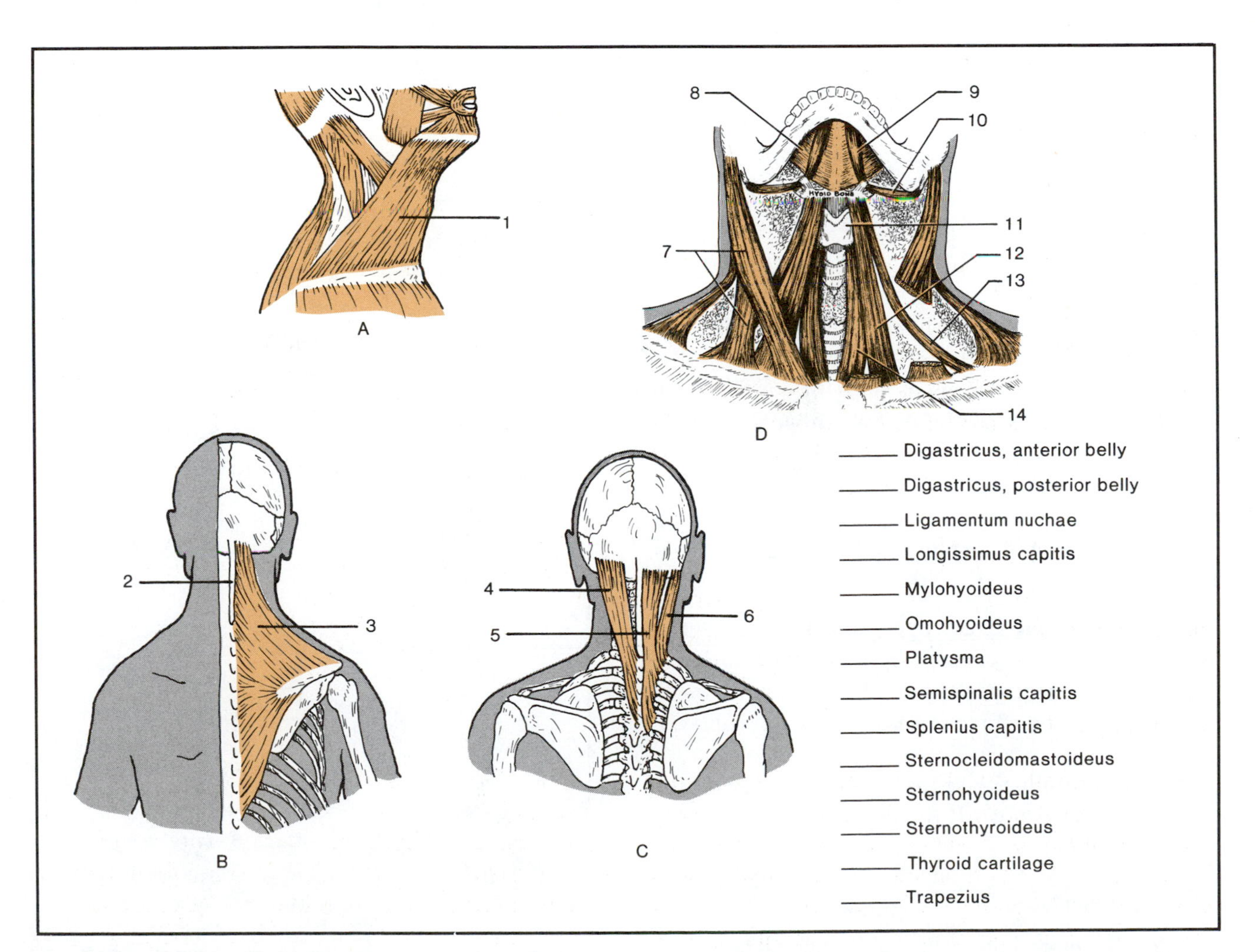

Figure 25.2 Neck muscles.

26 Trunk and Shoulder Muscles

In this study of the trunk and shoulder muscles, all of the surface and a majority of the deeper muscles will be studied. These muscles function primarily to move the shoulders, upper arms, spine, and head. Some of them assist in respiration.

Anterior Muscles

Removal of the skin and subcutaneous fat from the body will reveal muscles as shown in figure 26.1.

Surface Muscles

Pectoralis major This muscle is the thick fan-shaped one that occupies the upper quadrant of the chest. Its *origin* is on the clavicle, sternum, costal cartilages, and the aponeurosis of the external oblique. It *inserts* in the groove between the greater and lesser tubercles of the humerus.

Action: Adducts, flexes, and rotates the humerus medially.

Deltoideus This muscle is the principal muscle of the shoulder. It *originates* on the lateral third of the clavicle, the acromion, and the spine of the scapula. It *inserts* on the deltoid tuberosity of the humerus.

Action: Abduction of the arm when the entire muscle is activated; flexion, extension, and rotation (medial and lateral) takes place when only certain parts of the muscle are activated.

Serratus anterior The upper and lateral surfaces of the rib cage are covered by this muscle. It takes its *origin* on the upper eight or nine ribs and *inserts* on the anterior surface of the scapula near the vertebral border.

Action: Pulls the scapula forward, downward, and inward toward the chest wall.

Deeper Muscles

Pectoralis minor This muscle lies beneath the pectoralis major and is completely obscured by the

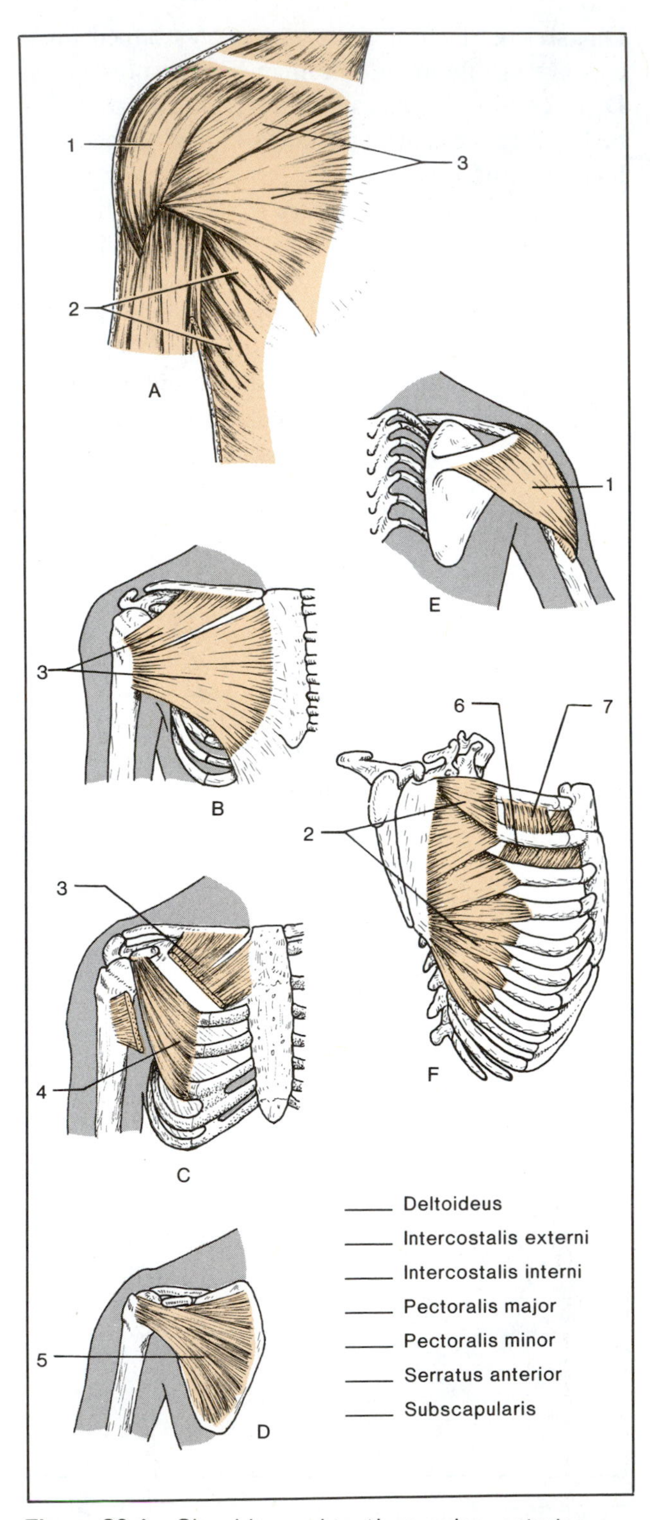

Figure 26.1 Shoulder and trunk muscles, anterior.

latter. It *arises* on the third, fourth, and fifth ribs, and *inserts* on the coracoid process of the scapula.

Action: Draws the scapula forward and downward with some rotation.

Intercostalis externi *(External intercostals)* Between the ribs on both sides are eleven pairs of short muscles, the external intercostals. The *origin* of each of these muscles is on the lower outer border of the upper rib; the *insertion* is on the upper outer border of the lower rib. Their fibers are directed obliquely forward on the front of the ribs. Label 7 in figure 26.1 illustrates one of these muscles.

Action: They pull the ribs closer to each other, causing them to be raised. Raising the ribs increases the volume of the thorax to cause inspiration of air in breathing.

Intercostalis interni *(Internal intercostals)* These antagonists of the external intercostals lie on the internal surface of the rib cage. The complete rib cage is lined with these muscles.

Each internal intercostal *arises* from the lower inner margin of the upper rib and *inserts* on the upper inner margin of the rib below. The fibers of these muscles are oriented in a direction that is opposite to the fibers of the external intercostals.

Action: Draw adjacent ribs closer together. This action has the effect of lowering the ribs and decreasing the volume of the thoracic cavity. Expiration of air from the lungs results.

Subscapularis Large triangular muscle that fills the subscapular fossa of the scapula. Some fibers *originate* on the vertebral (medial) margin; others are anchored to the lower two-thirds of the axillary margin of scapula. All fibers pass laterally to *insert* on the lesser tubercle of the humerus and the anterior portion of the joint capsule.

Action: Rotates the arm medially; assists in other directional movements of the arm, depending on position of the arm.

Assignment:

Label figure 26.1.

Posterior Muscles

The muscles of the upper back region that are revealed when the skin is removed are shown in illustration A, figure 26.2. The deeper muscles of the back and shoulder are shown in illustrations B, C, and D of figure 26.2.

Latissimus dorsi This large muscle of the back covers the lumbar area. It takes its *origin* in a broad aponeurosis that is attached to thoracic and lumbar vertebrae, the spine of the sacrum, iliac crest, and the lower ribs. Its *insertion* is on the intertubercular groove of the humerus.

Action: Extends, adducts, and rotates the arm medially; draws the shoulder downward and backward.

Infraspinatus This muscle derives its name from its position. It is attached to the inferior margin of the scapular spine. Although a portion of it can be seen in illustration A, its complete structure cannot be seen unless the deltoideus and trapezius are removed as in illustration C. Its *origin* occupies the infraspinous fossa. Its *insertion* is the middle facet of the greater tubercle of the humerus.

Action: Rotates the humerus laterally.

Teres major Of the three muscles that cover most of the scapula in illustration C, the teres major is the most inferior one. Note that, although the latissimus dorsi obscures part of it, a greater portion of it is visible in illustration A. It *originates* near the inferior angle of the scapula and *inserts* into the crest of the lesser tubercle of the humerus.

Action: Rotates the humerus medially and weakly adducts it.

Teres minor This small muscle lies between the infraspinatus and teres major. Only a small portion of it is visible in illustration A. It takes its *origin* from the lateral margin of the scapula and its *insertion* is on the lowest facet of the greater tubercle of the humerus.

Action: Rotates the arm laterally and weakly adducts it.

Levator scapulae This muscle lies under the trapezius in the neck region. It takes its *origin* on the transverse processes of the first four cervical vertebrae. Its *insertion* is on the upper portion of the vertebral border of the scapula.

Action: Raises the scapula and draws it medially. With the scapula in a fixed position, it can bend the neck laterally.

Supraspinatus This muscle is completely covered by the trapezius and deltoideus; thus, it cannot be seen in illustration A. It *arises* from the fossa above the scapular spine and *inserts* on the greater tubercle of the humerus.

Action: Assists the deltoid in abduction of the humerus.

Rhomboideus major and minor These two muscles lie beneath the trapezius. They are flat muscles that extend from the vertebral border of the scapula to the spine (see illustration C).

The rhomboideus minor is the smaller one. It occupies a position between the levator scapulae and the rhomboideus major. It *arises* from the lower part of the ligamentum nuchae and the first thoracic vertebra. It *inserts* on that part of the scapular vertebral margin where the scapular spine originates.

The rhomboideus major *arises* from the spinous processes of the second, third, fourth, and fifth thoracic vertebrae. It *inserts* below the rhomboideus minor on the vertebral border of the scapula.

Action: These two muscles act together to pull the scapula medially and slightly upward. The lower part of the major rotates the scapula to depress the lateral angle, assisting in adduction of the arm.

Sacrospinalis *(Erector spinae)* This long muscle extends over the back from the sacral region to the midshoulder region (see illustration B, figure 26.2). It consists of three portions: a lateral **iliocostalis,** an intermediate **longissimus,** and a medial **spinalis.** It *arises* from the lower and posterior portion of the sacrum, the iliac crest, and the lower two thoracic vertebrae. *Insertion* of the muscle is on the ribs and transverse processes of the vertebrae.

Action: This muscle is an extensor, pulling backward on the ribs and vertebrae to maintain erectness.

Assignment:

Label figure 26.2, and complete the Laboratory Report for this exercise.

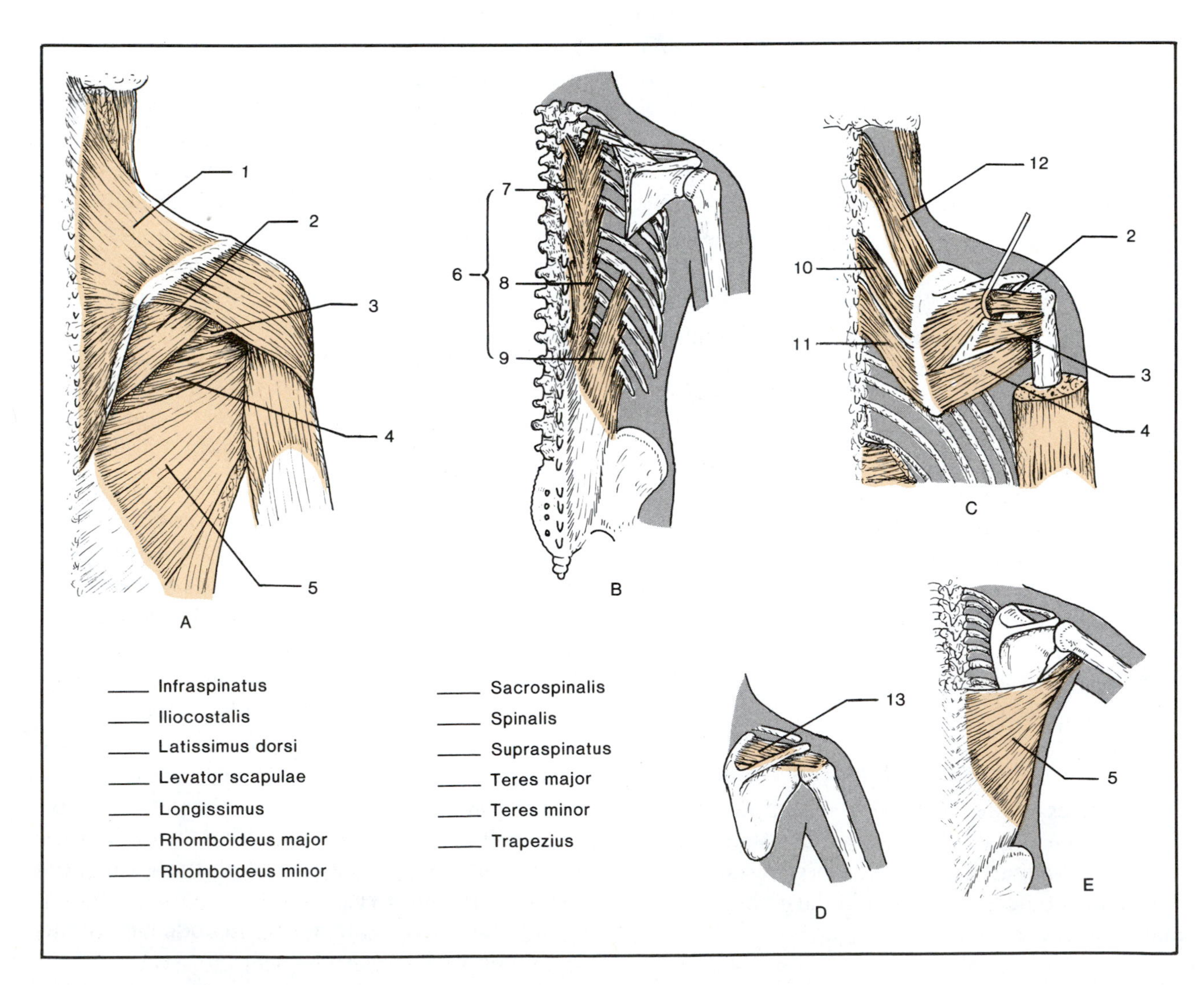

Figure 26.2 Shoulder and back muscles.

Arm Muscles 27

Muscles that move the upper arm are primarily muscles of the shoulder and trunk, which were studied in the last exercise. In this exercise the muscles of the arm that control movements of the forearm will be studied. One muscle, however, that was not mentioned in the last exercise that does move the upper arm is the **coracobrachialis.** It is a muscle of the upper arm and is illustrated in figure 27.1.

Coracobrachialis This muscle covers a portion of the upper medial surface of the humerus. It takes its *origin* on the apex of the coracoid process of the scapula. Its *insertion* is in the middle of the medial surface of the humerus.

Action: Carries the arm forward in flexion and adducts the arm.

Forearm Movements

The principal movers of the forearm are the *biceps brachii, brachialis, brachioradialis,* and *triceps brachii.* These muscles are also illustrated in figure 27.1. Note that incorporated into all the names of these upper arm muscles is the Latin term for the upper arm: *brachium.*

Biceps brachii This is the large muscle on the anterior surface of the upper arm that bulges when the forearm is flexed. Its *origin* consists of two tendinous heads: a medial tendon which is attached to the coracoid process, and a lateral tendon which fits into the intertubercular groove on the humerus. The latter tendon is attached to the supraglenoid tubercle of the scapula. At the lower end of the muscle the two heads unite to form a single tendinous *insertion* on the radial tuberosity of the radius.

Action: Flexion of the forearm; also, rolls the radius outward to supinate the hand.

Brachialis Immediately under the biceps brachii on the distal anterior portion of the humerus lies the brachialis. Its *origin* occupies the lower half of the humerus. Its *insertion* is attached to the front surface of the coronoid process of the ulna.

Action: Flexion of the forearm.

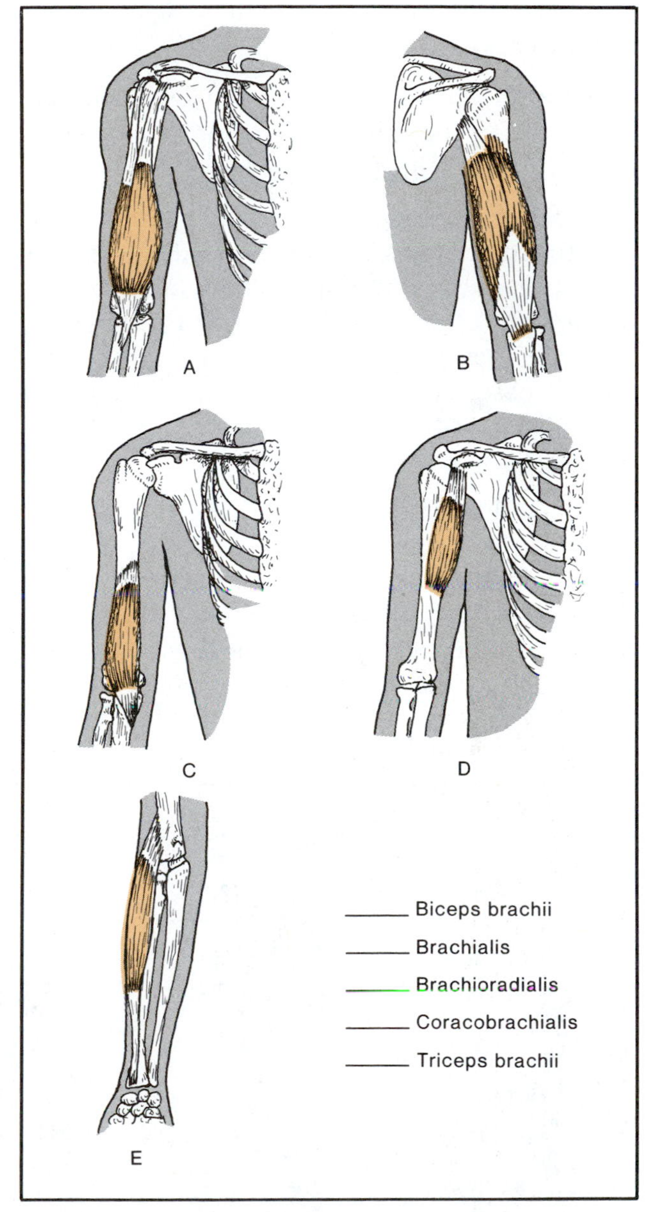

Figure 27.1 Arm muscles.

Brachioradialis This muscle is the most superficial muscle on the lateral (radial) side of the forearm. It *originates* above the lateral epicondyle of the humerus and *inserts* on the lateral surface of the radius slightly above the styloid process.

Action: Flexion of the forearm.

Triceps brachii The entire back surface of the upper arm is covered by this muscle. It has three heads of *origin.* A long head arises from the scapula, a lateral head from the posterior surface of the humerus, and a medial head from the surface below the radial groove. The tendinous *insertion* of the muscle is attached to the olecranon process of the ulna.

Action: Extension of the forearm; antagonist of the brachialis.

Assignment:

Identify the muscles in figure 27.1 by placing the correct letter in front of each name in the legend.

Hand Movements

Muscles of the arm that cause hand movements are illustrated in figures 27.2 and 27.3. All illustrations are of the right arm; thus, if the thumb points to the left side of the page the anterior aspect is being viewed; when pointing to the right, the posterior view is observed. Keeping this in mind will facilitate understanding the descriptions that follow.

Supination and Pronation

Two muscles can cause supination: the biceps brachii and the supinator.

The **supinator** is a short muscle near the elbow that *arises* from the lateral epicondyle of the humerus and the ridge of the ulna. It curves around the upper portion of the radius and *inserts* on the lateral edge of the radial tuberosity and the oblique line of the radius.

Pronation is achieved by the pronator teres and the pronator quadratus. The **pronator teres** is the upper one in illustration B, figure 27.2. It *arises* on the medial epicondyle of the humerus and *inserts* on the upper lateral surface of the radius.

The **pronator quadratus** is located in the wrist region. It *originates* on the distal portion of the ulna and *inserts* on the distal lateral portion of the radius.

Flexion of the Hand

Two muscles that flex the hand are shown in illustration C, figure 27.2.

The muscle that extends diagonally across the forearm is the **flexor carpi radialis.** It *arises* on the medial epicondyle of the humerus and *inserts* on the proximal portions of the second and third metacarpals.

The other muscle in illustration C is the **flexor carpi ulnaris.** Its *origin* is on the medial epicondyle of the humerus and the posterior surface (olecranon process) of the ulna. Its *insertion* consists of a tendon that attaches to the base of the fifth metacarpal.

While both muscles flex the hand, the radialis muscle causes abduction and the ulnaris muscle causes adduction of the hand.

Flexion of Fingers of the Hand

Three muscles (see illus. D, fig. 27.2, and illus. A, fig. 27.3) flex the fingers.

The broad muscle in illustration D that flexes all fingers except the thumb is the **flexor digitorum**

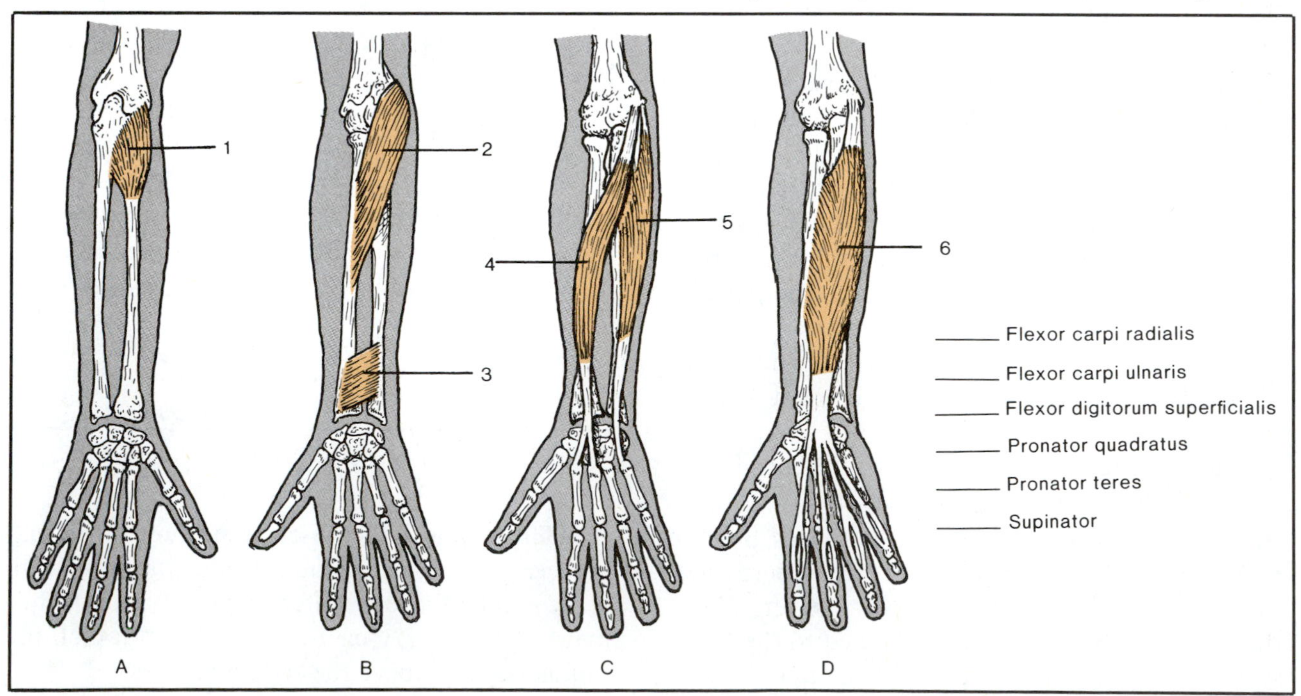

Figure 27.2 Forearm muscles.

superficialis. It *arises* on the humerus, ulna, and radius. Its *insertion* consists of tendons that are attached to the middle phalanges of the second, third, fourth, and fifth fingers.

The second muscle that flexes fingers is the **flexor digitorum profundus.** It is the larger muscle in illustration A, figure 27.3. It lies directly under the flexor digitorum superficialis. It *originates* on the ulna and the interosseous membrane between the radius and ulna. It *inserts* with four tendons on the distal phalanges of the second, third, fourth, and fifth fingers. This muscle flexes the distal portions of the fingers.

The **flexor pollicis longus** (Latin: *pollex,* thumb) is the smaller muscle in illustration A. Note that it *arises* on the radius, ulna, and interosseous membrane. Its *insertion* consists of a tendon that is anchored to the distal phalanx of the thumb. It flexes only the thumb.

Extension of Wrist and Hand

Three muscles extend the hand at the wrist. Illustrations B and D reveal them.

The **extensor carpi radialis longus** and **brevis** are shown in illustration B. The brevis muscle is medial to the longus. Both muscles *originate* on the humerus, with the longus taking a more proximal position. The longus *inserts* on the second metacarpal; the brevis on the middle metacarpal.

The **extensor carpi ulnaris** is the third muscle involved in extending the hand. It is the muscle on the medial edge of the arm in illustration D. It *arises* on the lateral epicondyle of the humerus and part of the ulna. It *inserts* on the fifth metacarpal.

Extension and Abduction of Fingers

Three muscles cause extension of the fingers; one abducts the thumb.

The **extensor digitorum communis** lies alongside the extensor carpi ulnaris; it is shown in illustration D. It *arises* from the lateral epicondyle of the humerus and *inserts* on the distal phalanges of fingers two through five. It extends all fingers except the thumb.

The **extensor pollicis longus, extensor pollicis brevis,** and **abductor pollicis** move the thumb. The longus muscle is the lower one in illustration C. It *arises* on both the ulna and radius. It *inserts* on the distal phalanx of the thumb. The brevis muscle, which lies superior to it, is not shown in illustration C. It *inserts* on the proximal phalanx of the thumb, and assists the longus in extending the thumb.

The abductor pollicis is the superior muscle shown in illustration C. It takes its *origin* on the interosseous membrane, and it *inserts* on the lateral portion of the first metacarpal and trapezium. Its only action is to abduct the thumb.

Assignment:

Label figures 27.2 and 27.3, and complete the Laboratory Report for this exercise..

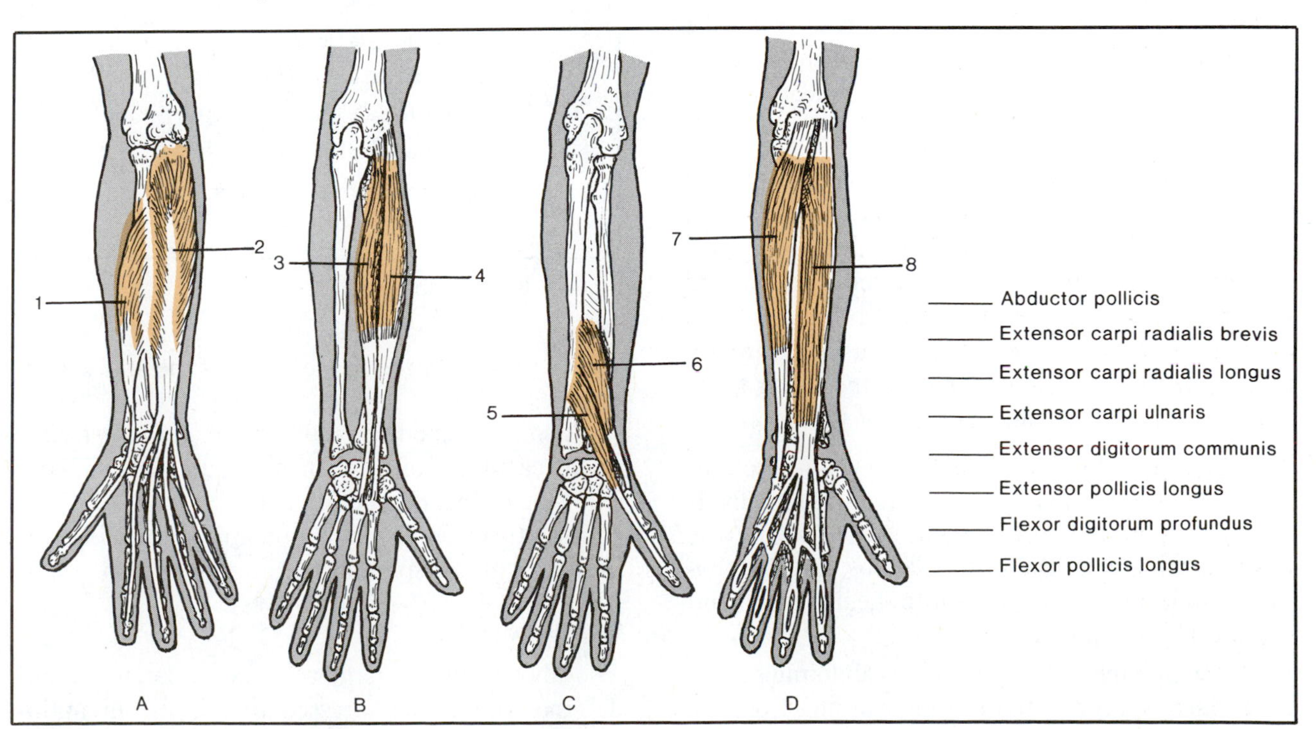

Figure 27.3 Forearm muscles.

28 Abdominal and Pelvic Muscles

For this study of the abdominal wall and pelvis, only two illustrations, involving eight muscles, will be labeled (figures 28.1 and 28.2).

Abdominal Muscles

The abdominal wall consists of four pairs of thin muscles. The illustration in figure 28.1 has portions of the right abdominal wall removed to reveal the nature of the laminations. Identify the following muscles in this illustration.

Obliquus externus *(External oblique)* This muscle is the most superficial layer on the abdominal wall. It is ensheathed by the **aponeurosis of the obliquus externus,** which terminates at the linea alba and inguinal ligament.

The muscle takes its *origin* on the external surfaces of the lower eight ribs. Although it appears to insert on the linea semilunaris, its actual *insertion* is the **linea alba** (white line), where fibers of the left and right aponeuroses interlace on the midline of the abdomen.

The lower border of each aponeurosis forms the **inguinal ligament,** which extends from the anterior spine of the ilium to the pubic tubercle. Label 5, figure 28.2 is of this ligament.

Obliquus internus *(Internal Oblique)* This muscle lies immediately under the external oblique, i.e., between the external oblique and the transversus abdominis. Its *origin* is on the lateral half of the inguinal ligament, the anterior two-thirds of the iliac crest, and the thoracolumbar fascia. Its *insertion* is on the costal cartilages of the lower three ribs, the linea alba, and the crest of the pubis.

Transversus abdominis The innermost muscle of the abdominal wall is the transversus abdominis. It *arises* on the inguinal ligament, the iliac crest, the costal cartilages of the lower six ribs, and the thoracolumbar fascia. It *inserts* into the linea alba and the crest of the pubis.

Rectus abdominis The right rectus abdominis is the long, narrow, segmented muscle running from the rib cage to the pubic bone. It is enclosed in a fibrous sheath formed by the aponeuroses of the above three muscles.

Its *origin* is on the pubic bone. Its *insertion* is on the cartilages of the fifth, sixth, and seventh ribs. The linea alba lies between the pair of muscles. Contraction of the rectus abdominis muscles aids in flexion of the spine in the lumbar region.

Collective Action

The above four abdominal muscles keep the abdominal organs compressed and assist in maintaining intraabdominal pressure.

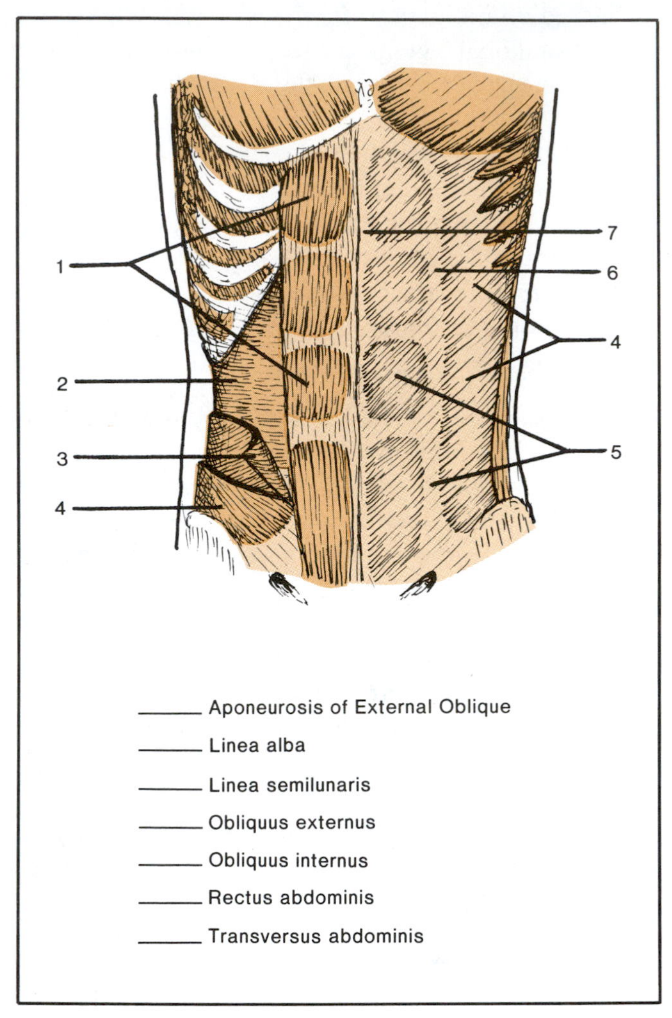

Figure 28.1 Abdominal muscles.

They act as antagonists to the diaphragm. When the latter contracts, they relax. When the diaphragm relaxes they contract to effect expiration of air from the lungs. They also assist in defecation, micturition, parturition, and vomiting. Flexion of the body at the lumbar region is also achieved by them.

Assignment:

Label figure 28.1.

Pelvic Muscles

The principal muscles of the pelvic region are shown in figure 28.2. They are the quadratus lumborum, psoas major, and iliacus.

Quadratus lumborum The muscle that extends from the iliac crest of the os coxa to the lowest rib in figure 28.2 is this muscle. Its *origin* is on the iliac crest, the iliolumbar ligament, and the transverse processes of the lower four lumbar vertebrae. Its *insertion* is on the inferior margin of the last rib and the transverse processes of the upper four lumbar vertebrae.

Acting together, these two muscles extend the spine at the lumbar vertebrae. Lateral flexion or abduction results when one acts independently of the other.

Psoas major This is the long muscle shown on the right side of the pelvic cavity in figure 28.2. Only the cut ends of the left muscle are shown in this illustration.

The psoas major *arises* from the sides of the bodies and transverse processes of the lumbar vertebrae. It *inserts* with the iliacus on the lesser trochanter of the femur.

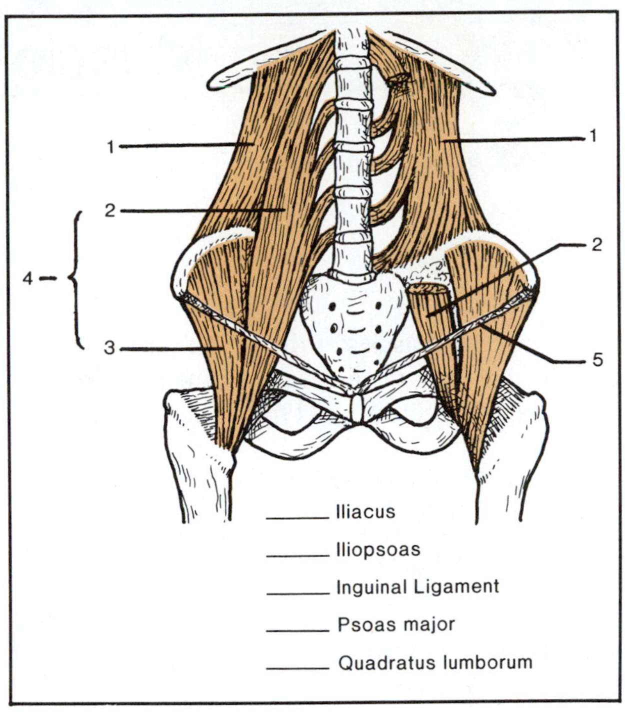

Figure 28.2 Pelvic muscles.

The two psoas majors work synergistically with the rectus abdominis muscles to flex the lumbar region of the vertebral column.

Iliacus This muscle extends from the iliac crest to the proximal end of the femur. Its *origin* is the whole iliac fossa. Its *insertion* is the lesser trochanter of the femur. The psoas major and iliacus are jointly referred to as the **iliopsoas** because of their intimate relationship at their insertion.

The iliacus works synergistically with the psoas major to flex the femur on the trunk.

Laboratory Report

Label figure 28.2 and answer the questions on combined Laboratory Report 28,29 that pertain to abdominal and pelvic muscles.

29 Leg Muscles

Twenty-seven muscles of the leg will be studied in this exercise. As in the case of arm muscles they are grouped according to type of movement.

Thigh Movements

Seven muscles that move the femur are shown in figure 29.1. All originate on a part of the pelvis.

Gluteus maximus This muscle of the buttock region is covered by a deep fascia, the *fascia latae,* that completely invests the thigh muscles. Emerging downward from the fascia lata is a broad tendon, the **iliotibial tract,** which is attached to the tibia.

This muscle *arises* on the ilium, sacrum, and coccyx; it *inserts* on the iliotibial tract and the posterior part of the femur. It causes extension and outward rotation of the femur.

Gluteus medius This muscle lies immediately under the gluteus maximus, covering a good portion of the ilium. Its *origin* is on the ilium and its *insertion* is on the lateral part of the greater trochanter. It causes abduction and medial rotation of the femur.

Gluteus minimus This is the smallest of the three gluteal muscles, and is located immediately under the gluteus medius. It, too, *arises* on the ilium; it *inserts* on the anterior border of the greater trochanter. The femur is abducted, rotated inward, and slightly flexed by this muscle.

Piriformis This small muscle takes its *origin* on the anterior surface of the sacrum and *inserts* on the upper border of the greater trochanter. It causes outward rotation, some abduction, and extension of the femur.

Adductor longus and Adductor brevis These two muscles are shown in illustration E. The adductor longus *originates* on the front of the pubis and *inserts* on the linea aspera of the femur.

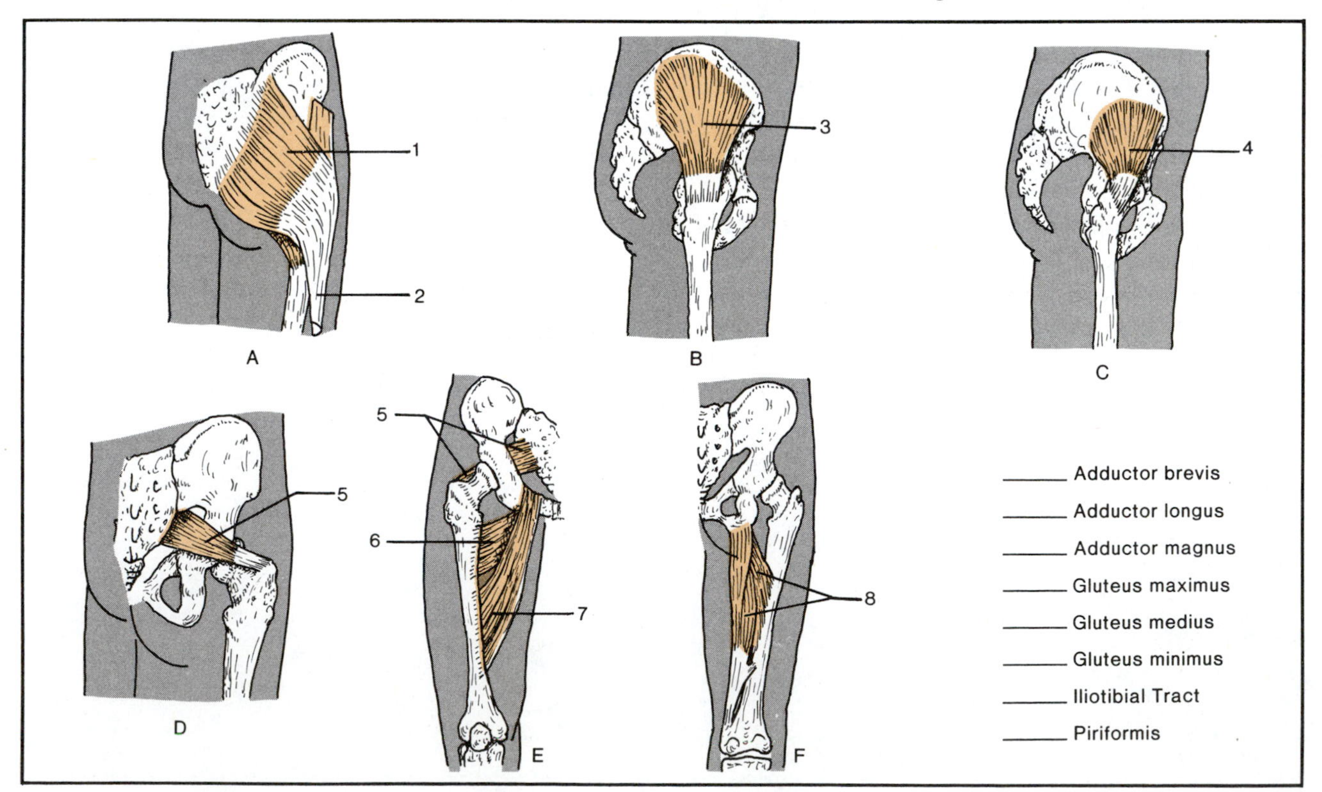

Figure 29.1 Muscles that move the femur.

The adductor brevis *arises* on the posterior side of the pubis and *inserts* on the femur above the longus muscle. In addition to adduction, these muscles flex and rotate (medially) the femur.

Adductor magnus This muscle is the strongest of the three adductors. It *arises* on the ischium and part of the pubis. It *inserts* on the linea aspera of the femur in the same region as the other two adductors. Its action is synergistic with the longus and brevis adductors.

Assignment:

Label figure 29.1.

Thigh and Lower Leg Movements

The muscles illustrated in figure 29.2 are primarily concerned with flexion and extension of the lower part of the leg. Most of them are anchored to some part of the os coxa.

Hamstrings Three muscles, the *biceps femoris, semitendinosus,* and *semimembranosus,* constitute a group of muscles on the back of the thigh known as hamstrings. They are grouped together in illustration A.

Biceps femoris This muscle occupies the most lateral position of the three hamstrings. It has two heads: one long, and the other short. The long head, which obscures the short one, *arises* on the ischial tuberosity. The short head *originates* on the linea aspera of the femur. *Insertion* of the muscle is on the head of the fibula and the lateral condyle of the tibia.

Semitendinosus Medial to the biceps femoris is this muscle, which *arises* on the ischial tuberosity and *inserts* on the upper end of the shaft of the tibia.

Semimembranosus This muscle occupies the most medial position of the three hamstrings. Its *origin* consists of a thick semimembranous tendon attached to the ischial tuberosity. Its *insertion* is primarily on the posterior medial part of the medial condyle of the tibia.

Collective Action: All of these muscles flex the calf upon the thigh. They also extend and rotate the thigh. Rotation by the biceps is outward; the other two muscles cause inward rotation.

Quadriceps femoris The large muscle that makes up the anterior portion of the thigh is the quadriceps femoris. It consists of four parts, which are shown in illustrations C and D, figure 29.2. Note that all of them are united into a common tendon that passes over the patella to *insert* on the tibia.

Rectus femoris This portion of the quadriceps occupies a superficial central position. It *arises* by two tendons: one from the anterior inferior iliac spine and the other from a groove just above the acetabulum. The lower portion of the muscle is a broad aponeurosis that terminates in the tendon of *insertion.*

Vastus lateralis This is the largest and lateral portion of the quadriceps femoris. It *arises* from the lateral lip of the linea aspera.

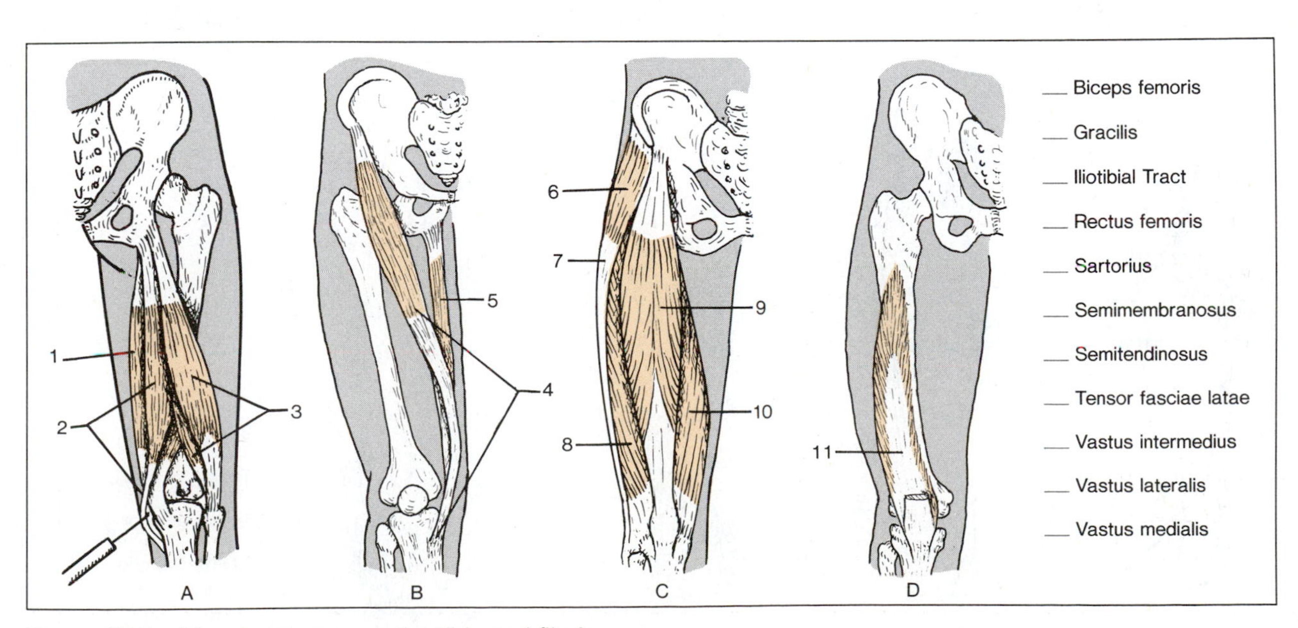

Figure 29.2 Muscles that move the tibia and fibula.

Vastus medialis This muscle occupies the medial position on the thigh. It *arises* from the linea aspera.

Vastus intermedius Since this muscle is completely obscured by the other three muscles, it is shown separately in illustration D. Its *origin* is on the front and lateral surfaces of the femur.

Collective Action: The entire quadriceps femoris extends (straightens) the leg. Flexion of the thigh, however, is achieved by the rectus femoris.

Sartorius This is the longest muscle shown in illustration B. It *arises* from the anterior superior spine of the ilium and *inserts* on the medial surface of the tibia. It flexes the calf on the thigh, the thigh upon the pelvis, and rotates the leg laterally.

Tensor fasciae latae This muscle is shown in illustration C, figure 29.2. It *arises* from the anterior outer lip of the iliac crest, the anterior superior spine, and from the deep surface of the fascia lata. It is *inserted* between the two layers of the iliotibial tract at about the junction of the middle and upper thirds of the thigh. It flexes and slightly rotates the thigh.

Gracilis This muscle is located on the medial surface of the thigh. It *arises* from the lower margin of the pubic bone and *inserts* on the medial surface of the tibia near the insertion of the sartorius. It adducts, flexes, and rotates the thigh medially.

Assignment:

Label figure 29.2.

Lower Leg and Foot Movements

Figures 29.3 and 29.4 illustrate most of the muscles of the lower leg that cause flexion and foot movements.

Triceps surae The large superficial muscle that covers the calf of the leg is the triceps surae. It consists of two parts that are shown in illustrations A and B, figure 29.3. They are united by a common **tendon of Achilles** that *inserts* on the calcaneous of the foot. Collectively, the two muscles cause plantar flexion (standing on tiptoe).

Gastrocnemius This outer portion of the triceps surae has two heads that *arise* from the posterior surfaces of the medial and lateral condyles of the femur. In addition to plantar flexion, this muscle can, independently, flex the calf on the thigh.

Soleus This inner portion *arises* on the heads of the fibula and tibia.

Tibialis anterior The anterior lateral portion of the tibia is covered by this muscle. It *arises* from the lateral condyle and upper two-thirds of the tibia. Its distal end is shaped into a long tendon that passes over the tarsus and *inserts* on the inferior surface of the first cuneiform and first metatarsal bones. It causes dorsiflexion and inversion of the foot.

Tibialis posterior This muscle is a deep one that lies on the posterior surfaces of the tibia and fibula. It *arises* from both these bones and the interosseous membrane that extends between them. It *inserts* on the inferior surfaces of the navicular, the

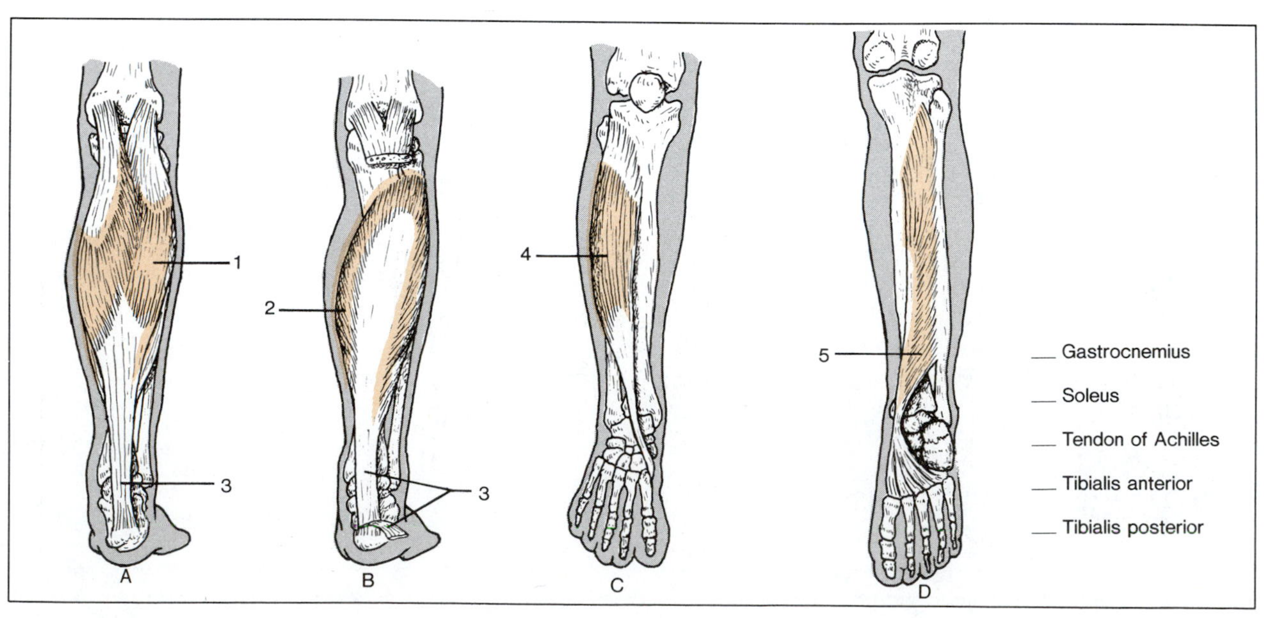

Figure 29.3 Lower leg muscles.

cuneiforms, the cuboid, and the second, third, and fourth metatarsals. It causes plantar flexion and inversion of the foot. It assists in maintenance of the longitudinal and transverse arches of the foot.

The Peroneus Muscles (Latin: *peroneus,* fibula) Three peroneus muscles are shown in illustrations A and B, figure 29.4. Note that all three of them *originate* on the fibula. The **peroneus longus** is the longest one and takes its origin on the head and upper two-thirds of the fibula. The **peroneus tertius** is the smallest one, and the **peroneus brevis** is the one of in-between size. Note that the longus *inserts* on the first metatarsal and second cuneiform bones. The brevis and tertius muscles *insert* at different points on the fifth metatarsal bone. The longus and brevis muscles cause plantar flexion and eversion of the foot. The peroneus tertius causes dorsiflexion and eversion of the foot.

Flexor Muscles Two flexor muscles of the foot are shown in illustration C, figure 29.4. The **flexor hallucis longus** (Latin: *hallux,* big toe) has its *origin* on the fibula and intermuscular septa. Its long distal tendon *inserts* on the distal phalanx of the great toe. It flexes the great toe.

The **flexor digitorum longus** is the longer muscle shown in illustration C that *originates* on the tibia and the fascia that covers the tibialis posterior. Its distal end divides into four tendons that *insert* on the bases of the distal phalanges of the second, third, fourth, and fifth toes. It flexes the distal phalanges of the four smaller toes.

Extensor Muscles Two extensors of the foot are shown in illustration D, figure 29.4. The **extensor digitorum longus** is the longer one that takes its *origin* on the lateral condyle of the tibia, part of the fibula, and part of the interosseous membrane. Its tendon of insertion divides into four parts that *insert* on the superior surfaces of the second and third phalanges of the four smaller toes. It extends the proximal phalanges of the four smaller toes. It also flexes and inverts the foot.

The **extensor hallucis longus** is the other extensor in illustration D. It takes its *origin* on the fibula and interosseous membrane. Its distal tendon *inserts* at the base of the distal phalanx of the great toe. This muscle extends the proximal phalanx of the great toe and aids in dorsiflexion of the foot.

Assignment:

Label figures 29.3 and 29.4.

Surface Muscles Review

Figure 29.5 has been included here in this exercise to summarize our study of the muscles of the body as a whole. Only the surface muscles are shown. To determine your present understanding of the human musculature, attempt to label these diagrams first by *not referring back* to previous illustrations. This type of self-testing will determine what additional study is needed.

Laboratory Report

Complete the last portion of combined Laboratory Report 28,29.

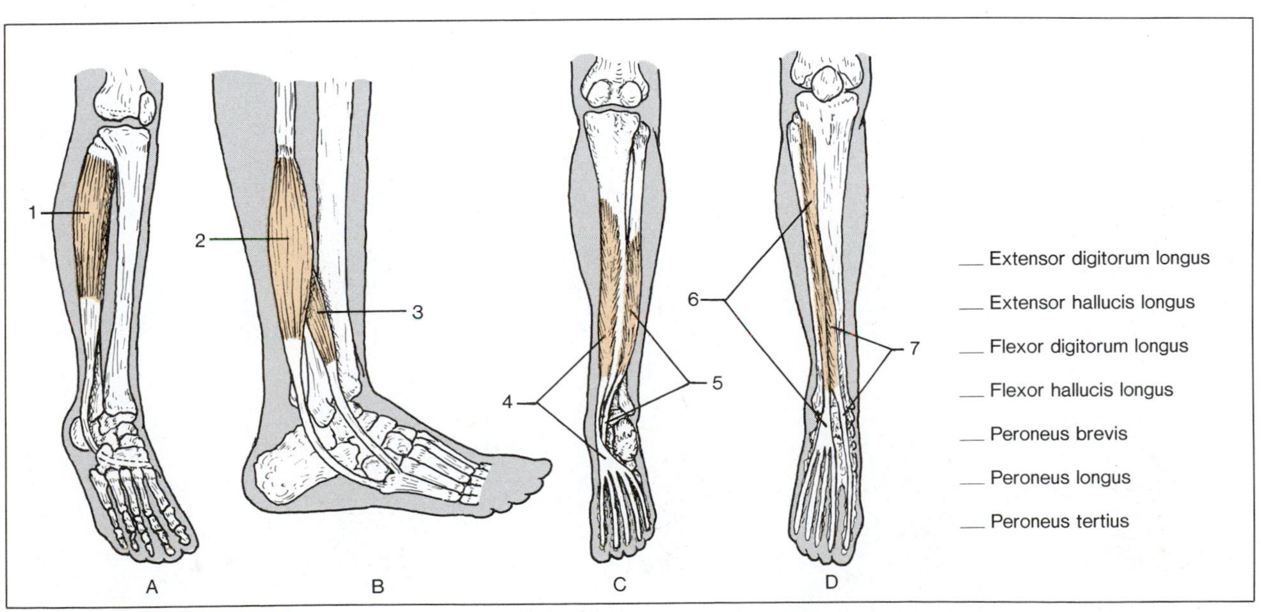

Figure 29.4 Lower leg muscles.

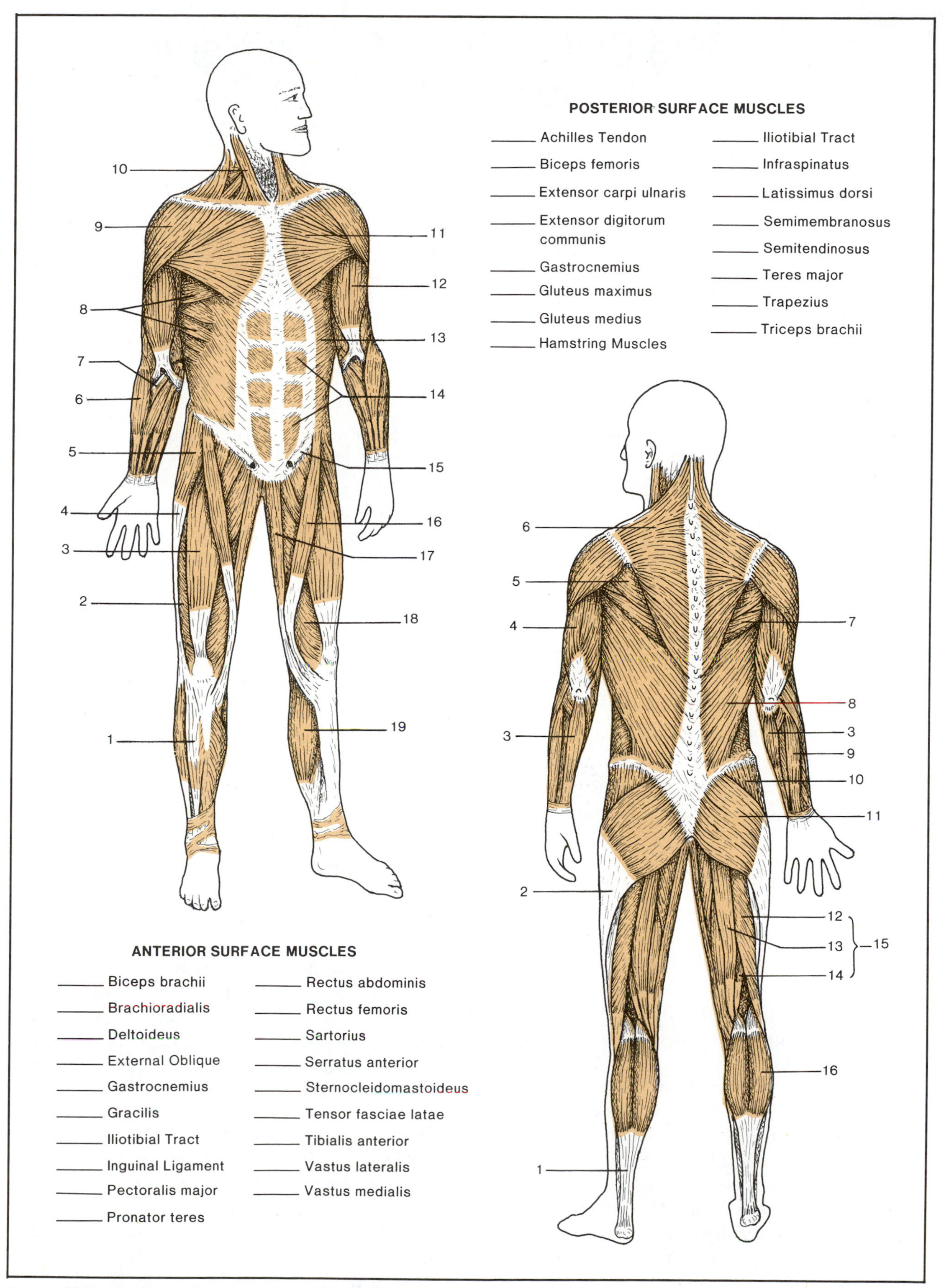

Figure 29.5 Major surface muscles of the body.

30 The Spinal Cord, Spinal Nerves, and Reflex Arcs

Automatic stereotyped responses to various types of stimuli enable animals to adjust quickly to adverse environmental changes. These automatic responses, generated by the nervous system, are called **reflexes.**

Reflexes that result in automatic regulation of body function can be either somatic or visceral. **Somatic reflexes** involve skeletal muscle responses, and **visceral reflexes** involve the adjustments of smooth muscle, cardiac muscle, and glands to stimuli.

The neural pathway utilized in performing a reflex is called a **reflex arc.** This pathway involves receptors, spinal nerves, the spinal cord, and effectors. It is the purpose of this exercise to study each of the components that are involved in both types of reflexes.

The Spinal Cord

The spinal cord is a downward extension of the medulla oblongata of the brain. Figure 30.1 reveals its gross structure, as seen posteriorly. To expose it, the posterior portions of the vertebrae and sacrum have been removed.

The spinal cord starts at the upper border of the atlas and terminates as the **conus medullaris** at the lower border of the first lumbar vertebra. In fetal life the spinal cord occupies the entire length of the vertebral canal (spinal cavity), but as the vertebral column continues to elongate, the spinal cord fails to lengthen with it; thus, the vertebral canal extends downward beyond the end of the spinal cord.

Extending downward from the conus medullaris is an aggregate of fibers called the **cauda equina** (horse's tail), which fills the lower vertebral canal. The innermost fiber of the cauda equina, which is located on the median line, is called the **filum terminale interna.** This delicate median prolongation of the conus medullaris becomes the **filum terminale externa** after it passes through the dural sac. The **dural sac** (dura mater) is continuous with the dura mater that surrounds the brain. Only a portion of it (label 25) is shown in figure 30.1. The remainder of the dura mater has been left out of this illustration for clarity.

The Spinal Nerves

The spinal nerves emerge from the spinal cord in pairs through the intervertebral foramina on each side of the spinal cavity. There are eight cervical, twelve thoracic, five lumbar, and five sacral pairs. These thirty pairs of nerves, plus one pair of coccygeal nerves, make a total of thirty-one pairs of spinal nerves.

Reference to the sectional view of the spinal cord in figure 30.1 reveals that each spinal nerve is connected to the spinal cord by structures called **anterior** and **posterior roots.** Note that the posterior root (label 21) has an enlargement, the **spinal ganglion,** which contains cell bodies of sensory neurons. Neuronal fibers from these roots pass through the outer **white matter** of the spinal cord into the inner **gray matter** where neuronal connections *(synapses)* are made.

Cervical Nerves The first cervical nerve (C_1) emerges from the spinal cord in the space between the base of the skull and the atlas. The eighth cervical nerve (C_8) emerges between the seventh cervical and first thoracic vertebrae. Note that the first four cervical nerves are united in the neck region to form a network called the **cervical plexus.** The remaining cervical nerves (C_5, C_6, C_7, and C_8) and the first thoracic nerve unite to form the **brachial plexus** in the shoulder region.

Thoracic *(Intercostal)* Nerves The first thoracic nerve (T_1) emerges through the intervertebral foramen between the first and second thoracic vertebrae. The twelfth thoracic nerve (T_{12}) emerges through the foramen between the twelfth thoracic and first lumbar vertebrae. Each of these nerves lies adjacent to the lower margin of the rib above it.

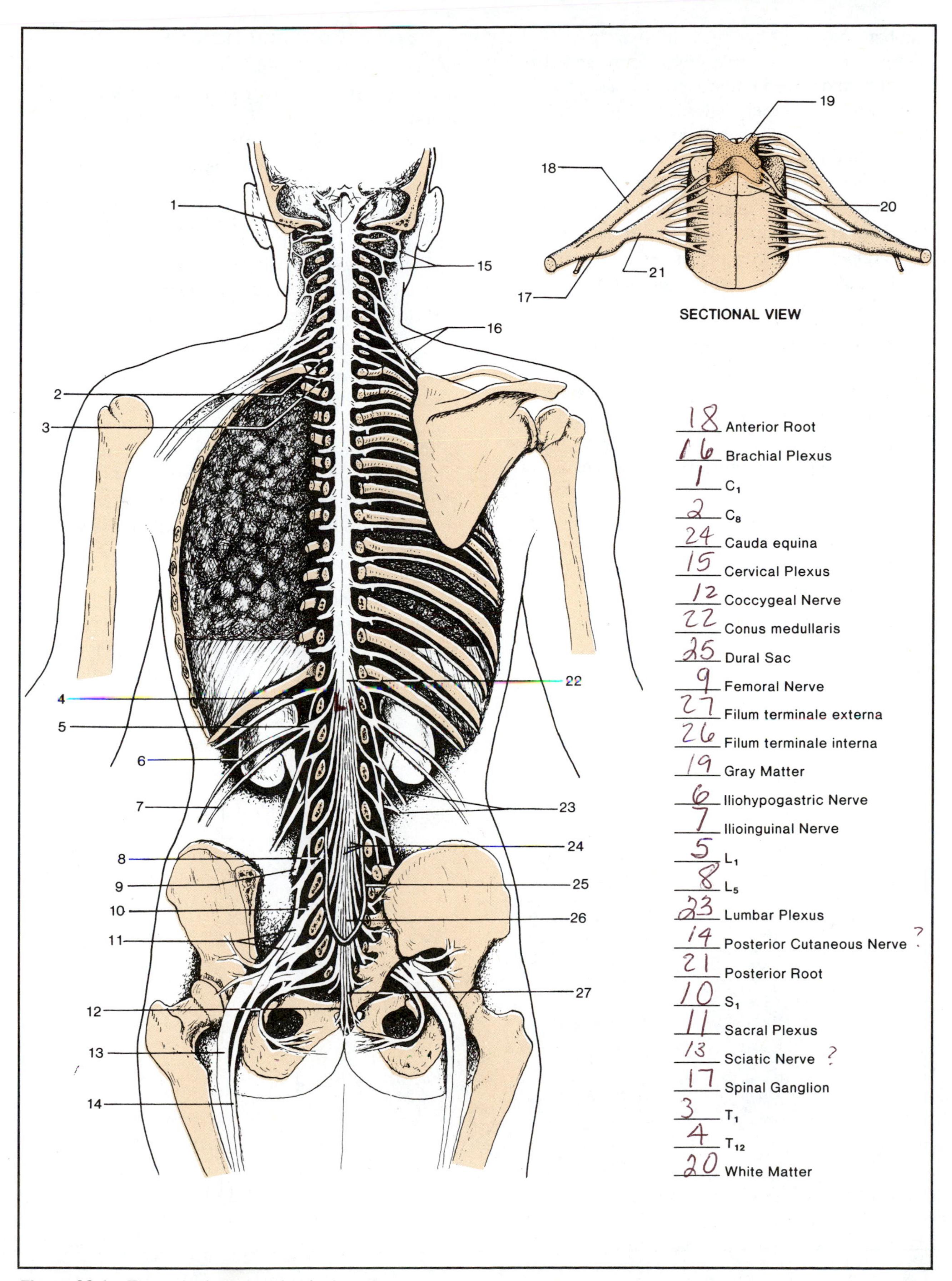

Figure 30.1 The spinal cord and spinal nerves.

Lumbar Nerves The first lumbar nerve (L_1) emerges from the intervertebral foramen between the first and second lumbar vertebrae. Close to where the nerve emerges from the spinal cavity, it divides to form an upper **iliohypogastric** nerve and a lower **ilioinguinal** nerve.

Note in figure 30.1 that L_1, L_2, L_3, and most of L_4 are united to form the **lumbar plexus** (label 23). The largest trunk emanating from this plexus is the **femoral** nerve, which innervates part of the leg. In reality, the femoral is formed by the union of L_2, L_3, and L_4.

Sacral and Coccygeal Nerves The remaining nerves of the cauda equina include five pairs of sacral and one pair of coccygeal nerves. A **sacral plexus** (label 11) is formed by the union of roots of L_4, L_5, and the first three sacral nerves (S_1, S_2, and S_3). The largest nerve that emerges from this plexus is the **sciatic** nerve which passes down into the leg. The smaller nerve that parallels the sciatic is the **posterior cutaneous** nerve of the leg. The **coccygeal** nerve (label 12) contains nerve fibers from the fourth and fifth sacral nerves (S_4 and S_5).

Assignment:

Label figure 30.1.

Spinal Cord and Nerve Protection

Injury to the spinal cord is medically grave in that once the spinal cord is traumatized or severed the damage is permanent. Fortunately, the bony protection of the spinal cavity and the tissues around the spinal cord afford considerable protection. The presence of areolar connective tissue, meninges, and cerebrospinal fluid act as a combined shock absorber to protect the vulnerable cord. The cross-sectional view shown in figure 30.2 illustrates how the cord and spinal nerves are surrounded by these protective layers.

Cord Structure Before examining the surrounding tissues, identify the following structures on the cord in figure 30.2. The most noticeable characteristic of the spinal cord is the pattern of the **gray matter,** which resembles the configuration of the outstretched wings of a swallow-tailed butterfly. Surrounding this darker material is the **white matter,** which consists primarily of myelinated nerve fibers.

Along the median line of the spinal cord are two fissures and a small canal. The **posterior median sulcus** is the upper fissure in figure 30.2. The **anterior median fissure** is the wider fissure on the anterior surface of the spinal cord. In the gray matter

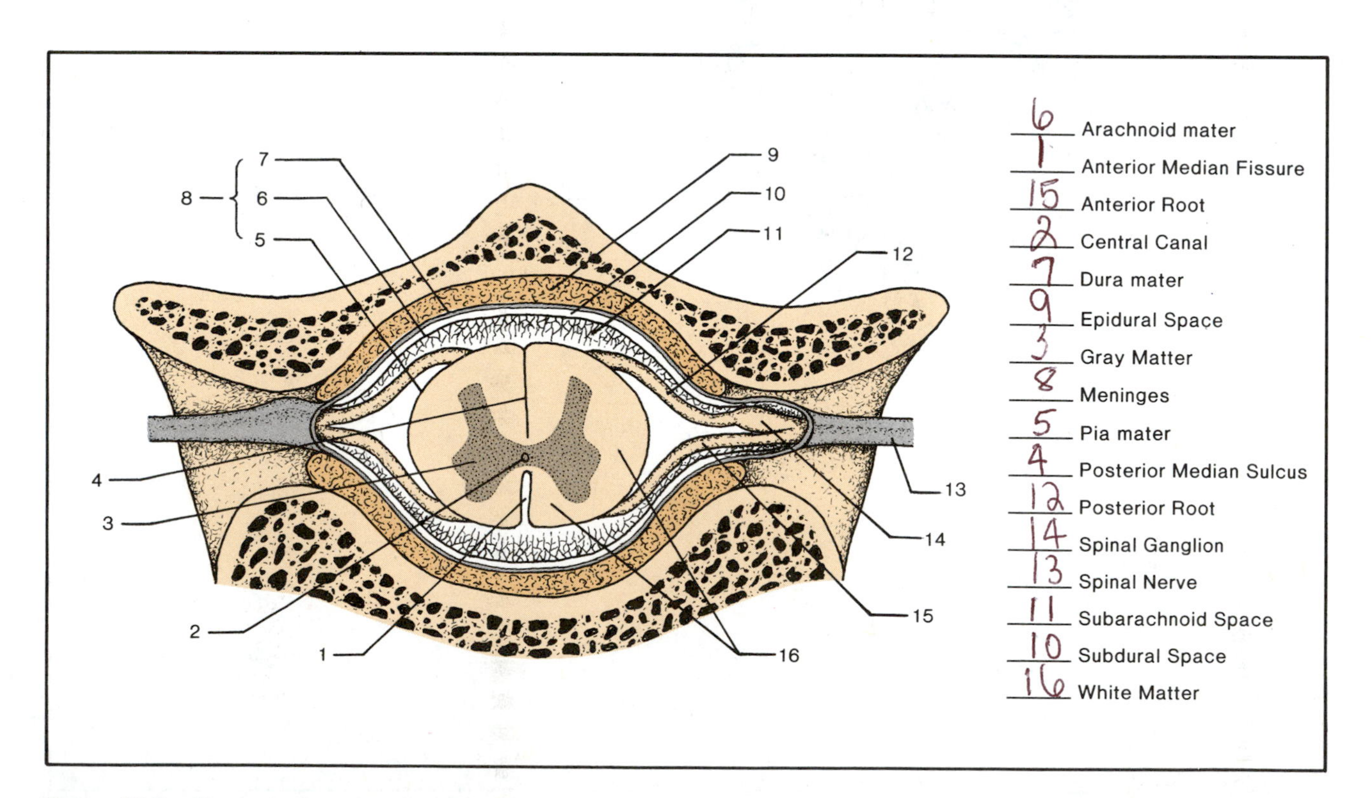

Figure 30.2 The spinal cord and meninges.

on the median line lies a tiny **central canal,** evidence of the tubular nature of the spinal cord. This canal is continuous with the ventricles of the brain and extends into the filum terminale. It is lined with ciliated ependymal cells.

Spinal Nerves Note how the spinal nerves (label 13) emerge from the vertebrae on each side of the spinal column. Identify the **posterior root** with its **spinal ganglion** and the **anterior root.**

The Meninges Surrounding the spinal cord (and brain) are three meninges (*meninx,* singular). The outermost membrane is the **dura mater** (label 7), which also forms a covering for the spinal nerves. A cutaway portion of the dura on each side in figure 30.2 reveals the inner spinal nerve roots. The dura mater is the toughest of the three meninges, consisting of fibrous connective tissue. Between the dura and the bony vertebrae is an **epidural space,** which contains areolar connective tissue and blood vessels.

The innermost meninx is the **pia mater.** It is a thin delicate membrane that is actually the outer surface of the spinal cord. Between the pia mater and dura mater is the third meninx, the **arachnoid mater.** Note that the outer surface of the arachnoid mater lies adjacent to the dura mater. The space between the dura and arachnoid mater is actually a potential cavity that is called the **subdural space.** The inner surface of the arachnoid mater has a delicate fibrous texture that forms a netlike support around the spinal cord. The space between the arachnoid mater and the pia mater is the **subarachnoid space.** This space is filled with cerebrospinal fluid that provides a protective fluid cushion around the spinal cord.

Assignment:

Label figure 30.2.

The Somatic Reflex Arc

A brief mention of the somatic reflex arc was made in Exercise 19. In that exercise our prime concern was with the structure of neuronal elements that make up the circuit. Figure 30.3 is very similar to figure 19.1, although figure 30.3 is simpler.

The principal components of this type of reflex arc are (1) a **receptor,** such as a neuromuscular spindle or a cutaneous end-organ that receives the stimulus; (2) a **sensory** (afferent) **neuron,** which carries impulses through a peripheral nerve and posterior root to the spinal cord; (3) an **interneuron** (association neuron), which forms synaptic connections between the sensory and motor neurons in the gray matter of the spinal cord; (4) a **motor** (efferent) **neuron,** which carries nerve impulses from the central nervous system through the ventral root to the effector via a peripheral nerve; and (5) an **effector** (either a muscle or a gland) that responds to the stimulus by contracting or secreting.

Reflex arcs may have more than one interneuron, or may be completely lacking in interneurons. If no interneuron is present, as is true of the knee jerk reflex, the reflex arc is said to be *monosynaptic.* When one or more interneurons exist, the reflex arc is classified as being *polysynaptic.*

The Visceral Reflex Arc

Muscular and glandular responses of the viscera to internal environmental changes are controlled by

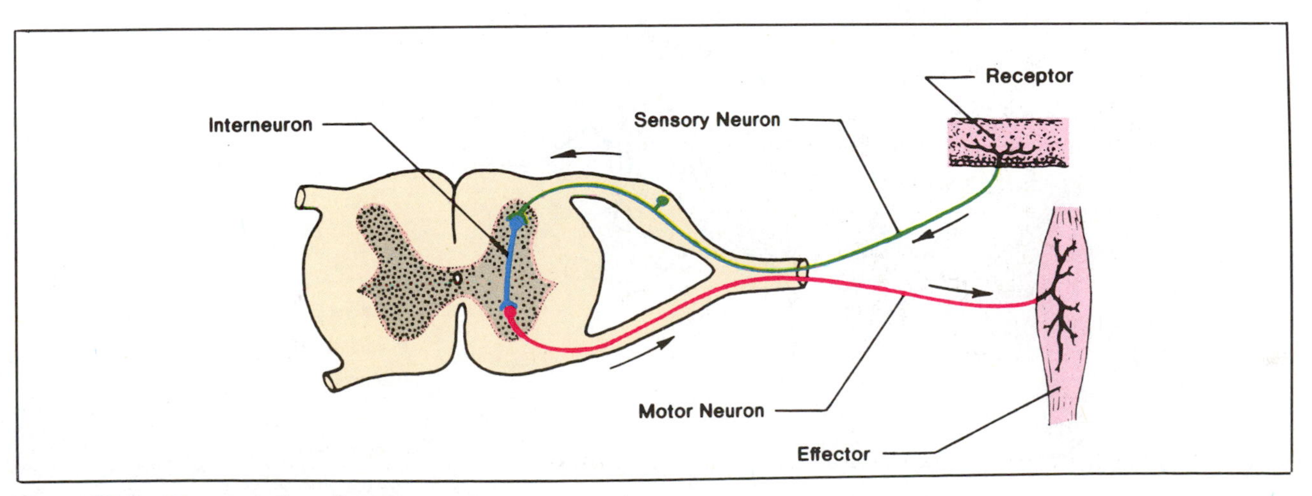

Figure 30.3 The somatic reflex arc.

the **autonomic nervous system.** The reflexes of this system follow a different pathway in the nervous system. Figure 30.4 illustrates the components of a visceral reflex arc.

The principal anatomical difference between somatic and visceral reflex arcs is that the latter type has two efferent neurons instead of one. These two neurons synapse outside of the central nervous system in an autonomic ganglion.

Two types of autonomic ganglia are shown in figure 30.4: vertebral and collateral. The **vertebral autonomic ganglia** are united to form a chain that lies along the vertebral column. The **collateral ganglia** are located further away from the central nervous system.

A visceral reflex originates with a stimulus acting on a receptor in the viscera. Impulses pass along the dendrite of a **visceral afferent neuron.** Note that the cell bodies of these neurons are located in the **spinal ganglion.** The axon of the visceral afferent neuron forms a synapse with the **preganglionic efferent neuron** in the spinal cord. Impulses in this neuron are conveyed to the **postganglionic efferent** at synaptic connections in either type of autonomic ganglion (note the short portions of cutoff postganglionic efferent neurons in the vertebral autonomic ganglia). From these ganglia the postganglionic efferent neuron carries the impulses to the organ innervated.

In this brief mention of the autonomic nervous system, the student is reminded that the system consists of two parts: the sympathetic and parasympathetic divisions. The *sympathetic* (thoracolumbar) *division* involves the spinal nerves of the thoracic and lumbar regions. The *parasympathetic* (craniosacral) *division* incorporates the cranial and sacral nerves. Most viscera of the body are innervated by nerves from both of these systems. This form of double innervation enables an organ to be stimulated by one system and inhibited by the other.

Assignment:

Label figure 30.4.

Laboratory Report

Complete the Laboratory Report for this exercise.

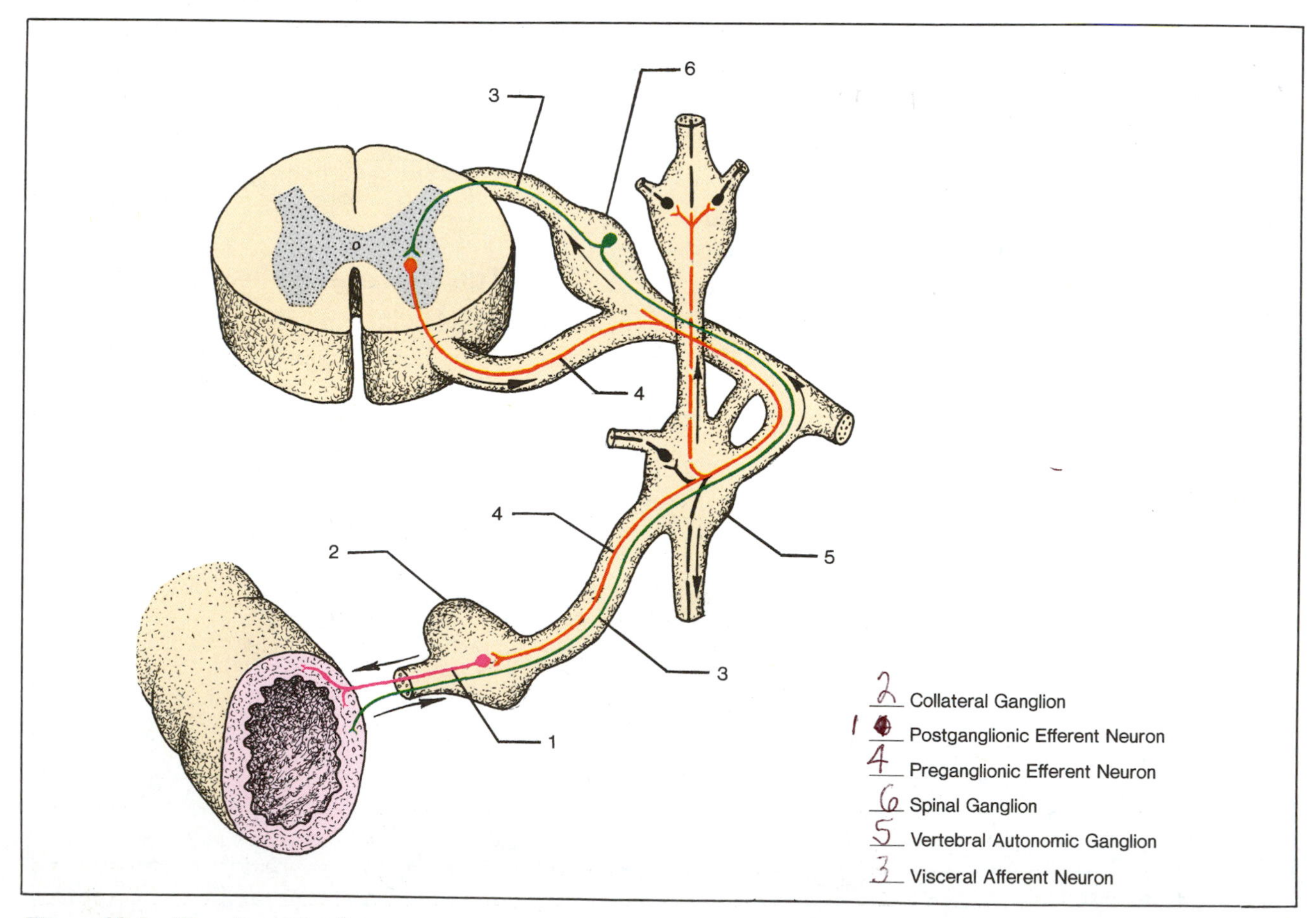

Figure 30.4 The visceral reflex arc.

Somatic Reflexes 31

We learned in Exercise 30 that somatic reflexes are automatic responses that occur in skeletal muscles when the proper stimulus is applied to the appropriate receptor. In this laboratory period we will study somatic reflexes, first in the frog, then in humans.

Frog Experiments

Somatic reflexes, whether in frogs or humans, have five characteristics that will be studied here on the frog. They are (1) function, (2) speed of reaction, (3) radiation, (4) inhibition, and (5) synaptic fatigue. A pithed frog will be used in all these tests.

Preparation of Frog

To destroy the brain of the frog in preparation for these tests, proceed as follows.

Rinse the frog under cool tap water and grip it in your left hand, as shown in illustration A, figure 31.1. Use your left index finger to force the frog's nose downward so that the head makes a sharp angle with the trunk.

Locate the transverse groove between the skull and the vertebral column by pressing down on the skin with your fingernail to mark its location. Now, force a sharp dissecting needle into the crevice, forcing it forward into the center of the skull through the foramen magnum.

Twist the needle from side to side to destroy the brain. This process of destroying the brain is called **single pithing** and the animal is referred to as a **spinal frog.**

After five or six minutes, test the effectiveness of the pithing by touching the cornea of each eye with the dissecting needle. If the lower eyelid on each eye does not rise to cover the eyeball, the pithing is complete. This test is valid *only* after approximately five minutes to allow recovery from a temporary state of neural shock to the entire spinal cord.

You can determine when neural shock of the spinal cord has ended by pinching one of the toes with forceps. If a withdrawal reflex occurs, spinal shock has ended. You are now ready to proceed with the various experiments.

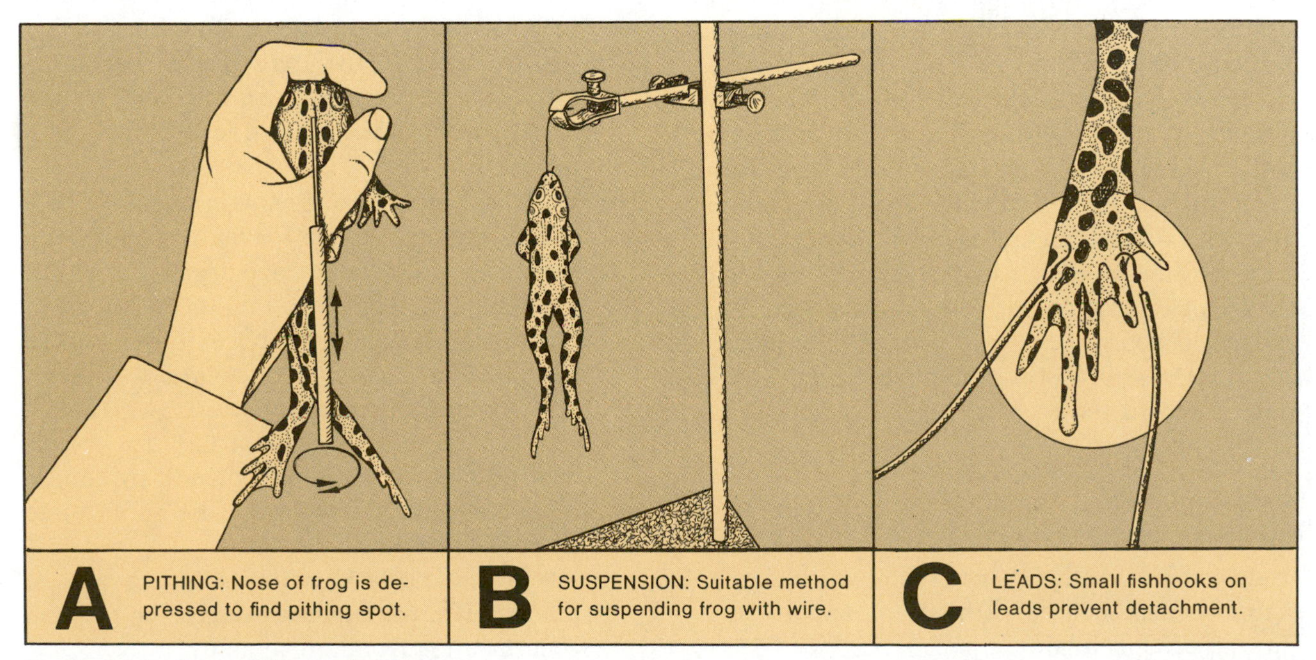

Figure 31.1 Frog pithing and setups.

Functional Nature

Muscle tone, movement, and coordinated action in a spinal frog would indicate that these physiological activities occur independently of cognition and are the result of reflex activity. Test for these phenomena as follows:

Materials:

spinal frog
ring stand and clamp
galvanized iron wire, pliers, forceps
filter paper squares (3 mm)
28% acetic acid
10% $NaHCO_3$ in squeeze bottle
squeeze bottle of tap water
beaker (250 ml size)

With the frog in a squatting position, gently draw out one of the hind legs. Is any pull exerted by the animal? When the foot is released, does it return to its previous position? Squeeze the muscles of the hind legs. Do they feel flaccid or firm? Does muscle tone seem to exist? Can the animal be induced to jump by prodding its posterior? Place the animal in a sink basin filled with water. Does it attempt to swim? Does it float or sink? Answer all these questions on the Laboratory Report.

Suspend the frog with a wire hook through the lower jaw as shown in illustration B, figure 31.1. With forceps, dip a small square of filter paper in acetic acid, shake off excess, and apply to the ventral surface of the thigh for about **10 seconds.** Is the frog able to remove the irritant? Do you think that the frog feels the burning sensation caused by the acid? Explain.

Wash the acid off the leg with 10% $NaHCO_3$, spraying it on the affected area with a squeeze bottle. Allow the liquid to drain from the leg into an empty beaker. Let stand for 1 or 2 minutes and rinse with water from squeeze bottle. Place another piece of filter paper and acid on another portion of the body (leg or abdomen). Observe reaction. What happens if you restrain the limb that attempts removal? Does any other adaptive behavior take place? Record these observations on the Laboratory Report.

Reaction Time

To determine whether or not the reaction time to a stimulus is influenced by the strength of the stimulus, proceed as follows:

Materials:

7 test tubes (20 mm dia.) in test tube rack
3 beakers (150 ml size)
Kimwipes
1 graduate (10 ml size)
1 squeeze bottle of tap water
1 squeeze bottle of 1% HCl
1 squeeze bottle of 1% $NaHCO_3$

1. Dispense 30 ml of 1% $NaHCO_3$ into a beaker.
2. Label seven test tubes: 1.0, .5, .4, .3, .2, .1, and .05%.
3. With a graduate, measure 10 ml of 1% HCl from the squeeze bottle, and pour it into the first tube.
4. Into the other six tubes, measure out the correct amounts of 1% HCl and tap water to make 10 ml each of the correct percentages of HCl. (Examples: 5 ml of 1% HCl and 5 ml of water in the .5% tube; 4 ml of 1% HCl and 6 ml of water in the .4% tube; etc.) All liquids are dispensed from the two squeeze bottles.
5. Pour the contents from the .05% tube into a clean beaker and allow the long toe of one of the hind legs to be immersed in it for **90 seconds.** Be sure to prevent other parts of the foot from touching the sides of the beaker. If the foot is withdrawn, record on the Laboratory Report the number of seconds that elapsed from the time of immersion to withdrawal. If there is no withdrawal, remove the beaker of .05% HCl at 90 seconds.
6. Bathe the foot in the beaker of 1% $NaHCO_3$ for **15 seconds** and rinse the foot by spraying with water. Use the third unused beaker to catch any water sprayed on the foot. Dry the foot off with Kimwipes and let rest for 2 or 3 minutes.

   Make two more tests with the .05% HCl and record an average for the three tests. Be sure to neutralize, wash, and dry between tests.
7. Pour the .05% HCl back into its test tube and empty the tube of .1% HCl into the beaker. Place the long toe into this solution, as previously, and again record the withdrawal time in seconds. After neutralization, bathing, drying, and resting the foot, repeat the test two times, as above, with this solution. Be sure not to exceed 90 seconds.
8. Repeat these procedures for all other tubes of diluted HCl, progressively increasing the acid concentration from low to high.

9. Plot the reaction times against the concentration on the graph provided on the Laboratory Report.

Reflex Radiation

If the foot of a spinal frog is subjected to increasing intensity of electrical stimulation, a phenomenon called *reflex radiation* occurs. To demonstrate it, one applies a tetanic stimulus, first at a low level, and then gradually increases it. Proceed as follows to observe what takes place.

Materials:

electronic stimulator
2-wire leads with small fishhooks soldered to the leads

1. Connect the jacks of the 2-wire leads to the output posts of the stimulator. (These leads have very small fishhooks soldered to one end to enable easy attachment to the skin. Handle them gingerly. They are sharp!)
2. Attach the hooks to the skin of the left foot as shown in illustration C, figure 31.1.
3. Set the stimulator to produce a minimum tetanic stimulus as follows: Voltage at 0.1 volt, Duration at 50 msec, Frequency at 50 pps, Stimulus switch at REGULAR, Polarity at NORMAL, and Output switch on BIphasic.
4. Press the Mode switch to REPEAT and look for a response. If no response, sequentially increase the voltage by doubling each time (.2, .4, .8, etc.) until a response occurs. Describe the response on the Laboratory Report.
5. Increase the voltage in steps of 2 or 3 volts at a time and note any changes that occur in the reflex pattern.

Reflex Inhibition

Just as it is possible to stifle a sneeze or prevent a knee jerk, it should be possible to override the reflex to acid with electrical stimulation. With the stimulator still hooked up to the left foot from the previous experiment, lower the toes of the right foot into a beaker of HCl of a concentration that produced a moderately fast reflex response. As the right foot is lowered into the acid, a moderate tetanizing stimulus is administered to the left foot. Adjust the voltage until the foot in the acid is inhibited. Record on the Laboratory Report the acid concentration and voltage that inhibited the reflex action.

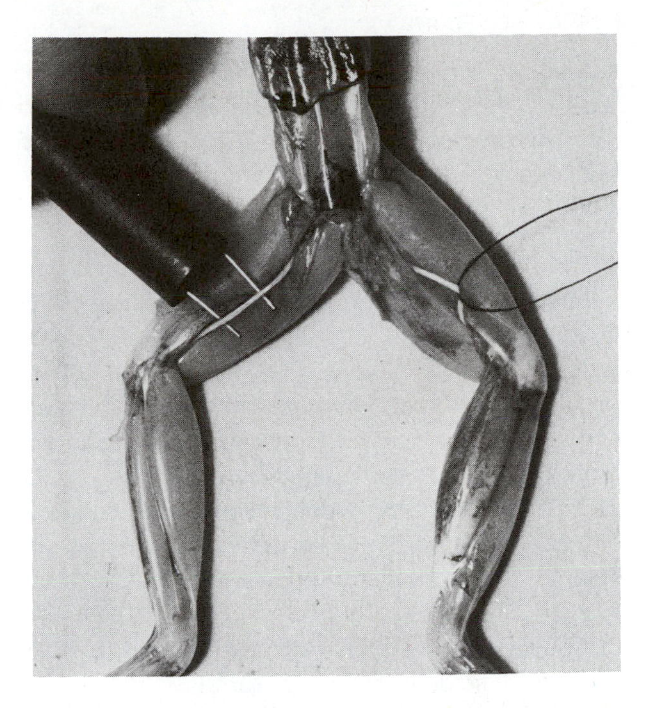

Figure 31.2 Setup for determining when synaptic fatigue occurs. Note string on nerve to prevent trauma when nerve is manipulated.

Synaptic Fatigue

If your specimen is still responding well to stimuli, it may be used in this final experiment. If it does not prove viable, replace it with a freshly pithed specimen for this last test.

In this experiment, it will be necessary to remove the skin from both hind legs and expose the sciatic nerves of both thighs (see figure 31.2). Stimulate the sciatic nerve of the left leg and note that muscle contraction occurs in both legs. Continue this stimulation until the muscles in the right leg fail to respond. Wait thirty seconds and stimulate again to make sure that no right leg muscle contraction occurs.

Now, stimulate the sciatic nerve in the right leg. Does this cause muscles in the right leg to contract? Where did the fatigue take place when the left sciatic nerve was overstimulated?

Reflexes in Medical Diagnosis

Reflex testing is a standard, useful clinical procedure employed by physicians in search of neurological pathology. Damage to intervertebral disks, tumors, polyneuritis, apoplexy, and many other conditions can be better understood with the aid of reflex studies. The diagnostic reflexes studied here are the ones most frequently employed by physicians.

Interpretation of reflex responses is often subjective and requires considerable experience on the part of the diagnostician. Our purpose in performing the tests here, obviously, is not to diagnose, but rather to test and observe, and to understand why a particular test is performed.

Two Types of Reflexes

Clinically, reflexes are categorized as being either deep or superficial. The **deep reflexes** include all reflexes that are elicited by a sharp tap on an appropriate tendon or muscle. They are also called *jerk, stretch,* or *myotatic* reflexes. The receptors (spindles) for these reflexes are located in the muscle, not in the tendon. When the tendon is tapped, the muscle is stretched; stretching the muscle, in turn, activates the muscle spindle, which triggers the reflex response.

Superficial reflexes are withdrawal reflexes elicited by noxious or tactile stimulation. They are also called *cutaneous reflexes.* Instead of percussion initiating these reflexes, the skin is stroked or scratched to induce a response.

Reflex Aberrations

Abnormal responses to stimuli may be diminished (hyporeflexia), exaggerated (hyperreflexia), or pathological. **Hyporeflexia** may be due to malnutrition, neuronal lesions, aging, deliberate relaxation, etc. **Hyperreflexia** is often accompanied by marked muscle tone due to the loss of inhibitory control by the motor cortex. Among other things, it may also be caused by strychnine poisoning. **Pathological reflexes** are reflex responses that occur in one or more muscles other than the muscle where the stimulus originates. It is these three deviations that we will look for in performing these tests. The following scale is used for evaluating each response.

- ++++ very brisk, hyperactive; often indicative of pathology; may be associated with clonus
- +++ brisker than average; may or may not be indicative of pathology
- ++ average; normal
- + somewhat diminished
- 0 no response

Materials:

reflex hammer

Reflexes of the Arm and Hand

Biceps Reflex This reflex is a deep reflex which causes flexion of the arm. It is elicited by holding the subject's elbow with the thumb pressed over the tendon of the biceps brachii, as shown in illustration A, figure 31.3. To produce the desired response, strike a sharp blow to the first digit of the thumb with the reflex hammer.

If the reflex is absent or diminished (0 or +), apply reinforcement by asking the subject to clench

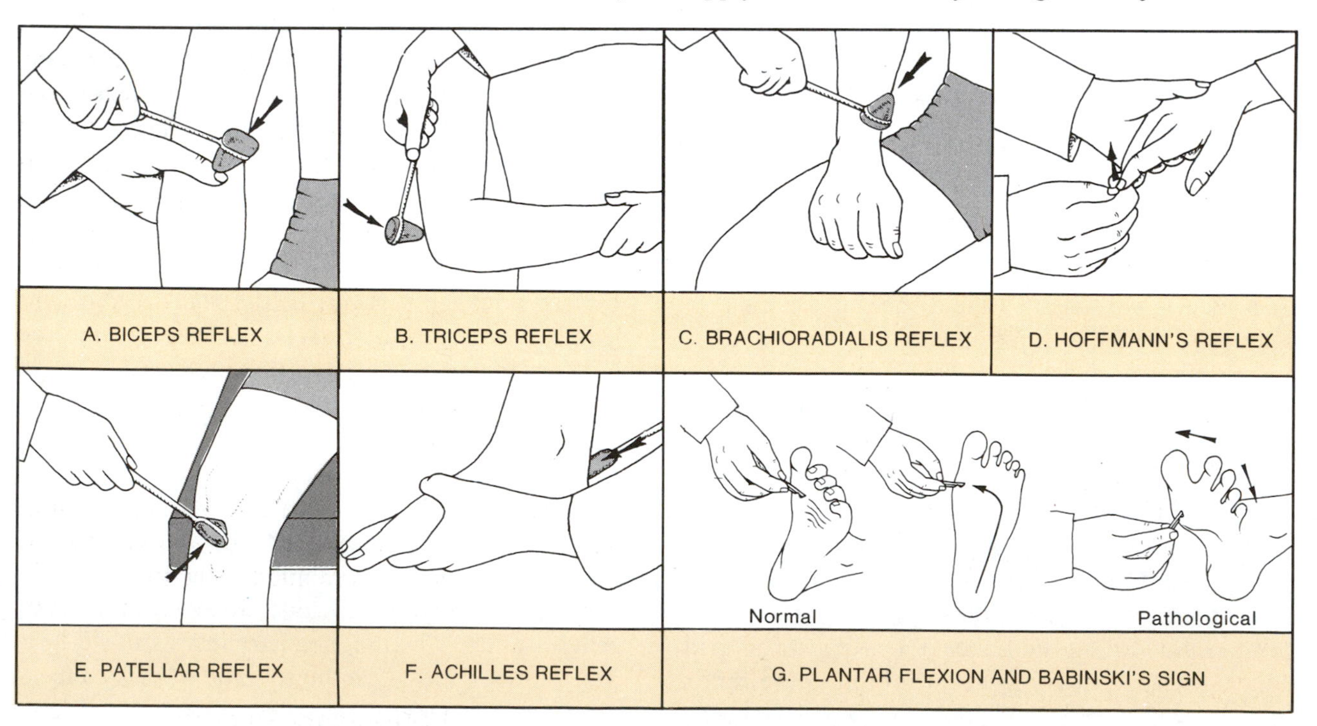

Figure 31.3 Methods used for producing seven types of somatic reflexes.

his/her teeth or squeeze his/her thigh with the other hand. Test both arms and record the degree of response on the Laboratory Report.

This reflex functions through C_5 and C_6 spinal nerves.

Triceps Reflex This deep reflex causes extension of the arm in normal individuals. To demonstrate this reflex, flex the arm at the elbow, holding the wrist as shown in illustration B, figure 31.3, with the palm facing the body. Strike the triceps brachii tendon above the elbow with the pointed end of the reflex hammer. Use the same reinforcement techniques as above if necessary. Test both arms and record the degree of response on the Laboratory Report.

This reflex functions through C_7 and C_8 spinal nerves.

Brachioradialis Reflex In normal individuals this reflex manifests itself in flexion and pronation of the forearm and flexion of the fingers. To demonstrate this reflex, direct the subject to rest the hand on the thigh in the position shown in illustration C, figure 31.3. Strike the forearm with the wide end of the reflex hammer about 1 inch above the end of the radius. The illustration reveals the approximate spot. Use reinforcement techniques, if necessary. The tendon that is struck here is for the brachioradialis muscle of the arm.

This reflex functions through the same spinal nerves (C_5 and C_6) as the biceps brachii reflex.

Hoffmann's Reflex This reflex is an example of a deep reflex in which the response does not occur only in the muscle that is stretched. As stated earlier, a positive reflex is classified as pathological. Hoffmann's reflex also exhibits characteristics of hyperreflexia.

This reflex is produced by flicking the terminal phalanx of the index finger upward, as shown in illustration D, figure 31.3. In individuals who have pyramidal tract lesions, the thumb of the hand is adducted and flexed, and other fingers exhibit twitchlike flexion. Such a broad field of reflex action resulting from a brief stretch of the flexor muscles of one finger manifests deep reflex hyperactivity and pathological characteristics.

Test both hands of the subject and record results on the Laboratory Report.

Reflexes of the Leg

Patellar Reflex This monosynaptic reflex is variously referred to as the *knee reflex, knee jerk,* or *quadriceps reflex.* To perform this test, the subject should be seated on the edge of a table with the leg suspended and somewhat flexed over the edge.

To elicit the typical response, strike the patellar tendon, which is just below the kneecap. The correct spot is shown in illustration E, figure 31.3.

If the response is negative, utilize the **Jendrassic maneuver** for facilitation. This technique involves the subject locking the fingers in front of the body and pulling each hand against each other isometrically.

A phenomenon called **clonus** is sometimes seen when inducing this reflex. Clonus is characterized by a succession of jerklike contractions that follow the normal response and persist for a period of time. This condition is a manifestation of hyperreflexia and indicates damage within the central nervous system.

Test both legs, recording the results on the Laboratory Report. This reflex functions through L_2, L_3, and L_4 spinal nerves.

Achilles Reflex This reflex is also referred to as the *ankle jerk.* It is characterized by plantar flexion when the Achilles tendon is struck a sharp blow. Stretching this tendon affects the muscle spindles in the triceps surae, causing it to contract. To perform this test, grip the foot with the left hand as shown in illustration F, figure 31.3, forcing it upward somewhat. With the subject relaxed, strike the Achilles tendon as shown. The reflex functions through S_1 and S_2 spinal nerves. Hyporeflexia here is often a sign of hypothyroidism.

Plantar Flexion *(Babinski's sign)* This reflex is the only one in this series that is of a superficial nature. Note in illustration G, figure 31.3, that the normal reaction to stroking the sole of the foot in an adult is plantar flexion. If dorsiflexion occurs, starting in the great toe and spreading to the other toes *(Babinski's sign),* it may be assumed that there is myelin damage to fibers in the pyramidal tracts. Incidentally, dorsiflexion is normal in infants, especially if they are asleep. Babinski's sign disappears in infants once myelinization of nerve fibers is complete. This reflex functions through S_1 and S_2 spinal nerves.

Perform this test on the bottom of the foot of a subject, using a hard object such as a key. Follow the pattern shown in the middle illustration.

Laboratory Report

Complete the Laboratory Report for this exercise.

32 Brain Anatomy: External

This study of the human brain will be done in conjunction with dissection of the sheep brain. The ample size and availability of sheep brains make them preferable to brains of most other animals; also, the anatomical similarities that exist between the human and sheep brains are considerably greater than their differences.

Preserved human brains will also be available for study in this laboratory period. Dissection, however, will be confined to the sheep brain. A discussion of the meninges will precede the sheep brain dissection.

The Meninges

It was observed in Exercise 30 that the spinal cord, as well as the brain, is surrounded by three membranes called *meninges*. Due to the importance of these protective structures it will be necessary to reemphasize certain anatomical characteristics of these membranes.

Dura mater The outermost meninx is the *dura mater,* which is a thick tough membrane made up of fibrous connective tissue. The lateral view of the opened skull in figure 32.1 reveals the dura mater lifted away from the exposed brain. On the median line of the skull it forms a large **sagittal sinus** (label 6) which collects blood from the surface of the brain. Note that a large number of cerebral veins empty into this sinus.

Extending downward between the two halves of the cerebrum is an extension of the dura mater, the **falx cerebri.** This structure can also be seen in the frontal section. At its anterior end, the falx cerebri is attached to the crista galli of the ethmoid bone.

Arachnoid mater Inferior to the dura mater lies the second meninx, the *arachnoid mater* (label 8). Between this delicate netlike membrane and the surface of the brain is the **subarachnoid space,** which contains cerebrospinal fluid. Note that the arachnoid mater has small projections of tissue, the **arachnoid granulations,** which extend up into the sagittal sinus. These granulations allow cerebro-

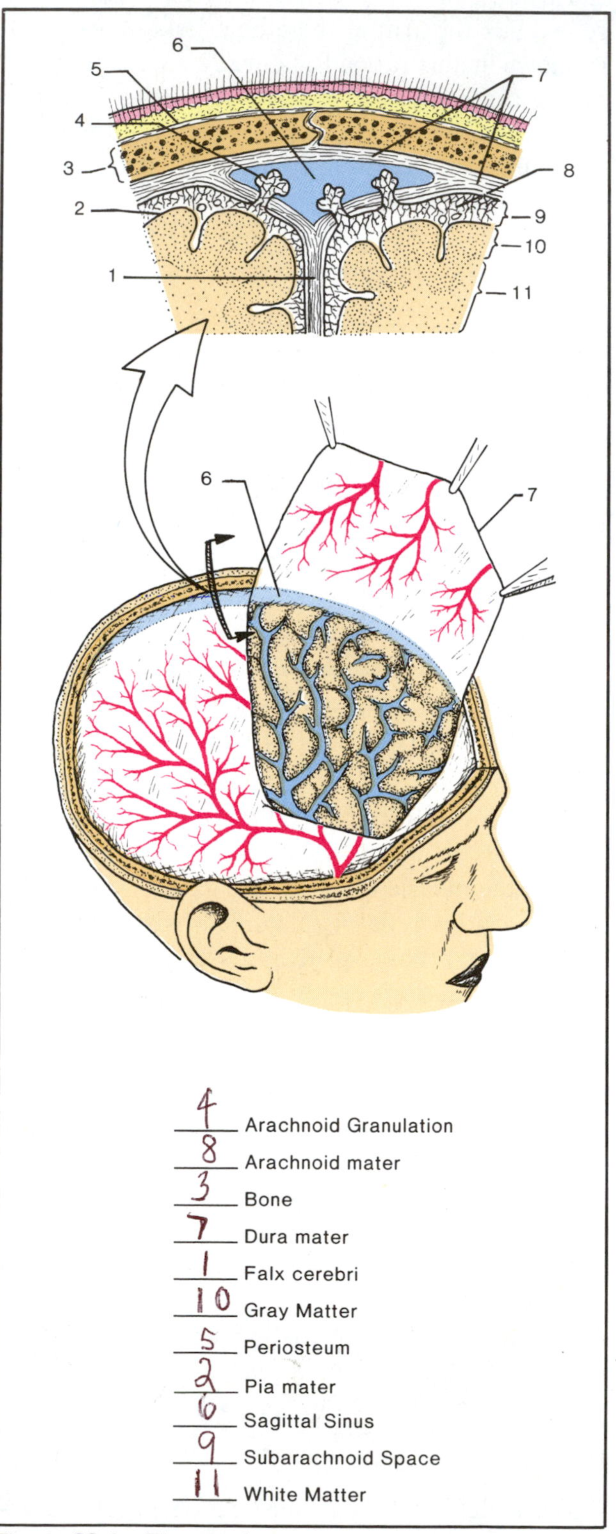

Figure 32.1 The meninges.

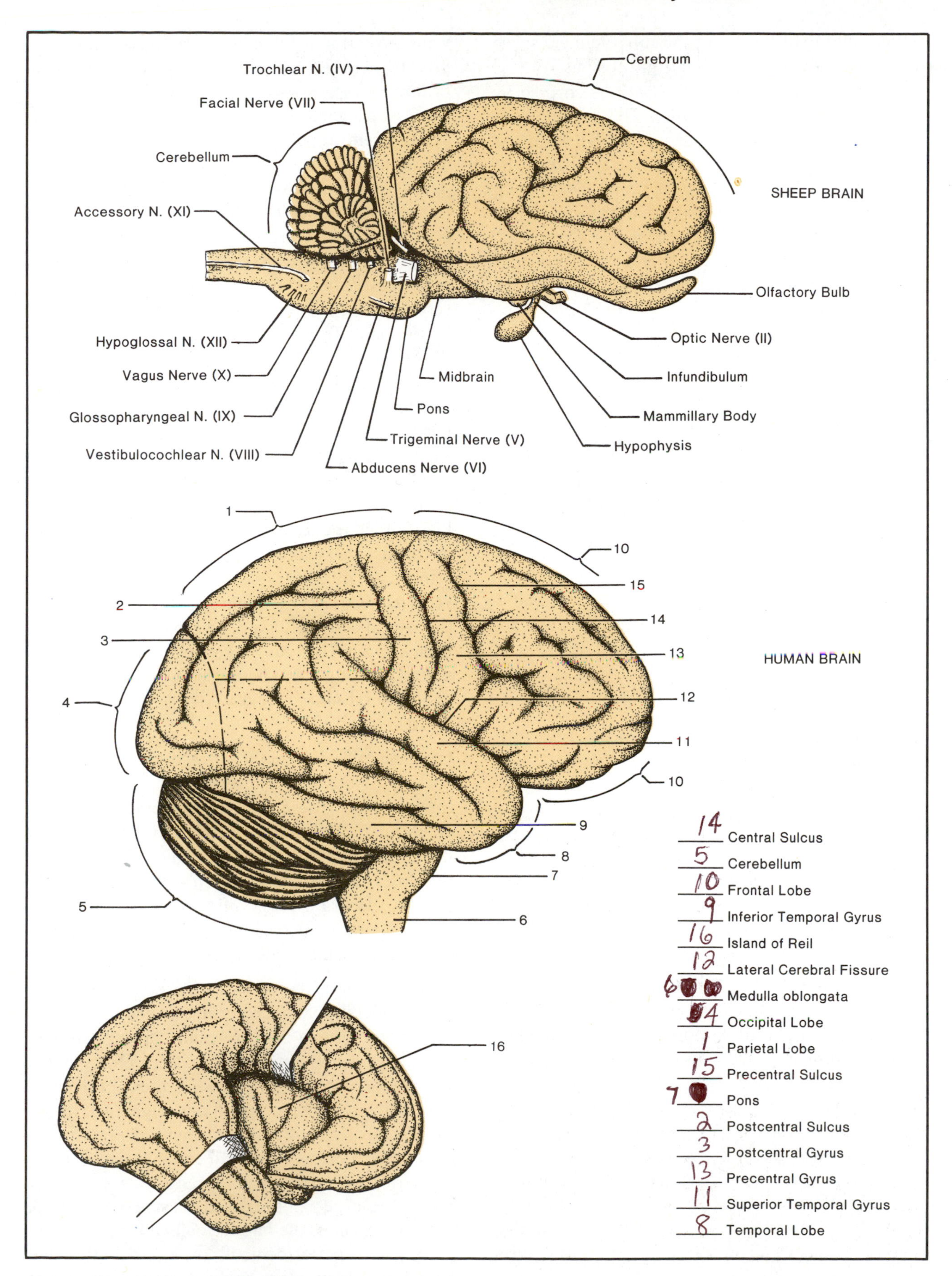

Figure 32.2 Lateral aspects of sheep and human brains.

spinal fluid to diffuse from the subarachnoid space into the venous blood of the sagittal sinus.

Pia mater The surface of the brain is covered by the third meninx, the *pia mater*. This membrane is very thin. Note that the surface of the brain has grooves, or **sulci,** which increase the surface of the **gray matter** (label 10). As in the case of the spinal cord, the gray matter is darker than the inner **white matter** because it contains cell bodies of neurons. The white matter consists of neuron fibers that extend from the gray matter to other parts of the brain and spinal cord.

Assignment:

Label figure 32.1.

Sheep Brain Dissection

As you dissect the sheep brain and identify the various structures on it, compare the sheep brain with the human brain. Note that the illustrations of comparable views of both brain types are on the same page or opposite pages. Once you have located a structure on the sheep brain, try to find a comparable structure on the human brain. Performing the dissection in this way will help you to learn human brain anatomy.

Materials:

preserved sheep brains
preserved human brains
human brain models
dissecting instruments
dissecting trays

Place a sheep brain on a dissecting tray and, by comparing it to figures 32.2 and 32.4, identify the **cerebrum, cerebellum, pons Varolii, midbrain,** and **medulla oblongata.** Study each of these five major divisions to identify the fissures, grooves, ridges, and other structures as follows.

Cerebral Hemispheres In both the sheep and man, the cerebrum is the predominant portion of the brain. Note that when observed from the dorsal or ventral aspect, the cerebrum is divided along the median line to form two **cerebral hemispheres** by the **longitudinal cerebral fissure.**

Observe that the surface of the cerebrum is covered with ridges and furrows of variable depth. The deeper furrows are called **fissures,** and the shallow ones, **sulci.** The ridges or convolutions are called **gyri.** The frontal section in figure 32.1 reveals how this infolding of the cerebral surface greatly increases the amount of gray matter of the brain.

The principal fissures of the human cerebrum are the **central sulcus** (fissure of Rolando), **lateral cerebral fissure** (fissure of Sylvius), and the **parieto-occipital fissure.** In figure 32.2 the central sulcus is label 14; the lateral cerebral fissure is label 12. The greater portion of the parieto-occipital fissure is seen on the medial surface of the cerebrum (label 1, figure 33.2). If the lateral cerebral fissure is retracted as shown in the lower illustration of figure 32.2, an area called the **island of Reil** is exposed.

Each half of the human cerebrum is divided into four lobes. The **frontal lobe** is the most anterior portion. Its posterior margin falls on the central sulcus and its inferior border consists of the lateral cerebral fissure. The **occipital lobe** is the most posterior lobe of the brain. The anterior margin of this lobe falls on the parieto-occipital fissure. Since most of this fissure is seen on the medial face of the cerebrum, an imaginary dotted line in figure 32.2 reveals its position. The **temporal lobe** lies inferior to the lateral cerebral fissure and extends back to the parieto-occipital fissure. An imaginary horizontal dotted line provides a border between the temporal and **parietal lobes.** Labels 1, 4, 8, and 10 are for these four lobes.

Cerebellum Examine the surface of the sheep cerebellum. Note that its surface is furrowed with sulci. How do these sulci differ from those of the cerebrum? The human cerebellum is constricted in the middle to form right and left hemispheres. Can the same be said for the sheep's cerebellum? This part of the brain plays an important part in the maintenance of posture and the coordination of complex muscular movements.

Midbrain Since the midbrain lies concealed under the cerebellum and cerebrum, expose its dorsal surface by forcing the cerebellum downward with the thumb as illustrated in figure 32.3. Observe that you have exposed five roundish bodies in this area: four bodies of the corpora quadrigemina and a single pineal gland on the median line.

The two pairs of rounded eminences that dominate this area constitute the *corpora quadrigemina.* The large pair are called the **superior colliculi;** the smaller ones, the **inferior colliculi.** In some animals, the superior colliculi are called **optic lobes** because of their close association with the optic

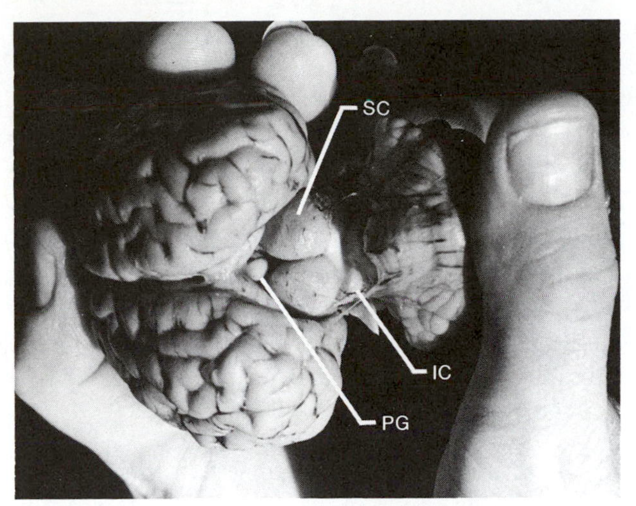

Figure 32.3 Structures on midbrain: SC—superior colliculus; IC—inferior colliculus; PG—pineal gland.

tracts. In animals, the colliculi are important analytical centers concerned with brightness and sound discrimination. Their functions in man are still obscure.

The **pineal gland** lies on the median line of the midbrain, anterior to the corpora quadrigemina. From an evolutionary point of view, there are indications that the pineal gland is a remnant of the third eye that exists in certain reptiles. This gland produces a hormone called *melatonin,* which in lower animals inhibits the estrus cycle. The exact role of melatonin in humans is not completely understood at this time; it may have something to do with the estrus cycle. It is probably significant that this structure is, histologically, glandular until puberty, and then becomes fibrous in nature after puberty.

Pons Varolii Locate this portion of the sheep's brain by examining its ventral surface and comparing it with figures 32.4 and 32.5. This part of the brain stem contains fibers that connect parts of the cerebellum and the medulla with the cerebrum. It also contains nuclei of the fifth, sixth, seventh, and eighth cranial nerves.

Medulla oblongata This portion of the brain stem is also known as the **spinal bulb.** It contains centers of gray matter that control the heart, respiration, and vasomotor reactions. The last four cranial nerves emerge from the medulla.

Assignment:

Label figure 32.2.

The Cranial Nerves

There are twelve pairs of nerves that emerge from various parts of the brain and pass through foramina of the skull to innervate parts of the head and trunk. Each pair has a name as well as a position number. Although most of these nerves contain both motor and sensory fibers *(mixed nerves),* a few contain only sensory fibers *(sensory nerves).* The sensory fibers have their cell bodies in ganglia outside of the brain; the cell bodies of the motor neurons, on the other hand, are situated within *nuclei* of the brain.

Compare your sheep brain with figures 32.2 and 32.4 to assist you in identifying each pair of nerves. As you proceed with the following study, avoid damaging the brain tissue with dissecting instruments. Keep in mind, also, that once a nerve is broken off it is difficult to determine where it was.

I Olfactory Nerve This cranial nerve contains sensory fibers for the sense of smell. Since it extends from the mucous membranes in the nose through the ethmoid bone to the olfactory bulb, it cannot be seen on your specimen. In the **olfactory bulb** synapses are made with fibers that extend inward to olfactory areas in the cerebrum. Note that on the human brain an **olfactory tract** (label 2, figure 32.5) extends between the bulb and the brain.

II Optic Nerve This sensory nerve functions in vision. It contains axon fibers from ganglion cells of the retina of the eye. Part of the fibers from each optic nerve cross over to the other side of the brain as they pass through the **optic chiasma.** From the optic chiasma the fibers pass through the **optic tracts** to the thalamus and finally to the visual areas of the cerebrum. Identify these structures on the sheep brain.

III Oculomotor Nerve This nerve emerges from the midbrain and supplies somatic nerve fibers to the *levator palpebrae* (eyelid muscle) and four of the extrinsic ocular muscles: *superior rectus, medial rectus, inferior rectus,* and *inferior oblique.* (The superior oblique and lateral rectus muscles are innervated by IV and VI, respectively.)

The oculomotor nerve also supplies parasympathetic fibers to the *constrictor pupillaris* of the iris and the *ciliaris muscle* of the ciliary body. These fibers constrict the iris and change the lens shape as needed in accommodation (focusing).

If the oculomotor nerves are not visible on your sheep brain, it may be that they are obscured by the hypophysis (pituitary gland). The sheep brain illustrated in figure 32.4 lacks this structure. To remove the hypophysis without damaging other structures, lift it up with a pair of forceps and cut the infundibulum with a scalpel or a pair of scissors.

IV Trochlear Nerve This nerve provides muscle sense and motor stimulation of the *superior oblique* muscle of the eye. As in the case of the oculomotor, this nerve also emerges from the midbrain. Because of its small diameter it is one of the more difficult nerves to locate.

V Trigeminal Nerve Just posterior to the trochlear nerve lies the largest cranial nerve, the trigeminal. Although it is a mixed nerve, its sensory functions are much more extensive than its motor functions. Because of its extensive innervation of parts of the mouth and face, it is described in detail on pages 151 through 153.

VI Abducens Nerve This small nerve provides innervation of the *lateral rectus* muscle of the eye. It is a mixed nerve in that it provides muscle sense as well as muscular contraction. On the human brain it emerges from the lower part of the pons near the medulla.

VII Facial Nerve This nerve consists of motor and sensory divisions. Label 15, figure 32.5, reveals the double nature of this nerve on the human brain. It innervates muscles of the face, salivary glands, and taste buds of the anterior two-thirds of the tongue.

VIII Vestibulocochlear *(Auditory)* Nerve This nerve goes to the inner ear. It has two branches: the *vestibular division,* which innervates the semicircular canals, and the *cochlear division,* which innervates the cochlea. The former branch functions in maintaining equilibrium; the cochlear portion is auditory in function. The vestibulocochlear nerve emerges just posterior to the facial nerve on the human brain.

IX Glossopharyngeal Nerve This mixed nerve emerges from the medulla posterior to the vestibulocochlear. It functions in reflexes of the heart, taste, and swallowing. Taste buds on the back of the tongue are innervated by some of its fibers. Efferent fibers innervate muscles of the pharynx (swallowing) and the parotid salivary glands (secretion).

X Vagus Nerve This cranial nerve, which originates on the medulla, exceeds all of the other cranial nerves in its extensive ramifications. In addition to supplying parts of the head and neck with nerves, it has branches that extend down into the chest and abdomen. It is a mixed nerve. Sensory fibers in this nerve go to the heart, external acoustic meatus, pharynx, larynx, and thoracic and abdominal viscera. Motor fibers pass to the pharynx, base of the tongue, larynx, and to the autonomic ganglia of

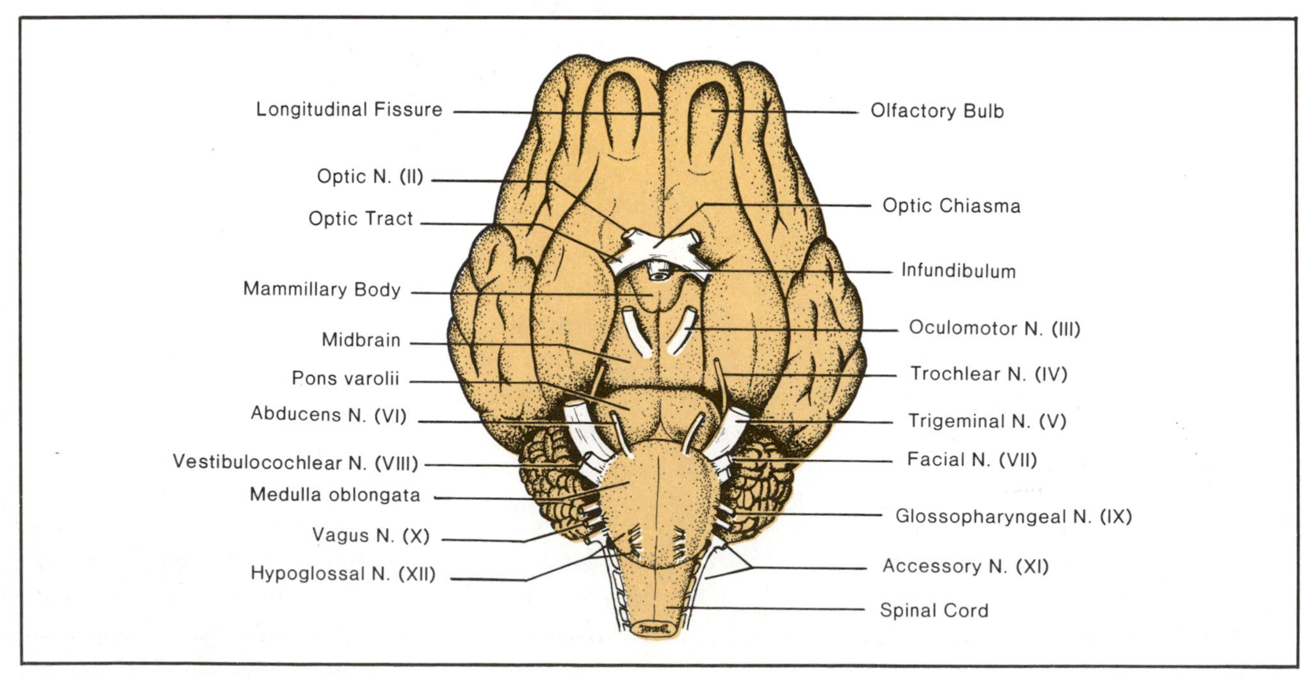

Figure 32.4 Ventral aspect of sheep brain.

thoracic and abdominal viscera. Many references will be made to this nerve in subsequent discussions of the physiology of the lungs, heart, and digestive organs.

XI Accessory Nerve Since this nerve emerges from both the brain and spinal cord it is sometimes called the *spinal accessory* nerve. Note that on both the sheep and human brains it parallels the lower part of the medulla and the spinal cord with branches going into both structures. On the human brain it appears to occupy a more posterior position than the twelfth nerve. Afferent and efferent spinal components innervate the sternocleidomastoid and trapezius muscles. The cranial portion of the nerve innervates the pharynx, upper larynx, uvula, and palate.

XII Hypoglossal Nerve This nerve emerges from the medulla to innervate several muscles of the tongue. It contains both afferent and efferent fibers. On the human brain in figure 32.5, it has the appearance of a cutoff tree trunk with roots extending into the medulla.

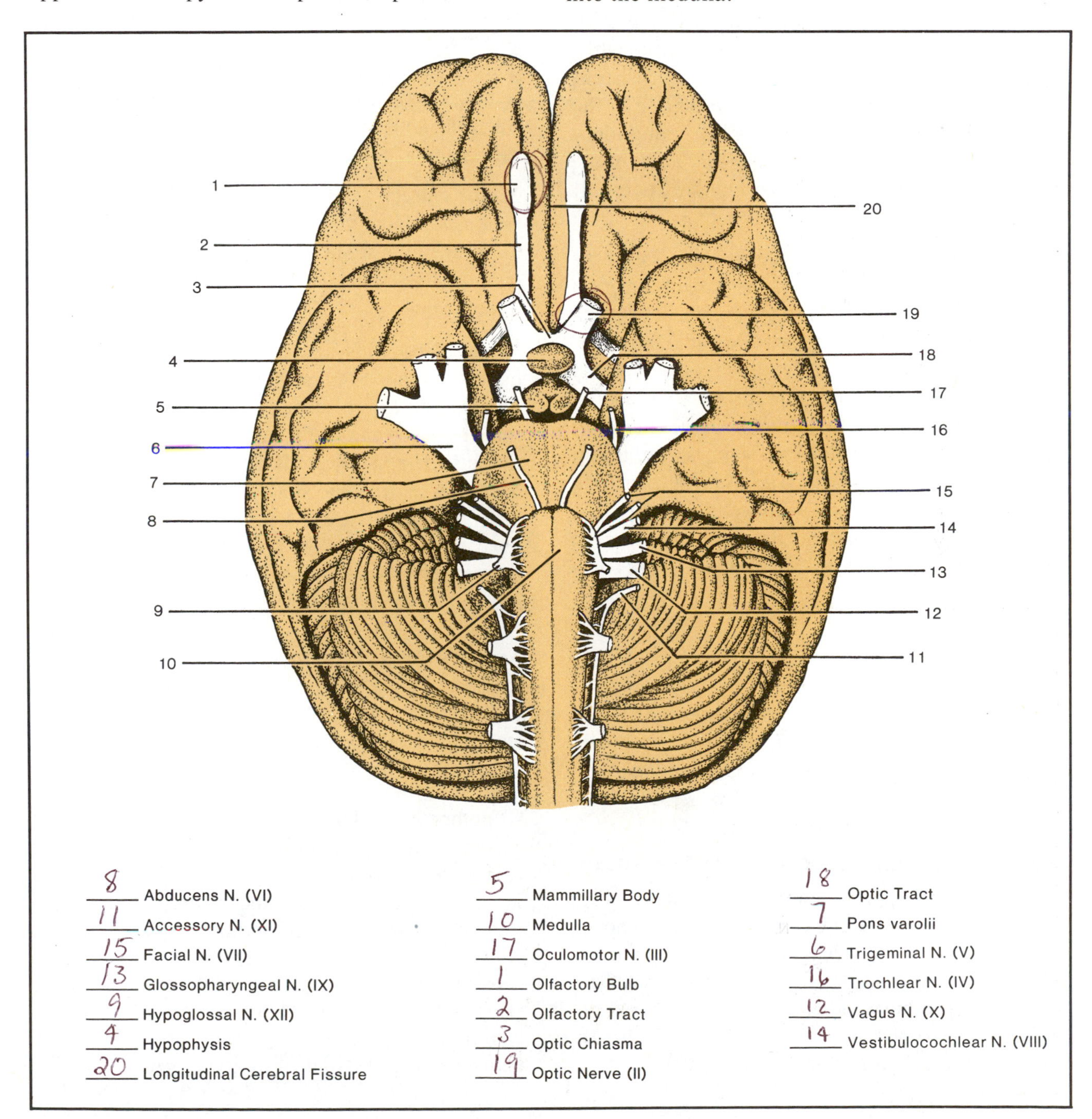

Figure 32.5 Inferior aspect of human brain.

The Hypothalamus

The ventral portion of the brain that includes the mammillary bodies, infundibulum, and part of the hypophysis is collectively called the *hypothalamus.* This portion of the brain is a part of the diencephalon which lies between the cerebral hemispheres and the midbrain. The entire diencephalon, including the hypothalamus, will be studied more in detail in the next exercise when the internal parts of the brain are revealed by dissection.

Mammillary Bodies This part of the hypothalamus lies superior to the hypophysis. Although it appears as a single body on the sheep, it consists of a pair of rounded eminences just posterior to the infundibulum on the human. The mammillary bodies receive fibers from the olfactory areas of the brain and ascending pathways; fibers exit from it to the thalamus and other brain nuclei.

Hypophysis This structure, which is also called the *pituitary gland,* is attached to the base of the brain by a stalk, the **infundibulum.** It consists of two distinctly different parts: an anterior adenohypophysis and a posterior neurohypophysis. Only the neurohypophysis and infundibulum are considered to be part of the hypothalamus because they both have the same embryological origin as the brain. The adenohypophysis originates as an outpouching of the pharynx and is quite different histologically. A close functional relationship exists between the hypothalamus, hypophysis, and the autonomic nervous system.

Assignment:

Label figure 32.5.

The remainder of this exercise pertains to the labeling of figures 32.6 and 32.7 and does not pertain to sheep brain dissection. To continue the sheep brain study, proceed to Exercise 33.

Functional Localization of the Cerebrum

Extensive experimental studies on monkeys, apes, and humans have resulted in considerable knowledge of the functional areas of the cerebrum. Figure 32.6 shows the positions of these various centers. Before attempting to identify each of these areas be sure you know the boundaries (sulci and fissures) of the cerebral lobes (figure 32.2).

Somatomotor Area This area occupies the surface of the precentral gyrus of the frontal lobe. Electrical stimulation of this portion of the cerebral cortex in a conscious human results in movement of specific muscular groups.

Premotor Area The large area anterior to the somatomotor area is the premotor area. It exerts control over the motor area.

Somatosensory Area This area is located on the postcentral gyrus of the parietal lobe. It functions to localize very precisely those points on the body where sensations of light touch and pressure originate. It also assists in determining organ position. Other sensations such as aching pain, crude touch, warmth, and cold are localized by the thalamus rather than this area.

Motor Speech Area This area is located in the frontal lobe just above the lateral cerebral fissure, anterior to the somatomotor area. It exerts control over the muscles of the larynx and tongue that produce speech.

Visual Area This area is located on the occipital lobe. Note that only a small portion of it is seen on the lateral aspect; the greater portion of this area is on the medial surface of the cerebrum. On this surface it extends anteriorly to the parieto-occipital fissure, becoming narrower as it approaches the fissure. This area receives impulses from the retina via the thalamus. Destruction of this region causes blindness, although light and dark are still discernible.

Auditory Area This area receives nerve impulses from the cochlea of the inner ear via the thalamus. It is responsible for hearing and speech understanding. It is located on the superior temporal gyrus, which borders on the lateral cerebral fissure.

Olfactory Area This area is located on the medial surface of the temporal lobe. The recognition of various odors occurs here. Tumors in this area cause individuals to experience nonexistent odors of various kinds, both pleasant and unpleasant.

Association Areas Adjacent to the somatosensory, visual, and auditory areas are association areas that lend meaning to what is felt, seen, or heard. These areas are lined regions in figure 32.6.

Common Integrative Area The integration of information from the above three association areas and the olfactory and taste centers is achieved by a small area which is called the common integrative or *gnostic* area. This area is located on the **angular gyrus,** which is positioned approximately midway between the three association areas.

Assignment:

Label figure 32.6.

The Trigeminal Nerve

Although a detailed study of all cranial nerves is precluded in an elementary anatomy course, the trigeminal nerve has been singled out for a thorough study here. This fifth cranial nerve is the largest one that innervates the head and is of particular medical-dental significance.

Figure 32.7 illustrates the distribution of this nerve. You will note that it has branches that supply the teeth, tongue, gums, forehead, eyes, nose, and lips. The following description identifies the various ganglia and branches.

On the left margin of the diagram is the severed end of the nerve at a point where it emerges from the brain. Between this cut end and the three main branches is an enlarged portion, the **Gasserian ganglion.** This ganglion contains the nerve cell bodies of the sensory fibers in the nerve. The three branches that extend from this ganglion are the *ophthalmic, maxillary,* and *mandibular nerves.* Locate the bony portion of the skull in figure 32.7 through which these three branches pass.

Ophthalmic Nerve The ophthalmic nerve is the upper branch which passes out of the cranium through the *superior orbital fissure.* It has three branches which innervate the lacrimal gland, the upper eyelid, and the skin of the nose, forehead, and scalp. One of the branches also supplies parts of the eye such as the *cornea, iris,* and the *ciliary body* (muscle attached to the lens of the eye). Superior to the lower branch of the ophthalmic is the **ciliary ganglion,** which contains nerve cell bodies that are incorporated into reflexes controlling the ciliary body. The ophthalmic nerve and its branches are not involved in dental anesthesia.

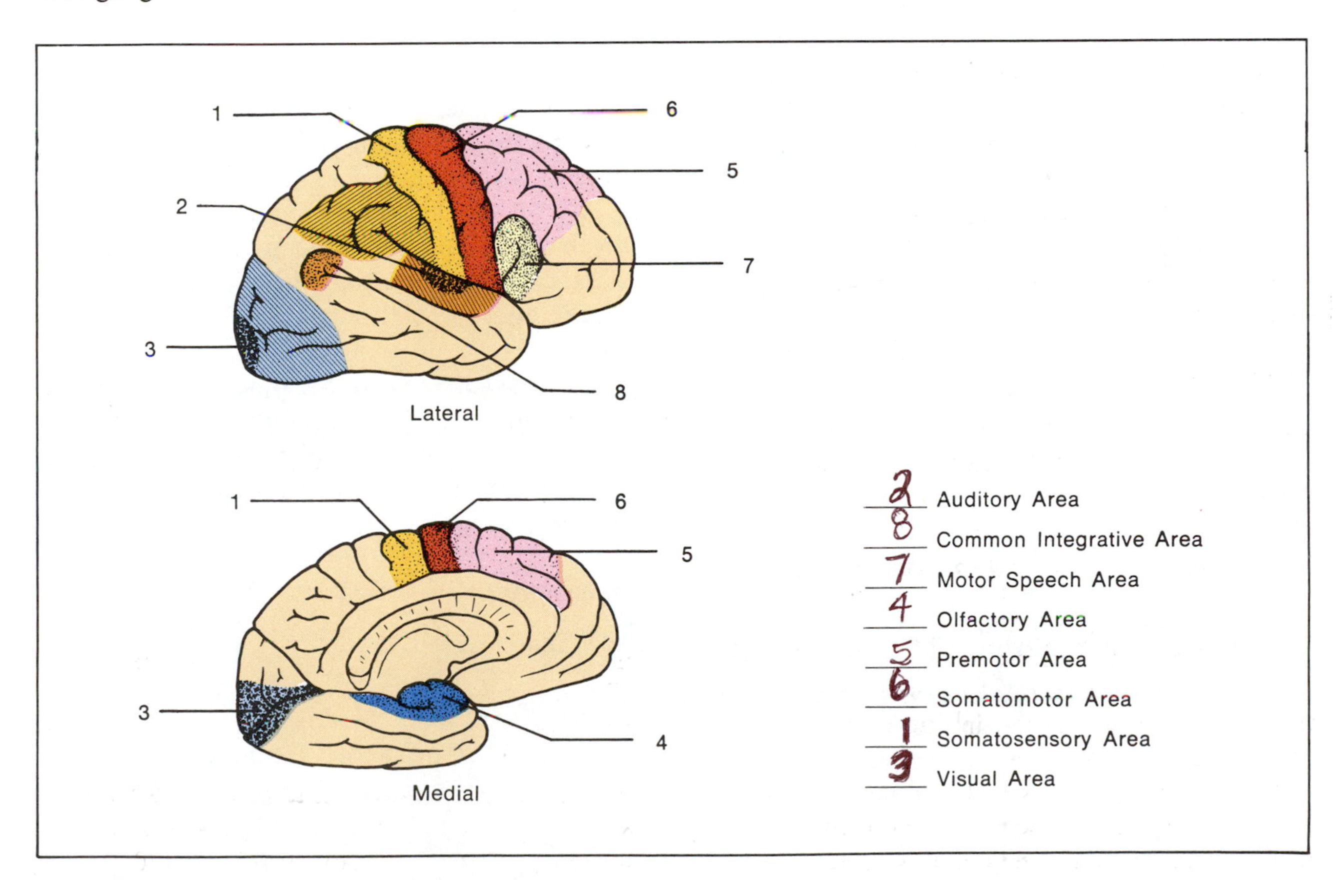

Figure 32.6 Functional areas of human brain.

Maxillary Nerve This nerve, also known as the *second division* of the trigeminal nerve, is a sensory nerve that provides innervation to the nose, upper lip, palate, maxillary sinus, and upper teeth. It is the branch that comes off just inferior to the ophthalmic nerve.

Note that the maxillary nerve has two short branches that extend downward to an oval body, the **sphenopalatine ganglion.** The major portion of the maxillary nerve becomes the **infraorbital nerve,** which passes through the **infraorbital nerve canal** of the maxilla. This canal is shown with dotted lines in figure 32.7. This nerve emerges from the maxilla through the **infraorbital foramen** to innervate the tissues of the nose and upper lip.

The upper teeth are innervated by three superior alveolar nerves. The **posterior superior alveolar nerve** is a branch of the maxillary nerve that enters the posterior surface of the upper jaw and innervates the molars. The **anterior superior alveolar nerve** is a branch of the infraorbital nerve that supplies the anterior teeth and bicuspids with nerve fibers. The anterior superior alveolar nerve forms a loop with the **middle superior alveolar nerve.** This latter nerve is shown clearly where a portion of the maxilla has been cut away. It contains fibers that pass to the bicuspids and first molar.

To desensitize all of the upper teeth on one side of the maxilla, a dentist can inject anesthetic near either the maxillary nerve or the sphenopalatine ganglion. Desensitization of the maxillary nerve is called a *second division nerve block.* This type of nerve block will affect the palate as well as the teeth because it involves fibers of the **anterior palatine nerve.** This latter nerve extends downward from the sphenopalatine ganglion to the soft tissues of the palate through the *greater palatine foramen.*

Mandibular Nerve The mandibular nerve, or *third division* of the trigeminal nerve, is the most inferior branch of this nerve. It innervates the teeth and gums of the mandible, the muscles of mastication, the anterior part of the tongue, the lower part of the face, and some skin areas on the side of the head.

Moving downward from the Gasserian ganglion, the first branch of the mandibular nerve that we encounter is the **nerve to the muscles of mastication.** This nerve has five branches with small identifying letters near their cut ends (**T** for temporalis, **M** for masseter, **EP** for external pterygoid, **B** for buccinator, and **IP** for internal pterygoid). The fibers in these nerves control the motor activities of these five muscles.

The largest branch of the mandibular nerve is the **inferior alveolar nerve,** which enters the ramus of the mandible through the **mandibular foramen** and supplies branches to all the teeth on one side of the mandible. At the mental foramen the inferior alveolar nerve becomes the **incisive nerve,** which innervates the anterior teeth. Emerging from the **mental foramen** is the **mental nerve,** which supplies the soft tissues of the chin and lower lip.

Desensitization of the mandibular teeth is generally achieved by injecting anesthetic near the mandibular foramen. This type of injection, called a *lower nerve block* or *third division nerve block,* is widely used in dentistry because the external and internal alveolar plates of the mandibular alveolar process are too dense to facilitate anesthesia by infiltration of the solid bone. The lower nerve block of one side of the mandible desensitizes all of the teeth on one side except for the anterior teeth. The fact that the anterior teeth receive nerve fibers from both the left and right incisive nerves prevents complete desensitization.

A branch of the mandibular nerve that emerges just above the inferior alveolar nerve and innervates the tongue is the **lingual nerve.** As in the case of the inferior alveolar nerve, it is sensory in function.

The **long buccal nerve** is another sensory branch of the mandibular nerve that innervates the buccal gum tissues of the molars and bicuspids. This nerve arises from a point which is just superior to the lingual nerve.

Assignment:

Label figure 32.7.

Laboratory Report

Complete the Laboratory Report for this exercise.

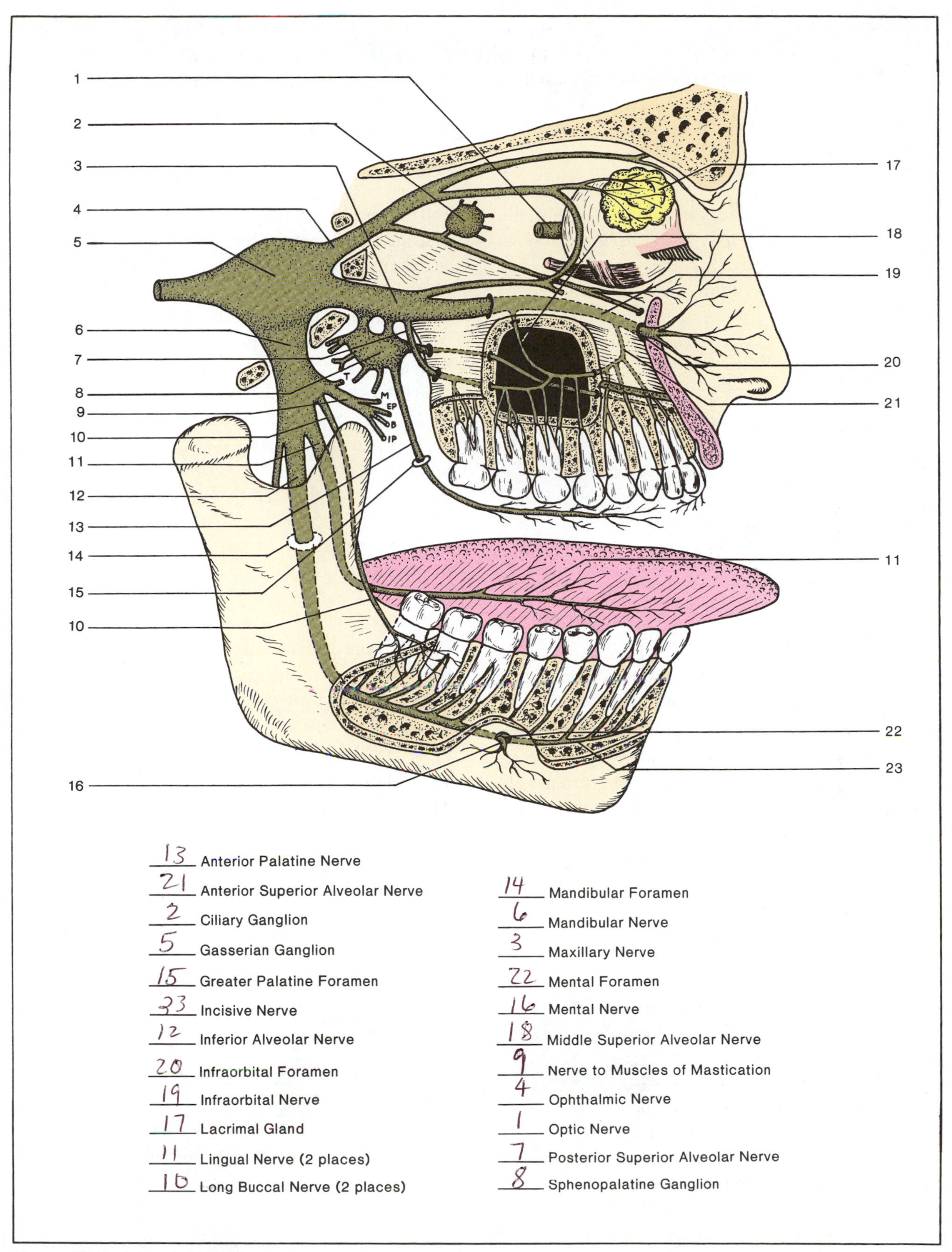

Figure 32.7 The trigeminal nerve.

33 Brain Anatomy: Internal

The interrelations of the various divisions of the brain can be seen only by studying sections such as those in figures 33.1 and 33.2. Bundles of interconnecting fibers unite the cerebral hemispheres, cerebellum, pons, and medulla to each other in a manner that enables the brain to function as an integrated whole. In this part of our brain study we will identify those important fiber tracts as well as brain cavities, meningeal spaces, and other related parts.

Materials:

- preserved sheep brains
- preserved human brains
- human brain models
- dissecting instruments
- long sharp knife
- dissecting tray

Midsagittal Section

With the preserved sheep brain resting ventral side down on a dissecting tray, place a long sharp meat-cutting knife (butcher knife) in the longitudinal cerebral fissure. Carefully slice through the tissue, cutting the brain into right and left halves on the midline. If only a scalpel is available, attempt to cut the tissue as smoothly as possible. Proceed to identify the structures on the sheep brain and their counterparts on the human brain as follows.

If the brain has been cut exactly on the midline, the only portion of the cerebrum that is cut is the **corpus callosum.** Locate this long band of white matter on the sheep brain by referring to figure 33.1. Also, compare your specimen with the human brain in figure 33.2. The corpus callosum is a commissural tract of fibers that connects the right and left cerebral hemispheres, correlating their functions. Note that the corpus callosum on the human brain is significantly thicker.

The Diencephalon

Between the corpus callosum of the cerebral hemisphere and the midbrain exists a section called the *diencephalon* or *interbrain.* This region includes the fornix, third ventricle, thalamus, intermediate mass, and hypothalamus.

Fornix Locate this structure on the sheep brain. This body contains fibers that are an integral part

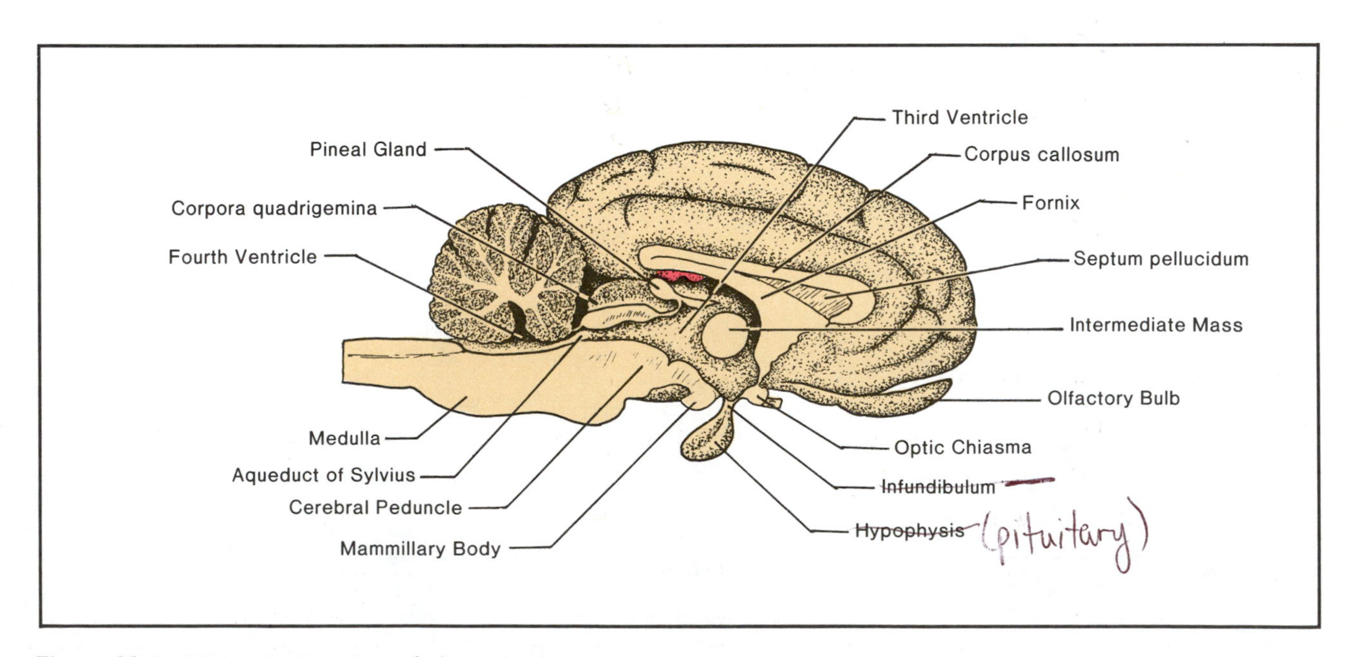

Figure 33.1 Midsagittal section of sheep brain.

of the olfactory mechanism of the brain, the *rhinencephalon.* Note how much larger, proportionately, it is on the sheep than on the human.

Intermediate Mass This oval area in the midsection of the diencephalon is the only part of the thalamus that can be seen in the midsagittal section. The **thalamus** is a large ovoid structure on the lateral walls of the third ventricle. It is an important relay station in which sensory pathways of the spinal cord and brain form synapses on their way to the cerebral cortex. The intermediate mass contains fibers that pass through the third ventricle, uniting both sides of the thalamus.

Hypothalamus This portion of the diencephalon extends about two centimeters up into the brain from the ventral surface. It has many nuclei that control various physiological functions of the body (temperature regulation, hypophysis control, coordination of the autonomic nervous system). The only part of the hypothalamus that is labeled in figure 33.2 is the left **mammillary body** (label 7).

Inferior and anterior to the mammillary body in figure 33.2 lie the **infundibulum** and **hypophysis.**

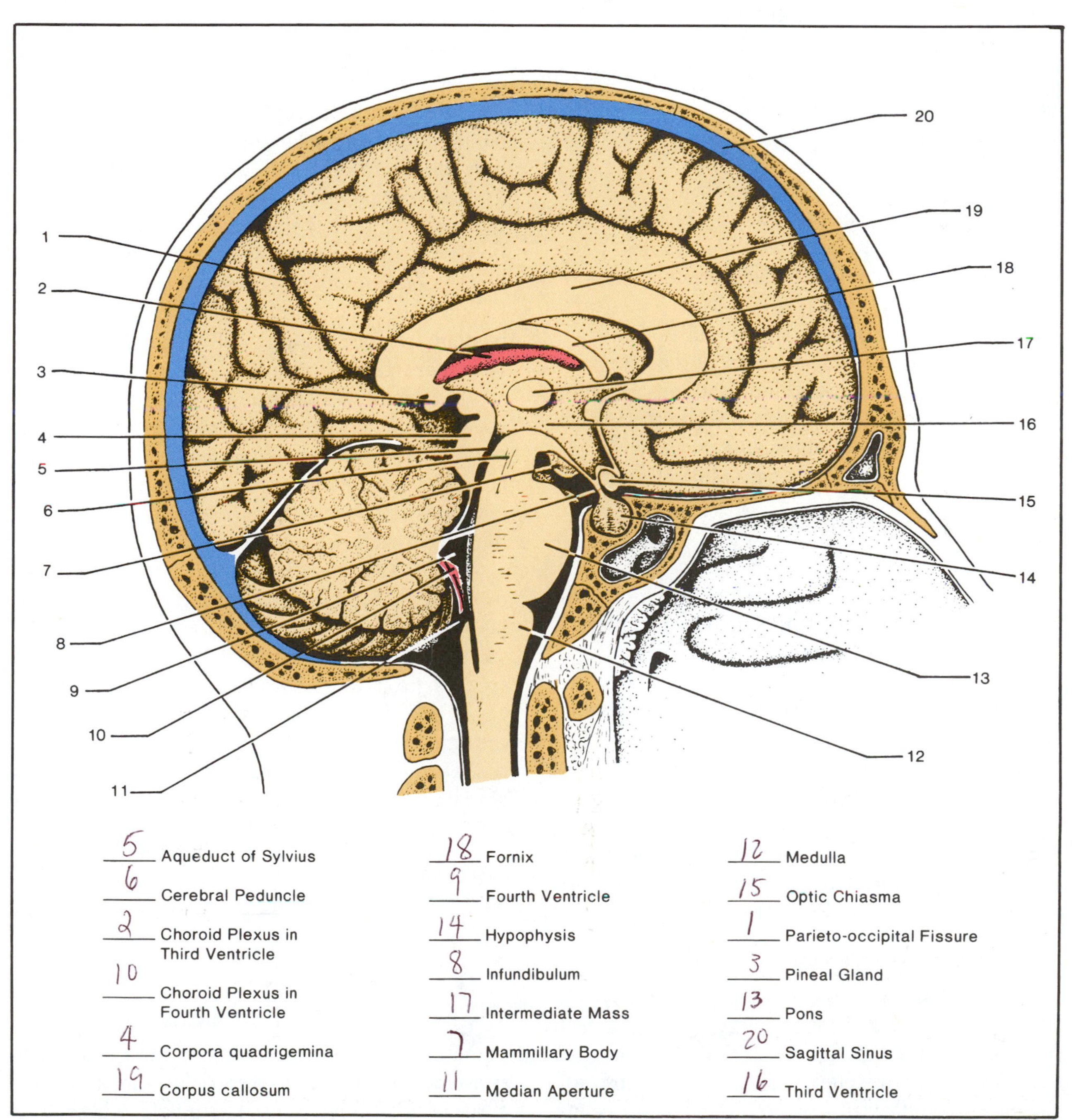

Figure 33.2 Midsagittal section of human brain.

Note how the latter lies encased in the bony recess of the *sella turcica.* The **optic chiasma** is the small round body just above the hypophysis, anterior to the infundibulum.

The Midbrain

Locate the following structures of the sheep's midbrain: **corpora quadrigemina, pineal gland, cerebral peduncle,** and **aqueduct of Sylvius.**

The *cerebral peduncles* consist of a pair of cylindrical bodies made up largely of ascending and descending fiber tracts that connect the cerebrum with the other three brain divisions.

The *aqueduct of Sylvius* is a duct that runs longitudinally through the midbrain, connecting the third and fourth ventricles.

The Cerebellum

Examine the cut surface of the cerebellum and note that gray matter exists near the outer surface. The cerebellum receives nerve impulses along cerebellar peduncles from motor and visual centers of the brain, the semicircular canals of the inner ears, and muscles of the body. Nerve impulses pass from this region to all the motor centers of the body wall to maintain posture, equilibrium, and muscle tonus. None of the activities of the cerebellum are of a conscious nature. It is significant that each hemisphere controls the muscles on its side of the body.

The Pons and Medulla

These two parts make up the greater portion of the brain stem. The pons, medulla, and midbrain constitute the entire *brain stem.* The pons is that portion of the brain stem between the midbrain and medulla. Note that it consists primarily of white matter (fiber tracts). Locate the pons and medulla on both the sheep and human brains.

Within the brain stem is a complex interlacement of nuclei and white fibers (gray and white matter) that constitutes the *reticular formation.* Nuclei of this formation generate a continuous flow of impulses to other parts of the brain to keep the cerebrum in a state of alert consciousness.

The Ventricles and Cerebrospinal Fluid

The brain has four cavities, or **ventricles,** which contain cerebrospinal fluid. There are two **lateral ventricles** situated in the lower medial portions of each cerebral hemisphere. Although they are not visible in a midsagittal section, the **septum pellucidum,** a thin membrane, may be seen which separates these two cavities. A frontal section of the brain, such as the one in figure 33.4, shows the position of the lateral ventricles (label 5). The **third** and **fourth ventricles,** however, are readily visible in figures 33.1 and 33.2. Note that the intermediate mass passes through the third ventricle. Within this ventricle is seen a **choroid plexus** (label 2, figure 33.2), which secretes cerebrospinal fluid. Each ventricle has its own choroid plexus.

Cerebrospinal Fluid Pathway The path of cerebrospinal fluid is indicated in figure 33.3. From the lateral ventricles the fluid enters the third ventricle through the **foramen of Monro.** Note in figure 33.3 that this foramen lies anterior to the fornix and choroid plexus of the third ventricle.

From the third ventricle the cerebrospinal fluid passes into the **aqueduct of Sylvius.** This duct conveys the cerebrospinal fluid to the fourth ventricle. The **choroid plexus of the fourth ventricle** lies on the posterior wall of this cavity.

From the fourth ventricle the cerebrospinal fluid exits into the subarachnoid space that surrounds the cerebellum through three foramina: one median aperture (foramen of Magendie), and two lateral apertures (foramina of Luschka).

In figure 33.3, the **median aperture** is the lower opening, which has two arrows leading from it. That part of the subarachnoid space that it empties into is the **cisterna cerebellomedullaris.** Since the **lateral apertures** are located on the lateral walls of the fourth ventricle, only one is shown in figure 33.3.

From the cisterna cerebellomedullaris the cerebrospinal fluid moves up into the **cisterna superior** (above the cerebellum) and finally into the subarachnoid space around the cerebral hemispheres. From the subarachnoid space this fluid escapes into the blood of the sagittal sinus through the **arachnoid granulations.** The cerebrospinal fluid passes down the subarachnoid space of the spinal cord on the posterior side and up along its anterior surface.

Assignment:

After identifying all the structures of the sheep brain, label figure 33.2.

Label figure 33.3.

Frontal Sections

To get a better understanding of the special relationships of the thalamus, ventricles, and other parts of the brain, it is necessary to study frontal sections of the brain. Once frontal sections of the sheep brain are studied, a frontal section of the human brain is relatively easy to understand. Two frontal sections of a sheep brain and one of the human brain are shown in figure 33.4.

Infundibular Section *(Sheep Brain)*

Place a whole sheep brain on a dissecting tray with the dorsal side down. Cut a frontal section with a long knife by starting at the point of the infundibulum. Cut through perpendicularly to the tray and to the longitudinal axis of the brain. Then, make a second cut parallel to the first cut ¼ inch further back. The latter cut should be through the center of the hypophysis.

Place this slice, *with the first cut upward,* on a piece of paper toweling for study. Illustration A, figure 33.4, is of this surface. Proceed to identify the following structures. You may need to refer back to figure 33.1 for reference.

Note the pattern of the **gray matter** of the cerebral cortex. Force a probe down into a sulcus. What meninx do you break through as the probe moves inward? What is the average thickness of the gray matter in the sulci of the dorsal part of the brain?

Locate the two triangular **lateral ventricles.** Probe into one of the ventricles to see if you can locate the **choroid plexus.** What is its function? Identify the **corpus callosum** and **fornix.**

Identify the **third ventricle,** which is situated on the median line under the fornix. Also, identify the **thalamus** and **intermediate mass.** Does the thalamus appear to consist of white or gray matter? What is the function of the white matter between the thalamus and cerebral cortex?

Hypophyseal Section *(Sheep Brain)*

Produce a section of the brain by cutting another ¼-inch slice off the posterior portion of the brain. Place this slice, with its *anterior face upward,* on paper toweling. This section through the hypophysis should look like illustration B, figure 33.4. Identify the structures in illustration B that are labeled.

Infundibular Section *(Human Brain)*

Illustration C, figure 33.4, reveals the close similarities of sheep and human brains, yet differences are apparent. To complete this study of the brain, identify the following structures.

Locate first the two triangular **lateral ventricles** and the **third ventricle.** Note how the **longitudinal cerebral fissure** extends downward to the **corpus callosum** which forms the upper wall of the lateral ventricles. Although no label exists for the **septum pellucidum,** can you find it? Identify the two masses of gray matter just below the septum pellucidum that constitute the **fornix.**

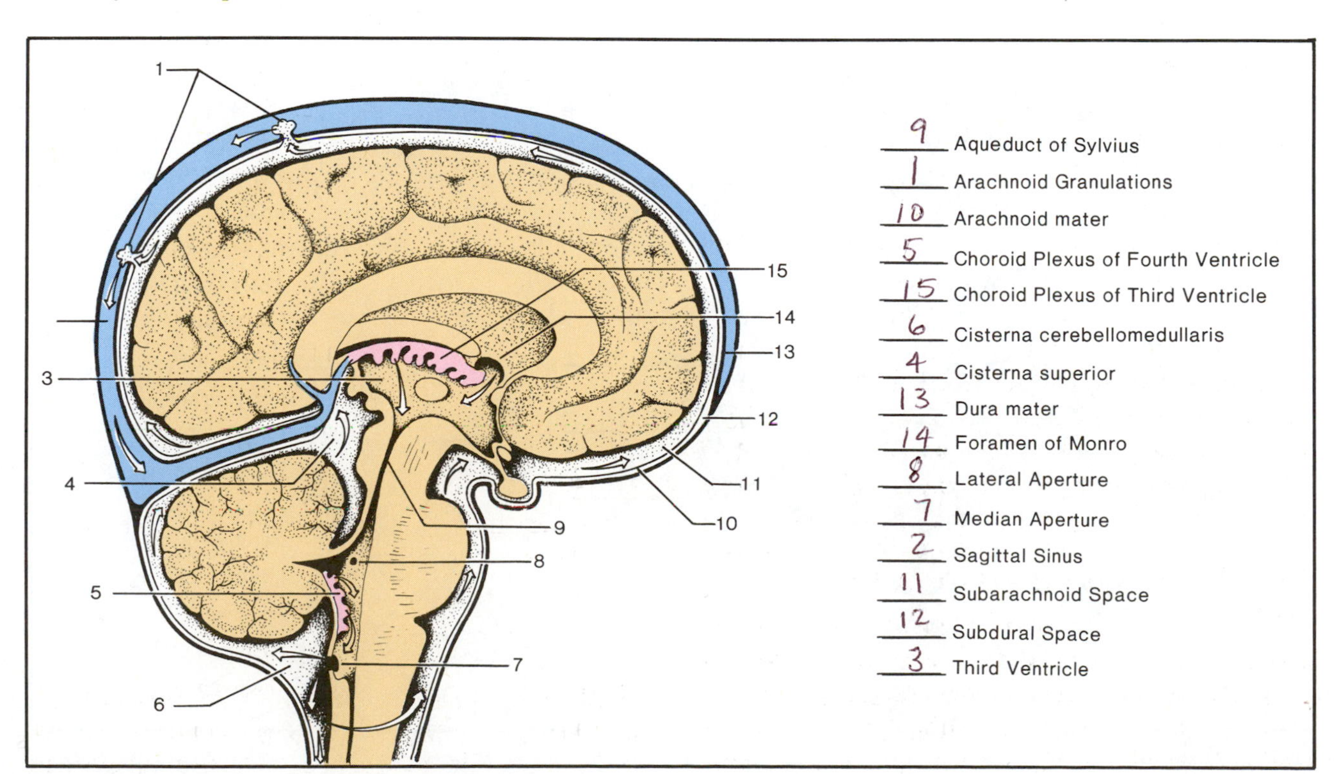

Figure 33.3 Origin and circulation of cerebrospinal fluid.

Identify the areas on each side of the third ventricle that make up the nuclei of the **hypothalamus.** Approximately eight separate masses of gray matter are seen here. What physiological functions are regulated by these nuclei?

Locate the two large areas on the walls of the lateral ventricles that constitute the **thalamus.** Note, also, how these two portions of the thalamus are united by the **intermediate mass** that passes through the third ventricle.

Labels 12, 13, 14, and 15 are masses of gray matter known collectively as the **basal ganglia.** The largest basal ganglia are the **putamen** and **globus pallidus** (labels 12 and 13). The latter is medial to the putamen. Together the putamen and globus pallidus form a triangular mass, the **lentiform nucleus.** These two ganglia exert a steadying effect on voluntary muscular movements.

The **caudate nuclei** are smaller basal ganglia located in the walls of the lateral ventricles superior to the thalamus. These nuclei have something to do with muscular coordination since surgically produced lesions in these centers can correct certain kinds of palsy.

Although some anatomists prefer to designate the thalamus and hypothalamus as basal ganglia, functionally, they differ considerably.

Assignment:

Label figure 33.4.

Laboratory Report

Complete the Laboratory Report for this exercise.

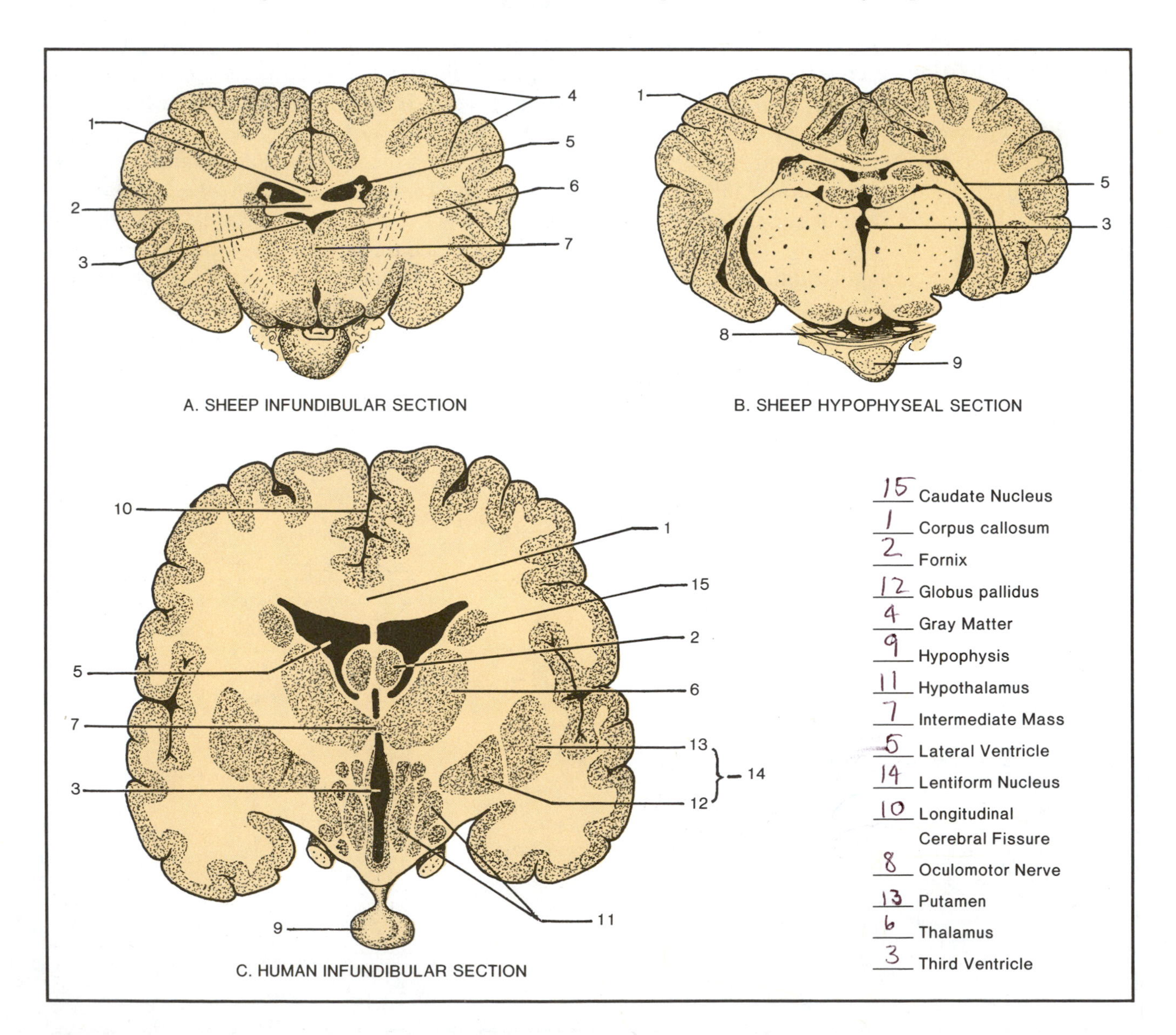

Figure 33.4 Frontal sections of sheep and human brains.

Anatomy of the Eye 34

In preparation for the study of eye function in the next exercise, our focus here will be on the anatomy of the eye through dissection, ophthalmoscopy, and microscopy. Before performing these studies, however, label figures 34.1, 34.2, and 34.3 that pertain to the lacrimal system, internal anatomy, and the extrinsic muscles.

The Lacrimal Apparatus

Each eye lies protected in a bony recess of the skull and is covered externally by the **eyelids.** Lining the eyelids and covering the exposed surface of the eyeball is a thin membrane, the **conjunctiva.** The conjunctival surfaces are kept constantly moist by protective secretions of the **lacrimal gland,** which lies between the upper eyelid and the eyeball.

Figure 34.1 reveals the external anatomy of the right eye, with particular emphasis on the lacrimal apparatus. Note that the lacrimal gland consists of two portions: a large **superior lacrimal gland** and a smaller **inferior lacrimal gland.** Secretions from these glands pass through the conjunctiva of the eyelid via six to twelve small ducts.

The flow of fluid from these ducts moves across the eyeball to the *medial canthus* (angle) of the eye, where the fluid enters two small orifices, the **lacrimal puncta.** These openings lead into the **superior** and **inferior lacrimal ducts:** the superior one is in the upper eyelid and the inferior one is in the lower eyelid. These two ducts empty directly into an enlarged cavity, the **lacrimal sac,** which, in turn, is drained by the **nasolacrimal duct.** The tears finally exit into the nasal cavity through the end of the nasolacrimal duct.

Two other structures, the caruncula and the plica semilunaris, are seen near the puncta in the medial canthus of the eye. The **caruncula** is a small red conical body which consists of a mound of skin containing sebaceous and sweat glands and a few small hairs. It is this structure that produces a whitish secretion which constantly collects in this region. Lateral to the caruncula is a curved fold of conjunctiva, the **plica semilunaris.** This structure contains some smooth muscle fibers; in the cat and many other animals it is more highly developed and is often referred to as the "third eyelid."

Assignment:

Label figure 34.1.

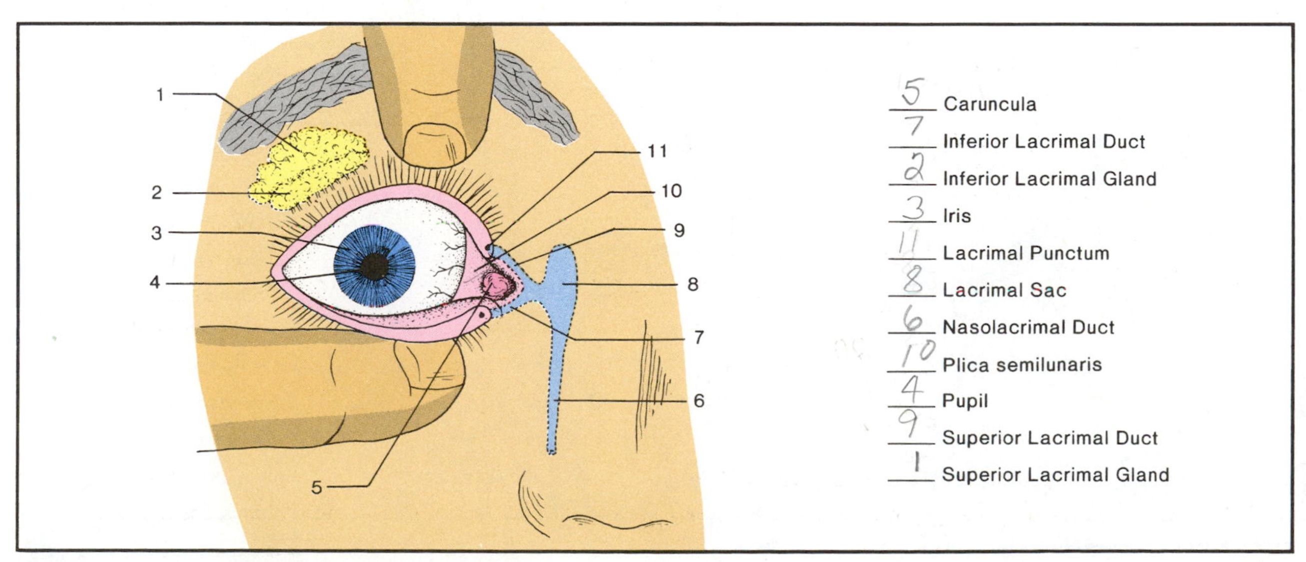

Figure 34.1 The lacrimal apparatus of the eye.

Internal Anatomy

Figure 34.2 is a horizontal section of the right eye as seen looking down upon it. Note that the wall of the eyeball consists of three layers: an outer **scleroid coat,** a middle **choroid coat,** and an inner **retina.** The continuity of the surface of the retina is interrupted only by two structures, the optic nerve and the fovea centralis.

The **optic nerve** (label 3) contains fibers leading from the rods and cones of the retina to the brain. That point on the retina where the optic nerve makes its entrance is lacking in rods and cones. The absence of light receptors at this spot makes it insensitive to light; thus, it is called the **blind spot.** It is also known as the **optic disk.**

Note that the optic nerve is wrapped in a sheath, the **dura mater** (an extension of the dura that surrounds the brain). To the right of the blind spot is a pit, the **fovea centralis,** which is the center of a round yellow spot, the **macula lutea.** The macula, which is only a half a millimeter in diameter, is composed entirely of cones and is that part of the eye where all critical vision occurs.

Light entering the eye is focused on the retina by the **lens,** an elliptical crystalline clear structure suspended in the eye by a **suspensory ligament.** Attached to this ligament is the **ciliary body,** a circular smooth muscle that can change the shape of the lens to focus on objects at different distances.

In front of the lens is a cavity that contains a watery fluid, the **aqueous humor.** Between the lens and the retina is a larger cavity which contains a more viscous, jellylike substance, the **vitreous body.** Immediately in front of the lens lies the circular colored portion of the eye, the **iris.** It consists of circular and radiating muscle fibers that can change the size of the **pupil** of the eye. The iris regulates the amount of light that enters the eye through the pupil.

Covering the anterior portion of the eye is the **cornea,** a clear, transparent structure that is an extension of the scleroid coat. It acts as a window to the eye, allowing light to enter.

Aqueous humor is constantly being renewed in the eye. The enlarged inset of the lens-iris area in figure 34.2 illustrates where the aqueous humor is produced (point A) and where it is reabsorbed (point B). At point A, which is in the **posterior chamber,** aqueous humor is produced by capillaries of the ciliary body. As this fluid is produced, it

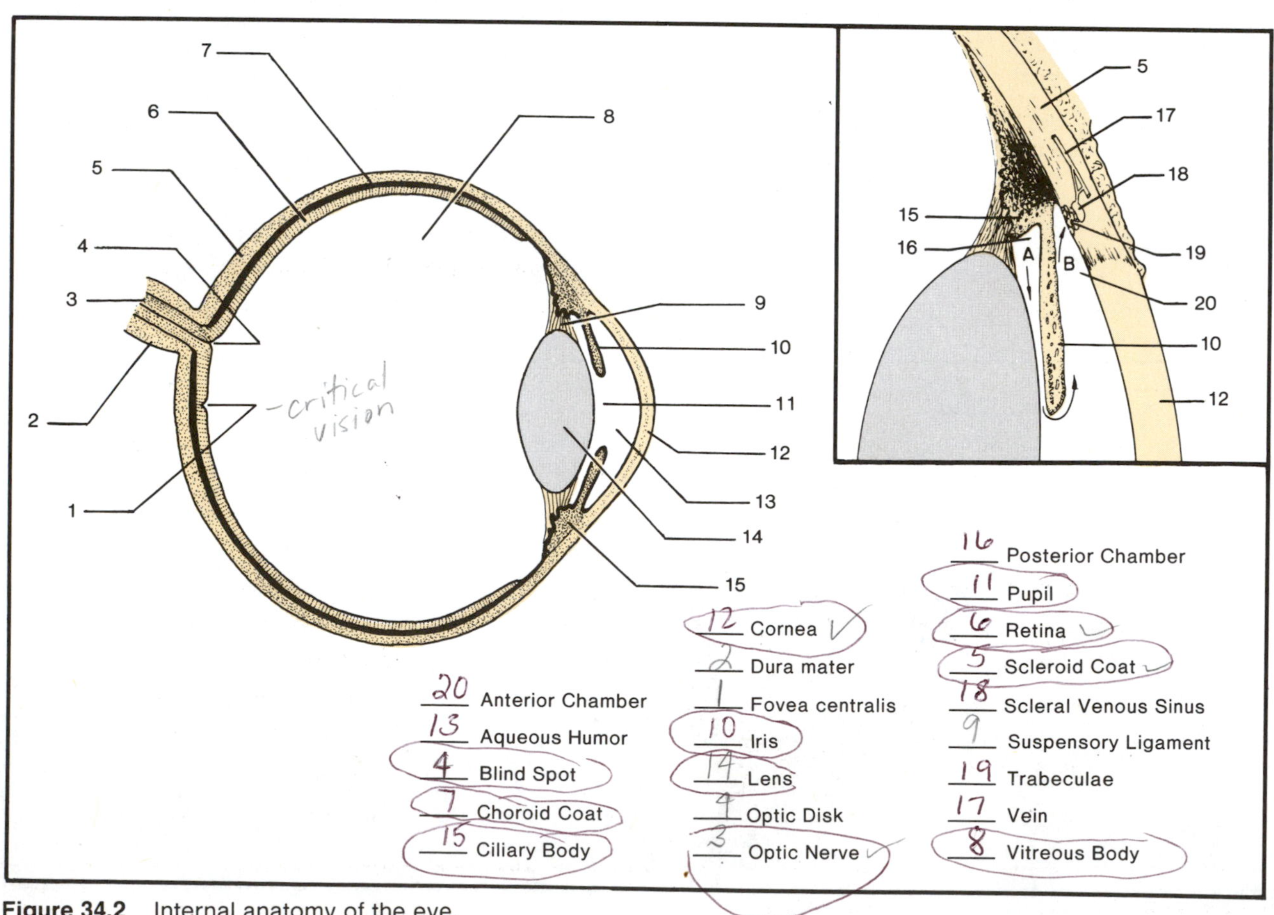

Figure 34.2 Internal anatomy of the eye.

passes through the pupil into the **anterior chamber** and is reabsorbed into minute spaces called **trabeculae** at the juncture of the cornea and the iris (point B). From the trabeculae the fluid passes to the **scleral venous sinus** (canal of Schlemm), which empties into the venous system.

Assignment:

Label figure 34.2.

The Ocular Muscles

The ability of each eye to be moved in all directions within its orbit is controlled by six *extrinsic* ocular muscles. Movement of the upper eyelid is controlled by a single muscle, the **superior levator palpebrae.**

Figure 34.3, which is a lateral view of the right eye, reveals the insertions of all seven of these muscles. The origins of these muscles are on various bony formations at the back of the eye orbit.

The muscle on the side of the eye that has a portion of it removed is the **lateral rectus.** Opposing it on the other side of the eye is the **medial rectus.** The muscle that inserts on top of the eye is the **superior rectus;** its antagonist on the bottom of the eye is the **inferior rectus.** Just above the stub of the lateral rectus is seen the insertion of the **superior oblique** which passes through a cartilaginous loop, the **trochlea.** Opposing the superior oblique is the **inferior oblique,** of which only a portion of the insertion can be seen.

Assignment:

Label figure 34.3.

Beef Eye Dissection

The nature of the tissues of the eye can best be studied by actual dissection of an eye. A beef eye will be used for this study. Figure 34.4 illustrates various steps that will be followed.

Materials:

beef eye, preferably fresh
dissecting instruments
dissecting tray

1. Examine the external surfaces of the eyeball. Identify the **optic nerve** and look for remnants of **extrinsic muscles** on the sides of the eyeball.
2. Note the shape of the **pupil.** Is it round or elliptical? Identify the **cornea.**
3. Holding the eyeball as shown in illustration 1, figure 34.4, make an incision into the scleroid coat with a sharp scalpel about one-quarter of an inch away from the cornea. Note how difficult it is to penetrate. Insert scissors into the incision and carefully cut all the way around the cornea, *taking care not to squeeze the fluid out of the eye.* Refer to illustration 2, figure 34.4.
4. Gently lift the front portion off the eye and place it on the tray with the inner surface upward. The lens usually remains attached to the vitreous body, as shown in illustration 3. If the eye is preserved (not fresh), the lens may remain in the anterior part of the eye.
5. Examine the inner surface of the anterior portion and identify the thickened, black circular **ciliary body.** What function does the black pigment perform?

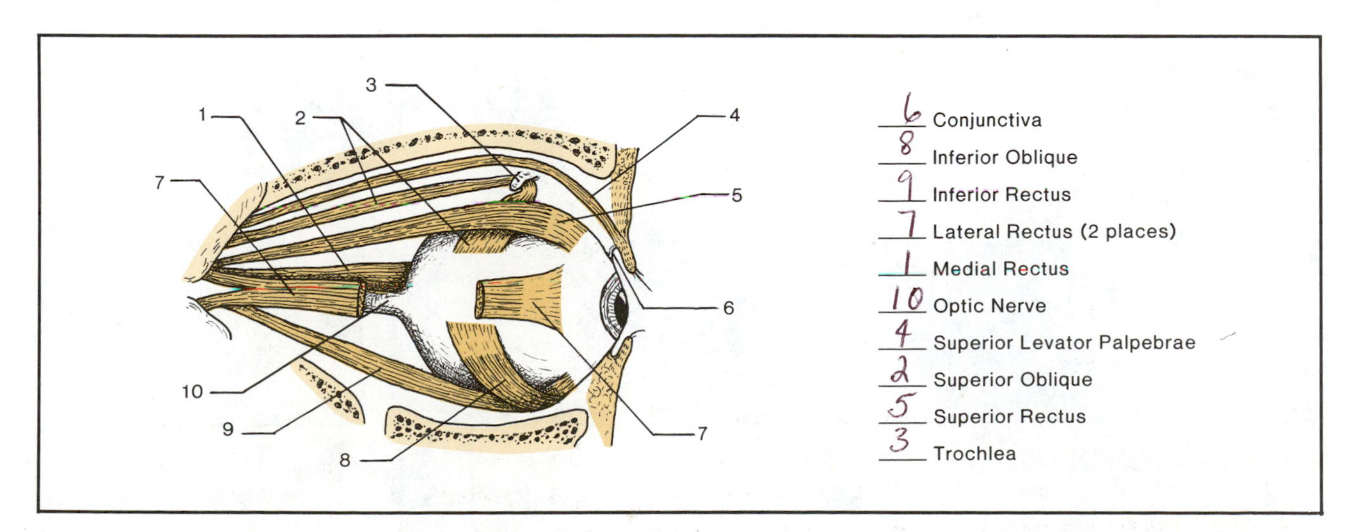

Figure 34.3 The extrinsic muscles of the eye.

6. Study the **iris** carefully. Can you distinguish **circular** and **radial** muscle fibers?
7. Is there any **aqueous humor** left between the cornea and iris? Compare its consistency with the **vitreous body.**
8. With a dissecting needle separate the lens from the vitreous body by working the needle around its perimeter as shown in illustration 3. Hold the lens up by its edge with a forceps and look through it at some distant object. What is unusual about the image?
9. Place the lens on printed matter. Are the letters magnified?
10. Compare the consistency of the lens at its center and near its circumference. Use a probe or forceps. Do you detect any differences?
11. Locate the **retina** at the back of the eye. It is a thin colorless inner coat which separates easily from the pigmented **choroid coat.** Now, locate the **blind spot.** This is the area where the retina is attached to the back of the eye.
12. Note the iridescent nature of a portion of the choroid coat. This reflective surface is the **tapetum lucidum.** It causes the eyes of animals to reflect light at night and appears to enhance night vision by reflecting some light back into the retina.
13. Answer all questions on the Laboratory Report that pertain to this dissection.

Ophthalmoscopy

Routine physical examinations invariably involve an examination of the **fundus,** or interior, of the eye. A careful examination of the fundus provides valid information about the general health of the patient. The retina of the eye is the only part of the body where relatively large blood vessels may be inspected without surgical intervention.

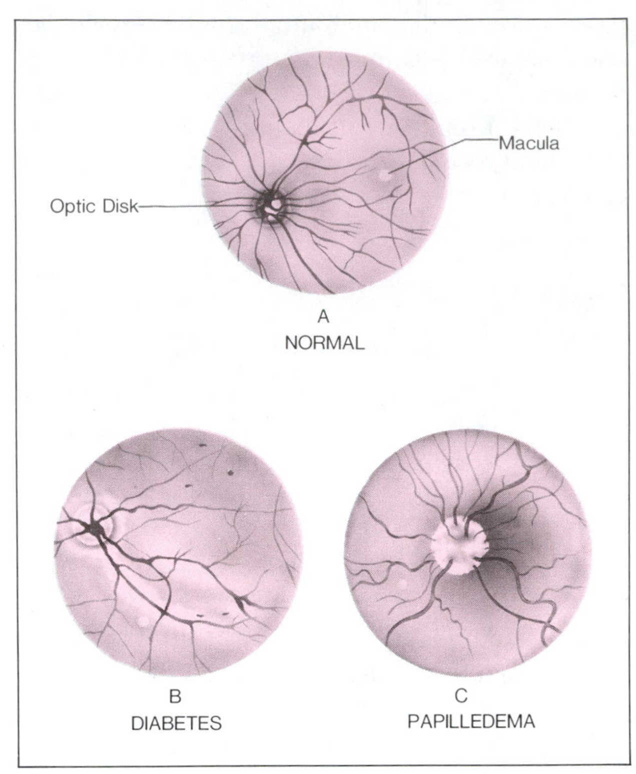

Figure 34.5 Normal and abnormal retinas as seen through an ophthalmoscope.

Figure 34.5 illustrates what one might expect to see in a normal eye (A); the eye of a diabetic (B); and the eye of one who has hypertension, a brain tumor, or meningitis (C). In *diabetic retinopathy* the peripheral vessels are generally irreg-

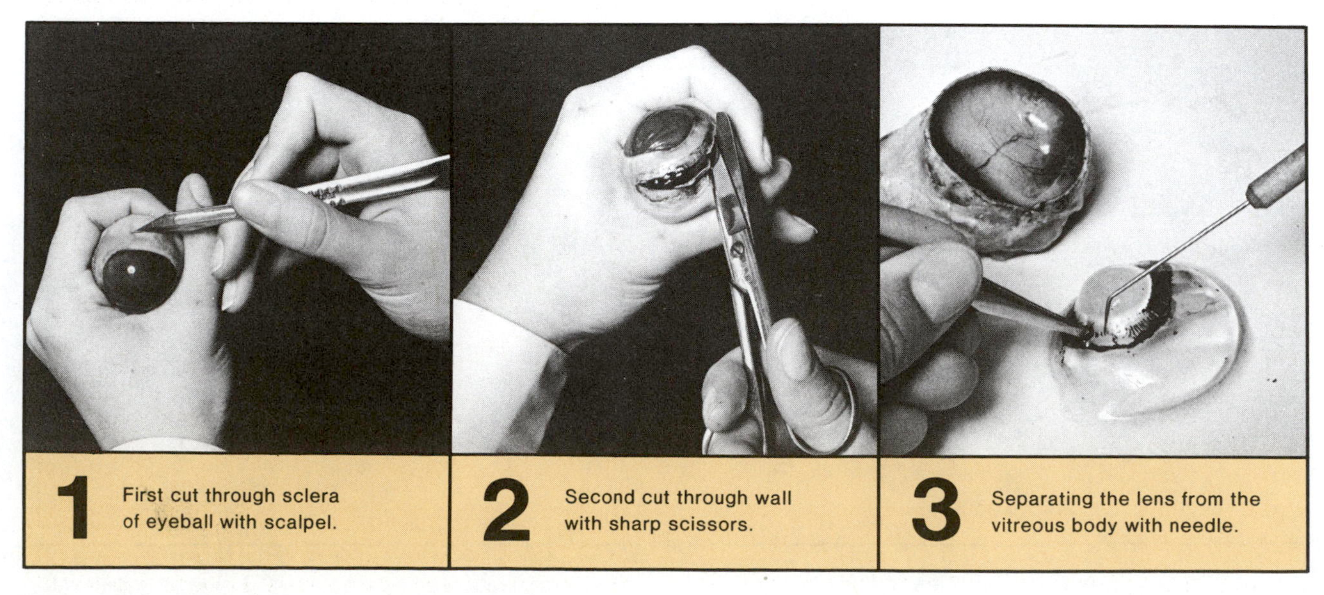

Figure 34.4 Three steps in beef eye dissection.

ular, fewer in number, and often show small hemorrhages. In *papilledema,* or choked disk, the optic disk is blurred, enlarged, and often hemorrhagic; the veins are also considerably enlarged. These are only a few examples of pathology showing up in the fundus.

The instrument that one uses for examining the fundus is called an **ophthalmoscope.** In this exercise you will have an opportunity to examine the interior of the eyes of your laboratory partner. Your partner, in turn, will examine your eyes. It will be necessary to make this examination in a darkened room. Before this examination can be made, however, it is essential that you understand the operation of the ophthalmoscope.

The Ophthalmoscope

Figure 34.6 reveals the construction of a Welch Allyn ophthalmoscope. Within its handle are two C-size batteries that power an illuminating bulb that is located in the viewing head. Note that at the top of the handle there is a rheostat control for regulating the intensity of the light source. Only after depressing the lock button can this control be rotated.

The head of the instrument has two rotatable notched disks: a large lens selection disk and a smaller aperture selection disk. A viewing aperture is located at the upper end of the head.

While looking through the viewing aperture, the examiner is able to select one of twenty-three small lenses by rotating the **lens selection disk** with the index finger held as shown in figure 34.7. Twelve of the lenses on this disk are positive (convex) and eleven of them are negative (concave). Numbers that appear in the **diopter window** indicate which lens is in place. The positive lenses are represented as black numbers; negative lens numbers are red. A black "0" in the window indicates that no lens is in place.

If the eyes of both subject and examiner are normal (*emmetropic*), no lens is needed. If the eye of the subject or examiner is farsighted (*hypermetropic*), the positive lenses will be selected. Nearsighted (*myopic*) eyes require the use of negative lenses. The degree of myopia or hypermetropia is indicated by the magnitude of the number. Since the eyes of both the examiner and subject affect the lens selection, the selection becomes entirely empirical.

The **aperture selection disk** enables the examiner to change the character of the light beam that

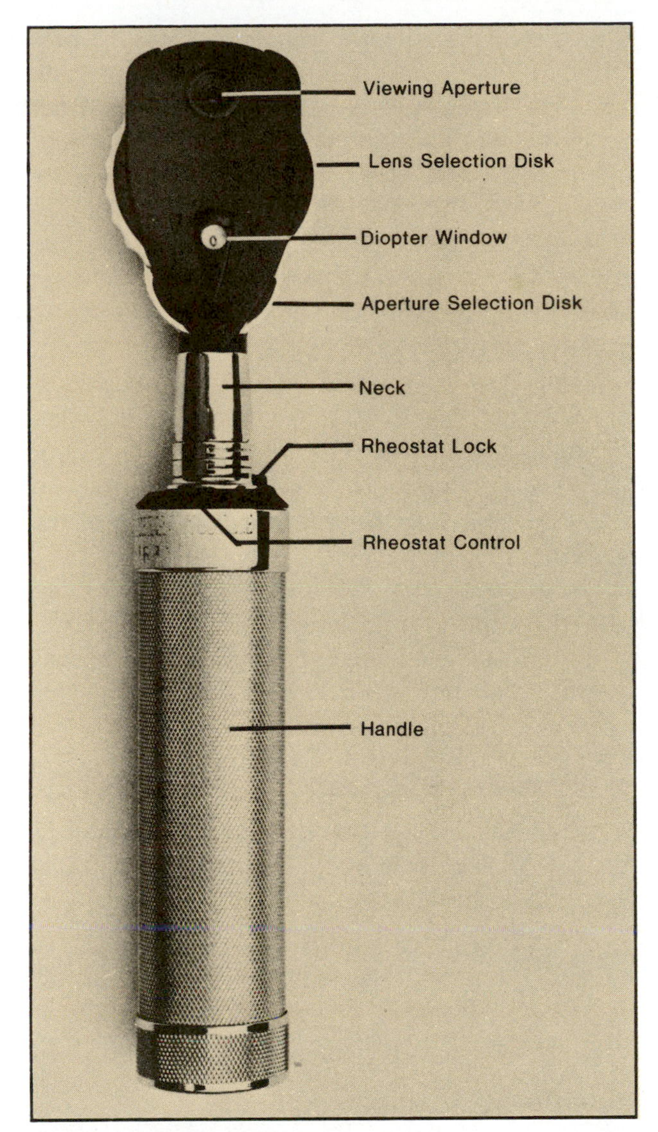

Figure 34.6 The ophthalmoscope.

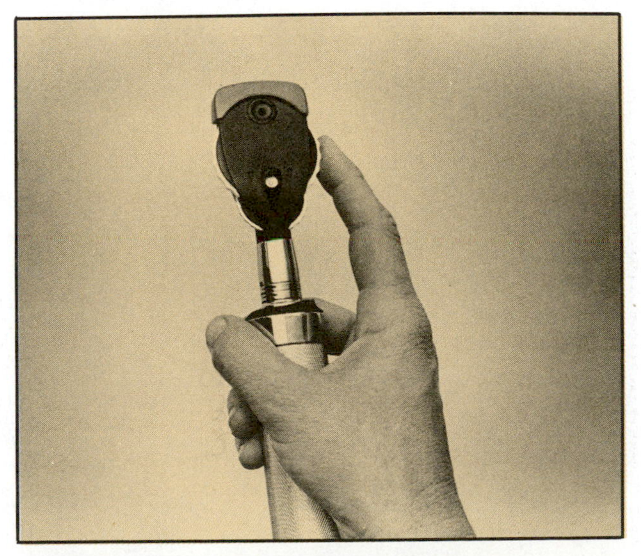

Figure 34.7 Viewing lenses of ophthalmoscope are rotated into position by moving the lens selection disk with index finger.

is projected into the eye. The light may be projected as a grid, straight line, white spot, or green spot. **The green spot is most frequently used** because it is less irritating to the eye of the subject, and it causes the blood vessels to show up more clearly.

Using an ophthalmoscope for the first time seems difficult for some and relatively easy for others. The examiner with emmetropic eyes has a definite advantage over one with myopia, astigmatism, or hypermetropia. Even for individuals with normal eyes, there is the troublesome reflection of light from the retina back into the examiner's eye. The best way to minimize this problem is to **direct the light beam toward the edge of the pupil** rather than through the center.

A precautionary statement *must* be heeded concerning light exposure: **Avoid subjecting an eye to more than one-minute light exposure.** After one-minute exposure, allow several minutes of rest for recovery of the retina.

Procedure

Prior to using the ophthalmoscope in the darkened room, it will be necessary for you to become completely familiar with its mechanics.

Materials:

ophthalmoscope
metric tape or ruler

Desk Study of Ophthalmoscope

Turn on the light source by depressing the lock button and rotating the rheostat control. Observe how the light intensity can be varied from dim to very bright. Rotate the aperture selection disk as you hold the ophthalmoscope about two inches away from your desktop. Keep the instrument parallel to the desk surface. How many different light patterns are there? Select the large white spot for the next phase of this study.

Rotate the lens selection disk until a black "5" appears in the diopter window. While peering through the viewing window at the print on this page, bring the letters into sharp focus. Have your lab partner measure with a metric tape the distance in millimeters between the ophthalmoscope and the printed page when the lettering is in sharp focus. Record this distance on the Laboratory Report. Do the same thing for the 10D, 20D, and 40D lenses, recording all measurements.

Now, set the lens selection disk on "0" and the light source on the green spot. The instrument is properly set now for the beginning of an eye examination. Read over the following procedure *in its entirety* before entering the darkened room.

The Examination

Figures 34.8 and 34.9 illustrate how the ophthalmoscope is held in different viewing situations. Note that the examiner uses the right hand when examining the right eye and the left when examining the left eye. Proceed as follows:

1. With the "0" in the diopter window and the light turned on, grasp the ophthalmoscope, as shown in figure 34.8. Note that the index finger rests on the lens selection disk.

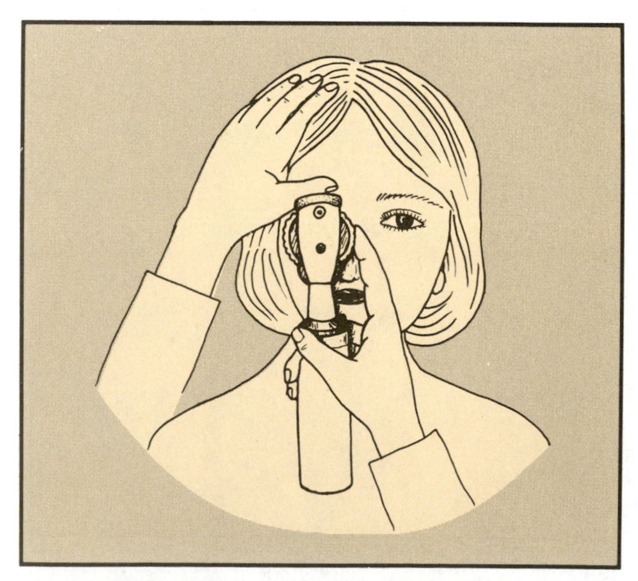

Figure 34.8 When examining the right eye, the ophthalmoscope is held with the right hand.

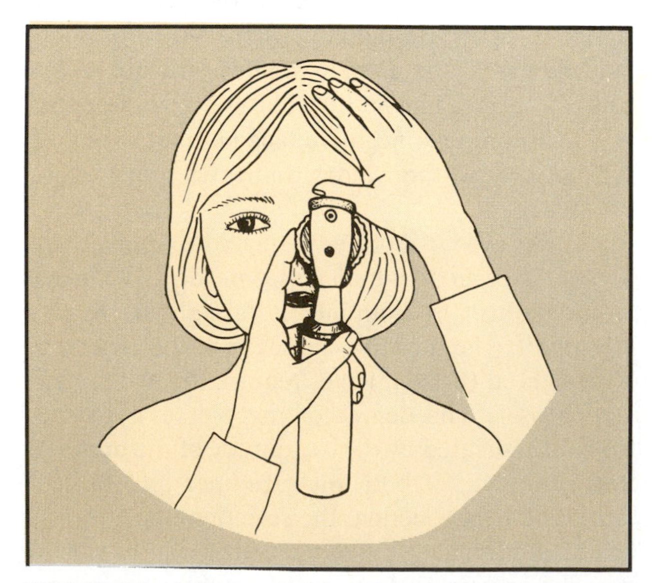

Figure 34.9 When examining the left eye, the ophthalmoscope is held with the left hand.

2. Place the viewing aperture in front of your right eye and steady the ophthalmoscope by resting the top of it against your eyebrow. See figure 34.10.

   Start viewing the subject's right eye at a distance of about 12 inches. Keep your subject on your right and instruct your subject to **look straight ahead at a fixed object** at eye level.
3. Direct the beam of light into the pupil and examine the lens and vitreous body. As you look through the pupil, a red **reflex** will be seen. If the image is not sharp, adjust the focus with the lens selection disk.
4. While keeping the pupil in focus, move in to within two inches of the subject, as illustrated in figure 34.11. The red reflex should become more pronounced. **Try to direct the light beam toward the edge of the pupil** rather than in the center to minimize reflection from the retina.
5. Search out the **optic disk** and adjust the focus, as necessary, to produce a sharp image. Observe how blood vessels radiate out from the optic disk. Follow one or more of them to the periphery.

   Locate the **macula.** It is situated about two disk diameters *laterally* from the optic disk. Note that it lacks blood vessels. It is easiest to observe when the subject is asked to look directly into the light, a position that **must be limited to one second only.**
6. Examine the entire retinal surface. Look for any irregularities, depressions, or protrusions from its surface. It is not unusual to see a roundish elevation, not unlike a pimple, on its surface.

   To examine the periphery, instruct the subject to (1) **look up** for examination of the superior retina, (2) **look down** for examination of the inferior retina, and (3) **look laterally and medially** for those respective areas.
7. **Caution:** Remember to limit the examinations to **one minute!** After this time limit is reached, examine the left eye, using the left hand to hold the ophthalmoscope and reversing the entire procedure.
8. Switch roles with your laboratory partner. When you have examined each other's eyes, examine the eyes of other students. Report any unusual conditions to your instructor.

Histological Studies

Most prepared slides available for laboratory study are made from either the rabbit or the monkey. Both animals are suitable. Proceed as follows to identify structures depicted in the photomicrographs of figures HA–15 and HA–16 of the Histology Atlas.

Materials:

prepared slides:

monkey eye, x.s. (H10.635, H10.64, or H10.76)

rabbit eye, sagittal section (H10.61 or 10.62)

The Cornea With the lowest powered objective on your microscope scan a cross section of the eye of

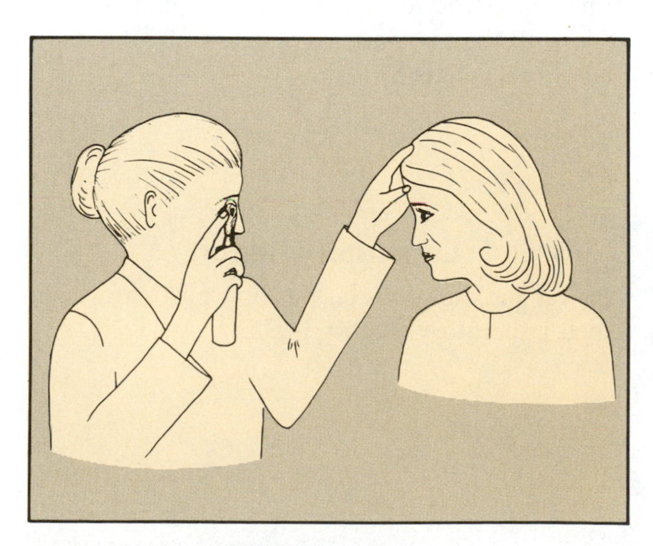

Figure 34.10 When viewing the lens and vitreous body, use this position, relative to subject.

Figure 34.11 When examining the retinal surfaces, move in close as illustrated here.

a rabbit or monkey until you find the cornea. Refer to illustrations A and B, figure HA–15. Note that the cornea consists of three layers: the epithelium, stroma, and endothelium. For fine detail, as in illustration B, use high-dry.

The **corneal epithelium** is the outer portion that consists of stratified squamous. While the outermost cells of this layer are flattened squamous cells, the basal cells are columnar.

The **corneal endothelium** is the innermost layer of cells, which is adjacent to the aqueous humor. This thin layer consists of low cuboidal cells.

The **corneal stroma** *(substantia propria)* comprises nine-tenths of the thickness of the cornea. It consists of collagenous fibrils, fibroblasts, and cementing substance. The fibrils are arranged in lamellae that run parallel to the corneal surface. The components of this stroma are held together by a mucopolysaccharide cement. The chemical structure and fibril arrangement contribute to the transparency of the cornea.

The Lens Note that the lens is enclosed in a homogenous elastic **capsule** to which the **suspensory ligament** is attached. The lens is formed from epithelial cells that become elongated to a fibrous shape (lens fibers), losing most of their organelles and nuclei. All that remains in a mature lens fiber are a few microtubules and clumps of free ribosomes. These cells are not entirely inert, and they persist throughout life.

The Ciliary Muscle This ring of smooth muscle tissue is a part of the body wall of the eye. Identify the **ciliary processes,** which form ridges on the ciliary muscle. The latter provide an anchor for the suspensory ligaments.

Body Wall of Eye Examine a section through the body wall at the back of the eye. Identify the sclera, choroid coat, and retina. Note that the **sclera** consists of closely packed collagenous fibers, elastic fibers, and fibroblasts.

Note, also, the large amount of pigment in the **choroid coat** that is produced by **melanocytes.** This is the layer that supplies nourishment to the retina and scleroid coat. Look for blood vessels.

The **retina** is the inner photosensitive layer of the body wall.

Retinal Layers The retina is composed of five main classes of neurons: the photoreceptors (rods and cones), bipolar cells, ganglion cells, horizontal cells, and amacrine cells. The first three form a direct pathway from the retina to the brain. The horizontal and amacrine cells form laterally directed pathways that modify and control the message being passed along the direct pathway.

Study a section of the retina and identify the various layers. Refer to figure HA–16 for reference. Any Turtox slide of monkey eye section is suitable for retinal study.

At the base of the retina, where it meets the choroid layer, are the **rods** and **cones.** The nuclei for these photoreceptors are located in the **outer nuclear layer.** To stimulate these receptors light must pass through all the other cell layers.

Nuclei of the amacrine and bipolar neurons are located in the **inner nuclear layer.** Dendrites of these association neurons make synaptic connections with axons of the rods and cones in the **outer synaptic layer.**

The nuclei of ganglion cells are located closest to the exposed surface of the retina: in a layer designated as the **ganglion cell layer.** The dendrites of ganglion cells make synaptic connections with amacrine and bipolar cells in the **inner synaptic layer.** Nonmyelinated axons of the ganglion cells fill up the **nerve fiber layer;** they converge at the optic disk to form the optic nerve. Some neuroglial cell nuclei can be seen in the nerve fiber layer.

The Fovea Centralis Study a slide that reveals the structure of the fovea (H10.635). Compare your slide with the photomicrographs in illustrations B and C in figure HA–16.

Laboratory Report

Answer the questions on the first portion of combined Laboratory Report 34,35.

Visual Tests 35

The various tests outlined in this exercise relate to the observation of normal conditions, as well as the detection of some of the more common types of abnormalities. By performing these tests you will learn much about the physiology of vision.

Materials:

- penlite (image suppression)
- 12″ ruler or tape measure (blind spot)
- 3″ × 5″ card and pins (Scheiner's experiment)
- Snellen eye charts (visual acuity)
- Ishihara or Stilling charts (color blindness)
- laboratory lamp (reflexes)

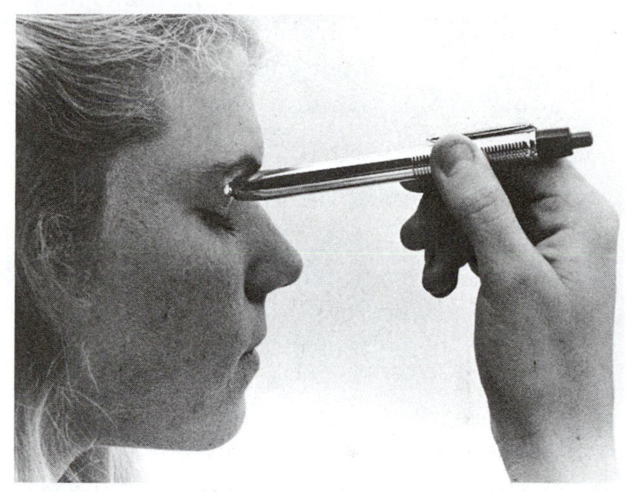

Figure 35.1 Proper position for penlite when observing the Purkinje tree.

Image Suppression
(The Purkinje Tree)

When you examined the fundus of the eye with an ophthalmoscope in the last laboratory exercise, you observed that blood vessels and capillaries were very much in evidence. These branching vessels lie very close to receptor cells and cast sharp shadows on most regions of the retina. You may wonder why these shadows do not show up in one's field of vision. The answer is that the brain has the capacity to suppress disturbing images of this sort that are always in the normal direct line of vision.

To illustrate that these blood vessels do cast an image on the retina and that they can be brought into one's field of vision, one need only illuminate the interior of the eye with rays of light that enter through the eyelid and sclera instead of through the pupil. Since light rays do not normally enter the eye in this manner, the brain will not suppress any shadows cast by these blood vessels on the retina. The branching image that one sees is called the *Purkinje tree.*

To perform this experiment on your own eye, proceed as follows: Hold a penlite with your right hand against the eyelid of your right eye at an angle of 45° to the right, as shown in figure 35.1. The eyelid may be open or closed. If the eyelid is open, look at a dimly lighted wall. If the eyelid is closed, do not face a brightly lighted window. While exposing the eye to the light, move the penlite from side to side to produce the Purkinje image. The amount of movement should be very slight, only a few millimeters from side to side. A most striking image can be produced if this experiment is performed in a dark closet. Describe the image on Laboratory Report 34,35.

The Blind Spot

The examination of the fundus in Exercise 34 clearly revealed the nature of the optic disk. It was noted that where the optic nerve enters the eyeball, there is an absence of rods and cones, which renders that part of the retina insensitive to light. The illustration in figure 35.2 can be used to detect the presence of this blind spot in each eye.

To test the right eye, close the left eye and stare at the plus sign with the right eye as this page is moved from about 18 inches toward the face. At first, both the plus sign and dot are seen simultaneously. Then, at a certain distance from the eye, the dot will disappear as it comes into focus on the optic disk of the retina.

Figure 35.2 The blind spot test.

Perform this test on both of your eyes. To test the left eye, it is necessary to look at the dot instead of the plus sign. Have your laboratory partner measure the distance from your eye to the test chart with a ruler or tape measure. Record the measurements on the chart on Laboratory Report 34,35.

Near Point Determination

The ability of the lens of the eye to produce a sharp image on the retina is, partially, a function of its elasticity. When focusing on close objects, the lens must be considerably more spherical, or convex, than when focusing on distant objects. To become more convex, the tension on the lens periphery is relaxed as the ciliary muscle is stimulated.

In an infant the ability of the eye to accommodate to distance is at its maximum. On the average, an infant's eye lens has a range of 17 diopters from distance sighting to near objects. As an individual gets older, the lens gradually becomes less elastic and the degree of accommodation diminishes. Between the ages of 45 and 50, one's accommodation is reduced to 2 diopters. Beyond 60 years of age, accommodation is nearly nonexistent. This condition is called **presbyopia.** This loss of lens elasticity is probably due to protein denaturation in the lens. It is for this reason that most individuals over 45 years of age find it necessary to acquire bifocal or trifocal glasses.

A measure of the elasticity of the lens in the eye can be made by determining the near point of the eye. The **near point** is the closest distance at which one can see an object in sharp focus. At 20 years of age the near point is approximately 3½″; at 30, 4½″; at 40, 6½″; at 50, 20½″; and at 60 it may be 33″.

To determine the near point in each eye, use the letter "T" at the beginning of this paragraph. Close one eye and move the page up to the eye until the letter becomes blurred; then move it away until you get a clear undistorted image. Have your laboratory partner measure the distance from your eye to the page with a ruler or tape measure. The closest distance at which the image is clear will be the near point. Test the other eye also, and record your results on the Laboratory Report.

Scheiner's Experiment Another method for determining the near point is with Scheiner's experiment. With this method one peers through two small pinholes in a card at an object such as a common pin. Figure 35.3 illustrates the method.

The pinholes should be in the center of the card and be no more than 2 mm apart, center to center. While peering through the two holes, it will be noted that the holes overlap so that it seems as if there is only a single opening. Holes that are too far apart, however, will appear as two distinct openings with an opaque bridge between them. Perform this experiment as follows:

1. Pierce two small pinholes in the center of a card that are no more than 2 mm apart, center to center.
2. Peer through the holes at a common pin that is held up at **arm's length.** If the two holes merge as a single opening, they are the correct distance apart. **It is important that the pin be viewed through the space where the two luminous circles overlap.**
3. With the pin still at arm's length, focus on a distant object. The pin should now appear double since the eye is focused on infinity.
4. Now focus on the pin and note that a single image is seen.
5. Move the pin slowly toward your eye, keeping it in focus until the image changes from single to double. **Be sure to keep the eye looking through the overlapping area of the two holes.**

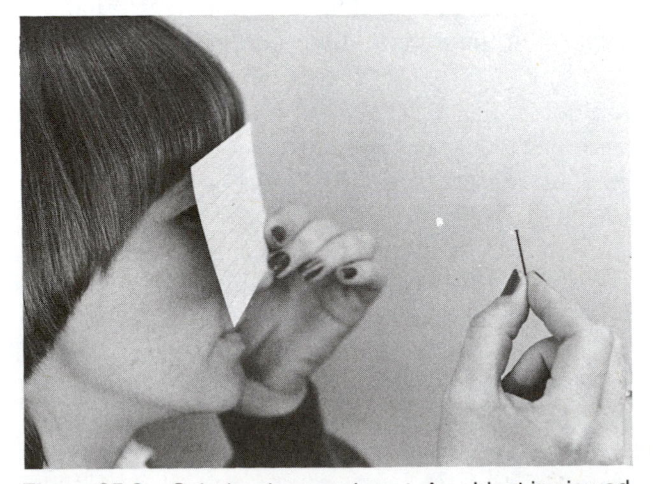

Figure 35.3 Scheiner's experiment. An object is viewed through two pinholes in a card.

The distance from the eye at which the image changes from single to double is the **near point.** Have your laboratory partner measure the distance with a ruler or tape measure.

6. Repeat this experiment several times to establish an average near point. How does this measurement compare with the first method used? Record the measurements on Laboratory Report 34,35.

Visual Acuity

A completely normal human eye lacking any form of deviation is able to differentiate, at a distance of ten meters, two points that are only one millimeter apart. Points that are less than a millimeter apart at this distance will be seen as a single spot. The size and proximity of cones in the fovea determine this degree of visual acuity.

If pinpoints of light from two different objects strike adjacent cones, only a single image will be seen. This is due to the fact that the brain lacks the ability to record the stimuli that it receives as separate entities. However, if the images fall on two cones separated by an unexcited cone, the brain recognizes two separate points. Because the diameter of a cone is approximately two micrometers (1/500 mm), the images on the fovea must be at least two micrometers apart. On the basis of this information it can be calculated that the brain can differentiate two pinpoints that enter the eye at an angle of 26 seconds. This is somewhat less than one-half a degree!

The Snellen eye chart (figure 35.4) has been developed with this information in mind. It is printed with letters of various sizes. When you stand at a certain distance from it, usually 20 feet, and are able to read the letters on a line designated to be read at 20 feet, you are said to have 20/20 vision. The ability to read these letters indicates that there are no aberrations of the lens or cornea to interfere with the angle of pinpoints of light reaching the retina of the eye. If you are only able to read the larger letters, such as those that should be read at 200 feet, you are said to have 20/200 vision.

At some place in the laboratory, a Snellen eye chart will be posted on the wall. A twenty-foot mark will also be designated on the floor. Working with your laboratory partner, test each other's eyes. Test one eye at a time, with the other eye covered. Record your test results on the Laboratory Report.

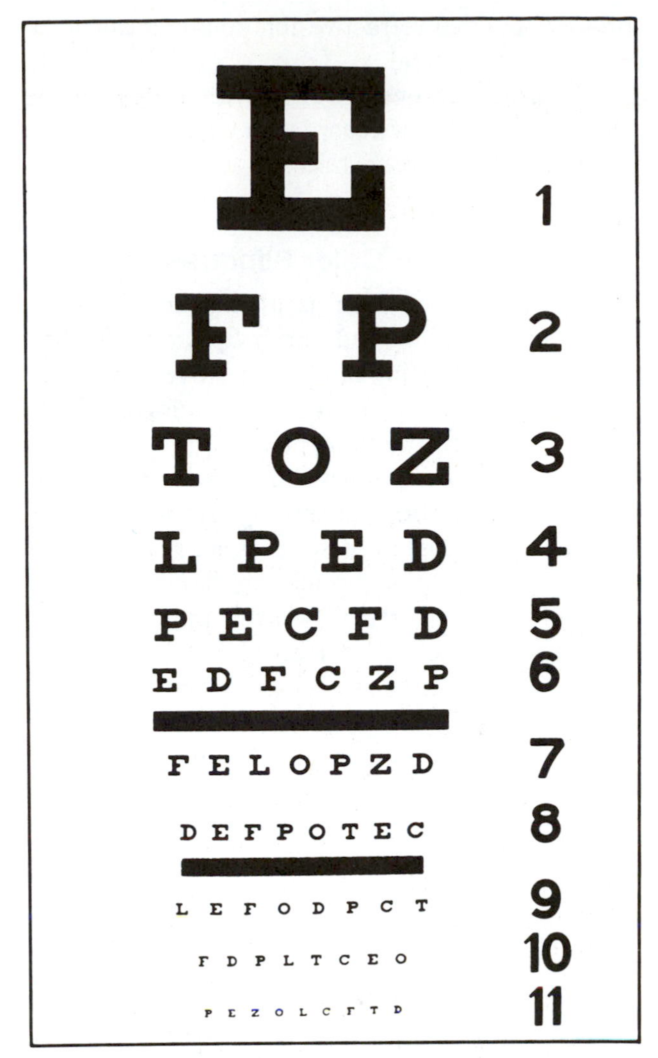

Figure 35.4 The Snellen eye chart.

Test for Astigmatism

If the lens of an eye has an uneven curvature on one of its surfaces, a condition called *astigmatism* exists. Figure 35.5 reveals the difference between normal and astigmatic lenses. Note that the astigmatic lens has a greater curvature on its left upper surface than on the lower left quadrant. This type of lens will cause greater bending of light rays as they pass through one axis of the lens than when passing through another axis. The image that one sees with this type of lens will be blurred in one axis and sharp in other axes.

To determine the presence of astigmatism, look at the center of the diagram in figure 35.6 and note if all radiating lines are in focus and have the same intensity of blackness. If all lines are sharp and equally black, no astigmatism exists. Of course, the presence of other refractive abnormalities can make this test impractical.

Look at the wheellike chart with each eye while closing the other and record your conclusions on the Laboratory Report.

Test for Color Blindness

The perception of color is a sensation achieved partly by the retina and partly by the brain. The receptors of the retina that are sensitive to color are the cones. According to the *Young-Helmholtz theory of color perception,* there are three different types of cones, each of which responds, maximally, to a different color. The three colors which cause maximum response in these cones are red, blue, and green. The degree of stimulation that each type of cone gets from a particular wavelength of light determines what color is perceived by the brain.

When the retina is exposed to red monochromatic light (wavelength of 610 nanometers), the red cones are stimulated at 75%, the green cones at 13%, and the blue cones, not at all. The ratio of stimulation for red is thus 75:13:0. When the brain receives this ratio of stimulation from the three types of cones, the interpretation is for red color.

When a monochromatic blue light (wavelength of 450 nanometers) strikes the retina, the red cones are not stimulated at all (0), the green cones are stimulated to a value of 14%, and the blue cones to a value of 86%. The ratio here of 0:14:86 is interpreted by the brain as blue. For green, the ratio is 50:85:15. Orange-yellow produces a ratio of 100:50:0. When exposed to white light, which has no specific wavelength since it is a mixture of all colors of the spectrum, the three types of cones are stimulated equally.

Color blindness is a sex-linked hereditary condition which affects 8% of the male population and 0.5% of females. The most common type is red-green color blindness, in which either the red or green cones are lacking. If red cones are lacking, a condition called **protanopia** exists. Individuals that have this condition see blue-greens and purplish-tinted reds as gray. A lack of green cones is designated as **deuteranopia.**

Although both protanopes and deuteranopes have difficulty differentiating reds and greens, their visual spectrums differ enough so that they can be diagnosed with color test charts. Illustration 4, figure 35.7, is a test plate for differentiating protanopes and deuteranopes. While a normal person would see the number 96 on this plate, a protanope would see only the number 6 and a deuteranope would see only the number 9.

Other than protanopia and deuteranopia, there are rarer forms of color vision deficiencies, such as blue-color weakness, yellow-color weakness, and total-color blindness. These forms of color vision deficiencies are not as well understood.

Using Ishihara Plates

In our color blindness test here we will use Ishihara's book of plates (*Ishihara's Tests for Colour Blindness,* Concise Edition). Procure one of these books from the supply table.

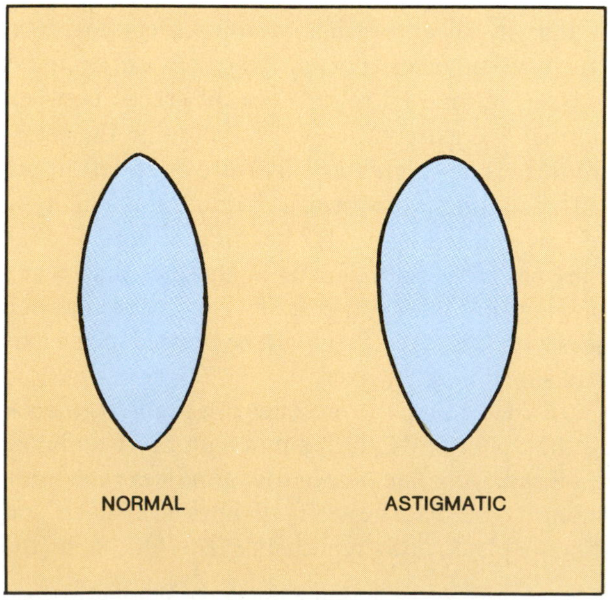

Figure 35.5 Normal and astigmatic lenses.

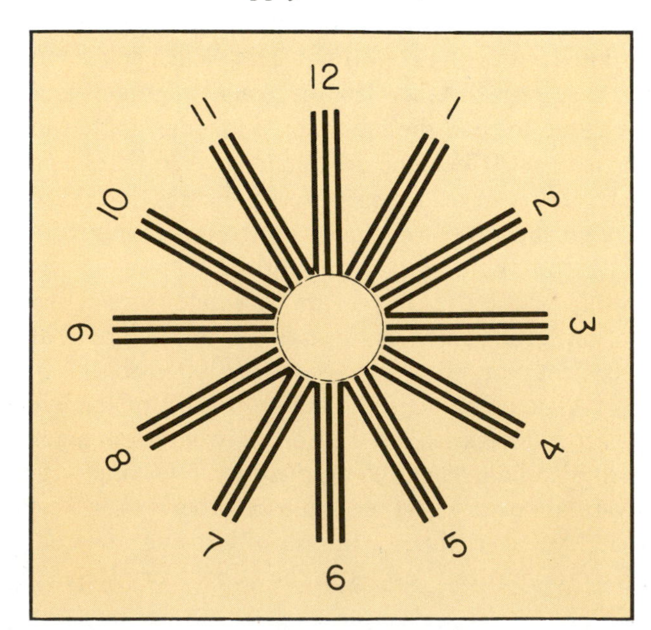

Figure 35.6 Astigmatism test chart.

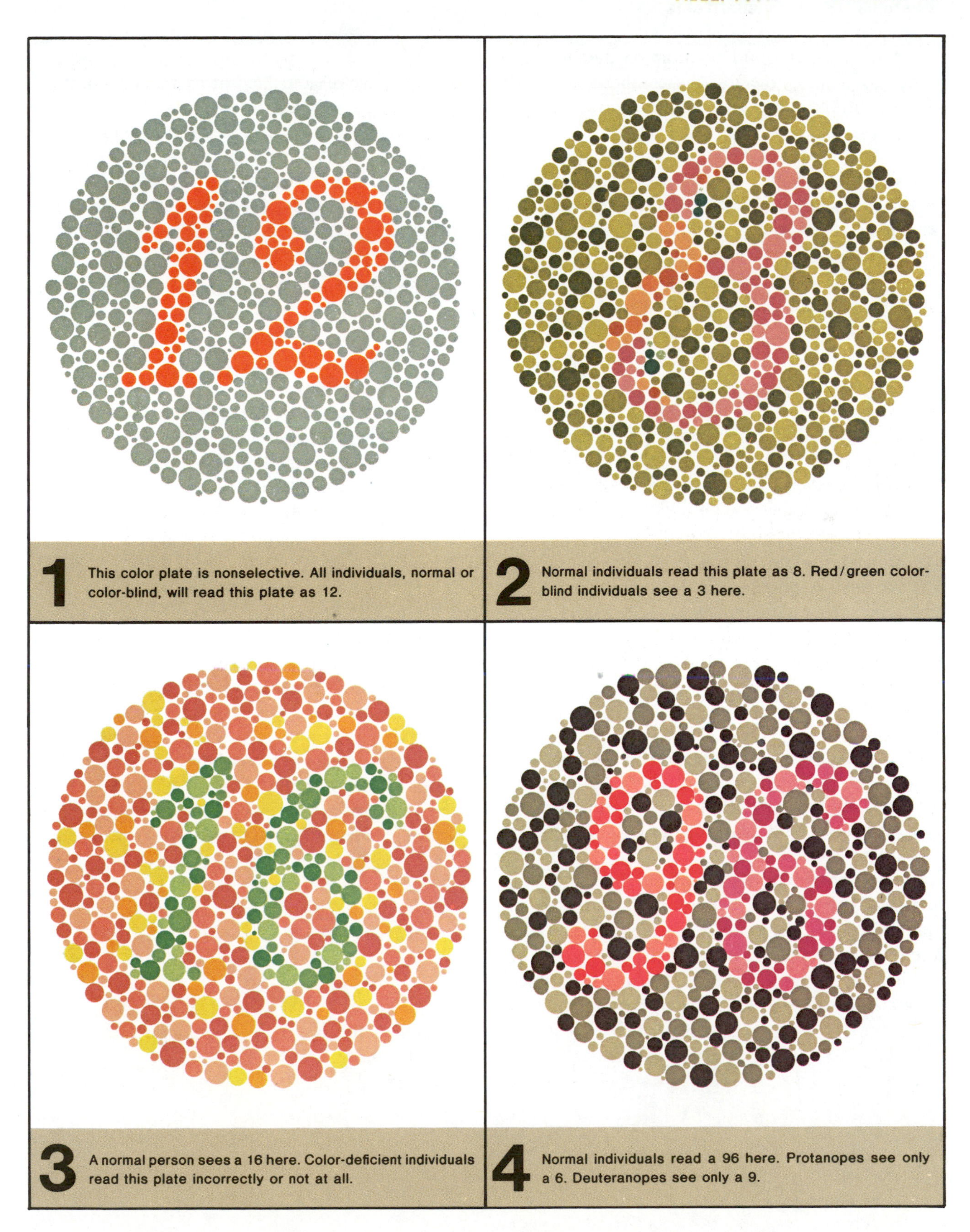

The above has been reproduced from *Ishihara's Tests for Colour Blindness* published by Kanehara & Co., Ltd., Tokyo, Japan, but tests for color blindness cannot be conducted with this material. For accurate testing, the original plates should be used.

Figure 35.7 Ishihara color test plates.

Working with your laboratory partner, test each other by holding the test plates about 30 inches away from the subject. If sunlight is available, use it.

Start with plate #1 and proceed consecutively through all 14 plates. The subject should respond with an answer for each plate within **3 seconds.** The examiner will record the responses on the chart on the Laboratory Report.

After all plates have been observed and recorded, compare the responses with the correct answers and make a statement at the bottom of the chart (conclusion).

Pupillary Reflexes

The tests and experiments performed so far have been concerned primarily with the role of the lens and retina in vision. The following experiments illustrate the roles of the pupil in adjustment of the eye to distance and light intensity.

Accommodation to Distance Three events take place when the eyes change their focus from a distant object to a close object:

1. Eye muscles react to achieve convergence of the eyes.
2. The lens becomes more convex.
3. A change occurs in the size of the pupil.

Working with your laboratory partner or a selected member of the class, perform the following experiment to observe what happens to the pupils when the eyes focus on a near object after looking at a distant object.

This experiment works best if the subject has **pale blue eyes.** If your laboratory partner does not qualify in this respect, request some other class member to be the subject and allow several other students to observe the results. Have the subject look at the wall on the side of the room opposite to windows. The eyes are now relaxed and focused at infinity.

While closely watching the pupils, place the printed page of your laboratory manual within six inches of the subject's face and ask the individual to focus on the print. Incidentally, the light intensity on the printed page and distant wall should be equal. Do the pupils remain the same size when looking at the printed page? Repeat the experiment several times and record your results on the Laboratory Report.

Accommodation to Light Intensity Sudden exposure of the retina to a bright light causes immediate reflex contraction of the pupil in direct proportion to the degree of light intensity. The pupil contracts to approximately 1.5 mm when the eye is exposed to intense light, and it enlarges to almost 10 mm in complete darkness. This approximates a total difference in pupillary area of about 40 times. In this reflex, impulses pass from the retina via the optic nerve through two centers in the brain and then return to the sphincter of the iris through the ciliary ganglion.

Perform this simple experiment to learn a little more about the pathway of this reflex. As in the previous experiment, select a subject with pale blue eyes so that the pupil size is more easily observed. Have the subject hold this laboratory manual, vertically, with the spiral binding close to the forehead and extending downward along the bridge of the nose. Now, position an unlighted laboratory lamp about six inches from the right eye. While watching the pupil of the left eye, turn on the lamp for one second and then turn it off. Make sure that no light spills over from the right side of the book. Did the pupil that was not exposed to light become smaller? What does this reaction tell us with respect to the pathways of the nerve impulses?

Laboratory Report

Complete the last half of combined Laboratory Report 34,35.

The Ear: Its Role in Hearing 36

Four aspects of the ear will be studied in this exercise: (1) its anatomical components that pertain to hearing; (2) the physiology of hearing; (3) hearing tests; and (4) histological studies. The vestibular apparatus, which functions in equilibrium, will be studied in Exercise 37. A brief statement pertaining to the characteristics of sound will precede our study of the ear. The sections on ear anatomy and the physiology of hearing should be completed prior to the hearing tests and microscopy.

Characteristics of Sound

Sound waves moving through the air are propagated in wave forms which possess characteristics relative to the vibratory motion that generates them. The ear is able to distinguish tones that differ in pitch, loudness, and quality. **Pitch** is determined by the frequency of vibration; **loudness** pertains to the intensity of the vibration; and **quality** relates to the nature of the vibrations as revealed by the wave shape.

Figure 36.1 illustrates several curves depicting both the shapes of sound waves and the characteristics of the vibrations which produce them. The sine curves A and B in illustration I differ in frequency, with B producing the tone of higher pitch. Curves A and C in the middle group are of the same frequency; they differ only in amplitude, or loudness. Although the amplitude of A is twice that of C, the loudness of A will be four times that of C. This is because the intensity (I) of sound is directly proportional to the square of the amplitude (a):

$$I = 2\pi^2 V f^2 a^2 d$$

V = velocity of wave propagation
f = frequency
d = density of medium

Waves A and D in illustration III differ in shape, D having some components of higher frequency that are not present in A. These curves represent sounds of different quality.

The ability of the ear to distinguish differences in pitch, amplitude, and quality depend on (1) the conduction and pressure amplification of sound waves through the ossicles of the middle ear; (2) the stimulation of receptor cells in the cochlea; and (3) the conveyance of action potentials in the cochlear nerve to the auditory centers in the brain for interpretation. The interference of any component of this system results in hearing loss. The types of hearing loss will be studied as hearing tests are performed. Let us now explore the anatomy of the ear as it functions in normal anatomy.

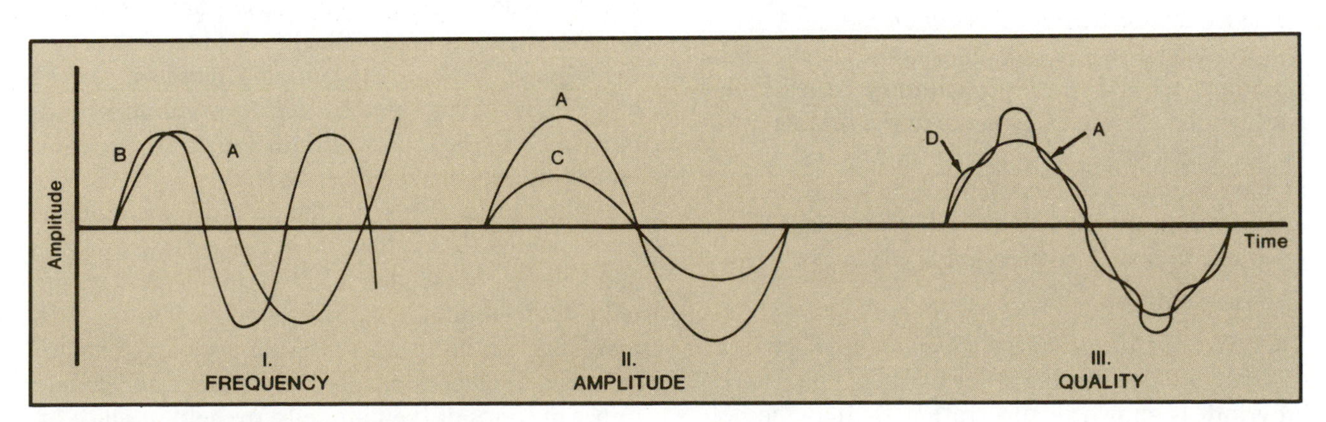

Figure 36.1 Differences in sound waves.

Components of the Ear

Anatomically, the human ear consists of three distinct divisions: the external ear, the middle ear, and the internal ear. Figure 36.2 is a diagrammatic representation of its various components.

The External Ear

The outer or external ear consists of two parts: the auricle and external auditory meatus. The **auricle,** or **pinna,** is the outer shell of skin and elastic cartilage that is attached to the side of the head. The **external auditory meatus** is a canal about one inch in length that extends from the auricle into the head through the temporal bone. The skin lining the canal contains some small hairs and modified apocrine sweat glands that produce a waxy secretion called *cerumen.* The inner end of this meatus terminates at the **tympanic membrane,** or eardrum. The auricle serves to collect and direct sound waves into the tympanic membrane through the meatus.

The Middle Ear

This division of the ear consists of a small cavity in the temporal bone between the tympanic membrane and the inner ear. It contains three small bones (*ossicles*) that are united to form a lever system. The outermost ossicle, which is attached to the tympanic membrane, is the **malleus,** a hammer or club-shaped bone. The middle bone, an anvil-shaped structure, is the **incus.** The innermost bone, which fits into the **oval window** of the inner ear, is stirrup-shaped and is called the **stapes.**

It is the role of these ossicles to transfer the forces from the eardrum to the cochlear fluids of the inner ear through the oval window. Although most sound waves reach the inner ear via the ossicles (*ossicular conduction*), some sounds, specifically very loud ones, reach the cochlea through the bones of the skull (*bone conduction*).

Leading downward from the middle ear to the nasopharynx is a duct, the **auditory** (Eustachian) **tube,** which allows air pressure in the middle ear to be equalized with the outside atmosphere. A valve at the nasopharynx end of the tube keeps the tube closed. Acts of yawning or swallowing cause it to open temporarily for pressure equalization.

The Internal Ear

This part of the ear consists of two labyrinths: the osseous and membranous labyrinths. The **osseous labyrinth** is shown in illustration B. It is the hollowed out portion of the temporal bone that contains an inner tubular structure of membranous tissue, the **membranous labyrinth.** The entire membranous labyrinth is shown in the next exercise (figure 37.1). Some portions of the membranous labyrinth are shown in cutaway portions of the osseous labyrinth in illustration B, figure 36.2.

Within the membranous labyrinth is a fluid, the **endolymph.** Between the membranous and osseous labyrinths is a different fluid, the **perilymph.** These fluids act as conduction media for the forces involved in hearing and maintaining equilibrium.

The osseous labyrinth consists of three semicircular canals, the vestibule, and the cochlea. The **vestibule** is that portion that has the oval window on its side into which the stapes fits. The three **semicircular canals** branch off the vestibule to one side, and the **cochlea,** which is shaped like a snail's shell, emerges from the other side.

On the osseous labyrinth are two nerves, which are branches of the eighth cranial, or vestibulocochlear, nerve. The **vestibular nerve** is the upper nerve branch, which passes from the sensory areas of the semicircular ducts, saccule, and utricle. The other branch is the **cochlear nerve,** which emerges from the cochlea.

The cochlea consists of a coiled, bony tube with three chambers extending along its full length. Illustration C in figure 36.2 reveals a cross section of the cochlear tube. The upper chamber, or scala vestibuli (label 16), is so-named because it is continuous with the vestibule. The lower larger chamber is the **scala tympani.** The **round window** (label 1) is a membrane-covered opening that is on the osseous wall of the scala tympani. Between these two chambers is a triangular cross section that represents the **cochlear duct.** This duct is bounded on its upper surface by the **vestibular membrane** and on its lower surface by the **basilar membrane.** On the upper surface of the basilar membrane lies the **organ of Corti,** which contains the receptor cells of hearing. All of these chambers contain fluid: perilymph in the scala vestibuli and scala tympani; endolymph in the cochlear duct.

Illustration D reveals the detailed structure of the organ of Corti. Note how the hair tips of the **hair cells** are embedded in a gelatinouslike flap, the **tectorial membrane.** Leading from each hair cell is a nerve fiber that passes through the basilar membrane and becomes part of the cochlear nerve. The upper margins of the hair cells are held in place by

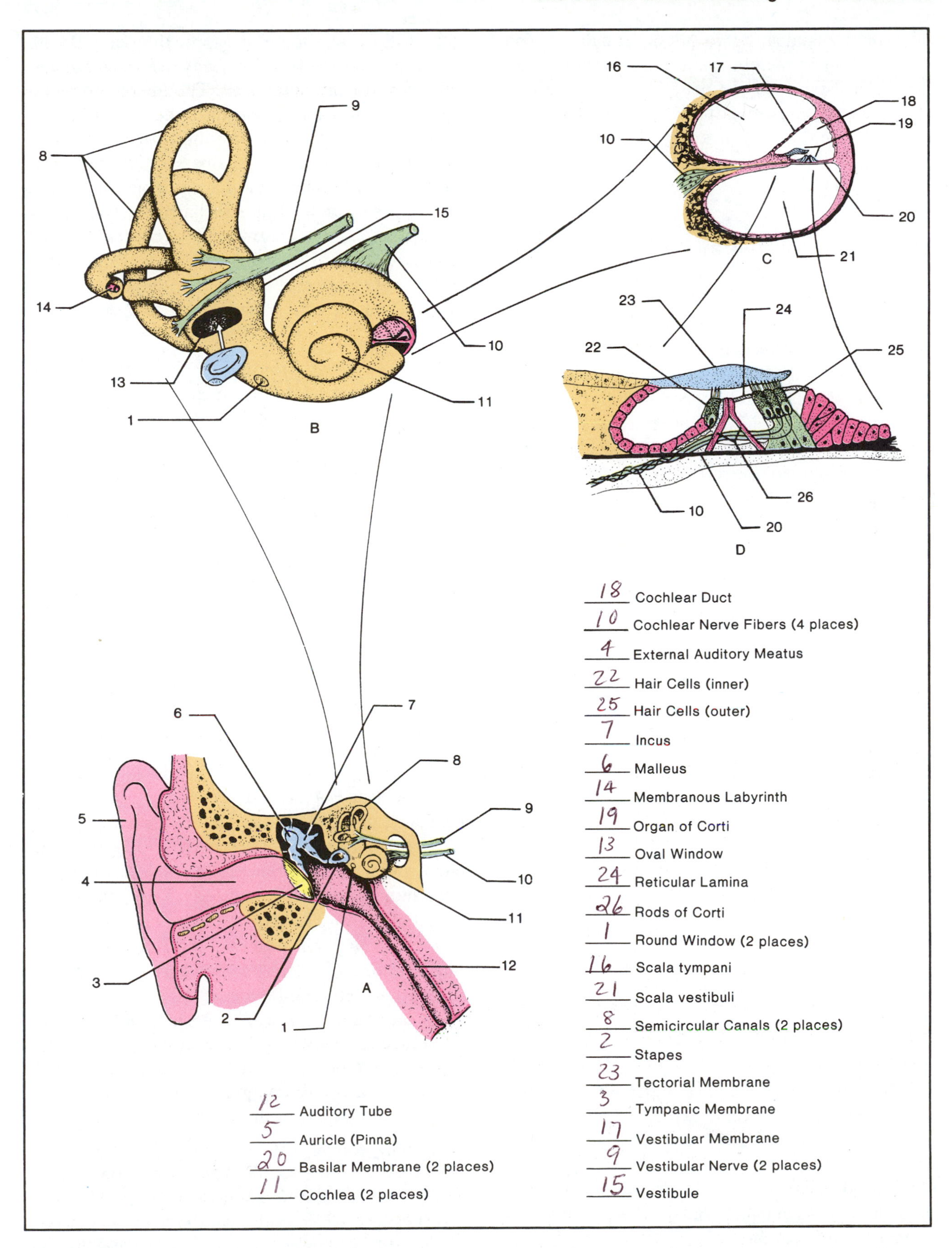

Figure 36.2 Anatomy of the ear.

the **reticular lamina.** Between the reticular lamina and the basilar membrane are reinforcing structures called the **rods of Corti.** The significance of all these structures must be taken into account in any plausible theory of hearing.

Assignment:

Label figure 36.2.

The Physiology of Hearing

As stated previously, hearing occurs when action potentials are received by the auditory centers of the brain. These action potentials, which pass along the cochlear nerve fibers of the eighth cranial nerve, are initiated by the hair cells in the organ of Corti. Activation of the hair cells depends on forces within the cochlear fluids, basilar membrane structure, secondary energy transfer, and endocochlear potential. A discussion of roles of each of these follows.

Role of Cochlear Fluids In figure 36.3 the cochlea has been uncoiled to reveal the relationships of its various chambers. When the stapes moves in and out of the oval window, sound wave energy moves through the scala vestibuli and into the scala tympani. Since the perilymph is incompressible, the elasticity of the round window membrane accommodates the pressure changes by moving in and out in synchrony with the movements of the stapes.

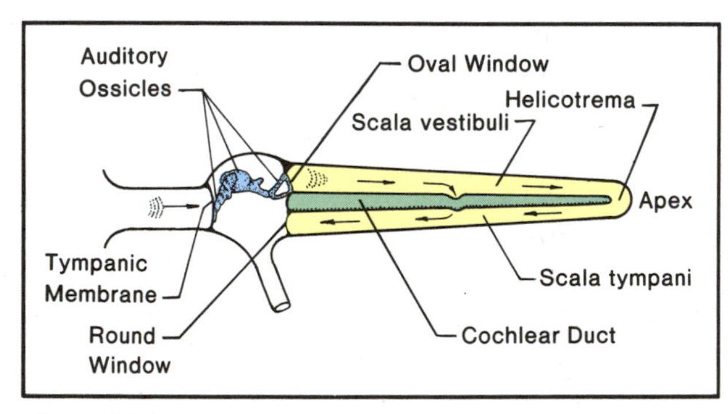

Figure 36.3 Pathway of sound wave transmission in the ear.

The energy of sound waves moving through the cochlea reaches the round window in two ways: (1) directly into the scala tympani through the helicotrema at the apex of the cochlea, and (2) through the flexible cochlear duct to the scala tympani. Arrows in figure 36.3 illustrate both routes of energy propagation.

Basilar Membrane Structure The first structure of the organ of Corti that reacts to vibrations of different sound frequencies is the basilar membrane. Within this membrane are approximately 20,000 fibers that project from the bony center of the cochlea toward the outer wall. The fibers closest to the stapes are short (0.04 mm) and stiff; they vibrate when stimulated by vibrations caused by high-frequency sounds. The fibers at the end of the cochlea near the helicotrema are long (0.5 mm), more limber, and vibrate in harmony with low-frequency sounds. These fibers are elastic reedlike structures that are not fixed at their distal ends. Because they are free at one end, they are able to vibrate like reeds of a harmonica to specific frequencies.

Secondary Energy Transfer Once fibers at a particular point on the basilar membrane are set into motion by a specific sound frequency, the rods of Corti transfer the energy from the basilar membrane to the reticular lamina, which supports the upper portion of the hair cells (see illustration D, figure 36.2). This secondary transfer of energy causes all these components to move as a unit, with the end result that the hairs of the hair cells are bent and stressed. The back and forth bending of the hairs, in turn, causes alternate changes in the electrical potential across the hair cell membrane. This alternating potential, known as the *receptor potential* of the hair cell, stimulates the nerve endings that are at the base of each hair cell, and an action potential in the cochlear nerve fibers is initiated.

Endocochlear Potential The sensitivity of the hair cells to depolarization is greatly enhanced by the chemical differences between the perilymph and endolymph. Perilymph has a high sodium-to-potassium ratio; endolymph has a high potassium-to-sodium ratio. These ionic differences result in an *endocochlear potential* of 80 millivolts. It is significant that the tops of the hair cells project through the reticular lamina into the endolymph of the scala media, but the lower portions of these cells lie bathed in perilymph. It is believed that this potential difference of 80 mv between the top and bottom of each hair cell greatly sensitizes the cell to slight movement of the hairs.

Frequency Range The variability in length and stiffness of the fibers within the basilar membrane enables the ear to differentiate sound waves as low as 30 cycles per second near the apex of the cochlea, and as high as 20,000 cps near the base of the cochlea. In between these two extremes, at various spots on the basilar membrane, the membrane responds to the in-between frequencies. This method of pitch localization by the basilar membrane is called the *place principle.*

Loudness Determination As noted in figure 36.1, the loudness of a sound is reflected in the sound wave. Increased loudness of sounds causes an increase in the amplitude of vibration of the basilar membrane. With an increase in basilar membrane vibration, more and more hair cells become stimulated, causing *spatial summation* of impulses; that is, more nerve fibers are carrying more impulses, and this is interpreted by the brain as increased loudness.

Assignment:

Answer the questions on the Laboratory Report that pertain to the physiology of hearing.

Hearing Tests

Although deafness may have many different origins, there are essentially two principal kinds of deafness: nerve and conduction deafness. If the cochlea or cochlear nerve is damaged, the condition is referred to as **nerve deafness.** Damage to the eardrum or ossicles, on the other hand, will result in **conduction deafness.** While conduction deafness can usually be remedied with surgery or hearing aids, nerve deafness cannot be corrected.

The three kinds of hearing tests outlined here are ones that are most frequently used. Each type of test serves a specific function. Perform those tests for which equipment is available.

Watch Tick Method

The use of a spring-wound pocket watch to screen patients for hearing loss has merit in simplicity. The inability of a patient to hear a ticking watch at prescribed distances from the ear can alert the examining physician to a hearing problem. Further tests using tuning forks or the audiometer can then be brought into play.

Materials:

spring-wound pocket watch
cotton plugs
meterstick or measuring tape

1. Seat the subject comfortably in a chair and plug one ear with cotton.
2. Instruct the subject to use the index finger to signal when sound is first heard as the watch approaches or when sound disappears as the watch is moved away from the ear.
3. With the subject looking straight ahead, place the watch at a position that is out of hearing range of the ear, usually about three feet. Keep the face of the watch parallel to the side of the head.
4. Move the watch toward the ear at **a speed of about ½ inch per second.** This is a very slow movement. The subject should signal when the **first tick** is heard.
5. Measure the distance with the meterstick or tape measure, record the information on the Laboratory Report, and repeat the test two or three times. On repeat tests change the approaches to the ear to prevent the factor of subject anticipation.
6. Reverse the procedure by starting close to the ear and receding from it. In this test, the subject signals the **last tick** that is heard.
7. Transfer the cotton plug to the other ear and follow the same procedure again.

Tuning Fork Methods

The distinction between nerve and conduction deafness can be readily made with two tuning fork methods: the **Rinne** and **Weber** tests. In one test the base of the tuning fork is applied to the mastoid process behind the ear; in the other test the fork is placed on the forehead. In both tests the tuning fork is set in vibration by bouncing the fork off the heel of the hand as shown in illustration 1, figure 36.4. Avoid striking hard surfaces, such as the tabletop. These tests may be performed independently or with the aid of an examiner.

Materials:

tuning fork
cotton earplugs

The Rinne Test This test enables you to identify the cause of hearing loss by placing the base of a tuning fork against the mastoid process. The method readily differentiates between nerve and conduction deafness.

1. Plug one ear with cotton.
2. Set a tuning fork in motion by striking it on the heel of the hand. Place it in front of the unplugged ear with the tine facing the ear, as shown in illustration 2, figure 36.4; 3 to 6 inches from the ear is adequate.

   If there is a minimum of hearing loss, the vibrating sound will be heard immediately. As the sound intensity diminishes, a point will be reached when the sound finally disappears.

At this instant, place the stem of the tuning fork against the mastoid process as in illustration 3, figure 36.4. **If the sound vibrations reappear, conduction deafness exists.**

If considerable hearing loss is present, and the tuning fork vibrations cannot be heard for long or at all, place the stem of the vibrating fork against the mastoid process and note if the sound can be heard through bone conduction. If the sound is loud through the bone, conduction deafness exists. **If no sound is heard, nerve deafness exists.**

3. Repeat the test on the other ear.
4. Record results on the Laboratory Report.

The Weber Test When the stem of a vibrating tuning fork is placed on the forehead of a person with normal hearing, the sound of the fork is heard at equal intensity in both ears. If an individual with **conduction deafness** is tested in this manner, **the sound will be heard louder in the deaf ear than in the normal ear.** The reason is that the deaf ear, which is normally not activated by sound waves through the air, is more acutely atuned to sound waves being conducted to the cochlea through bone.

On the other hand, if this test is performed on a person who has one normal ear and one ear with **nerve deafness,** the **sound will be more intense in the normal ear** than in the deaf one.

Place a vibrating tuning fork on your forehead and compare the sounds heard in each ear. Report your conclusions on the Laboratory Report.

Audiometry

The audiometer is an instrument used for determining hearing losses in the audible range of normal speech. It measures the ability of the ear to hear sounds in the **frequency range** of 125 to 8,000 cps. Although the hearing range of the human ear may be as broad as 30 to 20,000 cps, it is the 125–8,000 cps range that is important in hearing the spoken word.

This instrument also measures the **level of intensity** of sounds in the audible range that one can hear. The sensation of loudness experienced by the ear is not related in a simple way to the intensity of sound that strikes the tympanic membrane. While the range of sensitivity between a faint whisper and the loudest noise is one trillion times, the ear interprets this great difference as approximately a 10,000-fold change. What happens here is that the scale of intensity is compressed by the action of the tympanic membrane, ossicles, and cochlear duct, enabling the ear to have a much broader sensitivity range.

Because of this extreme range in sound intensities, it is necessary to express the intensity in terms of the logarithms of their actual intensities. The basic unit is called a **bel** (after Alexander Bell). Sounds that differ one bel in intensity differ by ten times. A **decibel** is one-tenth of a bel. It is in decibel units that hearing is measured, mainly because this is the smallest unit that the human ear is able to differentiate.

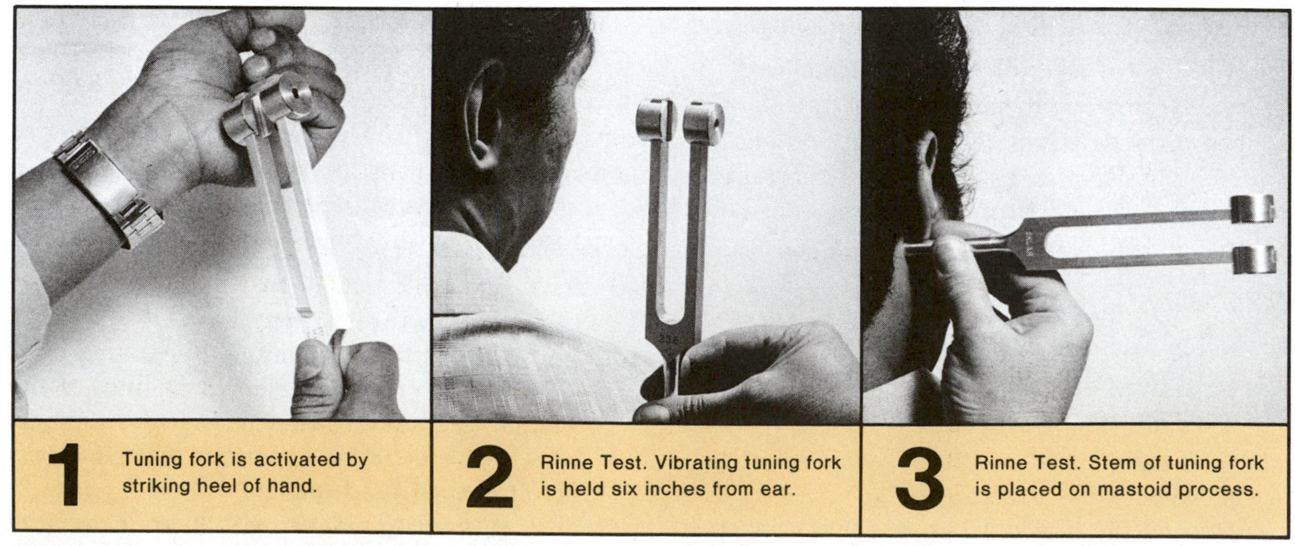

Figure 36.4 Tuning fork manipulation in hearing tests.

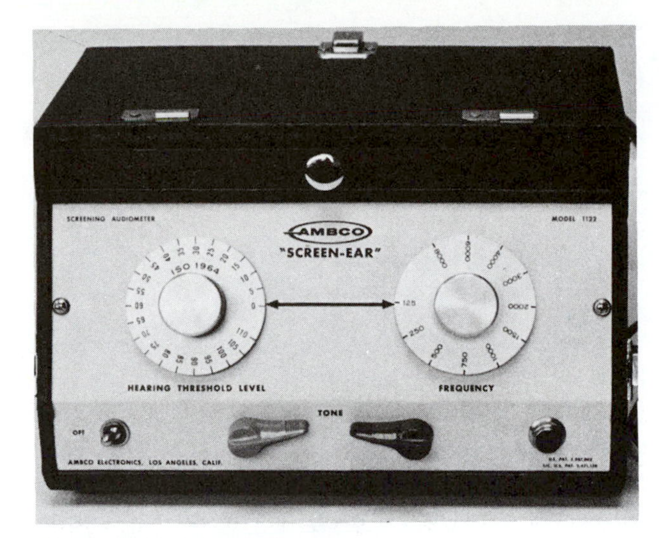

Figure 36.5 The audiometer.

The audiometer is, essentially, an electronic oscillator with earphones. As shown in figure 36.5, it has one control that regulates the frequency and another control that regulates the intensity, or loudness. Two separate switches, or tone controls, near the bottom are used to release the tone into the earphones; the left switch (blue) is for the left earphone and the right (red) one is for the right earphone.

The Hearing Level Control is calibrated so that the zero intensity level of sound at each frequency is the loudness that can be barely heard by a normal person. The numbers on this control represent decibels. If it is necessary to adjust this control to 30 at 125 cps frequency, it means that the patient has a hearing loss of 30 decibels.

In this experiment, work with your laboratory partner to plot an audiogram on the Laboratory Report for each ear. Proceed as follows:

Materials:

audiometer
red and blue pencils

1. Place the headset securely on the subject's head so that the ear cushions make good contact over the ears. Make sure that the ear cushions are not obstructed by clothing, earrings, etc. **Important:** Make sure that the earphone with a red cord is placed over the right ear.
2. To enable the subject to become familiar with the tone, set the Frequency Control on 1,000 cps and the Hearing Level Control at 50 db; then, depress the right (red) Tone Control so that the subject can hear the tone in the **right ear.**
3. Rotate the Frequency Control throughout all frequencies at this 50 db level to let the subject hear the frequencies prior to testing. Depress the Tone Control *only* when the dial is set on each frequency.
4. Return the Frequency Control to 1,000 cps and, while depressing the red Tone Control, rotate the Hearing Level Control counterclockwise slowly until you reach the point where the subject is just barely able to recognize the tone. If, for instance, the subject can hear the tone at 30 decibels, but not hear the tone at 25 decibels, the subject's threshold is established at 30 decibels for that frequency setting. Have the subject signal with a hand signal when the tone is heard.
5. Record this measurement on the audiogram chart of the Laboratory Report by marking a red "O" on the vertical line for 1,000 cps, where it intersects the line for the decibel value. (In the above example, it would be where the 30 decibel line intersects the 1,000 cps line.)
6. Now rotate the Frequency Control to 2,000 cps following the above procedures. After testing at this frequency, test the right ear at the remaining higher frequencies up to 8,000 cps. Finally, finish the tests on the right ear by testing at frequencies of 250 and 500 cps.
7. Test the **left ear,** following steps 2 through 6 above. Use the blue Tone Control switch for testing the left ear. Record each threshold level with an "X" on the appropriate frequency line with a blue pencil.
8. Connect the points on the chart with appropriately colored lines to produce a hearing graph for the ear.

Histological Study

Examine slides H10.53 and H10.54 to study the parts of the cochlea and crista ampullaris shown in figure HA–17 of the Histology Atlas. Both of these slides are made from guinea pig tissue. In particular, attempt to identify all the structures of the organ of Corti that are shown in illustration C. Note in illustration B that the cupula is obscured by the collapsed wall of the ampulla.

Complete the Laboratory Report for this exercise.

37 The Ear: Its Role in Equilibrium

The semicircular ducts, saccule, and utricle function in maintaining equilibrium. These three components are combined into a unit called the *vestibular apparatus.*

In addition to the vestibular apparatus, three other factors play important roles in maintaining equilibrium. They are (1) visual recognition of horizon position, (2) proprioceptor sensations in joint capsules, and (3) cutaneous sensations through certain exteroceptors. Sensations from these four sources are continuously subconsciously synthesized at the cortical level to provide the correct reflex responses necessary to maintain equilibrium.

In this exercise we will evaluate the relative significance of visual, proprioceptive, and vestibular sensations in postural control. Some clinical tests for evaluating static and dynamic equilibrium mechanisms will also be explored.

Two types of equilibrium are identifiable: static and dynamic. **Static equilibrium** pertains to the effect of gravity on receptors. **Dynamic equilibrium** compensates for angular movements of the body in different directions. The physiology of each type follows.

Static Equilibrium

The sensory receptors of the vestibular apparatus that respond to gravitational forces are located in the saccule and utricle. Note in figure 37.1 that the **saccule** connects directly to the cochlear duct and the **utricle** lies at the base of the semicircular canals. These two structures, being a part of the entire membranous labyrinth, contain **endolymph.**

On the inner walls of the saccule and utricle lie two sensory structures, the **maculae.** Each macula consists of **hair cells,** a **gelatinous matrix,** and calcium carbonate crystals called **otoconia.** The otoconia are embedded in the gelatinous matrix. Fibers of the vestibular nerve carry impulses from the hair cells to the brain.

When the head is tilted slightly in any direction, the otoconia are pulled by gravitational forces, causing the cilia of the hair cells to be bent by the action of the gelatinous matrix against them. Bending of the cilia in one direction causes impulse traffic in the vestibular nerve fibers to increase markedly; bending in another direction decreases the impulse traffic to a point of no reaction at all. This system is so sensitive to malequilibrium that

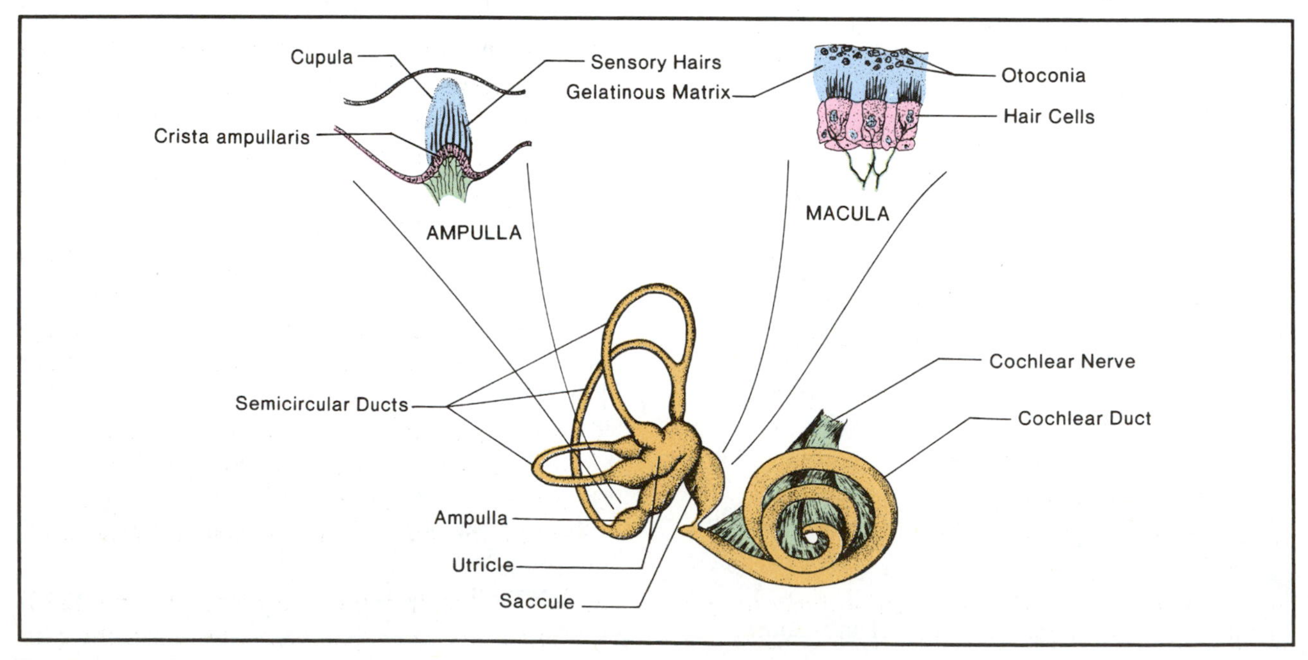

Figure 37.1 The membranous labyrinth.

as little as ½° shifting of the head in any direction from vertical is detectable.

Simply tilting the head, however, does not bring about a sense of malequilibrium. Fortunately, proprioceptors in the joints of the neck transmit inhibitory signals to the brain stem that neutralize the effects of the vestibular receptors. It is only when the whole body becomes disoriented that the hair-cell initiated impulses are not inhibited.

Balancing Test One of the simplest tests for determining the integrity of this static equilibrium mechanism is to have the subject stand perfectly still with eyes closed. Work with your partner to test each other's sense of balance. If there is any damage to this system, the subject will waver and tend to fall. Individuals with long-term damage, however, are often able to stand fairly well due to well-developed proprioceptive mechanisms.

Dynamic Equilibrium

The receptor hair cells for dynamic equilibrium are contained in the **ampullae** of the **semicircular ducts.** These hair cells are arranged in a crest, the **crista ampullaris,** within each ampulla. The hair tufts, in turn, are embedded in a gelatinous mass to form a structure called the **cupula.** Identify all these structures in figure 37.1.

The three semicircular ducts are arranged at right angles to each other in three different planes. When the head moves in any direction, the semicircular duct in the plane of directional movement moves relative to the endolymph within it; in other words, the fluid is stationary as the duct moves.

This movement causes the cupula to act as a trapdoor, moving in either direction due to the force of the endolymph. The hair cells in this structure act much like the hair cells of the macula, in that bending in one direction produces increased action potentials and bending in the other direction is inhibitory. Nerve impulses along the vestibular nerve reflexly excite the appropriate muscles to maintain equilibrium.

Nystagmus

The principal function of the semicircular ducts is to maintain equilibrium *in the initial stages* of angular or rotational movement. As soon as one begins to change direction, the semicircular ducts trigger reflexes to the proper muscles in anticipation of malequilibrium. A reflex movement, called nystagmus, occurs during this movement to produce fixed momentary images on the retina.

Nystagmus is characterized by rapid and slow jerky movements of the eyes as the body is rotated. It is caused by reflexes transmitted through the vestibular nuclei, the cerebellum, and the medial longitudinal fasciculus to the ocular nuclei. *Its function is to permit fixed images rather than blurred ones during head movements.*

If the head is slowly rotated to the right, the eyes will, at first, move slowly to the left. This reflex eye movement is the **slow component** of nystagmus. As soon as the eyes have moved as far to the left as they possibly can, they quickly shift to the right, producing what is called the **fast component** of nystagmus. The first slow component is initiated by the vestibular apparatus; the second fast component involves the brain stem.

Group Demonstration

Proceed as follows to demonstrate nystagmus. Work in teams of five students, using one member as a subject. The subject, seated in a swivel chair, will be rotated by the other four members. By using several individuals to anchor and rotate the chair, the subject will be accorded maximum safety. Prior to beginning the test, the **chair's back should be securely tightened** to prevent backward tilting.

Materials:

swivel-type armchair

1. Have the subject sit in the chair, cross-legged and gripping the arms of the chair.
2. The other four members of the team will form a circle around the chair, each member placing one foot against the base of the chair to prevent chair base movement.
3. Revolve the subject to the right (**clockwise**) and observe the eye movement as the subject gazes straight ahead during the rotation. Make ten revolutions at a rate of one revolution per second.
4. Allow the subject to remain seated for one or two minutes after the test to regain stability.
5. Repeat the procedure with another subject, but rotate this individual in a **counterclockwise** direction.
6. With the remaining three team members, follow the same procedure, but have the subjects keep their eyes closed during rotation.

Immediately upon stopping the chair, have the subjects open their eyes for observation.

7. Record all observations on the Laboratory Report.

Caution: The next test must not be performed here without special permission. It can induce nausea and vomiting!

Ice Water Test *(Caloric Stimulation)*

A simple test that may be used to determine whether or not the vestibular apparatus of one ear is functioning properly is to perfuse the external ear canal with a small amount of ice water. If the vestibular apparatus is intact and functional, the subject will experience a discomforting sensation of rotation, and *nystagmus will immediately be initiated.* If the subject experiences no discomfort or nystagmus, it may be assumed that damage to the vestibular apparatus or the vestibular nerve has occurred.

The physiological explanation of this reaction is that the cold water increases the density of the endolymph in a portion of the semicircular canals. With some of the endolymph heavier than other endolymph, convection currents are created in the semicircular canals that cause a sense of malequilibrium.

This test is often used to test for vestibular nerve damage resulting from overdosage of certain antibiotics such as streptomycin.

Proprioceptive Influences

To observe the relative roles of proprioceptors, vision, and vestibular reflexes in equilibrium, perform the following experiments. Retain the same five-member teams, as previously.

Materials:

swivel-type armchair
pencil
blindfold

At-Rest Reactions

Direct the subject to sit in the swivel chair and perform the following simple acts to establish norms for comparison.

1. **With eyes closed,** have the subject place the heel of the right foot on the toes of the left foot.
2. **With eyes closed,** have the subject bring the index finger of the right hand to the tip of the nose from an extended arm position.

   Question: How do proprioceptors in the appendages function to accomplish these two feats? Record your conclusions on the Laboratory Report.
3. Next, have the subject, **with eyes open,** raise his or her hand from the right knee vertically and forward to point his or her finger at the eraser of a pencil held about two feet directly in front of the subject. Have the subject repeat this pointing six times, once per second on command of the person holding the pencil. After each pointing, the hand is returned to rest on the right knee.
4. Have the subject repeat step 3 **with eyes closed.** How close does the subject approach the eraser with eyes closed? Report results on the Laboratory Report.

Effects of Rotation

The role of the eyes in kinesthetic efficiency will now be determined in tests similar to those above. Proceed as follows:

1. Direct the subject to sit cross-legged and keep the eyes open. Revolve the chair at one revolution per second for ten full turns. Halt the chair in exactly the same starting position.

   Ask the subject to point to the pencil eraser held in the same position as in the previous test. As before, the subject points to the eraser one time per second for a total of six times. After each pointing, the hand is returned to rest on the right knee. Record observations on the Laboratory Report.
2. Blindfold the subject, orient the chair to the same starting position, and, by feel, show the subject where the pencil eraser is located.
3. Now, rotate the subject for ten full turns, stopping the chair at the same starting point. As soon as the chair has come to rest, ask the subject to point to the spot where he or she thinks the eraser is, **but don't let subject touch the eraser.** Repeat the pointing a total of six times. Remember, the pencil eraser must be held in the same position as previously, and must not be touched.
4. Repeat the entire procedure with another subject, but reverse the direction of rotation.

Laboratory Report

Complete the Laboratory Report for this exercise.

HISTOLOGY ATLAS

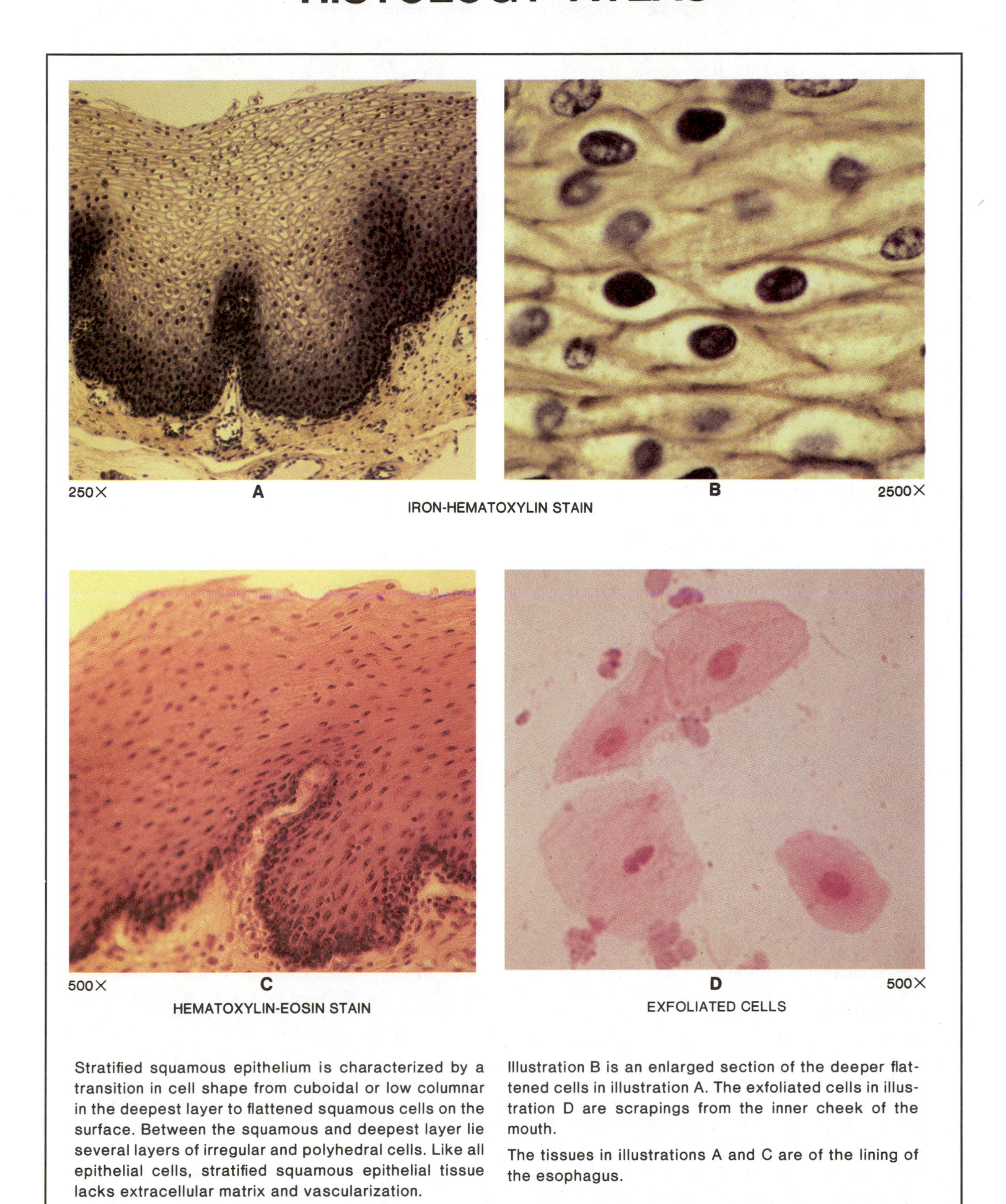

Stratified squamous epithelium is characterized by a transition in cell shape from cuboidal or low columnar in the deepest layer to flattened squamous cells on the surface. Between the squamous and deepest layer lie several layers of irregular and polyhedral cells. Like all epithelial cells, stratified squamous epithelial tissue lacks extracellular matrix and vascularization.

Illustration B is an enlarged section of the deeper flattened cells in illustration A. The exfoliated cells in illustration D are scrapings from the inner cheek of the mouth.

The tissues in illustrations A and C are of the lining of the esophagus.

Figure HA-1 Stratified squamous epithelium.

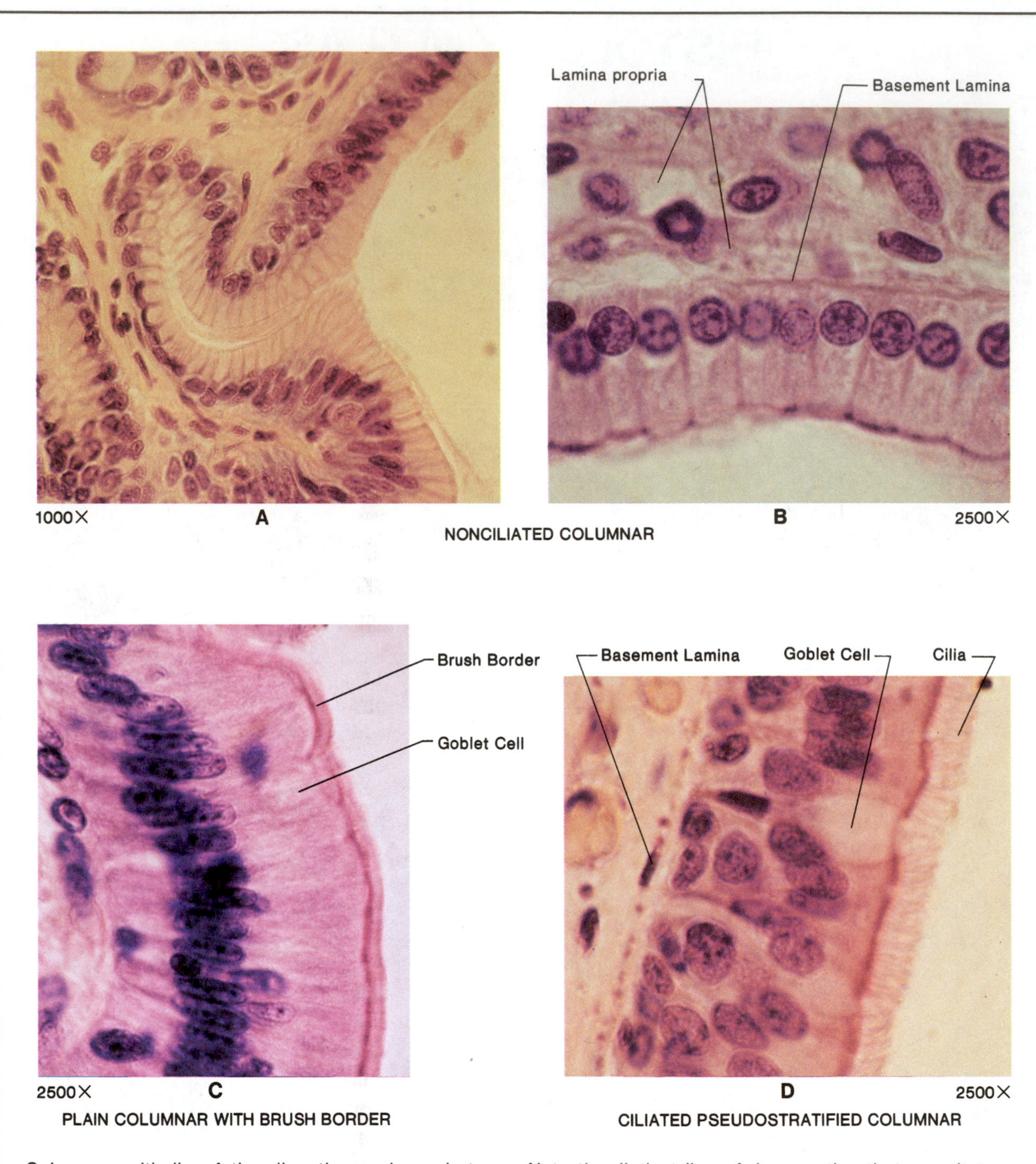

Columnar epithelia of the digestive and respiratory tracts exhibit large numbers of goblet cells that produce mucus. Although only illustrations C and D have them labeled, they can also be seen in illustration A.

The distinct differences between cilia and a brush border are seen in illustrations C and D. While cilia form from centrioles, brush borders are modified microvilli. Another modification of microvilli are cilialike structures called stereocilia. These nonmotile organelles are seen on columnar cells that line the vas deferens (see figure HA-34).

Note the distinct line of demarcation that constitutes the basement lamina in illustrations B and D. This thin layer between the epithelial cells and the lamina propria consists of a colloidal complex of protein, polysaccharide, and reticular fibers.

The lamina propria, to which all epithelial tissues are connected, consists of connective tissue, vascular and lymphatic channels, lymphocytes, plasma cells, eosinophils, and mast cells.

Figure HA-2 Columnar epithelium.

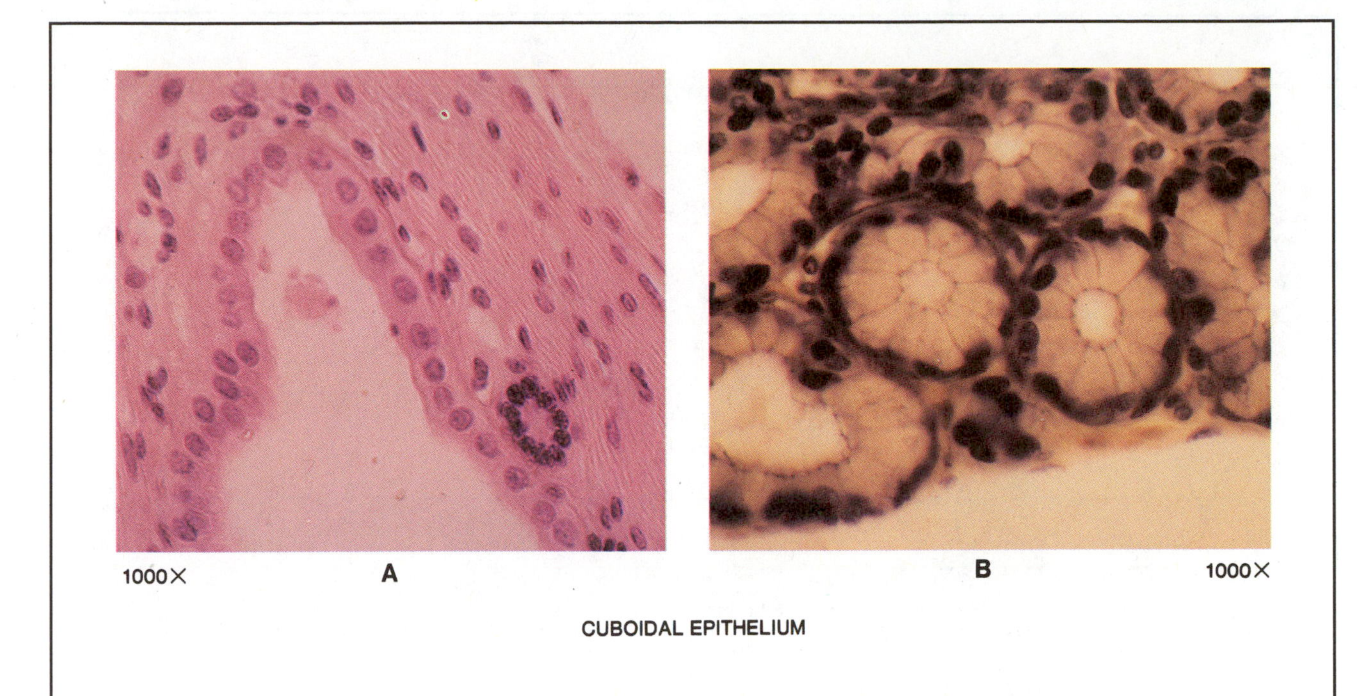

CUBOIDAL EPITHELIUM

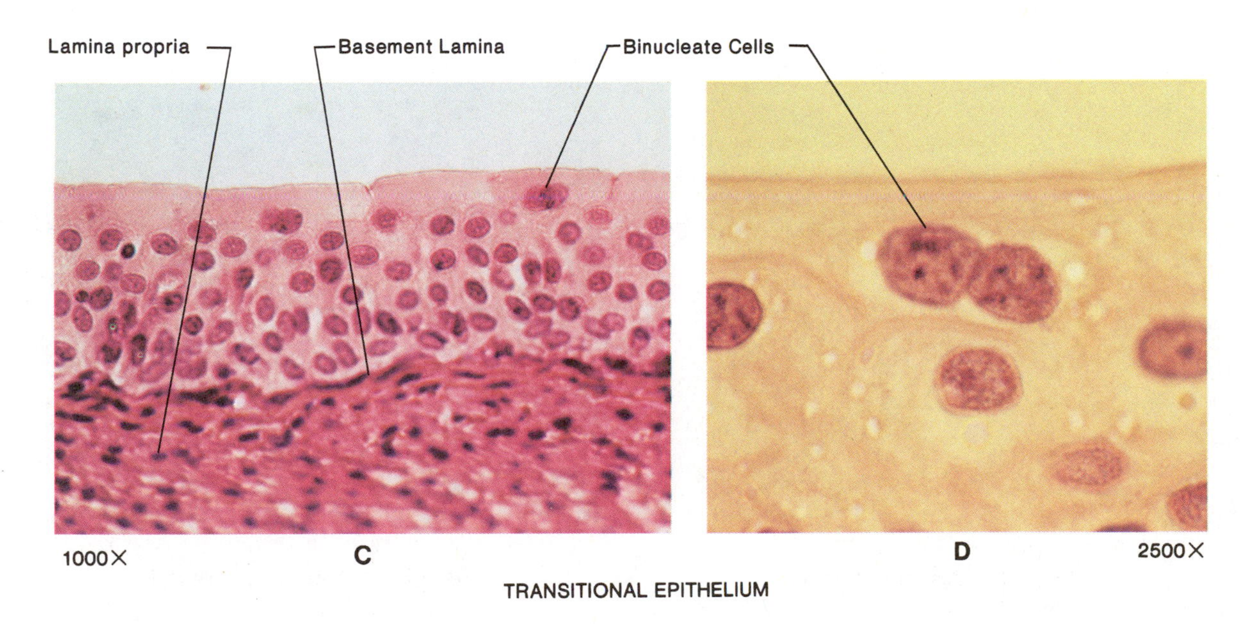

TRANSITIONAL EPITHELIUM

While cuboidal cells are usually thought of as having a squarish appearance, as in illustration A, they often take on a pyramidal structure when observed surrounding the lumen of a duct or a small gland as in illustration B. Cuboidal epithelia may serve both absorptive and secretory functions, as in the case of tubules in the kidney.

Transitional epithelium is a stratified epithelium whose surface cells do not fall into squamous, cuboidal, or columnar categories. Note that the surface cells are dome-shaped and often binucleate, while the basal cells are more like stratified columnar cells. Between the surface cells and basal cells can be seen layers of pear-shaped cells in a loose configuration. This type of tissue is seen in the wall of the urinary bladder, the urethra, and certain places in the kidneys. The loose nature of the cells makes it desirable in places where organ distention demands elasticity of tissue.

Figure HA-3 Cuboidal and transitional epithelium.

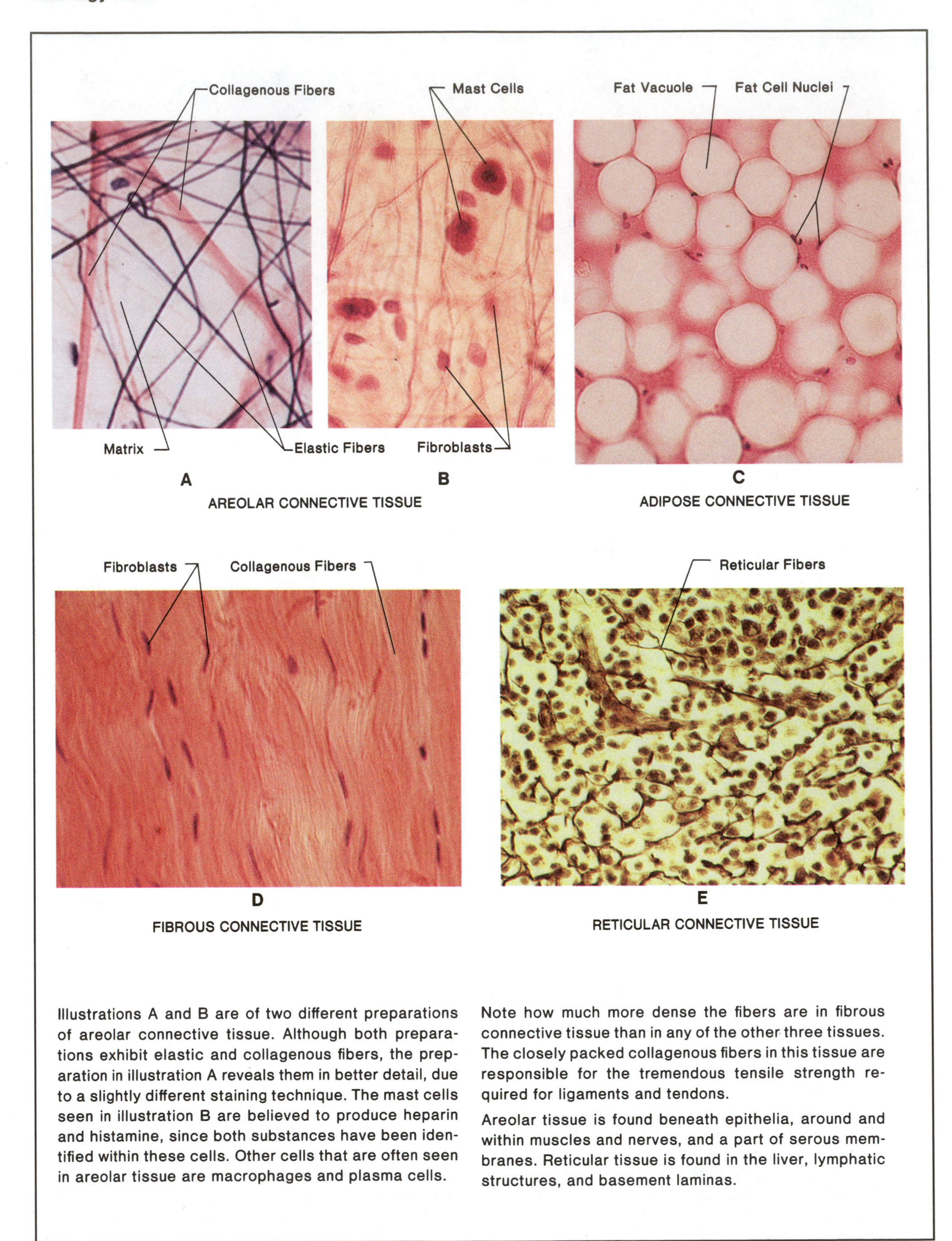

Illustrations A and B are of two different preparations of areolar connective tissue. Although both preparations exhibit elastic and collagenous fibers, the preparation in illustration A reveals them in better detail, due to a slightly different staining technique. The mast cells seen in illustration B are believed to produce heparin and histamine, since both substances have been identified within these cells. Other cells that are often seen in areolar tissue are macrophages and plasma cells.

Note how much more dense the fibers are in fibrous connective tissue than in any of the other three tissues. The closely packed collagenous fibers in this tissue are responsible for the tremendous tensile strength required for ligaments and tendons.

Areolar tissue is found beneath epithelia, around and within muscles and nerves, and a part of serous membranes. Reticular tissue is found in the liver, lymphatic structures, and basement laminas.

Figure HA-4 Connective tissues (1000×).

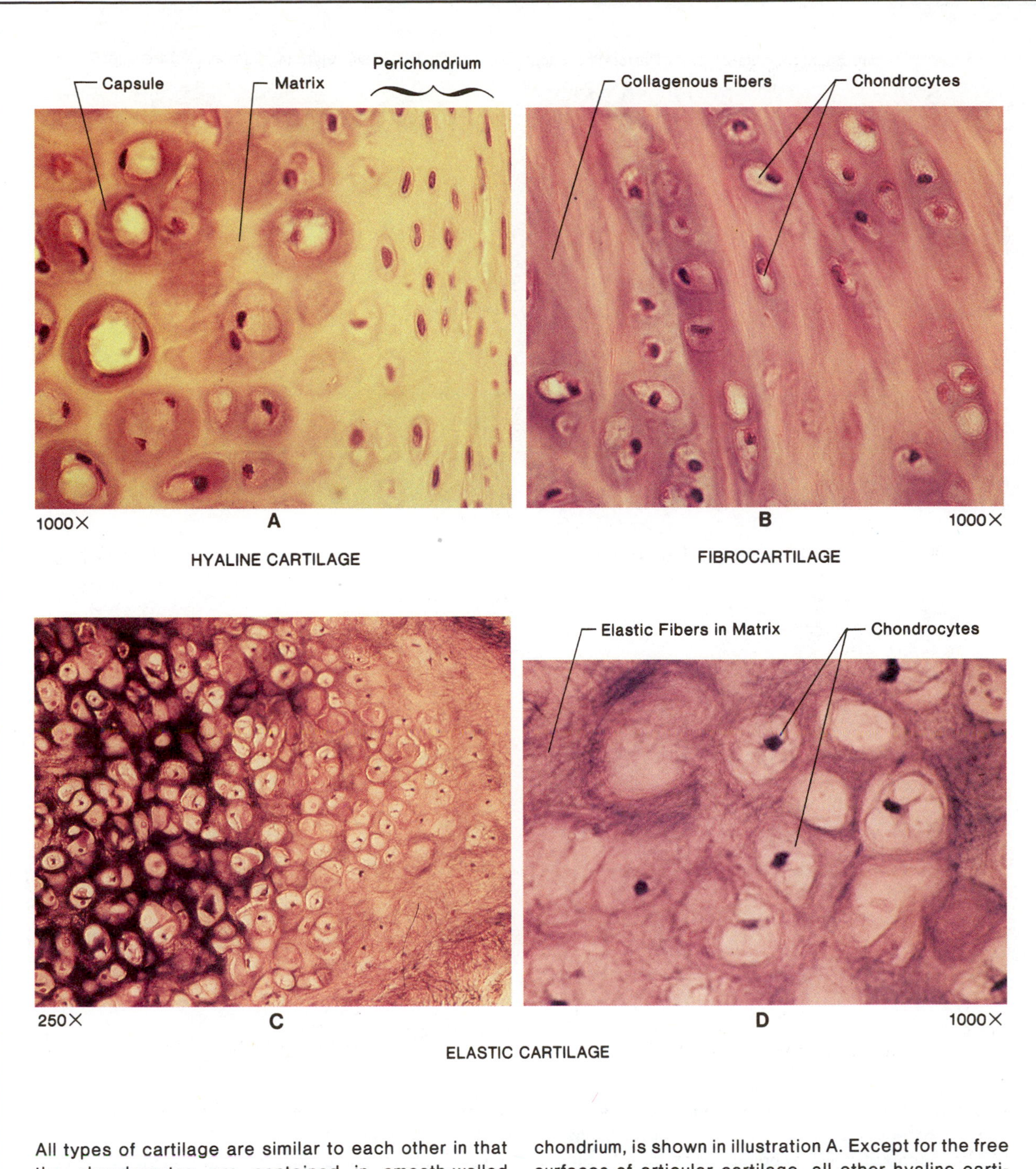

All types of cartilage are similar to each other in that the chondrocytes are contained in smooth-walled spaces called lacunae. Observe that the walls of the lacunae are quite dense and stain darker than surrounding matrix. These walls are referred to as capsules. Note that the chondrocytes appear to have large vacuoles in them. This apparent vacuolation is an artifact of tissue preparation due to the leaching out of fat and glycogen by solvents used in the staining process. Note that a dense layer of connective tissue, the perichondrium, is shown in illustration A. Except for the free surfaces of articular cartilage, all other hyaline cartilage structures are invested with a perichondrium.

The firmness of the matrix in hyaline cartilage is due to the presence of condroitin sulfate and collagen. Fibrocartilage is much stronger than hyaline cartilage because of the preponderance of collagenous fibers in its matrix. The presence of elastic fibers as well as collagenous fibers in elastic cartilage makes this type of cartilage more flexible.

Figure HA-5 Types of cartilage.

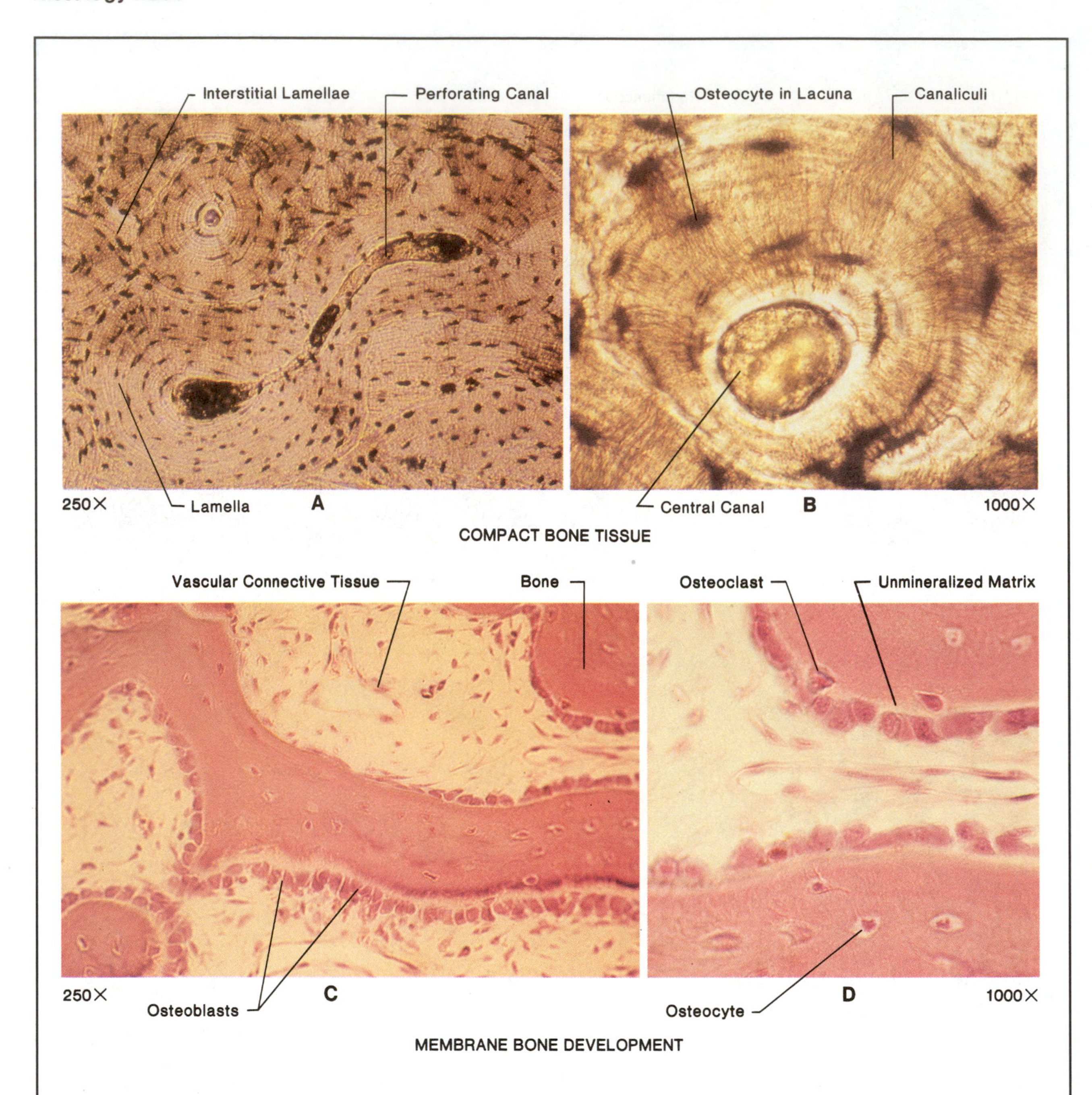

Bone formation has two origins: (1) from cartilage and (2) from osteogenic mesenchymal connective tissue. Compact bone of the long bones forms from cartilage; bones of the skull, on the other hand, develop as shown in illustrations C and D. The latter are often referred to as "membrane bones."

Note the loci of the different types of bone cells in the above illustrations. Illustrations C and D reveal that osteoblasts in membranous bone formation are seen on the leading edge of the forming bone. The first stage is the secretion of matrix by the osteoblasts. Reticular fibers are then added to the matrix by surrounding mesenchymal cells. Finally, mineralization occurs due to osteoblastic activity.

Osteoclasts are larger multinucleated cells that are involved in shaping bone structure by bone resorption. These cells are surrounded by a clear area (Howship's lacuna), evidence of mineral resorption. The osteoclast seen in illustration D is in an early stage of development; thus, Howship's lacuna is not very large. Once a bone cell becomes completely surrounded by bone it is referred to as an osteocyte. Nourishment in mature compact bone reaches osteocytes through canaliculi.

Figure HA-6 Bone histology.

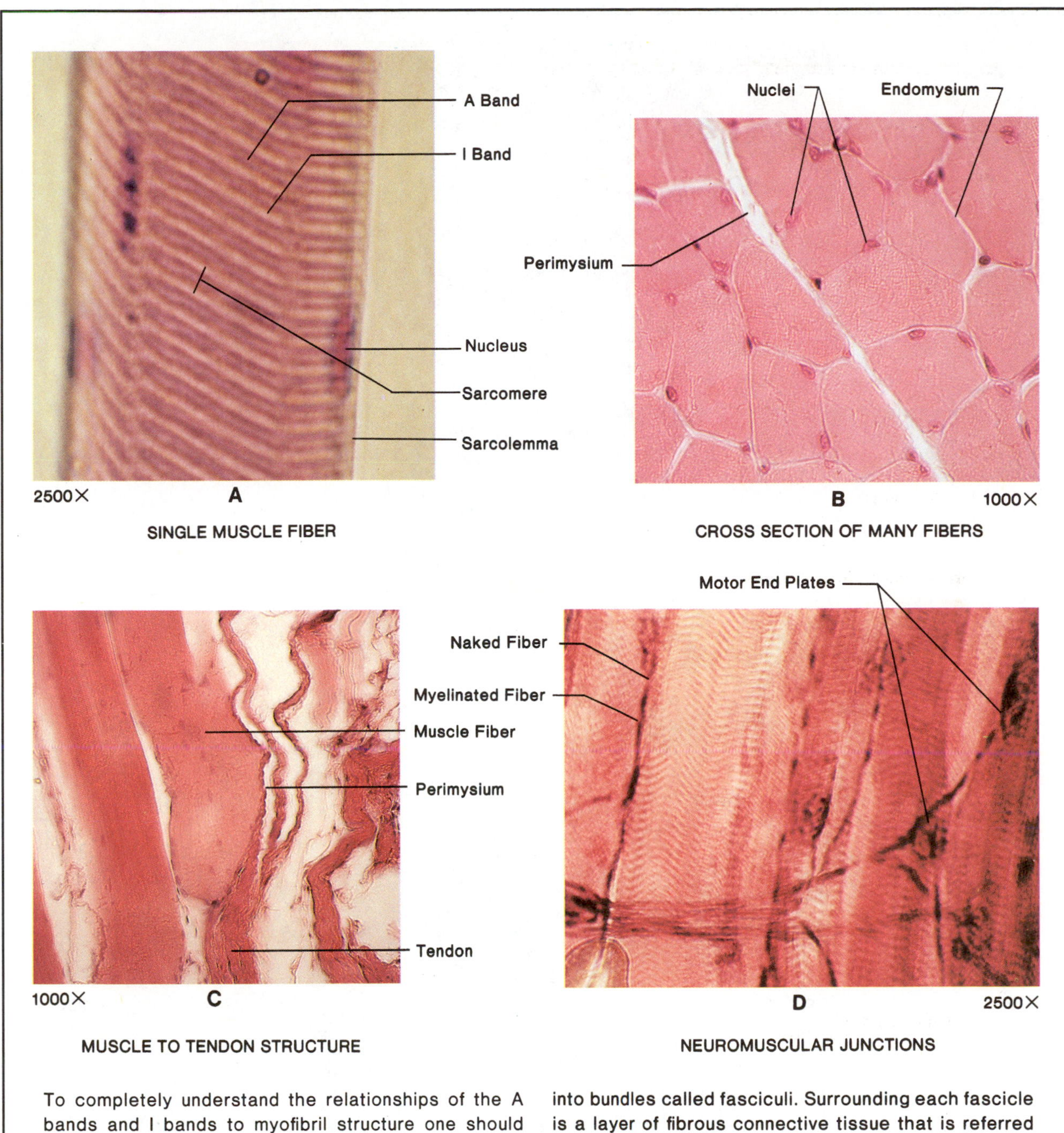

To completely understand the relationships of the A bands and I bands to myofibril structure one should compare illustration A with figure 21.1. Much detail of myofibril structure can only be seen with the electron microscope.

A significant characteristic of skeletal muscle cells is that they are multinucleated, or syncytial. Note that the nuclei are elongated and situated near the sarcolemma of the cell. The peripheral location of the nuclei shows up well in illustration B.

Illustration B reveals that muscle fibers are separated from each other by endomysium, and grouped together into bundles called fasciculi. Surrounding each fascicle is a layer of fibrous connective tissue that is referred to as the perimysium.

Illustration C illustrates how the fibrous connective tissue of the endomysium, perimysium, and epimysium is continuous with the tendons which attach muscles to bone. The connective tissue of tendons, in turn, is continuous with the periosteum of bone.

Note in illustration D how the myelinated motor nerve fibers lose their myelin sheaths and become "naked" where they join the motor end plates. Compare with figure 20.1.

Figure HA-7 Skeletal muscle microstructure.

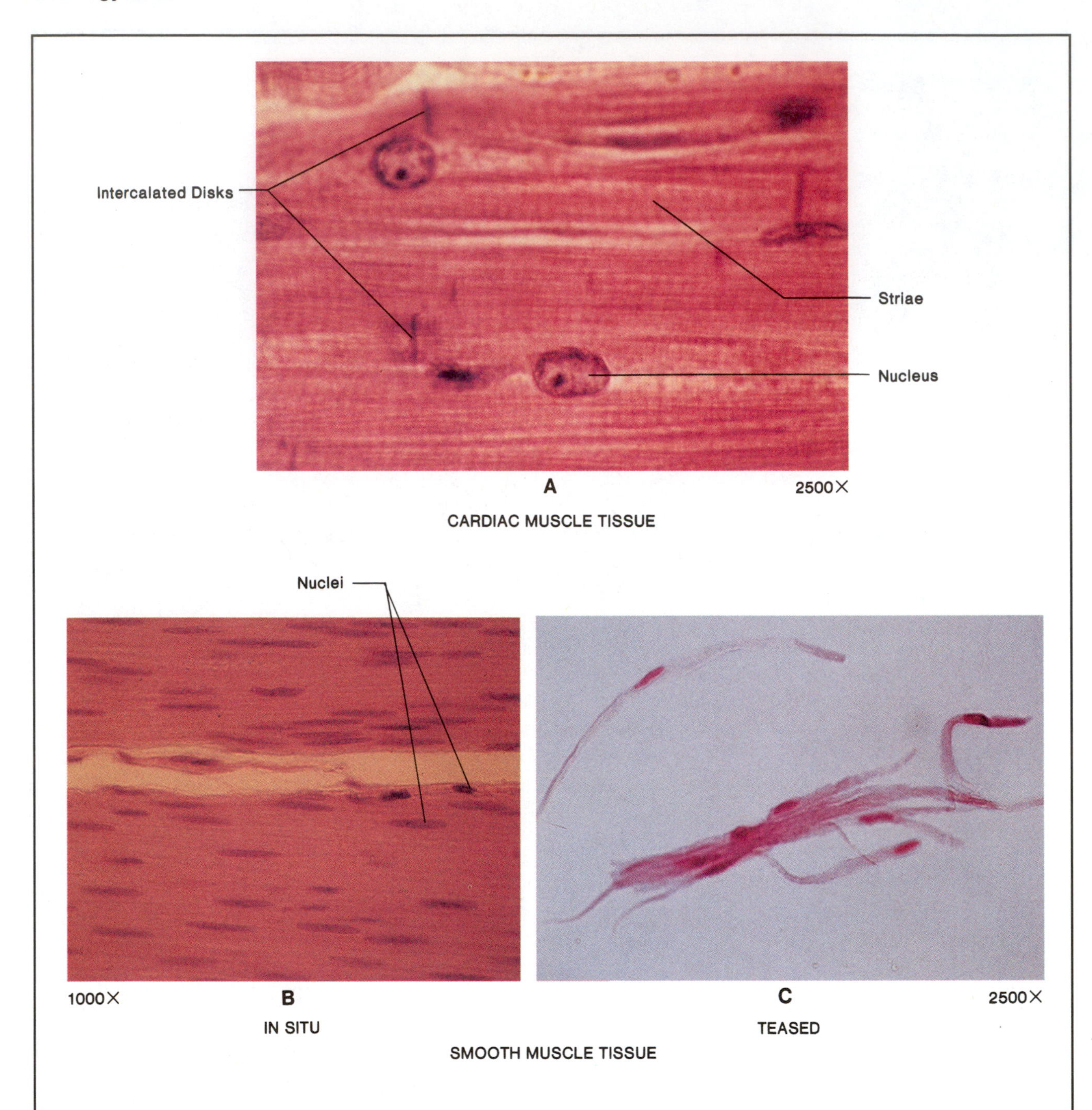

The distinct contrast of the striae in the cardiac muscle preparation in illustration A is evidence that these fibers are in a state of complete relaxation; fibers prepared from tissue in contraction generally do not reveal striae as distinctly. Although this tissue may appear to be syncytial, it is not. Cardiac fibers have only one nucleus per fiber with intercalated disks forming the limits for each unit.

The smooth muscle tissue seen in illustration B is of a portion of the intestinal wall. Due to the closely packed nature of the cells in such structures it is very difficult to see individual cells. Illustration C reveals how cells of this type of tissue appear when separated with a probe prior to staining. Each cell has a single nucleus and lacks the cross-striations seen in skeletal and cardiac muscle tissue.

Figure HA-8 Cardiac and smooth muscle tissue.

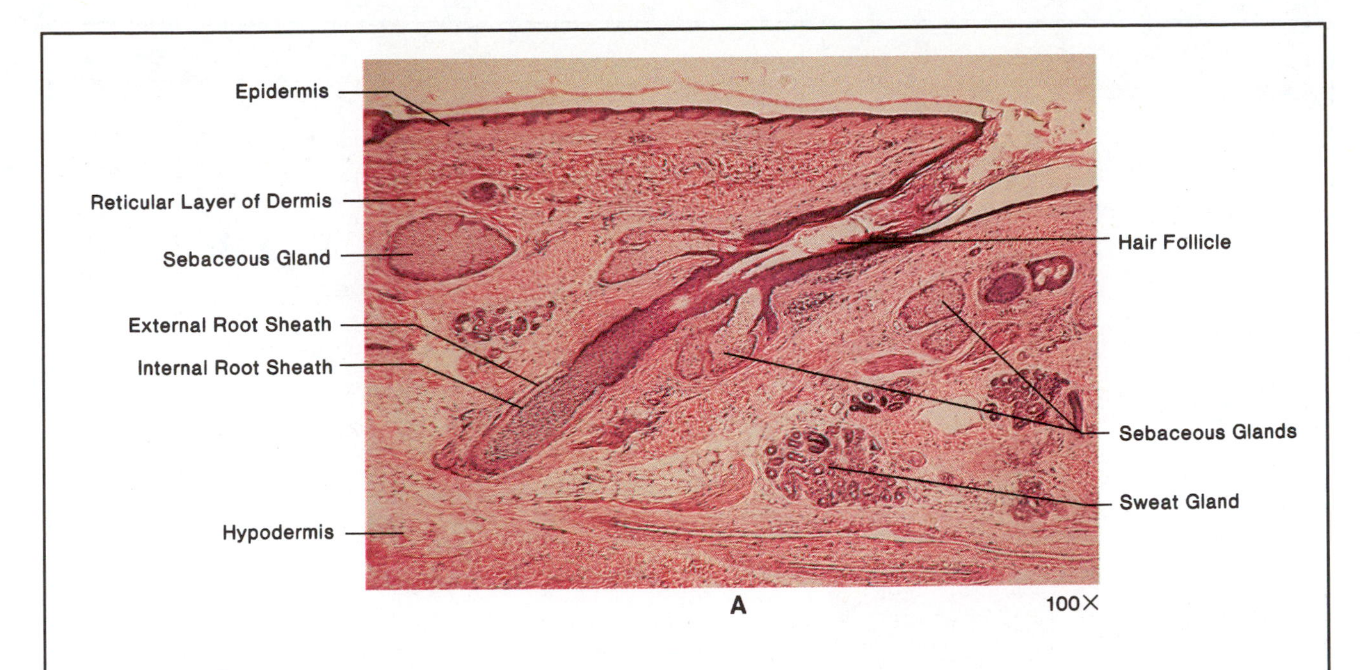

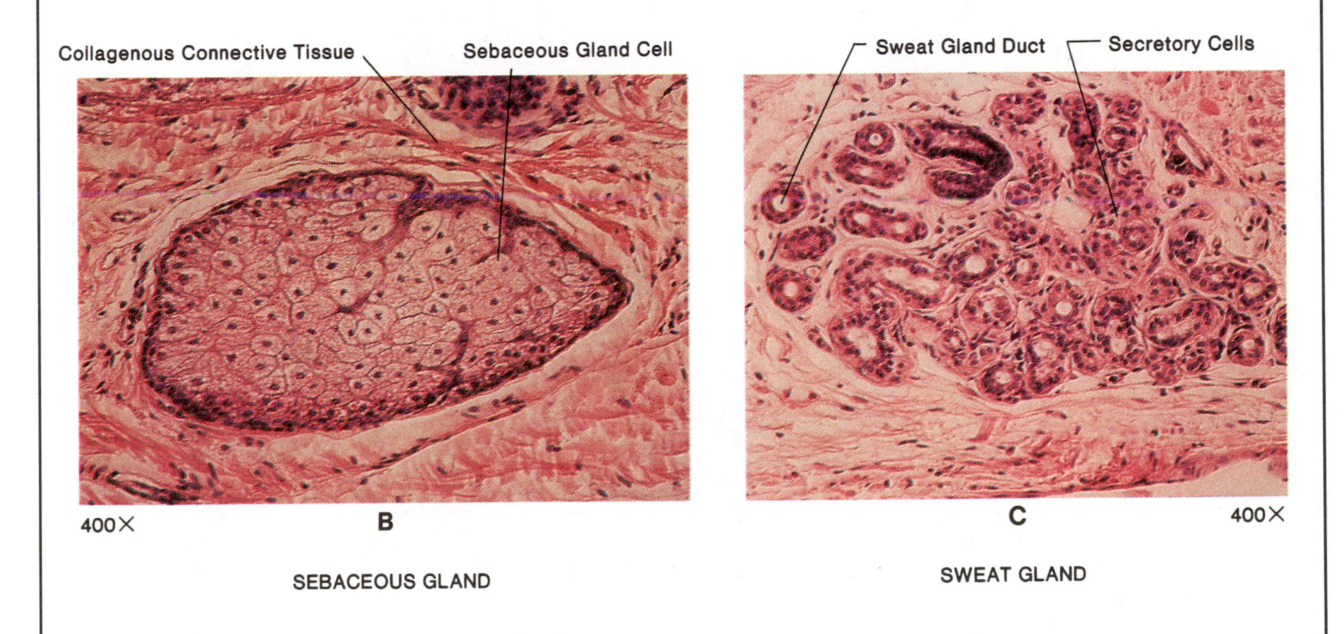

SEBACEOUS GLAND

SWEAT GLAND

The relationship of the sebaceous and sweat glands to other structures of the skin is illustrated on this page. The low magnification of illustration A reveals that the epidermis is a very thin layer as compared to the dermis. Except for the palms and soles of the feet, the epidermis is usually only 0.1 mm thick. The dermis or corium, which underlies the epidermis, varies from 0.3 to 4.0 mm thick in different parts of the body. Note that both the sebaceous and sweat glands are located in the reticular layer of the dermis.

Observe that the sweat glands are tubulor-alveolar structures lined with cuboidal or columnar epithelium. The sebaceous glands, on the other hand, consist of rounded masses of cells, cuboidal at the periphery, and polygonal in the center. The secretion of sebaceous glands is formed by the breaking down of the central cells into an oily complex which is forced out into the space between the follicle and hair shaft. Although not shown in illustration A, the sweat glands open out directly onto the surface of the skin.

Figure HA-9 Scalp histology.

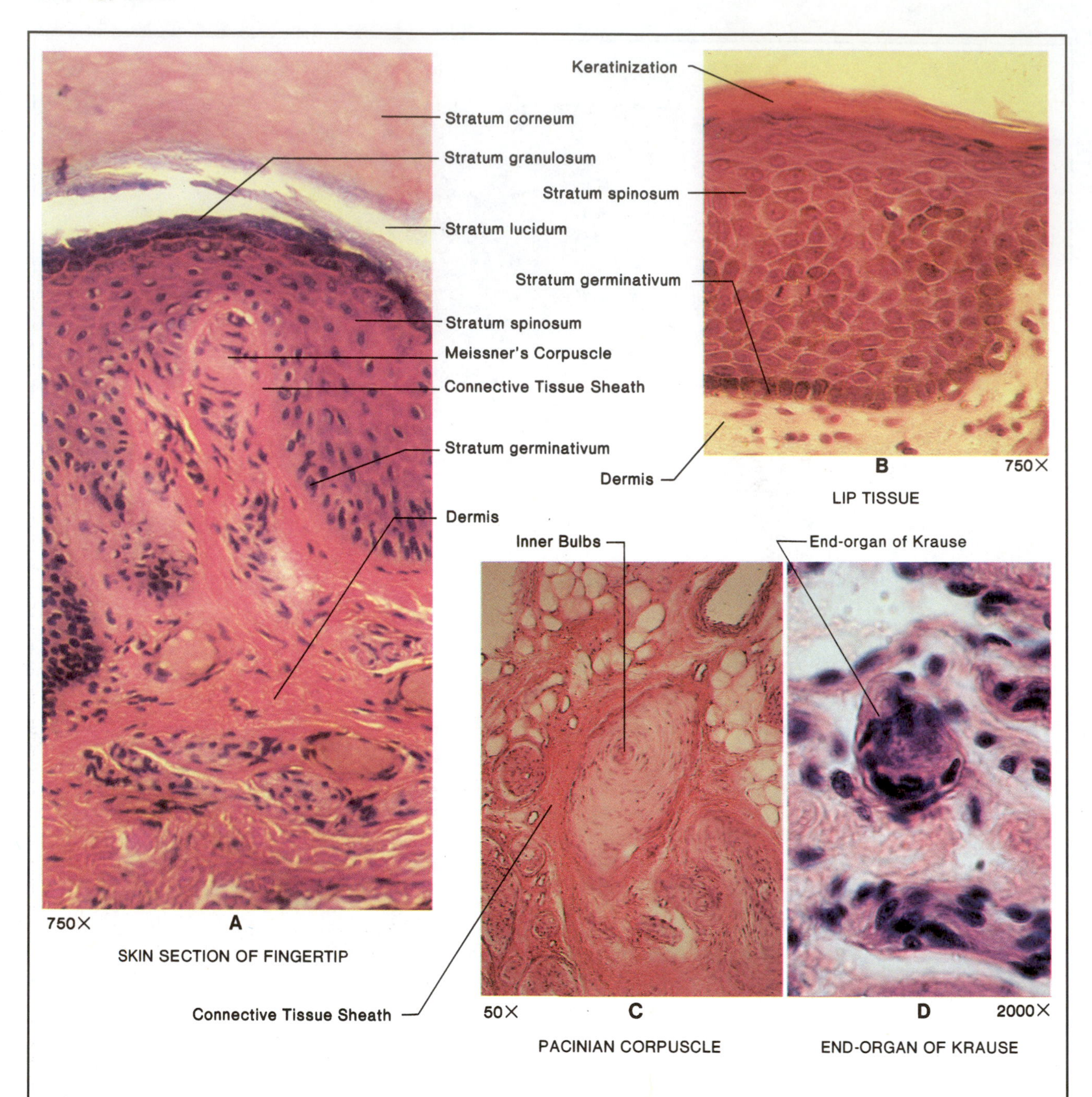

A comparison of the skin of the scalp (figure HA-9), a fingertip (illustration A), and the lip (illustration B) reveals that the integument differs in construction in different regions of the body. These differences are due to the presence or absence of hair and the degree of keratinization. The fingertips, palms, and soles of the feet have epidermal layers with a thick stratum corneum and no hair follicles. Skin on the lips has some keratinization, but to a much lesser degree than the fingertips.

Three receptors of the skin are illustrated here: Meissner's, pacinian, and Krause corpuscles. Meissner's corpuscle, which is seen in illustration A, is located in the papillary layer of the dermis. Note that it is encased in a sheath of connective tissue. These receptors are sensors of discriminative touch.

Pacinian corpuscles (illustration C) consist of laminated collagenous material and several inner bulbs. Pressure on the lamina triggers impulses in nerve endings in the bulbs. They are so large that they can be seen without magnification.

End-organs of Krause are small corpuscles that are sensitive to cold temperatures. They, too, are quite numerous in the skin.

Figure HA-10 Skin structure.

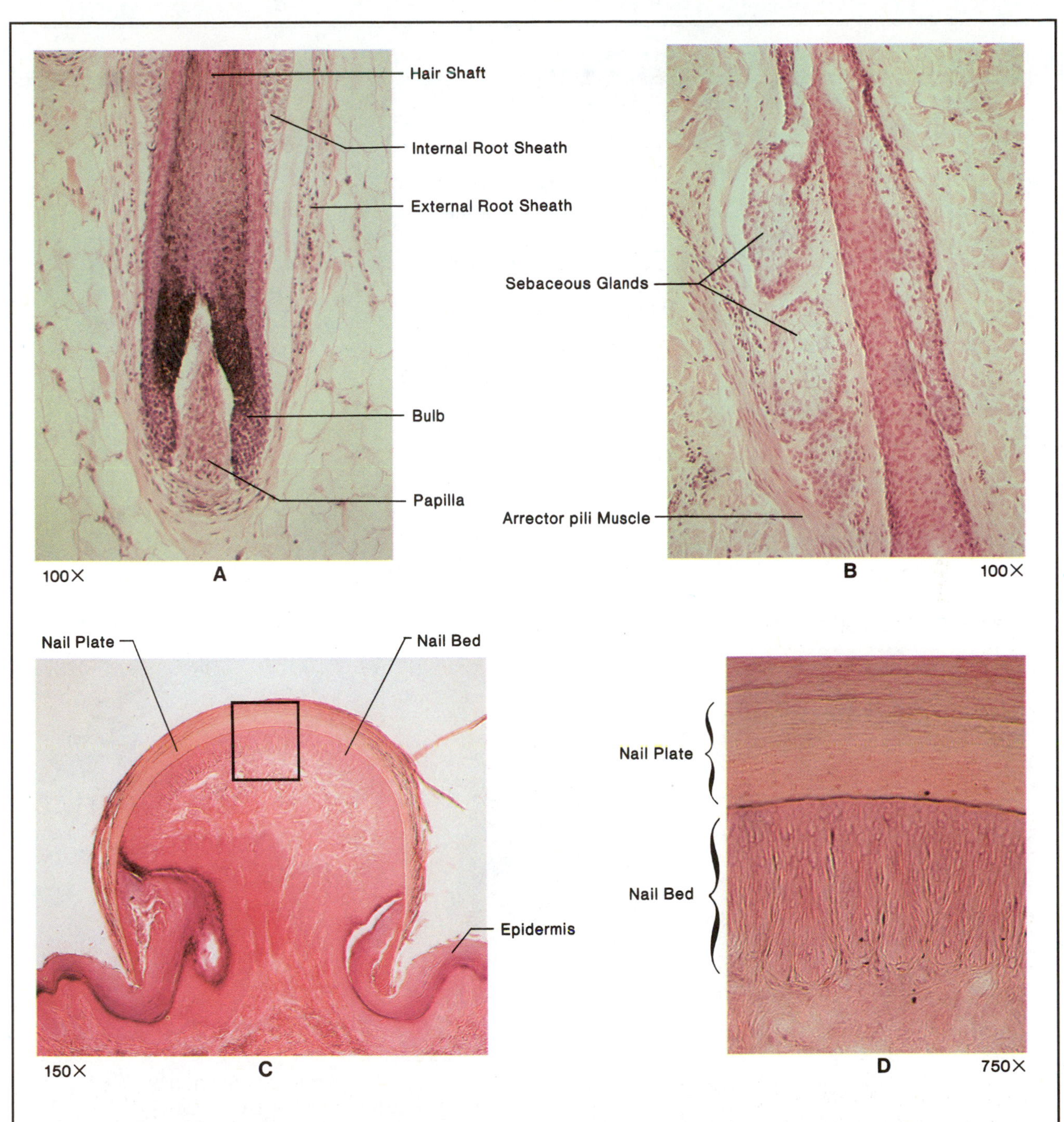

Hair consists of horny threads that are derived from the epidermis. Its shaft is enveloped by a tube, the follicle. Three components of the hair follicle (illustration A) are the internal root sheath, an external root sheath, and a connective tissue sheath. The two inner sheaths are derived from the epidermis; the outer connective tissue sheath is formed from the dermis.

Illustration B reveals the relationship of the arrector pili muscle to the sebaceous glands. These small muscles, which consist of smooth muscle tissue, play a role in forcing sebum out into the hair follicle.

Nails are a modification of the epidermis. The fingernail in illustration C is from an infant. Microscopically, the body of the nail consists of several layers of flattened, clear cells that take the place of the stratum corneum of skin. These cells are harder than stratum corneum layers and have shrunken nuclei.

The nail bed under the nail plate is the epidermis, but lacks the stratum granulosum and stratum lucidum; only the deeper epidermal layers are present. Illustration D reveals its structure.

Figure HA-11 Hair and nail structure of the integument.

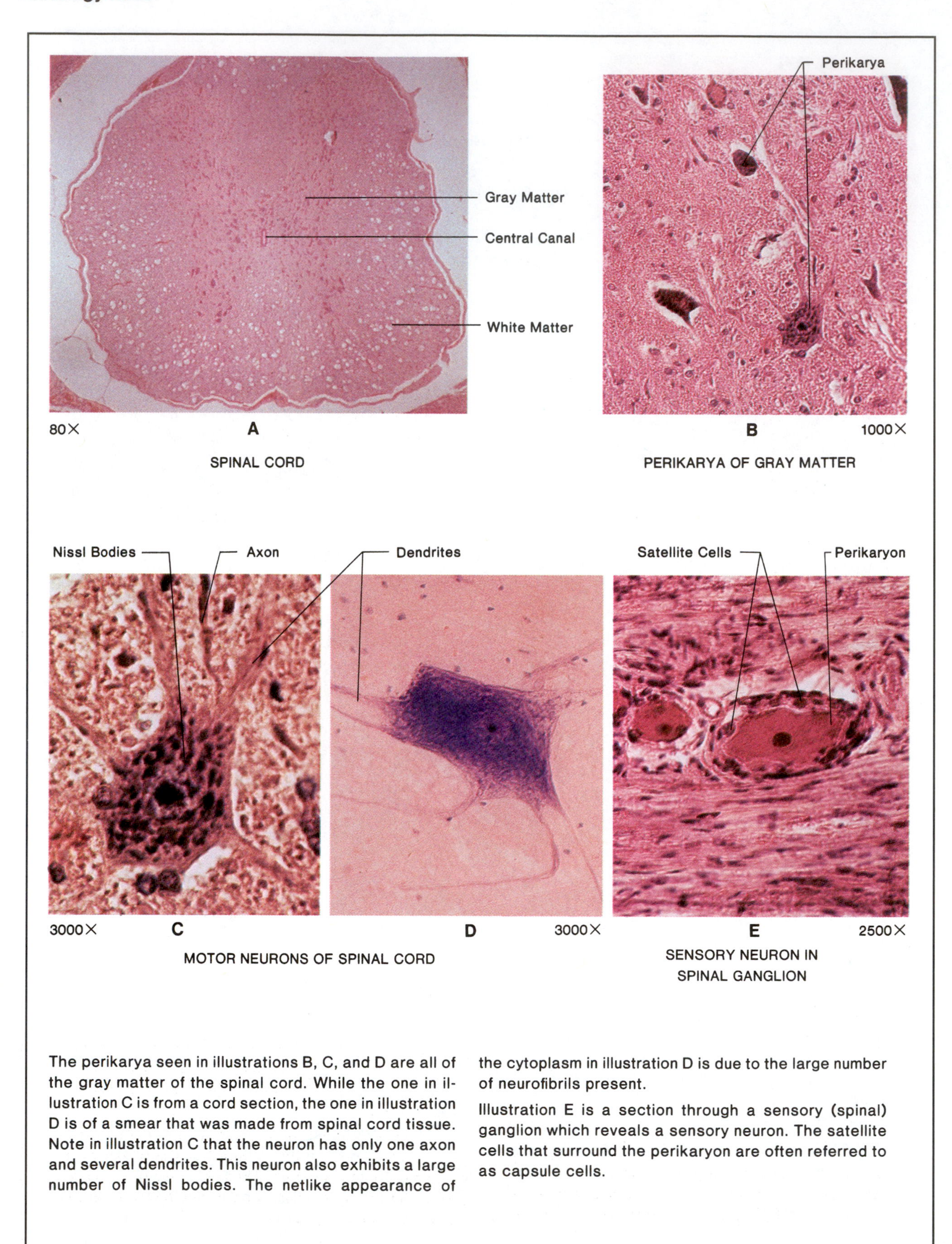

The perikarya seen in illustrations B, C, and D are all of the gray matter of the spinal cord. While the one in illustration C is from a cord section, the one in illustration D is of a smear that was made from spinal cord tissue. Note in illustration C that the neuron has only one axon and several dendrites. This neuron also exhibits a large number of Nissl bodies. The netlike appearance of the cytoplasm in illustration D is due to the large number of neurofibrils present.

Illustration E is a section through a sensory (spinal) ganglion which reveals a sensory neuron. The satellite cells that surround the perikaryon are often referred to as capsule cells.

Figure HA-12 Neurons of the spinal cord and spinal ganglia.

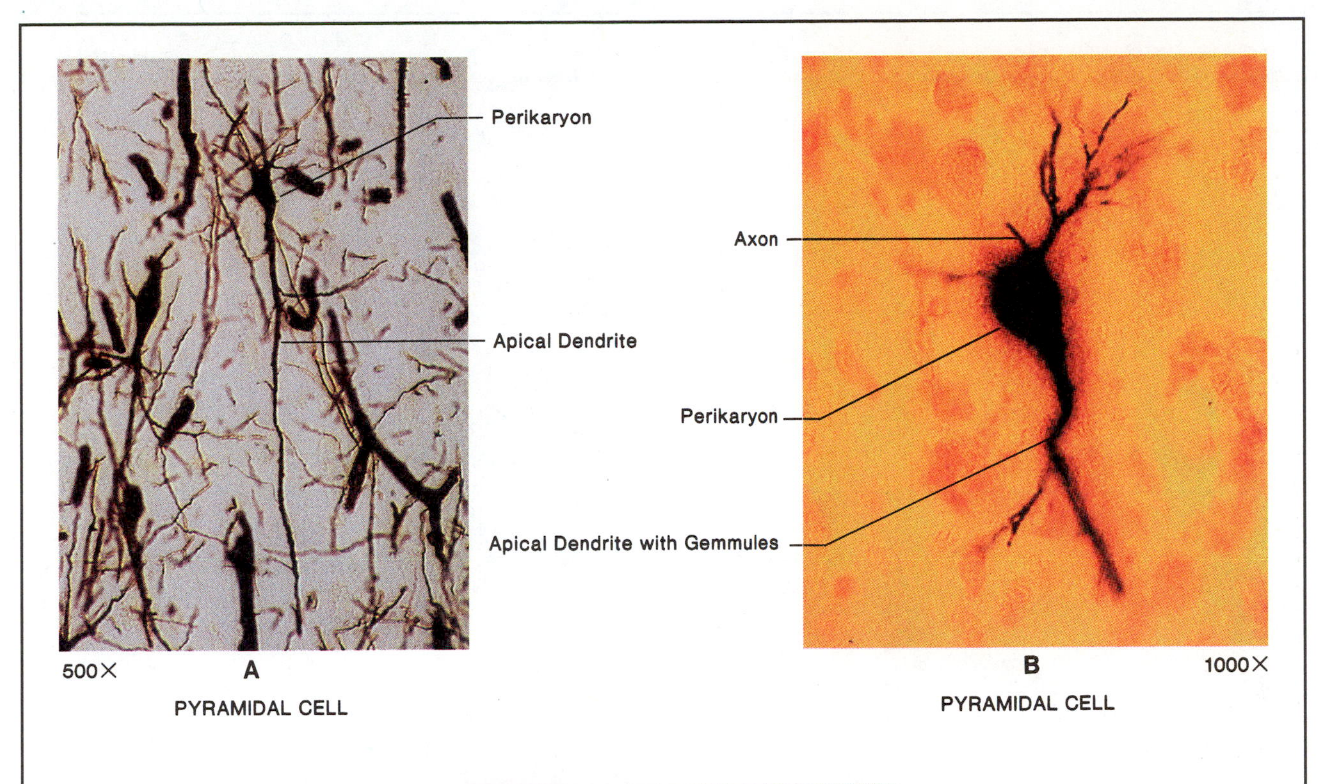

PYRAMIDAL CELL

PYRAMIDAL CELL

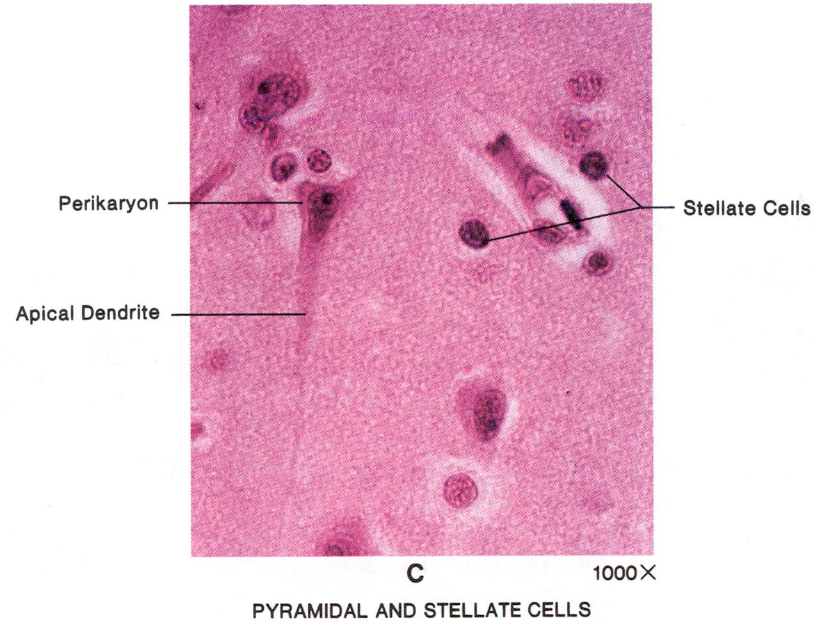

PYRAMIDAL AND STELLATE CELLS

The principal nerve cells of the cerebrum are pyramidal cells. Note in each of the above photomicrographs that a long apical dendrite extends from the perikaryon. The apical dendrite is always oriented toward the surface of the cerebrum. In illustration B there is evidence of "gemmules" on the surface of the dendrite. Gemmules are small processes that greatly increase the surface area of dendrites, allowing large neurons to receive as many as 100,000 separate axon terminals or synapses.

Note the difference in size between the axon and dendrites in illustration B. The axon is much smaller in diameter and has a smoother surface due to the absence of gemmules.

The stellate cells shown in illustration C are association neurons that provide connections between pyramidal cells of the cerebrum. These interneurons are also present in the cerebellum, where they provide linkage between the Purkinje cells.

Figure HA-13 Neurons of the cerebrum.

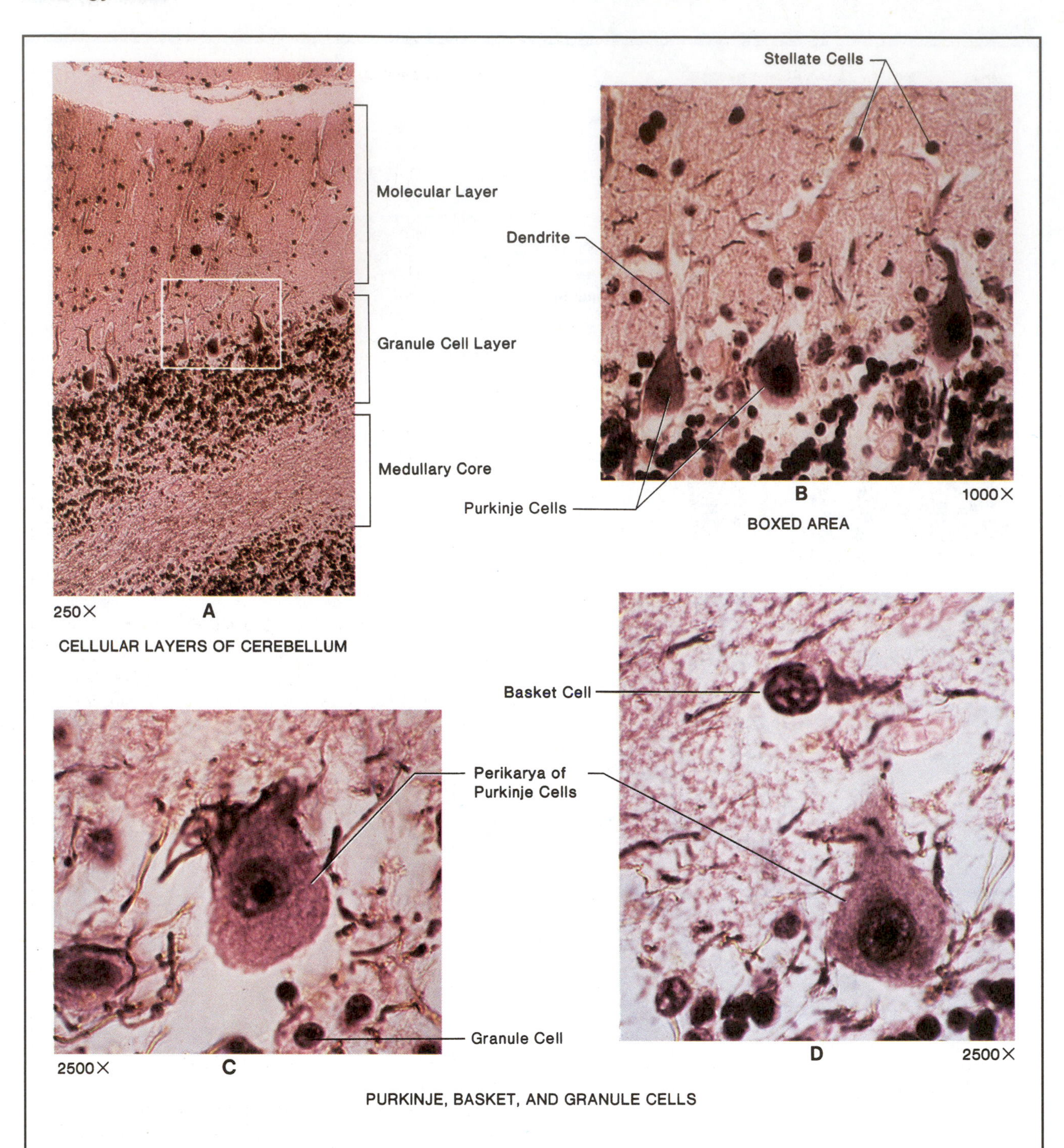

Illustration A is a section through the cerebellum that depicts an outermost molecular layer, a middle granule cell layer, and an inner medullary core.

The most distinctive neurons of the cerebellum are the Purkinje cells. Note in illustration A that these neurons form a layer deep within the molecular layer near the edge of the granule cell layer. Each flask-shaped cell has a thick dendrite that is directed toward the cerebellar cortex. A short distance out from the perikaryon the dendrite branches out into two thick branches, which, in turn, divide many times to arborize out into the molecular layer. The axon of each Purkinje cell extends from the perikaryon through the granular and medullary layers, and finally, through collaterals, reenters the molecular layer to contact other Purkinje cells.

Stellate, basket, and granule cells are all multipolar association neurons. Note the proximity of the perikarya of the basket cells to the Purkinje cells.

Figure HA-14 Neurons of the cerebellum.

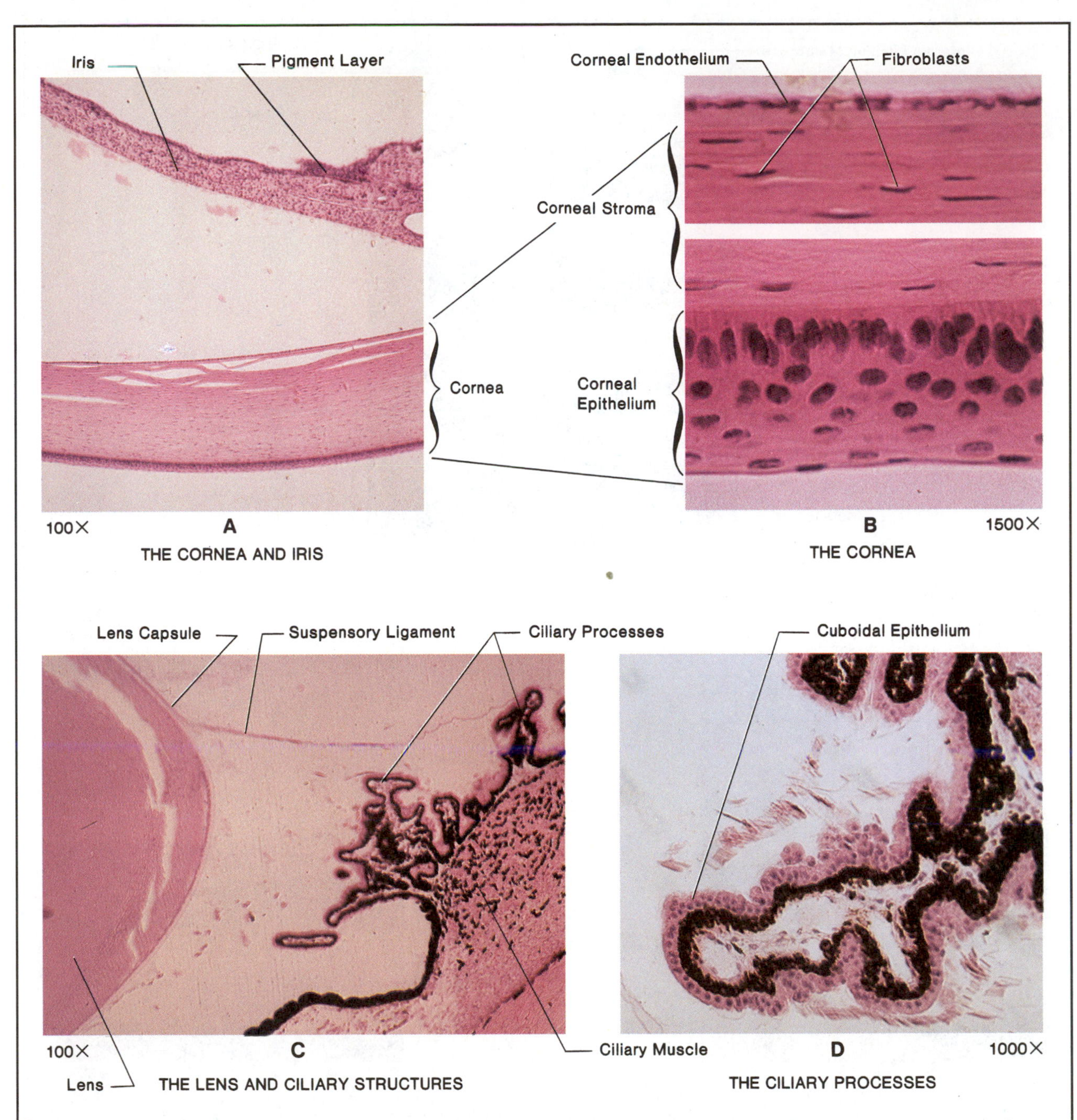

Illustration A reveals the relationship of the iris to the cornea. If a portion of the lens were shown in this photomicrograph, it would be above the iris, since the iris lies between the lens and the cornea.

Illustration B is a compacted enlargement of the cornea, in which many of the middle layers have been removed so that all significant layers can be seen. Note that the outer corneal epithelium consists of stratified squamous epithelium, and that the inner corneal endothelium consists of a single layer of low cuboidal cells. Between these two layers is the corneal stroma (*substantia propria*), which makes up nine-tenths of the thickness of the cornea. This transparent stroma is made up of collagenous fibrils and fibroblasts, all held together with a mucopolysaccharide cement.

Note in illustration C that the lens is encased in an elastic capsule to which the suspensory ligament is attached. The lens is formed from epithelial cells that become elongated and lose their nuclei. Note, also, that the ciliary processes shown in illustration D are covered with cuboidal epithelial cells. It is these cells that produce the aqueous humor that fills the space between the lens and the cornea.

Figure HA-15 The cornea, lens, iris, and ciliary structures.

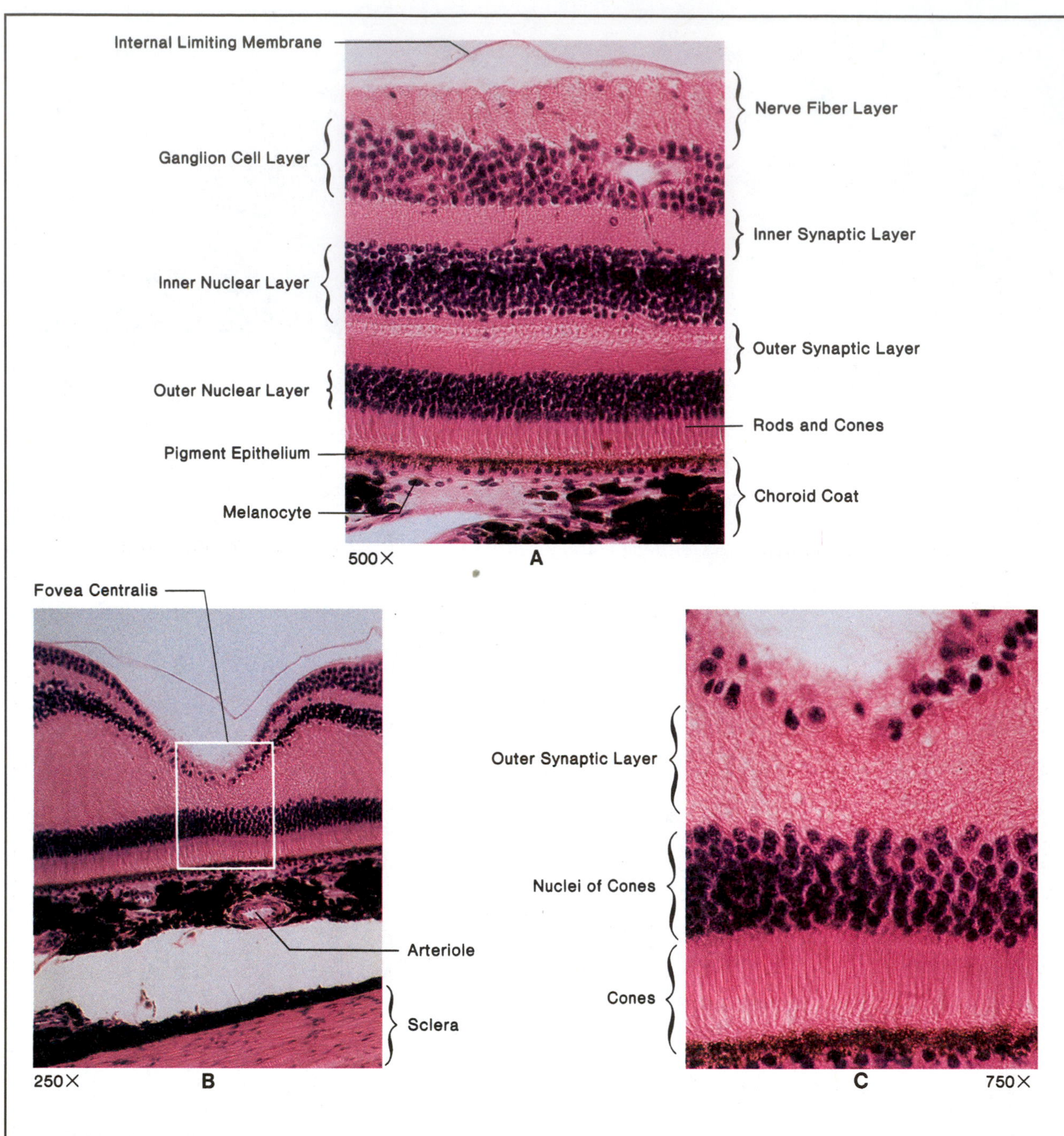

Illustration A reveals the various neuronal layers of the retina. Note that the photoreceptors (rods and cones) are in the deepest layer next to the choroid coat. Light, striking the retina, must pass through all these layers to reach the receptors.

The nuclei seen in the outer nuclear layer are of the rods and cones. The outer synaptic (*plexiform*) layer is where the synapses are made between axons of the rods and cones and the dendrites of bipolar cells and the processes of horizontal cells. The inner nuclear layer contains nuclei of bipolar neurons and association neurons (horizontal and amacrine cells). The inner synaptic (*plexiform*) layer is the place where synaptic connections are made between the cells of the inner nuclear layer and the ganglion cells. The nerve fiber layer consists of nonmyelinated axons of the ganglion cells. These fibers converge on the optic disk to form the optic nerve. Illustrations B and C reveal the nature of the fovea centralis. Note that the receptors in the fovea consist only of cones and that the inner synaptic and nerve fiber layers are lacking.

Figure HA-16 The retina of the eye.

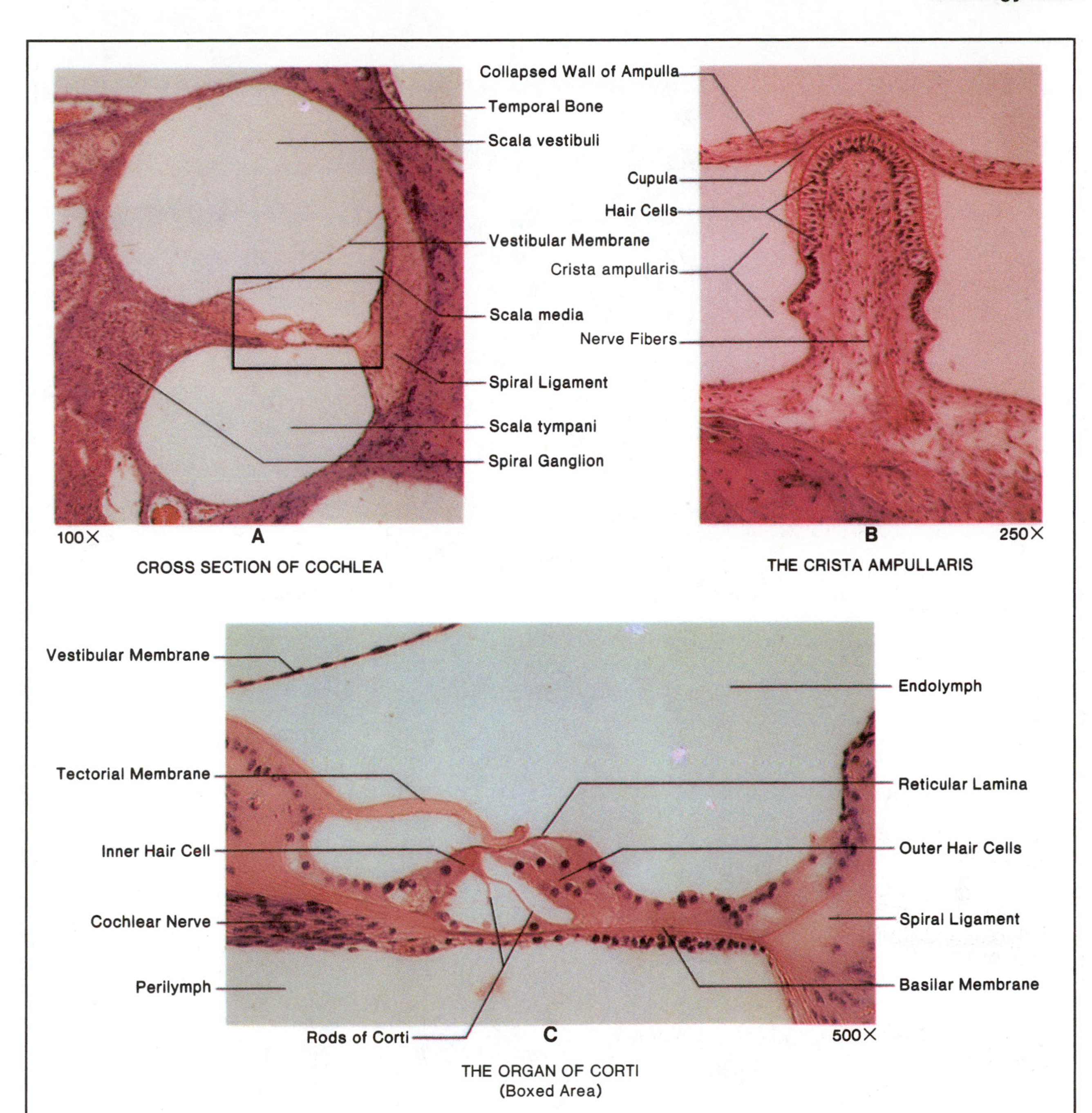

The cochlea of the ear lies within the temporal bone. Along its outer wall is seen a spiral ligament (illustration A) that is formed from periosteum of the bone. Note that fibers of the spiral ligament are continuous with the basilar membrane. Observe that the cochlea has three chambers: scala vestibuli, scala tympani, and scala media. The cochlear duct is the flexible portion of the cochlea that is bounded on one side by the vestibular membrane and on the other side by the basilar membrane. Endolymph is present in the cochlear duct; perilymph fills the scala vestibuli and scala tympani.

The organ of Corti shown in illustration C is an enlargement of the boxed area in illustration A. Note how the gelatinous flap of the tectorial membrane lies in contact with hairs that emerge from the hair cells. Discrimination of sound frequencies results from nerve impulses sent to the brain via the cochlear nerve in conjunction with vibrations in the endolymph and basilar membrane.

The obliteration of the cupula in illustration B occurred during slide preparation. Other structures in the crista ampullaris are normal.

Figure HA-17 The cochlea and crista ampullaris.

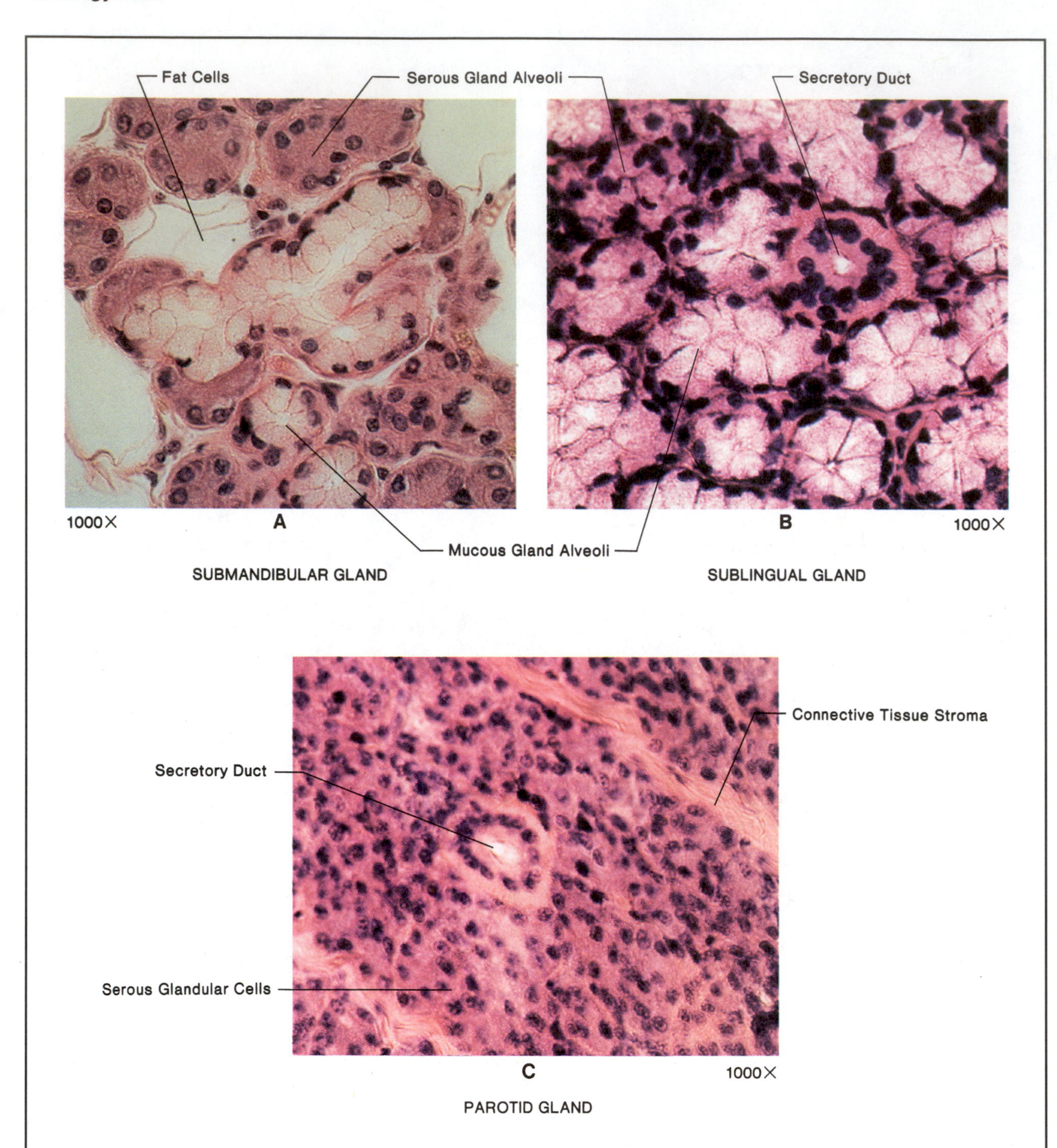

Whether a salivary gland produces mucus or the less viscous serous fluid is easily determined by microscopic examination of its tissue. Note in illustrations A and B that mucus-producing cells are much paler and larger than cells that produce serous fluid. The presence of both types of secretory cells in the submandibular and sublingual glands indicates that these glands secrete a mixture of mucus and serous fluid into the oral cavity. Although the submandibular gland is primarily serous, the sublingual produces mostly mucus. The absence of mucous acini in the parotid glands indicates that only serous fluid is produced by these glands.

Figure HA-18 Histology of major salivary glands.

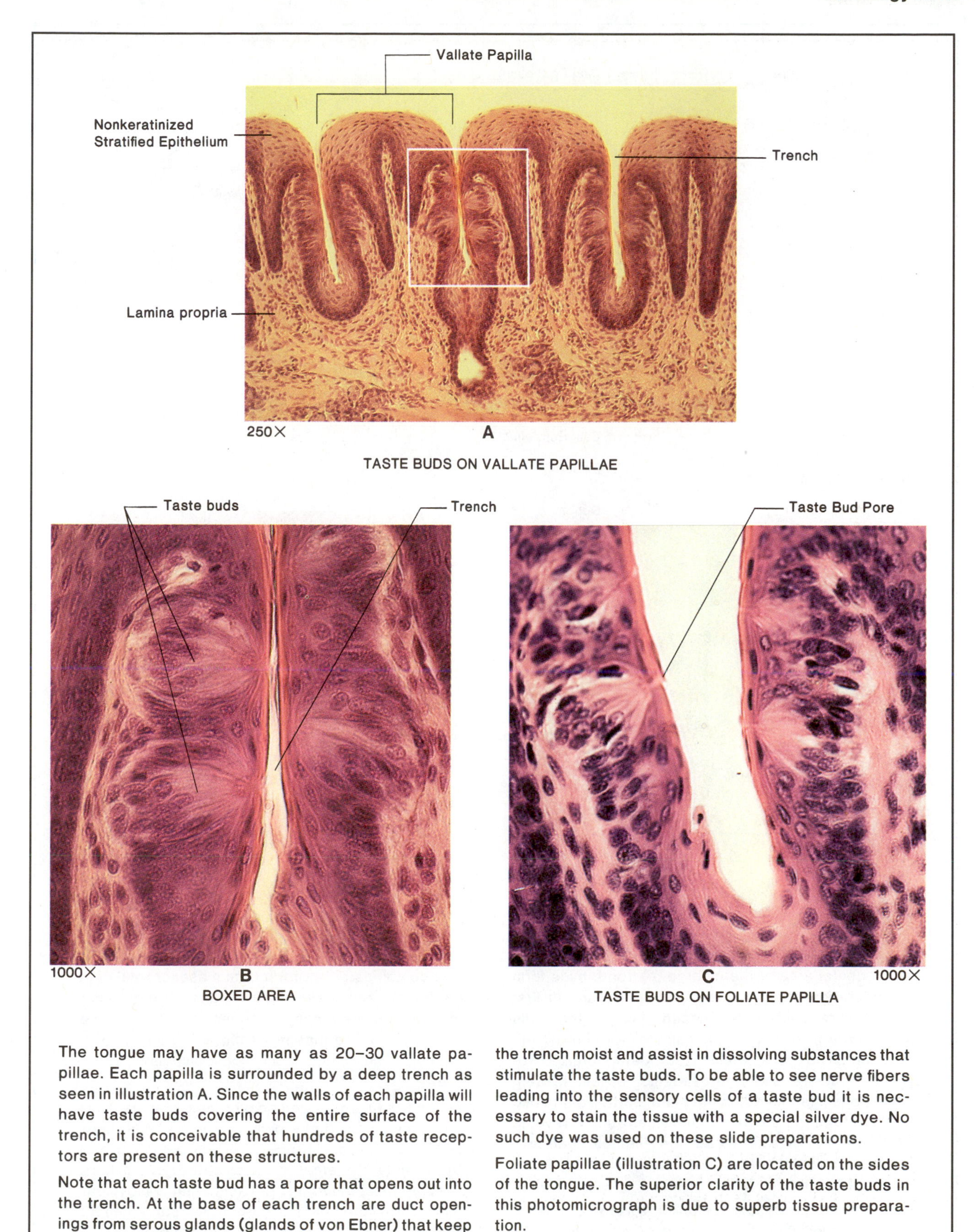

The tongue may have as many as 20–30 vallate papillae. Each papilla is surrounded by a deep trench as seen in illustration A. Since the walls of each papilla will have taste buds covering the entire surface of the trench, it is conceivable that hundreds of taste receptors are present on these structures.

Note that each taste bud has a pore that opens out into the trench. At the base of each trench are duct openings from serous glands (glands of von Ebner) that keep the trench moist and assist in dissolving substances that stimulate the taste buds. To be able to see nerve fibers leading into the sensory cells of a taste bud it is necessary to stain the tissue with a special silver dye. No such dye was used on these slide preparations.

Foliate papillae (illustration C) are located on the sides of the tongue. The superior clarity of the taste buds in this photomicrograph is due to superb tissue preparation.

Figure HA-19 Taste buds.

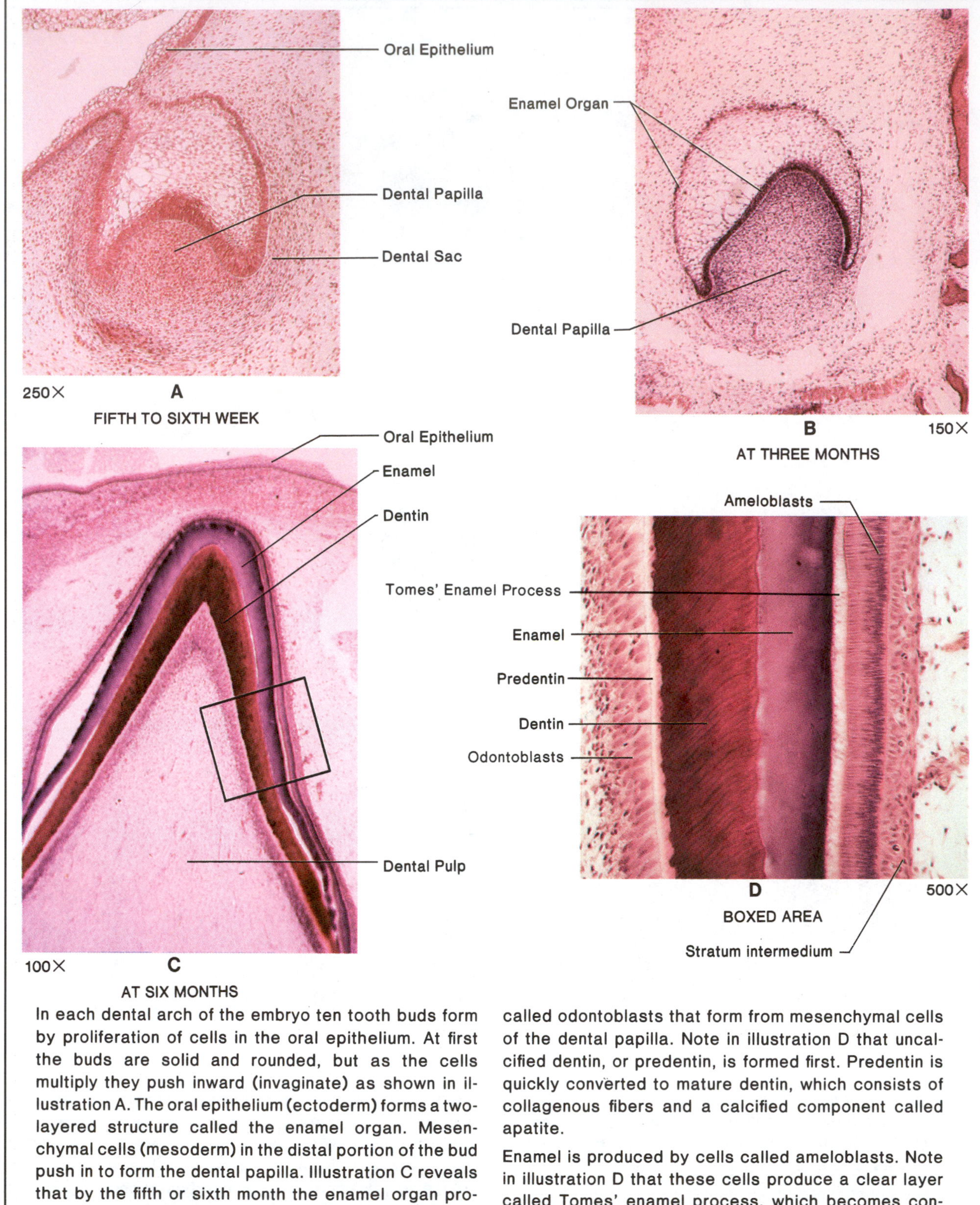

In each dental arch of the embryo ten tooth buds form by proliferation of cells in the oral epithelium. At first the buds are solid and rounded, but as the cells multiply they push inward (invaginate) as shown in illustration A. The oral epithelium (ectoderm) forms a two-layered structure called the enamel organ. Mesenchymal cells (mesoderm) in the distal portion of the bud push in to form the dental papilla. Illustration C reveals that by the fifth or sixth month the enamel organ produces the enamel of the tooth, and the cells of the dental papilla differentiate to form dentin.

Dentin is laid down just before the appearance of enamel. It is produced by a layer of elongated cells called odontoblasts that form from mesenchymal cells of the dental papilla. Note in illustration D that uncalcified dentin, or predentin, is formed first. Predentin is quickly converted to mature dentin, which consists of collagenous fibers and a calcified component called apatite.

Enamel is produced by cells called ameloblasts. Note in illustration D that these cells produce a clear layer called Tomes' enamel process, which becomes converted to mature enamel by calcification with mineral salts. As new enamel forms, the ameloblasts move outward away from the dentin and odontoblasts. When the tooth erupts, the ameloblasts are sloughed off.

Figure HA-20 Embryological stages in tooth development.

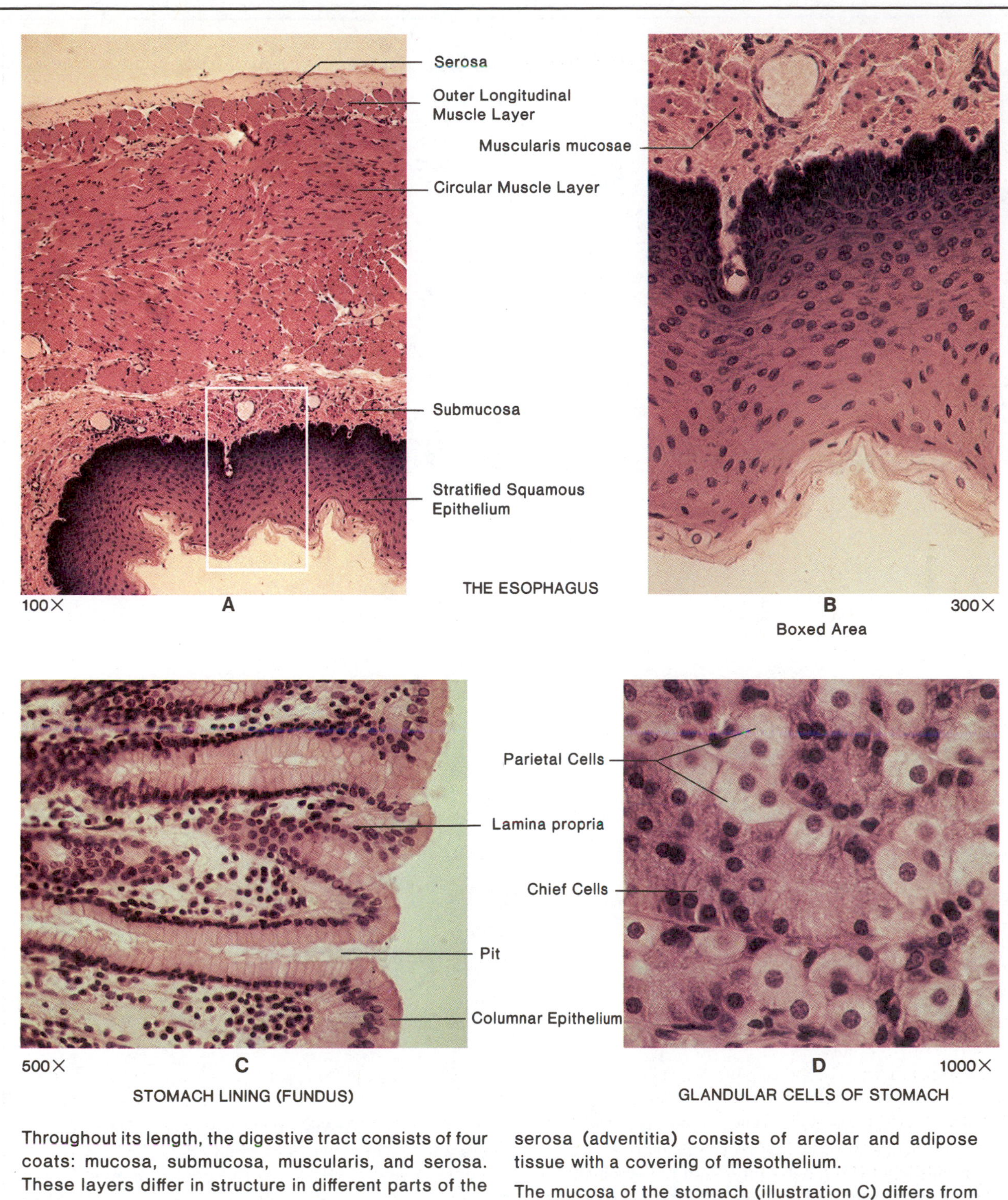

Throughout its length, the digestive tract consists of four coats: mucosa, submucosa, muscularis, and serosa. These layers differ in structure in different parts of the tract.

Note that the mucosa of the esophagus consists of stratified squamous epithelium, similar to what is seen in the oral cavity. The submucosa is a layer of areolar tissue that contains mucous glands, nerves, blood vessels, and some muscle tissue. The muscularis layer consists of longitudinal and circular fibers. The serosa (adventitia) consists of areolar and adipose tissue with a covering of mesothelium.

The mucosa of the stomach (illustration C) differs from the esophagus in that it consists of simple columnar epithelial cells. The surface of the mucosa contains numerous pits. Deep in the pits lie glands that contain parietal and chief cells. The chief cells are serous cells that produce the precursor of pepsin; the parietal cells produce the antecedent of hydrochloric acid and the intrinsic factor gastrin.

Figure HA-21 The esophagus and stomach.

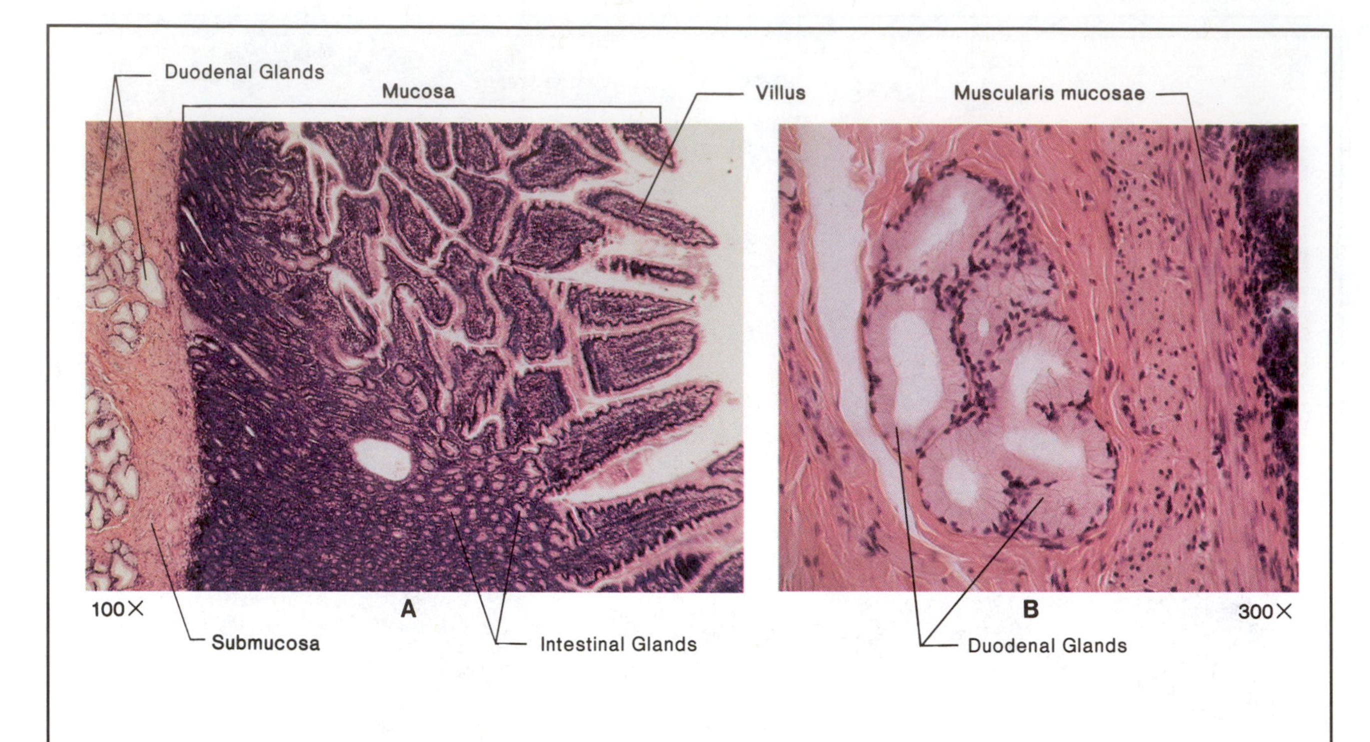

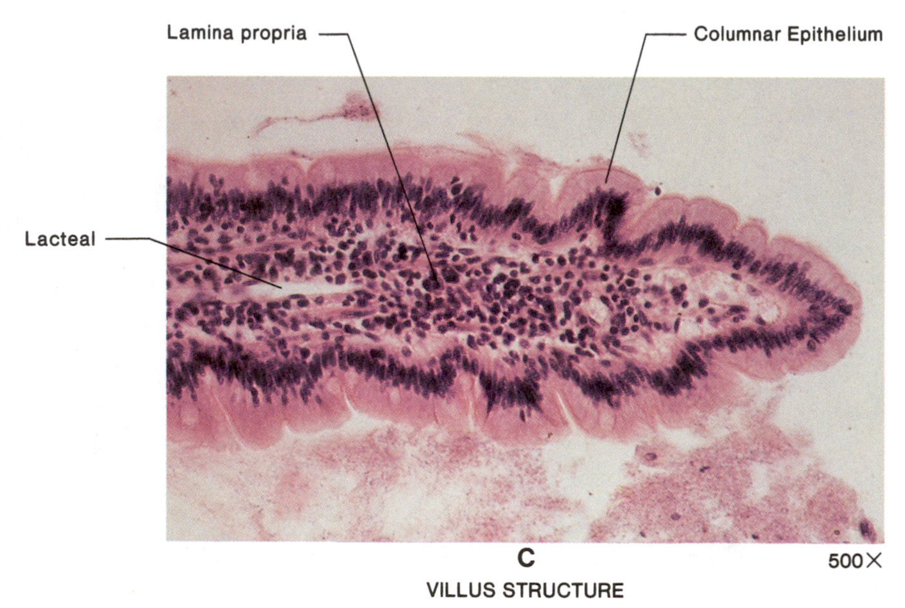

VILLUS STRUCTURE

The mucosa of the duodenum consists of the epithelium and the lamina propria. Within the lamina propria are numerous intestinal glands (crypts of Lieberkuhn). The entire mucosal surface is covered with millions of small fingerlike projections called villi.

Where the lamina propria meets the muscularis mucosae, the mucosa ends. Note the large number of mucus-secreting duodenal (Brunner's) glands (illustration B) that are located in the submucosa.

Illustration C reveals the structure of a single villus. The entire epithelium consists of simple columnar cells interspersed with goblet cells. The core of the villus is the lamina propria, which contains loose connective tissue, smooth muscle fibers, blood vessels, and a lymphatic vessel, the lacteal. The muscularis layer (not shown in illustration C) contains an inner circular layer of smooth muscle tissue, and an outer longitudinal layer of smooth fibers. The outer surface of the duodenum is covered by the serosa.

Figure HA-22 The duodenum.

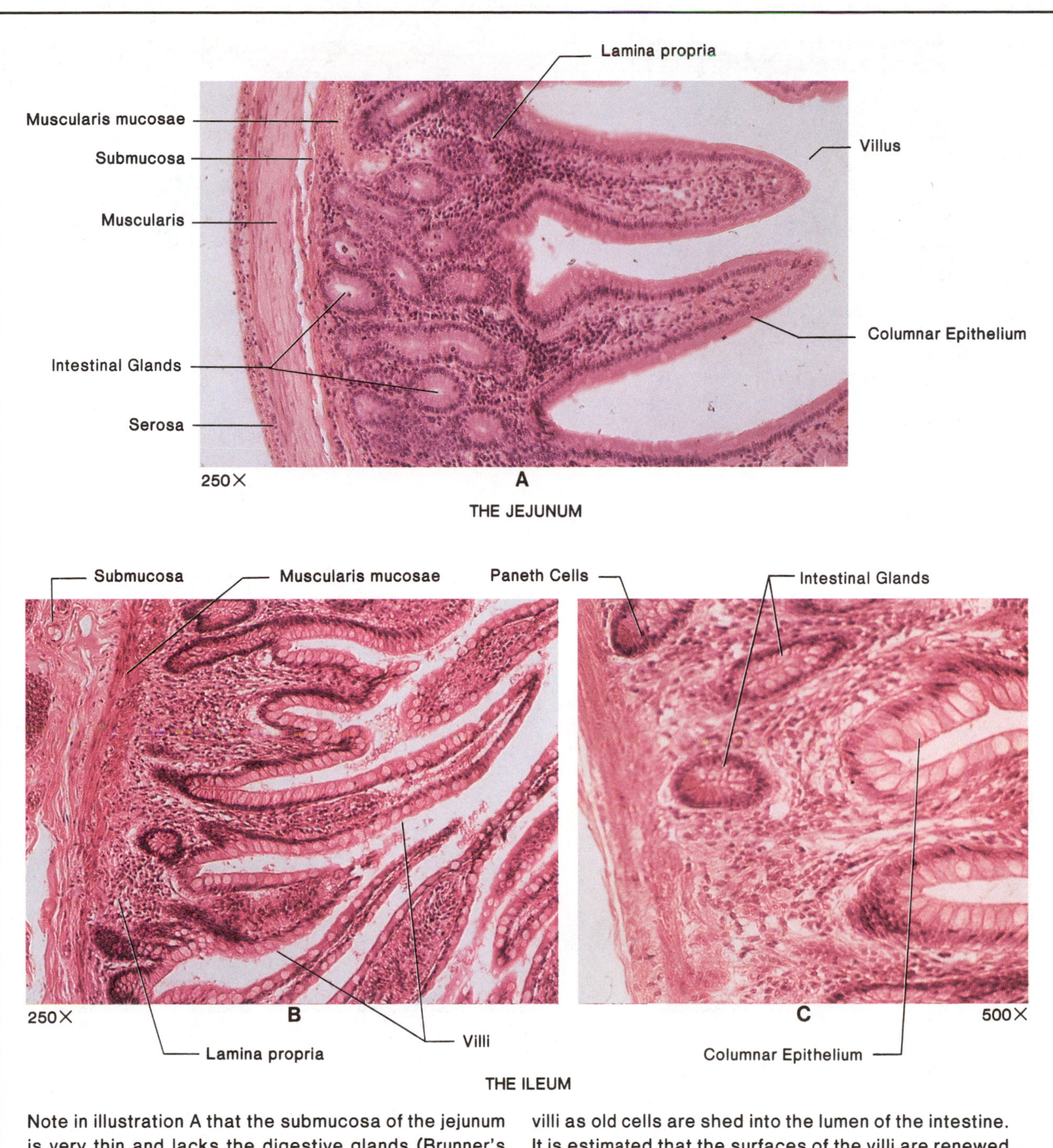

Note in illustration A that the submucosa of the jejunum is very thin and lacks the digestive glands (Brunner's duodenal) that are present in the duodenum. The mucosa, however, contains the same intestinal glands (crypts of Lieberkühn) that are present in the duodenum.

Observe that the intestinal glands in illustration C have two kinds of cells: Paneth cells and simple columnar. Although the Paneth cells produce no digestive enzymes, they may have something to do with the production of lysozyme, an antibacterial enzyme. The columnar cells, which resemble the columnar cells of the epithelium, provide new cells for the surfaces of the villi as old cells are shed into the lumen of the intestine. It is estimated that the surfaces of the villi are renewed every few days in the human intestine.

The villi of the ileum tend to be shorter and more club-shaped than those of the jejunum. Villi of the jejunum are often forked instead of leaflike as in illustration A.

The epithelial cells of the mucosa are of two types: columnar absorbing cells and goblet cells. The absorbing cells are of the tall columnar variety with nuclei at the bases of the cells. These cells exhibit distinct brush borders (figure HA-2). The goblet cells produce protective mucus.

Figure HA-23 The jejunum and ileum.

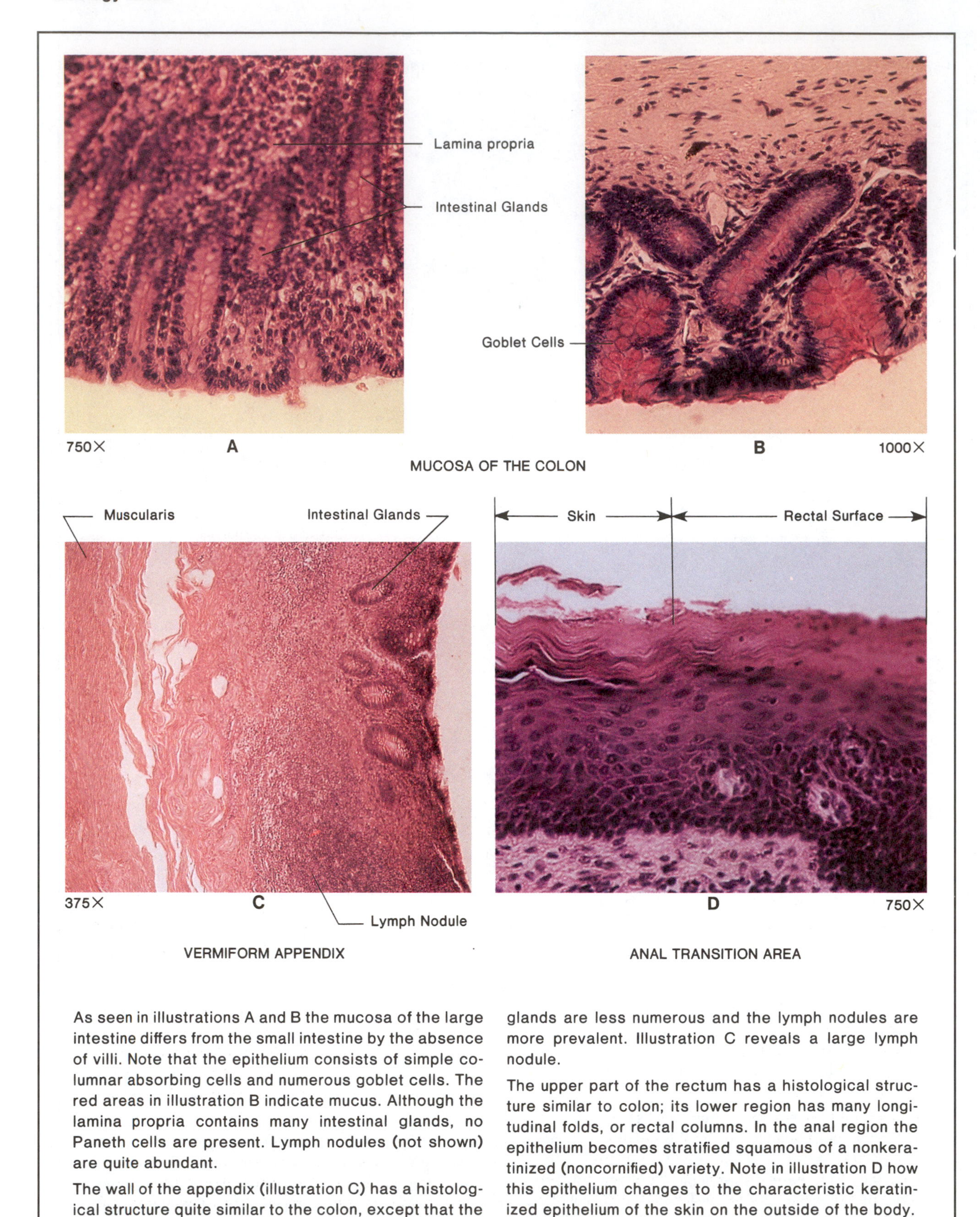

As seen in illustrations A and B the mucosa of the large intestine differs from the small intestine by the absence of villi. Note that the epithelium consists of simple columnar absorbing cells and numerous goblet cells. The red areas in illustration B indicate mucus. Although the lamina propria contains many intestinal glands, no Paneth cells are present. Lymph nodules (not shown) are quite abundant.

The wall of the appendix (illustration C) has a histological structure quite similar to the colon, except that the glands are less numerous and the lymph nodules are more prevalent. Illustration C reveals a large lymph nodule.

The upper part of the rectum has a histological structure similar to colon; its lower region has many longitudinal folds, or rectal columns. In the anal region the epithelium becomes stratified squamous of a nonkeratinized (noncornified) variety. Note in illustration D how this epithelium changes to the characteristic keratinized epithelium of the skin on the outside of the body.

Figure HA-24 The lower digestive tract.

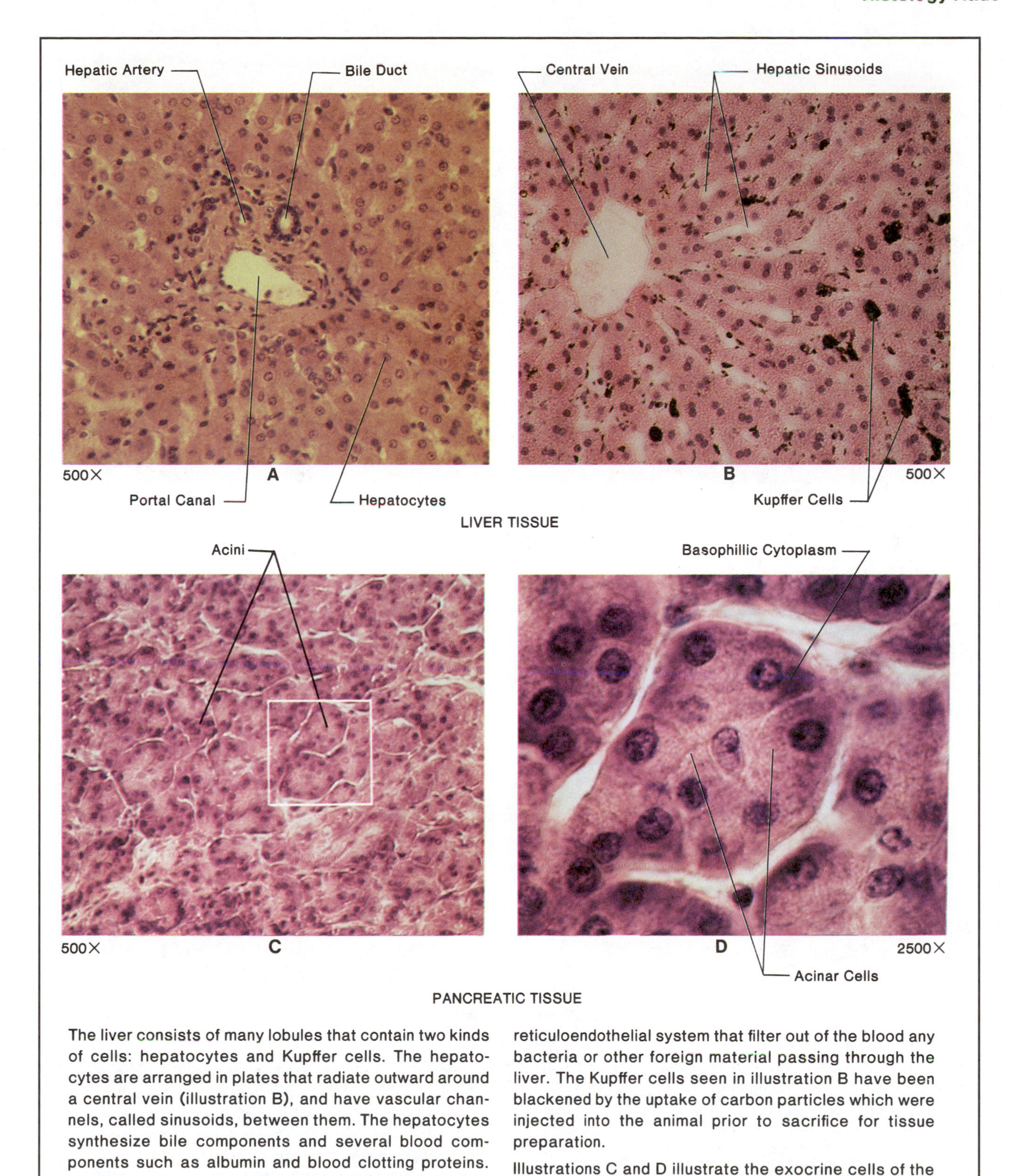

The liver consists of many lobules that contain two kinds of cells: hepatocytes and Kupffer cells. The hepatocytes are arranged in plates that radiate outward around a central vein (illustration B), and have vascular channels, called sinusoids, between them. The hepatocytes synthesize bile components and several blood components such as albumin and blood clotting proteins. The secretions of these cells drain into the blood in the central vein of the lobule by way of the sinusoids. The central veins of each liver lobule, in turn, empty into a portal canal which lies between the lobules. The Kupffer cells are stationary phagocytic cells of the reticuloendothelial system that filter out of the blood any bacteria or other foreign material passing through the liver. The Kupffer cells seen in illustration B have been blackened by the uptake of carbon particles which were injected into the animal prior to sacrifice for tissue preparation.

Illustrations C and D illustrate the exocrine cells of the pancreas. Note how the pyramid-shaped cells are arranged in alveoli (acini). The basophilic portion of the cells is specialized for the production of zymogen, an inactive protein precursor.

Figure HA-25 Liver and pancreas histology.

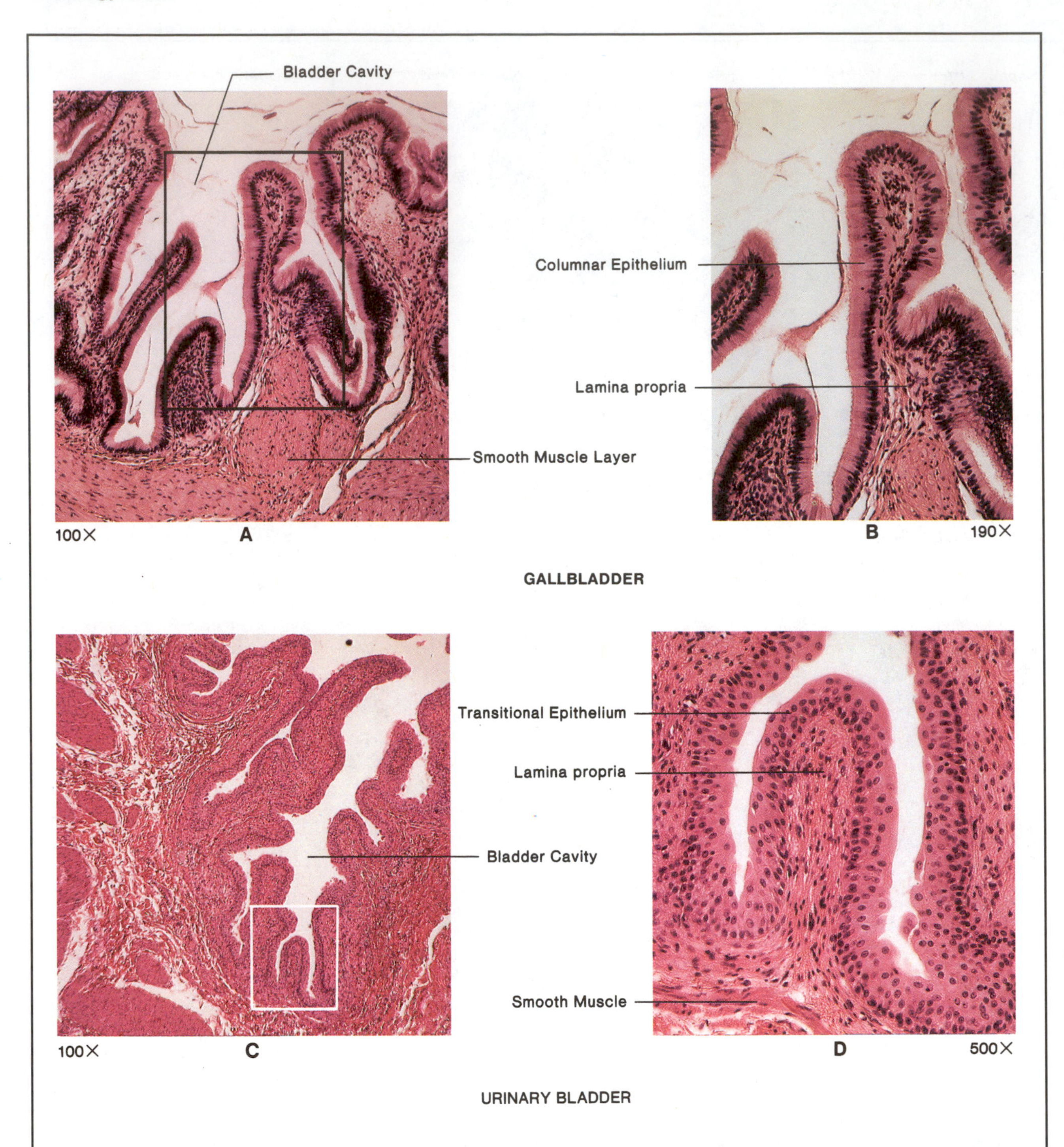

Although the gallbladder and urinary bladders are organs of two different systems, they have similarities as well as distinct differences when compared side by side.

At first glance the considerable infolding of the mucosa of both bladders is evident. As both bladders fill up with their contents (bile or urine), the extensive infolding allows for expansion to contain the fluids that pour into them.

Close examination of the inner epithelial linings, however, reveals distinct differences. Note that the epithelium of the gallbladder consists of a single layer of columnar cells, and the urinary bladder mucosa is much thicker, being made up of transitional epithelium. These differences are as one would expect: columnar cells in most of the digestive tract and transitional epithelium in the urinary system.

Figure HA-26 The gall and urinary bladders.

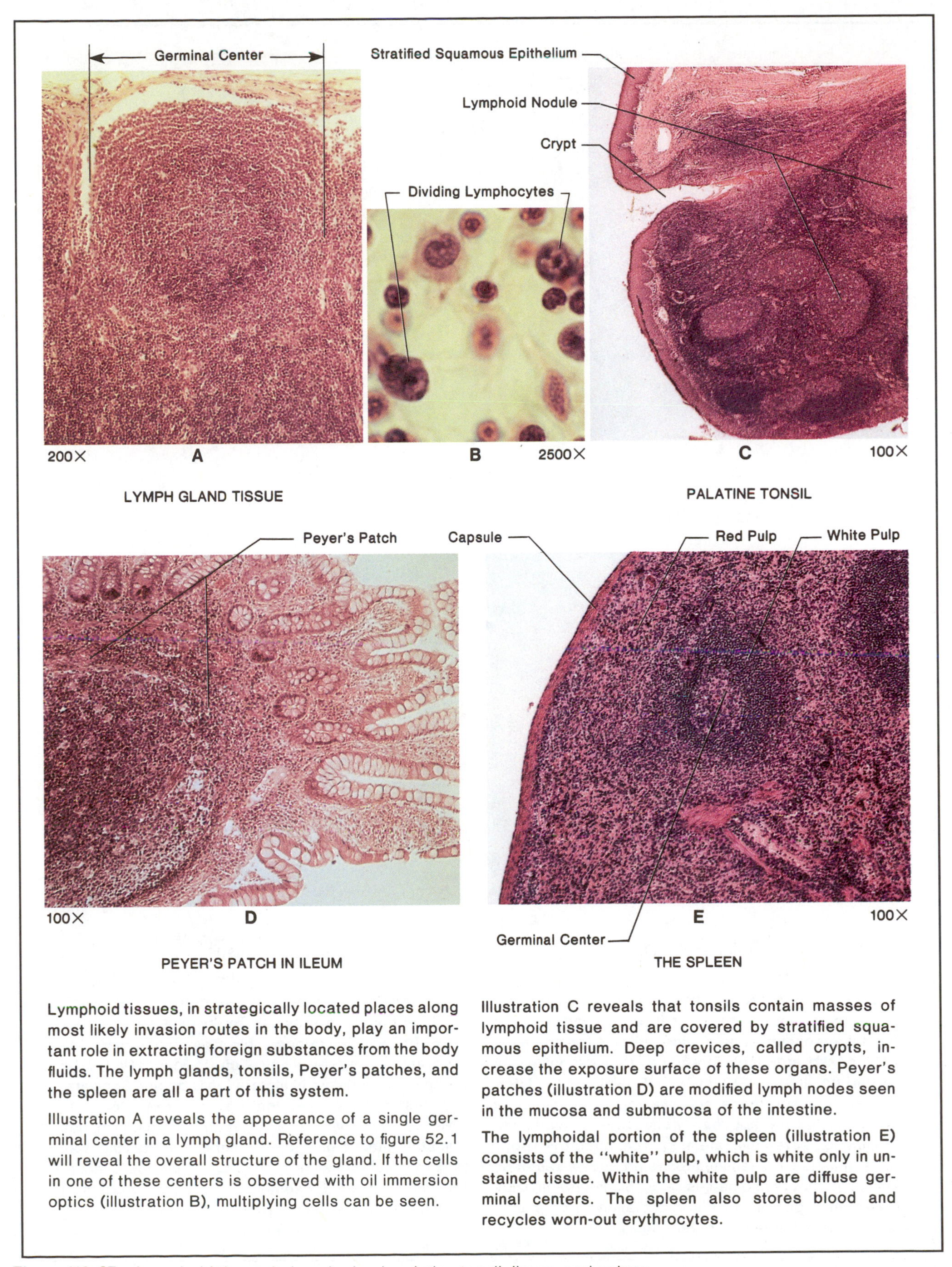

Lymphoid tissues, in strategically located places along most likely invasion routes in the body, play an important role in extracting foreign substances from the body fluids. The lymph glands, tonsils, Peyer's patches, and the spleen are all a part of this system.

Illustration A reveals the appearance of a single germinal center in a lymph gland. Reference to figure 52.1 will reveal the overall structure of the gland. If the cells in one of these centers is observed with oil immersion optics (illustration B), multiplying cells can be seen.

Illustration C reveals that tonsils contain masses of lymphoid tissue and are covered by stratified squamous epithelium. Deep crevices, called crypts, increase the exposure surface of these organs. Peyer's patches (illustration D) are modified lymph nodes seen in the mucosa and submucosa of the intestine.

The lymphoidal portion of the spleen (illustration E) consists of the "white" pulp, which is white only in unstained tissue. Within the white pulp are diffuse germinal centers. The spleen also stores blood and recycles worn-out erythrocytes.

Figure HA-27 Lymphoid tissue in lymph gland, palatine tonsil, ileum, and spleen.

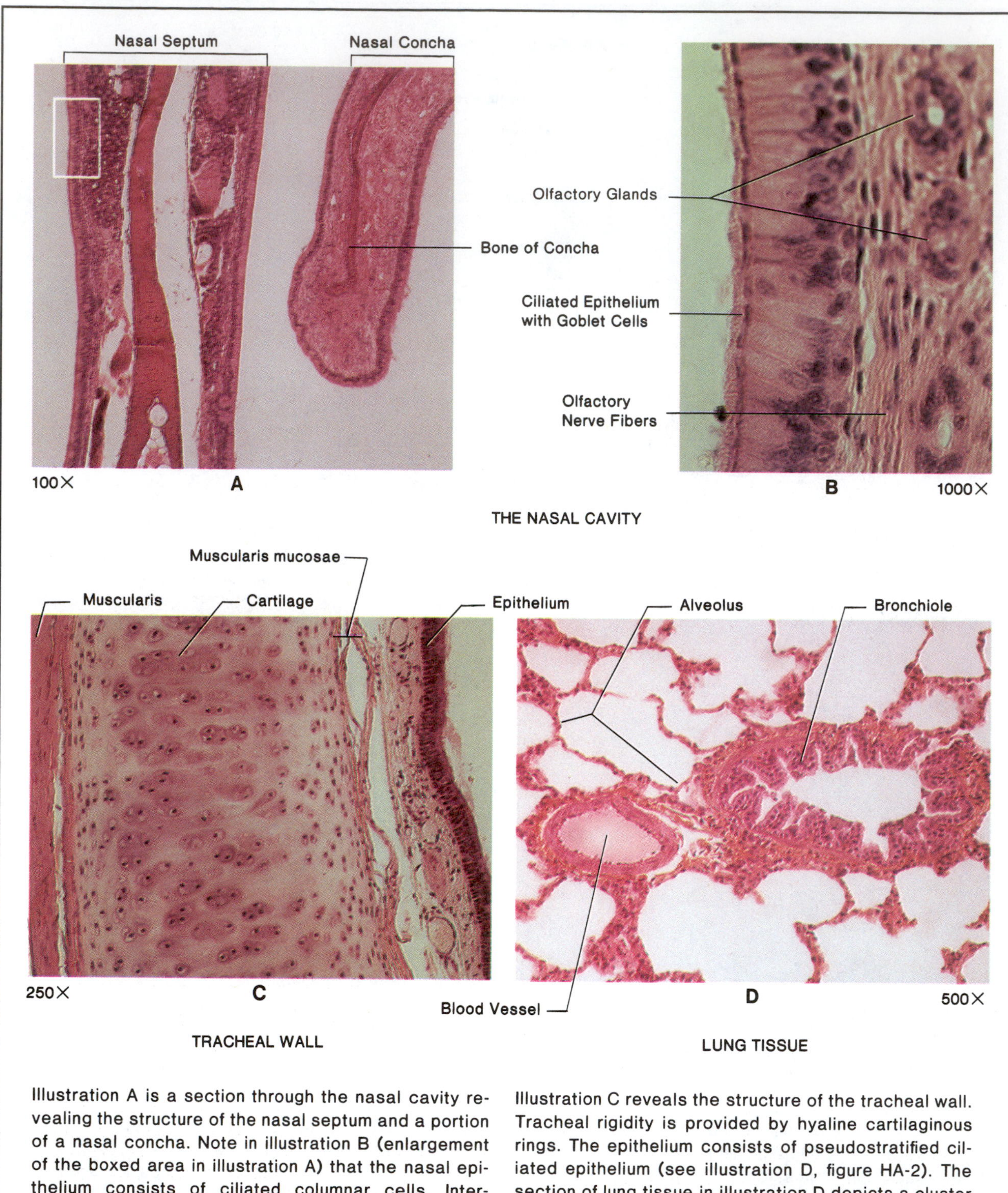

Illustration A is a section through the nasal cavity revealing the structure of the nasal septum and a portion of a nasal concha. Note in illustration B (enlargement of the boxed area in illustration A) that the nasal epithelium consists of ciliated columnar cells. Interspersed between these columnar cells are olfactory receptors that are not readily visible here due to the absence of silver dye staining. The olfactory (Bowman's) glands in the lamina propria produce a secretion made up of mucus and serous fluid. This secretion helps to keep the epithelium moist and dissolves molecules that stimulate the olfactory receptors.

Illustration C reveals the structure of the tracheal wall. Tracheal rigidity is provided by hyaline cartilaginous rings. The epithelium consists of pseudostratified ciliated epithelium (see illustration D, figure HA-2). The section of lung tissue in illustration D depicts a cluster of alveoli, a bronchiole, and a small artery. Note that several alveoli open into a larger atrium, which empties ultimately into a bronchiole. Observe that the inner wall of a bronchiole consists of cuboidal or low columnar epithelium. Cilia are present in the proximal portion of bronchioles and absent distally.

Figure HA-28 Histology of respiratory structures.

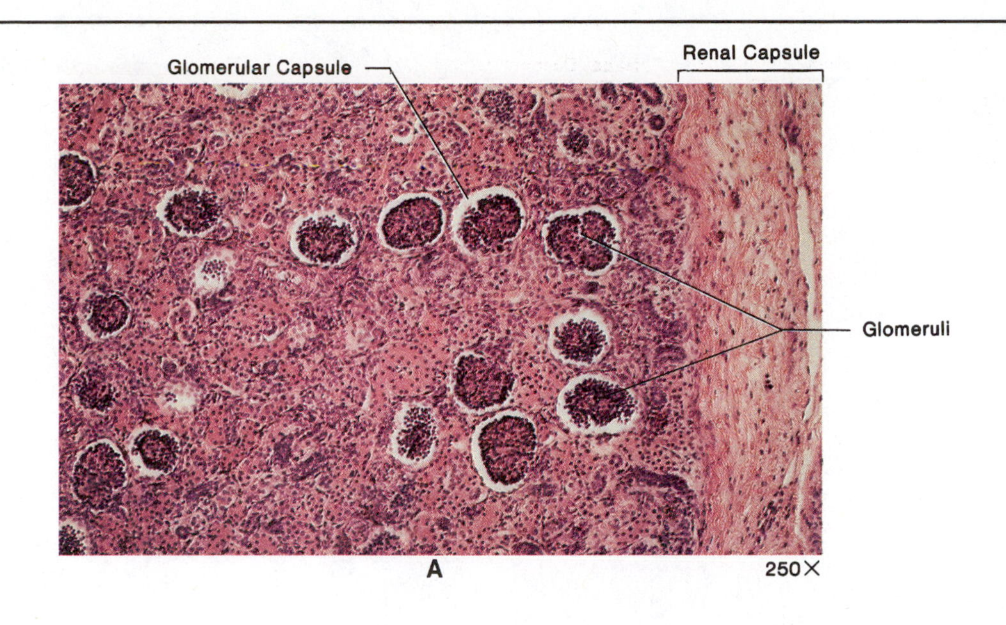

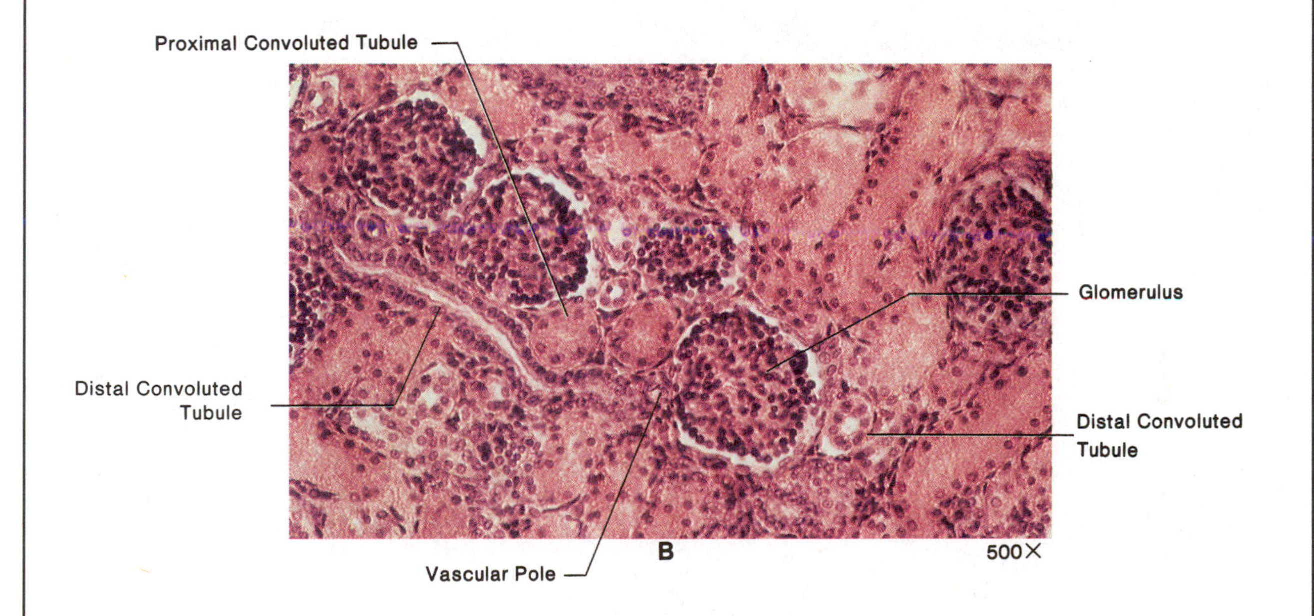

Illustration A reveals the structure of the outermost portion of the cortex of the kidney. The entire organ is enclosed by a thick fibrous capsule.

Blood enters each glomerulus through its vascular pole (shown in illustration B). High vascular pressure in the glomeruli results in the production of a glomerular filtrate consisting of water, glucose, amino acids, and other substances. This glomerular filtrate is collected by the glomerular capsule and passes down the length of the nephron collecting tubule, being altered in composition as it approaches the calyx of the kidney. Note that the proximal convoluted tubules can be differentiated from the distal convoluted tubules by the size of the cuboidal cells that comprise their walls: large in the proximal tubule and small in the distal tubule. Eighty percent of water absorption occurs in the proximal convoluted tubules.

Figure HA-29 Histology of the cortex of the kidney.

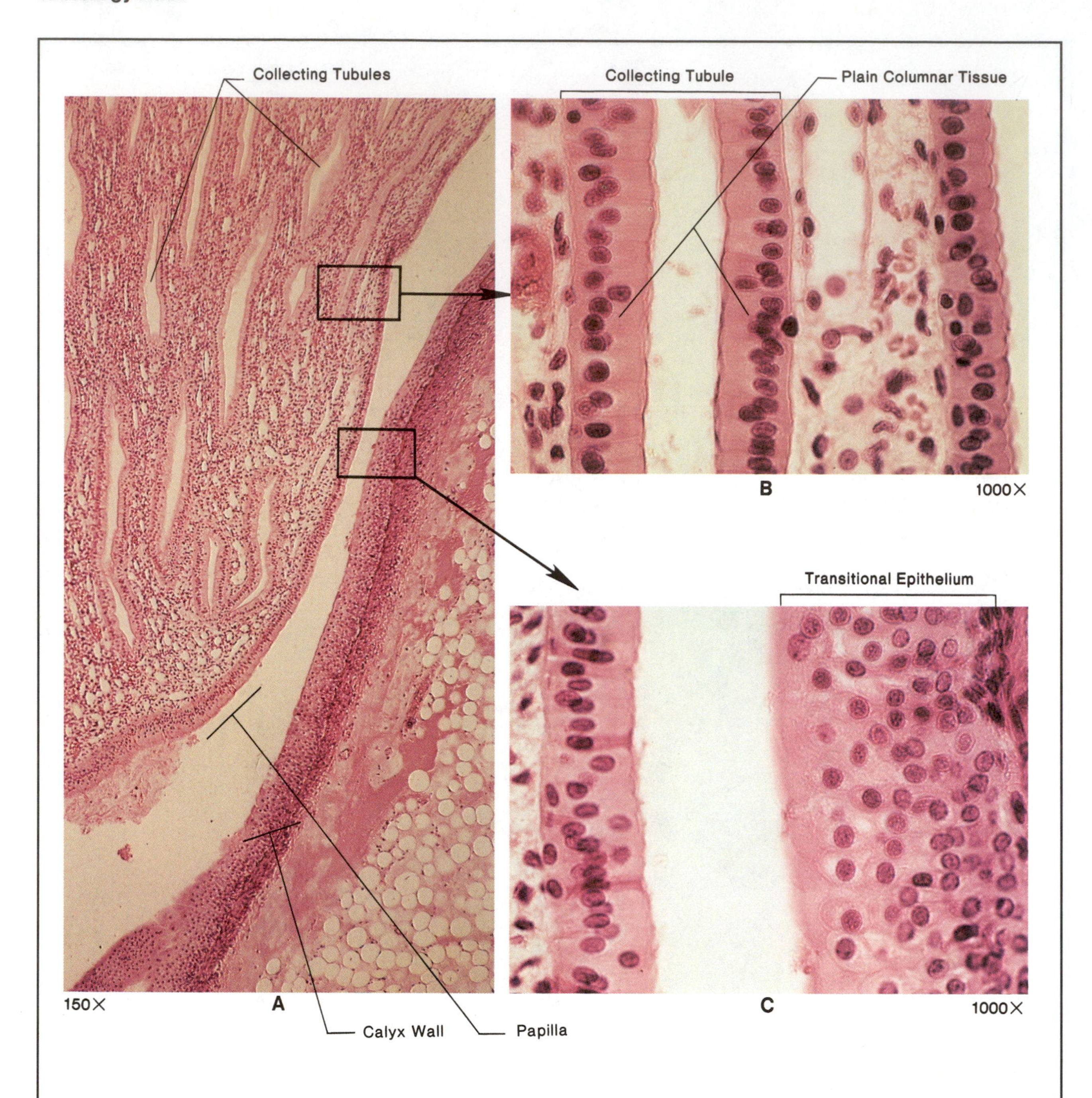

Illustration A portrays a portion of the tip of a pyramid and the wall of the calyx. Illustrations B and C reveal the histological nature of these two structures. The collecting tubules, which collect urine from several nephrons, are lined with distinct low columnar epithelium. Examination of these cells with oil immersion optics will reveal the existence of a distinct brush border. Both absorption and secretion takes place through these columnar cells.

Note in illustration C that the wall of the calyx is lined with transitional epithelium. This same type of tissue is seen lining the ureters and the urinary bladder (see illustrations C and D, figure HA-26).

Figure HA-30 Histology of the papilla and calyx of the kidney.

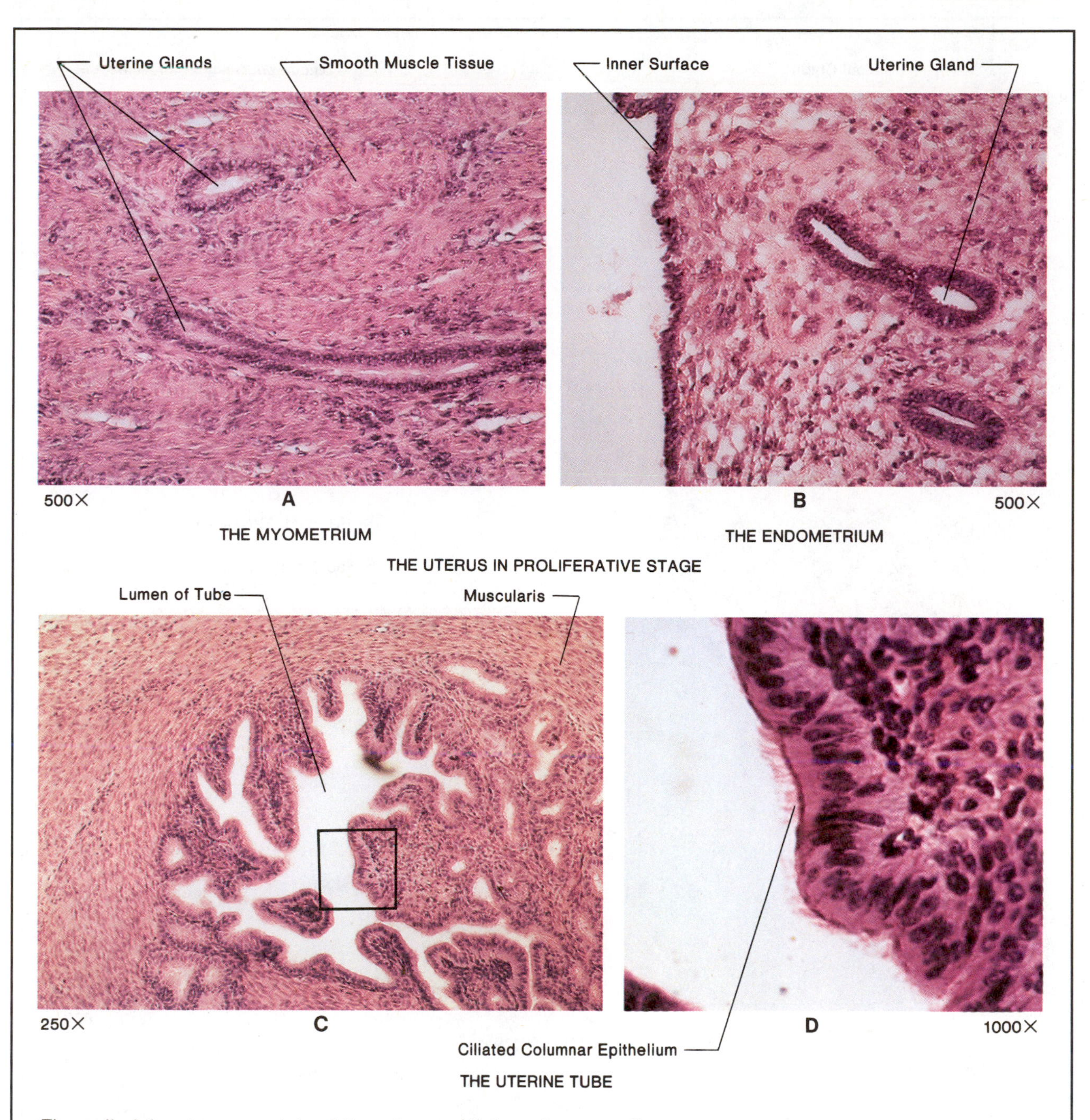

The wall of the uterus consists of three layers: (1) the endometrium, which corresponds to the mucosa and submucosa; (2) the myometrium, or muscularis; and (3) the perimetrium, a typical serous membrane. The myometrium (illustration A), which forms three-fourths of the uterine wall, consists of three layers of smooth muscle fibers: an inner longitudinal layer, a middle circular layer, and an outer oblique layer. Some glands extend into it from the endometrium. The endometrial surface cells are columnar and partially ciliated. During the proliferative stage (immediately after menstruation) the endometrium undergoes regeneration of the columnar cells and increased growth of the mucosal glands. Vascularity of the tissue also becomes more pronounced. The proliferative stage terminates at the 13th or 14th day of the menstrual cycle.

The uterine tube (illustration C) consists of an inner mucosa, a middle muscularis, and an outer serosa. The mucosa epithelium consists of a mixture of ciliated and nonciliated columnar cells (see illustration D). The muscularis consists of inner circular and outer longitudinal layers of smooth muscle fibers. There is no muscularis mucosae as seen in the digestive tract.

Figure HA-31 The uterus and uterine tube.

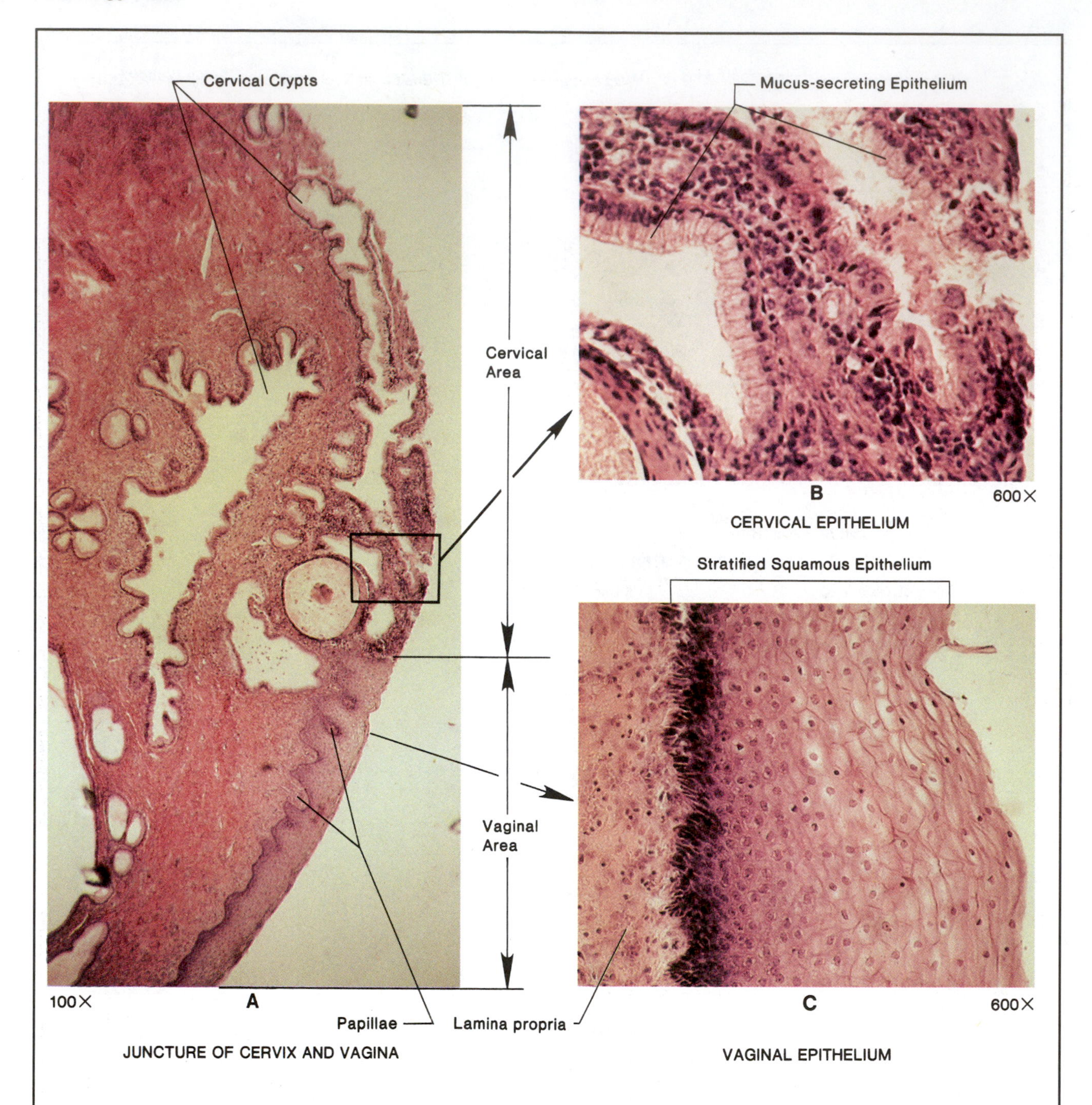

Illustration A shows the contrasting histological differences that exist between the epithelia of the vagina and cervix. Note in illustration C that the vaginal epithelium consists of nonkeratinized stratified squamous epithelium. Examination of the lamina propria of the vaginal mucosa with high-dry or oil immersion optics will reveal the presence of large numbers of lymphocytes. Note in illustration A that the epithelium has a large number of papillae.

The most distinguishing characteristic of the cervical portion of the uterus is the presence of approximately one hundred mucus-secreting cervical crypts. As indicated in illustration B, the epithelium of these crypts consists, primarily, of plain columnar cells; some ciliated cells are also present, however. Approximately 20 to 60 mg of mucus are produced daily. During ovulation mucus production increases to over 700 mg daily. Mucus plays an important role in fertility since it is the first secretion met by the sperm entering the female tract. Occasionally the exit of a crypt will become occluded, causing a cyst to form. The large round structure near the box in illustration A is such a structure.

Figure HA-32 The cervix and vagina.

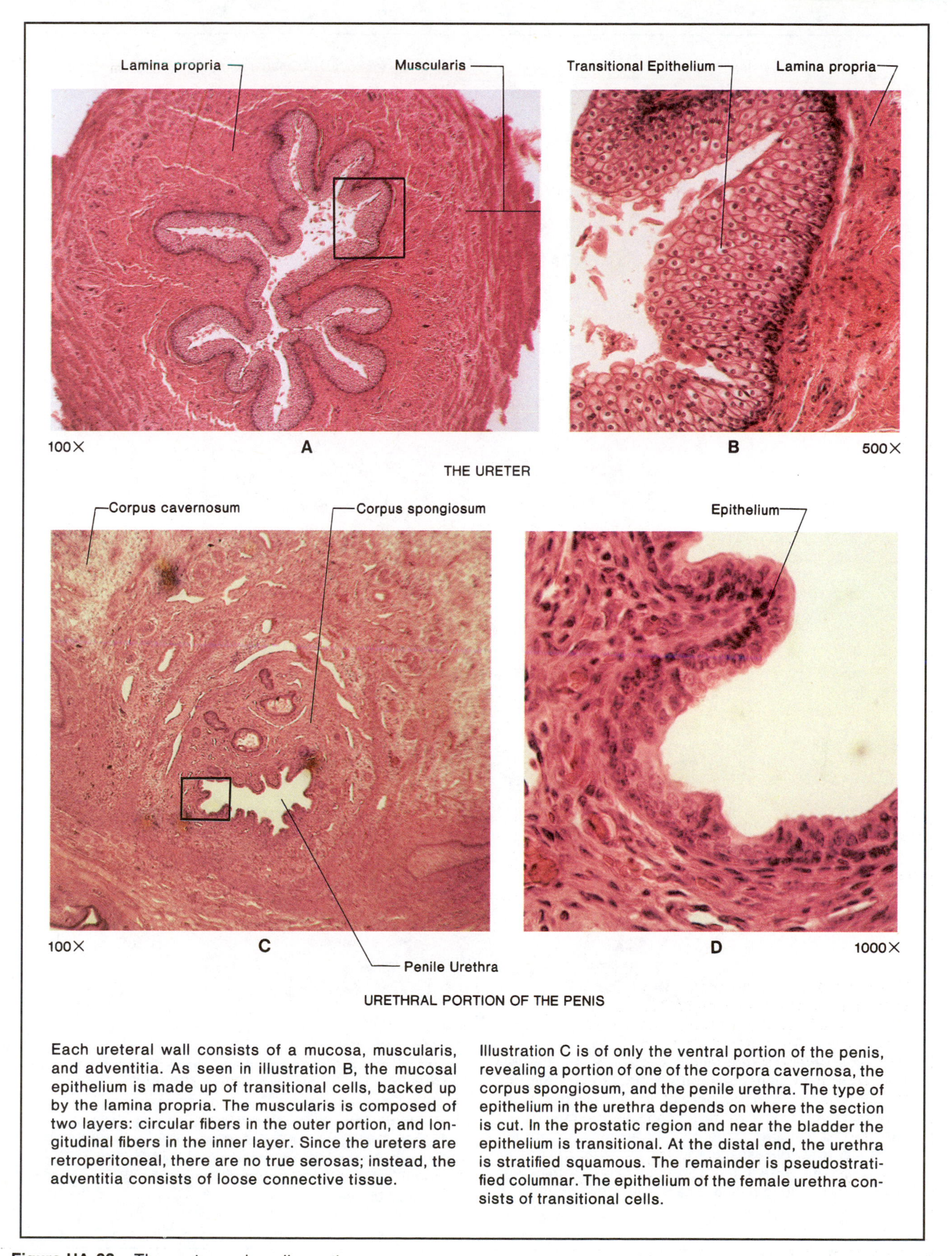

Each ureteral wall consists of a mucosa, muscularis, and adventitia. As seen in illustration B, the mucosal epithelium is made up of transitional cells, backed up by the lamina propria. The muscularis is composed of two layers: circular fibers in the outer portion, and longitudinal fibers in the inner layer. Since the ureters are retroperitoneal, there are no true serosas; instead, the adventitia consists of loose connective tissue.

Illustration C is of only the ventral portion of the penis, revealing a portion of one of the corpora cavernosa, the corpus spongiosum, and the penile urethra. The type of epithelium in the urethra depends on where the section is cut. In the prostatic region and near the bladder the epithelium is transitional. At the distal end, the urethra is stratified squamous. The remainder is pseudostratified columnar. The epithelium of the female urethra consists of transitional cells.

Figure HA-33 The ureter and penile urethra.

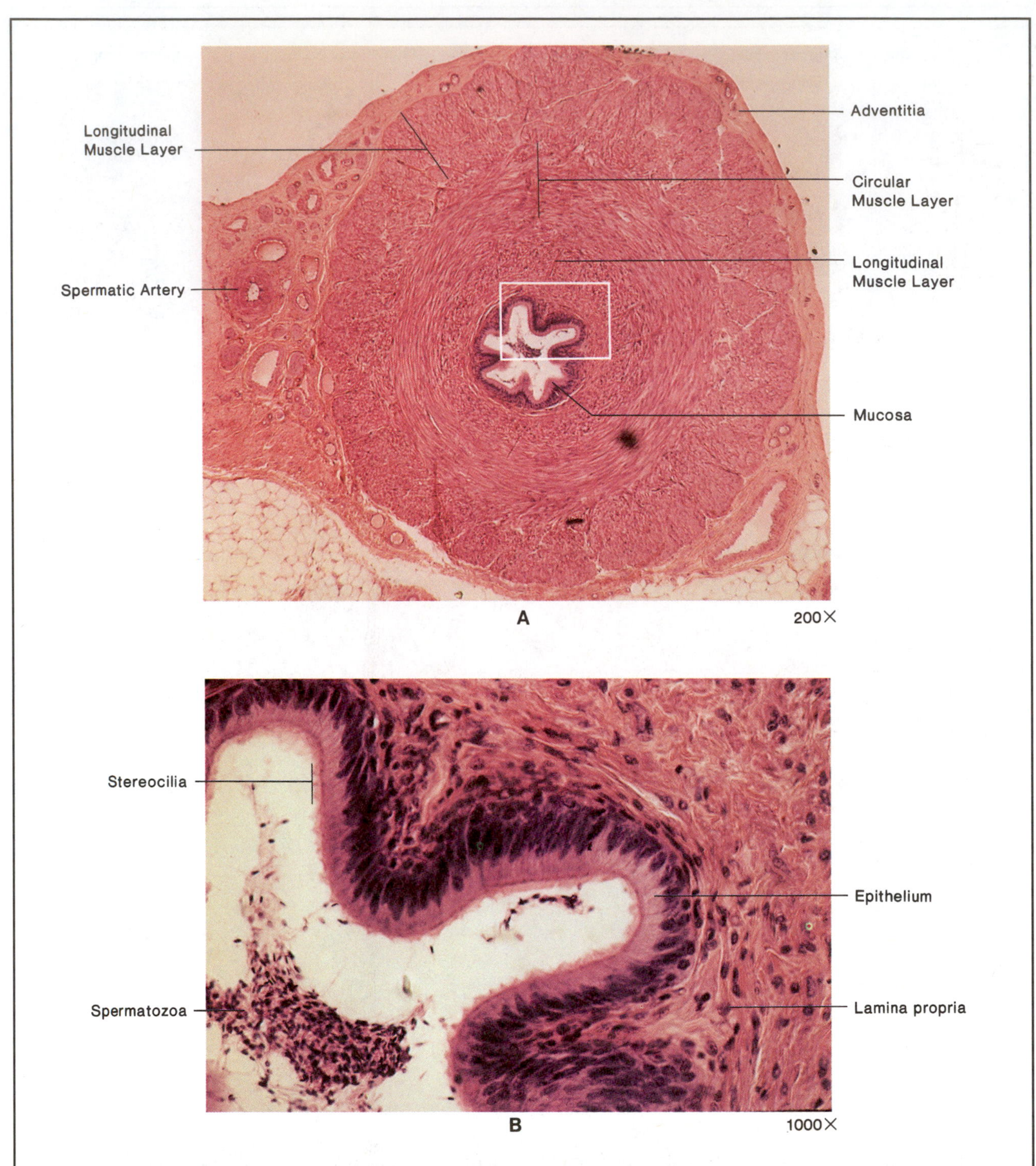

The vas deferens (spermatic duct) is a thick-walled muscular tube that transports spermatozoa from the epididymis to the ejaculatory duct where fluid from the seminal vesicles joins the spermatozoa (see figure 87.1). The wall has three distinct areas: an outer adventitia, a middle muscularis (three-layered), and an inner mucosa. Note that the muscularis is very thick with two layers of longitudinal fibers and a middle layer of circular fibers. Peristaltic waves produced by these muscles help to propel the spermatozoa during ejaculation.

The mucosa, consisting of the lamina propria and epithelium, is unique in that the columnar cells that make up the epithelium have stereocilia. These structures are nonmotile enlarged microvilli that cover the exposed surfaces of the cells.

Figure HA-34 The vas deferens.

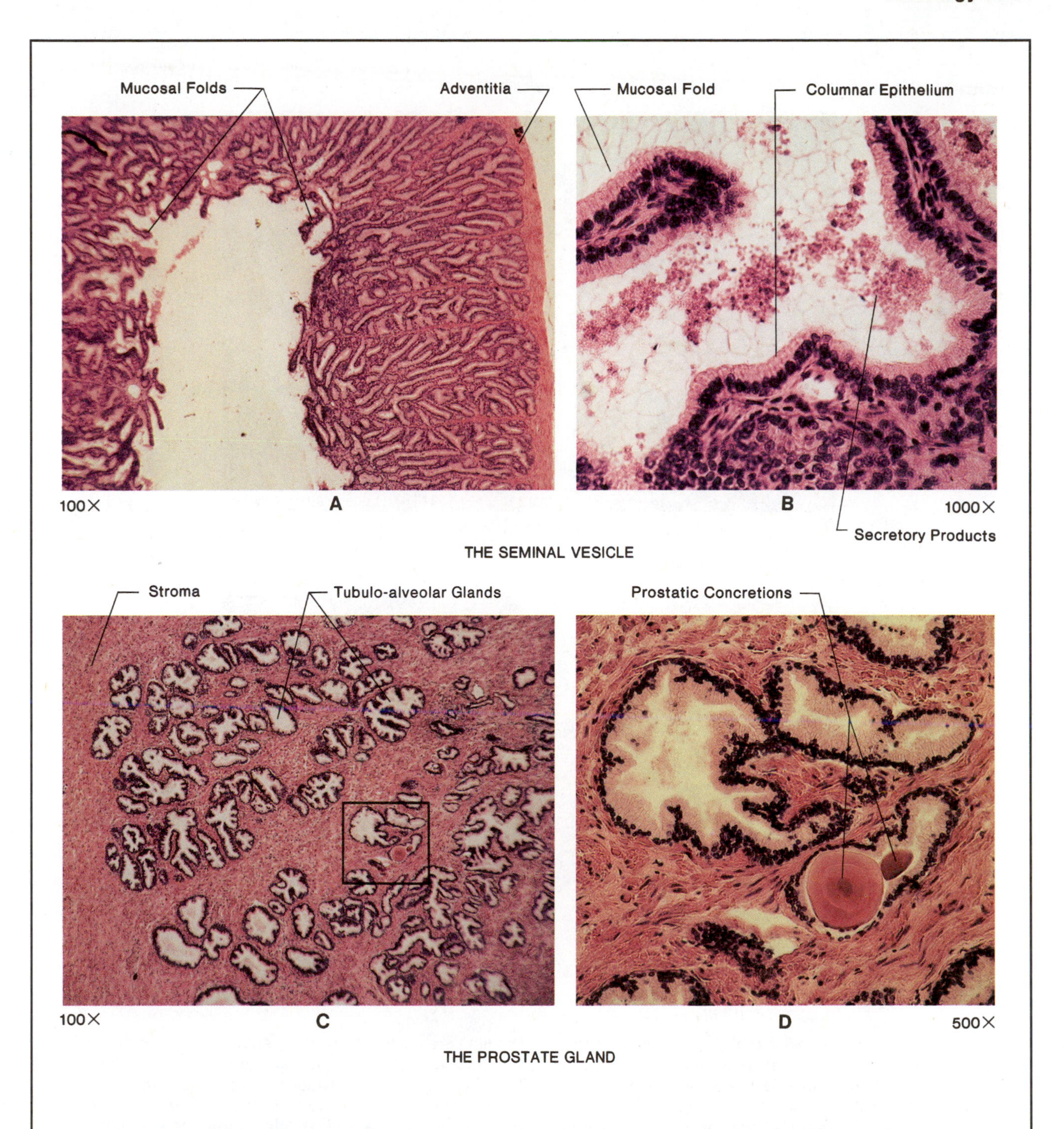

The mucosa of the paired seminal vesicles is unique in its extensive ramification of folds as seen in illustration A. Note in illustration B that the epithelium consists of nonciliated columnar cells. The seminal vesicles contribute a thick yellowish fluid that is alkaline and rich in fructose, a source of energy for the spermatozoa. It also contains coagulating enzymes that cause the seminal fluid to congeal in the female tract after it is deposited. The prostate gland consists of thirty to fifty tubulo-alveolar glands. These glands exhibit considerable infolding to accommodate distention during storage of prostatic fluid. The secretory cells of the epithelium consist of columnar cells superimposed on a few flattened cells. The stroma (50% of the gland) is made up of equal parts of smooth muscle fibers and fibrous tissue. The prostatic concretions that are shown in illustration D are lamellar bodies that often become calcified in older men.

Figure HA-35 Seminal vesicle and prostate gland.

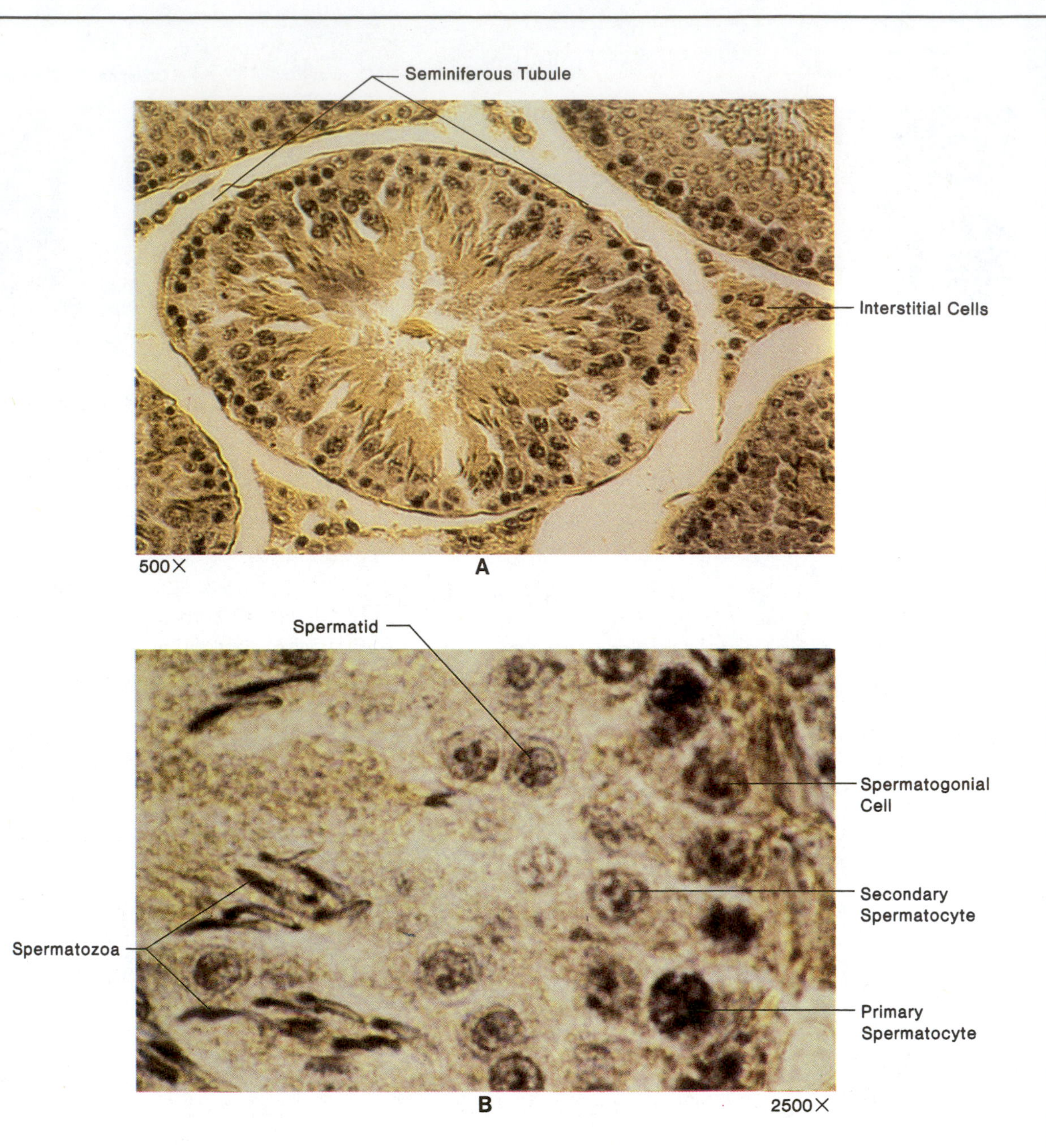

The production of spermatozoa, known as spermatogenesis, occurs in the seminiferous tubules of the testes. Illustration A reveals the structure of a single tubule. Between the tubules can be seen the interstitial cells that secrete testosterone into the circulatory system.

All spermatozoa originate from the primordial germ cells, or spermatogonia, that are located at the periphery of the seminiferous tubule. These cells have a diploid complement of 46 chromosomes. The first step in this process is the synapsis (union) of the homologous chromosomes to form tetrads in the primary spermatocyte. Note the proximity of the primary spermatocyte to the spermatogonial cells in illustration B.

The next step is for the primary spermatocyte to divide, meiotically, to produce two secondary spermatocytes. This meiotic division produces two cells (secondary spermatocytes) that have a haploid number (23) of chromosomes. Note in illustration B that the secondary spermatocytes are closer to the lumen of the seminiferous tubule. Finally, the two secondary spermatocytes divide meiotically, again, to produce a total of four spermatids which contain 23 chromosomes each. The spermatids continue to develop into mature spermatozoa.

Figure HA-36 Spermatogenesis.

Blood Cell Counts 38

Three types of blood cell counts are described here in this exercise. The principal purpose in performing these counts is to learn as much as possible about blood cells in the limited amount of time provided by a few laboratory periods: becoming a skilled technician should not be anticipated. The blood to be used in these three experiments will be your own. Only a drop at a time will be needed. It will be necessary for you to work with your laboratory partner.

Before starting this experiment, take a few moments and carefully review the sanitary precautions recommended in the Introduction. All the experiments in this hematology unit are safe to perform, *if proper precautions are observed.*

The Differential WBC Count

Figure 38.1 illustrates the various leukocytes, blood platelets, and erythrocytes that one encounters when a blood smear is stained with Wright's stain. Figure 38.2 illustrates the various steps that one uses for making a slide that can be used to study stained blood cells.

Note in figure 38.1 that there are five basic types of leukocytes. Three of them, the neutrophils, eosinophils, and basophils, have conspicuous granules in their cytoplasm; these leukocytes are designated as **granulocytes.** The monocytes and lymphocytes lack cytoplasmic granules and are called **agranulocytes.**

A *differential WBC count* is performed on a Wright's stained slide, using an oil immersion objective to determine the percentages of each type of leukocyte. The clinical value of such a slide can be considerable.

The normal percentage ranges for the various leukocytes are **neutrophils** 50–70%, **lymphocytes** 20–30%, **monocytes** 2–6%, **eosinophils** 1–5%, and

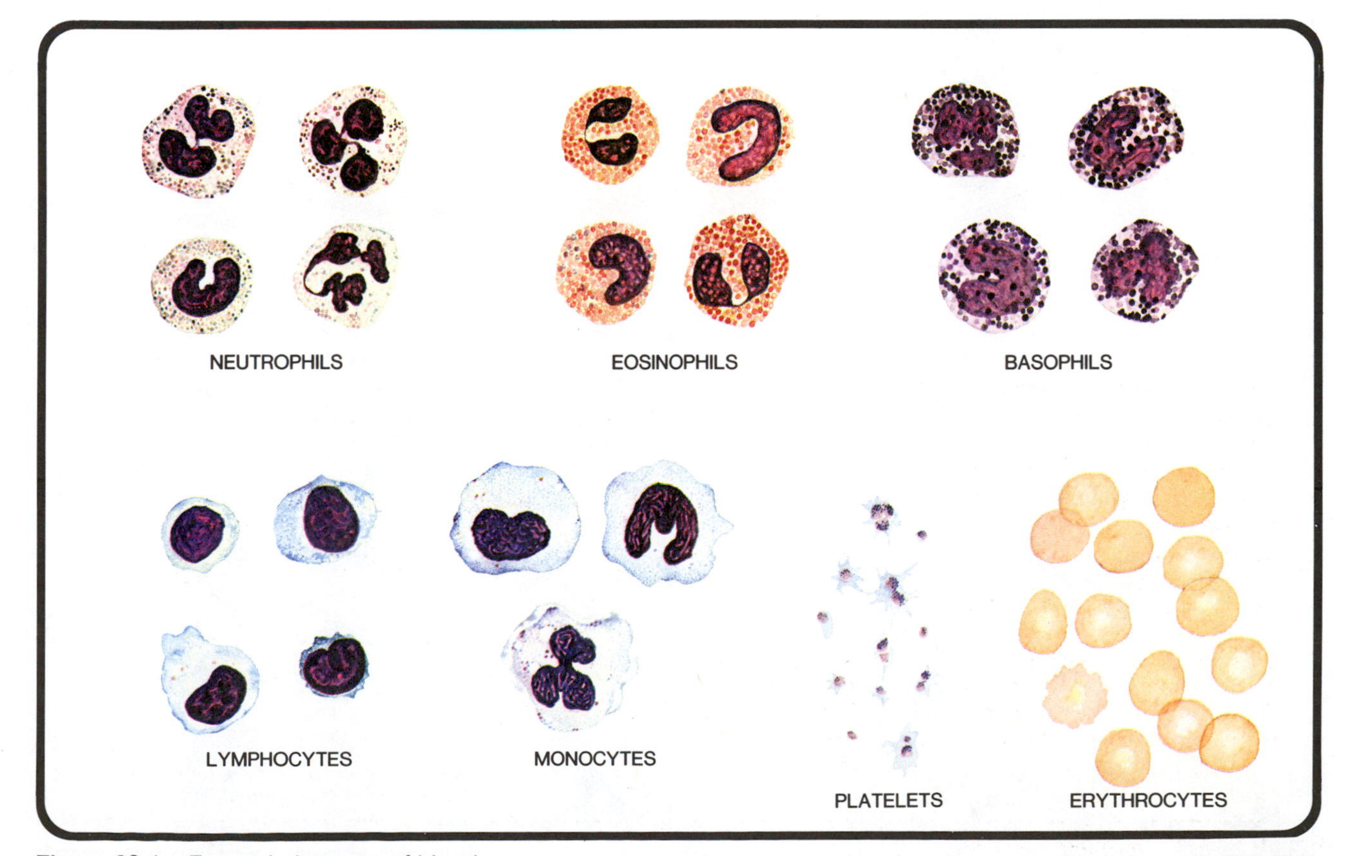

Figure 38.1 Formed elements of blood. K. P. Talaro

basophils 0.5–1%. High or low counts in certain categories may indicate the presence of certain pathological conditions.

High neutrophil counts are referred to as **neutrophilia** and may indicate localized infections such as appendicitis, or abscesses in some part of the body. Decreased numbers of neutrophils, on the other hand, are designated as **neutropenia,** which may be evidence of typhoid fever, undulant fever, or influenza.

Eosinophilia (high eosinophil count) may indicate allergic conditions or invasions by certain parasitic roundworms such as *Trichinella spiralis,* or pork worm. Counts of eosinophils may rise as much as 50% in cases of trichinosis.

High lymphocyte counts, or **lymphocytosis,** may be present in whooping cough or some viral infections.

To determine the percentages of each type of leukocyte it is necessary to count and record each type of cell in a total count of 100 white blood cells. After identifying and tallying this many leukocytes you should become quite familiar with each type of cell.

Preparation of Slide

Two of the most difficult steps in preparing a good slide are (1) putting the right size drop of blood on the slide, and (2) spreading the blood *evenly* over the slide. Take special care with these two steps.

Materials:

clean microscope slides (3 or 4 per student)
sterile disposable lancets
sterile absorbent cotton
Wright's stain
dropping bottle of distilled water
70% alcohol (for disinfection)
wax pencil and bibulous paper

1. Clean 3 or 4 slides with soap and water. Keep flat surfaces *free of fingerprints* after they have been cleaned.
2. After scrubbing *middle finger* with alcohol, stick with lancet, and place a drop of blood ¾ inch from one end of slide. **Drop size should be approximately ⅛-inch diameter,** no more, no less.
3. Spread the blood on the slide, using another clean slide as a spreader. Note in illustration 3, figure 38.2, that **blood is dragged over the**

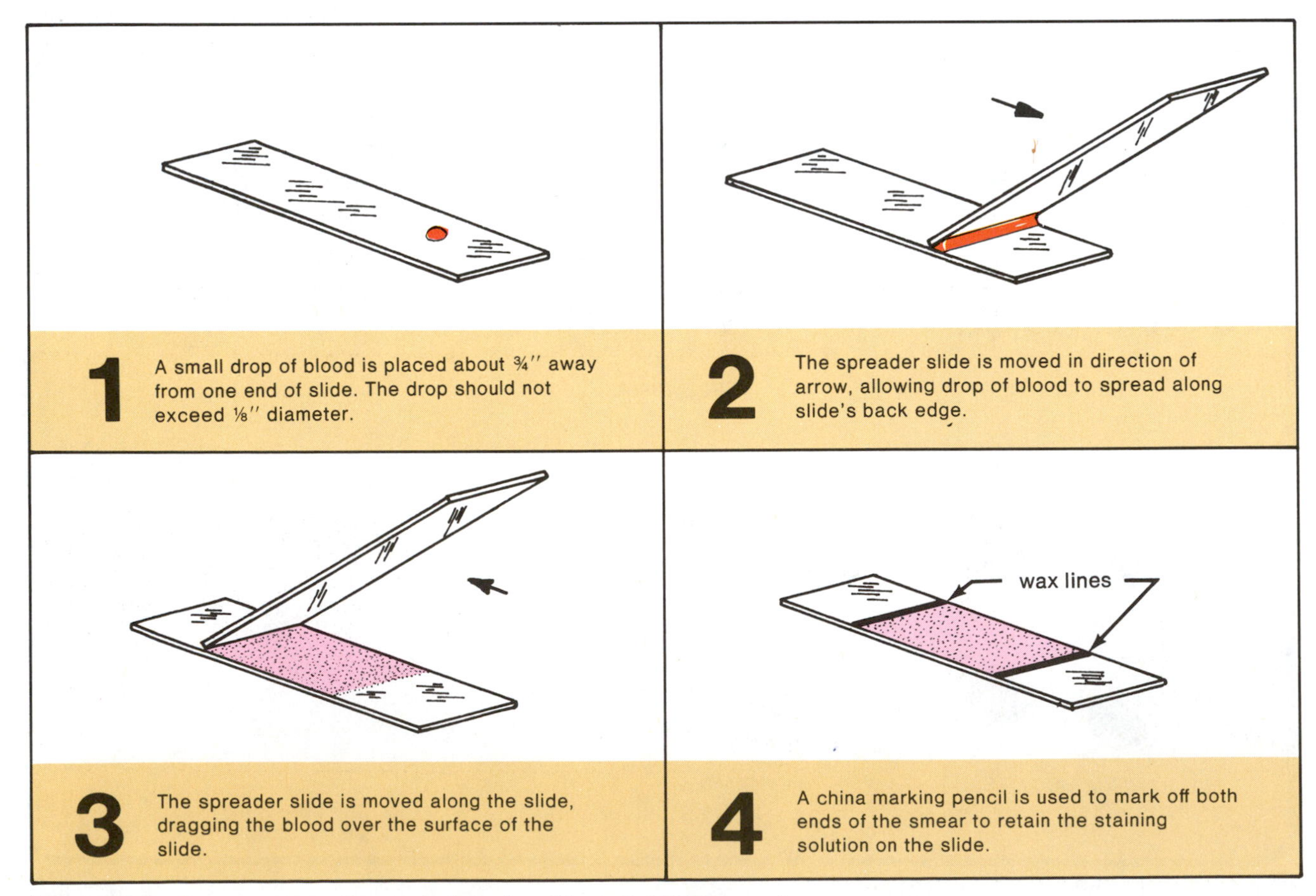

Figure 38.2 Smear preparation technique for differential WBC count.

slide, not pushed. Hold the spreader slide at an angle somewhat greater than 45° to get a smear of proper thickness.

4. With a wax pencil, draw a line on each side of the smear to confine the stain.
5. Cover the blood smear with Wright's stain, **counting the drops** as you add them. Stain for **4 minutes** and then add the same number of drops of distilled water to the stain and let stand for **another 10 minutes.** Blow gently on the mixture every few minutes to keep the solutions mixed.
6. Gently wash off the slide under running water for **30 seconds** and shake off the excess. Blot dry with bibulous paper.

Performing the Cell Count

As soon as the slide is completely dry, scan it with the low-power objective to find the area of the slide that has the best distribution of cells. Avoid the excessively dense areas. Place a drop of oil near the edge of the smear in the selected area and examine with the oil immersion objective.

Remove your Laboratory Report sheet from the back of the manual and record each type of leukocyte encountered. Follow the path indicated in figure 38.3. For identification of each type of cell, refer to figure 38.1.

Total White Blood Cell Count

While the differential count provides us with the relative percentages of different types of leukocytes, it alone cannot give us a true picture of the extent of infection. One must also know the total number of WBCs per cubic millimeter of blood.

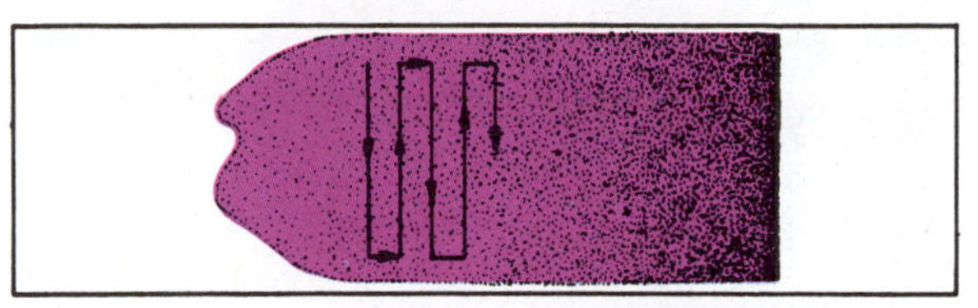

Figure 38.3 Examination path for differential count.

Although the number of white blood cells may vary with the time of day, exercise, and other factors, a range of 5,000 to 9,000 WBCs per cubic millimeter is considered normal. If an individual were to have an abnormally high neutrophil percentage and a total count of 17,000 WBCs, the presence of an infection of some sort would be highly probable.

To determine the number of leukocytes in a cubic millimeter of blood, one must dilute the blood and count the WBCs on a specialized slide called a **hemacytometer.** Figures 38.4 and 38.5 reveal various steps in preparation for this count.

Note in figure 38.4 that blood is drawn up into a special pipette and then diluted in the pipette with a weak acid solution. After shaking the pipette to mix the acid and blood, a small amount of diluted blood is allowed to flow under the cover glass of the hemacytometer. The count of white blood cells is made with the low-power microscope objective. Proceed as follows.

Preparation of Hemacytometer

Working with your laboratory partner, assist each other to prepare a "charged" hemacytometer as follows:

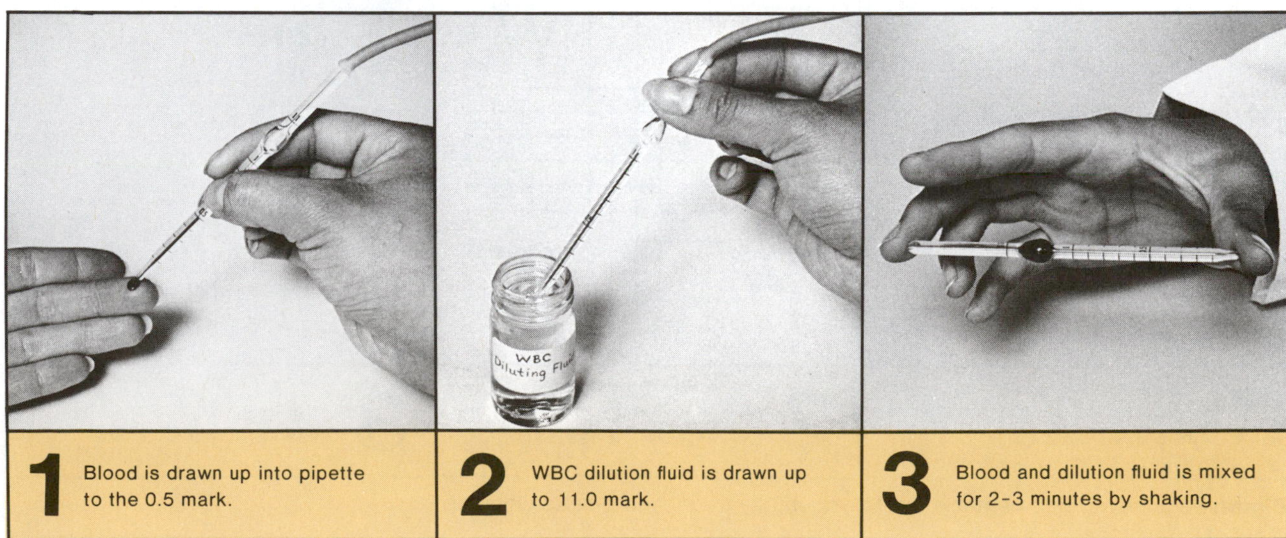

Figure 38.4 Dilution and mixing procedures for WBC blood count.

Materials:

hemacytometer and cover glass
WBC diluting pipette and rubber tubing
WBC diluting fluid
mechanical hand counter
pipette cleaning solutions
cotton, alcohol, lancets, clean cloth

1. Wash the hemacytometer and cover glass with soap and water, rinse well, and dry with a clean cloth or Kimwipes.
2. Produce a free flow of blood, wipe away the first drop and draw the blood up into the diluting pipette to the 0.5 mark. See illustration 1, figure 38.4. Slight excess over 0.5 mark can be withdrawn with blotting paper.

   If the blood goes substantially past the 0.5 mark, discharge the blood, wash the pipette in the four cleansing solutions (illustration 3, figure 38.5), and start over. The ideal way is to draw up the blood *exactly* to the mark on the first attempt.
3. As shown in illustration 2, figure 38.4, draw the WBC diluting fluid up into the pipette until it reaches the 11.0 mark.
4. Place your thumb over the tip of the pipette, slip off the tubing, and place your third finger over the other end (illustration 3, figure 38.4).
5. Mix the blood and diluting fluid in the pipette for **2–3 minutes** by holding it as shown in illustration 3, figure 38.4. The pipette should be held *parallel to the tabletop* and move through a 90° arc, with the wrist held rigidly.
6. Discharge one-third of the bulb fluid from the pipette by allowing it to drop onto a piece of paper toweling.
7. While holding the pipette as shown in illustration 1, figure 38.5, deposit a **tiny drop** on the polished surface of the counting chamber next to the edge of the cover glass. **Do not let the tip of the pipette touch the polished surface for more than an instant.** If it is left there too long, the chamber will overfill.

   A properly filled chamber will have diluted blood filling only the space between the cover glass and counting chamber. No fluid should run down into the moat.
8. Charge the other side if the first side was overfilled.

Performing the Count

Place the hemacytometer on the microscope stage and bring the grid lines into focus under the **low-power** (10×) objective. Use the coarse adjustment knob and reduce the lighting somewhat to make both the cells and lines visible.

Locate one of the "W" (white) sections shown in illustration 2, figure 38.5. Since the diluting fluid contains acid, all erythrocytes have been destroyed; only the leukocytes will show up as very small dots.

Do the cells seem to be evenly distributed? If not, charge the other half of the counting chamber

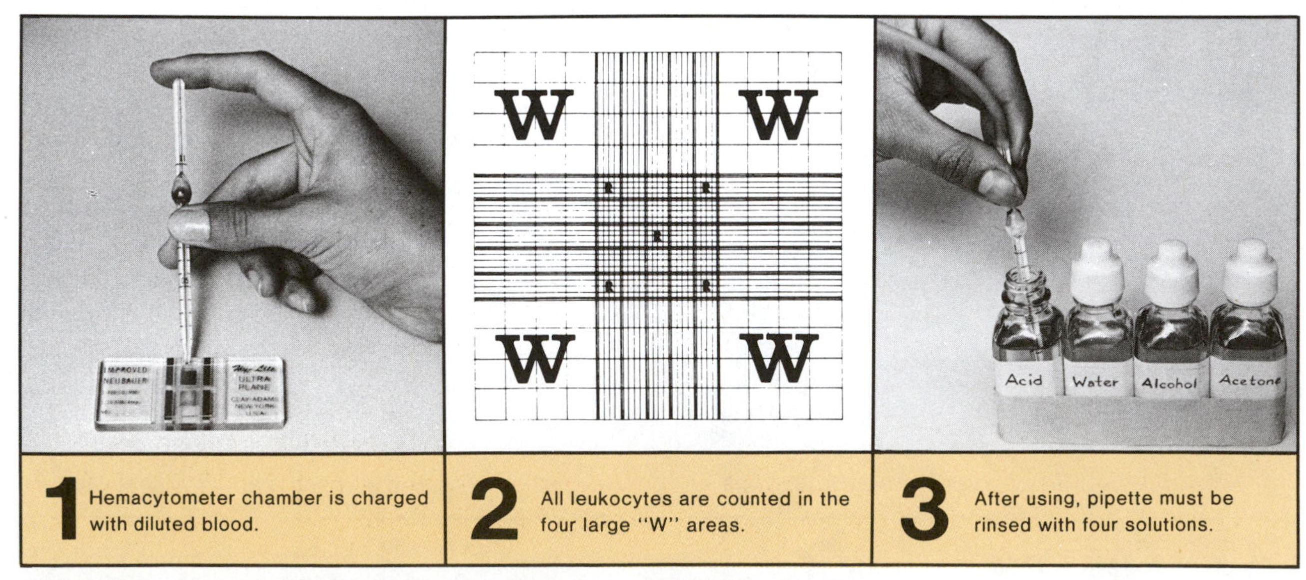

Figure 38.5 Charging hemacytometer, counting areas, and pipette cleaning.

after further mixing. If the other chamber had been previously charged unsuccessfully by overflooding, wash off the hemacytometer and cover glass, shake the pipette for 2–3 minutes, and recharge it.

Count all the cells in the four "W" areas, using a mechanical hand counter. To avoid overcounting of cells at the boundaries, **count the cells that touch the lines on the left and top sides only.** Cells that touch the boundary lines on the right and bottom sides should be ignored. This applies to the boundaries of each entire "W" area.

Discharge the contents of the pipette and rinse it out by sequentially flushing with the following fluids: acid, water, alcohol, and acetone. See illustration 3, figure 38.5.

Calculations

To determine the number of leukocytes per cubic millimeter, **multiply the total number of cells counted in the four "W" areas by 50.** The factor of 50 is the product of the volume correction factor and dilution factor, or

$$2.5 \times 20 = 50.$$

The volume correction factor of 2.5 is arrived at in this way: Each "W" area is exactly one square millimeter by 0.1 mm deep. Therefore, the volume of each "W" section is 0.1 cu mm. Since four "W" sections are counted, the total amount of diluted blood that is examined is 0.4 cu mm. And, since we are concerned with the number of cells in one cu mm instead of 0.4 cu mm, we must multiply our count by 2.5 derived from dividing 1.0 by 0.4.

Record your results on the Laboratory Report.

Red Blood Cell Count

Normal RBC counts for adult males average around 5,400,000 (±600,000) cells per cubic millimeter. The normal average for women is 4,600,000 (±500,000) per cu mm. The difference in sexes does not exist prior to puberty. At high altitudes, higher values will be normal for all individuals.

Low and high RBC counts result in anemia or polycythemia. Although low RBC counts do result in anemia, individuals can be anemic and still have normal RBC counts. If the cells are small, or if they lack sufficient hemoglobin, anemia will result. Thus, **anemia** may be defined, simply, as a condition in which the oxygen-carrying capacity of the blood is reduced.

Polycythemia, a condition characterized by above normal RBC counts, may be due to living at high altitudes (*physiological polycythemia*) or red marrow malignancy (*polycythemia vera*). In physiological polycythemia, counts may run as high as 8,000,000 per cu mm. In polycythemia vera, counts of 10–11 million are not uncommon.

Although the general procedures for the RBC count are essentially the same as for counting white blood cells, the diluting fluid, pipette, and mathematics are different. The RBC pipette differs from the WBC pipette in that the former has 101 scribed above the bulb instead of 11. The RBC diluting fluid may be one of several isotonic solutions such as physiological saline or Hayem's solution. To perform a red blood cell count, follow this procedure:

Materials:

RBC diluting pipette
rubber tubing and mouth piece
RBC diluting fluid
other supplies used for WBC count

1. Draw the blood up to the 0.5 mark and the diluting fluid up to the 101 mark. Mixing is performed in the same manner as for the WBC count.
2. Charge the chamber using the same procedures as for the WBC count.
3. Count all the cells in the five "R" areas (see illustration 2, figure 38.5). Use the **high-power objective** and observe the same rules as for the WBC count pertaining to line counts.
4. Multiply the total count of five areas by 10,000 (dilution factor is 200, volume correction factor is 50).
5. Clean the pipette with acid, water, alcohol, and acetone.

Laboratory Report

Record your results on part B and Table II of Laboratory Report 38–42.

39 Hemoglobin Percentage Measurement

Since hemoglobin is the essential oxygen-carrying ingredient of erythrocytes, its quantity in the blood determines whether or not a person is anemic. We learned in Exercise 38 that one may have a normal RBC count and still be anemic if the erythrocytes are smaller than normal. Such a condition, which is called *microcytic anemia,* results in one having an insufficient amount of hemoglobin. In other cases, a normal count in which the erythrocytes have a low hemoglobin percentage (*hypochromic cells*) may also result in anemia. Thus, it is apparent that the amount of hemoglobin present in a unit volume of blood is the critical factor in determining whether or not an individual is anemic.

Determination of the hemoglobin content of blood can be made by various methods. One of the oldest methods and, incidentally, a most inaccurate one, is the Tallqvist scale. In this technique a piece of blotting paper that has been saturated with a drop of blood is compared with a color chart to determine the percentage of hemoglobin. Very few, if any, physicians utilize this method today. A rapid, accurate method is to insert a cuvette of blood into a photocolorimeter that is calibrated for blood samples. Such a method, however, requires a rather expensive piece of electronic equipment. It also requires a considerable quantity of blood.

A relatively inexpensive instrument for hemoglobin determinations is the *hemoglobinometer.* Such a device compares a hemolyzed sample of blood with a color standard. Figures 39.1 through 39.4 illustrate the major steps to follow in using the American Optical hemoglobinometer (*Hb-Meter*). It is this piece of equipment that we will use in this test in determining hemoglobin content of a blood sample.

The hemoglobin content of blood is expressed in terms of grams per 100 ml of blood. Three different standards are used, depending on the community where the test is performed. For our purposes we will use the 15.6 gms/100 ml as standard. For adult males, the normal on this scale is 14.9 ± 1.5 gms/100 ml; for adult females, 13.7 ± 1.5; for children at birth, 21.5 ± 3; for children at 1 year, 12 ± 1.5; and for children at 4 years, 13 ± 1.5 gms/100 ml. A conversion scale exists on the side of the A/O Hb-Meter which allows one to determine hemoglobin percentages from the grams Hb/100 ml. Proceed as follows to determine your own hemoglobin percentage.

Materials:

American Optical Hb-Meter
hemolysis applicators
lancets
cotton
alcohol

1. Disassemble the blood chamber by pulling the two pieces of glass from the metal clip. Note that one piece of glass has an H-shaped moat cut into it. This piece will receive the blood. The other piece of glass has two flat surfaces and serves as a cover plate.
2. Clean both pieces of glass with alcohol and Kimwipes. Handle by edges to keep clean.
3. Reassemble the glass plates in the clip so that the grooves on the moat plate face the cover plate. The moat plate should be inserted only halfway to provide an exposed surface to receive the drop of blood. See figures 39.1 and 39.2.
4. Disinfect and puncture the finger with a disposable lancet.
5. Place a drop of blood on the exposed surface of the moat plate, as shown in figure 39.1.
6. Hemolyze the blood on the plate by mixing the blood with the pointed end of a hemolysis applicator as shown in figure 39.2. It will take 30–45 seconds for all the red blood cells to rupture. Complete hemolysis has occurred when the blood loses its cloudy appearance and becomes a transparent red liquid.
7. Push the moat plate in flush with the cover

plate and insert the sample into the side of the instrument, as in figure 39.3.

8. Place the eyepiece to your eye with the left hand in such a manner that the left thumb rests on the light switch button on the bottom of the hemoglobinometer.
9. While pressing the light button with the left thumb, move the slide button on the side of the instrument back and forth with the right index finger until the two halves of the split field match. The index mark on the slide knob indicates the grams of hemoglobin per 100 ml of blood. Read the percent hemoglobin on the 15.6 scale.

Laboratory Report

Record your results on Table II of the Laboratory Report 38–42.

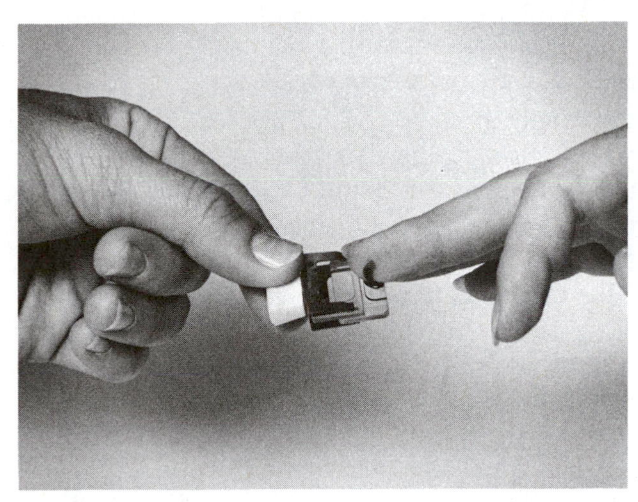

Figure 39.1 A fresh drop of blood is added to the moat plate of the blood chamber assembly. The blood must flow freely. Avoid squeezing the finger excessively.

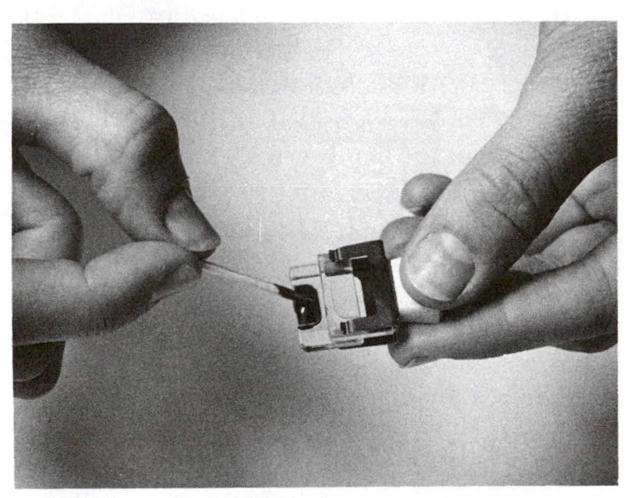

Figure 39.2 Blood sample is hemolyzed on moat plate with a wooden hemolysis applicator. Thirty-five to forty-five seconds are required for complete hemolysis.

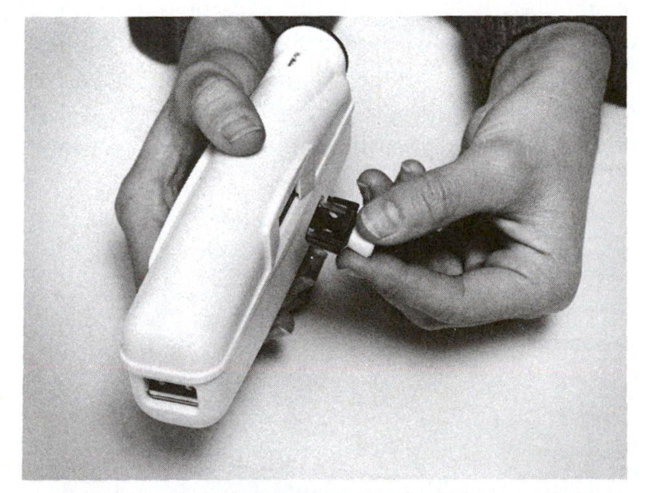

Figure 39.3 Charged blood chamber is inserted into slot of hemoglobinometer. Before insertion, the unit should be tested to make sure the batteries are active.

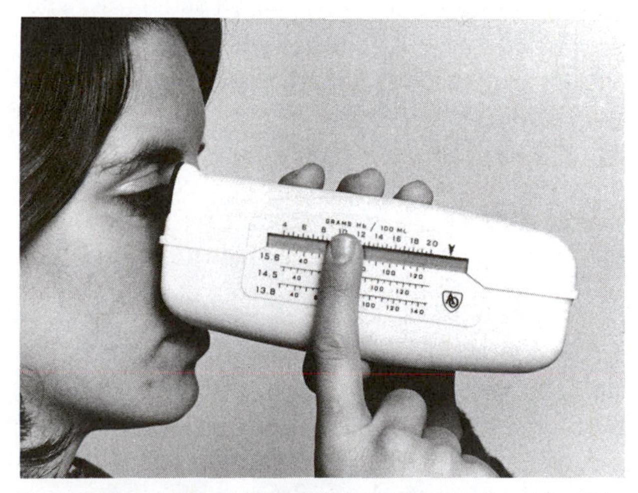

Figure 39.4 Blood sample is analyzed by moving slide button with right index finger. When the two colors match in density, the grams/100 ml is read on the scale.

40 Packed Red Cell Volume

In Exercises 38 and 39, we employed two different tests for determining the presence or absence of anemia: (1) the RBC count and (2) the hemoglobinometer. We have seen that anemia may be due to a dilution of the number of erythrocytes or to a deficiency of hemoglobin. It was observed that the most significant factor is the amount of hemoglobin that is present in a given volume of blood. A third method that can be used in anemia diagnosis is to determine the volume percentage of blood that is occupied by the red blood cells that have been packed by centrifugation. This method is called the **VPRC,** or volume of packed red cells.

The VPRC is determined by centrifuging a blood sample in a special centrifuge tube called a *hematocrit,* or in a special type of capillary tubing. The centrifuge must be specifically designed to accommodate the hematocrit or capillary tubing. The speed of the centrifuge and the radius of its head must fall within rigid specification standards. The mathematical relationship of centrifuge head radius and speed is as follows:

$$\text{RPM} = \frac{202{,}146{,}700}{\text{r (cm)}} .$$

In this exercise, we will utilize a micro method for determining the VPRC. As illustrated in figure 40.1, only a drop of blood is needed to perform the test. Capillary tubing, which has been heparinized, is used to collect the blood sample. It is centrifuged at high speed for only four minutes. (Macro methods that utilize hematocrits usually centrifuge the blood for 30 minutes at a much slower speed.) After centrifugation, the percentage of total volume occupied by the packed red blood cells is read on a special tube reader (figure 40.6), or directly on the head of the centrifuge, depending on the type of centrifuge.

An interesting relationship exists between grams of hemoglobin per 100 ml and the VPRC: *the VPRC is usually three times the gm Hb/100 ml.* In men, the normal VPRC range is between 40% and 54%, with 47% as average. In women, the normal range is between 37% and 47%, with 42% as average. The great advantages of this method over the two previous methods are (1) simplicity, (2) speed, and (3) high degree of accuracy. Although this micro method is not as accurate as macro methods, it still functions in an accuracy

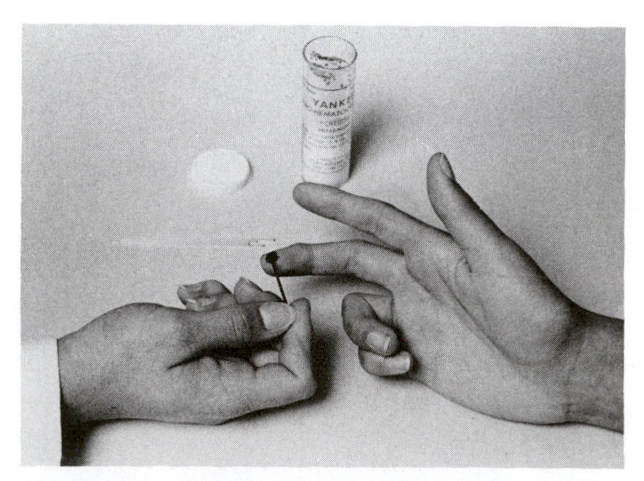

Figure 40.1 Blood is drawn up into heparinized capillary tube for hematocrit determination.

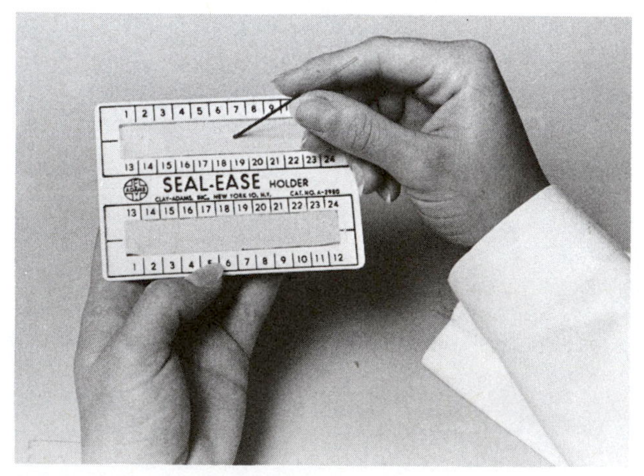

Figure 40.2 The end of the capillary tube is sealed with clay.

range of ±2% for most blood samples. Proceed as follows to determine the VPRC of your own blood.

Materials:

lancets
cotton
alcohol
micro-hematocrit centrifuge
tube reader
heparinized capillary tubes
Clay-Adams Seal-Ease

1. Produce a free flow of blood on the finger. Wipe away the first drop of blood.
2. Place the marked end (red) of the capillary tube into the drop of blood and allow the blood to be drawn up about two-thirds of the way into the tube. Holding the open end of the tube downward from the blood source will cause the tube to fill more rapidly.
3. Seal the blood end of the tube with Seal-Ease. See figure 40.2.
4. Place the tube into the centrifuge with the sealed end against the ring of rubber at the circumference. Load the centrifuge with even number of tubes (2, 4, 6, etc.) to properly balance the load.
5. Secure the inside cover with a wrench (figure 40.4) and fasten down the outside cover.
6. Turn on the centrifuge, setting the timer for **four minutes.**
7. Determine the VPRC by placing the tube in the mechanical tube reader. Instructions for reading are on the instrument.

Laboratory Report

Record your results on Table II, Part E, of Laboratory Report 38–42.

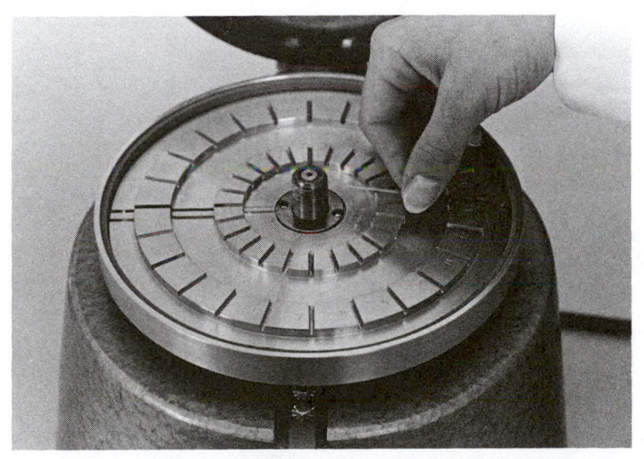

Figure 40.3 The capillary tubes are placed in the centrifuge with the sealed end toward the perimeter.

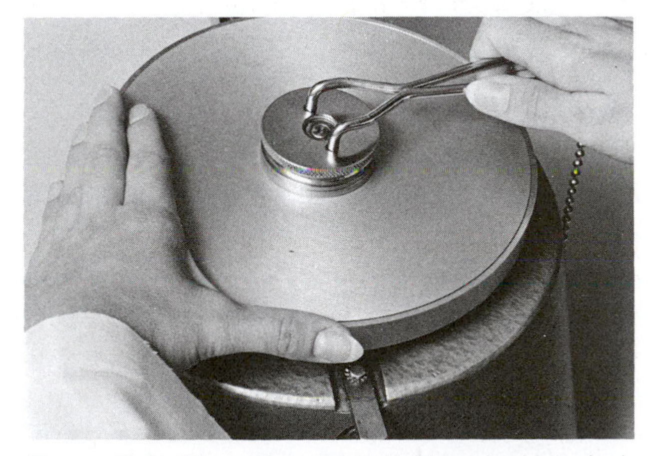

Figure 40.4 The safety lid is tightened with a lock wrench.

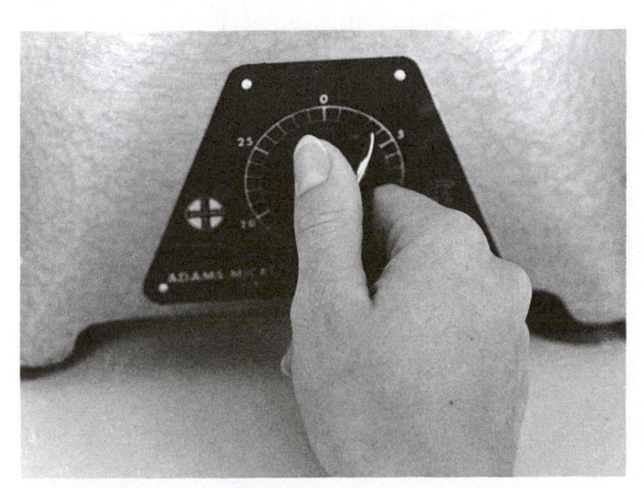

Figure 40.5 The timer is set for four minutes by turning the dial clockwise.

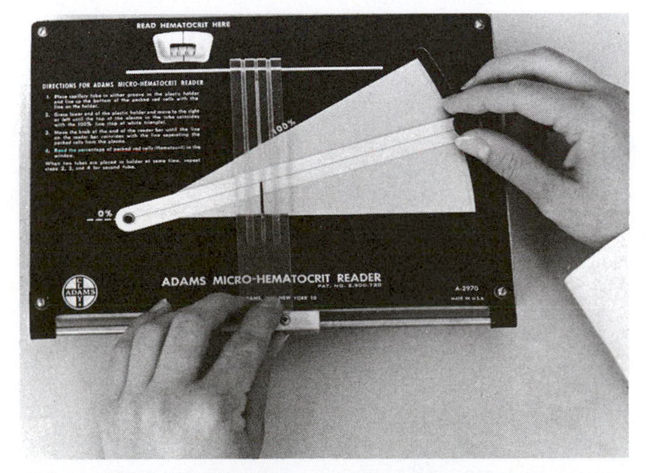

Figure 40.6 The capillary tube is placed in the tube reader to determine the hematocrit.

41 Blood Typing

The surfaces of red blood cells possess various antigenic molecules that have been designated by letters such as A, B, C, D, E, c, d, e, M, N, etc. Although the exact chemical composition of all these substances is undetermined, some of them appear to be carbohydrate residues (oligosaccharides).

The presence or absence of these various substances determines the type of blood possessed by an individual. Since the chemical makeup of cells is genetic, an individual's blood type is the same in old age as it is at birth. It never changes. The only factors that we are concerned with here in this exercise are the A, B, and D(Rh) antigens since they are most commonly involved in transfusion reactions.

To determine an individual's blood type, drops of blood typing sera are added to suspensions of red blood cells to detect the presence of **agglutination** (clumping) of the cells. ABO typing may be performed at room temperature with saline suspensions of red blood cells as shown in figure 41.1. Rh typing (D factor), on the other hand, requires higher temperatures (around 50° C) and whole blood instead of diluted blood. For ABO typing, the diluted blood procedure is preferable. For convenience, however, the warming box method may be used for combined ABO and Rh typing.

ABO Blood Typing

Materials:

- small vial (10 mm diam. × 50 mm long)
- disposable lancets (*B-D Microlance, Sera-sharp,* etc.)
- 70% alcohol and cotton
- wax pencil and microscope slides
- typing sera (anti-A and anti-B)
- applicators or toothpicks
- saline solution (0.85%)
- 1 ml pipettes

1. Mark a slide down the middle with a marking pencil, dividing the slide into two halves (see figure 41.1). Write *anti-A* on the left side and *anti-B* on the right side.
2. Pour approximately 1 ml of saline solution into a small vial or test tube.
3. Scrub the middle finger with a piece of cotton saturated with 70% alcohol and pierce it with a sterile disposable lancet. Allow two or three drops of blood to mix with the saline by holding the finger over the end of the vial and washing it with the saline by inverting the tube several times.
4. Place a drop of this red cell suspension on each side of the slide.
5. Add a drop of anti-A serum to the left side of the slide and a drop of anti-B serum to the right side. *Do not contaminate the tips of the serum pipettes with the material on the slide.*
6. After mixing each side of the slide with *separate* applicators or toothpicks, look for agglutination. The slide should be held about 6 inches above an illuminated white background and rocked gently for two or three minutes. Record your results on the Laboratory Report as of three minutes.

Combined ABO and Rh Typing

As stated, Rh typing must be performed with heat on blood that has not been diluted with saline. A warming box such as the one in figure 41.2 is essential in this procedure. In performing this test, two factors are of considerable importance: first, only a small amount of blood must be used (a drop of about 3 mm diam. on the slide) and second, proper agitation must be executed. The agglutination that occurs in this antibody-antigen reaction results in finer clumps; therefore, closer examination is essential. If the agitation is not properly performed, agglutination may not be as apparent as it should be. In this combined method, we will use the

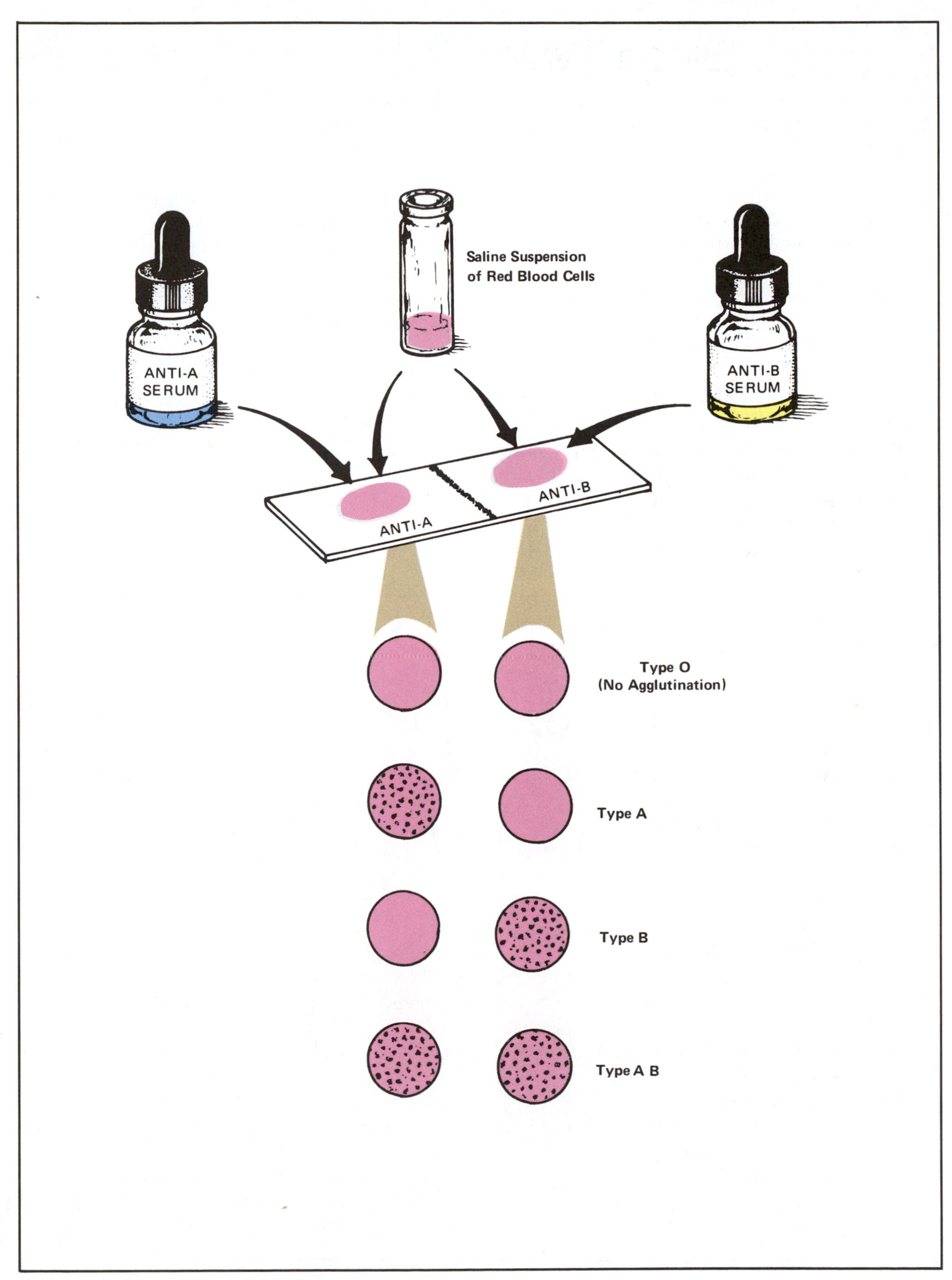

Figure 41.1 Blood typing ABO groups.

whole blood for the ABO typing also. Although this method works out satisfactorily as a classroom demonstration for the ABO groups, it is *not as reliable* as the previous method in which saline and room temperature are used, *and should not be used clinically.*

Materials:

- slide warming box with a special marked slide
- anti-A, anti-B, and anti-D typing sera
- applicators or toothpicks
- 70% alcohol
- cotton
- disposable sterile lancets (*B-D Microlance, Sera-sharp,* etc.)

1. Scrub the middle finger with a piece of cotton saturated with 70% alcohol and pierce it with a sterile disposable lancet. Place a small drop in each of three squares on the marked slide on the warming box. To get the proper proportion of serum to blood, do not use a drop larger than 3 mm diameter on the slide.
2. Add a drop of anti-D serum to the blood in the anti-D square, mix with a toothpick, and note the time. **Only two minutes should be allowed for agglutination.**
3. Add a drop of anti-B serum to the anti-B square and a drop of anti-A serum to the anti-A square. Mix the sera and blood in both squares with *separate fresh* toothpicks.
4. Agitate the mixtures on the slide by slowly rocking the box back and forth on its pivot. At the end of two minutes, examine the anti-D square carefully for agglutination. If no agglutination is apparent, consider the blood to be Rh negative. By this time, the ABO type can also be determined.

Laboratory Report

Complete Laboratory Report 38–42.

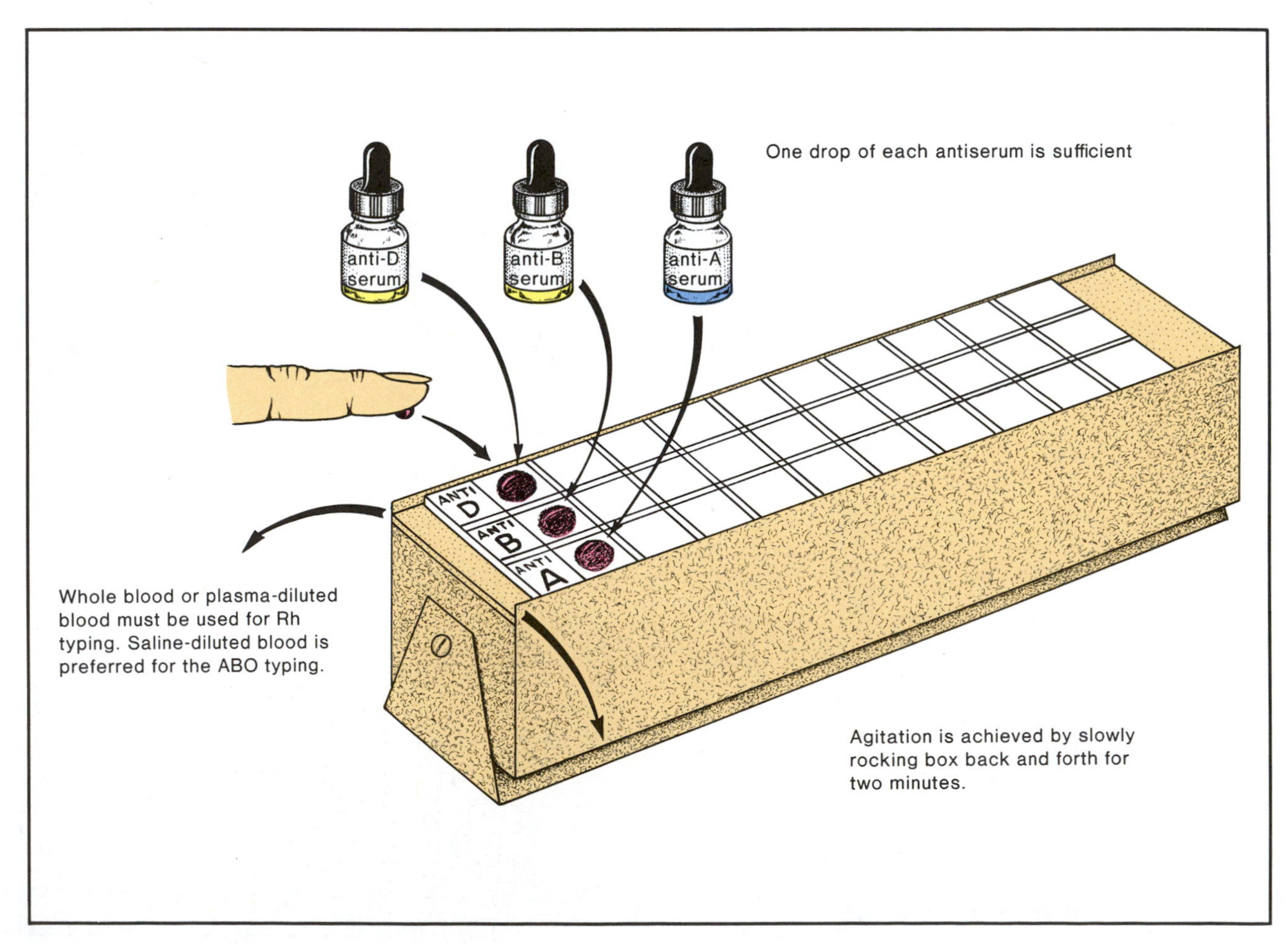

Figure 41.2 Blood typing with warming box.

Coagulation Time 42

The coagulation of blood is a complex phenomenon involving over thirty substances. The majority of these substances inhibit coagulation and are called *anticoagulants;* the remainder, which promote coagulation, are designated as *procoagulants.* Whether or not the blood will coagulate depends on which group predominates in a given situation. Normally, the anticoagulants predominate, but when a vessel is ruptured, the procoagulants in the affected area assume control, causing a clot (**fibrin**) to form in a relatively short period of time.

Of the various methods that have been devised to determine the rate of blood coagulation, the one outlined here is the simplest and most reliable. Figures 42.1 through 42.3 illustrate the general procedure. Proceed as follows:

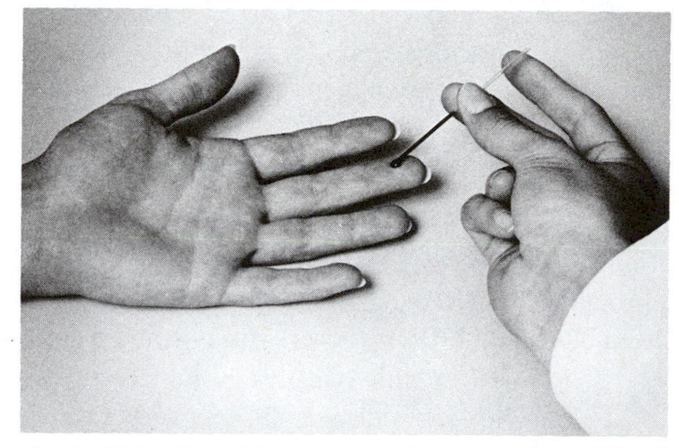

Figure 42.1 Blood is drawn up into a nonheparinized capillary tube.

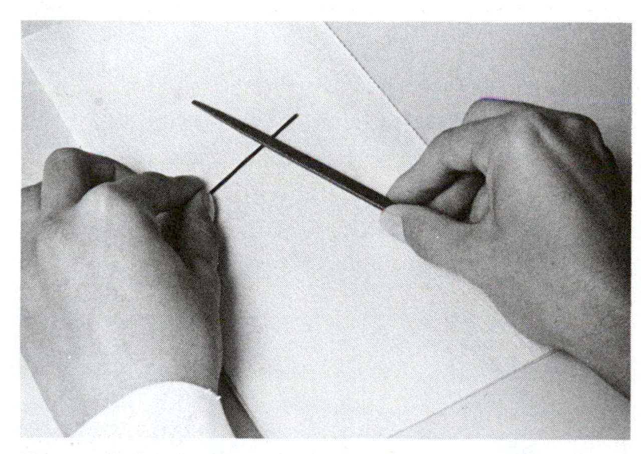

Figure 42.2 At one-minute intervals, sections of the tube are filed and broken off for the test.

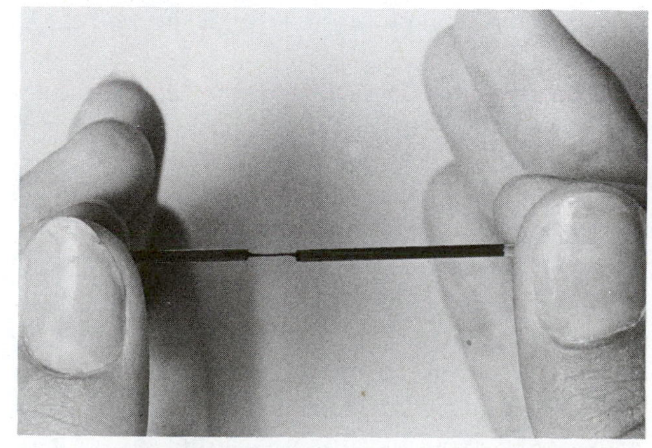

Figure 42.3 When the blood has coagulated, strands of fibrin will extend between broken ends of tube.

Materials:

lancets, cotton, alcohol
capillary tubes (0.5 mm diameter)
3-cornered file

1. Puncture the finger to expose a free flow of blood. **Record the time.** Place one end of the capillary tube into the drop of blood. Hold the tube so that the other end is **lower** than the drop of blood so that the force of gravity will aid the capillary action.
2. At **one-minute intervals,** break off small portions of the tubing by scratching the glass with a file first. **Important:** Separate the broken ends slowly and gently while looking for coagulation. Coagulation has occurred when threads of fibrin span the gap between the broken ends. **Record the time** as that time from which the blood first appeared on the finger to the formation of fibrin.

Laboratory Report

Record your results on Table II, Part E, of Laboratory Report 38–42.

43 Anatomy of the Heart

In this study of the anatomy of the human heart we will use a sheep heart for dissection and comparison. Before beginning the dissection, however, study figures 43.1 and 43.2.

Internal Anatomy

Figure 43.1 reveals the internal anatomy of the human heart. Red colored vessels carry oxygenated blood and blue ones carry deoxygenated blood.

Chambers of the Heart

Note that the heart has four chambers: two small upper atria and two larger ventricles. The atria receive all the blood that enters the heart and the ventricles pump it out. Note that the **right atrium** in the frontal section is on the left side of the illustration and the **left atrium** is on the right. Observe, also, that although the **right ventricle** is somewhat larger than the **left ventricle,** the left ventricle has a thicker wall. Separating the two ventricles is a partition, the **ventricular septum.**

Wall of the Heart

The wall of the heart consists of three layers: the myocardium, the endocardium, and the epicardium. The **myocardium** is the muscular portion of the wall that is composed of cardiac muscle tissue. Note in the enlarged section of the wall that the myocardium makes up the bulk of the wall thickness.

Lining the inner surface of the heart is the **endocardium.** It is a thin serous membrane that is continuous with the endothelial lining of the arteries and veins. An infection of this membrane is called *endocarditis.*

Attached to the outer surface of the heart is another serous membrane, the **epicardium,** or **visceral pericardium.** The production of serous fluid by this membrane on the outer surface of the heart enables the heart to move freely within the pericardial sac. Note that the **pericardial sac,** or **parietal pericardium,** consists of two layers: an inner serous layer and an outer fibrous layer. Between the heart and the parietal pericardium is the **pericardial cavity** (label 19). Its dimension has been exaggerated here.

Vessels of the Heart

The major vessels of the heart are the two venae cavae, the pulmonary artery, the four pulmonary veins, and the aorta. The **superior vena cava** is the upper blue vessel on the right atrium; it conveys deoxygenated blood to the right atrium from the head and arms. The **inferior vena cava** is the lower blue vessel that empties deoxygenated blood from the trunk and legs into the right atrium. From the right atrium blood passes to the right ventricle, where it leaves the heart through the **pulmonary artery.** This artery branches into the **right** and **left pulmonary arteries,** which carry blood to the lungs.

Blood is drained from the lungs by means of the four **pulmonary veins** (four small red vessels), which carry it to the left atrium. This blood passes into the left ventricle and finally out through the **aorta.** The initial emergence of the aorta is obscured, but it can be seen as the large red vessel at the top of the heart. The aorta carries oxygenated blood to the systemic circulation.

Note what appears to be a short vessel between the pulmonary artery and the aorta. It is called the **ligamentum arteriosum.** During prenatal life this ligament is a functional blood vessel, the *ductus arteriosus,* that allows blood to pass from the pulmonary artery to the aorta.

Valves of the Heart

The heart has two atrioventricular and two semilunar valves. Between the right atrium and the right ventricle is the **tricuspid valve.** Between the left atrium and the left ventricle is the **bicuspid,** or **mitral, valve.** The differences between these two valves are seen in the superior sectional view of figure 43.1. Note that the tricuspid valve has three flaps or cusps, and the bicuspid valve has only two cusps. Observe, also, that the edges of the cusps

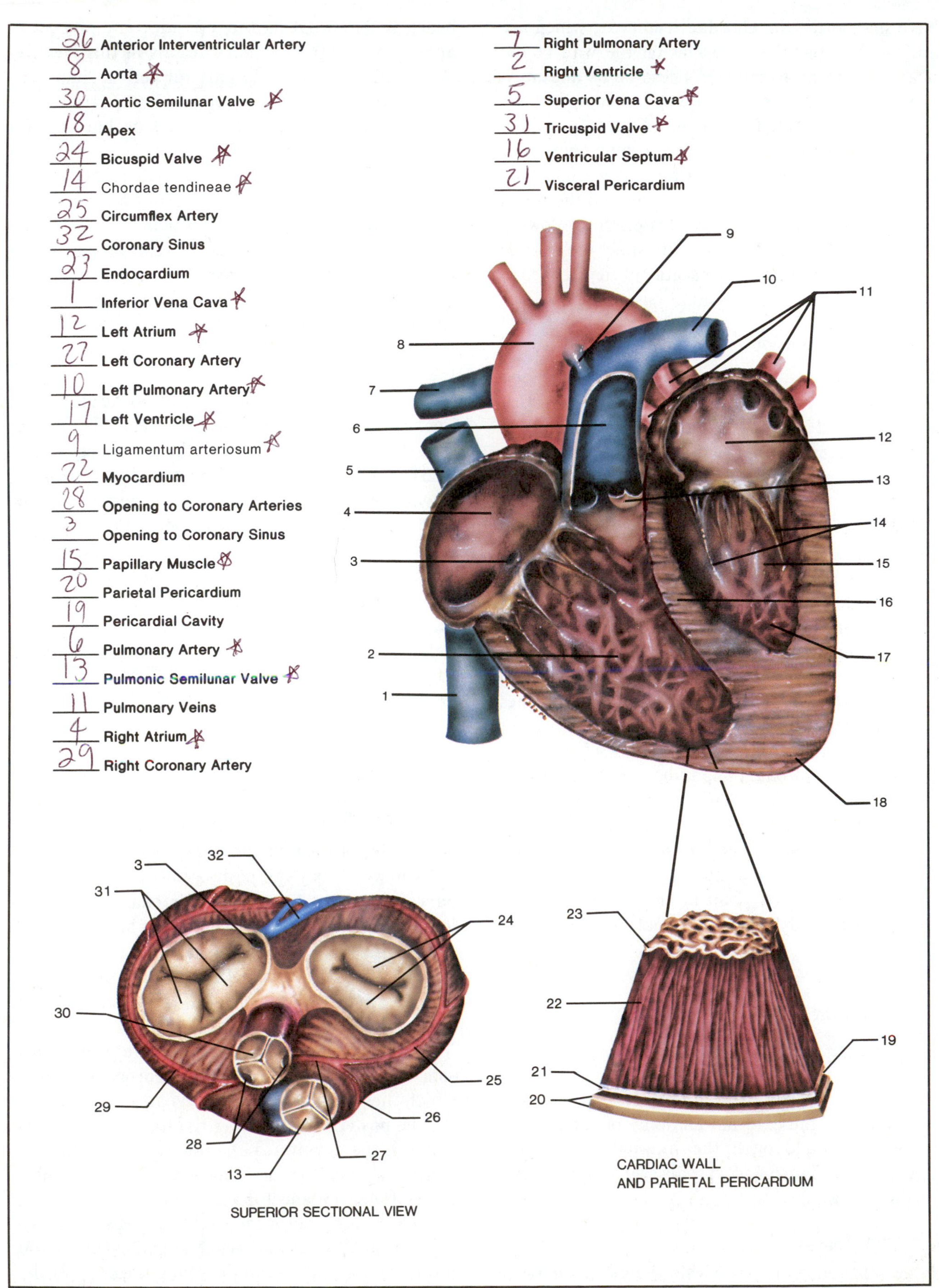

Figure 43.1 Internal anatomy of the heart.

have fine cords, the **chordae tendineae,** which are anchored to **papillary muscles** on the wall of the heart (see frontal section). These cords and muscles prevent the cusps from being forced up into the atria during systole (ventricular contraction).

The other two valves are the **pulmonic semilunar** and **aortic semilunar valves.** They are located at the bases of the pulmonary artery and the aorta. They prevent blood in those vessels from flowing back into the heart during diastole (relaxation phase). Note in the superior sectional view that each of these valves has three small cusps.

Coronary Circulation

The vessels that supply the heart muscle with blood comprise the *coronary circulatory system.* Oxygenated blood in the aorta passes into right and left coronary arteries through two small openings at a point just superior to the aortic semilunar valve. The **openings to the coronary arteries** are seen in the wall of the aorta in the superior sectional view in figure 43.1. The **left coronary artery** has two principal branches: a **circumflex artery** that passes around the heart in the left atrioventricular sulcus and the **anterior interventricular artery,** which lies in the interventricular sulcus on the anterior surface of the heart. The **right coronary artery** lies in the right atrioventricular sulcus and has branches that supply the posterior and anterior surfaces of the ventricular muscle.

Once the coronary blood has been relieved of its oxygen and nutrients it is picked up by the various veins that parallel the arteries and empty into the **coronary sinus** (label 2, posterior aspect, figure 43.2). Blood in the coronary sinus is emptied into the right atrium. Its point of entry can be seen in the superior and frontal sectional views of figure 43.1.

Assignment:

Label figure 43.1.

External Anatomy

The study of the external anatomy of the heart is relatively simple once the internal anatomy is understood. Figure 43.2 reveals two external views of the heart.

Anterior Aspect

Note that the heart lies within the **mediastinum** at a slight tilt to the left so that the **apex,** or tip of the heart, is somewhat on the left side. The obscured appearance of the coronary vessels in this view is due to the fact that the **parietal pericardium** lies intact over the organ. The clarity of the structures on the posterior view is due to the fact that the parietal pericardium has been removed.

Differentiating the right ventricle from the left ventricle is best determined by locating the **interventricular sulcus** first. This sulcus is the slight depression over the ventricular septum that contains the **anterior interventricular artery** and the **great cardiac vein.** The yellow material seen in the sulcus consists of fatty deposits. The left ventricle is on the left side of the interventricular sulcus, and the right ventricle is to the right of it.

Observe that the aorta forms an **aortic arch** over the heart and descends behind the heart. That part that is near the heart is called the **ascending aorta.** The part that passes down behind the heart is called the **descending aorta.** Only a short portion of the descending aorta is seen at the bottom of the illustration.

Locate the right and left pulmonary arteries which emerge from under the aortic arch. Also, identify the ligamentum arteriosum between the pulmonary artery and the aorta. The superior vena cava shows up well near the top of the right atrium. Only a short portion of the inferior vena cava is seen at the bottom of this view. The **right coronary artery** and the **small cardiac vein** are visible in the atrioventricular sulcus.

Posterior Aspect

The posterior view of the heart in figure 43.2 reveals more clearly the position of the four pulmonary veins. The two right pulmonary veins are located near the venae cavae. Note how close together the venae cavae are located. The appearance of a single blue vessel lying below the aorta is actually the site of division of the pulmonary artery into right and left branches.

The following coronary arteries are seen on this side of the heart: (1) the **right coronary artery,** which lies in the right atrioventricular sulcus; (2) the **posterior descending right coronary artery,** which is the downward extension of the right coronary artery; (3) the **posterior interventricular artery** (label 1); and (4) the **circumflex artery.**

The major coronary veins on this side are (1) the **middle cardiac vein,** which parallels the posterior descending right coronary artery; (2) the **left posterior ventricular vein** (label 3); and (3) the

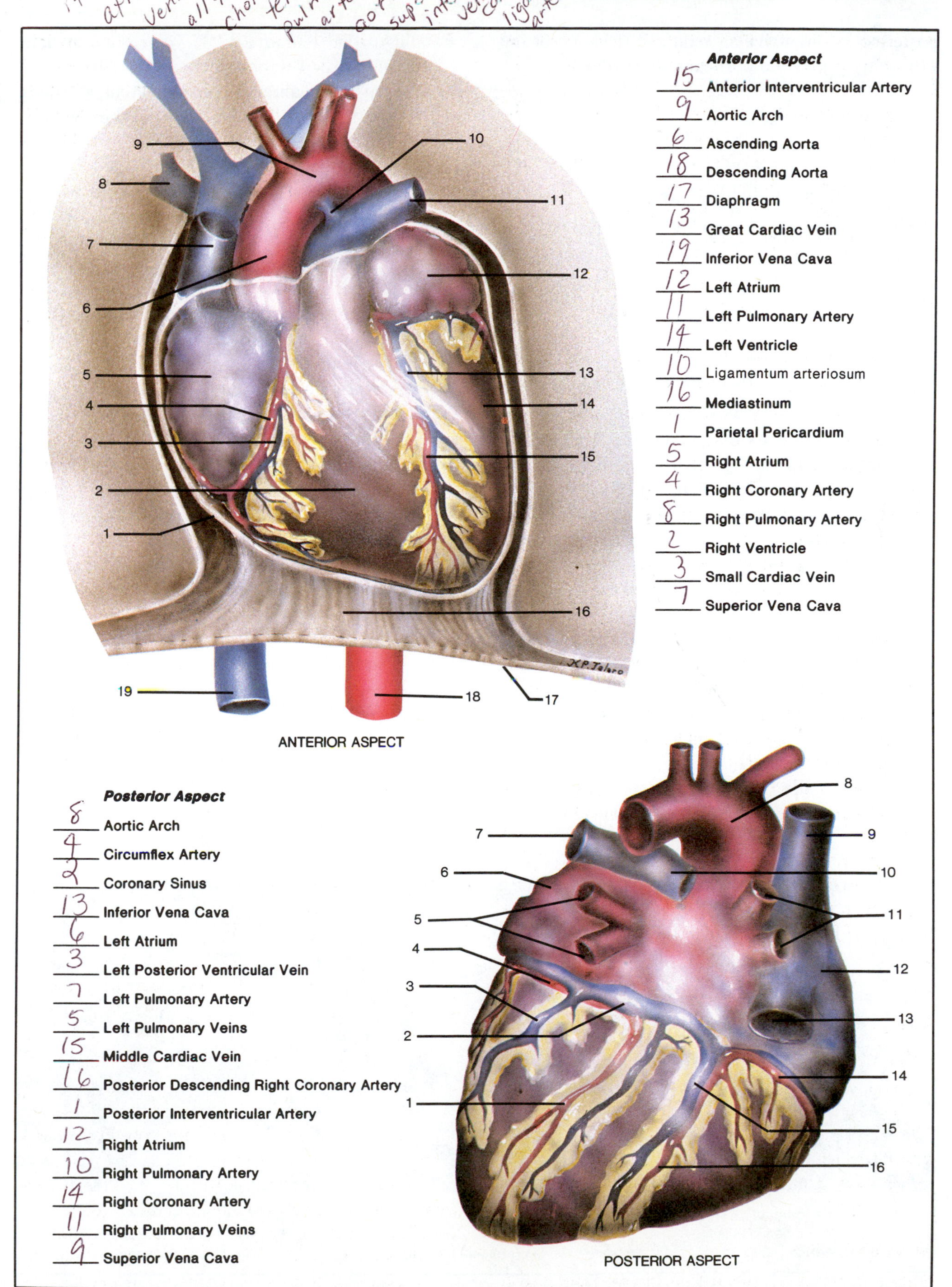

Figure 43.2 External anatomy of the heart.

posterior interventricular vein. All these coronary veins empty into the **coronary sinus** (label 2).

Assignment:

Label figure 43.2.

Sheep Heart Dissection

While dissecting the sheep heart, attempt to identify as many structures as possible that are shown in figures 43.1 and 43.2.

Materials:

sheep heart, fresh or preserved
dissecting instruments and tray

1. Rinse the heart with cold water to remove excess preservative or blood. Allow water to flow through the large vessels to irrigate any blood clots out of its chambers.
2. Look for evidence of the **parietal pericardium.** This fibroserous membrane is usually absent from laboratory specimens, but there may be remnants of it attached to the large blood vessels of the heart.
3. Attempt to isolate the **visceral pericardium** (*epicardium*) from the outer surface of the heart. Since it consists of only one layer of squamous cells, it is very thin. With a sharp scalpel try to peel a small portion of it away from the myocardium.
4. Identify the **right** and **left ventricles** by squeezing the walls of the heart as shown in illustration 1, figure 43.3. The right ventricle will have the thinner wall.
5. Identify the **anterior interventricular sulcus** (between the two ventricles), which is usually covered with fatty tissue. Carefully trim away the fat in this sulcus to expose the **anterior interventricular artery** and the **great cardiac vein.**
6. Locate the **right** and **left atria** of the heart. The right atrium is seen on the left side of illustration 1, figure 43.3.
7. Identify the **aorta,** which is the large vessel just to the right of the right atrium as seen in illustration 1, figure 43.3. Carefully peel away some of the fat around the aorta to expose the **ligamentum arteriosum.**
8. Locate the **pulmonary artery,** which is the large vessel between the aorta and the left atrium as seen when looking at the anterior side of the heart. If the vessel is of sufficient length, trace it to where it divides into the **right** and **left pulmonary arteries.**
9. Examine the posterior surface of the heart. It should appear as in illustration 2, figure 43.3. Note that only the right atrium and two ventricles can be seen on this side. The left atrium is obscured by fat from this aspect. Look for the four thin-walled **pulmonary veins** that are embedded in the fat. Probe into these vessels and you will see that they lead into the left atrium.
10. Locate the **superior vena cava,** which is attached to the upper part of the right atrium.

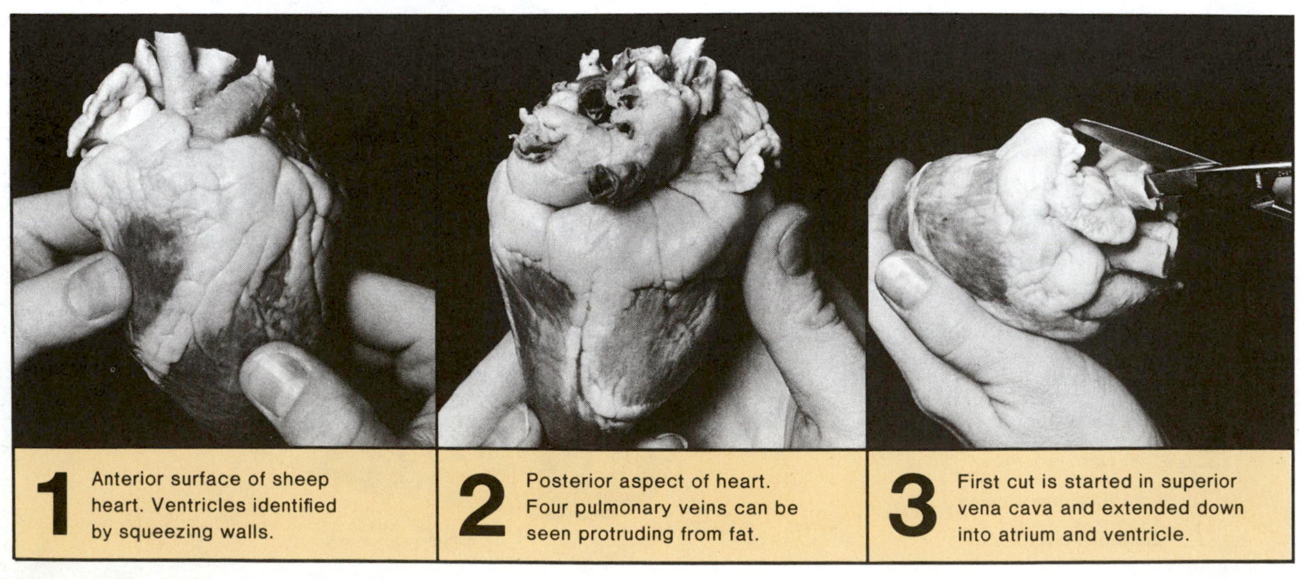

Figure 43.3 Sheep heart dissection.

Insert one blade of your dissecting scissors into this vessel (illustration 3, figure 43.3), and cut through it into the atrium to expose the **tricuspid valve** between the right atrium and right ventricle. Don't cut into the ventricle at this time.

11. Fill the right ventricle with water, pouring it in through the tricuspid valve. Gently squeeze the walls of the ventricle to note the closing action of the valve's cusps.
12. Drain the water from the heart and continue the cut with scissors from the right atrium through the tricuspid valve down to the apex of the heart.
13. Open the heart and flush it again with cold water. Examine the interior. The open heart should look like illustration 1, figure 43.4.
14. Examine the interior wall of the right atrium. This inner surface has ridges, giving it a comblike appearance; thus, it is called **pectinate muscle** (*pecten,* comb). Between the inferior vena cava and the tricuspid valve you should see the **opening to the coronary sinus.** It is through this opening that blood of the coronary circulation is returned to the venous circulation. This opening can be seen in illustration 1, figure 43.4.
15. Insert a probe under the cusps of the tricuspid valve. Are you able to see three flaps?
16. Locate the **papillary muscles** and **chordae tendineae.** How many papillary muscles do you see in the right ventricle? Identify the **moderator band,** which is a reinforcement cord between the ventricular septum and the ventricular wall. Its presence prevents excessive stress from occurring in the myocardium of the right ventricle.
17. With scissors, cut the right ventricular wall up along its lower margin parallel to the anterior interventricular sulcus to the pulmonary artery. Continue the cut through the exit of the right ventricle into the pulmonary artery. Spread the cut surfaces of this new incision to expose the **pulmonic semilunar valve.** Wash the area with cold water to dispel blood clots.
18. Insert one blade of your scissors into the left atrium, as shown in illustration 2, figure 43.4. Cut through the atrium into the left ventricle. Also, cut from the left ventricle into the aorta, slitting this vessel longitudinally.
19. Examine the **bicuspid (mitral) valve,** which lies between the left atrium and left ventricle. With a probe, identify the two cusps.
20. Examine the pouches of the **aortic semilunar valve.** Compare the number of pouches of this valve with the pulmonic valve.
21. Look for the two **openings to the coronary arteries,** which are in the walls of the aorta just above the aortic semilunar valve. Force a blunt probe into each hole. Note how they lead into the two coronary arteries.

Laboratory Report

Complete the Laboratory Report for this exercise.

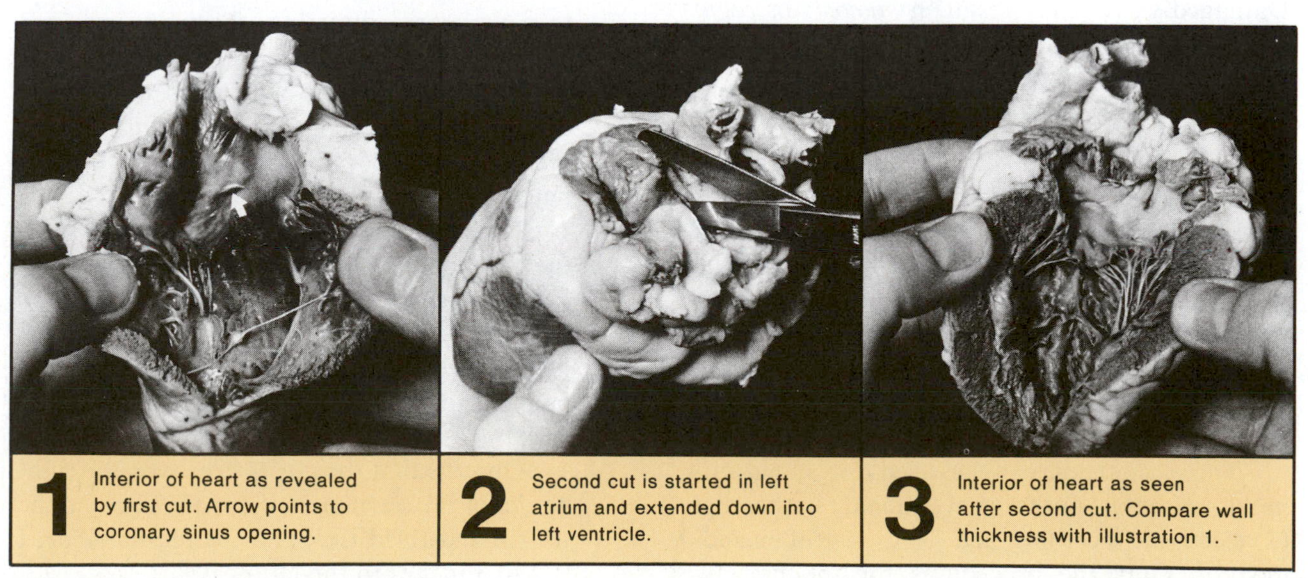

1 Interior of heart as revealed by first cut. Arrow points to coronary sinus opening.

2 Second cut is started in left atrium and extended down into left ventricle.

3 Interior of heart as seen after second cut. Compare wall thickness with illustration 1.

Figure 43.4 Sheep heart dissection.

44 Cardiovascular Sounds

Cardiovascular sounds are created by myocardial contraction and valvular movement. Three distinct sounds can be detected.

The **first sound** (lub) occurs simultaneously with the contraction of the ventricular myocardium and the closure of the mitral and tricuspid valves.

The **second sound** (dup) follows the first one after a brief pause. It differs from the first sound in that it has a snapping quality of higher pitch and shorter duration. This sound is caused by the virtually simultaneous closure of the aortic and pulmonic valves at the end of ventricular systole.

The **third sound** is low pitched and is loudest near the apex of the heart. It is more difficult to detect, but can be heard most easily when the subject is recumbent. It appears to be caused by the vibration of the ventricular walls and the AV valves during systole.

Abnormal heart sounds are collectively referred to as **murmurs.** They are usually due to damaged valves. Mitral valve damage due to rheumatic fever is a most frequent cause of heart murmur; the aortic valve is next in susceptibility to damage. Damaged valves may result in *stenosis* or *regurgitation* of blood flow.

Examination Procedures

Listening to sounds of the body is called **auscultation.** For maximum results in monitoring valvular sounds it is necessary to be familiar with four specific spots on the thorax. Illustration 1, figure 44.1, shows their locations. Note that these auscultatory areas *do not coincide* with the anatomic locations of the various valves; this is because valvular sounds are projected to different spots on the rib cage.

Three methods can be used for monitoring heart sounds: (1) the conventional stethoscope, (2) electronic recording, and (3) the use of an audio monitor. Only the procedures for the first two methods are presented. The audio monitor, if available, will be used only as a demonstration. Since noise is distracting in these experiments, it is imperative that all students keep voices low.

Audio Monitor Demonstration *(by Instructor)*

Using a student as a subject in recumbent position, the instructor will use an audio monitor to demonstrate heart sounds to the class. If a sensitive unit, such as the Grass AM7, is used, it will not be necessary to bare the subject's chest.

When attempting to demonstrate second sound splitting, hand signals should be used instead of verbalization. The subject can be instructed to inspire when the instructor's hand is raised and expire when the hand is lowered.

Stethoscope Auscultation

Work with your laboratory partner to auscultate each other's heart with a stethoscope. Use the following procedure:

Materials:

stethoscope
alcohol and absorbent cotton

1. Clean and disinfect the earpieces of a stethoscope with cotton saturated in alcohol. Fit them to your ears, directing them inward and upward.
2. Try first to maximize the **first sound,** by locating the stethoscope on the *tricuspid* and *mitral* auscultatory areas. See illustration 1, figure 44.1.

   Note that the *mitral area* coincides with the apex of the heart (fifth rib), and the *tricuspid area* is about 2 to 3 inches medial to this spot.
3. Now, move the stethoscope to the aortic and pulmonic areas to hear the **second sound** more clearly.

   Observe that the *aortic area* is on the right border of the sternum in the space between the second and third ribs, and the *pulmonic area* is 2 to 3 inches to the left of it.
4. While listening to the second sound, attempt to detect the **splitting of the second sound** while

the subject is inhaling. Use hand signals instead of verbalization to communicate with subject. Have the subject inspire when you raise your hand and expire when you lower your hand.

This characteristic of the second sound during inspiration is normal. It is due to the fact that during inspiration more venous blood is forced into the right side of the heart, causing delayed closure of the pulmonic valve during systole. Result: aortic and pulmonic closure sounds are heard separately.

5. After listening to the sounds with the subject sitting erect, listen to them with the subject supine.
6. Can you detect the **third sound**?
7. Is there any evidence of **murmurs**?
8. Compare the sounds before and after exercise, recording your observations on the Laboratory Report. For exercise, leave the room to run up and down the stairs, or jog for a short distance.
9. Before ending the examination, check all the auscultatory areas in figure 44.1 to make sure that you have identified all valvular sounds.

Electronic Recording *(Unigraph Setup)*

Produce a phonocardiogram of the heart sounds of a supine subject, using these procedures:

Materials:

Unigraph, Trans/Med 6605 adapter
Statham P23AA pressure transducer

1. Attach a Trans/Med 6605 to the input end of the Unigraph.
2. Attach the Statham transducer and microphone jack to the 6605 adapter. Refer to illustration 3, figure 44.1.
3. Turn on the Unigraph and set the following controls: c.c. at STBY, heat control at two o'clock position, speed control at slow position, Gain at 1 MV/CM, and Mode on TRANS.
4. Place the microphone against one of the auscultatory areas and start the chart moving by placing the c.c. lever at Chart On. Adjust the temperature control knob to get a good recording. Center the tracing with centering control knob and adjust sensitivity to produce at least 5 mm stylus displacement.
5. Move speed control lever to high speed position.
6. Make recordings over each auscultatory area. Be sure to mark chart to identify areas.
7. Attempt to record splitting of second sound, using same techniques used with stethoscope.
8. Attach phonocardiograms to Laboratory Report.
9. Deactivate all controls on Unigraph at end of experiment.

Laboratory Report

Complete the Laboratory Report for this exercise.

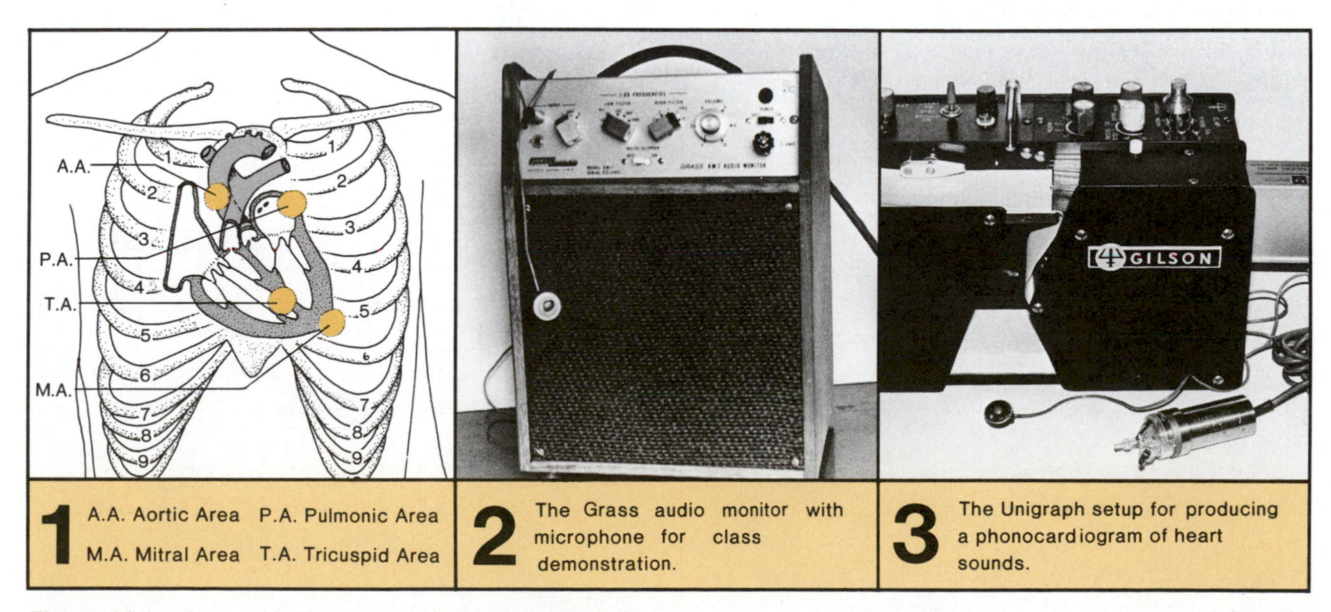

1 A.A. Aortic Area P.A. Pulmonic Area M.A. Mitral Area T.A. Tricuspid Area

2 The Grass audio monitor with microphone for class demonstration.

3 The Unigraph setup for producing a phonocardiogram of heart sounds.

Figure 44.1 Auscultatory areas and two electronic setups.

45 Electrocardiogram Monitoring: Using Chart Recorders

In this laboratory period we will learn the procedures that are involved in making an electrocardiogram (ECG), utilizing a chart recorder such as the Gilson Unigraph or the Narco Physiograph®. Separate instructions will be provided for both types of equipment. If the Cardiocomp (Exercise 46) is to be used instead of one of these chart recorders, it will be necessary for you to read over the applications provided here in this exercise since this background information is lacking in Exercise 46.

The ECG Wave Form

Myocardial contractions in the heart originate with depolarization of the sinoatrial (SA) node of the right atrium. As the myocardium of the right atrium is depolarized and contracts in response to depolarization of the SA node, the atrioventricular (AV) node is activated, causing it to send a depolarization wave via the atrioventricular bundle and conduction myofibers to the ventricular myocardium. The electrical potential changes that result from this depolarization and repolarization can be monitored to produce a record called an **electrocardiogram,** or **ECG.**

To produce such a record it is necessary to attach a minimum of two and as many as ten electrodes to different portions of the body to record the differences in electrical potential that occur. If one electrode is placed slightly above the heart and to the right, and another is placed slightly below the heart to the left, one can record the wave form at its maximum potential.

Figure 45.1 reveals a typical ECG as recorded on a Unigraph, and figure 45.2 illustrates the appearance of an individual normal ECG wave form. The initial depolarization of the SA node, which causes atrial contraction, manifests itself as the **P wave.** This depolarization is immediately followed by repolarization of the atria. Atrial repolarization, not usually seen with surface electrodes, can be demonstrated with implanted electrodes; such a wave is designated as the *TA wave.*

Depolarization of the ventricles via the atrioventricular bundle and conduction myofibers results in the production of the **QRS complex.** This depolarization causes ventricular systole.

As soon as depolarization of the ventricular muscle is completed, repolarization takes place, producing the **T wave.** Some ECG wave forms show an additional wave occurring after the T wave. It is designated as the **U wave.** This small wave is attributed to the gradual repolarization of the papillary muscles.

The final result of this depolarization wave is that the myocardium of the ventricles contracts as

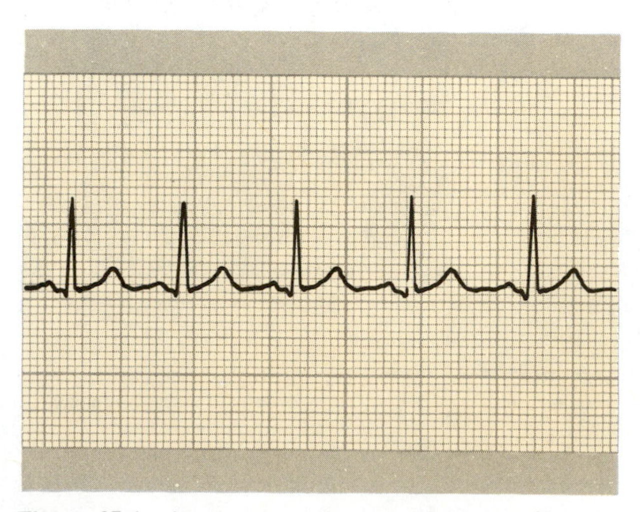

Figure 45.1 An electrocardiogram (Unigraph tracing).

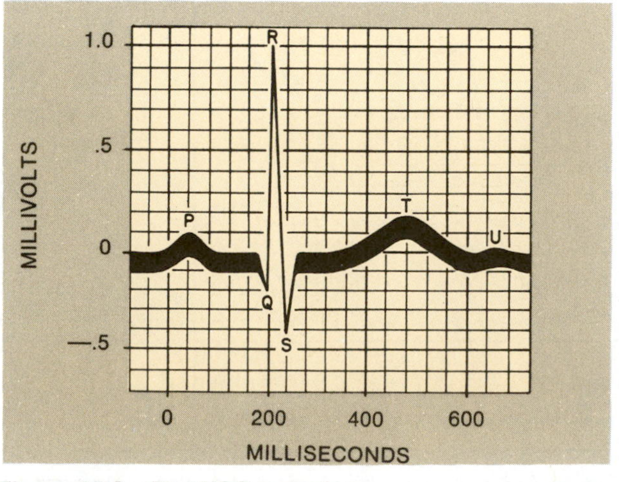

Figure 45.2 The ECG wave form.

a unit, forcing blood out through the pulmonary artery and aorta. Repolarization of the affected tissues quickly follows in preparation for the next contraction cycle.

The value of the ECG to the cardiologist is immeasurable. Almost all serious abnormalities of the heart muscle can be detected by analyzing the contours of the different waves. Interpretation of the ECG wave forms to determine the nature of heart damage is a highly developed technical skill involving vector analysis and goes considerably beyond the scope of this course; however, figure 45.4 illustrates three examples of abnormal ECGs in which heart damage has occurred.

Electrodes and Leads

As indicated earlier, as few as two and as many as ten electrodes can be used to produce an ECG. An understanding of Einthoven's triangle (figure 45.3) will clarify the significance of the different electrode positions and the meaning of leads I, II, and III. Einthoven was the first person to produce an electrocardiogram around the turn of the nineteenth century.

The number of **electrodes** used for the Unigraph setup is three: the two wrists and the left ankle (figure 45.6). For the Physiograph® setup, five electrodes will be used (figure 45.5): the four appendages and one on the chest over the heart. The right leg acts as a ground in the five electrode setup.

Three leads, designated as leads I, II, and III, are derived from these electrode hookups. A **lead,** as designated by Einthoven, *is the electrical potential in the fluids around the heart between two electrodes during depolarization.* On the Physiograph® and Cardiocomp setups you will have an opportunity to make lead selections.

Note in figure 45.3 that the apexes of the upper part of the triangle represent the points at which the two arms connect electrically with the fluids around the heart. The lower apex is the point at which the left leg (+) connects electrically with the pericardial fluids. **Lead I** records the electrical potential between the right (−) and left (+) arm electrodes. **Lead II** records the potential between the right arm (−) and the left leg (+). **Lead III** records the potential between the left arm (−) and the left leg (+). These potentials are recorded graphically on the ECG in millivolts.

When we put leads I, II, and III together on one diagram, as in figure 45.3, we see that the three leads form a triangle around the heart, known as

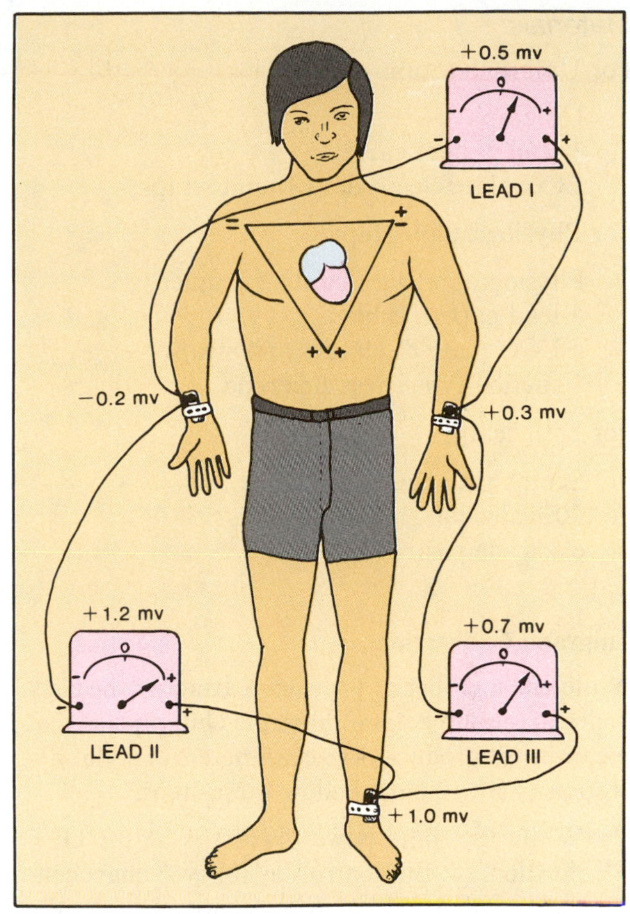

Figure 45.3 In the conventional three-lead hookup, the two arms and left leg form apexes of a triangle (Einthoven's) surrounding the heart. Note that the sum of the voltages in leads I and III equals the voltage in lead II (Einthoven's Law).

Einthoven's triangle. The significance of this triangle is that if the potentials of two of the leads are recorded, the third one can be determined mathematically. Or, stated another way: *the sum of the voltages in leads I and III equals the voltage of lead II.* This principle is known as **Einthoven's law:**

Lead II = Lead I + Lead III.

Procedure

Students will work in teams of three or four with one individual being the subject, and the others making the electrode hookups and operating the controls. It will be necessary to calibrate the recorder before recording the ECG; separate calibration instructions must be followed for the type of recorder being used. Failure in this experiment usually results from (1) poor skin contact, (2) damaged electrode cables, or (3) incorrect hookup.

Materials:

For Unigraph setup:

Unigraph
3-lead patient cable
ECG plate electrodes (3) and straps

For Physiograph® setup:

Physiograph® with cardiac coupler
5-lead patient cable
4 ECG plate electrodes and straps
1 suction-type chest electrode

For both setups:

Scotchbrite pad
70% alcohol
electrode paste

Unigraph Calibration

While one member of your team attaches the electrodes to the subject, calibrate the Unigraph so that one millivolt produces two centimeters of stylus deflection (1 mv/2 cm). Follow these steps:

1. While the main power switch is at the OFF position, set the controls as follows: chart control switch at STBY, speed control lever at the slow position, gain selector knob at 1 MV/CM, sensitivity knob completely counterclockwise, mode selector control at CC-Cal, and the Hi Filter and Mean switches on NORM.
2. Turn on the power switch and rotate the stylus heat control to the two o'clock position.
3. Place the chart control switch on Chart On and observe the trace and center the stylus, using the centering control.
4. Increase the sensitivity by rotating the sensitivity control clockwise one-half turn.
5. Depress the 1 MV button, hold it down for about 2 seconds, and then release it. The upward deflection should measure 1 cm. The downward deflection should equal the upward deflection (figure 45.8). If the travel is not 1 cm, adjust the sensitivity control and depress the 1 MV button until exactly 1 cm is achieved.
6. Place the chart control switch on STBY and label the chart with the date and subject's name.
7. Place the mode selector control on ECG and move the speed selector lever to the high speed position (25 mm per second). The Unigraph is now ready to accept the patient cable.

Physiograph® Calibration

While one member of your team attaches the electrodes to the subject, calibrate the channel sensitivity of the Physiograph® so that one millivolt produces two centimeters of pen deflection (1 mv/2 cm). Follow these steps:

1. Place the Time Constant switch of the cardiac coupler in the 3.2 position.
2. Place the Gain switch on the coupler to the X2 position.
3. Place the Lead Selector Control in the calibrate (Cal) position.

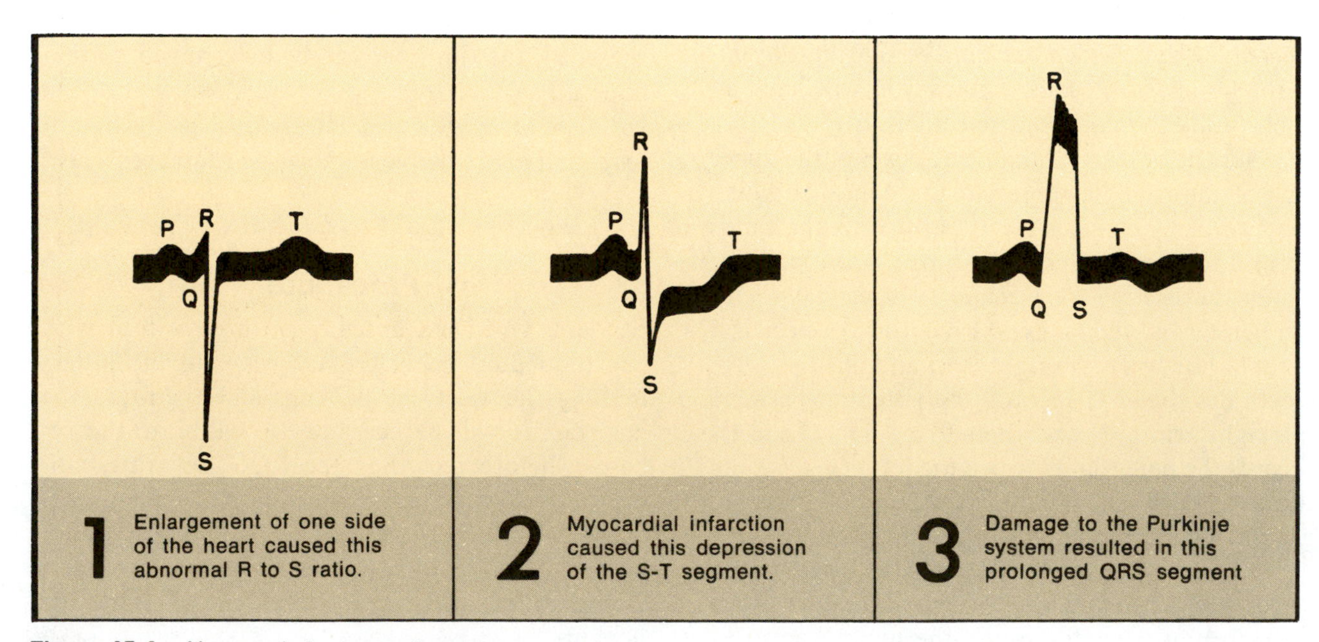

Figure 45.4 Abnormal electrocardiograms.

4. Set the outer knob of the channel Amplifier Sensitivity Control on the 10 MV/CM position, and rotate the inner sensitivity knob completely clockwise until it clicks.
5. Start the chart moving at 0.25 cm/sec, and lower the pens onto the paper. Adjust the pen to the center of its arc with the Position Control.
6. Place the channel amplifier record button into the ON position.
7. Place the CAL toggle switch on the cardiac coupler into the 1 MV position. Observe that the channel recording pen will deflect *upward* 2 centimeters.
8. Hold the CAL switch in the 1 MV position until the recording pen returns to the center line, then release the switch. Observe that upon release, the channel recording pen will now deflect *downward* 2 cm.
9. If the upward and downward deflections are not *exactly* 2 centimeters, make adjustments with the amplifier sensitivity controls.
10. Stop recording. The channel is now calibrated.

Preparation of Subject

While the recorder is being calibrated, the electrodes should be attached to the subject by other members of the team. Although it is not absolutely essential, it is desirable to have the individual recumbent on a comfortable cot.

Note in figure 45.6 that the two wrists and left ankle will be used for electrode attachment if the Unigraph is used. As shown in figure 45.5, both wrists, both ankles, and the heart region will have electrodes attached if the Physiograph® is to be used. Proceed as follows to attach the electrodes to the subject:

1. To achieve good electrode contact on the wrists and ankles, rub the contact areas with a Scotchbrite pad, disinfect the skin with 70% alcohol, and add a little electrode paste to the flat contact surface of each electrode.
2. Attach each of the plate electrodes with a rubber strap, as shown in figure 45.5.
3. If you are using the five electrode setup, affix the suction-type electrode over the heart region, using an ample amount of electrode paste on the contact surface.
4. Attach the leads of the patient cable to the electrodes as follows:

   Unigraph Setup: Attach the black unnumbered lead to the left ankle; one of the numbered leads to the right wrist; and the other numbered lead to the left wrist.

   Physiograph® Setup: Attach each lead according to its label designation: LL, left leg; RL, right leg; LA, left arm; RA, right arm; and C, cardiac.
5. Make certain that the leads are securely connected to the electrode binding posts by *firmly* tightening the knobs on the binding posts. If

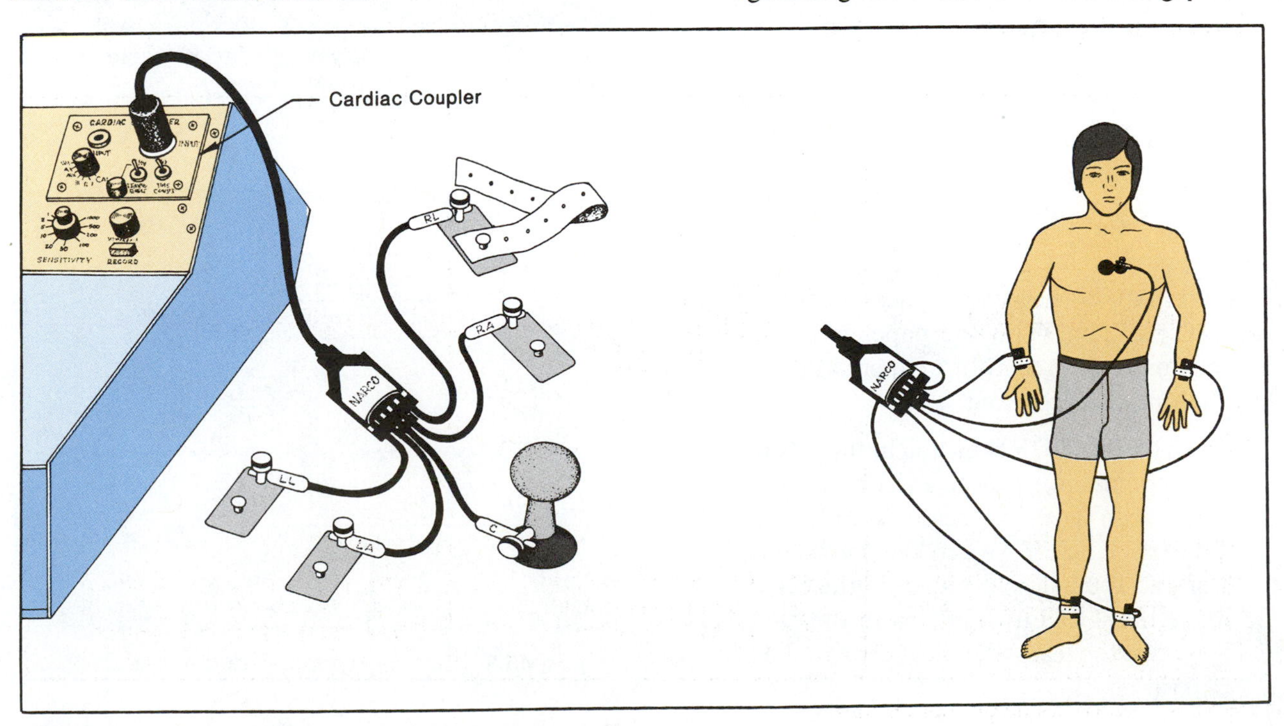

Figure 45.5 Physiograph® setup for ECG monitoring.

the patient cable has alligator clamps instead of straight jacks, take care to see that the clamps can't slip off easily.

6. If the subject is sitting in a chair instead of lying down, **make sure that the chair is not metallic.**
7. Connect the free end of the patient cable to the appropriate socket on the Unigraph or Physiograph® cardiac coupler.
8. The subject is now ready for monitoring.

Monitoring

Tell the subject to relax and be still. Turn on the recorder and observe the trace. Follow the steps outlined below for the type of recorder you are using.

Unigraph Adjust the stylus heat control so that optimum recording is achieved.

1. Compare the ECG with figures 45.1 and 45.2 to see if the QRS spike is up or down (positive or negative). If it is negative (downward), reverse the leads to the two wrists.
2. After recording for about 30 seconds, stop the paper and study the ECG. Does the wave form appear normal? Determine the pulse rate by counting the spikes between two margin lines and multiplying by twenty.
3. Disconnect the leads to the electrodes and have the subject exercise for 2–5 minutes by leaving the room to run up and down stairs or jog outside the building.
4. After reconnecting the electrodes, record for another 30 seconds and compare this ECG with the previous one. Save the record for attachment to the Laboratory Report.

Physiograph® Place the lead selector control in the LEAD I position. With the paper moving, lower the pens onto the recording paper. Make certain that the timer is activated.

1. Place the RECORD button in the ON position and record one sheet of ECG activity.
2. If the pen goes off scale, simply press the TRACE RESET switch on the cardiac coupler and the pen will return to the center line.
3. Place the RECORD button in the OFF position, turn the lead selector control to the LEAD II position.

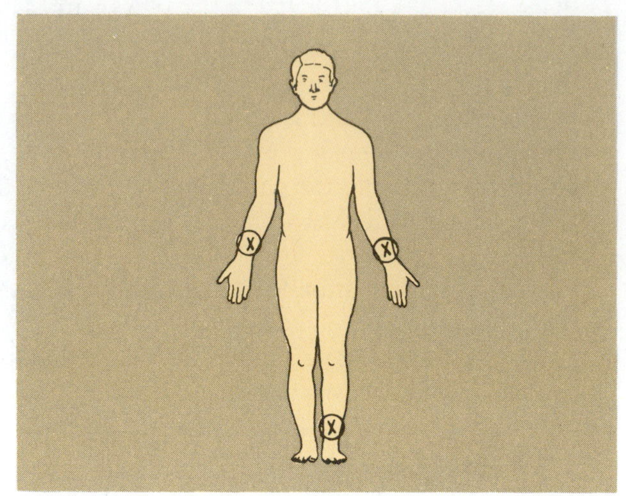

Figure 45.6 Three-lead hookup used for Unigraph setup.

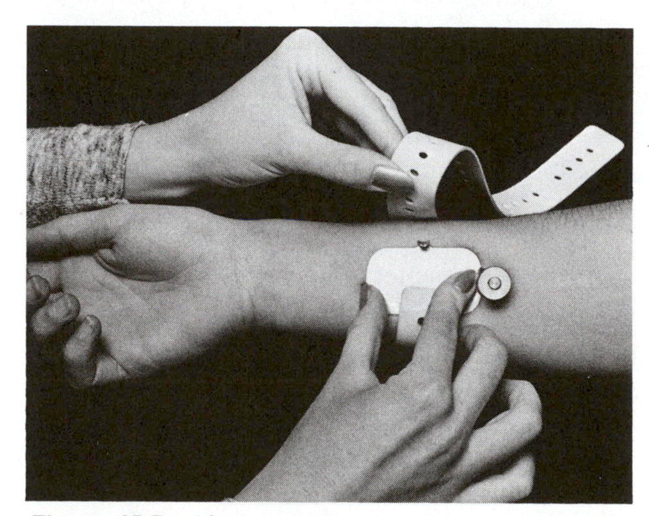

Figure 45.7 After applying electrode paste to electrode, it is strapped in place.

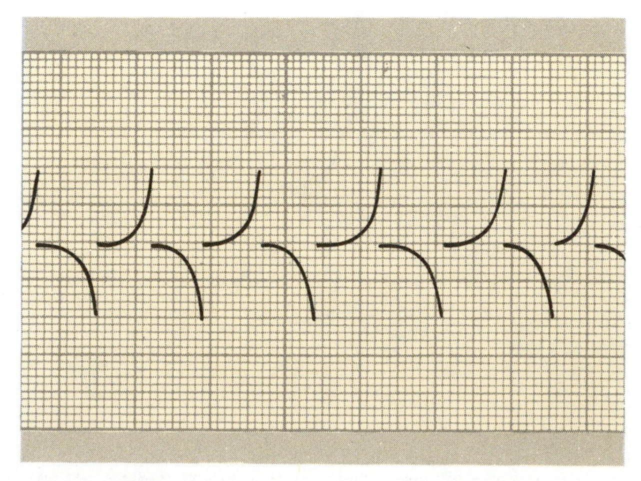

Figure 45.8 Sample recording of calibration of Unigraph for ECG.

4. Place the RECORD button back into the ON position and record another sheet of ECG activity.
5. Repeat step 4 above for the LEAD III position on the lead selector control.
6. Stop the recording and compare your results with the sample tracings in figure 45.9.

Laboratory Report

Complete the Laboratory Report for this exercise.

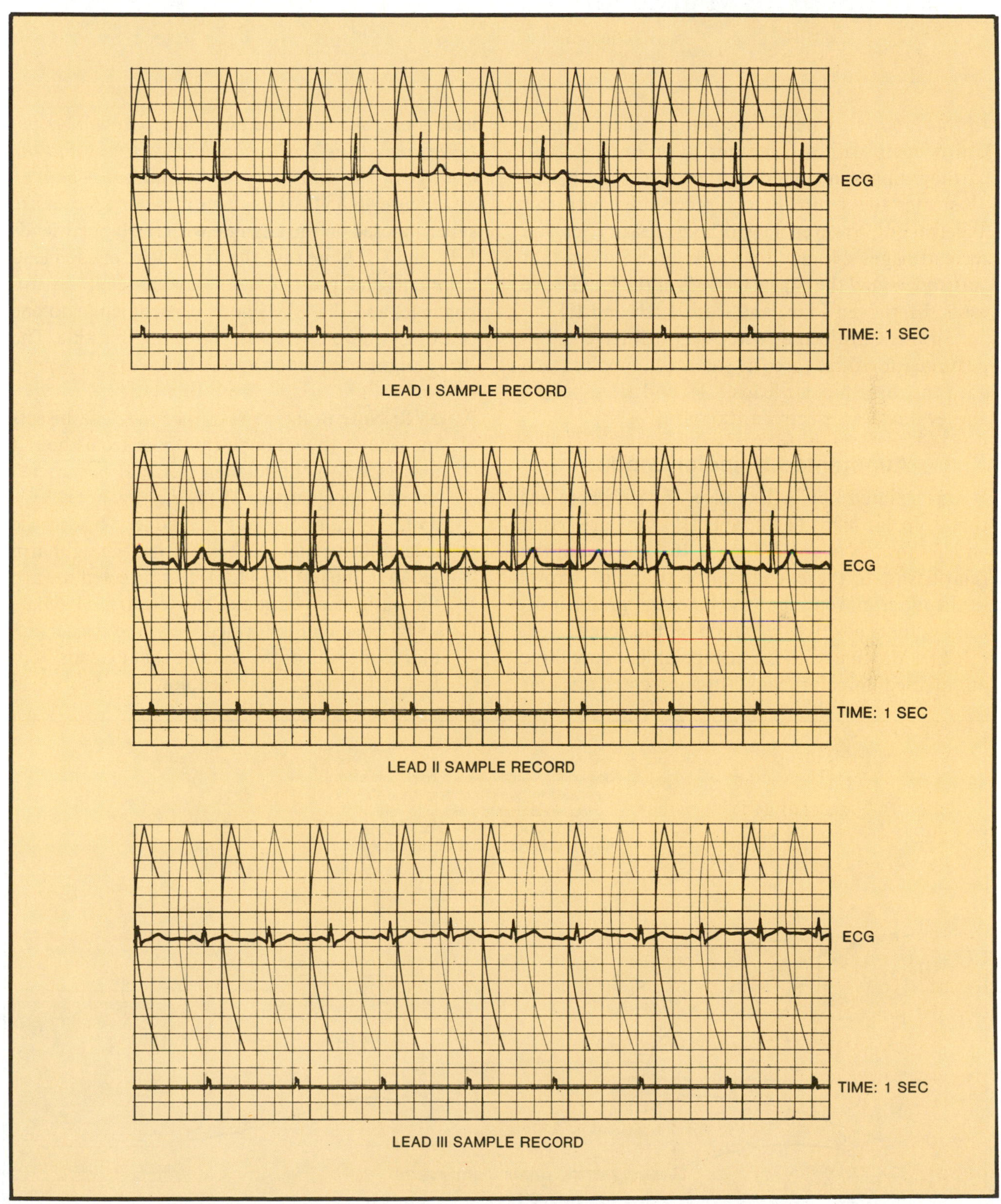

Figure 45.9 Physiograph® electrocardiogram sample recordings at different lead settings.

46 Electrocardiogram Monitoring: Using Computerized Hardware

In this laboratory period we will use the Intelitool Cardiocomp with a computer to produce electrocardiograms of members of the class. Figure 46.1 illustrates the setup. Before proceeding with this experiment, however, it will be necessary for you to read pages 240 and 241 in Exercise 45 so that you understand the characteristics of the ECG wave form, Einthoven's law, and leads I, II, and III.

In this Cardiocomp setup you will have an opportunity to utilize three augmented unipolar leads with the three bipolar leads I, II, and III to record and evaluate an electrocardiogram.

Augmented Unipolar Leads

It was pointed out in Exercise 45 that there are three bipolar leads (I, II, and III) that are based on Einthoven's triangle. Figure 46.2 illustrates the relationship of these leads to the triangle. Arrows in the diagram represent the potential gradient direction for a positive deflection on the screen.

In addition to these three bipolar leads are three augmented unipolar leads. An **augmented unipolar lead** is one in which the *electrical potential is between one of the positively charged limbs and the average of the other two limbs that are negative.*

Figure 46.3 illustrates how these three leads are superimposed on the triangle with the bipolar leads I, II, and III. Note that the arrowhead of each lead points to the positive (+) electrode of an appendage, and that the line is perpendicular to one side of Einthoven's triangle, bisecting that side. The three augmented leads are as follows:

aVR: Note in figure 46.3 that this lead bisects the side of the triangle that goes from the left arm (LA) to the left leg (LL). It is directed toward the electrode of the right arm (RA).
aVL: This unipolar augmented lead bisects the side of the triangle that goes from the right arm (RA) to the left leg (LL). It is directed toward the positive electrode of the left arm (LA).
aVF: This lead is formed by a line perpendicular to the side of the triangle that extends from RA to LA and is directed downward to LL.

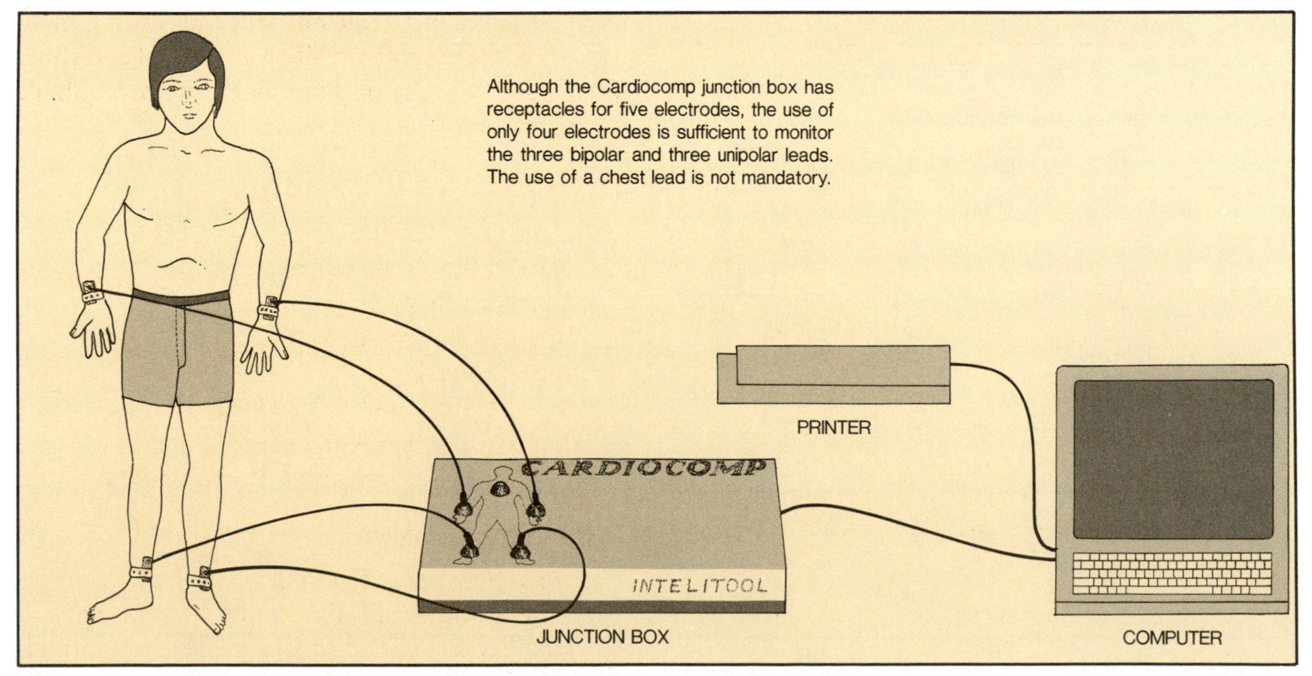

Figure 46.1 Cardiocomp setup for ECG monitoring.

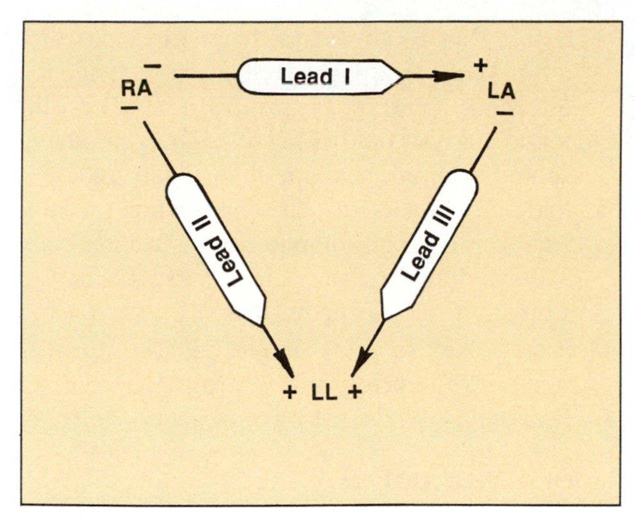

Figure 46.2 Bipolar lead orientation according to Einthoven's triangle.

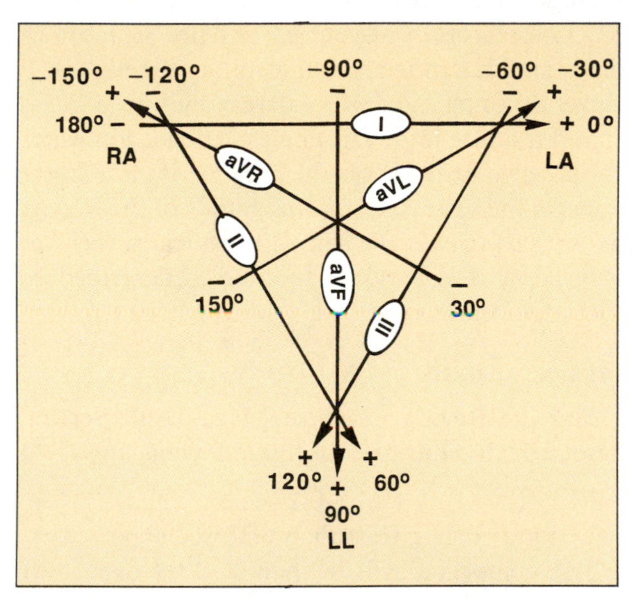

Figure 46.3 Diagram illustrating the relationship of bipolar to unipolar augmented leads.

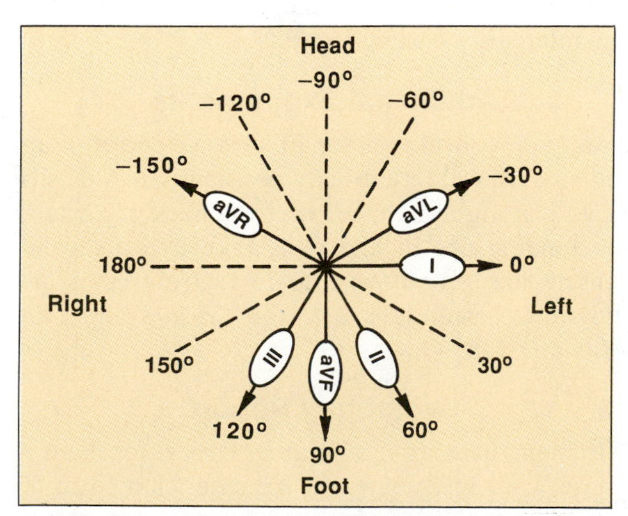

Figure 46.4 Directional orientation of the six most commonly used leads in ECG monitoring.

Figure 46.4 is an attempt to clarify the directional nature of the six leads. Note that by using six leads, the direction of the electrical potential through the heart can be monitored, radially, in thirty degree segments. By scanning the six leads, a cardiologist can determine the direction of the potential *by simply noting in which lead the R wave has the strongest deflection.* If the strongest deflection happens to be in lead II, as is usually the case, one knows that the potential is in the direction from the right shoulder to the left foot, or about 60 degrees. Another reason for using six leads instead of only three bipolar leads is that there is less chance of missing some unusual cardiac event.

Chest Leads

Note in figure 46.2 that only four electrodes are used in our experiment and that the Cardiocomp has one receptacle for a chest electrode. In clinical practice most cardiologists rely on six standardized chest leads in addition to the ones studied here. While chest leads are invaluable to the cardiologist, they reveal nothing about the wave form or arrhythmias that cannot be determined with the bipolar and unipolar augmented leads; thus, for simplicity in our situation, we will not use a chest electrode.

Procedure

Working together in groups of two or more students, proceed as follows to do a series of ECG measurements on all members of the class, using the Cardiocomp. While one student is being monitored, another student will be wired up for testing. Note that two Cardiocomp manuals are available for reference.

Materials:

computer and printer
Cardiocomp junction box
Cardiocomp program diskette
blank initialized diskette
4 electrode wires
4 flat-plate electrodes and electrode gel
Scotchbrite pad and alcohol swabs
Cardiocomp User and *Lab Manuals*

Electrode Attachment

1. To achieve good electrode contact on the wrists and ankles, rub the contact areas with a Scotchbrite pad, disinfect the skin with an alcohol swab, and add a little electrode gel to the flat contact of the surface of each electrode.

2. Attach each of the plate electrodes to the limbs with the rubber strap and snap the wires to the electrodes. Be sure to match the labels to the limbs.
3. Plug the electrode wires into the proper receptacles of the Cardiocomp junction box.
4. Have the subject lie down and avoid movement. **It is very important that you arrange the wires parallel to the body, separate from each other and away from any power cables to avoid "noise."**

Six-Lead Data Acquisition

Now that the subject is properly hooked up to the electrodes and in a relaxed, horizontal position, proceed to monitor the ECG as follows:

1. Insert the Cardiocomp program diskette into drive A and the data diskette into drive B and turn on the computer. If you are using an Apple IIe, IIc, or IIGS, be sure to make the caps lock key active (in the down position).
2. Touch any key and the Cardiocomp program will take over. After you have described the system configuration with the proper answers, the Main Menu, which appears as below, will appear on the screen. Command key reminder is also shown.

( ) Electrocardiogram
( ) Review–Screen Analysis
( ) Interval Analysis
( ) Vectorgram
( ) Save Current Data
( ) Load Old Data File
( ) Disk Commands

(No Data in Memory)

Note: The Cardiocomp software knows that Review–Screen Analysis, Interval Analysis, Vectorgram, and Save Current Data modes require experimental data. If no data has been put into the system, the (No Data in Memory) line will flash.

3. Select **Electrocardiogram** from the Main Menu. The Electrocardiogram Menu will appear as follows on the screen:

ELECTROCARDIOGRAM MENU

( ) General Purpose ECG
( ) Automatic Change ECG
( ) Run Exam 'Strip'
( ) Simultaneous ECG–Leads I,II,III
( ) Return to Main Menu (ESC)

(No Data in Memory)

4. Select **Run Exam 'Strip'** from Electrocardiogram Menu. In this mode you must provide the following answers: (1) six frames, (2) no interval analysis, and (3) save each exam, using subject's initials. Do not change settings.
5. Press **any key** to start the exam when the subject is ready. The computer will automatically acquire data for the six leads. Press **S** key to stop any trace of current frame.
6. Press **P** key to print. Begin hooking up next subject while printing is in progress.
7. Run the exam for the next subject.

Review–Screen Analysis

Select **Review–Screen Analysis** from the Main Menu and **Full Screen Review–Analyze** from the Review–Screen Analysis Menu.

Look carefully at your ECG strip. Scan the six leads and determine which lead has the **highest R wave.** By using figure 46.4 determine the angle of that particular lead. The angle of the lead in which the R wave deflection is the greatest is close to the angle of the R wave. If the magnitude of the R wave deflection is about the same in two leads, then the R wave axis is between the angles described by those two leads.

Interval Analysis

Using the Review–Analyze Menu (Full Screen) determine the duration of the following:

- PR interval ___ (0.12–0.2 sec expected)
- QT interval ___ (less than 0.38 sec expected)
- QRS complex ___ (less than 0.10 sec expected)

Measure the interval between R waves of several heartbeats. Add them up and divide by the number of intervals to determine the ventricular rate. Are the intervals the same length?

Additional Experiments

You may wish to monitor ECG wave forms of an individual in different body positions, such as sitting up straight, hunched over the back of a chair, or standing up. In any case, take care to avoid muscle noise. If you wish to do a cardiovectorcardiogram consult Lab 2 of the *Cardiocomp Lab Manual,* page 38.

Laboratory Report

Although there is no Laboratory Report for this exercise, your instructor will indicate what form of report is required.

Pulse Monitoring 47

When the ventricles of the heart undergo systole, a surge of blood flows into the arterial tree, which manifests itself as the **pulse** in the extremities. Since the rate and strength of the pulse indicate cardiovascular function, physicians always monitor the pulse in routine medical examinations. This physiological parameter is usually determined simply by placing the fingers over one of the patient's arteries in the wrist or neck region.

In this exercise we will utilize a photoelectric plethysmograph on the index finger to record the pulse. The recording produced with this type of transducer is called a **plethysmogram** (figure 47.2). As indicated in figures 47.1 and 47.3 procedures will be provided for both the Unigraph and Physiograph®.

The details of construction and manner of operation of the plethysmograph were described in Exercise 16. It operates on the principle of a light beam that is projected into the tissues of the finger. When the beam is altered by a surge of blood passing through the finger, a photoresistor is activated by the scattered light to produce a signal that can be recorded.

In addition to determining the pulse rate, this type of transducer has merit in detecting indirect and relative blood pressure differences, but not precise blood pressures. The reason that precise blood pressures cannot be determined with this type of transducer is that it is very difficult to calibrate recorders in such a way that the volume of blood surging through the finger at any given moment represents an exact blood pressure. This limitation, however, does not prevent this type of transducer from being a useful tool; as we shall see in this ex-

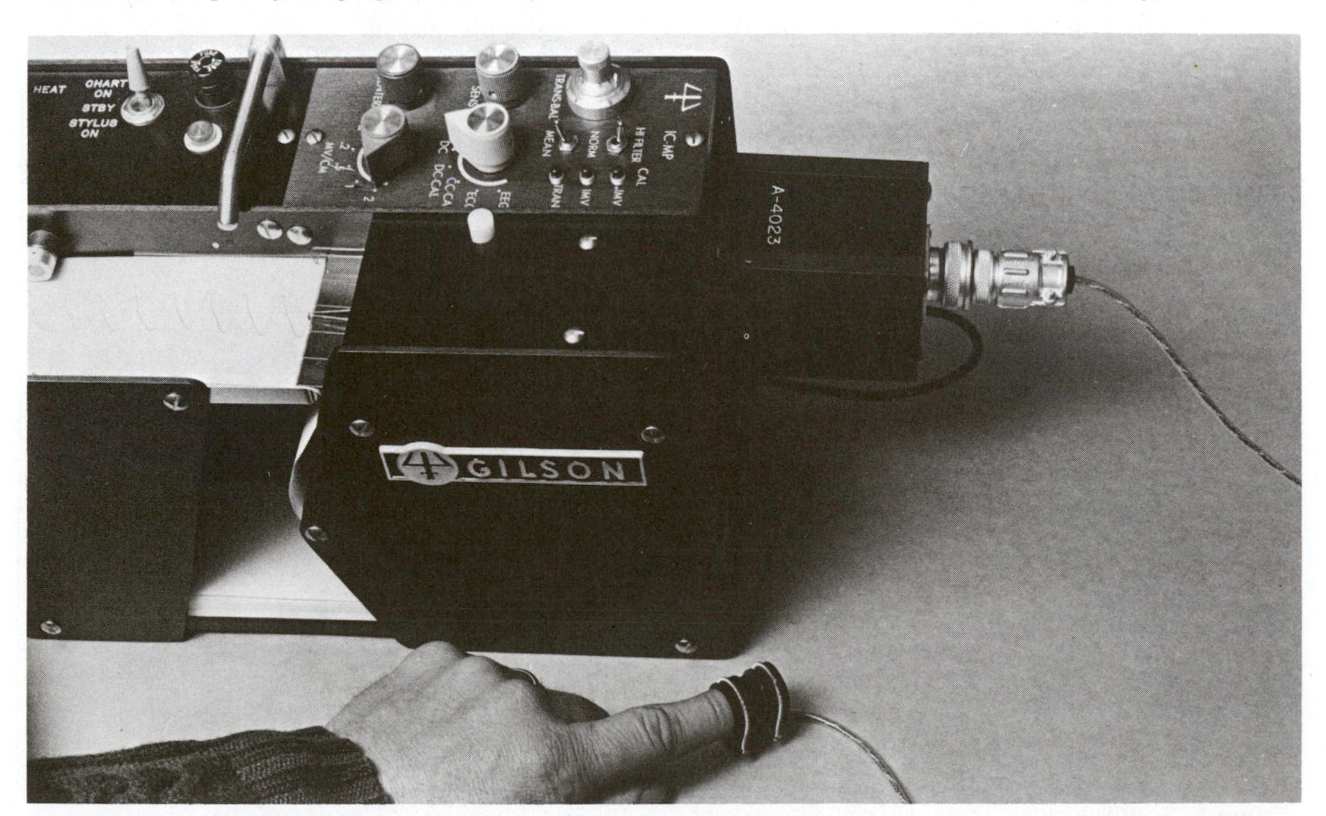

Figure 47.1 Pulse monitoring with Unigraph.

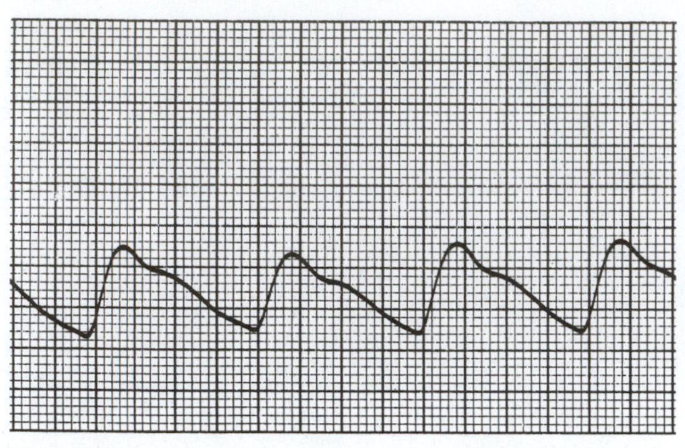

Figure 47.2 A plethysmogram (Unigraph tracing).

periment, much can be learned about the mechanics of blood pressure with it.

In this experiment we hope to accomplish the following: (1) determine the exact pulse rate; (2) explore some of the factors that affect the pulse rate; (3) identify the dicrotic notch and its significance; and (4) determine the range of pulse rates among class members. Proceed as follows:

Materials:

For Unigraph setup:

Unigraph
A4023 adapter (Gilson)
Photoresistor pulse pickup (Gilson T4020)

For Physiograph® setup:

Physiograph® with transducer coupler
pulse transducer (Narco 705–0050)

Preliminary Preparations

Prepare the subject and make the preliminary adjustments on the recorder according to the following suggestions.

The Subject

1. Place the subject in a prone position on top of a laboratory table or on a cot, with both arms lying relaxed, parallel to the body.
2. Attach the pulse transducer to the index finger of one hand. Be sure that the light source faces the pad of the fingertip. This hand should be near the edge of the table so that it can be lowered toward the floor, or extended upward toward the ceiling.
3. Make certain that the subject is comfortable and relaxed. The laboratory must be free of distracting stimuli.

Unigraph Setup

1. Attach the A4023 adapter to the receiving end of the Unigraph. Make sure that the locknut is tightened securely and that the phone jack is also plugged in.

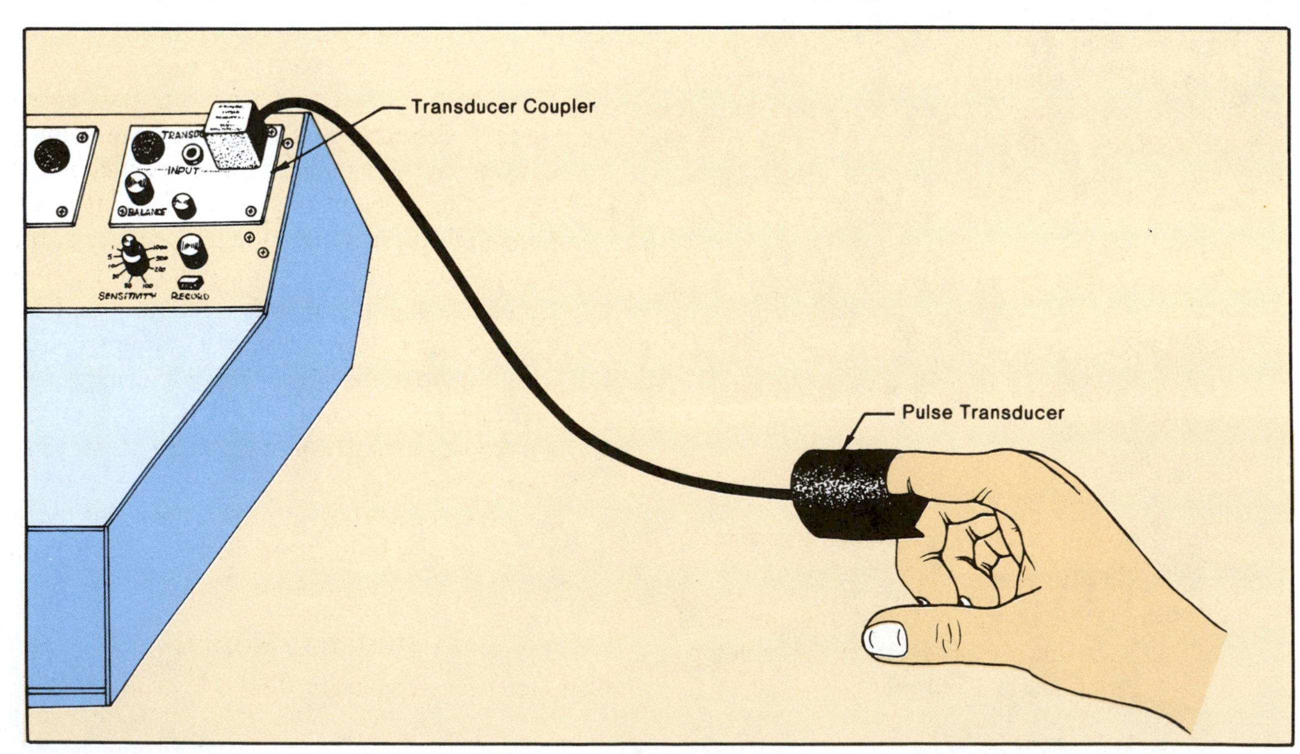

Figure 47.3 Pulse monitoring setup with Physiograph®.

2. Secure the jack of the pulse pickup to the adapter. Plug in the power cord.
3. Set the controls as follows: speed control lever at slow position, stylus heat control at two o'clock position, gain control at 2 MV/CM, sensitivity control completely counterclockwise, and Mode selector on DC.
4. Turn on the power switch, and set the chart control switch at Stylus On. Wait about 15 seconds for the stylus to warm up and then place the chart control switch at Chart On.
5. Observe the trace.
6. Adjust the centering control to get the trace near the center of the paper.
7. Increase the sensitivity by turning the sensitivity control clockwise until you get approximately 2 cm pen deflection. If this amount of deflection cannot be achieved, turn the sensitivity control down again and increase the gain to 1 MV/CM. Now, increase the sensitivity with the sensitivity control again to get the desired deflection.
8. Proceed to the instructions under Investigative Procedure.

Physiograph® Setup

1. Plug the end of the transducer cable into the proper socket on the transducer coupler.
2. Turn on the power switch and set the paper speed at 0.5 cm/sec.
3. Start the chart paper moving and lower the pens to the recording paper. Make certain that the timer is activated.
4. Use the position control to position the pen so that it is recording one centimeter below the center line.
5. Place the RECORD button in the ON position.
6. Adjust the amplifier sensitivity control to produce pulse waves that are approximately 2 centimeters high.

Investigative Procedure

Now that the pulse of the subject is being recorded, proceed as follows with your observations:

1. With the subject lying still, and the hand with the pulse transducer lying flat on the tabletop, record the pulse for **2 minutes.**
2. Stop the recording and calculate the pulse rate. (On the Unigraph chart, the space between two margin marks represents 30 seconds; thus, it is necessary to count the spikes and fractions thereof between two marks and multiply by two.)
3. Refer to illustration A, figure SR–2, Appendix D, for a Physiograph® sample record.
4. Resume recording, and have the subject slowly raise his or her arm to a fully vertical position above the body, hold that position for **15 seconds,** and then slowly return the arm to its original position on the table. Allow the pulse to stabilize. Refer to illustration B, figure SR–2 for sample Physiograph® record.
5. Now, have the subject slowly lower his or her arm over the edge of the table until it hangs down vertically. Hold this position for **15 seconds,** and then slowly return the arm to its former resting position. Allow the pulse to stabilize. Refer to illustration C, figure SR–2 for sample record.
6. Have the subject take a deep breath and hold it for **15 seconds,** before exhaling and resuming normal breathing. Refer to illustration A, figure SR–3 for sample record.
7. Locate the brachial artery below the biceps muscle. Occlude this vessel for **15–20 seconds,** and then release the pressure. Allow the pulse to stabilize. Refer to illustration B, figure SR–3 for sample record.
8. Increase the speed to 2.5 cm/sec (fast speed on the Unigraph) and record for **one minute.** Do you see a pronounced **dicrotic notch** on the descending slope of the curve? This interruption of the curve is due to sudden closure of the aortic valve of the heart during diastole.
9. Return the speed to 0.25 cm/sec (slow on Unigraph), and mildly frighten or startle the subject. This can be done with an unexpected, sudden handclap or other loud noise. The stimulus must be totally unanticipated. Refer to illustration C, figure SR–3 for sample record.

Terminate the recording after sufficient sample recordings have been made for all members of your team.

Laboratory Report

Complete the Laboratory Report for this exercise.

48 Blood Pressure Monitoring

The contractions of the ventricles of the heart exert a propelling force on the blood which manifests itself as blood pressure in vessels throughout the body. Since the heart acts as a pump that forces out spurts of blood at approximately seventy times a minute while at rest, the pressure during a short interval of time will fluctuate up and down.

The force on the walls of the blood vessels is greatest during contraction and is called the **systolic pressure.** When the heart relaxes (diastole) and fills up with blood in preparation for another contraction, the pressure falls to its lowest value. The pressure during this phase is called the **diastolic pressure.**

There are several different methods that one might use for determining blood pressure. Probably the most sensitive and precise way is to insert a hollow needle into a vessel, which allows the force of the blood pressure to act on a pressure transducer. The signal from such a device can be fed into an amplifier and recorder for monitoring. The impracticality of this method for us in this laboratory precludes its use here.

Two methods will be outlined here for student participation: (1) the conventional stethoscope-pressure cuff method used by medical personnel in routine diagnoses, and (2) electronic recording with a Physiograph® or Unigraph, using a pressure cuff. A third method, using an audio monitor, may be used for instructor demonstration. Your instructor will indicate which procedures will be used.

A Demonstration

When one takes the blood pressure with a sphygmomanometer and stethoscope, one listens for the sound of blood rushing through a partially occluded brachial artery in the arm. These sounds, which are called **Korotkoff** (Korotkow) **sounds,** can readily be demonstrated with an audio monitor, utilizing a setup similar to the one in figure 48.1.

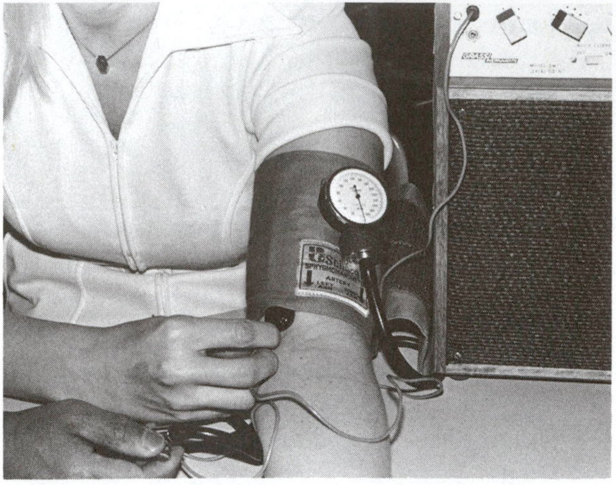

Figure 48.1 Auscultation of Korotkoff sounds with an audio monitor is performed by slipping a microphone under the cuff over brachial artery.

Materials:

- microphone
- audio monitor
- pressure cuff (sphygmomanometer)

To demonstrate these sounds, the instructor will wrap a pressure cuff around the upper arm of a subject, insert a microphone under the cuff over the brachial artery, and demonstrate the Korotkoff sounds that are listened for in determining systolic and diastolic blood pressures. Pressure will be built up in the cuff by pumping enough air into it to a level where all blood flow is shut off in the brachial artery.

While watching the pressure gauge, air will be released from the cuff slowly until the Korotkoff sounds are first heard by the blood rushing through a partially occluded artery. The pressure at which the sounds are first heard is called the **systolic pressure.** The pressure at which the sounds disappear is called the **diastolic pressure.** By correlating the sounds heard through the audio monitor with the pressure gauge, a large number of students can, in a short period of time, quickly grasp the concept.

Stethoscope Method

Students, working in pairs, will take each other's blood pressure. To determine the broad range of "normal" blood pressures, individual systolic pressures will be recorded on the chalkboard. From this tabulation a median systolic pressure for the class will be determined. The effects of exercise on blood pressure will also be studied.

Materials:

stethoscope
sphygmomanometer

1. Wrap the cuff around the right upper arm. The method of attachment will depend on the type of equipment. Your instructor will show you the best way to use the specific type that is available in your laboratory. See figure 48.2.

   Make certain that the subject's forearm rests comfortably on the table.
2. Close the metering valve on the neck of the rubber bulb. *Don't twist it so tight that you won't be able to open it!*
3. Pump air into the sleeve by squeezing the bulb in your right hand. Watch the pressure gauge. Allow the pressure to rise to about 180 mm Hg and continue to hold the metering valve closed.
4. Position the bell of the stethoscope just below the cuff at a point that is *midway between the epicondyles of the humerus.* This is the lowest extremity of the brachial artery. About 2.5 cm distal to this point, the artery bifurcates to become the radial and ulnar arteries. See figure 48.3.

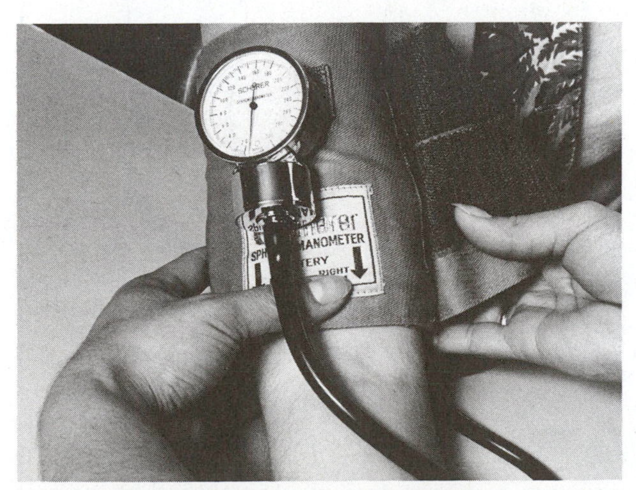

Figure 48.2 Sphygmomanometer cuff is wrapped around the upper arm, keeping lower margin of cuff above line through the epicondyles of the humerus.

5. Slowly release the valve so that the pressure goes down slowly. Listen carefully as you watch the pressure fall. When you just begin to hear the Korotkoff sounds, note the pressure on the gauge. This is the systolic pressure.
6. Continue listening as the pressure falls. Just when you are unable to hear the Korotkoff sounds anymore, note the pressure on the gauge. This is the diastolic pressure.
7. Record the systolic pressure on the chalkboard and the systolic/diastolic pressures on the Laboratory Report.
8. Repeat two or three times to see if you get consistent results.
9. Now, have the individual do some exercise, such as running up and down stairs a few times. Measure the blood pressure again and record your results on the Laboratory Report.

Blood Pressure Recording

A chart recording of blood pressure can be made with either a Unigraph or Physiograph®. The Physiograph® setup is illustrated in figure 48.4. If a Unigraph is used, one needs a Statham pressure transducer similar to the one shown in figure 56.2. Procedures are provided here for using either type of recorder.

In this portion of the exercise we will (1) record systolic and diastolic pressures; (2) determine mean arterial and mean pulse pressures; (3) observe the effects of postural changes on arterial pressure; and (4) study the effects of physical exertion on arterial pressure.

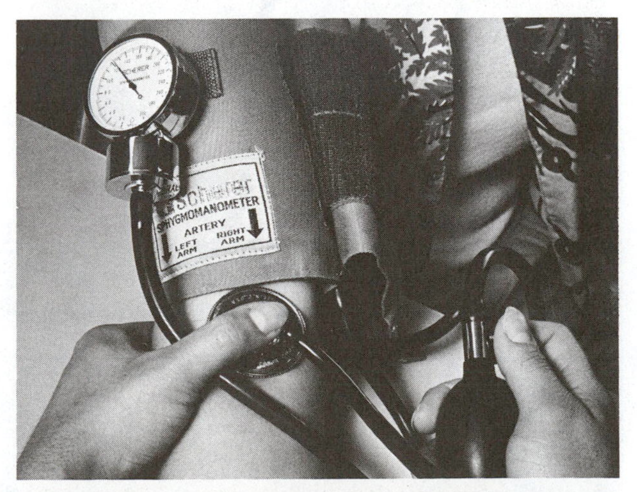

Figure 48.3 The inflated cuff shuts off the flow of blood in the brachial artery. Korotkoff sounds are listened for as pressure is gradually lowered.

Materials:

For Unigraph setup:

Unigraph
adult pressure cuff
Statham T-P231D pressure transducer
transducer stand and clamp for transducer

For Physiograph® setup:

Physiograph® with ESG coupler
adult pressure cuff (Narco PN 712–0016)

Preliminary Preparations

As has been true for all previous experiments, separate instructions will be provided here for the Physiograph® and the Unigraph. One difference between the two setups is that the Gilson setup utilizes a separate pressure transducer, but the Physiograph® doesn't need one; instead, the ESG coupler of the Physiograph® has a pressure transducer built into it.

Unigraph Setup Refer to figure 56.2 to see how the Statham pressure transducer is mounted to a transducer stand with a transducer clamp. The transducer is seen in the background with a hose attached to it. If no support is available for this unit, it can be placed on its side on the table; make sure, however, that it is in a spot where it cannot roll off onto the floor.

1. Insert the jack of the pressure transducer cord into the receptacle of the Unigraph. No adapter is necessary.
2. Place the pressure cuff on the upper right arm, securing it tightly in place.
3. Attach the open end of the tube from the cuff to the fitting on the pressure transducer.
4. Test the cuff by pumping some air into it and noting if the pressure holds when the metering valve is closed on the bulb. Release the pressure for the comfort of the subject while other adjustments are made.
5. Plug in the Unigraph and turn on the power. Set the speed control lever in the slow position and the stylus heat control knob in the two o'clock position.
6. Set the gain at 1 MV/CM, and turn the sensitivity control completely counterclockwise to its lowest value. Set the mode on TRANS.
7. **Calibrate the Statham transducer** by depressing the TRANS button and simultaneously adjusting the sensitivity knob to get 2 cm deflection. The calibrated line represents 100 mm Hg.
8. Proceed to the Investigative Procedure portion of this experiment.

Physiograph® Setup Refer to figure 48.4 to see how the various components of this setup are hooked up. Note that the pressure bulb tube is affixed to the

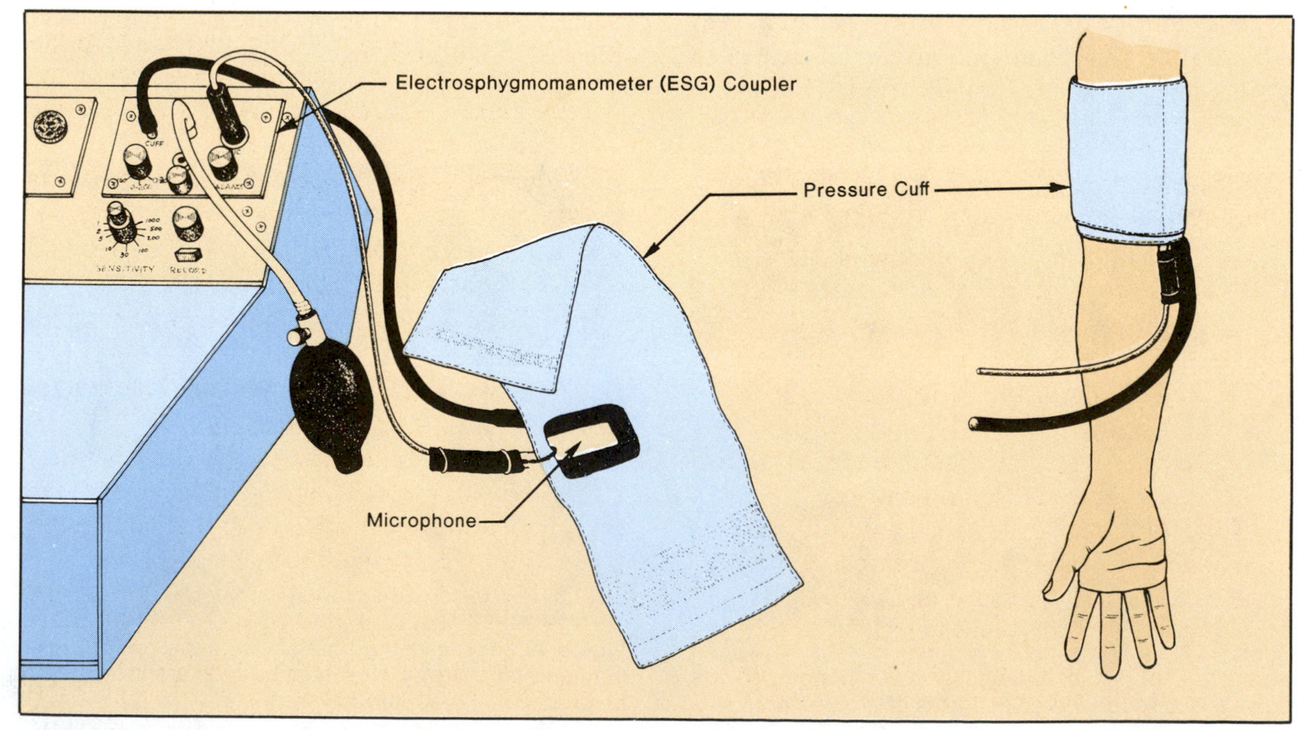

Figure 48.4 Blood pressure monitoring setup with Physiograph®.

"Bulb" fitting of the ESG coupler, the pressure cuff tube to the "Cuff" fitting on the coupler, and the microphone cable is inserted into the "Mic" receptacle of the ESG coupler. Before any tests can be run it will be necessary to balance and calibrate the Physiograph®. Proceed as follows:

1. **Balance the ESG coupler** as follows:
   a. Start the chart moving at 0.25 cm/sec and lower the pens onto the paper.
   b. Keep the amplifier RECORD button in the OFF position.
   c. Orient the recording pen to the centerline by using the position control knob.
   d. Start recording by pressing the RECORD button to the ON position (down).
   e. With the balance control knob on the ESG coupler, return the recording pen to the centerline established in step c above.
   f. Check the balance by placing the amplifier RECORD button in the OFF position. If the system is balanced, the pen will remain on the centerline.
   g. If the pen does not remain on the centerline, repeat steps e and f until balancing is achieved.
2. **Calibrate the coupler** to produce 2.5 cm of pen deflection which is equivalent to 100 mm Hg. Figure 48.5 is a sample recording of this procedure. Proceed as follows:
   a. Start recording by pressing the RECORD button to the ON position.
   b. With the position control knob, set the recording pen on a baseline that is exactly 2.5 cm below the centerline; i.e., 2.5 cm above the "0" pressure line.
   c. While depressing the "100 mm Hg CAL" button on the coupler, rotate the inner knob of the amplifier sensitivity control until the pen is moved back to the center-line (i.e., 2.5 cm above the 0 pressure line).
   d. Release the CAL button and see if it returns exactly to the 0 pressure line.
   e. Repeat steps c and d several times to make sure that exact calibration exists.

   Note: The ESG coupler is now calibrated so that each block on the recording paper represents 20 mm Hg pressure, and each centimeter of pen deflection is equivalent to 40 mm Hg.
3. From this point on *do not touch* the balance control knob, the position control knob, or the inner knob of the amplifier sensitivity control. Any readjustments of these controls will produce inaccurate results.

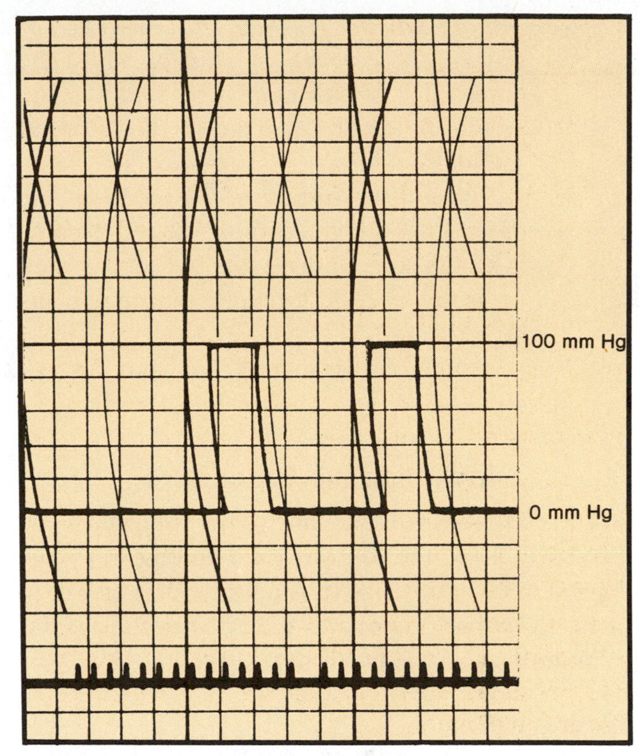

Figure 48.5 Pressure calibration: Zero pressure line "0" is calibrated so that it is 2.5 cm below the centerline. Each block on recording paper represents 20 mm Hg; thus, the centerline represents 100 mm Hg.

4. Place the pressure cuff on the subject's right arm in such a way that the microphone is positioned over the brachial artery. Secure the cuff tightly around the upper arm with the Velcro surfaces holding it closed.
5. **Calibrate the cuff microphone** sensitivity as follows:
   a. Start the chart moving at 0.5 cm/sec and lower the pens to the paper.
   b. Place the RECORD button in the ON position.
   c. Close the valve on the hand bulb by rotating it clockwise.
   d. With the subject's arm on the table in a relaxed position, pump up the cuff until the recording pen indicates approximately 160 mm Hg (4 cm of pen deflection).
   e. Release the bulb valve *slightly* so that the recorded pressure begins to fall at a rate of approximately 10 mm Hg/sec. Note the Korotkoff sounds on the recording.

f. *Taking care not to touch the inner knob of the amplifier sensitivity control,* adjust the outer knob of this control so that you get an amplitude of 1–1.5 cm on the recorded sound.

Usually, a setting of 10 or 20 on the outer knob produces good sound recordings. On some individuals the setting may be as low as 5 or 2; on others it may be as high as 100.

6. You are now ready to perform your investigation.

Investigative Procedure

Four recordings will be made of the subject's blood pressure under the following conditions: (1) while seated in a chair; (2) after standing for 4 or 5 minutes; (3) while prone; and (4) after exercising for 5 minutes. Use the following steps.

Subject in Chair

1. Start the paper advance at 0.5 cm/sec, and lower the pens onto the chart. Activate the timer.
2. Press the amplifier RECORD button to the ON position.
3. Tighten the pressure bulb valve (turn fully clockwise) and inflate the cuff until you are recording a pressure of approximately 40 mm Hg above the expected systolic pressure. Usually, 160–180 mm Hg is sufficient.
4. Release the hand bulb valve slightly so that the recorded pressure begins to fall at a rate of approximately 10 mm Hg/sec.
5. Watch for the appearance of the **first** and **last Korotkoff sounds** on the recording. These pressures can be considered to be the "normal" systolic and diastolic pressures of your subject.
6. Terminate the recording and refer to Appendix D, illustration A, figure SR–4 for Physiograph® sample record.

Subject Standing Up

Allow the subject to stand up for 4 or 5 minutes and repeat steps 1 through 6 above while the individual is still standing. Refer to illustration B, figure SR–4 for sample record.

Subject Reclining

Allow the subject to lie on a laboratory table or cot for 4 or 5 minutes, and then follow steps 1 through 6 again to record the blood pressure in this position. Refer to illustration C, figure SR–4 for sample record.

After Exercise

Remove the cuff from the subject and allow him or her to perform 5 minutes of strenuous exercise, such as running, jumping, or climbing stairs.

After exercising, seat the subject in a chair and *immediately* attach the cuff. Record the systolic and diastolic pressures, using the same procedures as above. Remember, it will be necessary to inflate the cuff to a higher pressure (240 mm Hg) to occlude the brachial artery. Refer to illustration A, figure SR–5 for Physiograph® sample record.

Make a series of recordings one minute apart until the pressure has returned to what it was in the original (seated) test. Refer to illustrations B and C, figure SR–5 for sample records.

Laboratory Report

Complete the first portion of Laboratory Report 48,49.

Peripheral Circulation Control (Frog) 49

The arterial and venous divisions of the circulatory system are united by an intricate network of **capillaries** in the tissues. Capillaries are short thin-walled vessels of approximately 9 micrometers in diameter. They are so numerous that all body cells are no more than two to three cells away from one of these vessels.

The combined diameter of all the capillaries causes the blood to slow down as it passes through them. This slow flow of blood is of great importance because it allows sufficient time for the exchange of materials between the blood and tissue cells. The pathway of exchange is as follows:

$$\text{capillary blood} \rightleftarrows \text{tissue fluid} \rightleftarrows \text{cells}$$

Whether or not blood enters a capillary is determined by the size of its preceding arteriole and the action of its **precapillary sphincter muscle.** The arteriole has the capacity of vasodilation and vasoconstriction to affect the amount of blood flow through it. The precapillary sphincter, which guards the entrance to the capillary, can also restrict or permit blood flow. Both the arterioles and sphincters are under neural control of the vasomotor center of the medulla oblongata through the autonomic nervous system. Although the vasomotor center is influenced by many factors, carbon dioxide concentration of the blood is of major importance.

Locally, carbon dioxide, histamine, and epinephrine may influence the volume of blood flow. High CO_2 concentrations in certain tissues will result in increased blood flow in these tissues with a decrease of blood in those tissues that lack the high CO_2 concentration. Increased epinephrine levels in the blood cause vasodilation in some areas (muscles, heart, lungs) and vasoconstriction in other areas. Histamine, a hormone that is present in large quantities in allergic reactions, causes extensive va-

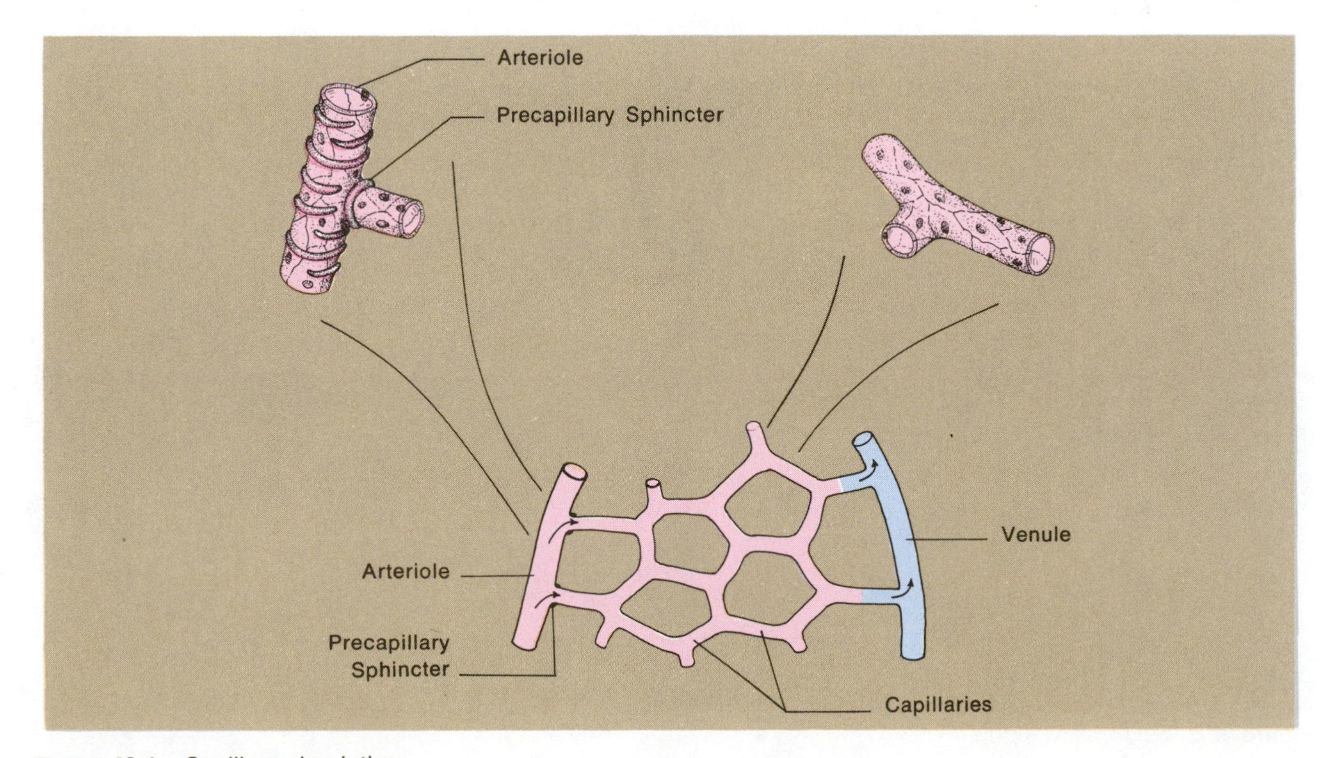

Figure 49.1 Capillary circulation.

sodilation in peripheral circulation. Such reactions may cause a precipitous drop in blood pressure, and death.

In this exercise we will study some of these factors that influence capillary circulation. A frog will be strapped to a board so that the capillaries in the webbing between its toes can be studied under a microscope.

Materials:

small frog, frog board, cloth straps (4″ × 12″)
string, pins, rubber bands
epinephrine (1:1000)
histamine (1:10,000)

1. Obtain a frog from the stock table and strap it to a frog board as illustrated in figure 49.2. Wrapping a piece of cloth around the body and head and tying it securely with string should hold the animal in place.
2. Pin the webbing of the foot over the hole in the board. Keep the webbing moistened with water.
3. If necessary, secure the board to the microscope stage with a large rubber band.
4. Position the foot webbing over the light source and focus on it with the 10× objective. Observe the blood flowing through the vessels. Locate an **arteriole,** with its characteristic pulsating, rapid flow, and a **venule,** with its slightly slower, steady flow. Also, identify a **capillary,** with its smaller diameter and blood cells moving through it slowly and in single file.
5. Study a capillary more closely to see if you can locate the juncture of the capillary and arteriole, which is the site of the **precapillary sphincter.** You probably won't be able to see the sphincter, but you can observe the irregular flow of blood cells into the capillary, which results partly from the sphincter muscle's action.
6. Now, remove most of the water from the frog's foot by blotting it with paper toweling and add several drops of 1:10,000 **histamine** solution. Observe the change in blood flow and record your observations on the Laboratory Report.
7. Wash the foot with water, blot it, and then add several drops of 1:1000 **epinephrine** solution. Observe any change in blood flow and record your results on the Laboratory Report.
8. Wash the foot again, remove the frog from the board, and return it to the stock table. Clean the microscope stage, if necessary.

Laboratory Report

Complete the last portion of combined Laboratory Report 48,49.

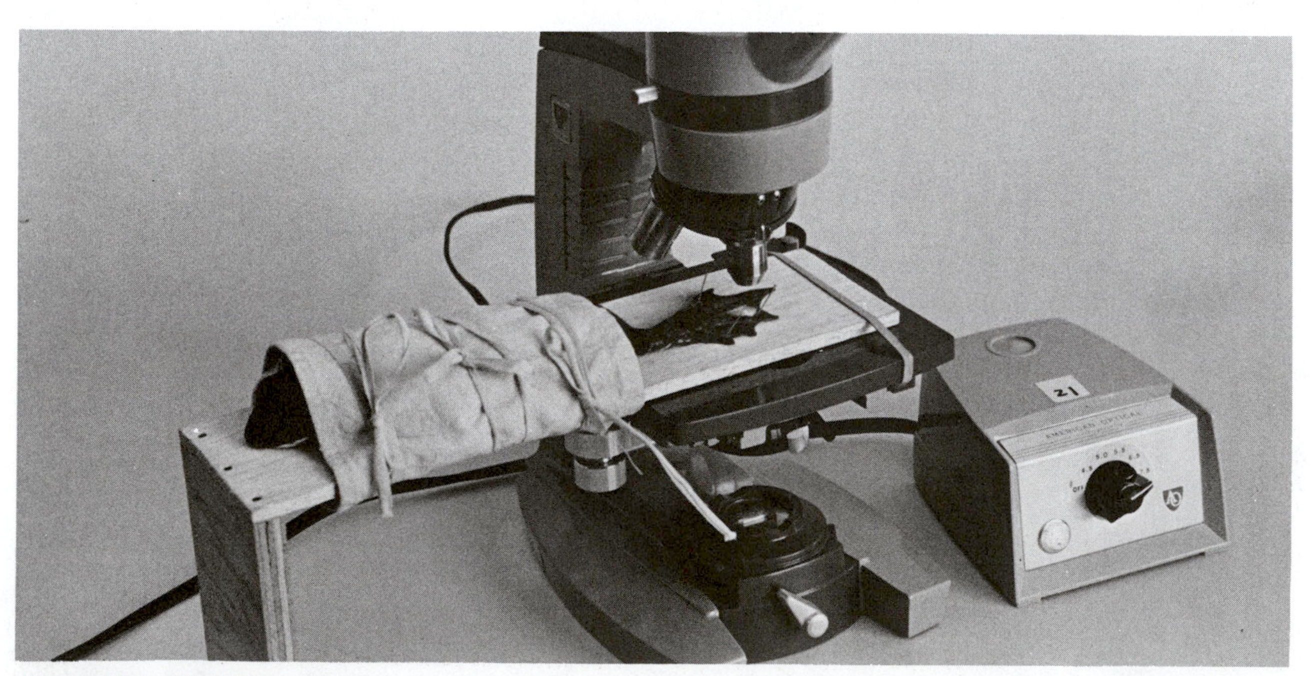

Figure 49.2 Blood flow setup.

The Arteries and Veins 50

The principal arteries and veins of the circulatory system will be studied in this exercise. Ideally, the various blood vessels would be sought out on a cadaver. If no cadaver is used, only the circulatory plan in figure 50.1 will be worked out, and the labels identified in figures 50.2 and 50.3.

The Circulatory Plan

Figure 50.1 is an incomplete flow diagram of the heart and major regions of the body. If you can complete this diagram by filling in the proper blood vessels and providing correct labels, you can assume that you understand the overall plan of the circulatory system as described below. As you read the following description, complete the diagram.

Blood enters the right atrium (left side of illustration) from the **superior** and **inferior venae cavae,** which have collected blood from all parts of the body. This blood is dark colored because it is low in oxygen and high in carbon dioxide content. From the right atrium the blood passes to the right ventricle. When the heart contracts, blood leaves the right ventricle through the **pulmonary artery** to the lungs, where it picks up oxygen and gives off carbon dioxide. The blood leaves the lungs by way of the **pulmonary veins** (a single vessel on the diagram) and passes back to the left atrium of the heart. Blood in the pulmonary veins is brightly colored due to its high oxygen content. The pulmonary arteries, veins, and lung capillaries constitute the *pulmonary system.*

From the left atrium the blood passes to the left ventricle. When the heart contracts, blood leaves the left ventricle through the **aortic arch.** This blood, which is rich in oxygen, passes to all parts of the body. The aortic arch has branches that go to the head and arms. (For simplicity, only one blood vessel is shown passing to these regions.)

Passing downward on the right side of the illustration, the aortic arch becomes the **aorta,** which has branches going to the liver, digestive organs, kidneys, pelvis, and legs. The branch that enters the liver is the **hepatic artery.** The intestines are supplied by the **superior mesenteric artery,** and the kidneys receive blood through the **renal arteries.** The pelvis and legs are supplied by several arteries, but only one is shown for simplicity.

The blood leaving most of the organs of the trunk (including the pelvis and legs) empties directly into a large collecting vein, the inferior vena

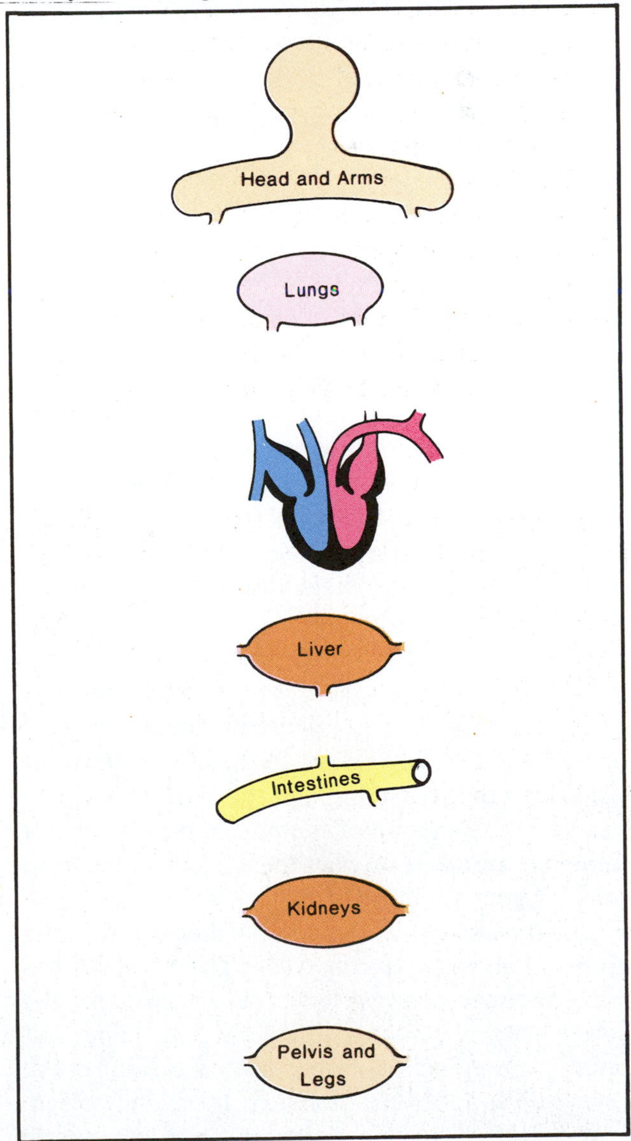

Figure 50.1 The circulatory plan.

cava. The **renal veins** from the kidneys and the **hepatic vein** from the liver empty directly into the inferior vena cava.

Note that blood from the intestines does not go directly into the vena cava; instead, all blood from this region passes to the liver by way of the **portal vein.** This route of the blood from the intestines to the liver is called the *portal circulation.*

Assignment:

After drawing in all the necessary blood vessels, color the oxygenated vessels **red** and the deoxygenated ones **blue.** Provide arrows to show direction of blood flow and label the vessels.

The Arterial Division

The principal arteries of the body are shown in the left-hand illustration of figure 50.2. Starting at the heart, blood leaves it through the large curved artery, the **aortic arch.** From the upper surface of this large vessel emerge the left subclavian, left common carotid, and the innominate artery.

The **left subclavian** artery is the artery that passes from the aortic arch into the left shoulder behind the clavicle. In the armpit (*axilla*) the subclavian becomes the **axillary** artery. The axillary, in turn, becomes the **brachial** artery in the upper arm. This latter artery divides into the **radial** and **ulnar** arteries that follow the radial and ulnar bones of the forearm.

The **innominate** artery is the short vessel coming off the aortic arch on the right side of the body. It gives rise to two arteries: the **right common carotid** that goes to the right side of the head, and the **right subclavian** that passes behind the right clavicle.

The third branch of the aortic arch is the **left common carotid.** It supplies the left side of the head with oxygenated blood.

As it passes down through the thorax, the aortic arch becomes the **aorta.** Although the aorta has many branches leading to various organs, only the larger ones are shown in figure 50.2. The first branch, emerging just below the heart, is the **celiac** artery. It supplies the stomach, liver, and spleen. Just below the celiac is the **superior mesenteric** artery, which supplies most of the small intestine and part of the large intestine. Inferior to the superior mesenteric are a pair of **renal** arteries that supply the kidneys. The single branch of the aorta just below the renals is the **inferior mesenteric** artery, which supplies part of the large intestine and rectum with blood.

In the lumbar region the aorta divides into the right and left **common iliac** arteries. Each common iliac passes downward a short distance and then divides into a smaller inner branch, the **internal iliac** (*hypogastric*) artery, and a larger branch, the **external iliac** artery, which continues on down into the leg. The external iliac becomes the **femoral** artery in the upper three-fourths of the thigh. In the general region of the origin of the femoral artery, the **deep femoral** branches off of it. This artery courses backward and downward along the medial surface of the femur. In the knee region the femoral becomes the **popliteal** artery. Just below the knee the popliteal divides into the **posterior tibial** and **anterior tibial** arteries.

Assignment:

Label the arterial portion of figure 50.2.

The Venous Division

Starting at the heart in figure 50.2, we see the **superior vena cava** emptying into the upper part of the right atrium and the **inferior vena cava** leading into the lower part of the right atrium.

In the neck region are seen four veins: two medial **internal jugular** veins and two smaller lateral **external jugulars.** The internal jugulars empty into the innominate veins and the externals empty into the subclavian veins. The **innominate** veins are the short veins that empty into the superior vena cava.

Three veins, the cephalic, basilic, and brachial, collect blood in the upper arm. The **cephalic** vein courses along the lateral aspect of the arm and empties into the **subclavian.** The **basilic** lies on the medial side of the arm, and the **brachial** lies along the posterior surface of the humerus. Both the basilic and brachial empty into a short **axillary** vein that becomes the subclavian in the shoulder region. Between the basilic and cephalic veins in the elbow

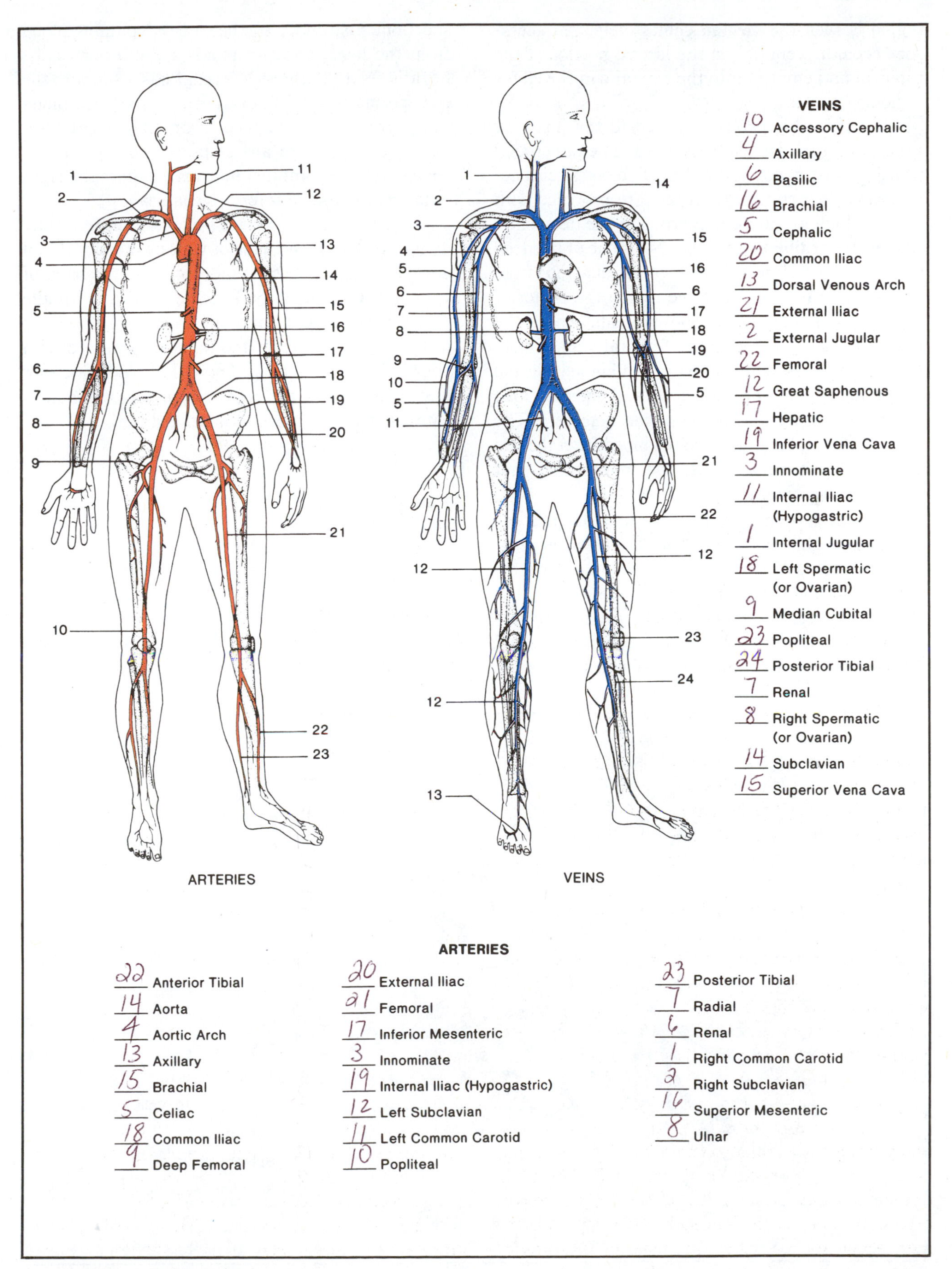

Figure 50.2 Arteries and veins.

region is seen the **median cubital** vein. An **accessory cephalic** vein lies on the lateral portion of the forearm and empties into the cephalic in the elbow region.

Blood in the legs is returned to the heart by superficial and deep sets of veins. The superficial veins are just beneath the skin. The deep veins accompany the arteries. Both sets are provided with valves which are more numerous in the deep ones. The **posterior tibial** vein (label 24) is one of the deep veins that lies behind the tibia. It collects blood from the calf and foot. In the knee region the posterior tibial vein becomes the **popliteal** vein. Above the knee this vessel becomes the **femoral** vein, which, in turn, empties into the **external iliac** vein in the upper leg region. The external and **internal iliac** (label 11) empty into the **common iliac** vein. The inferior vena cava, thus, receives blood from the two common iliacs, renal veins, and many others not shown in this diagram.

A large superficial vein of the leg, the **great saphenous,** originates from the **dorsal venous arch** on the superior surface of the foot, and enters the femoral vein at the top of the thigh.

Although many short veins empty into the inferior vena cava, only the hepatic, renals, and spermatics are shown in figure 50.2. The **hepatic** is the short one just under the heart that contains blood from the liver. The two **renals** are inferior to the hepatic. Note that the left renal has a branch, the **left spermatic** (or **left ovarian**), that carries blood away from the left testis or ovary. The **right spermatic** (or **right ovarian**) empties into the inferior vena cava at a point just inferior to where the right renal vein joins the inferior vena cava.

The presence of valves in veins can be vividly demonstrated on the back of your own hand. If you apply pressure to a prominent vein on the back of your left hand near the knuckles with the middle finger of your right hand and then strip the blood in the vein away from the point of pressure with your index finger, you will note that blood does not flow back toward the point of pressure. This indicates that valves prevent backward flow.

Assignment:

Label the veins in figure 50.2.

Portal Circulation

Figure 50.3 reveals the venous system that comprises the human portal circulation. All the veins

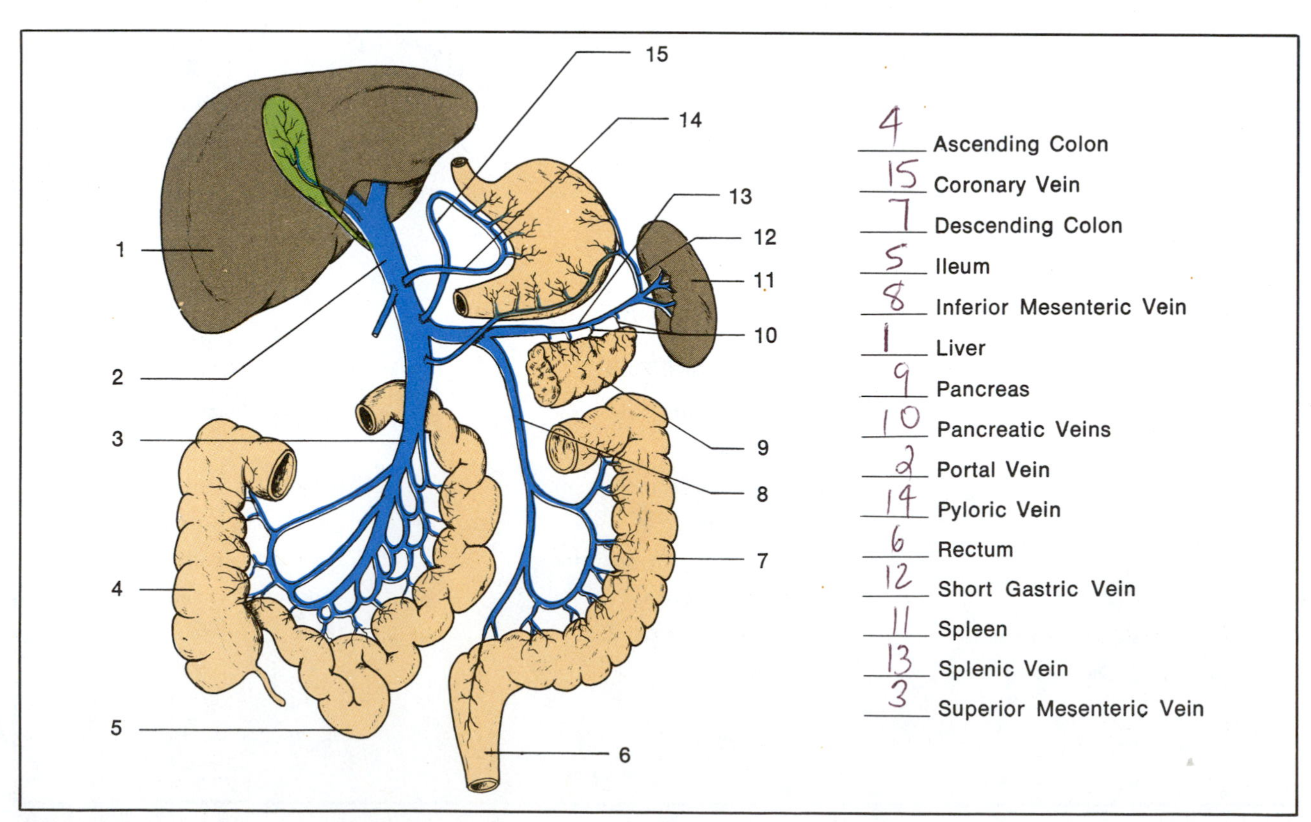

Figure 50.3 Portal circulation, human.

shown here from the stomach, spleen, pancreas, and intestine drain into the **portal** vein (label 2) which empties into the liver.

The large vessel that collects blood from the ascending colon (label 4) and the ileum (label 5) is the **superior mesenteric** vein. It empties directly into the portal vein. Blood from the descending colon (label 7) and rectum is collected by the **inferior mesenteric,** which empties into the **splenic,** or **lienal,** vein. The latter blood vessel parallels the length of the pancreas, receiving blood from the pancreas via several short **pancreatic** veins. Near the spleen, the splenic vein also receives blood from the stomach through the **short gastric** vein.

From the stomach a venous loop empties blood from the medial surface of the stomach into the portal vein. The upper portion of the loop is the **coronary** vein; the lower portion is the **pyloric vein.**

Assignment:

Label figure 50.3.

51 Fetal Circulation

During human embryological development the circulatory system is necessarily somewhat different from that after birth. The fact that the lungs are nonfunctional before birth dictates that less blood should go to the lungs. Also, because the food and oxygen supply must come from the mother by way of the placenta, blood vessels to and from the placenta through the umbilical cord must be present and functional.

Figure 51.1 illustrates, in a diagrammatic way, the circulatory pathway through the fetal heart. It is the purpose of this short exercise to trace the blood through the veins and arteries of figure 51.1 so that you understand the pathway of the blood prior to birth and comprehend the changes that must occur after birth.

In the lower left-hand corner of the illustration is seen a large vessel with two smaller vessels wrapped around it. The large vessel is the **umbilical vein,** which carries blood from the placenta of the mother. It contains all the nutrients and oxygen required by the developing fetus. The two small vessels wrapped around the umbilical vein are the **umbilical arteries.** These arteries return blood laden with carbon dioxide and wastes to the placenta. Near the liver of the fetus the umbilical vein narrows to become a short narrow vessel, the **ductus venosus.** At the entrance to the ductus venosus is a **sphincter muscle** that closes off that vessel when the umbilical cord is severed at birth. This helps prevent excessive loss of blood from the infant. The ductus venosus empties into the **inferior vena cava** where rich oxygenated blood from the placenta is mixed with unoxygenated blood. Most of this mixed blood in the inferior vena cava flows through the right atrium directly into the left atrium through an opening, the **foramen ovale.** Observe that this opening has a flaplike valve over it. Some blood in the right atrium, however, does flow into the right ventricle.

When the heart contracts, blood leaves the right ventricle through the pulmonary artery, which, in turn, divides to form the **right** and **left pulmonary arteries.** At the same time, oxygenated blood leaves the left ventricle through the ascending aorta to supply blood to the head, arms, trunk, and legs.

Note that between the pulmonary arteries and the aortic arch is a short vessel called the **ductus arteriosus.** This vessel transfers a considerable amount of blood directly into the aorta instead of allowing it to go to the lungs; enough blood passes to the lungs through the pulmonary arteries, however, to supply the needs of growing lung tissue.

The remainder of the circulatory system resembles the adult plan in that the aorta passes down through the body and divides into two common iliac arteries; each common iliac, in turn, divides to form an external iliac and a hypogastric (internal iliac) artery. Unlike the adult circulatory system, the hypogastric arteries in the fetus are much larger than the external iliacs.

Note that fetal blood is returned to the placenta via the **hypogastric arteries** that pass along each side of the bladder. From the bladder these arteries pass upward along the anterior abdominal body wall to entwine around the umbilical vein to become the **umbilical arteries.**

Once the fetus is separated from the placenta at birth, changes occur in the heart, veins, and arteries to provide a new route for the blood. As stated, one of the first things that happens is the closure of the umbilical vein by the sphincter muscle to help in the prevention of blood loss. Coagulation of blood in this vessel, as well as ligature by the physician, also assists in this respect. Within five days after birth the umbilical vein becomes the *round ligament* (ligamentum teres), which extends from the umbilicus to the liver in the adult.

As soon as the infant begins to breathe, more blood is drawn to the lungs via the pulmonary arteries. The abandonment of the path through the ductus arteriosus causes this vessel to collapse and begin its gradual transformation to connective tissue of the *ligamentum arteriosum.*

With the increase in blood volume and blood pressure in the left atrium, due to a greater blood flow from the lungs, the flap on the foramen ovale closes this opening. Eventually, connective tissue permanently seals off this valve.

The ductus venosus becomes a fibrous band, the *ligamentum venosum* of the liver. Those parts of the hypogastric arteries that lie along the bladder form into fibrous cords *(lateral umbilical ligaments)* and vesical arteries. The lateral umbilical ligaments extend anteriorly and upward from the bladder to the inner abdominal wall. The superior, middle, and inferior vesical arteries supply blood to the urinary bladder.

Laboratory Report

Label figure 51.1 and answer the questions on Laboratory Report 51,52 that pertain to this exercise.

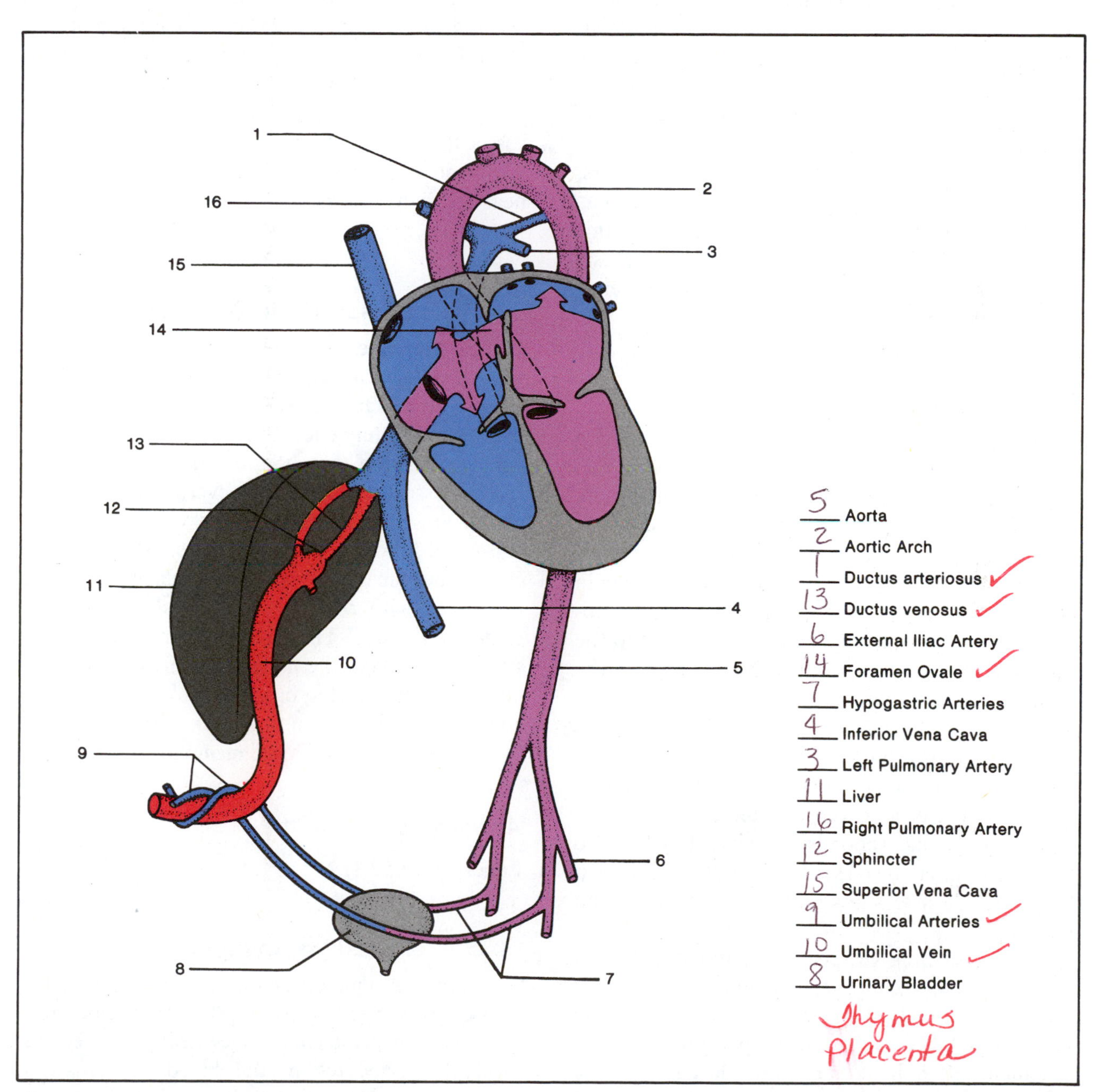

Figure 51.1 Fetal circulation.

52 The Lymphatic System and the Immune Response

The lymphatic system consists of a drainage system of lymphatic vessels that returns tissue fluid from the interstitial cell spaces to the blood. On its journey back to the blood, the lymph in these vessels is cleansed of bacteria and other foreign material by macrophages (reticuloendothelial system) in the lymph nodes. In addition to macrophages, the lymph nodes are packed with lymphocytes that play a leading role in cellular and humoral immunity. In this exercise we will explore the genesis and physiology of lymphocytes as well as the overall anatomy of the lymphatic system.

Lymph Pathway

The smallest vessels of the lymphatic system are the **lymph capillaries.** These tiny vessels have closed ends, are microscopic, and are situated among the cells of the various tissues of the body. The lymph capillaries unite to form the **lymphatic vessels** (*lymphatics*) which are the visible vessels shown in the arms and legs in figure 52.1. The irregular appearance of the walls of these vessels is due to the fact that they contain many valves to restrict the backward flow of lymph.

As lymph moves toward the center of the body through the lymphatics, it eventually passes through one or more of the small oval **lymph nodes.** Note that these nodes are clustered in the neck, groin, axillae, and abdominal cavity; they are also found in smaller clusters in other parts of the body. They vary in size from that of a pinhead to an almond.

Two collecting vessels, the thoracic and right lymphatic ducts, collect the lymph from different regions of the body. The largest one is the **thoracic duct,** which collects all the lymph from the legs, abdomen, left half of the thorax, left side of the head, and left arm. Its lower extremity consists of a saclike enlargement, the **cisterna chyli.** Lymph from the intestines passes through lymphatics in the mesentery to the cisterna chyli. The lymph from this region contains a great deal of fat and is usually referred to as **chyle.** The thoracic duct empties into the left subclavian vein near the left internal jugular vein. The **right lymphatic duct** is a short vessel that drains lymph from the right arm and right side of the head into the right subclavian vein near the right internal jugular.

Lymph Node Structure

Each lymph node has a slight depression on one side, the **hilum,** through which blood vessels and the **efferent lymphatic vessel** emerge. **Afferent lymphatic vessels** may enter the lymph node at various points. Although each lymph node has only one efferent vessel, it may have several afferent lymphatic vessels. Surrounding the entire node is a **capsule** of fibrous connective tissue. Just inside the capsule lies the **cortex,** which consists of (1) sinuses reinforced with reticular tissue and (2) **germinal centers** *(nodules).* The germinal centers represent structural units of the node. They arise from small nests of lymphocytes or lymphoblasts and reach about one millimeter in diameter. The central portion of a lymph node is the **medulla.**

Assignment:

Label figure 52.1.

Genesis of Lymphocytes

The term *lymphocyte* usually refers to a family of cells characterized by the absence of specific granules and the presence of a centrally located nucleus. Lymphocytes in the blood are small to medium in size (4–10 μm); those in lymph are usually larger (up to 15 μm). Although all lymphocytes are morphologically similar, there are considerable physiological differences between individual cells. In addition to being present in blood, lymph, and lymph nodes, they are also seen in the bone marrow, thymus, spleen, and lymphoid masses associated with the digestive, respiratory, and urinary passages.

Two different kinds of lymphocytes account for what are known as humoral and cellular immunity. *Humoral immunity* is that type of immunity which is achieved through circulating **antibodies.** *Cellular immunity* is immunity in which *sensitized lymphocytes* (**lymphoblasts**) act against foreign

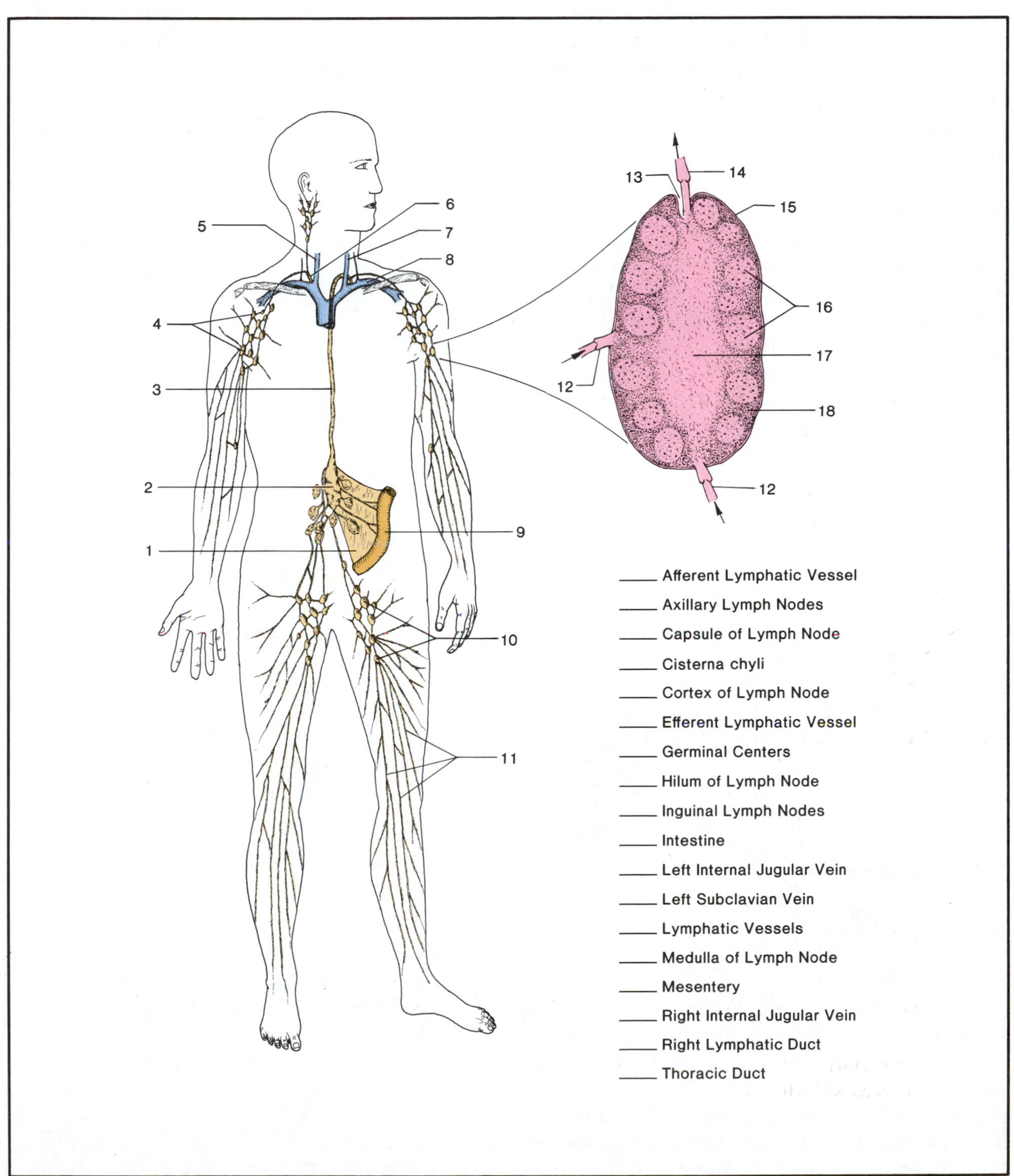

Figure 52.1 The lymphatic system.

cells; these sensitized cells are responsible for tissue rejection in organ transplants and grafts. Both types of immunity are *acquired,* since they are not present at birth and since they develop in the presence of foreign biological agents.

The lymphocytes that are responsible for antibody formation are called **B-lymphocytes,** or B-cells (*B* for bursa). The lymphocytes that are responsible for cellular immunity are **T-lymphocytes,** or T-cells (*T* for thymus).

Both types of lymphocytes originate from **lymphocytic stem cells** in the bone marrow before birth and shortly after birth. The stem cells are released into the blood and follow one of two pathways for "processing," as illustrated in figure 52.2. The cells that settle in the thymus for modification become T-cells; the remaining cells are processed in some other unknown area to become B-cells. Birds possess a *bursa of Fabricius* on the hindgut, where B-cells are formed; in man an equivalent of this organ is being sought. At present, it is speculated that lymphoid tissue in the fetal liver or even in the bone marrow may be responsible for B-cell processing.

Once the T-lymphocytes have matured in the thymus, they leave the thymus via the blood and colonize in specific loci of the lymph nodes and spleen. T-cells have a long life. In the presence of antigens they are transformed to **lymphoblasts,** which are responsible for cellular immunity.

The stem cells that mature in the B-cell processing tissues leave these tissues through the blood as B-lymphocytes and colonize in specific regions of the lymph nodes and spleen also. When these B-cells come in contact with the appropriate antigens they are transformed into **plasma cells** which produce the antibodies of humoral immunity.

Laboratory Assignment

Materials:

prepared slides of lymph node (H13.11), the palatine tonsil (H5.22), ileum (H5.58), and spleen (H13.18)

1. Examine a prepared slide of a lymph node and compare it with illustration A, figure HA–27. Locate a **germinal center.**
2. Examine the **lymphocytes** under high-dry or oil immersion to identify dividing lymphocytes as seen in illustration B, figure HA–27.
3. Examine slides of the palatine tonsil, ileum, and spleen to locate lymphoidal tissue.
4. Answer all questions on the Laboratory Report.

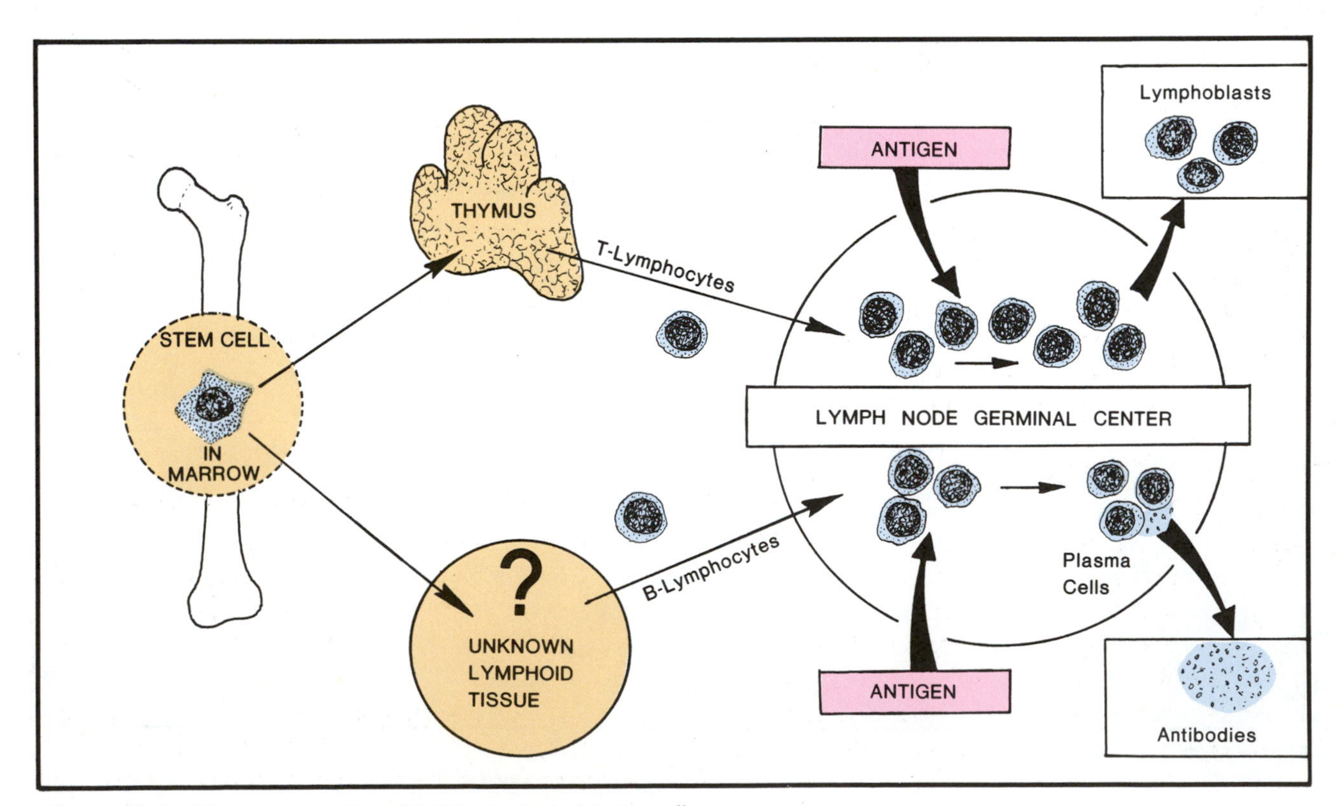

Figure 52.2 The genesis of lymphoblasts and plasma cells.

The Respiratory Organs

53

Both gross and microscopic anatomy of the respiratory system will be studied in this exercise. Sheep and frog materials will be available for certain observations. Before any laboratory dissections are performed, however, it is desirable that figures 53.1, 53.2, and 53.3 be labeled.

Materials:

- models of median section of head and larynx
- frog, pithed
- sheep pluck
- dissecting instruments and trays
- prepared slides of rat trachea and lung tissue
- Ringer's solution
- bone or autopsy saw

The Lungs, Trachea, and Larynx

Figure 53.1 illustrates the respiratory passages from the larynx to the lungs. Note that the larynx consists of three cartilages: an upper **epiglottis,** a large middle **thyroid cartilage,** and a smaller **cricoid cartilage.** Figure 53.2 reveals these cartilages in greater detail.

Below the larynx extends the **trachea** to a point in the center of the thorax where it divides to form two short **bronchi.** Note that both the trachea and bronchi are reinforced with rings of cartilage of the hyaline type. Each bronchus divides further into many smaller tubes called **bronchioles.** At the terminus of each bronchiole is a cluster of tiny sacs, the **alveoli,** where gas exchange with the blood takes place. Each lung is made up of thousands of these sacs.

Free movement of the lungs in the thoracic cavity is facilitated by the pleural membranes. Covering each lung is a **pulmonary pleura** and attached to the thoracic wall is a **parietal pleura.** Between these two pleurae is a potential cavity, the **pleural** (*intrapleural*) **cavity.** Normally, the lungs are firmly pressed against the body wall with little or no space between the two pleurae. The bottom of the lungs rests against the muscular **diaphragm,** the principal muscle of respiration. The parietal pleura is attached to the diaphragm also.

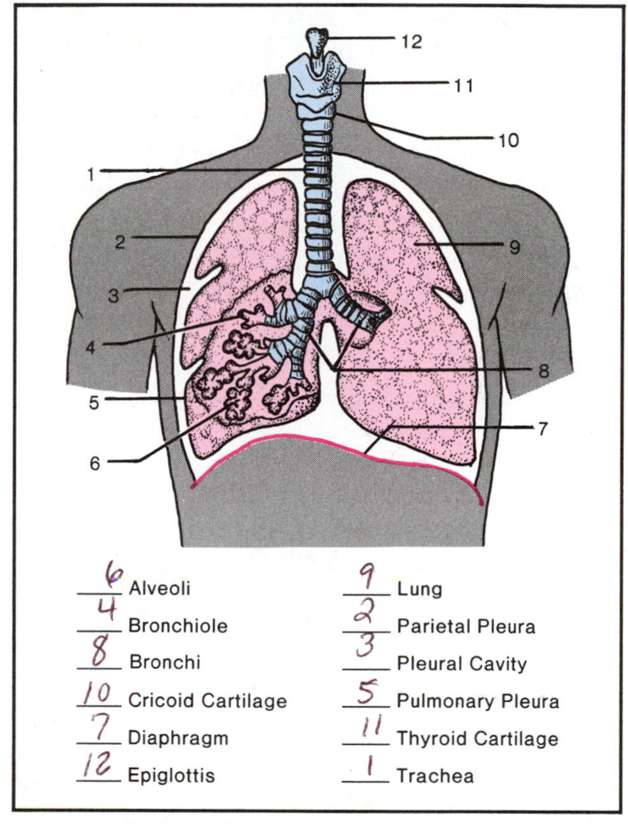

Figure 53.1 Respiratory tract.

Assignment:

Label figure 53.1.

The Upper Respiratory Tract

In addition to breathing, the upper respiratory tract also functions in eating and speech; thus, a study of the respiratory organs in this region necessarily includes organs concerned with these other activities. Figure 53.2 is of this area.

The two principal cavities of the head are the nasal and oral cavities. The upper **nasal cavity,** which serves as the passageway for air, is separated from the lower **oral cavity** by the **palate.** The anterior portion of the palate is reinforced with bone and is called the **hard palate.** The posterior part, which terminates in a fingerlike projection, the **uvula,** is the **soft palate.**

The oral cavity consists of two parts: the **oral cavity proper,** the larger cavity that contains the tongue; and the **oral vestibule** which is between the lips and teeth.

During breathing, air enters the nasal cavity through the nostrils, or **nares.** The lateral walls of this cavity have three pairs of fleshy lobes, the superior, middle, and inferior **nasal conchae.** These lobes serve to warm the air as it enters the body.

Above and behind the soft palate is the **pharynx.** That part of the pharynx posterior to the tongue, where swallowing is initiated, is the **oropharynx.** Above it is the **nasopharynx.** Two openings to the **auditory** *(Eustachian)* **tubes** are seen on the walls of the nasopharynx.

After air passes through the nasal cavity, nasopharynx, and oropharynx, it passes through the larynx and trachea to the lungs. The cartilages (labels 8, 10, and 11) of the larynx vary considerably in size and histology. The upper **epiglottis** consists of elastic cartilage; it helps to prevent food from entering the respiratory passages during swallowing. The **thyroid cartilage,** which consists of hyaline cartilage, forms the side walls of the larynx and the protuberance known as the "Adam's apple." Inferior to the thyroid cartilage is the **cricoid cartilage,** which consists of hyaline cartilage. Note that on the inner wall of the larynx is depicted a **vocal fold** which is, essentially, the true vocal cord of the larynx. The paired vocal cords are actuated by muscles through the **arytenoid cartilages** (label 5, figure 53.3) to produce sounds. One of these vocal cords exists on each side of the larynx. The vocal folds should not be confused with the

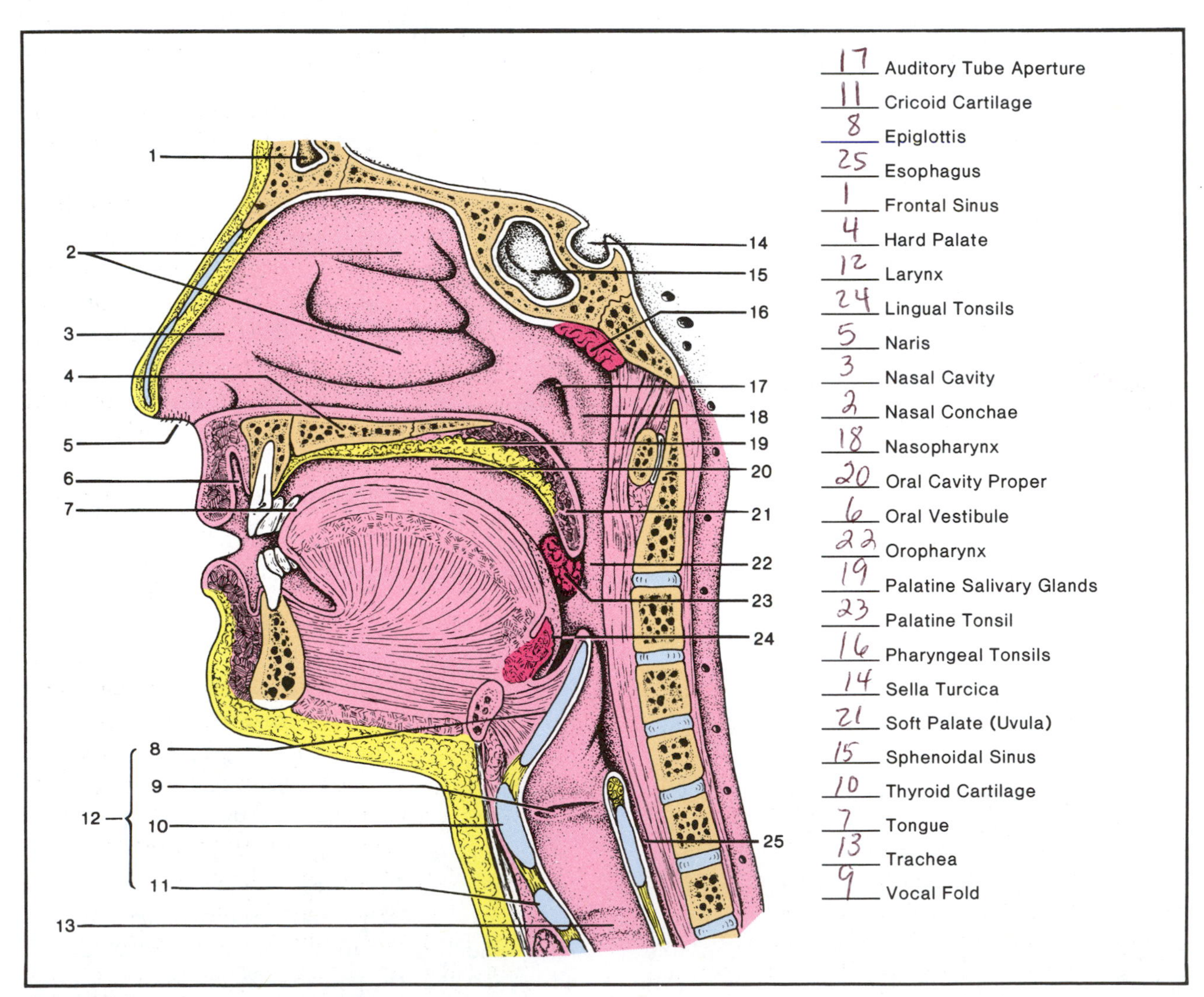

Figure 53.2 The upper respiratory passages.

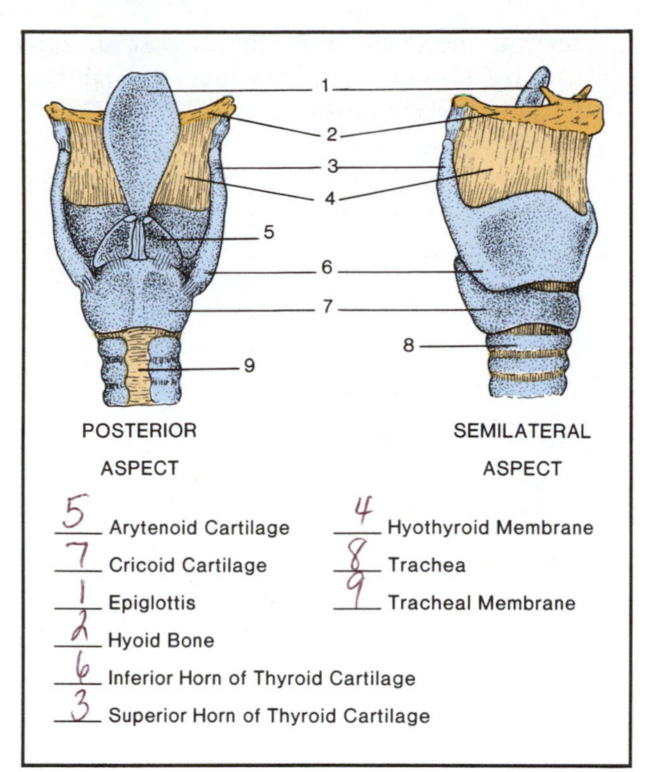

Figure 53.3 Human larynx.

ventricular folds (false vocal cords), which lie superior to the vocal folds. Posterior to the larynx and trachea lies the **esophagus,** which carries food from the pharynx to the stomach.

In several areas of the oral cavity and nasopharynx are seen islands of lymphoidal tissue called **tonsils.** The **pharyngeal tonsils,** or *adenoids,* are situated on the roof of the nasopharynx; the **lingual tonsils** are on the posterior inferior portion of the tongue; and the **palatine tonsils** are on each side of the tongue on the lateral walls of the oropharynx. These masses of tissue are as important as the lymph nodes of the body in protection against infection. The palatine tonsils are ones most frequently removed by surgical methods.

Assignment:

Label figures 53.2 and 53.3.

Study models of the head and larynx. Be able to identify all structures.

Sheep Pluck Dissection

A "sheep pluck" consists of the trachea, lungs, larynx, and heart of a sheep as removed during routine slaughter. Since it is fresh material rather than formalin-preserved, it is more lifelike than the organs of embalmed tissues. If plucks are available, proceed as follows to dissect one.

Materials:

sheep plucks, less the liver
dissecting instruments
drinking straws or serological pipettes

1. Lay out a fresh sheep pluck on a tray. Identify the major organs, such as the **lungs, larynx, trachea, diaphragm** remnants, and **heart.**
2. Examine the larynx more closely. Can you identify the **epiglottis, thyroid,** and **cricoid cartilages?** Look into the larynx. Can you see the **vocal folds?**
3. Cut a cross section through the upper portion of the trachea and examine a sectioned cartilaginous ring. Is it similar to the cat?
4. Force your index finger down into the trachea, noting the smooth and slimy nature of the inner lining. What kind of tissue lines the trachea that makes it so smooth and slippery?
5. Note that each lung is divided into lobes. How many lobes make up the right lung? The left lung? Compare with the human lung (figure 53.1).
6. Rub your fingers over the surface of the lung. What membrane on the surface of the lung makes it so smooth?
7. Free the connective tissue around the **pulmonary artery** and expose its branches leading into the lung. Also, locate the **pulmonary veins** that empty into the left atrium. Can you find the **ligamentum arteriosum,** which is between the pulmonary artery and aorta?
8. With a sharp scalpel, free the trachea from the lung tissue and trace it down to where it divides into the **bronchi.**
9. With a pair of scissors, cut off the trachea at the level of the top of the heart.
10. Now, cut the trachea down its posterior surface on the median line with a pair of scissors. The posterior surface of the trachea is opposite the heart. Extend this cut down to where it branches into the two primary bronchi. (Observe that the upper right lobe has a separate bronchus leading into it. This bronchus branches off some distance above where the primary bronchi divide from the trachea.)
11. Continue opening up the respiratory tree until you get down deep into the center of the lung. Note the extensive branching.
12. Insert a plastic straw or a serological pipette into one of the bronchioles and blow into the lung. If a pipette is used, **put the mouthpiece**

of the pipette into the bronchiole and blow on the small end. Note the great expansive capability of the lung.

13. Cut off a lobe of the lung and examine the cut surface. Note the sponginess of the tissue.
14. Record all your observations on the Laboratory Report.

Important: Wash your hands with soap and water at the end of the period.

Frog Lung Observation

To study the nature of living lung tissue in an animal, we will do a microscopic study of the lung of a frog shortly after pithing and dissection. A frog lung is less complex than the human lung, but it is made of essentially the same kind of tissue. Proceed as follows:

Materials:

frog, recently double pithed
dissecting instruments and dissecting trays
dissecting microscope
Ringer's solution in wash bottle

1. Pin the frog, ventral side up, in a wax-bottomed dissecting pan.
2. With your scissors, make an incision through the skin along the midline of the abdomen.
3. Make transverse cuts in both directions at each end of the incision and lay back the flaps of skin.

   Now, carefully make an incision through the right or left side of the abdomen over the lung and parallel to the midline. Take care not to cut too deeply. The inflated lung should now be visible.
4. With a probe, gently lift the lung out through the incision. Do not perforate the lung! From this point keep the lung moist with Ringer's solution.
5. If the lungs of your frog are not inflated, insert a medicine dropper into the slitlike glottis on the floor of the oral cavity and blow air into them by squeezing the bulb. Deflated lungs may be the result of excessive squeezing during pithing.
6. Observe the shape and general appearance of the frog lung. Note that it is basically saclike.
7. Place the frog under a dissecting microscope and examine the lungs. Locate the network of ridges on the inner walls. The thin-walled regions between the ridges represent the **alveoli.** Look carefully to see if you can detect blood cells moving slowly across the alveolar surface. Careful examination of the lungs may reveal the presence of parasitic worms. They are quite common in frogs.
8. When you have completed this study, dispose of the frog, as directed, and clean the pan and instruments.

Histological Study

Prepared slides of the nasal cavity, trachea, and lung tissue will be available for study. If Turtox slides are used, the tissues will probably be from monkey or human organs. Tissues from other mammals, such as the mouse or rabbit, may also be available.

Use figure HA–28 in the Histology Atlas for reference, and follow these suggestions in making your study.

Materials:

prepared slides of nasal septum (H6.1), trachea (H6.41), and lung tissue (H6.51)

Nasal Septum Scan this slide first with low power to locate the nasal septum and nasal concha. Refer to illustration A, figure HA–28, for reference. Identify the bony tissue and nasal epithelium.

Study the ciliated epithelium and the tissues beneath it with high-dry magnification. Look for the **olfactory** (Bowman's) **glands** (illustration B) which are located in the lamina propria. The secretion of these glands helps to keep the epithelial surface moist, and facilitates the solution of substances that stimulate the olfactory receptors. Observe that a large number of unmyelinated **olfactory fibers** lie between the glands and the ciliated epithelium. These fibers carry nerve impulses from olfactory receptors that are dispersed among the cells of the epithelium.

Trachea Identify the hyaline cartilage, ciliated epithelium, lamina propria, and muscularis mucosae. Look for goblet cells on the epithelial layer. Refer to illustration C, figure HA–28.

Lung Tissue Examine with the high-dry objective. Look for the thin-walled **alveoli,** a **bronchiole,** and blood vessels. Consult illustration D for reference.

Laboratory Report

Complete the Laboratory Report for this exercise.

Hyperventilation and Rebreathing 54

Activity of the respiratory muscles is controlled by the respiratory center of the medulla oblongata. The amounts of CO_2, H^+, and O_2 in the blood and cerebrospinal fluid determine whether these muscles are activated or inhibited. Of the three, the concentration of H^+ ion is most important; O_2 content is least important. Carbon dioxide assists in lowering the pH by forming carbonic acid.

The pH of blood and tissue fluid during normal breathing is around 7.4. During forced deep breathing (hyperventilation), the pH may be elevated to 7.6 as carbon dioxide is blown off. This decreased H^+ ion concentration depresses the respiratory center, lessening the desire for increased alveolar ventilation.

Hyperventilation in individuals can cause a drop in blood pressure, dizziness, and unconsciousness. All these symptoms are due to the washing out of CO_2 from the blood, causing *alkalosis*. This carbon dioxide depletion can be quickly restored by **rebreathing** into a paper bag.

In this experiment we will study the physiological effects of these two phenomena with and without recording equipment. The recording portion of the experiment can be performed with either a Gilson Duograph or Narco Physiograph.®

Nonrecording Experiment

Proceed as follows to observe the characteristics of hyperventilation, and to compensate for it with proper rebreathing techniques.

Materials:

paper bag

1. Breathe very deeply at a rate of about fifteen inspirations per minute for one or two minutes. Do not hurry the rate; concentrate on breathing deeply. Observe that the following symptoms occur:
   - It will become increasingly difficult to breathe deeply.
   - A feeling of dizziness will develop.

   Two events probably account for this: first, the blood pressure drops due to dilation of the splanchnic vessels which lessens blood to the brain; and second, the reduced CO_2 content of the blood causes vasoconstriction of the cerebral blood vessels, further reducing the blood supply.
2. Now, place a paper bag over your nose and mouth, holding it tightly to your face and breathing deeply into it for about **3 minutes.** Note how much easier it is to breathe into the bag than into free air. Why is this?
3. Allow your breathing rate to return to normal and breathe, normally, for 3 or 4 minutes. At the end of a normal inspiration, *without deep inhaling,* pinch your nose shut with the fingers of one hand, and hold your breath as long as you can. Do this three times, timing each duration carefully. Record these times and calculate the average on the Laboratory Report.
4. Now, breathe deeply, as in step 1, for **2 minutes,** and then hold your breath as long as you can. Record your results on the Laboratory Report.
5. Exercise by running in place for **2 minutes,** and then hold your breath as long as you can. Record your results on the Laboratory Report.

Recording Experiment

To produce a record of ventilation during hyperventilation and rebreathing, it will be necessary to use a bellows transducer as shown in figure 54.1.

Materials:

For Physiograph® setup:

Physiograph® with transducer coupler
bellows pneumograph transducer (PN 705–0190)
cable for bellows pneumograph transducer

For Duograph setup:

Duograph
Statham T–P231D pressure transducer
Gilson T–4030 chest bellows
stand and clamp for transducer

For both setups:

nose clamp
paper bag

Physiograph® Setup

Set up one channel, using the transducer coupler and bellows, as shown in figure 54.1.

1. Connect the bellows pneumograph transducer to the transducer coupler with a transducer cable. Be sure to match up the yellow dots on the connectors.
2. Balance the recording channel according to the instructions in Appendix C.
3. Attach the bellows pneumograph to the subject as follows:
   a. With the channel amplifier RECORD button in the OFF position, and the valve at the end of the rubber bellows open (turned counterclockwise), place the transducer on the subject's chest and fasten it with the attached leather strap.
   b. Locate the bellows as high as possible where it will get the greatest chest movement during breathing. Note that the transducer cable should be positioned over the shoulder to help support the weight of the transducer.
4. Start the paper advance at 0.25 cm/sec and lower the pens to the recording paper. Activate the timer.
5. Position the recording pen with the position control to a point that is 2 cm below the center line.
6. Close the valve on the bellows, while pinching the bellows slightly. The valve is closed by rotating it completely clockwise.
7. Place the channel amplifier RECORD button in the down (ON) position.
8. Starting at the 1000 position, rotate the outer channel amplifier sensitivity control knob to higher sensitivities (lower numbers) until normal shallow ventilation by the subject gives pen deflections of approximately 2 centimeters. Proceed to perform the following investigations.

Normal Respiration With the subject seated and relaxed, start the paper moving at 0.25 cm/sec, lower the pens, and activate the timer.

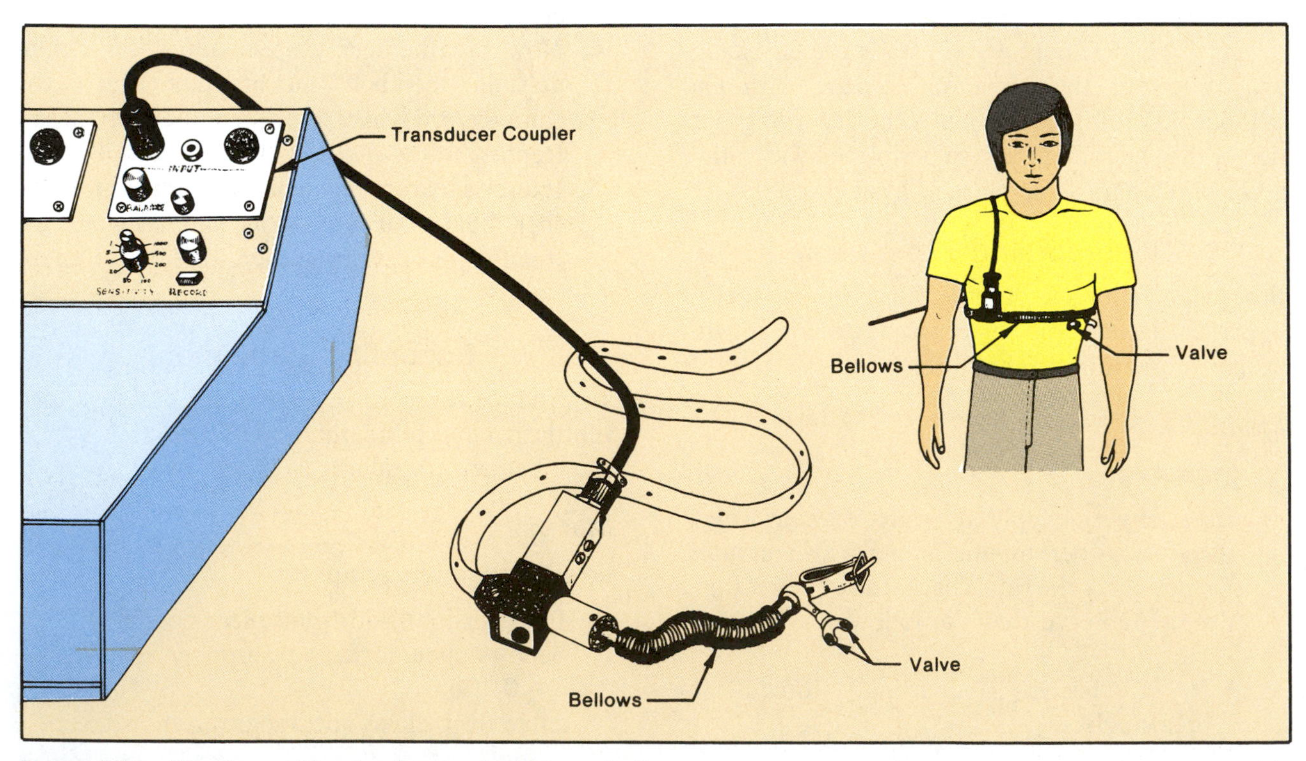

Figure 54.1 Physiograph® setup for monitoring respiration.

1. Place the amplifier RECORD button in the ON position.
2. Record one page of normal respiratory activity.
3. Increase the paper speed to 5 cm/sec and record several pages of normal respiratory activity.
4. Compare your tracings with sample records A and B in figure SR–6 of Appendix D.

Rebreathing Experiment Place a nose clamp on the subject and proceed as follows to note the effects of rebreathing into a paper bag.

1. Set the paper speed at 0.25 cm/sec and continue recording.
2. Record about a half page of normal breathing, and then place a paper bag tightly over the subject's mouth.
3. Allow the subject to breathe into and out of the bag until visible changes occur in the respiratory rate and depth.
4. Remove the bag and allow the respiratory rate to return to normal.
5. Compare your tracings with sample records C and D of figure SR–6, Appendix D.

Hyperventilation Continue recording at 0.25 cm/sec and have the subject hyperventilate by inhaling and exhaling as deeply as possible at a rate of one respiratory movement per second.

1. Have the subject continue hyperventilating for 1½ to 2 minutes or **until the subject begins to feel severe discomfort.**
2. Tell the subject to stop the deep breathing and allow the respiratory movements to normalize.
3. During normalization, look for changes that occur in rate or amplitude of respiratory movements as compared with prehyperventilation recordings.
4. Terminate the recordings.
5. Compare your tracings with the Physiograph® sample records in figure SR–7, Appendix D.

Duograph Setup

Prepare the subject and Duograph following this procedure:

1. Mount the Statham pressure transducer to the transducer stand with a clamp, and plug the jack from the transducer into the receptacle of channel 1 on the Duograph. Figure 56.2 illustrates the transducer and stand in the background.
2. Attach the chest bellows to the subject at the highest point where the greatest respiratory movements occur.
3. Attach the tubing from the bellows to the Statham transducer.
4. Activate the Duograph at slow speed noting the trace.
5. Adjust the excursion for normal breathing at about 1 centimeter. You are now ready to perform the following investigations.

Normal Respiration With the subject seated and relaxed, make the following recordings of normal respiratory activity.

1. For about 1½ to 2 minutes record normal breathing at the slow speed (2.5 mm/sec).
2. Increase the speed to fast (25 mm/sec) and record for another 1½ to 2 minutes.

Rebreathing Experiment Place a nose clamp on the subject and proceed as follows to note the effects of rebreathing into a paper bag.

1. Return the paper speed to slow, and continue recording.
2. For about 30 seconds, record normal ventilation and then place a paper bag tightly over the subject's mouth.
3. Allow the subject to breathe into and out of the bag until visible changes occur in the respiratory rate and depth.
4. Remove the bag and allow the respiratory rate to return to normal.

Hyperventilation Continue recording at slow speed, and have the subject hyperventilate by inhaling and exhaling, as deeply as possible, at a rate of one respiratory movement per second.

1. Have the subject continue hyperventilating for 1½ to 2 minutes, or **until the subject begins to feel severe discomfort.**
2. Tell the subject to stop the deep breathing and continue the recording as the respiratory movements normalize.
3. During normalization, look for changes that occur in the rate or amplitude of respiratory movements as compared with prehyperventilation recordings.
4. Terminate the recording.

Laboratory Report

Complete the Laboratory Report for this exercise.

55 The Diving Reflex

Many aquatic, air breathing animals have evolutionary adaptations that enable them to remain submerged under water for quite long periods of time. The **diving reflex** (or dive reflex) is one of these adaptations. In these animals there are prompt cardiovascular and respiratory adjustments that are triggered by partial entry of water into the air passages. These adjustments include a change in the heart rate, the shunting of blood from less essential tissues, and a cessation of breathing. In humans, these changes are less pronounced, but some adjustments are made. It is this series of changes that have accounted for the survival of some individuals, usually young children, that have fallen through the ice and remained under water for unusually long periods of time. Several of these near-drowning incidents have occurred in the past few years, reviving considerable interest in this phenomenon.

In an attempt to observe the existence of this reflex in the human, we will monitor the changes in respiration and pulse while the face of a subject is immersed in very cold water. Needless to say, a subject for this experiment should be one who has no aversion to facial submersion.

The equipment setup for this experiment requires two-channel recording. A chest bellows will be used to record breathing, and a pulse transducer will be used to record the heart rate (pulse). If this experiment is to be performed during the same laboratory period as the last exercise, then all you have to do is add the pulse transducer to the setup you used for the hyperventilation study.

The immersion maneuver may take place at your laboratory station, if large pans are available; or it may be made in a sink at a perimeter counter. The recording equipment will have to be set up wherever the immersion is to take place. As with previous experiments, separate instructions will be provided for Gilson and Narco equipment.

Several variables will be studied in this experiment, such as the effect of temperature changes, and the effect of hyperventilation prior to immersion. Proceed as follows:

Materials:

For Physiograph® setup:

- Physiograph® with two transducer couplers
- bellows pneumograph transducer
- pulse transducer (Narco 705–0050)

For Duograph setup:

- Duograph
- Statham T–P231D pressure transducer
- Gilson T–4030 chest bellows
- stand and clamp for transducer
- pulse pickup (Gilson T4020)
- A–4023 adapter (Gilson)

For both setups:

- pan for water
- ice, towel, thermometer

Duograph Setup

1. Attach the A–4023 adapter to the receptacle for channel # 1 of the Duograph. Make sure that the locknut is tightened securely, and that the phone jack of the pulse pickup is plugged into it.
2. Attach the finger pulse pickup and chest bellows to the subject. Keep the chain of the bellows as high into the armpits as possible.
3. Attach the tubing from the bellows to the Statham transducer, and plug the jack from the transducer into the receptacle for channel # 2 of the Duograph.
4. Make all connections in such a way that immersion movement does not tug on the cables.
5. While the subject is being prepared, fill the pan or sink with tap water, adjusting the temperature to 70° F.
6. Set the speed control lever in the slow position, and the stylus heat control knobs of both channels in the two o'clock position.
7. Set the Gain for both channels at 2 MV/CM, and turn the sensitivity control knobs of both channels completely counterclockwise (lowest values).
8. Set the Mode for channel # 1 on DC and for channel # 2 on TRANS.
9. Activate both channels on the Duograph to observe the traces. Adjust the excursion for tidal breathing at about 1 centimeter, using the sensitivity and gain controls. Adjust the pulse amplitude to about 2 cm.
10. Establish a reference by recording for at least

one minute before immersion. During this time, the subject will breathe regularly (tidal inspirations) without speaking or hyperventilating. **Once suitable sensitivity and gain are obtained, do not readjust them during the experiment.**

11. To further establish reference values, direct the subject to hold his or her breath for **one minute** after a tidal inspiration, without prior deep breathing or hyperventilating. Allow the subject to recover for 2 to 3 minutes by normal breathing.
12. After recovery, have the subject completely submerge his or her face into the water. Depress the event button at the exact moment of immersion, holding it down during the full time of immersion, and releasing it as soon as the head is raised from the water. **Immersion should last 15 to 30 seconds.** Use a large towel to protect the subject's clothing from getting wet. Repeat the procedure.
13. Chill the water to 60° F with ice cubes and repeat step 8. Do not leave ice floating on the water. Repeat the procedure.
14. Have the subject hyperventilate vigorously for **30 seconds,** and hold his or her breath **for as long as possible.** This is done without immersion.
15. When regular breathing has resumed, have the subject hyperventilate again for **30 seconds,** and then submerge his or her face **for as long as possible.** Be sure to depress the event button during immersion.
16. Disconnect the subject, put away the apparatus, clean the work area, and return all materials to their proper places. Analyze the tracing as a team and label it with additional information, as needed.

Physiograph® Setup

In the hyperventilation experiment (Exercise 54) one channel was set up for the bellows pneumograph transducer. For this experiment all that is necessary is to attach the pulse transducer to a second transducer coupler, and all the equipment you need is in place. Proceed as follows:

1. Attach the finger pulse transducer and chest bellows to the subject. As in the previous experiment, locate the bellows high enough to get maximum excursion during breathing.
2. Make all connections in such a way that immersion movement does not tug on the cables.
3. While the subject is being prepared, fill the pan or sink with tap water, adjusting the temperature to 70° F.
4. Turn on the power switch and set the paper speed at 0.25 cm/sec.
5. Start the chart paper moving and lower the pens to the paper. Make certain that the timer is activated.
6. Use the position controls on both channels to position the pens so that they are recording about one centimeter below the center line.
7. Place the RECORD button in the ON position.
8. Adjust the amplifier sensitivity controls so that the pulse wave amplitude is 2 cm and the tidal ventilation amplitude is 1 cm.
9. Establish a reference by recording for at least **one minute** before immersion. During this time, the subject will breathe regularly (tidal inspirations) without speaking or hyperventilating. **Do not readjust the amplifier sensitivity controls from this point on in the experiment.**
10. To further establish reference values, direct the subject to hold his or her breath for **one minute** after a tidal inspiration, without prior deep breathing or hyperventilating. Allow the subject to recover for 2 to 3 minutes by normal breathing.
11. After recovery, have the subject completely submerge his or her face into the water. Mark the chart at this point. **Immersion should last 15 to 30 seconds.** Use a large towel to protect the subject's clothing from getting wet. Repeat the procedure.
12. Chill the water to 60° F with ice cubes and repeat step 10. Do not leave ice floating on the water. Repeat the procedure.
13. Have the subject hyperventilate vigorously for **30 seconds,** and hold his or her breath for **as long as possible.**
14. When regular breathing has resumed, have the subject hyperventilate again for **30 seconds,** and then submerge his or her face for **as long as possible.** Be sure to mark the chart.
15. Terminate the recording, disconnect the subject, put away the apparatus, clean the work area, and return all materials to their proper places. Analyze the tracing as a team and label it with additional information, as needed.

Laboratory Report

Complete the first portion of combined Laboratory Report 55,56.

56 The Valsalva Maneuver

Antonio Valsalva, an Italian anatomist, discovered in 1723 that the middle ear could be filled with air if one expires forcibly against closed mouth and nostrils. This action became known as *the Valsalva maneuver.* The term also signifies attempting to expire air from the lungs against a closed glottis. In this maneuver, the abdominal muscles and internal intercostal muscles greatly increase intraabdominal and intrathoracic pressures. This momentary increase in intrathoracic pressure affects both the pulse and blood pressure.

Figure 56.1 illustrates what happens to the blood pressure and pulse in this maneuver. Note that the blood pressure goes up at the start of the straining; then, it falls a short time later. After the maneuver, the blood pressure goes up considerably more and the heart rate slows somewhat.

The initial rise in blood pressure is caused by the addition of intrathoracic pressure to the pressure within the aorta. The subsequent drop in blood pressure is caused by decreased cardiac output resulting from reduced venous return. The reduced venous return is caused by the restricted flow of blood into the right side of the heart through the compressed venae cavae. The final surge in increased cardiac output is a compensation for the reduced prior output.

The Valsalva maneuver is more than just a stunt to observe in the laboratory. It has real merit in the diagnosis of two pathological conditions.

First of all, patients that suffer from *autonomic insufficiency* exhibit no pulse changes during this maneuver. The cause of this condition is not well understood.

Secondly, patients with **hyperaldosteronism** fail to show pulse rate changes and blood pressure rise after the maneuver. This test can be used, diagnostically, on patients suspected of having aldosterone-secreting tumors. Once tumors are removed, normal responses to the Valsalva maneuver occur.

Students will work in teams of four or five to monitor the blood pressure and pulse of one of the team members during this maneuver. The equipment setups are similar to the last experiment, except that the bellows will be replaced with a pulse transducer. Figure 56.2 illustrates the Duograph setup.

Caution: Although the Valsalva maneuver is perfectly safe for normal, healthy individuals, any team member who has a history of heart disease should not be used as a subject.

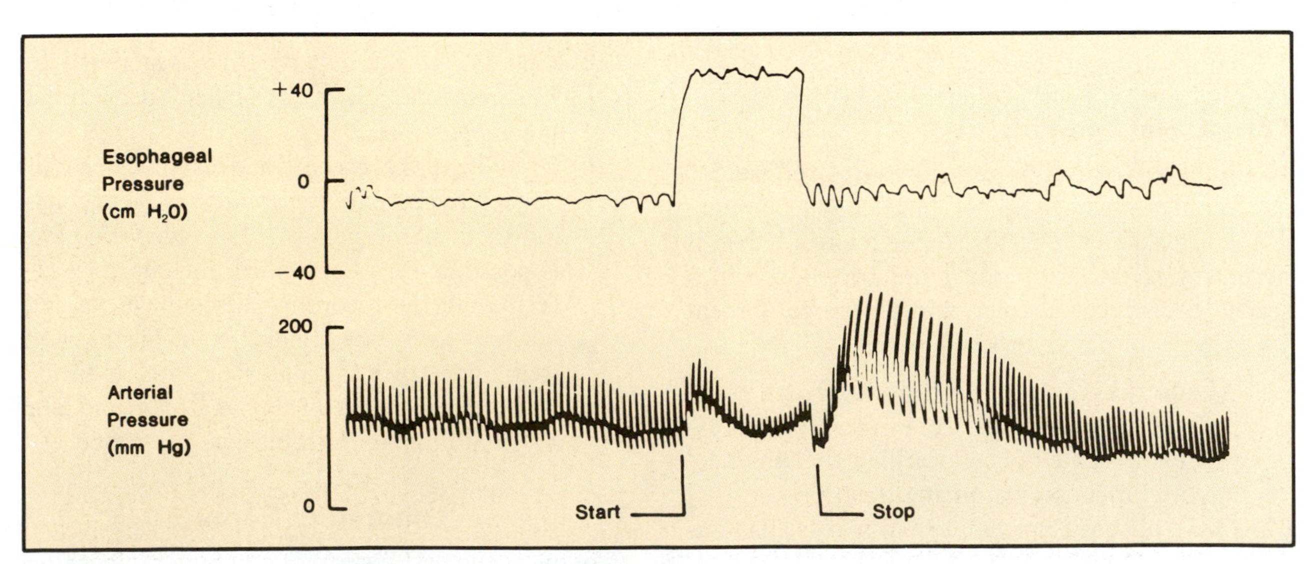

Figure 56.1 Blood pressure and pulse response in Valsalva's maneuver.

Materials:

For Duograph setup:

Duograph
Statham T–P231D pressure transducer
adult arm cuff
stand and clamp for T–P231D transducer
Gilson T–4020 finger pulse pickup
Gilson A–4023 adapter for pulse pickup

For Physiograph® setup:

Physiograph® with transducer and ESG couplers
adult pressure cuff (Narco PN 712–0016)

Duograph Setup

In the last experiment a chest bellows and pulse transducer were used. In this experiment the chest bellows will be replaced with an arm cuff, which works through the Statham pressure transducer. Set up the experiment on the basis of what you learned in the last exercise and as illustrated in figure 56.2. Make the following hookups and adjustments on the Duograph:

1. Place the cuff on the upper right arm, securing it tightly in place. Attach the tubing from the cuff to the Statham transducer.
2. Test the cuff by pumping some air into it and noting if the pressure holds when the metering valve is closed on the bulb. Release the pressure for the comfort of the subject while other adjustments are made.
3. Insert the subject's index finger into the finger pulse pickup.
4. Plug in the Duograph, and turn on the power switch.
5. Set the speed control lever in the slow position, and the stylus heat control knobs of both channels in the two o'clock position.
6. Set the Gain for channel # 1 at 2 MV/CM, and for channel # 2 at 1 MV/CM, and turn the sensitivity control knobs of both channels completely counterclockwise (lowest values).
7. Set the Mode for channel # 1 on DC and for channel # 2 on TRANS.
8. Calibrate the Statham transducer on channel # 2 by depressing the TRANS button, and, simultaneously, adjusting the sensitivity knob to get 2 cm deflection. The calibration line represents 100 mm Hg pressure.
9. Adjust the sensitivity on channel # 1 so that you get 1.5 to 2 cm deflection on the pulse.
10. Pump air into the cuff and establish a baseline for **one minute.**
11. Have the subject take a deep breath and perform a Valsalva maneuver, exerting as much internal pressure as possible. Just as the maneuver is begun, press down on the event

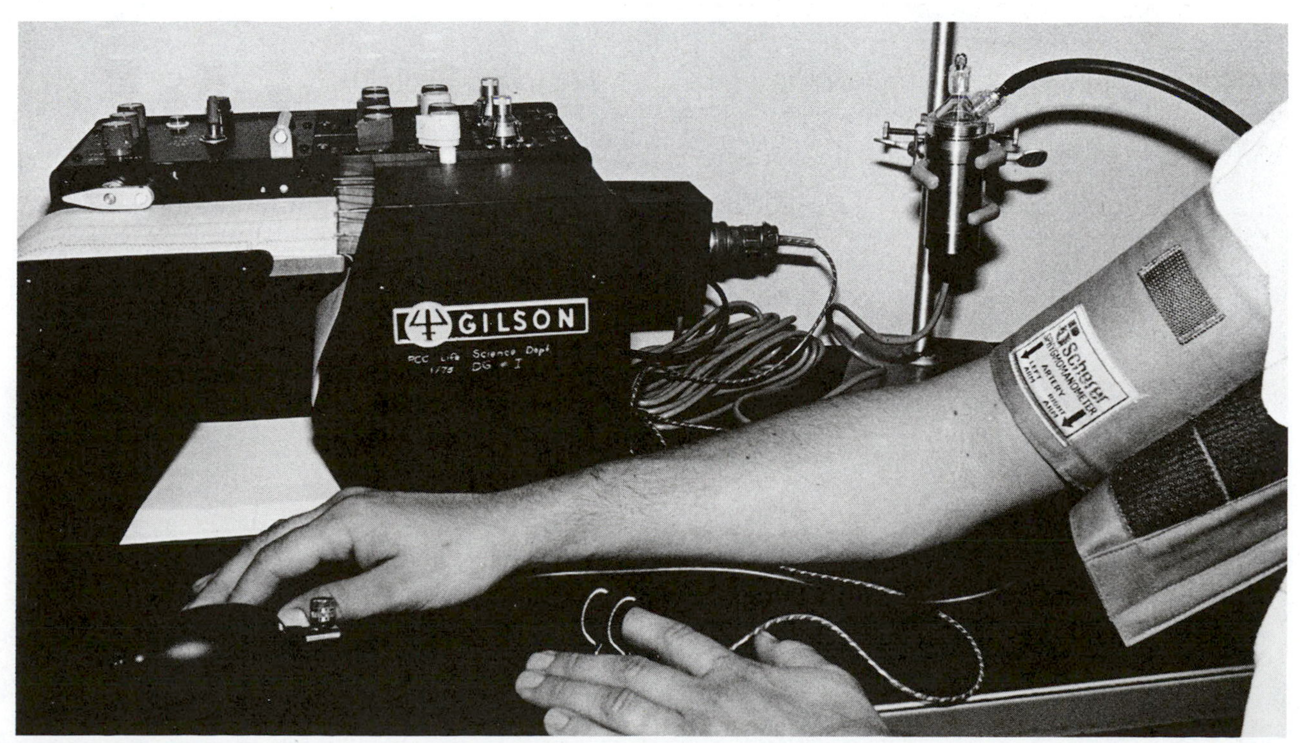

Figure 56.2 Monitoring pulse and blood pressure in Valsalva's maneuver.

marker button and hold it down until the subject ends the maneuver.

12. Repeat this maneuver several times to provide enough chart material for each team member.
13. Repeat the experiment using other team members as subjects.

Physiograph® Setup

In the last experiment a chest bellows and pulse transducer were used. In this experiment the chest bellows will be replaced with an arm cuff; thus, an ESG coupler on the Physiograph® will be used instead of a transducer coupler. Set up the experiment on the basis of what you learned in the last experiment, and what you did in Exercise 47. Make the following hookups and adjustments:

1. Refer back to figure 47.4 to see how the pressure cuff is hooked up to the ESG coupler.
2. Turn on the power switch, and set the paper speed at 0.25 cm/sec.
3. **Balance and calibrate the ESG coupler** according to the instructions on page 255.
4. From this point on *do not touch* the balance and position control knobs.
5. Place the cuff on the upper right arm, making certain that the microphone is positioned over the brachial artery.
6. **Calibrate the cuff microphone** according to the instructions on page 255.
7. Insert the subject's index finger into the finger pulse pickup, and plug the jack of the pulse transducer into the transducer channel.
8. Start the chart paper moving again, lower the pens to the paper, and make certain the timer is activated.
9. Adjust the amplifier sensitivity controls on the transducer channel so that the pulse wave amplitude is 1.5 to 2 cm.
10. Pump air into the cuff and establish a baseline for **one minute.**
11. Have the subject take a deep breath and perform a Valsalva maneuver, exerting as much internal pressure as possible.
12. Repeat this maneuver several times to provide enough chart material for each team member.
13. Repeat the experiment, using other team members as subjects.

Laboratory Report

Complete the last portion of combined Laboratory Report 55,56.

Spirometry: Lung Capacities 57

The volume of air that moves in and out of the lungs during breathing is measured with an apparatus called a **spirometer.** Two types are available: the hand-held (Propper) type (figure 57.2) and the tank-type recording spirometer (figure 58.1). When a recording spirometer is used, a **spirogram** similar to figure 57.1 can be made. The Propper spirometer is designed, primarily, for screening measurements of vital capacity.

In this exercise, we will use the Propper spirometer to determine individual respiratory volumes. Although this convenient device cannot measure inhalation volumes, it is possible to determine most of the essential lung capacities. While most members of the class are working with Propper spirometers, some students will be working with the recording spirometer to do the experiment in Exercise 58.

Materials:

Propper spirometer and disposable mouthpieces (Ward's Nat'l Science Est., Monterey, CA, Cat. No. 14W5070)
70% alcohol
cotton

Tidal Volume (TV)

The amount of air that moves in and out of the lungs during a normal respiratory cycle is called the *tidal volume.* Although sex, age, and weight determine this and other capacities, the average normal tidal volume is around 500 ml. Proceed as follows to determine your tidal volume.

1. Swab the stem of the spirometer with 70% alcohol and place a disposable mouthpiece over the stem.
2. Rotate the dial of the spirometer to zero as illustrated in figure 57.2.
3. After three normal breaths, expire three times into the spirometer while inhaling through the nose. Do not exhale forcibly. **Always hold the spirometer with the dial upward.**
4. Divide the total volume of the three breaths by 3. This is your tidal volume. Record this value on the Laboratory Report.

Minute Respiratory Volume (MRV)

Your *minute respiratory volume* is the amount of tidal air that passes in and out of the lungs in one

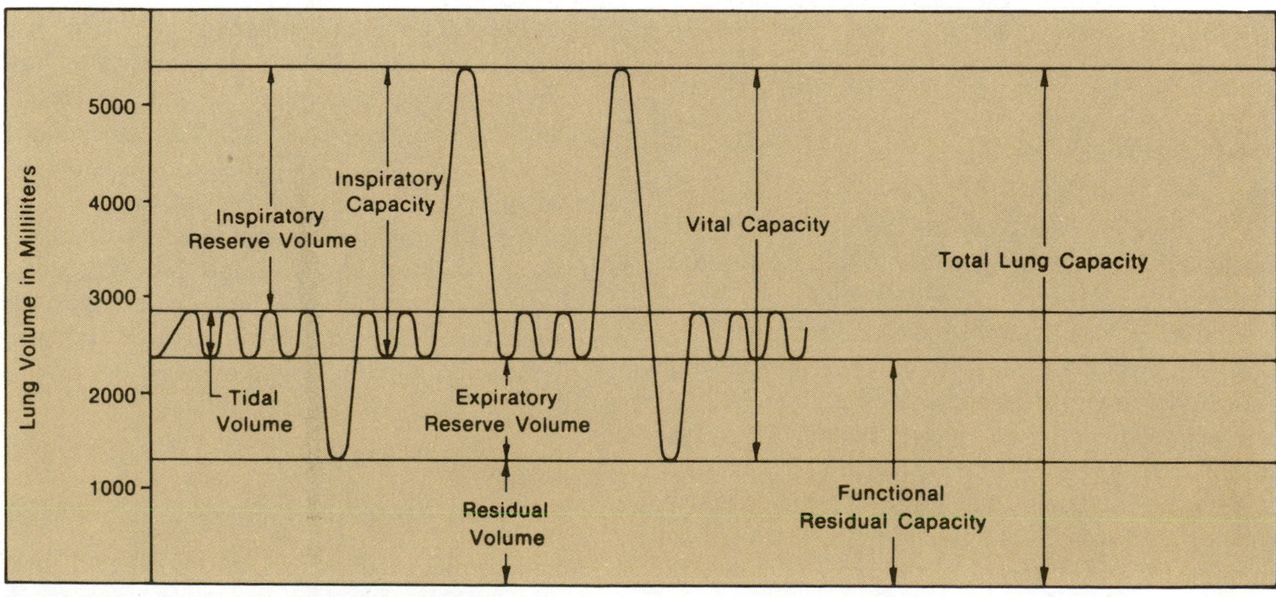

Figure 57.1 Spirogram of lung capacities.

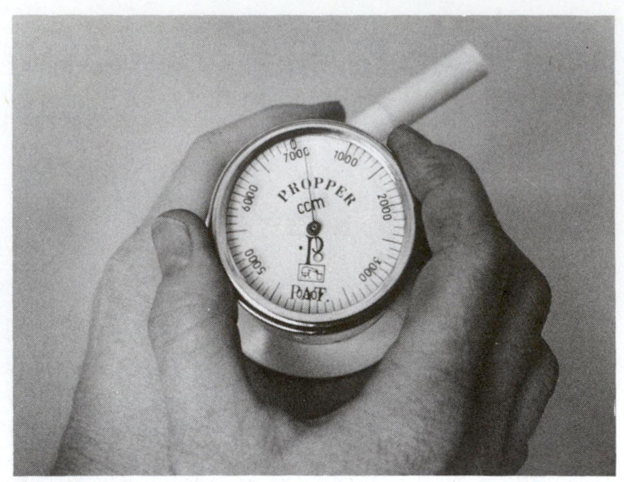

Figure 57.2 Dial face of spirometer is rotated to zero prior to measuring exhalations.

minute. To determine this value, count your respirations for one minute and multiply this by your tidal volume. Record this volume on the Laboratory Report.

Expiratory Reserve Volume (ERV)

The amount of air that one can expire beyond the tidal volume is called the *expiratory reserve volume.* It is usually around 1,100 ml.

To determine this volume, set the spirometer dial on 1,000 first. After making three normal expirations, expel all the air you can from your lungs through the spirometer. Subtract 1,000 from the reading on the dial to determine the exact volume. Record the results on the Laboratory Report.

Vital Capacity (VC)

If we add the tidal, expiratory reserve, and inspiratory reserve volumes, we arrive at the total functional or *vital capacity* of the lungs. This value is determined by directing an individual to take as deep a breath as possible and exhaling all the air possible. Although the average vital capacity for men and women is around 4,500 ml, age, height, and sex do affect this volume appreciably. Even the established norms can vary as much as 20% and still be considered normal. Tables of normal values for men and women are given in Appendix A.

Set the spirometer dial on zero. After taking two or three deep breaths and exhaling completely after each inspiration, take one final deep breath and exhale all the air through the spirometer. A slow, even, forced exhalation is optimum.

Repeat two or more times to see if you get approximately the same readings. Your VC should be within 100 ml each time. Record the average on the Laboratory Report. Consult tables V and VI in Appendix A for predicted (normal) values.

Inspiratory Capacity (IC)

If you take a deep breath to your maximum capacity after emptying your lungs of tidal air, you will have reached your maximum *inspiratory capacity.* This volume is usually around 3,000 ml. Note from the spirogram that this volume is the sum of the tidal and inspiratory reserve volumes.

Since this type of spirometer cannot record inhalations, it will be necessary to calculate this volume, using the following formula:

$$IC = VC - ERV.$$

Inspiratory Reserve Volume (IRV)

This is the amount of air that can be drawn into the lungs in a maximal inspiration after filling the lungs with tidal air. Since the IRV is the inspiratory capacity less the tidal volume, make this subtraction and record your results on the Laboratory Report.

Residual Volume (RV)

The volume of air in the lungs that cannot be forcibly expelled is the *residual volume.* No matter how hard one attempts to empty one's lungs, a certain amount, usually around 1,200 ml, will remain trapped in the tissues. The magnitude of the residual volume is often significant in the diagnosis of pulmonary impairment disorders. Although it cannot be determined by simple ordinary spirometric methods, it can be done by washing all the nitrogen from the lungs with pure oxygen and measuring the volume of nitrogen expelled.

Laboratory Report

After recording your lung capacities on Laboratory Report 57,58, evaluate your results.

Spirometry: The FEV_T Test 58

Impairment of pulmonary function in the form of asthma, emphysema, and cardiac insufficiency (left-sided heart failure) can be detected with a spirometer. The symptom common to all these conditions is shortness of breath, or *dyspnea.* The lung capacity that is pertinent in these conditions is the vital capacity.

In several types of severe pulmonary impairment, the vital capacities may exhibit nearly normal vital capacities; however, if the *rate* of expiration is recorded on a kymograph and timed, the extent of pulmonary damage becomes quite apparent. This test is called the *timed vital capacity* or *forced expiratory volume* (FEV_T). Figure 58.1 illustrates a spirometer (Collins Recording Vitalometer) which will be used in this experiment. The record produced on the kymograph in such a test is revealed in figure 58.2 and is called an *expirogram.*

To perform a timed vital capacity test, one makes a maximum inspiration and expels all the air from the lungs into the spirometer as fast as possible. The moving kymograph drum has chart paper on it that is calibrated vertically in liters and horizontally in seconds. A pen produces a record on the chart paper which reveals how much air is expelled per unit of time.

Approximately 95% of a normal person's vital capacity can be expelled within the first three seconds. From a diagnostic standpoint, however, the *percentage of vital capacity that is expelled during*

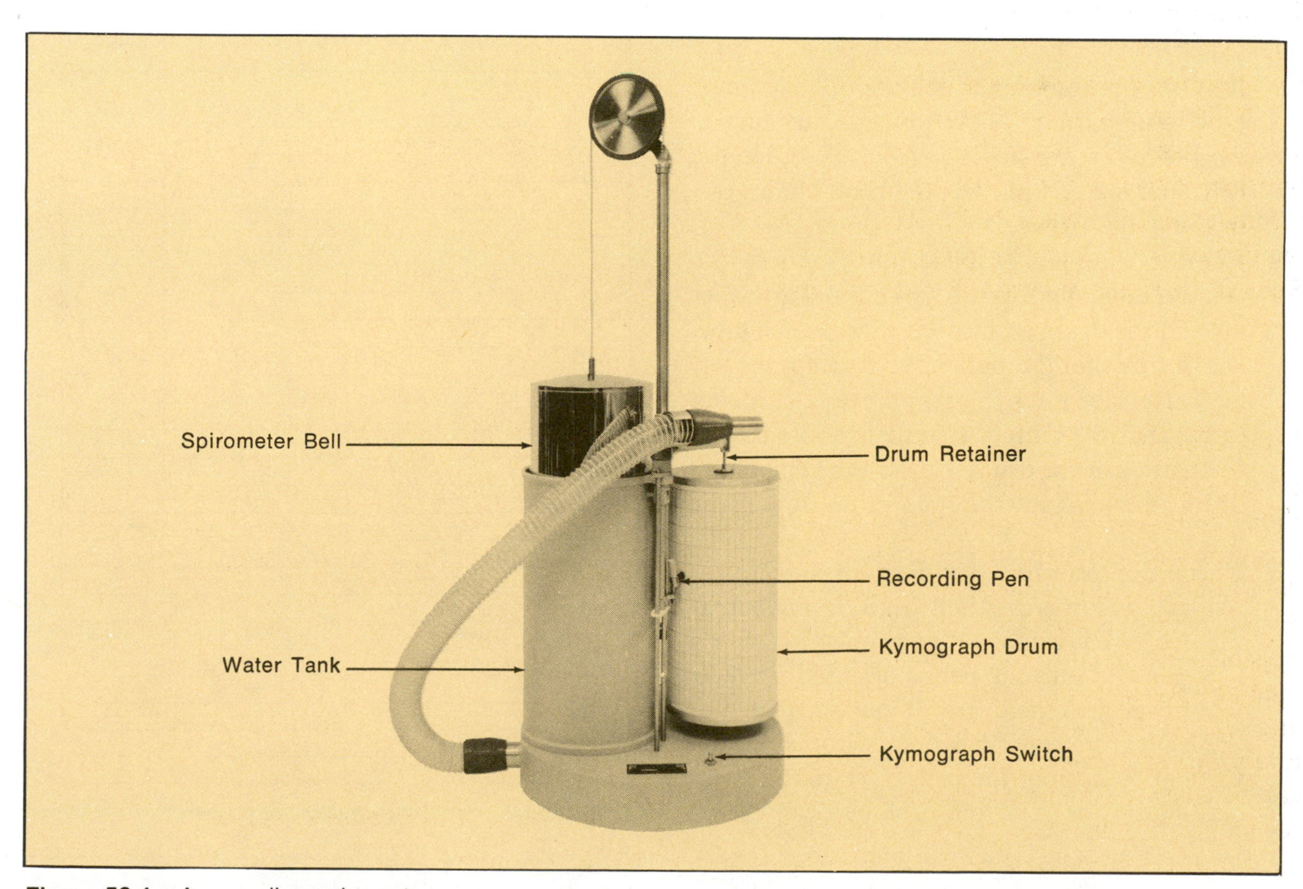

Figure 58.1 A recording spirometer.

the first second is of paramount importance. An individual with no pulmonary impairment should be able to expel 75% of the lung's total capacity during the first second. Individuals with emphysema and asthma, however, will expel a much lower percentage due to "air entrapment." Given sufficient time to exhale, they might be able to do much better.

In this experiment, each member of the class will have an opportunity to determine his or her FEV_T. Disposable mouthpieces will be available for each person. Follow this procedure:

Materials:

spirometer (Collins Recording Vitalometer)
disposable mouthpieces (Collins #P612)
kymograph paper (Collins #P629)
noseclip
Scotch tape
dividers
ruler

Preparation of Equipment

Prior to performing this test, the equipment should be readied as follows:

1. Remove the spirometer bell and fill the water tank to within 1½″ of the top. Before filling *be sure to close the drain petcock.* Replace the bell, making certain that the bead chain rests in the pulley groove.
2. Remove the kymograph drum by lifting the drum retainer first. Wrap a piece of chart paper around the drum and fasten with Scotch tape. Make sure that the right edge overlaps the left edge. Replace the kymograph drum, checking to see that the bottom spindle is in the hole in the bottom of the drum. Lower the kymograph drum retainer into the hole in the top of the drum.
3. Check the recording pen. Remove the protective cap which prevents it from drying out. If the pen is dry, remove it and replace it with a new pen. Be sure the pen is oriented properly with respect to the drum. It can be rotated to the correct marking position. Also, check the vertical position of the pen to see that it is recording on the zero line. The pen can be ad-

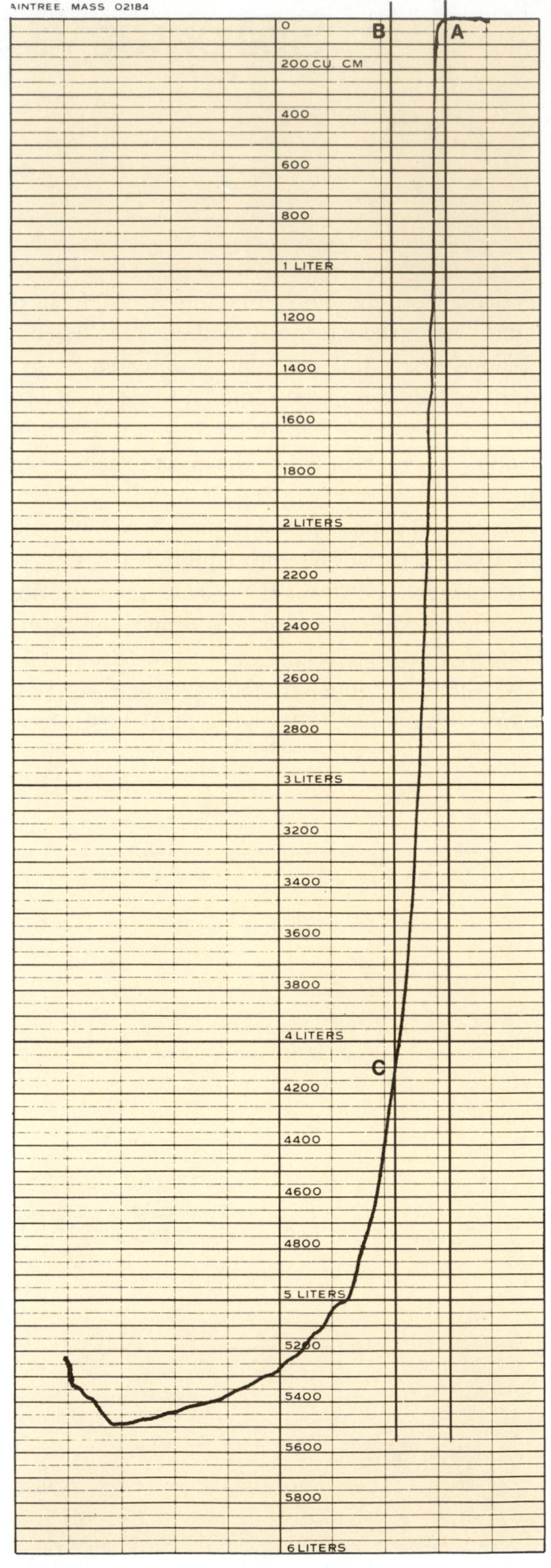

Figure 58.2 An expirogram.

justed vertically within its socket by simply pushing or pulling on it. Major adjustments should be made by loosening the setscrew.

4. Plug the electric cord into the electrical outlet. *This apparatus must not be used in outlets that lack a ground circuit.* The three-pronged plug must fit into a three-holed receptacle. If an adapter is used, be sure to make the necessary ground hookup. Electricity and water can be lethal!

Performing the Test

1. Before exhaling into the mouthpiece, check out the following items to make sure everything is ready.
   - Bell is in lowered position so that the recording pen is exactly on 0.
   - A clean sheet of paper is on the kymograph drum.
   - A sterile mouthpiece has been inserted in the end of the breathing tube.
2. Apply a noseclip to the subject to prevent leakage through the nose. Allow the subject to hold the tube.
3. Before turning on the instrument explain to the subject that (1) he or she will be taking as deep a breath as possible, (2) he or she will expel as much air as possible into the mouthpiece, *as rapidly and completely as possible,* and (3) inhalation should take place *before* the mouthpiece is placed in his or her mouth.
4. Turn on the kymograph and tell the subject to perform the test as described. As the subject blows into the tube, encourage him or her to *push, push, push* to get all the air out.
5. Turn off the kymograph.
6. After recording as many tracings as desired, remove the paper from the drum.

Analysis of Tracing

To determine the FEV_T from this expirogram, proceed as follows:

1. Draw a vertical line through the starting point of exhalation (see point A, figure 58.2).
2. Set a pair of dividers to the distance between two vertical lines. This distance represents one second.
3. Place one point of the dividers on point A and establish point B to the left of point A. Draw a parallel vertical line through point B. This line intersects the expiratory curve at point C.

   The one second volume (FEV_1) or one second timed vital capacity is represented by the distance B–C. This volume is read directly from the volume markings. In figure 58.2, it is 4,100 ml. The total vital capacity is also read directly from the volume markings. It is 5,500 ml.
4. Next, correct the recorded total vital capacity for body temperature, ambient pressure, and water saturation (BTPS). To do this, consult table IV in Appendix A for the conversion factor and multiply the recorded vital capacity by this factor.

   Example: If the temperature in the spirometer is 25 °C, the conversion factor is 1.075. Thus:

   $$5500 \text{ ml} \times 1.075 = 5913 \text{ ml.}$$

5. Also, convert the one second timed vital capacity (FEV_1) in the same manner:

   $$4100 \text{ ml} \times 1.075 = 4408 \text{ ml.}$$

6. Divide the FEV_1 by the total vital capacity:

   $$\frac{4408}{5913} = 75\%.$$

7. Determine how the subject's vital capacity compares with the predicted (normal) values in tables V and VI of Appendix A.

   Example: Individual in the above test was a 26-year-old male, 6'3" (184 cm) tall. From table VI his predicted vital capacity is 4,545 ml.

$$\frac{\text{Actual VC}}{\text{Predicted VC}} \times 100\% = \frac{5913}{4545} \times 100 = 131\%.$$

Laboratory Report

Complete the last portion of Laboratory Report 57,58.

59 Spirometry: Using Computerized Hardware

In this laboratory period we will use the Intelitool Spirocomp and computer to perform all of the spirometry experiments that are performed in Exercises 57 and 58. Figure 59.2 illustrates the setup. By utilizing the Spirocomp program we will be able to measure and graph the following:

- expiratory reserve volume (ERV)
- tidal volume (TV)
- measured vital capacity (mVC)

The following respiratory volumes can also be computed and graphed:

- inspiratory reserve volume (IRV)
- computed vital capacity (cVC)
- predicted vital capacity (pVC)

And finally, we will be able to determine forced expiratory volumes (FEV_T) for one, two, and three seconds.

The principal advantage of this computerized setup over the methods used in Exercises 57 and 58 is that all of the calculations are made automatically from data that is fed into the program. In addition, the program allows one to compute and average the values of up to 30 subjects. Respiratory efficiency comparisons of smokers with nonsmokers, athletes with nonathletes, and males with females are also easily made.

In performing these experiments you will be working with a laboratory partner. While one acts as the subject, breathing into the spirometer, the other person can run the computer and prompt the subject as to the procedures.

Note in the materials list that the Spirocomp manuals will be available for reference purposes. Page number references to certain sections in the Spirocomp manuals are made to assist you in cross-referencing certain procedures. (Example: *SUM, p. 9* in step 5 under Equipment Hookup. SUM stands for *Spirocomp User Manual,* SLM refers to *Spriocomp Lab Manual.*) Proceed as follows:

Materials:

computer with monitor
printer
Phipps & Bird wet spirometer
Spirocomp scale arm and interface box
computer game port to transducer cable
Spirocomp program diskette
Spirocomp User Manual
Spirocomp Lab Manual
blank diskette for saving data
valve assembly (Spirocomp)
4 reusable mouthpieces
noseclip
stopwatch or watch with second hand
Lysol disinfectant (or Clorox)
beaker of 70% alcohol
container of mild soap solution large enough to immerse one-way valve assembly

Equipment Hookup

The Spirocomp scale arm and interface box will already be attached to the spirometer post, so all you have to do is hook up the components, fill the spirometer, and calibrate the Spirocomp prior to starting any tests. Proceed as follows:

1. Fill the spirometer to a little over the fill line with fresh tap water, and add Lysol or Clorox to the water for disinfection.
2. Plug in the computer and printer power cords to *grounded* 110-volt electrical outlets.
3. *With the computer turned off,* connect the transducer to computer cable between the interface box and the game port of the computer.
4. Insert the program diskette and turn on the computer. Strike any key and the program will take over. If you are using Apple IIe, IIc, or IIGS, *be sure that the caps lock key is depressed.*
5. After you have described the system configuration by answering the appropriate questions (SUM, p. 9), you will be faced with the Main Menu which looks like this:

MAIN MENU

( ) Run Spirocomp
( ) Data Review (Graph)
( ) Data Review (Spreadsheet) [Edit]
( ) Save Current Data
( ) Load Old Data File
( ) Disk Commands

No Data in Memory

Command Keys: ↑(I) Moves cursor up
↓(M) Moves cursor down
RETURN Selects

6. Place all four mouthpieces into a beaker of 70% alcohol.
7. Select Run Spirocomp in the Main Menu, which leads you automatically into calibration.
8. **Calibrate** the Spirocomp transducer (SUM, p. 26) with the following steps:
   a. Grasp the chain of the pulley and lift it up so that the weight side, not the bell float side, is raised.
   b. While holding up the chain, turn the knob on the transducer until the computer reads 0 (±5).
   c. Lay the chain back down over the pulley, and align the plastic pointer (with rubber pointer guide right behind it) to 0 on the printed scale. Check the computer to make sure it still reads 0 (+5). Once zeroed in, press any key.

Note: If the chain slips at any time during data acquisition, be sure to recalibrate the transducer before continuing.

Experimental Procedure

Now that all components are active, proceed with the experiments under the Run Spriocomp menu. Figure 59.1 illustrates the various respiration capacities that will be measured. (Reminder: SUM = *Spirocomp User Manual;* SLM = *Spirocomp Lab Manual.*)

Tidal Volume

Note in figure 59.1 that the *tidal volume* is the amount of air that moves in and out of the lungs during a normal respiratory cycle. Although sex, age, and weight determine this value, the average normal tidal volume is around 500 ml. To get reliable results it is important that the subject be rested, not excited, and breathing in normal fashion. Proceed as follows:

1. Seat the subject in a straight high-backed chair and allow him or her to rest for five minutes before starting the test. The subject should not be allowed to observe the computer monitor during the test.
2. Swish the valve assembly in 70% alcohol, wipe it and a mouthpiece off, and insert the mouthpiece into the valve assembly.
3. Dip the noseclip in 70% alcohol, wipe it off, and place it on the nose of the subject.
4. Enter the requested subject data. This data will be used by the software to calculate the predicted vital capacity for a subject of a given gender, age, and height.

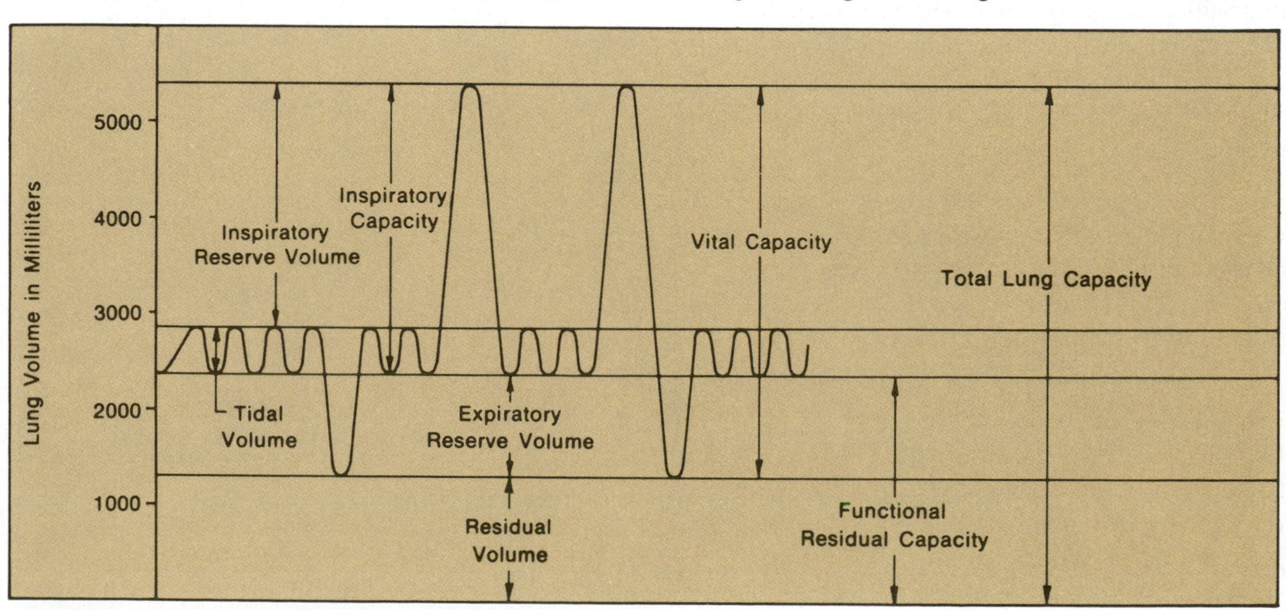

Figure 59.1 Spirogram of lung capacities.

Note: If you don't know your height in centimeters, enter it in inches, followed by an inch symbol ("). The software will automatically make the conversion. (Example: 69" will be converted to 175.)

5. Disconnect the valve assembly from the hose (SLM, p. 26), and empty the bell of the spirometer. This process can be hastened by *gently* pressing down on the bell float.
6. Have the subject breathe normally into the spirometer for one or two cycles. The valve system should cause the spirometer to fill on exhalation; fresh air from the room will be drawn in during inhalation. Incidentally, the valves will work best if they are wet.

   Note: If the valves are not working correctly, refer to pages 3 and 4 of the SUM.
7. *During an inhalation,* press the **T** key and follow the instructions on the screen (SLM, p. 27).

The software will read the changes in the spirometer for three cycles, automatically stop, and graph the average tidal volume. If you feel that you did not get good results, repeat the reading of the tidal volume before you continue.

Minute Respiratory Volume

Once you have determined your tidal volume, the next logical step is to calculate your minute respiratory volume. The *minute respiratory volume* is the amount of tidal air that passes in and out of the lungs in one minute. To determine this volume, count your respirations for one minute, using a stopwatch or any watch with a second hand, and multiply this by your tidal volume. A normal respiratory rate is usually between 12 to 15 cycles per minute. Example:

$$\text{MRV} = 12 \times 500$$
$$= 6000 \text{ ml per minute.}$$

Expiratory Reserve Volume

The amount of air that one can expire beyond the tidal volume is the *expiratory reserve volume.* It is usually around 1,100 ml. Note its position on the spirogram in figure 59.1. To measure this capacity proceed as follows:

1. Empty the spirometer bell float (step 5, above).
2. Have the subject breathe normally into the spirometer for one or two cycles.
3. *During an inhalation,* press the **E** key and follow the instructions on the screen, which says:

   BREATHE NORMAL CYCLES
4. *After two normal cycles* the software will prompt the subject to:

   STOP AFTER NORM EXHALE

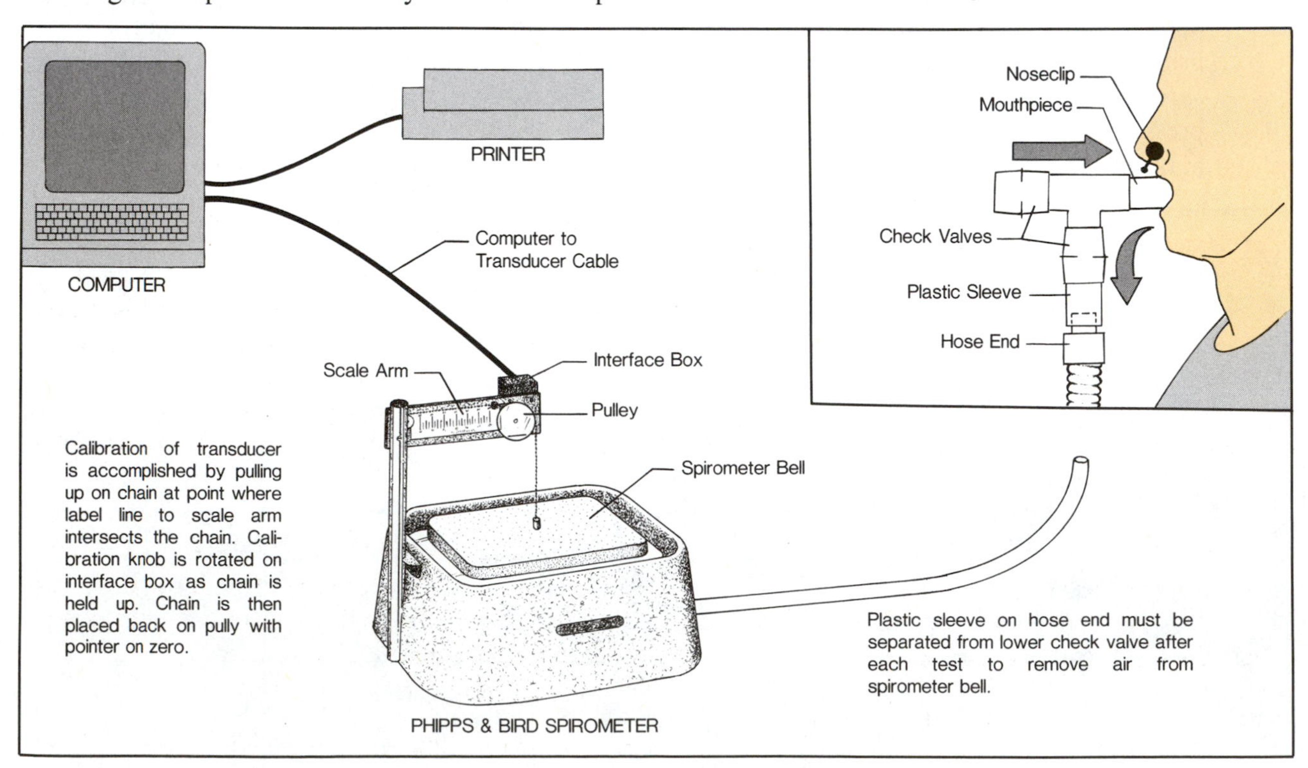

Figure 59.2 The Spirocomp setup for monitoring respiratory volumes.

Meaning: at the end of the next (third) exhale of a normal tidal cycle, stop all air movement and wait for the next prompt, which is:

EXHALE FULLY (bell will sound here)

Meaning: empty your lungs as forcefully and completely as possible. **It is also very important that you do not inhale at this point at all.**

As the subject holds his or her breath at this point for a few seconds, the software will calculate the expiratory reserve volume and graph it on the screen. If you feel that your results are not correct, repeat the reading of the expiratory reserve volume before you continue (SLM, p. 28).

Vital Capacity and FEV_T

Note in figure 59.1 that the *vital capacity* is the sum of the tidal, expiratory reserve, and inspiratory reserve volumes. This value is determined by directing an individual to take as deep a breath as possible and exhaling all the air possible. The average vital capacity for men and women is around 4,500 ml.

While measuring the vital capacity you will also be able to determine *timed vital capacity,* also referred to as the *forced expiratory volume$_T$ (FEV_T).*

If one takes a deep breath and expels it *as fast as possible,* the FEV_T of that individual is determined. An individual with no respiratory impairment should be able to expire 95% of his or her vital capacity within three seconds; however, it is the percentage of vital capacity that is expelled within the first second that is of paramount importance. **An individual with no pulmonary impairment should be able to expel 75 % of his or her vital capacity within one second.** Individuals with emphysema and asthma will have a much lower percentage due to air entrapment.

Proceed as follows to determine vital capacity and FEV_T.

1. Empty the spirometer bell float.
2. Press the **V** key and follow the instructions in the message window, which are:

   INHALE MAX THEN
   PRESS **V** EXHALE MAX

   Meaning: First inhale maximally and then simultaneously press the **V** key and exhale as forcefully, rapidly, and completely as possible into the spirometer.

   It is important that the subject keep trying to squeeze out every little bit of air until the software stops and graphs the vital capacity on the screen. If you feel that your results are not satisfactory, repeat the reading before you continue.
3. Read the vital capacity and FEV_1 on the screen (SUM, p. 23).

Evaluation and Next Subject

The subject record is now complete. The software has calculated the subject's inspiratory reserve volume [IR = VC-(TV + ERV)] and graphed it. The software has also calculated the predicted vital capacity and displayed the FEV_T.

To print out the results, press the **P** key. If you wish to add the next subject to the group record, press the **N** key and repeat the steps above for that individual.

If you are finished with the group record, press the **ESC** key. If you pressed the ESC key and you still want to add to the group record, simply select Run Spirocomp and press the **A** key for ADD to the current group data.

Compare your volumes and capacities to the following approximate norms:

	TV	ERV	IRV	VC
Males	500 ml	1,000 ml	3,000 ml	4,500 ml
Females	400 ml	750 ml	2,250 ml	3,400 ml

Compare your vital capacity to the vital capacity predicted for your age, height, and gender. Compare the computer-predicted values with tables V and VI in the Appendix.

Compare different groups to each other (smoker vs. nonsmoker, etc.). Are there differences? Explain.

Laboratory Report

Since there is no Laboratory Report for this exercise, your instructor will indicate how this experiment is to be reported.

60 Anatomy of the Digestive System

The study of the alimentary tract, liver, and pancreas will be pursued in this exercise. Although no animal will be dissected for this study, extensive assignments will be made of the tissues of organs of this system.

The Alimentary Canal

Figure 60.1 is a simplified illustration of the digestive system. The alimentary canal, which is about thirty feet in length, has been foreshortened in the intestines for clarity. The following text, which pertains to this illustration, is related to the passage of food through its full length.

Materials:

manikin

When food is taken into the oral cavity, it is chewed and mixed with saliva that is secreted by many glands of the mouth. The most prominent of these glands are the parotid, submandibular, and sublingual glands. The **parotid glands** are located in the cheeks in front of the ears, one on each side of the head. The **sublingual glands** are located under the tongue and are the most anterior ones of the three pairs. The **submandibular** glands are situated posterior to the sublinguals, just inside of the body of the mandible. (Figure 60.4 shows the position of these glands more precisely.)

After the food has been completely mixed with saliva it passes to the **stomach** by way of a long tube, the **esophagus.** The food is moved along the esophagus by wavelike constrictions called *peristaltic waves.* These constrictions originate in the **oropharynx,** which is the cavity at the top of the esophagus, posterior to the tongue. The upper opening of the stomach through which the food enters is the **cardiac valve.** The upper rounded portion, or **fundus,** of the stomach holds the bulk of the food to be digested. The lower portion, or **pyloric region,** is smaller in diameter, more active, and accomplishes most of the digestion that occurs in the stomach. That region between the fundus and the pyloric portion is the **body.** After the food has been acted upon by the various enzymes of the gastric fluid, it is forced into the small intestine through the **pyloric valve** of the stomach.

The *small intestine* is approximately 23 feet long and consists of three parts: the duodenum, jejunum, and ileum. The first 10 to 12 inches make up the **duodenum.** The **jejunum** comprises the next 7 or 8 feet, and the last coiled portion is the **ileum.** Complete digestion and absorption of food takes place in the small intestine.

Indigestible food and water pass from the ileum into the large intestine, or **colon,** through the **ileocecal valve.** This valve is shown in a cutaway section. The large intestine has four sections: the ascending, transverse, descending, and sigmoid colons. The **ascending colon** is the portion of the large intestine that the ileum empties into. At its lower end is an enlarged compartment or pouch, the **cecum,** which has a narrow tube extending down from it which is the **appendix.** The ascending colon ascends on the right side of the abdomen until it reaches the undersurface of the liver where it bends abruptly to the left, becoming the **transverse colon.** The **descending colon** passes down the left side of the abdomen where it changes direction again, becoming the **sigmoid colon.** The last portion of the alimentary canal is the **rectum,** which is about five inches long and terminates with an opening, the **anus.**

Leading downward from the inferior surface of the **liver** is the **hepatic duct.** This duct joins the **cystic duct,** which connects with the round saclike **gallbladder.** Bile, which is produced in the liver, passes down the hepatic duct and up the cystic duct to the gallbladder where it is stored until needed. When the gallbladder contracts, bile is forced down the cystic duct into the **common bile duct,** which extends from the juncture of the cystic and hepatic ducts to the intestine.

Between the duodenum and the stomach lies another gland, the **pancreas.** Its duct, the **pancreatic duct,** joins the common bile duct and empties into the duodenum.

Assignment:

Label figure 60.1. Disassemble the manikin and identify all of these structures.

Answer the questions on the Laboratory Report pertaining to this part of the exercise.

Oral Anatomy

A typical normal mouth is illustrated in figure 60.2. To provide maximum exposure of the oral structures, the lips (*labia*) have been retracted away from the teeth, and the cheeks (*buccae*) have been cut. The lips are flexible folds which meet laterally at the *angle* of the mouth where they are continuous with the cheeks. The lining of the lips, cheeks, and other oral surfaces consists of mucous membrane, the *mucosa.* Near the median line of the mouth on the inner surface of the lips, the mucosa is thickened to form folds, the **labial frenula,** or *frena.* Of the two frenula, the upper one is usually stronger. The delicate mucosa that covers the neck of each tooth is the **gingiva.**

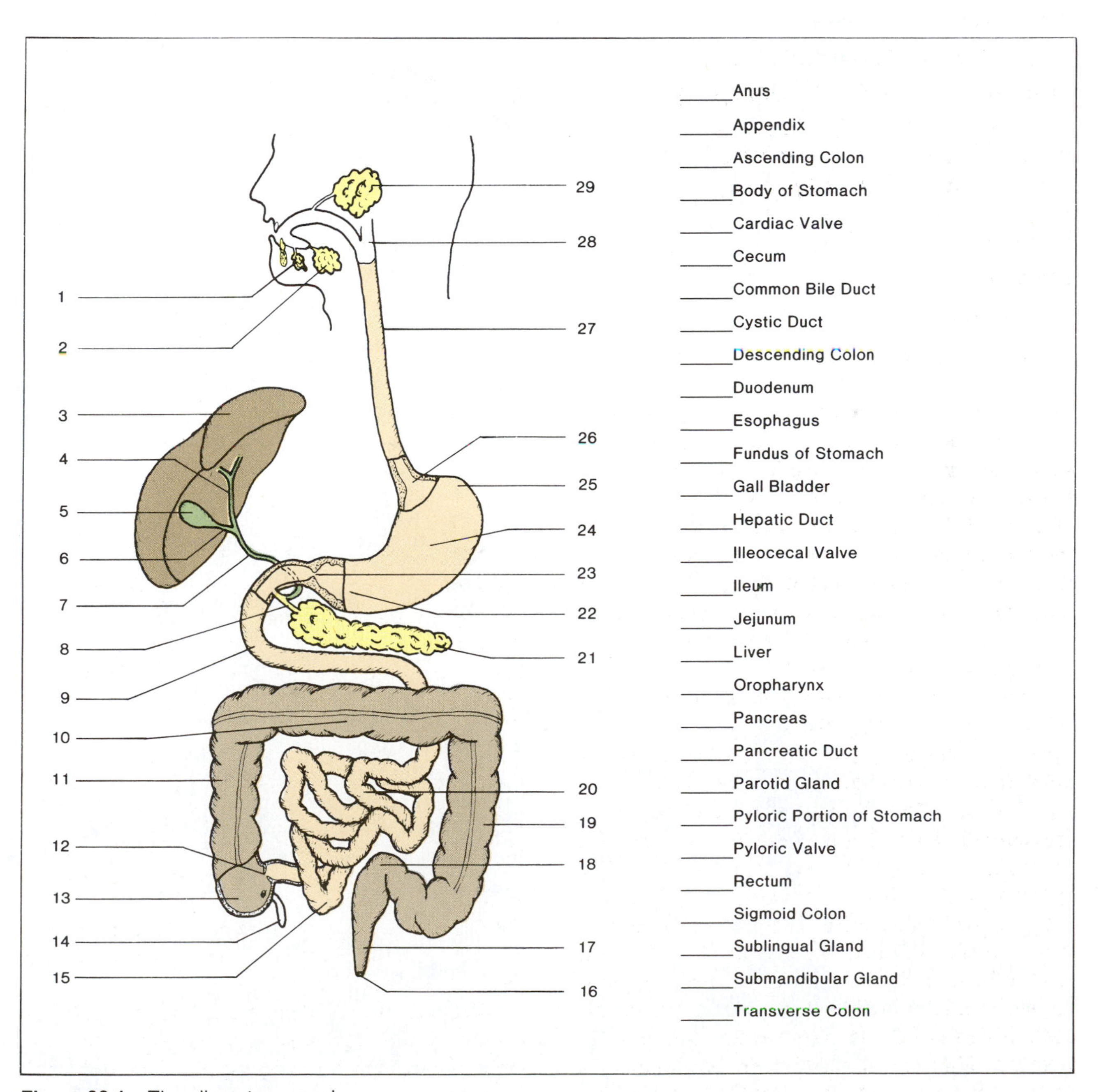

Figure 60.1 The alimentary canal.

The hard and soft palates are distinguishable as differently shaded areas on the roof of the mouth. The lighter shaded area is the **hard palate.** Posterior to it is a darker area, the **soft palate.** The **uvula** is a part of the soft palate. It is the fingerlike projection which extends downward over the back of the tongue. It varies considerably in size and shape in different individuals.

On each side of the tongue at the back of the mouth are the **palatine tonsils.** Each tonsil lies in a recess bounded anteriorly by a membrane, the **glossopalatine arch,** and posteriorly by a membrane, the **pharyngopalatine arch.** The tonsils consist of lymphoid tissue covered by epithelium. The epithelial covering of these structures dips inward into the lymphoid tissue forming glandlike pits called **tonsillar crypts** (label 11, figure 60.3). These crypts connect with channels that course through the lymphoid tissue of the tonsil. If they are infected, however, their protective function is impaired and they may actually serve as foci of infection. Inflammation of the palatine tonsils is called *tonsillitis.* Enlargement of the tonsils tends to obstruct the throat cavity and interfere with the passage of air to the lungs.

Assignment:

Label figure 60.2.

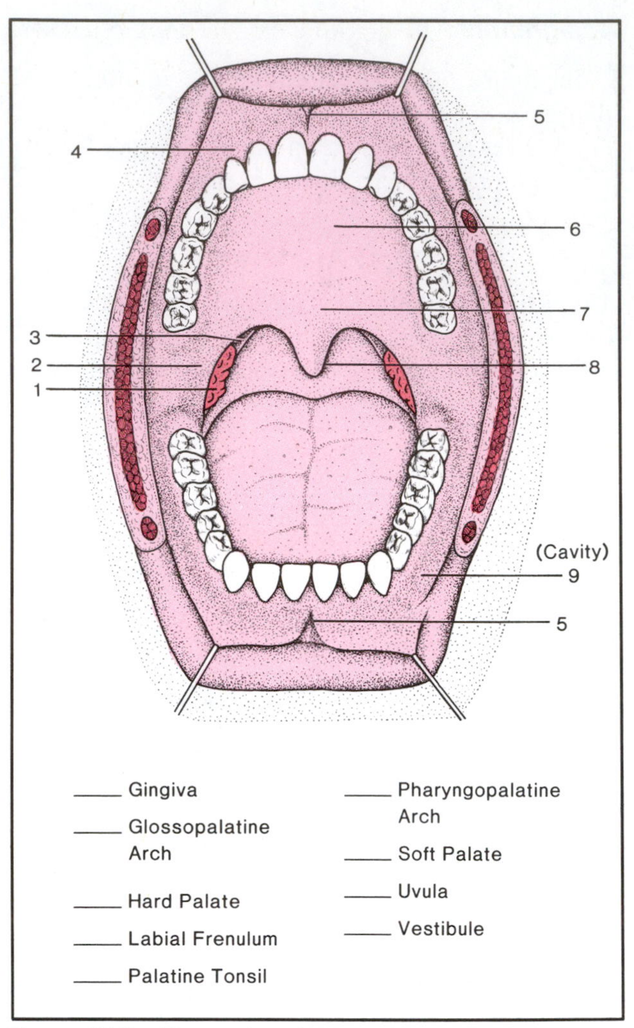

Figure 60.2 The oral cavity.

The Tongue

Figure 60.3 shows the tongue and adjacent structures. It is a mobile mass of striated muscle completely covered with mucous membrane. The tongue is subdivided into three parts: the apex, body, and root. The **apex** of the tongue is the most anterior tip which rests against the inside surfaces of the front teeth. The **body** (*corpus*) is the bulk of the tongue which extends posteriorly from the apex to the root. The body is that part of the tongue which is visible by simple inspection: i.e., without the aid of a mirror. The posterior border of the body is arbitrarily located somewhere anterior to the tonsillar material of the tongue. Extending down the median line of the body is a groove, the **central** (*median*) **sulcus.** The **root** of the tongue is the most posterior portion. Its surface is primarily covered by the **lingual tonsil.**

Extending upward from the posterior margin of the root of the tongue is the **epiglottis.** On each side of the tongue are seen the oval **palatine tonsils.**

The dorsum of the tongue is covered with several kinds of projections called *papillae.* The cutout section is an enlarged portion of its surface to reveal the anatomical differences between these papillae. The majority of the projections have tapered points and are called **filiform papillae.** These projections are sensitive to touch and give the dorsum a rough texture. This roughness provides friction for the handling of food. Scattered among the filiform papillae are the larger rounded **fungiform papillae.** Two fungiform papillae are shown in the sectioned portion. A third type of papilla is the donut-shaped **vallate** (*circumvallate*) **papilla,** which is the largest of the three kinds. They are arranged in a "V" near the posterior margin of the dorsum. Although the exact number may vary in individuals, eight vallate papillae are shown.

The fungiform and vallate papillae contain **taste buds,** the receptors for taste. The circular furrow of the vallate papilla in the cutout section reveals the presence of these receptors. Comparatively speaking, the vallate papillae contain many more taste buds than the fungiform papillae.

A fourth type of papilla is the **foliate papilla.** These projections exist as vertical rows of folds of mucosa on each side of the tongue, posteriorly. A few taste buds are also scattered among these papillae.

Assignment:

Label figure 60.3.

The Salivary Glands

The salivary glands are grouped according to size. The largest glands, of which there are three pairs, are referred to as the *major salivary glands.* They are the parotids, submandibulars, and sublinguals. The smaller glands, which average only 2–5 mm in diameter, are the *minor salivary glands.*

Major Salivary Glands

Figure 60.4 illustrates the relative positions of the three major salivary glands as seen on the left side of the face. Since all of these glands are paired, it should be kept in mind that the other side of the face has another set of these glands.

The largest gland, which lies under the skin of the cheek in front of the ear, is the **parotid gland.** Note that it lies between the skin layer and the *masseter muscle.* Leading from this gland is the **parotid** (*Stensen's*) **duct** which passes over the masseter and through the *buccinator muscle* into the oral cavity. The drainage opening of this duct usually exits near the upper second molar. The secretion of the parotid glands is a clear watery fluid which has a cleansing action in the mouth. It contains the digestive enzyme *salivary amylase,* which splits starch molecules into disaccharides (double sugar). The presence of sour (acid) substances in the mouth causes the gland to increase its secretion.

Inside the arch of the mandible lies the **submandibular** (*submaxillary*) **gland.** In figure 60.4 the mandible has been cut away to reveal two cut surfaces. Note that the lower margin of the submandibular gland extends down somewhat below the inferior border of the mandible. Also, note that a flat muscle, the *mylohyoid,* extends somewhat into

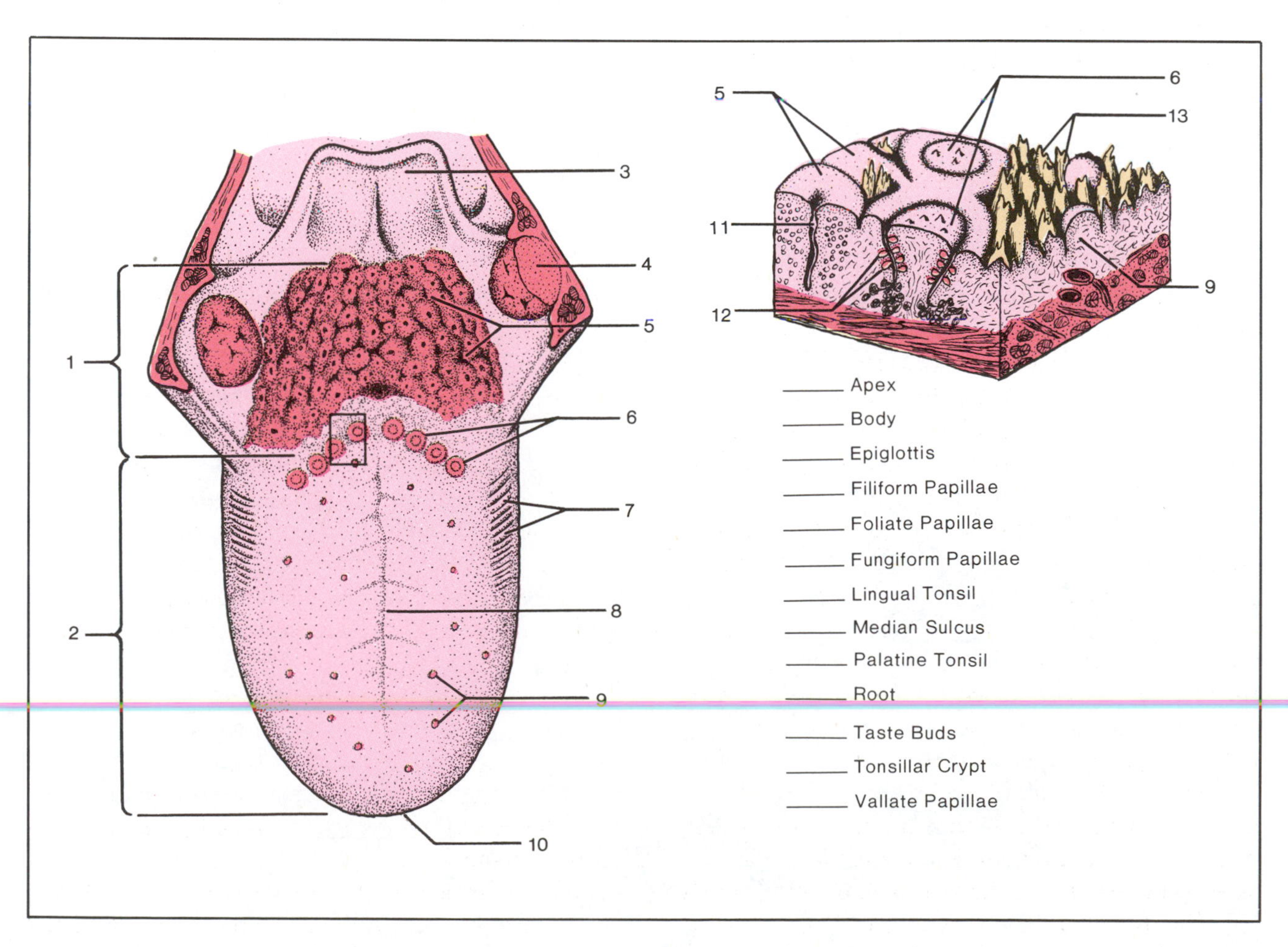

Figure 60.3 Tongue anatomy.

the gland. The secretions of this gland empty into the oral cavity through the **submandibular** (*Wharton's*) **duct.** The **opening of the submandibular duct** is located under the tongue near the lingual frenulum. The **lingual frenulum** is a mucosal fold on the median line between the tongue and the floor of the mouth. It is shown in figure 60.4 as a triangular membrane. The secretion of the submandibulars is quite similar in consistency to that of the parotid glands, except that it is slightly more viscous due to the presence of some mucin. This ingredient of saliva aids in holding the food together in a bolus. Bland substances such as bread and milk stimulate this gland.

The **sublingual gland** is the smallest of the three major salivary glands. As its name would imply, it is located under the tongue in the floor of the mouth. It is encased in a fold of mucosa, the **sublingual fold,** under the tongue. This fold is shown in figure 60.5 (label 5). Drainage of this gland is through several **lesser sublingual ducts** (*ducts of Rivinus*). The number of openings is variable in different individuals. The gland differs from the other two in that it is primarily a mucous gland. The high mucin content of its secretion lends a certain degree of ropiness to it.

Minor Salivary Glands

There are four principal groups of these smaller salivary glands in the mouth: the palatine, lingual, buccal, and labial glands. In illustration A, figure 60.5, the palatine mucosa has been removed to reveal the closely packed nature of the **palatine glands.** These glands, which are about 2–4 mm in diameter, almost completely cover the roof of the oral cavity. Where the mucosa has been removed near the lower third molar we see that the glands occur even next to the teeth. The glands in this particular area are called **retromolar glands.**

Illustration B in figure 60.5 shows the oral cavity with the tongue pointed upward and the inferior surface partially removed. The gland exposed in this area of the tongue is the **anterior lingual gland.**

The **buccal glands** cover the majority of the inner surface of the cheeks, and the **labial glands** are found under the inner surface of the lips. All of the minor glands except for the palatine glands produce salivary amylase. The palatine glands are primarily mucous glands.

Assignment:

Label figures 60.4 and 60.5.

The Teeth

Every individual develops two sets of teeth during the first twenty-one years of life. The first set, which begins to appear at approximately six months of age, are the *deciduous* teeth. Various other terms such as *primary, baby,* or *milk* teeth are also used in reference to these teeth. The second set of teeth, known as *permanent* or *secondary* teeth, begins to

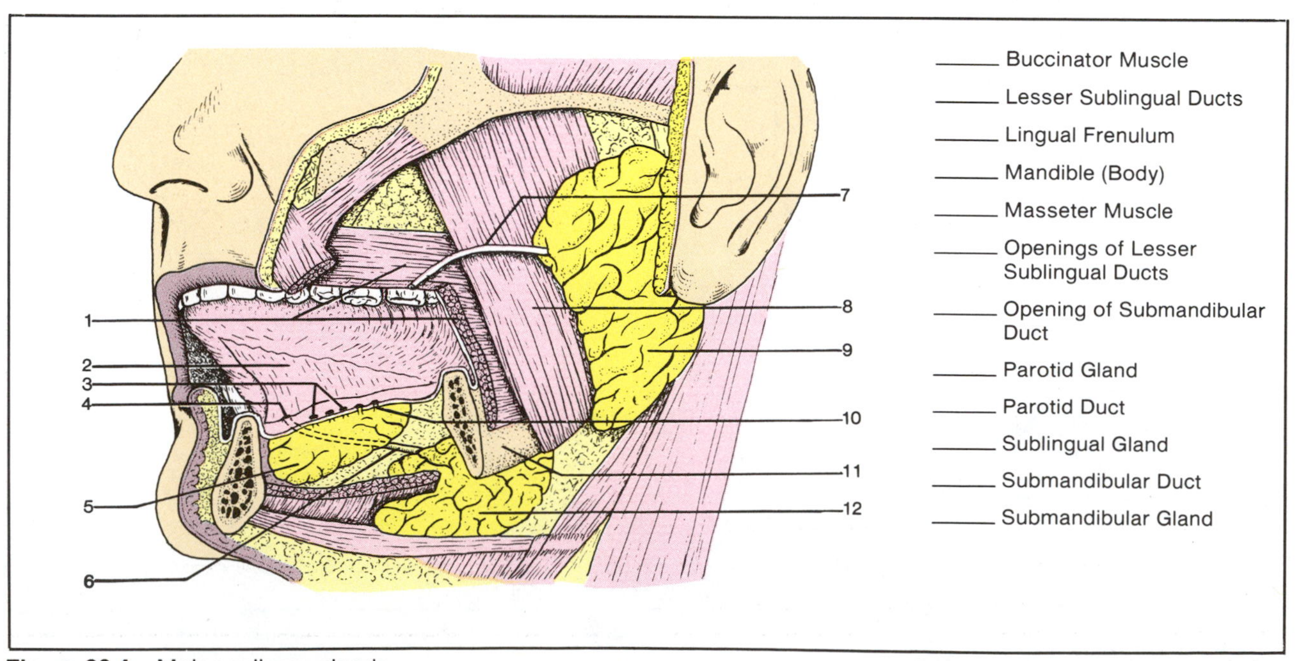

Figure 60.4 Major salivary glands.

appear when the child is about six years of age. As a result of normal growth from the sixth year on, the jaws enlarge, the secondary teeth begin to exert pressure on the primary teeth, and *exfoliation,* or shedding, of the deciduous dentition occurs.

The Deciduous Teeth

The deciduous teeth number twenty in all—five in each quadrant of the jaws. Normally, all twenty have erupted by the time the child is two years old. Starting with the first tooth at the median line, they are named as follows: **central incisor, lateral incisor, cuspid, first molar,** and **second molar.** The naming of the lower teeth follows the same sequence.

Once the two-year-old child has all of the deciduous teeth, no visible change in the teeth occurs until around the sixth year. At this time the first permanent molars begin to erupt.

The Permanent Teeth

For approximately five years (seventh to twelfth year) the child will have a *mixed dentition,* consisting of both deciduous and permanent teeth. As the submerged permanent teeth enlarge in the tissues, the roots of the deciduous teeth undergo *resorption.* This removal of the underpinnings of the deciduous teeth results eventually in exfoliation.

Figure 60.6 illustrates the permanent dentition. Note that there are sixteen teeth in one half of the mouth—thus, a total of thirty-two teeth. Naming them in sequence from the median line of the mouth they are **central incisor, lateral incisor, cuspid, first bicuspid, second bicuspid, first molar, second molar,** and **third molar.** The third molar is also called the *wisdom tooth.*

Comparisons of the teeth in figure 60.6 reveal that the anterior teeth have single roots and the posterior teeth may have several roots. Of the bicuspids, the only ones that have two roots (*bifurcated*) are the maxillary first bicuspids. Although all of the maxillary molars are shown with three roots (*trifurcated*) there may be considerable variability, particularly with respect to the third molar. The roots of the mandibular molars are generally bifurcated. The longest roots are seen on the maxillary cuspids.

Assignment:

Identify the teeth of skulls that are available in the laboratory.

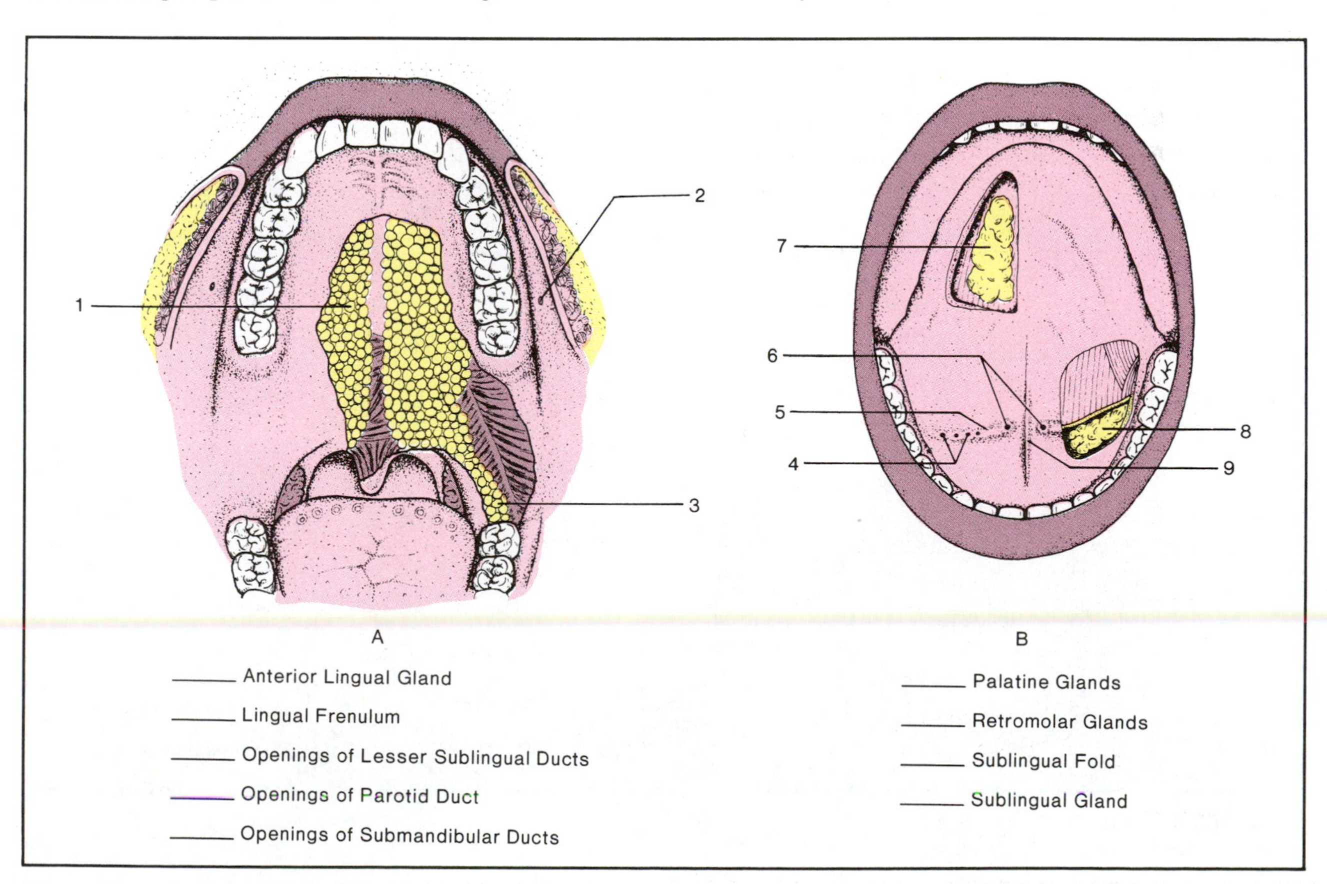

Figure 60.5 Minor salivary glands.

Tooth Anatomy

The anatomy of an individual tooth is shown in figure 60.7. Longitudinally, it is divided into two portions, the crown and root. The line where these two parts meet is the **cervical line** or *cemento-enamel juncture.*

The dentist sees the crown from two aspects: the anatomical and clinical crowns. The **anatomical crown** is the portion of the tooth that is covered with enamel. The **clinical crown,** on the other hand, is the portion of the crown that is exposed in the mouth. The structure and physical condition of the soft tissues around the neck of the tooth will determine the size of the clinical crown.

The tooth is composed of four tissues: the enamel, dentin, pulp, and cementum. **Enamel** is the most densely mineralized and hardest material in the body. Ninety-six percent of enamel is mineral. The remaining 4% is a carbohydrate-protein complex. Calcium and phosphorus make up over 50% of the chemical structure of enamel. Microscopically, this tissue is made up of very fine rods or prisms that lie approximately perpendicular to the outer surface of the crown. It has been estimated that the upper molars may have as many as twelve million of these small prisms per tooth. The hardness of enamel enables the tooth to withstand the abrasive action of one tooth against another.

Dentin is the material which makes up the bulk of the tooth and lies beneath the enamel. It is not as hard or brittle as the enamel and resembles bone in composition and hardness. This tissue is produced by a layer of **odontoblasts** (label 15) which lies in the outer margin of the pulp cavity.

Cementum is the hard dental tissue that covers the anatomical root of the tooth. It is a modified bone tissue, somewhat resembling osseous tissue. It is produced by cells called *cementocytes,* which are very similar to osteocytes. The primary function of the cementum is to provide attachment of the tooth to surrounding tissues in the alveolus.

The cavity that occupies the central portion of the tooth is called the **pulp cavity.** Extending down into the roots, this cavity becomes the **root canals.** Where this chamber extends up into the cusps of the crown, it forms the **pulpal horns.** The tissue of the pulp is essentially loose connective tissue. It consists of fibroblasts, intercellular ground substance, and white fibers. Permeating this tissue are blood vessels, lymphatic vessels, and nerve fibers. The functions of the pulp are to (1) provide nourishment for the living cells of the tooth, (2) provide some sensation to the tooth, and (3) produce dentin. The blood vessels and nerve fibers that enter the

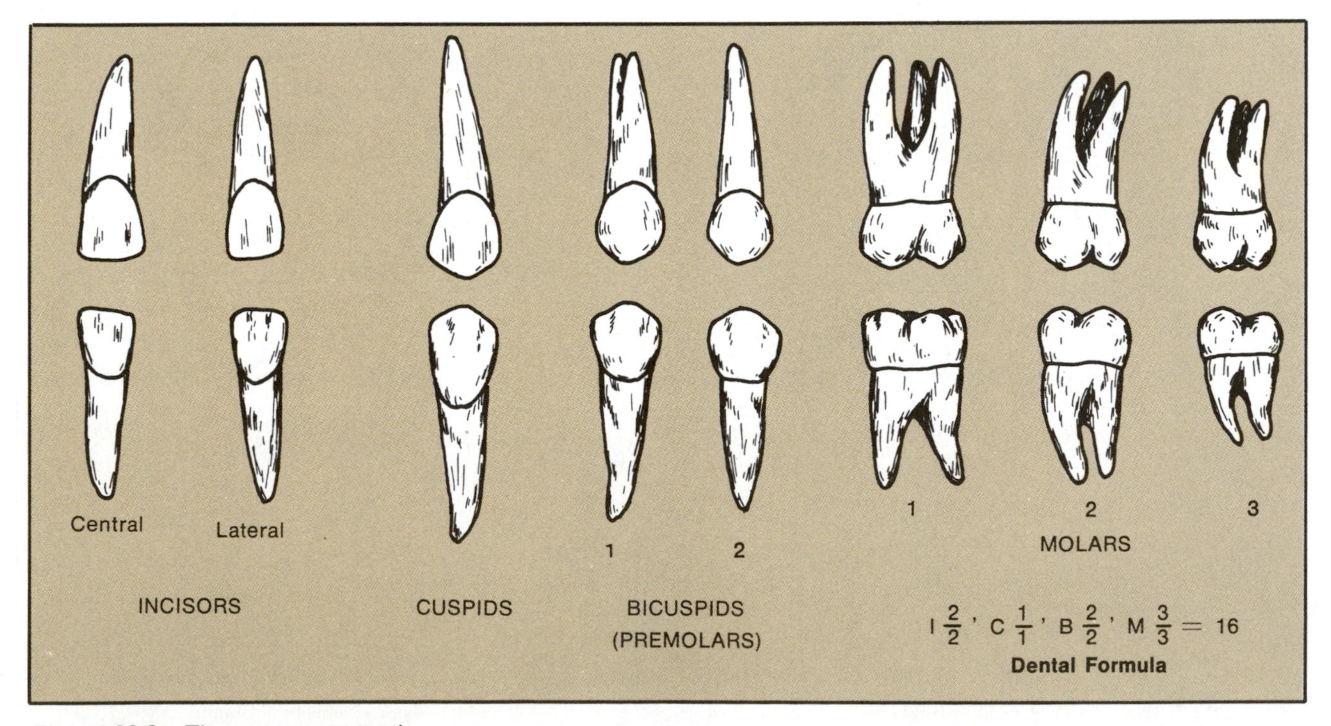

Figure 60.6 The permanent teeth.

pulp cavity do so through openings, the **apical foramina,** in the tips of the roots. Inflammation of the pulp, or *pulpitis,* may result in the destruction of the blood vessels and nerves producing a dead or *devitalized* tooth.

Between the cementum and the alveolar bone lies a vascular layer of connective tissue, the **periodontal membrane.** On the left side of figure 60.7, it is shown pulled away from the root. One of the principal functions of this membrane is tooth retention. The membrane consists of bundles of fibers that extend from the cementum to the alveolar bone, firmly holding the tooth in place. Its presence also acts as a cushion to reduce the trauma of occlusal action. Another very important function of this membrane is tooth sensitivity to touch. This sensitivity is due to the presence of free nerve ending receptors in the membrane. The slightest touch at the surface of the tooth is transmitted to these nerve endings through the medium of the periodontal membrane. Even if the apical parts of the membrane are removed, as in root tip resection, the sense of touch is not impaired.

To demonstrate the depth of the **gingival sulcus,** a probe is seen on the right side of figure 60.7. This crevice is frequently a site for bacterial putrefaction and the origin of *gingivitis.* The upper edge of the gingiva is called the **gingival margin.** The portion of the gingiva that lies over the alveolar bone is called the **alveolar mucosa.**

Assignment:

Label figure 60.7.

Histological Studies

Make a systematic histological study of the organs of the digestive system, using the Histology Atlas for reference.

Materials:

Prepared slides: taste buds (H5.120, H5.121), salivary glands (H5.131, H5.133, H5.134), esophagus (5.311), stomach (H5.415), duodenum, jejunum, and ileum (H5.58), colon (H5.613), appendix (H5.65), anal canal (H5.67), and tooth embryology (H5.1193, H5.1184)

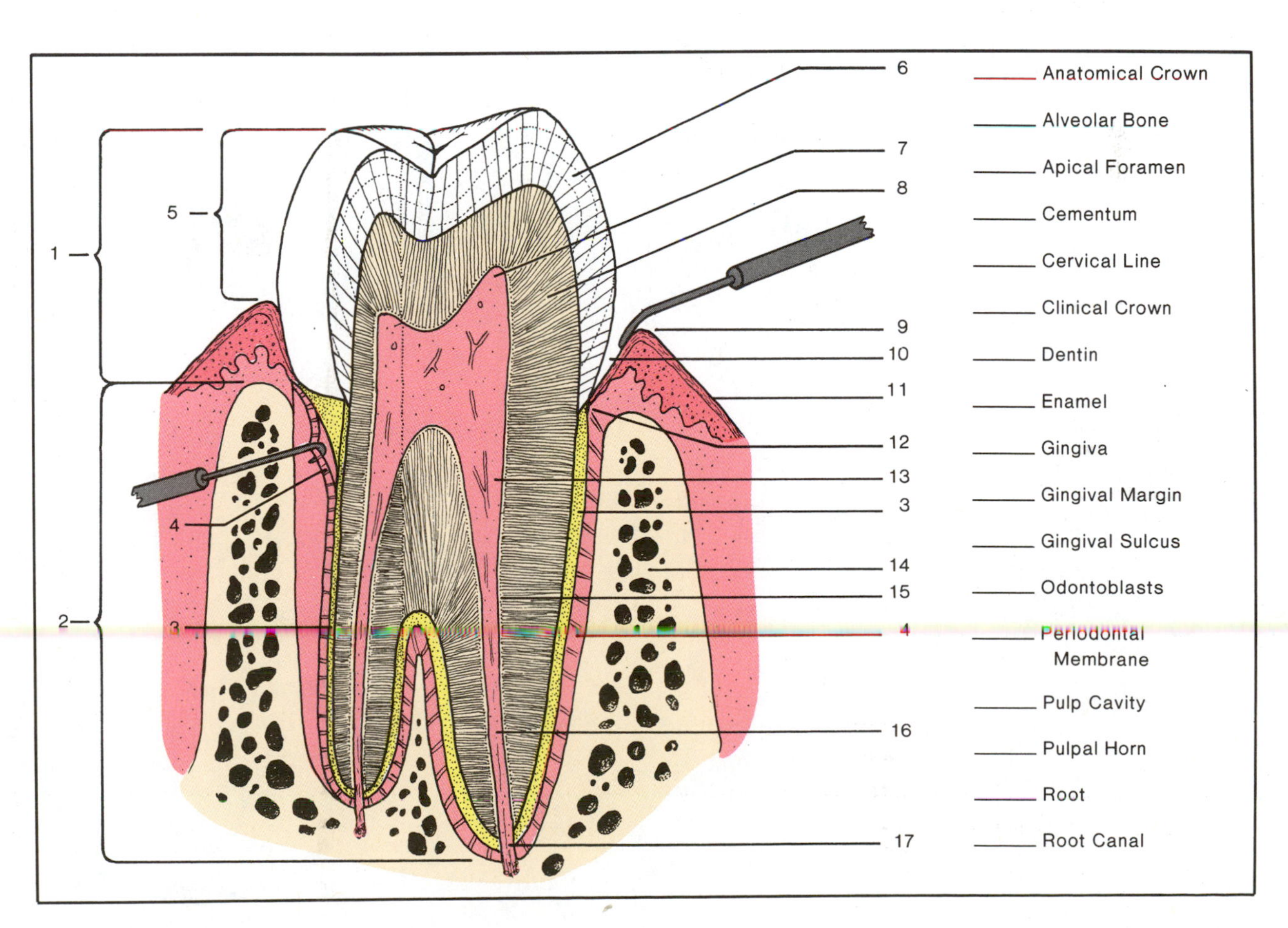

Figure 60.7 Tooth anatomy.

Salivary Glands Examine slides of the three major salivary glands, using figure HA–18 to note the histological differences between them.

Taste Buds Examine slides of vallate or foliate papillae. Identify the structures shown on figure HA–19.

Tooth Embryology Read the legend on figure HA–20 and examine slides to identify the structures shown in figure HA–20. Note in particular what cells give rise to dentin and enamel.

Esophagus Identify the four coats that make up the wall of the esophagus. Note in particular the type of epithelial tissue that lines the esophagus. Use figure HA–21 for reference.

Stomach Wall Examine a slide of the fundus of the stomach and compare it to illustrations C and D of figure HA–21. Study the deep cells within the pits and differentiate the **parietal** and **chief cells.** How do these cells differ in function? How does the mucosa of the stomach differ from the mucosa of the esophagus?

Small Intestine Study a slide that has all three sections of the small intestine. Use figures HA–22 and HA–23 for reference. Note the differences that exist between the duodenal and intestinal glands. How do these glands differ in function? What is the function of the **lacteal** in a villus? Look for lymphoidal tissue (**Peyer's patches**) in the ileum (see illustration D, Figure HA–27). This tissue makes up the gut-associated lymphoid tissue (GALT) of the digestive tract, an important part of the immune system.

Large Intestine Examine a slide of the colon and refer to illustrations A and B, figure HA–24. What is lacking on the mucosa of the colon that is seen in the small intestine? Identify all structures.

Appendix Refer to illustration C, figure HA–24 while studying a slide of the appendix. Compare its structure with the colon. What function would lymph nodules perform here?

Liver and Pancreas When examining slides of the liver and pancreas, read the legend on figure HA–25 which explains the significance of the structures labeled in the four illustrations.

Gallbladder Refer to illustrations A and B, figure HA–26, while studying a slide of the gallbladder. How does the mucosa of the gallbladder resemble the small intestine? How does it differ?

Anal Canal Determine what kind of tissue makes up the mucosa in this region. Consult illustration D, figure HA–24.

Laboratory Report

Complete the Laboratory Report for this exercise.

The Chemistry of Hydrolysis 61

The basic nutrients of the body consist of small quantities of vitamins and minerals plus large amounts of carbohydrates, fats, and proteins. While vitamins and minerals play a vital role in all physiological activities, it is the carbohydrates, fats, and proteins that provide the energy and raw materials for growth and tissue repair.

Food, as it is normally ingested, consists of large complex molecules of carbohydrates, fats, and proteins that cannot pass through the intestinal wall unless they are converted to smaller molecules by digestion in the stomach and intestines. The conversion of these large molecules to smaller ones is accomplished by digestive enzymes. All digestive enzymes split these molecules by hydrolysis.

Hydrolysis is a process whereby the larger molecules are split into smaller units by combining with water. Although hydrolysis without enzymes will occur automatically at body temperature, the process is extremely slow; the catalytic action of digestive enzymes, on the other hand, greatly hastens the process. The end result of this action is to reduce carbohydrates to monosaccharides, fats to fatty acids and glycerol, and proteins to amino acids. The large size of some food molecules requires that different enzymes work at different levels to produce intermediate size molecules before the end products are produced.

Digestive enzymes are produced by the salivary glands, the stomach wall, the intestinal wall, and the pancreas. Since the pancreas produces all three of the basic types of digestive enzymes (proteases, lipases, and carbohydrases), we will focus our attention here on this gland.

An enzyme extract from the pancreas of a freshly killed rat will be analyzed for the presence of the three types. Figure 61.1 reveals the steps that will be employed for extracting the pancreatic juice. Since enzymes are relatively unstable and easily denatured by certain adverse environmental conditions, it will be essential that chemical purity, cleanliness, and temperature control be maintained.

Once the pancreatic juice has been extracted, tests will be made for evidence of carbohydrase, lipase, and protease activity. Small amounts of the pancreatic juice will be added to substrates of starch, fat, and protein to observe the hydrolytic action of the various enzymes. Color changes that occur in each test will indicate the breaking of bonds within the large molecules to produce smaller molecules. Positive and negative test controls will be used for comparisons with the actual tests.

The class will be divided up into groups of four students. While two of the students are dissecting the rat, the other two will set up the necessary supplies and equipment for the extraction and assay procedures.

Extraction Procedure

Remove the pancreas from a freshly killed rat and produce the pancreatic extract using the following procedures.

Materials:

- freshly killed rat
- refrigerated centrifuge
- centrifuge tubes
- balance
- Sorvall blender
- beaker, 30 ml size
- Erlenmeyer flask, 125 ml size
- graduated cylinder, 100 ml size
- crushed ice
- cold mammalian Ringer's solution
- dissecting pan (wax bottom)
- dissecting instruments
- dissecting pins
- Parafilm

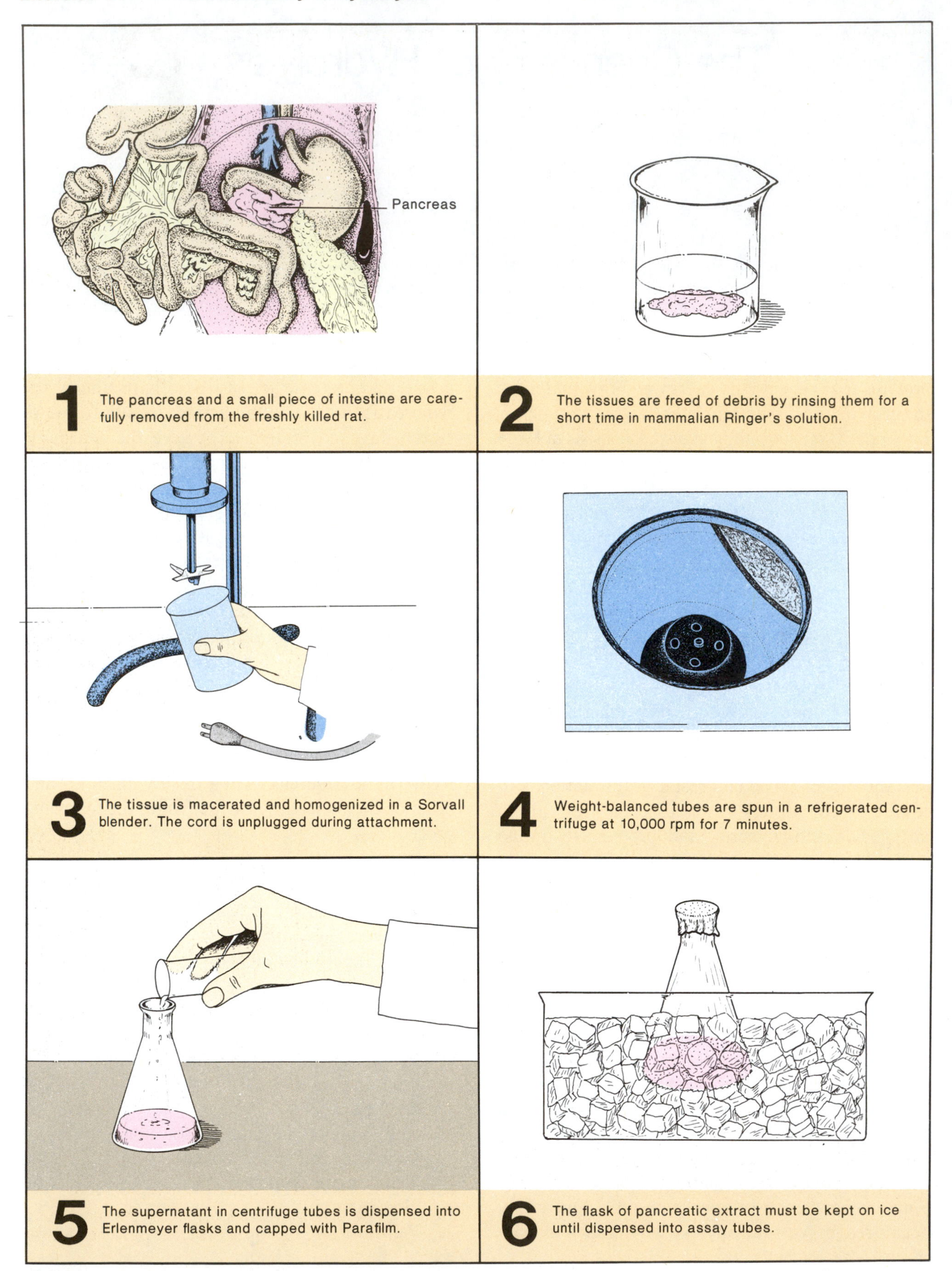

Figure 61.1 Procedure for extracting pancreatic enzymes.

1. Follow the steps outlined in figures 61.2 through 61.5 to open up the abdominal cavity of a freshly killed rat and remove the pancreas. Place the gland into a beaker containing a small amount (about 20 ml) of cold mammalian Ringer's solution. Snip off a small piece of the intestine and add that, too.
2. After washing the tissues in the Ringer's solution, macerate and homogenize them in a blender. The tissues of several groups should be blended together to get ample bulk.
3. Distribute the blended mixture into evenly balanced centrifuge tubes and spin the tubes in a refrigerated centrifuge (Beckman or other) for 7 minutes at 10,000 rpm. Tubes must be balanced by weighing carefully and adjusting the contents until they are equal.
4. Decant supernatant from the tubes into Erlenmeyer flasks for each group. The supernatant will be a milky pink mixture of digestive enzymes.
5. Place the flask of pancreatic extract into a container of ice to prevent enzyme deterioration.

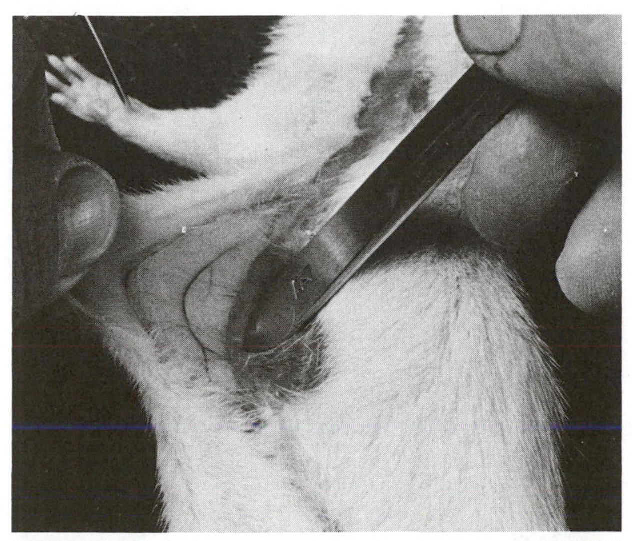

Figure 61.2 After pinning down the feet and cutting through the skin, separate the skin from the musculature with the scalpel handle.

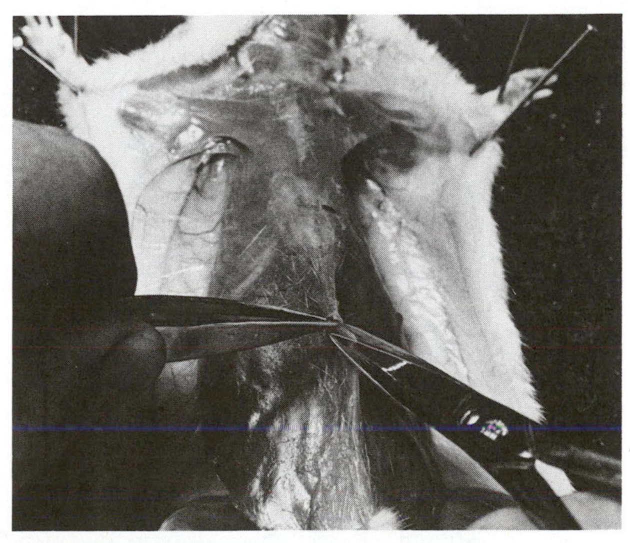

Figure 61.3 Hold the musculature up with the forceps while cutting through the wall. Be careful not to damage any organs.

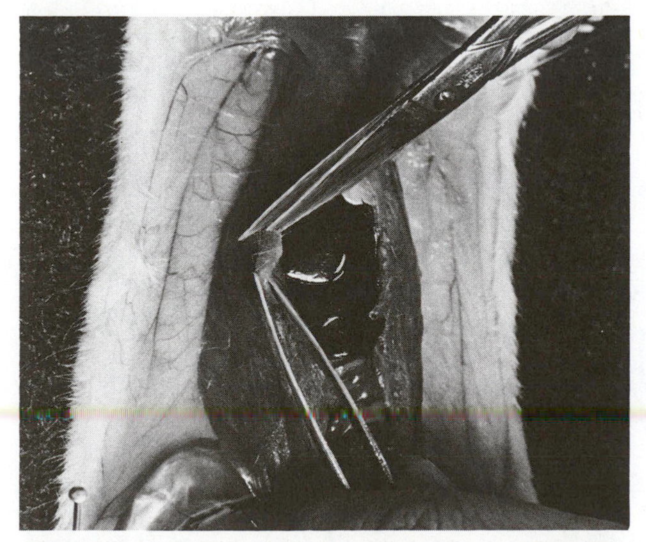

Figure 61.4 As the muscular body wall is opened, pull flaps of muscle wall laterally and pin them down to keep the cavity open.

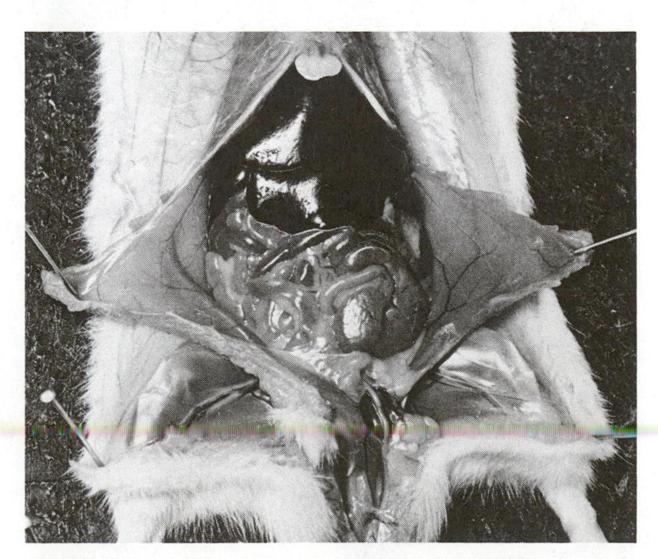

Figure 61.5 When the abdominal organs have been exposed, carefully expose the pancreas by lifting the liver with forceps.

Enzyme Assay

Figure 61.6 illustrates the number of test tubes that will be set up to assay the pancreatic extract for carbohydrase, protease, and lipase. The detection of the presence of each enzyme will depend on a specific color test after the tubes have been incubated in a 38° C water bath for one hour. Tube 1 in each series contains pancreatic extract, a substrate, and necessary buffering agents. The last tube in each series is a **positive test control** that contains the enzyme we are testing for in tube 1. All other tubes in each series should give negative results because of certain deficiencies. Note that tube 2 in each series contains the same ingredients as tube 1 except for two significant variables: (1) the tube is placed on ice while the others are incubated and (2) the pancreatic extract is not added until after one hour on ice. This tube is designated as our **zero time control.**

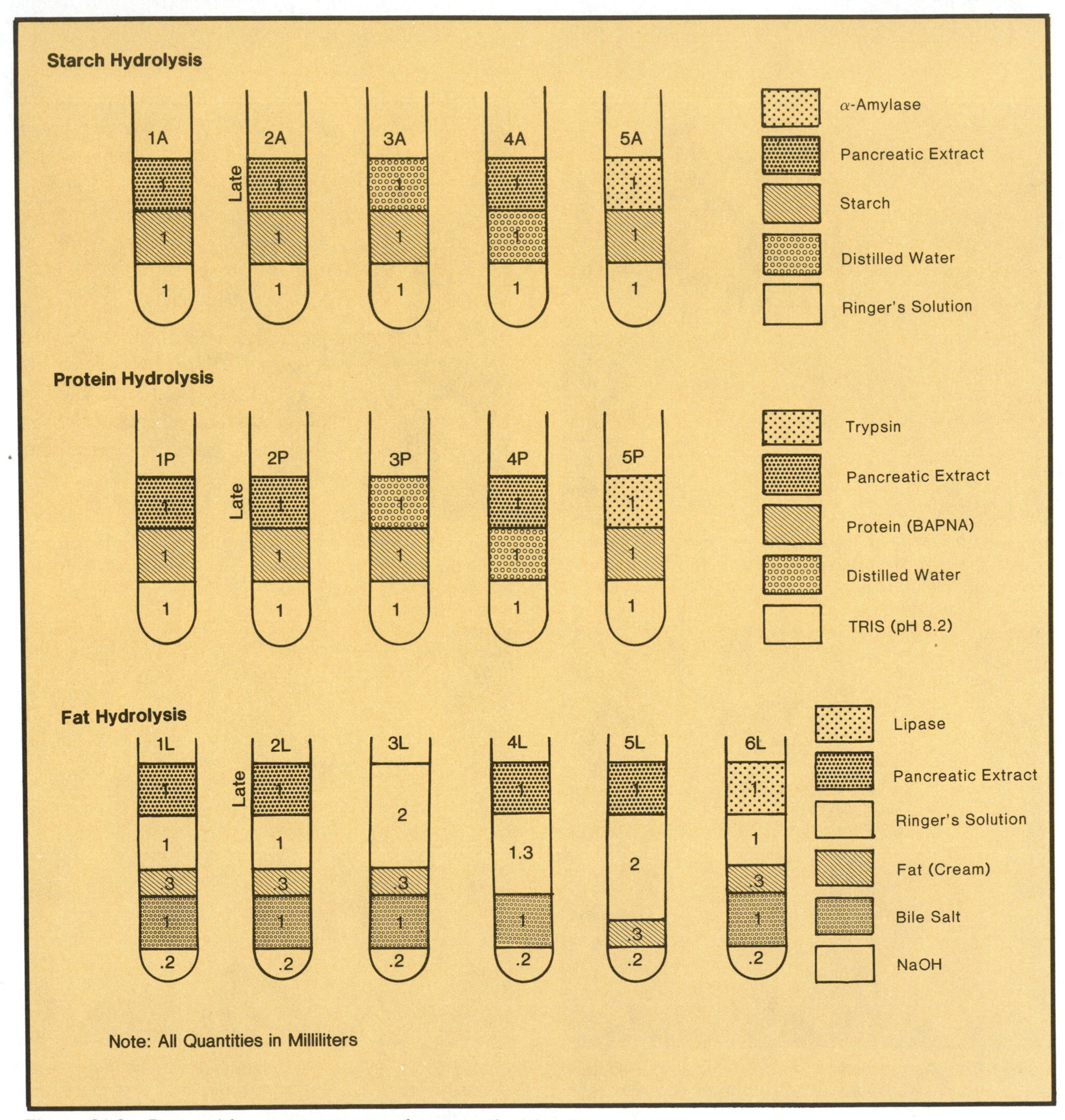

Figure 61.6 Protocol for enzyme assay of pancreatic extract.

All four members of each team will work cooperatively to set up these 16 tubes. After the tubes have been in the water baths for one hour, 1 ml of pancreatic juice will be added to the three zero time tubes. Various tests will then be performed to detect the presence of hydrolysis.

Materials:

water baths (38° C)
Vortex mixer
pipetting devices
test tube rack (wire type)
test tubes (16 mm × 150 mm), 16 per group
hot plates
pipettes, 1 ml, 5 ml, and 10 ml sizes
graduated cylinder (10 ml size)
bromthymol blue pH color standards
china marking pencil
photocolorimeter (Spectronic 20)
spot plates

solutions in dropping bottles

IKI solution	2% sodium taurocholate (a bile salt)
Benedict's solution	
Barfoed's solution	soluble starch, 0.1%
bromthymol blue	cream, unhomogenized
amylase, 1%	albumin, 0.1%
trypsin, 1%	NaCl, 0.2%
lipase, 1%	NaOH, 0.1N
TRIS (pH 8.2)	BAPNA solution

Carbohydrase

Carbohydrate in plants is stored, primarily, in the form of starch. Starch is composed of 98% amylose and 2% amylopectin. Amylose is a straight chain polysaccharide of glucose units attached by **α-1,4-glucosidic bonds.** Amylopectin is a branching polysaccharide with side chains attached to the main chain by **α-1,6-glucosidic bonds.** These linkages are illustrated in figure 61.7.

The digestion of starch is catalyzed by *α-amylase,* an enzyme present in saliva, pancreatic juice, and intestinal juice. This enzyme randomly attacks the α-1,4 bonds. An *α-1,6-glucosidase* is necessary to break the α-1,6 bond. The action of α-amylase on starch results in the formation of molecules of maltose and isomaltose. These two molecules are converted to glucose by the enzymes *maltase* and *isomaltase* in the epithelial cells of the small intestine.

Set up the five tube test series for α-amylase according to the following protocol:

1. With a china marking pencil, label the test tubes 1A through 5A (*A* for amylase).
2. With a 5 ml or 10 ml pipette, deliver 1 ml of **mammalian Ringer's solution** to each of the five tubes. Use a mechanical pipetting device such as the one illustrated in figure 61.10.
3. Flush out the pipette with water and deliver 1 ml of **distilled water** to tubes 3 and 4.

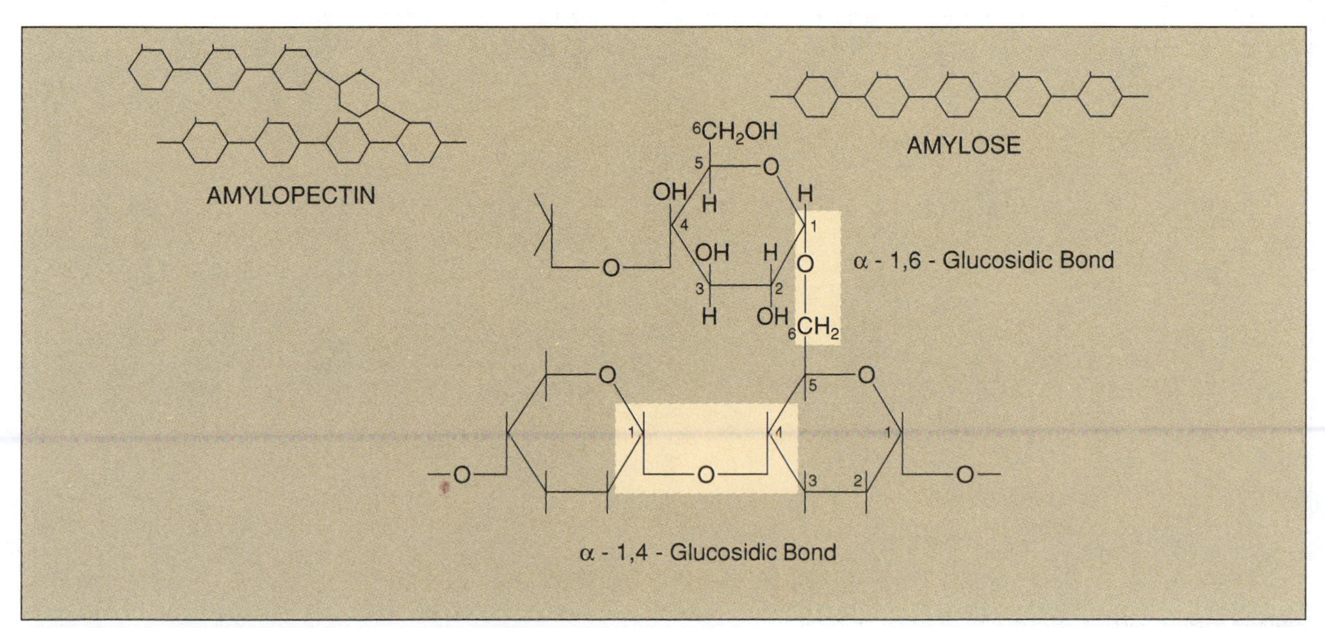

Figure 61.7 Carbohydrate linkage where hydrolysis occurs.

4. Using the same pipette, deliver 1 ml of **starch solution** to tubes 1, 2, 3, and 5. Note that tube 4 does not get any starch.
5. With a fresh 5 ml pipette, deliver 1 ml of **pancreatic juice** to tubes 1 and 4.
6. With a fresh 1 ml pipette, deliver 1 ml of **α-amylase** to tube 5.
7. Mix each tube on a Vortex mixer for a few seconds (see figure 61.11).
8. Remove tube 2 from the series, cover it with Parafilm, and place it in a rack that is immersed in ice water.
9. Cover the other four tubes with Parafilm and place the test tube rack with these tubes in a 38° C water bath for **one hour.**

Protease

The dietary proteins are derived, primarily, from meats and vegetables. These proteins consist of long chains of amino acids that are connected to each other by **peptide linkages.** Note in figure 61.8 that when a peptide linkage is hydrolyzed, the carbon of one amino acid takes on OH^- to form a carboxyl group (COOH), and H^+ is added to the other amino acid to form an amine (NH_2) group.

Protein digestion begins in the stomach with the action of *pepsin* and is completed in the small intestine. Pancreatic juice contains *trypsin, chymotrypsin,* and *carboxypolypeptidase,* which convert proteoses, peptones, and polypeptides to polypeptides and amino acids. *Amino polypeptidase* and *dipeptidase* within the epithelial cells of the small intestine hydrolyze the final peptide linkages as small polypeptides and dipeptides pass through the intestinal mucosa into the portal blood.

Since there are considerable physical differences between the linkages of different amino acids, there need be, and are, a multiplicity of proteolytic enzymes; no single enzyme can possibly digest protein all the way to its amino acids.

In our assay of pancreatic juice, we will use a synthetic peptide substrate that is designated as BAPNA (benzoyl D-L p-arginine nitroaniline). If our pancreatic extract contains trypsin, this protein will be hydrolyzed to release a yellow aniline dye, giving visual evidence of hydrolysis. Trypsin will be used in our positive test control.

1. With a china marking pencil, label five test tubes 1P through 5P (*P* for protease).
2. With a 5 ml or 10 ml pipette, deliver 1 ml of **pH 8.2 buffered solution (TRIS)** to each of the five tubes.
3. Flush out the pipette with water and deliver 1 ml of **distilled water** to tubes 3 and 4.
4. Using the same pipette, deliver 1 ml of **BAPNA solution** to tubes 1, 2, 3, and 5. Tube 4, which lacks this protein, will be a negative test control.
5. With a fresh 5 ml pipette, deliver 1 ml of **pancreatic extract** to tubes 1 and 4.

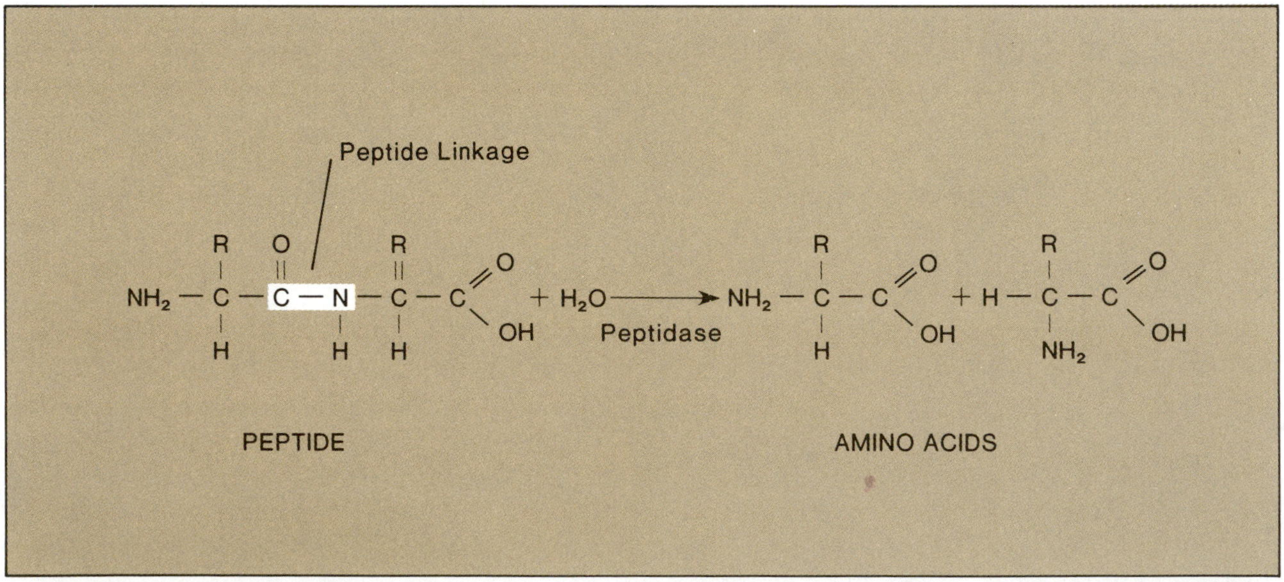

Figure 61.8 Protein hydrolysis.

6. With a fresh 1 ml pipette, deliver 1 ml of **trypsin** to tube 5.
7. Mix each tube on a Vortex mixer for a few seconds (see figure 61.11).
8. Remove tube 2 from the series, cover it with Parafilm, and place it in a rack that is immersed in ice water.
9. Cover the other four tubes with Parafilm and place the test tube rack with these tubes in a 38° C water bath for **one hour.**

Lipase

The hydrolysis of fats to glycerol and fatty acids is catalyzed by *lipase.* Although this enzyme is present in both gastric and pancreatic secretions, nearly all fat digestion occurs in the small intestine.

Lipase hydrolysis differs from carbohydrase and protease hydrolysis in that it requires an emulsifying agent to hasten the process. Bile salts, which are produced by the liver, act like detergents to break up large fat globules into smaller globules, greatly increasing the surface of fat particles. Since lipases are only able to work on the surface of fat globules, the increased surface area greatly enhances the speed of hydrolysis.

The most common fats of the diet are neutral fats, or **triglycerides.** As illustrated in figure 61.9 each triglyceride molecule is composed of a glycerol nucleus and three fatty acids. Although the final end products of triglyceride hydrolysis are fatty acids and glycerol, there are intermediate products of **monoglycerides** and **diglycerides.** Diagrammatically, the process looks like this:

E → F → Γ → | + Ξ

TRIGLYCERIDE DIGLYCERIDE MONOGLYCERIDE GLYCEROL FATTY ACIDS

Set up six test tubes to assay pancreatic extract for the presence of lipase.

1. With a china marking pencil, label six test tubes 1L through 6L (*L* for lipase).
2. Using a 1 ml pipette, deliver 0.2 ml of **0.1N NaOH** to each of the tubes. Use a mechanical pipetting device such as the one in figure 61.10.
3. With a 5 ml pipette, deliver 1 ml of **mammalian Ringer's solution** to tubes 1, 2, and 6; 2 ml to tubes 3 and 5; and 1.3 ml to tube 4.
4. Flush out the pipette with water and deliver 1 ml of **bile salt solution** to each of the tubes except tube 5.
5. Flush out the pipette with water again and deliver 0.3 ml of **cream** to all tubes except tube 4.
6. With a fresh 5 ml pipette, deliver 1 ml of **pancreatic juice** to tubes 1, 4, and 5.
7. Deliver 1 ml of **lipase** to tube 6, using a fresh 1 ml pipette.
8. Mix each tube on a Vortex mixer for a few seconds.
9. Add 2 drops of **bromthymol blue** to each tube. Check the color against a bromthymol blue

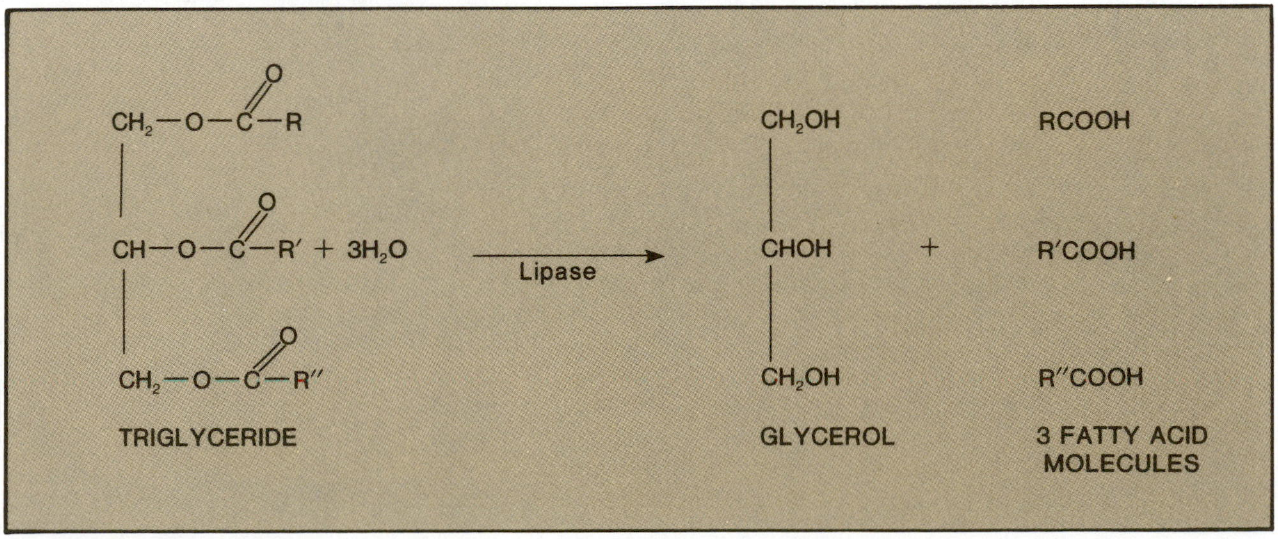

$CH_2-O-C(=O)-R$

$CH-O-C(=O)-R' + 3H_2O$ —Lipase→ CH_2OH, $CHOH$, CH_2OH + $RCOOH$, $R'COOH$, $R''COOH$

$CH_2-O-C(=O)-R''$

TRIGLYCERIDE — GLYCEROL — 3 FATTY ACID MOLECULES

Figure 61.9 Fat hydrolysis.

standard of pH 7.2. If the correct color does not develop, add NaOH, drop by drop, slowly, until the tube reaches the correct color of pH 7.2.

10. Remove tube 2 from the series, cover it with Parafilm, and place it in a rack that is immersed in ice water.
11. Cover the other five tubes with Parafilm and place the test tube rack with these tubes in a 38° C water bath for **one hour.**

Evaluation of Tests

Remove the tubes from both water baths after one hour and proceed as follows to complete the experiment:

1. To tube 2 of each series add 1 ml of pancreatic juice and mix for a few seconds on a Vortex mixer. Place each #2 tube in place within its series.
2. **Carbohydrase Test:** Remove a sample of test solution from each of the five tubes (1A through 5A) and place in separate depressions of a spot plate. Add 2 drops of IKI solution to each of the solutions on the plate. If starch is present due to lack of hydrolysis, the spot solution will be blue. If hydrolysis has occurred, test remaining solution in tube with Barfoed's and Benedict's tests to determine degree of digestion. See Appendix C for these tests. Record your results on the Laboratory Report.
3. **Protease Test:** Since hydrolysis produces a yellow color in the test tube, record the color of each tube on the Laboratory Report. If a Spectronic 20 photocolorimeter is set up and calibrated at 410 μm, take a reading for each tube and record the optical densities (OD) in the table on the Laboratory Report. Be sure to use appropriate cuvettes for the contents of each tube. (Before taking any readings, the photocolorimeter should be standardized with a blank made up of 3 ml of BAPNA and 0.1 ml of 0.001 M HCl.)
4. **Lipase Test:** Since fat hydrolysis results in a lowering of pH due to the formation of fatty acids, the bromthymol blue in the tubes will change from blue to yellow. Compare the six tubes with bromthymol blue color standard to determine the pH changes that have occurred.

Laboratory Report

Complete the first portion of combined Laboratory Report 61,62.

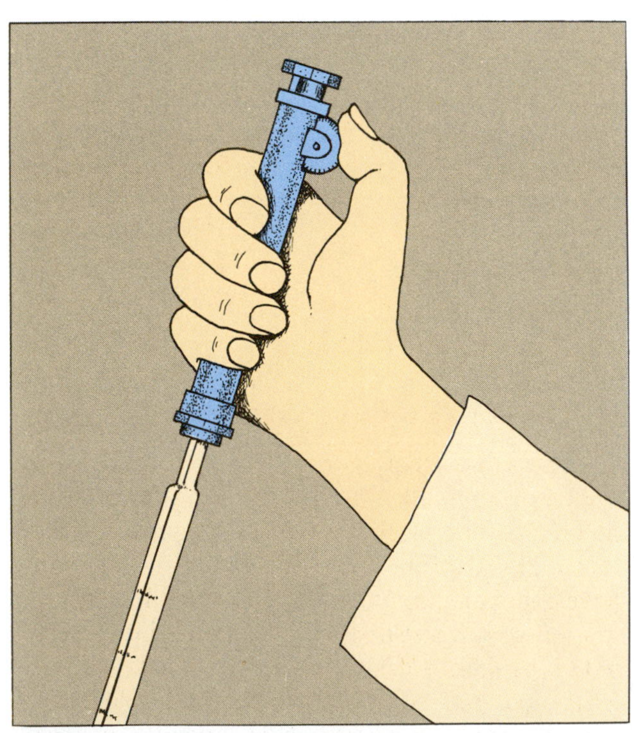

Figure 61.10 The volume of deliveries with pipetting device is controlled with the thumb.

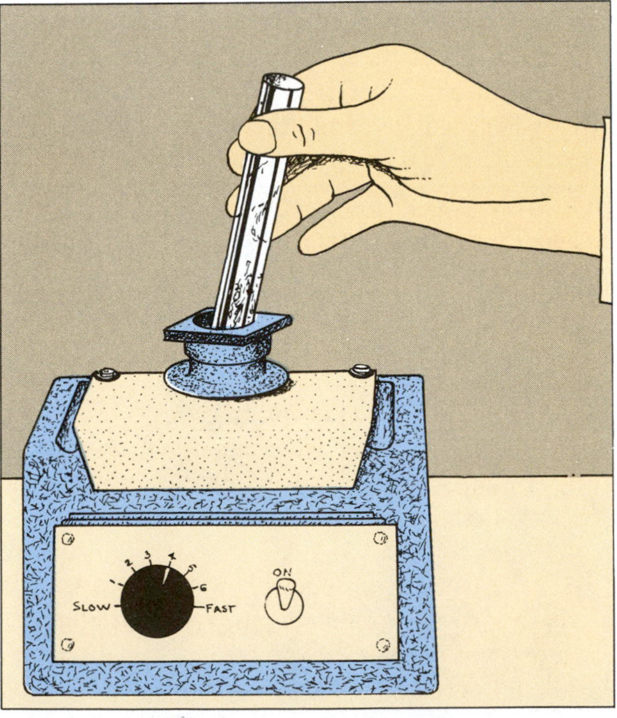

Figure 61.11 Each tube in the assay procedure must be mixed on the Vortex mixer.

Factors Affecting Hydrolysis 62

Enzymes are highly complex protein molecules of somewhat unstable nature due to the presence of weak hydrogen bonds. Many factors influence the rate at which hydrolysis occurs. Temperature and hydrogen ion concentration are probably the most important factors in this respect. It is the influence of these two conditions that will be studied in this exercise.

Salivary amylase (ptyalin) has been selected for this study. This enzyme hydrolyzes the starch amylose to produce soluble starch, maltose, dextrin, achrodextrins, and erythrodextrin. When iodine is used as an indicator of amylase activity, starch and soluble starch produce a blue color and erythrodextrin yields a red color. To detect the presence of maltose, Benedict's solution is used. With heat, maltose is a reducing sugar causing the reduction of soluble cupric to insoluble red cuprous oxide in Benedict's solution.

In this exercise, the saliva of one individual will be used. The saliva will be diluted to 10% with distilled water. To determine the effect of temperature on the rate of hydrolysis, iodine solution (IKI) will be added to a series of ten tubes (figure 62.1) containing starch and saliva. By adding the iodine at one-minute intervals to successive tubes, it is possible to determine, by color, when the hydrolysis of starch has been completed. For temperature comparisons, two water baths will be used. One will be at 20° C and the other will be at 37° C. To determine the effects of hydrogen ion concentration on the action of amylase, we will use buffered solutions of pH 5, 7, and 9.

The dispensing of solutions in this exercise may be performed with medicine droppers, graduates, or serological pipettes. If the student is unfamiliar with the proper techniques in using serological pipettes without mechanical pipetting devices, the latter should be used.

If laboratory time is limited, it might be necessary for the instructor to divide the class into thirds that will perform only a portion of the entire experiment. After each group completes its portion, the results can be tabulated on the chalkboard so that all students can record results for the complete experiment. The best arrangement is for students to work in pairs. The following assignments will work well for a two-hour laboratory period:

Group	Assignment
A (⅓ of Class)	Controls, 20° C, Boiling
B (⅓ of Class)	Controls, 37° C, Boiling
C (⅓ of Class)	Controls, pH Effects

Saliva Preparation

Prior to performing this experiment, the instructor will prepare a 10% saliva solution for the entire class as follows:

Materials:

- 50 ml graduate
- 250 ml graduated beaker
- paraffin
- 12 dropping bottles with labels

Place a small piece (¼″ cube) of paraffin under the tongue and allow it to soften for a few minutes before starting to chew it. As it is chewed, expectorate all saliva into a 50 ml graduate. When you have collected around 15 to 20 ml of saliva, fill the graduate with distilled water and pour into a graduated beaker. Add additional distilled water to make up a 10% solution. Dispense in labeled dropping bottles. For a class of 24 students, you will need 12 bottles.

Controls

A set of five control tubes are needed for color comparisons with test results. Tubes 1 and 2 will be used for detection of starch. Tubes 3, 4, and 5 will be used for the detection of the presence of maltose. Table 62.1 shows the ingredients in each tube and the significance of test results. Proceed as follows to prepare your set of controls.

Materials:

10% saliva solution
0.1% starch solution
1% maltose solution
Benedict's solution (in dropping bottle)
IKI solution (in dropping bottle)
test tubes (13 mm diam. × 100 mm long)
test tube racks (Wassermann type)
medicine droppers
250 ml beaker
electric hot plate
1 ml sterile pipettes
pipetting device

1. Label five clean test tubes *1, 2, 3, 4,* and *5.*
2. Fill the five tubes with the following reagents. If 1 ml pipettes are used, use a different pipette for each reagent.

   Tube 1: 1 ml starch solution and two drops of IKI solution.
   Tube 2: 1 ml distilled water and two drops of IKI solution.
   Tube 3: 1 ml starch solution, 2 drops saliva solution, and 5 drops Benedict's solution.
   Tube 4: 1 ml distilled water, 2 drops saliva solution, and 5 drops Benedict's solution.
   Tube 5: 1 ml maltose solution and 5 drops Benedict's solution.

3. Place tubes 3, 4, and 5 into a beaker of warm water and boil on an electric hot plate for **5 minutes** to bring about the desired color changes in tubes 3 and 5.
4. Set up five tubes in a test tube rack to be used for color comparisons. Tube 1 will be your positive starch control. Tubes 3 and 5 will be your positive sugar controls. Tubes 2 and 4 will be the negative controls for starch and sugar.

Effect of Temperature

(20° C and 37° C)

In this portion of the experiment we will determine the effect of room temperature (20° C) and body temperature (37° C) on the rate at which amylase acts. Note in figure 62.1 that tubes of starch and saliva are given two drops of IKI solution at one-minute intervals to observe how long it takes for hydrolysis to occur at a given temperature. When IKI is added to the first tube, the solution will be blue, indicating no hydrolysis. At some point along the way, however, the blue color will disappear, indicating complete hydrolysis. Note that the materials list is for performing the test at only one temperature.

Materials:

10 serological test tubes (13 mm diam. × 100 mm)
test tube rack (Wassermann type)
water bath (20° C or 37° C)
10% saliva solution
0.1% starch solution
china marking pencil
thermometer (Centigrade scale)
1 ml serological pipettes or 10 ml graduate pipetting device

Table 62.1. Control Tube Contents and Functions.

Tube Number	Carbohydrate	Saliva	Reagent	Test Results
1	Starch	–	IKI	Positive for Starch
2	–	–	IKI	Negative for Starch
3	Starch	2 Drops	Benedict's	Positive for Sugar Positive for Hydrolysis
4	–	2 Drops	Benedict's	Negative for Sugar
5	Maltose	–	Benedict's	Positive for Sugar

1. Label 10 test tubes 1 through 10 with a china marking pencil and arrange them sequentially in the front row of the test tube rack.
2. Dispense 1 ml of starch solution to each test tube, using either a 1 ml pipette or graduate. For accuracy, the pipette is preferred.
3. Place the rack of tubes and bottle of saliva in the water bath (20° C or 37° C).
4. Insert a clean thermometer into the bottle of saliva. When the temperature of the saliva has reached the temperature of the water bath (3–5 minutes), proceed to the next step.
5. Lift the bottle of saliva from the water bath, **record the time,** and put **one drop** of saliva into each test tube, starting with tube 1. Leave rack of tubes in bath. Be sure that the drop falls directly into the starch solution without touching the sides of the tube. Mix by agitation.
6. **Exactly one minute** after tube 1 has received saliva, add 2 drops of IKI solution to it and mix again. Note the color. **One minute later** add two drops of IKI to tube 2 and repeat this process every minute until all ten tubes have received IKI solution.
7. Compare the color of the tubes with the control tubes and determine the time required to digest the starch. Record your results on the Laboratory Report.
8. Wash the insides of all test tubes with soap and water and rinse thoroughly.

Effect of Boiling

In this portion of the exercise, we will compare the action of amylase that has been boiled for a few seconds with unheated amylase. Figure 62.2 illustrates the procedure.

Materials:

3 test tubes
test tube holder
Bunsen burner
10% saliva solution
0.1% starch solution
IKI solution

1. Label three test tubes *1, 2,* and *3.*
2. Dispense 1 ml of saliva solution to tubes 1 and 2, and 1 ml of distilled water to tube 3.
3. Heat tube 1 over a Bunsen flame so that it comes to a boil for a few seconds. Use a test tube holder.
4. Add 1 ml of starch solution to each of the three tubes and place them in the 37° C water bath for 5 minutes.
5. Remove the tubes from the water bath and test for the presence of starch by adding 2 drops of IKI to each tube. Record the results on the Laboratory Report.
6. Wash and rinse the test tubes.

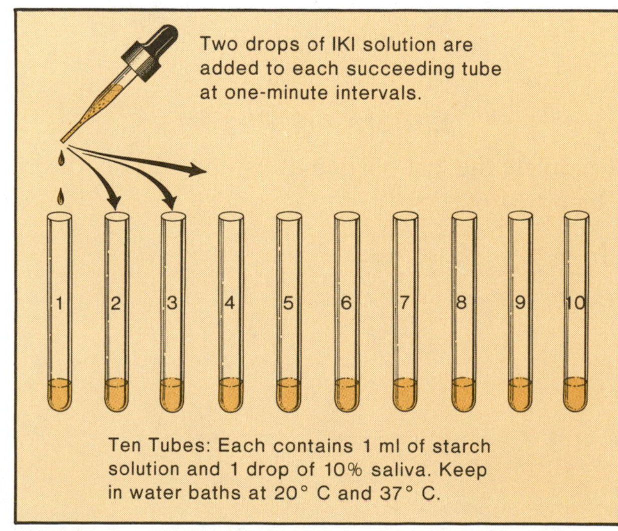

Figure 62.1 Amylase activity at 20° C and 37° C.

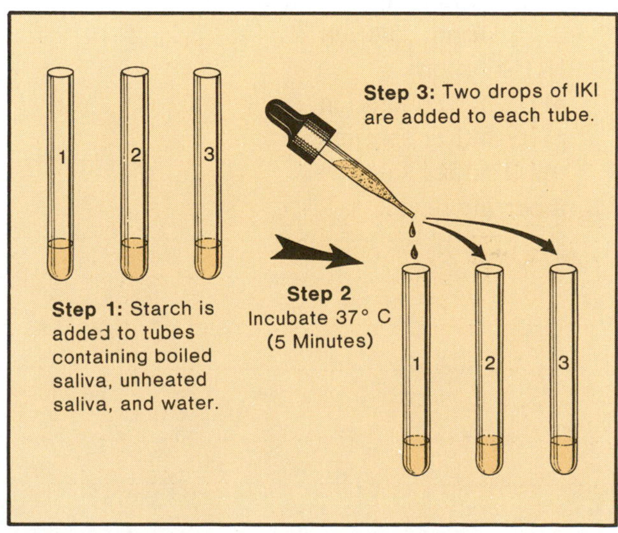

Figure 62.2 Effect of boiling on amylase.

Effect of pH on Enzyme Activity

To determine the effect of different hydrogen ion concentrations on amylase activity, the enzyme, starch, and buffered solutions will be incubated at 37° C. The general procedure is illustrated in figure 62.3.

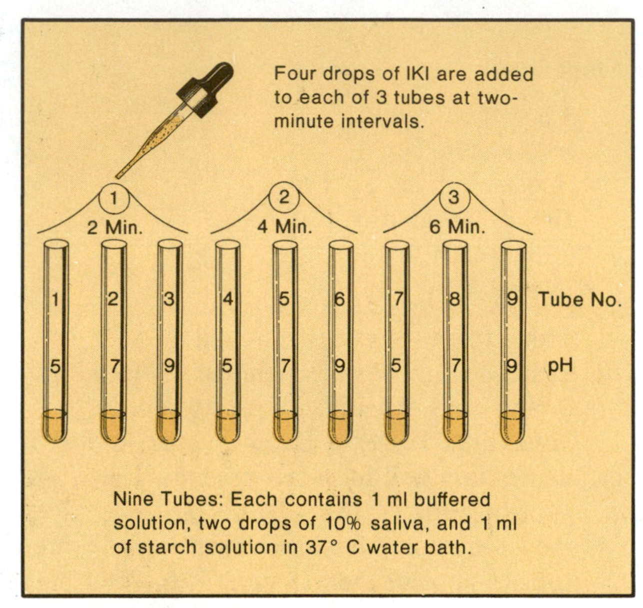

Figure 62.3 Effect of pH on amylase activity.

Materials:

- test tube rack (Wassermann type)
- 12 test tubes (13 mm diam. × 100 mm)
- 1 ml pipettes
- 10 ml graduate
- water bath (37° C)
- thermometer
- 10% saliva solution
- 0.1% starch solution
- IKI solution
- pH 5 buffer solution
- pH 7 buffer solution
- pH 9 buffer solution
- pipetting device

1. Label nine test tubes according to figure 62.3. Note that the tubes are arranged in three groups. Each group has three tubes of pH 5, 7, and 9 arranged in sequence.
2. With a 1 ml pipette, deliver 1 ml of pH 5 buffer solution to tubes 1, 4, and 7. Then, deliver 1 ml of pH 7 solution to tubes 2, 5, and 8, using a fresh pipette. With another fresh pipette, deliver 1 ml of pH 9 solution to tubes 3, 6, and 9.
3. Add two drops of saliva solution to each of the nine tubes containing buffered solutions.
4. Fill the three other empty tubes about two-thirds full of starch solution (total of approximately 15 ml). Place a clean thermometer in one of the three starch tubes and insert the whole rack of 12 tubes into the 37° C water bath.
5. When the thermometer in the starch tube reaches 37° C, pipette 1 ml of warm starch solution to each of the nine tubes. Use the starch from the 3 tubes in the rack. **Record the time** that the starch was added to tube 1.
6. **Two minutes** after tubes 1, 2, and 3 received starch, add 4 drops of IKI solution to each of these three tubes. At **four minutes** from the original recorded time, add 4 drops of IKI solution to tubes 4, 5, and 6. At **six minutes** from the original recorded time, add 4 drops of IKI solution to tubes 7, 8, and 9.
7. Remove the tubes from the water bath and compare them with the controls. Record your observations on the Laboratory Report.

Laboratory Report

Complete the last portion of combined Laboratory Report 61,62.

Anatomy of the Urinary System 63

This study consists of three parts: (1) illustration labeling, (2) sheep kidney dissection, and (3) microscopic examination of kidney tissue. As with previous exercises, the labeling should precede the laboratory activities.

Organs of the Urinary System

Figure 63.1 illustrates the components of the urinary system with its principal blood supply. It consists of two kidneys, two ureters, the urinary bladder, and a urethra. The **kidneys** are somewhat bean-shaped, of dark brown color, and located behind the peritoneum. The right kidney is usually positioned somewhat lower than the left one, probably because of its displacement by the liver. Each kidney is supplied blood through a **renal artery** that is a branch off the **abdominal aorta.** Blood leaves each kidney through a **renal vein** that empties into the **inferior vena cava.**

Urine passes from each kidney to the **urinary bladder** through a **ureter.** The upper end of each ureter is enlarged to form a funnellike **pelvis.** The lower end of each ureter enters the posterior surface of the bladder.

Leading from the urinary bladder to the exterior is a short tube, the **urethra.** In the male the urethra is about 20 centimeters long; in females it is approximately 4 centimeters in length.

The exit of urine from the bladder is called *micturition.* The passage of urine from the bladder is controlled by two sphincter muscles, the sphincter vesicae and the sphincter urethrae. The **sphincter vesicae** is a smooth muscle sphincter that is near the exit of the bladder. When approximately 300 ml of urine has accumulated in the bladder, the muscular walls of the bladder are stretched sufficiently to initiate a parasympathetic reflex that causes the bladder wall to contract. These contractions force urine past the sphincter vesicae into the urethra above the **sphincter urethrae.** This second sphincter, which is located approximately 1–3 centimeters below the sphincter vesicae, consists of skeletal muscle fibers and is voluntarily controlled. The presence of urine in the urethra above this sphincter creates the desire to micturate; however, since the valve is under voluntary control, micturition can be inhibited. When both sphincters are relaxed, urine passes from the body.

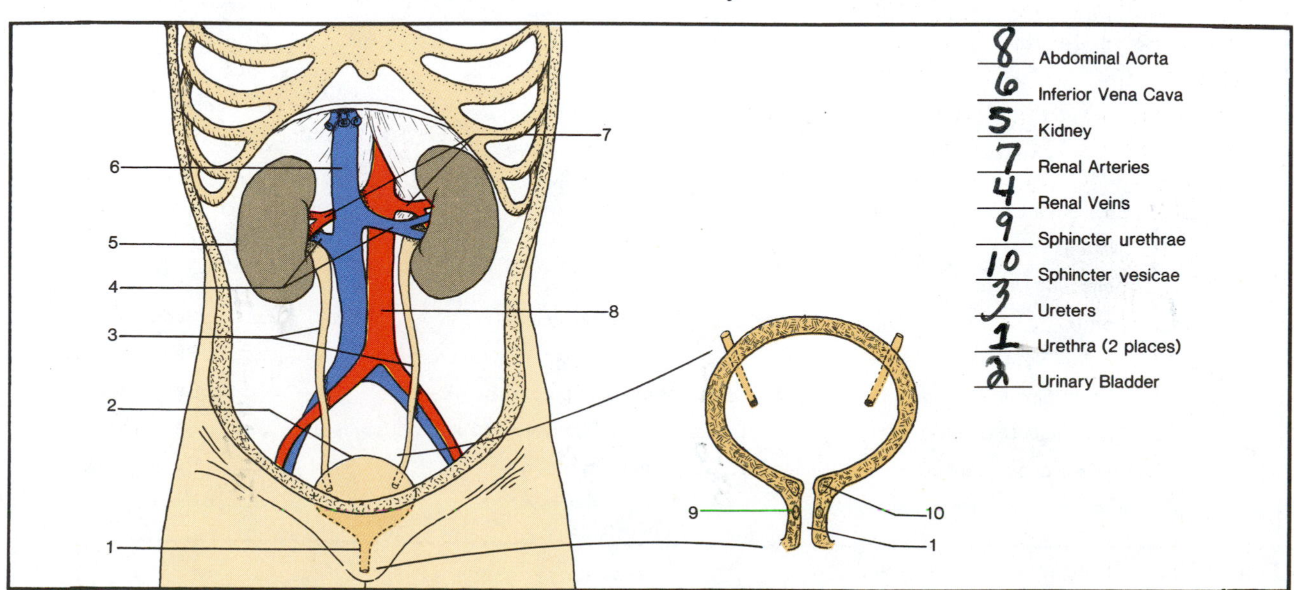

Figure 63.1 The urinary system.

Assignment:

Label figure 63.1.

Kidney Anatomy

Figure 63.2 reveals a frontal section of a human kidney. Its outer surface is covered with a thin fibrous **renal capsule** (label 8). In addition to this thin covering, the kidney is provided support and protection by a fatty capsule which completely encases it. The latter is not shown in figures 63.1 or 63.2.

Immediately under the capsule is the **cortex** of the kidney. The cortex is reddish brown due to its great blood supply. The lighter inner portion is called the **medulla.** The medulla is divided into cone-shaped **renal pyramids.** Nine of these pyramids are seen in figure 63.2. Cortical tissue, in the form of **renal columns,** extends down between the pyramids. Each renal pyramid terminates as a **renal papilla,** which projects into a calyx. The **calyces** are short tubes that receive urine from the renal papillae; they empty into the large funnellike **renal pelvis.**

Nephrons

The basic functioning unit of the kidney is the *nephron.* The enlarged section in figure 63.2 reveals two nephrons. Figure 63.3 illustrates a single nephron in greater detail. It has been estimated that there are around one million nephrons in each

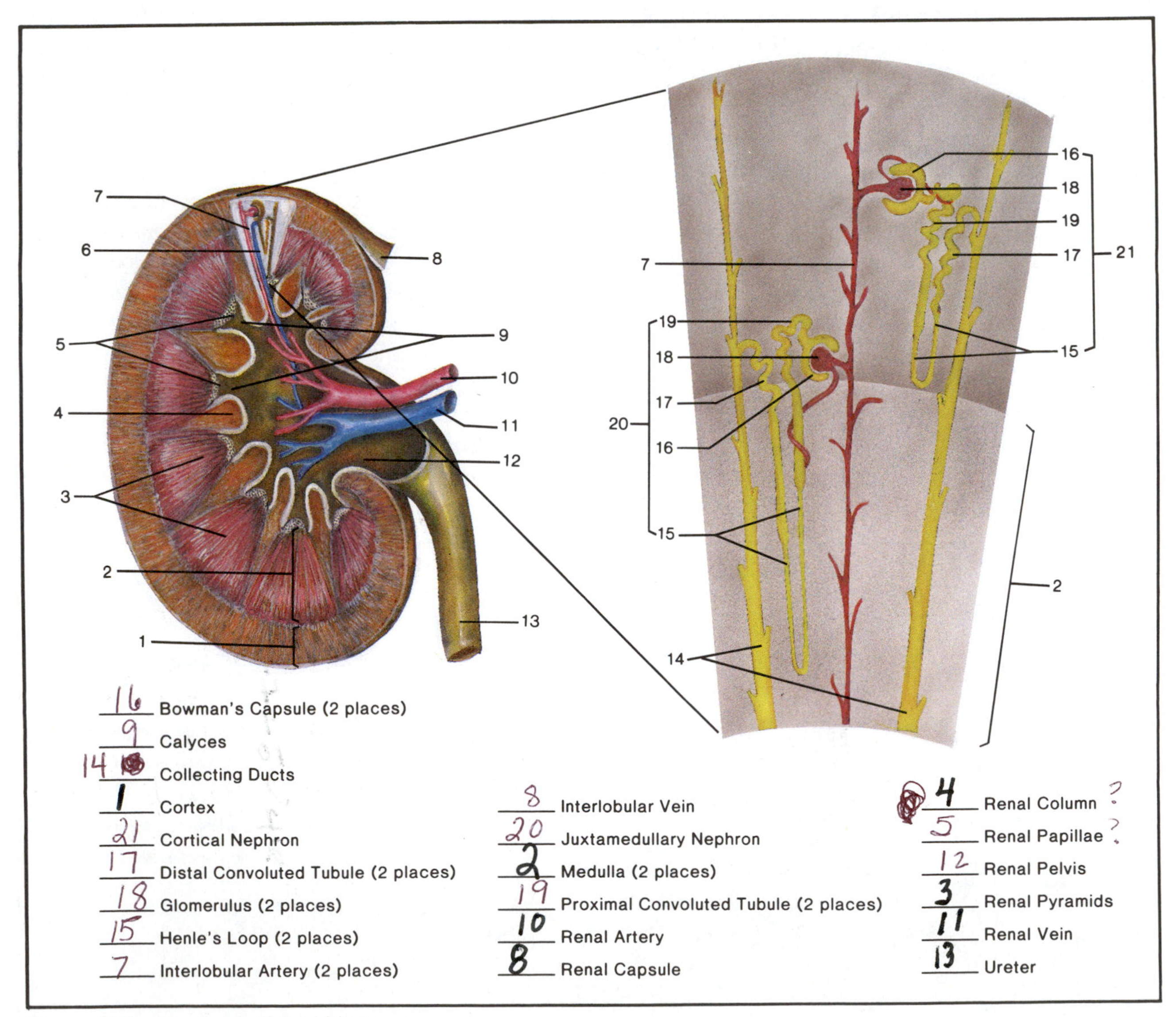

Figure 63.2 Anatomy of the kidney.

kidney. Approximately 80% of the nephrons are located in the cortex; these are designated as **cortical nephrons.** The remainder, or **juxtamedullary nephrons** (label 20, figure 63.2), are located partially in the cortex and partially in the medulla.

The formation of urine by the nephron results from three physiological activities that occur in different regions of the nephron: (1) filtration, (2) reabsorption, and (3) secretion. Because of the differing functions of various regions of the nephron, the fluid that first forms at the beginning of the nephron is quite unlike the urine that enters the calyces of the kidney.

Note that each nephron consists of an enlarged end, the **renal corpuscle,** and a long tubule which empties eventually into the calyx of the kidney. Each renal corpuscle has two parts: an inner tuft of capillaries, the **glomerulus,** and an outer double-walled caplike structure, the **glomerular** (*Bowman's*) **capsule.** Blood that enters the kidney through the **renal artery** reaches each nephron through an **interlobular artery** (label 2, figure 63.3). A short **afferent arteriole** conveys blood into the glomerulus from the interlobular artery. Blood exits from the glomerulus through the **efferent arteriole,** which has a much smaller diameter than the afferent vessel.

The high intraglomerular blood pressure forces a highly dilute fluid, the **glomerular filtrate,** to pass into the glomerular capsule. This fluid consists of glucose, amino acids, urea, salts, and a great deal of water.

This filtrate passes next through the **proximal convoluted tubule** (label 8), the **descending limb of Henle's loop** (label 12), the **ascending limb of Henle's loop,** and the **distal convoluted tubule** (label 10) before emptying into the large **collecting duct** on the right side of the diagram. A single collecting duct may have several nephrons emptying into it.

As the filtrate moves through these various regions, water, glucose, amino acids, and other substances are reabsorbed back into the blood of the **peritubular capillaries** that enmesh the entire route. Eighty percent of the water is reabsorbed through the walls of the proximal convoluted tubules. Water reabsorption is facilitated by the *antidiuretic hormone* of the posterior pituitary and *aldosterone* of the adrenal cortex. Cells lining the collecting ducts alter urine composition by secreting ammonia, uric acid, and other substances into the lumen of the duct.

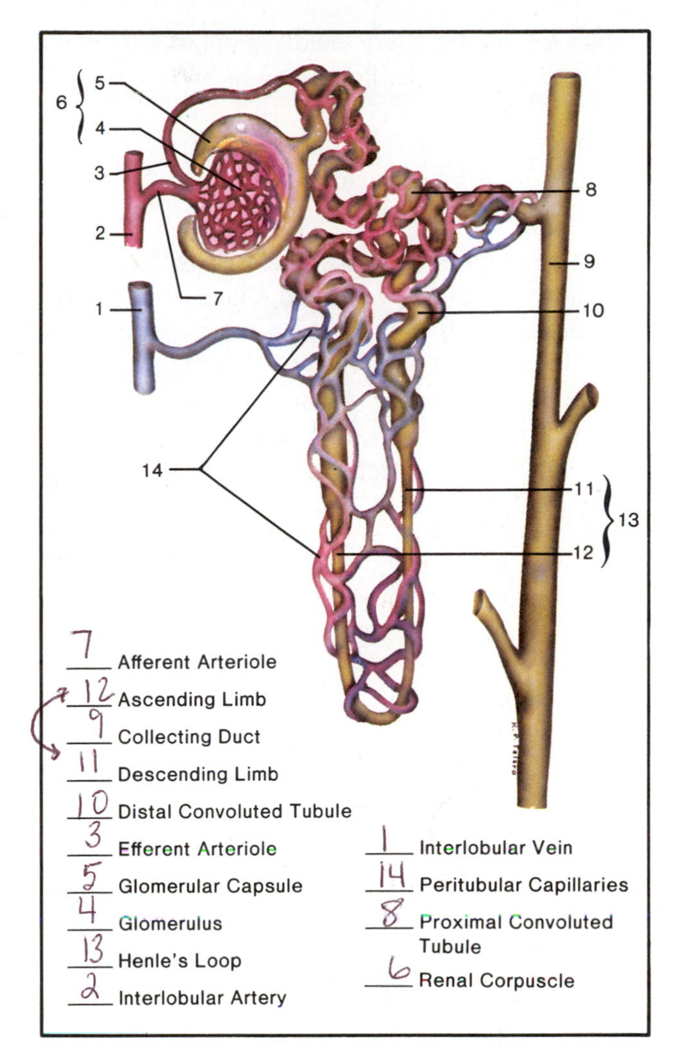

Figure 63.3 The nephron.

Assignment:

Label figures 63.2 and 63.3.

Histological Study

Materials:

Prepared slides: kidney tissue (H9.11, H9.115, or H9.14), ureter (H9.16), and urinary bladder (H9.21)

Examine slides of the **kidney cortex, kidney medulla, ureter,** and **urinary bladder,** and compare them to figures HA–29, HA–30, HA–26 (illustrations C and D), and HA–33 (illustrations A and B). Note the differences in kinds of tissue seen in the tubules, calyx, urinary bladder, and ureters.

Sheep Kidney Dissection

Dissection of the sheep kidney may be combined with the cat dissection to provide greater anatomical clarity. Although the injected cat kidney may illustrate better the blood supply, the fresh sheep kidney will reveal more vividly some of the lifelike characteristics of the kidney.

Materials:

dissecting kit, tray, long knife
fresh sheep kidneys

1. If the kidneys are still encased in fat, peel it off carefully. As you lift the fat away from the kidney, look carefully for the **adrenal gland,** which should be embedded in the fat near one end of the kidney. Remove the adrenal gland from the fat and cut it in half. Note that the gland has a distinct outer **cortex** and inner **medulla.**
2. Probe into the surface of the kidney with a sharp dissecting needle to see if you can differentiate the **capsule** from the underlying tissue.
3. With a long knife, slice the kidney longitudinally to produce a frontal section similar to figure 63.2. Wash out the cut halves with running water.
4. Identify all the structures seen in the kidney section of figure 63.2.

Laboratory Report

Complete the Laboratory Report for this exercise.

Urine: Composition and Tests 64

Urine is a highly complex aqueous solution of organic and inorganic substances. Urea, uric acid, creatinine, sodium chloride, ammonia, and water are its principal ingredients. Most substances are either waste products of cellular metabolism or products derived directly from certain foods that are eaten.

Our principal focus in this laboratory session will be with the detection of abnormal substances in urine which may have pathological significance. Some tests will be of physical properties; most tests will be for chemical substances. A total of ten specific examinations will be made.

The chemical tests will be of two types: (1) classical *in vitro* methods and (2) test strip methods. In most laboratories the test strip method will be used because of its simplicity and reliability. Proceed as follows to test a urine sample.

Materials:

test strips: *Albustix, Clinistix, Ketostix, Hemastix, Bili-Labstix*
other items: listed under each test

Collection of Specimen

Urine should be collected in a clean container and stored in a cool place until testing. If the sample is to be collected just prior to lab testing, plastic disposable containers will be made available on the supply table. The best time for collecting a urine sample is three hours after a meal. First samples taken in the morning are least likely to reveal abnormal substances in the urine.

Appearance of Test Specimen

The visual appearance of a urine sample is the first important thing to consider. Its color and turbidity can provide clues as to evidence of pathology.

Color Normal urine will vary from light straw to amber color. The color of normal urine is due to a pigment called *urochrome,* which is the end product of hemoglobin breakdown:

Hemoglobin ⟶ Hematin ⟶ Bilirubin
Urochrome ⟵ Urochromogen ⟵ (Bilirubin)

The following are deviations from normal color that have pathological implications:

Milky: pus, bacteria, fat, or chyle.

Reddish amber: urobilinogen or porphyrin. Urobilinogen is produced in the intestine by the action of bacteria on bile pigment. Porphyrin may be evidence of liver cirrhosis, jaundice, Addison's disease, and other conditions.

Brownish yellow or green: bile pigments. Yellow foam is definite evidence of bile pigments.

Red to smoky brown: blood and blood pigments.

Carrots, beets, rhubarb, and certain drugs may color the urine, yet have no pathological significance. Carrots may cause increased yellow color due to carotene; beets cause reddening; and rhubarb may cause urine to become brown.

Evaluate your urine sample according to the above criteria and record the information on the Laboratory Report.

Transparency (*Cloudiness*) A fresh sample of normal urine should be clear, but may become cloudy after standing for a while. Cloudy urine may be evidence of phosphates, urates, pus, mucus, bacteria, epithelial cells, fat, and chyle. Phosphates disappear with the addition of dilute acetic acid and urates will dissipate with heat. Other causes of turbidity can be analyzed by microscopic examination.

After shaking your sample, determine the degree of cloudiness and record it on the Laboratory Report.

Specific Gravity

The specific gravity of a 24-hour specimen of normal urine will be between 1.015 and 1.025. Single urine specimens may range from 1.002 to 1.030. The more solids in solution, the higher will be the specific gravity. The greater the volume of urine in a 24-hour specimen, the lower will be its specific gravity. A low specific gravity will be present in chronic nephritis and *diabetes insipidus.* A high specific gravity may indicate *diabetes mellitus,* fever, and acute nephritis.

Materials:

urinometer (cylinder and hydrometer)
thermometer
filter paper

1. Fill the urinometer cylinder three-fourths full of well-mixed urine. Remove any foam on the surface with a piece of filter paper.
2. Insert the hydrometer into the urine and read the graduation on the stem at the level of the bottom of the meniscus.
3. Take the temperature of the urine and compensate for the temperature by adding .001 to the specific gravity for each 3° C above 25° C, and subtracting the same amount for each degree below 25° C. Record this **adjusted specific gravity** on the Laboratory Report.
4. Wash the cylinder and hydrometer with soap and water after completing this test.

Hydrogen Ion Concentration

Although freshly voided urine is usually acid (around pH 6), the normal range is between 4.8 and 7.5. The pH will vary with the time of day and diet. Twenty-four-hour specimens are less acid than fresh specimens and may become alkaline after standing due to bacterial decomposition of urea. High acidity is present in acidosis, fevers, and high protein diets. Excess alkalinity may be due to urine retention in the bladder, chronic cystitis, anemia, obstructing gastric ulcers, and alkaline therapy. The simplest way to determine pH is to use pH indicator paper strips.

Materials:

pH indicator paper strip (*pHydrion* or *nitrazine* papers)

1. Dip a strip of pH paper into the urine three consecutive times and shake off the excess liquid.
2. After **one minute,** compare the color with color chart. Record observations on the Laboratory Report.

Protein Detection

Although the large size of protein molecules normally prevents their presence in normal urine, certain conditions can allow them to filter through. Excessive muscular exertion, prolonged cold baths, and excessive protein ingestion may result in *physiologic albuminuria. Pathologic albuminuria,* on

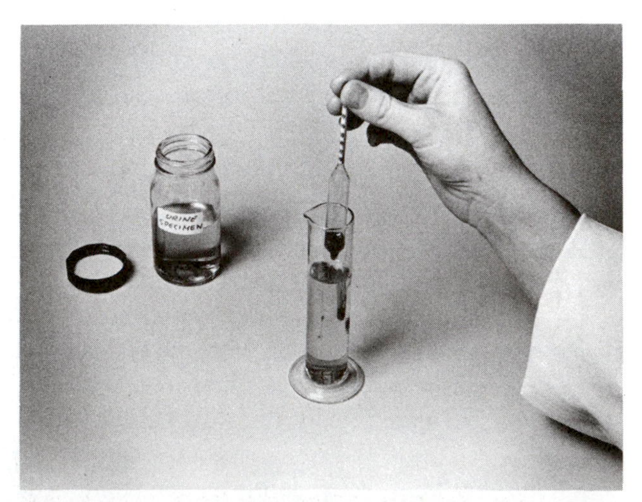

Figure 64.1 Specific gravity. The hydrometer is lowered into the urine after the foam has been removed with a piece of filter paper. The reading is taken from the bottom of the meniscus.

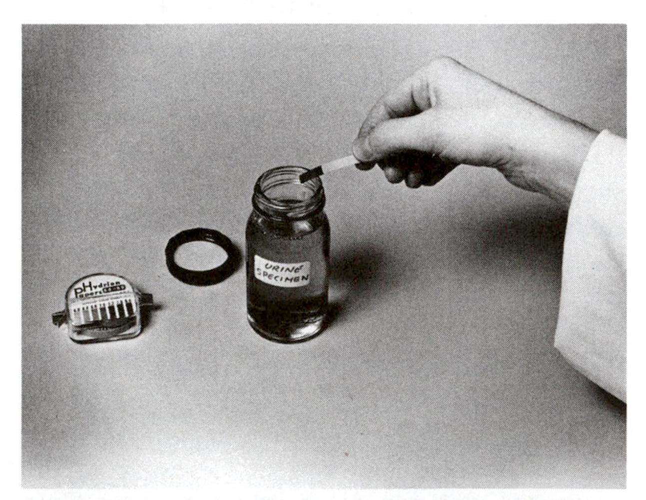

Figure 64.2 Determination of pH. Only fresh urine samples should be tested because pH tends to rise with aging of urine. Read the pH one minute after paper has been dipped.

the other hand, exists when albumin in the urine results from kidney congestion, toxemia of pregnancy, febrile diseases, and anemias.

Detection of albumin may be accomplished by precipitation with chemicals, heat, or both. In our test here we will use *Exton's method,* which employs sulfosalicylic acid and heat.

Materials:

Exton's reagent
test tubes, one for each urine sample
test tube holder
Bunsen burner
pipettes (5 ml size) or graduates

1. Pour or pipette equal volumes of urine and Exton's reagent into a test tube. Approximately 3 ml of each should suffice.
2. Shake the tube from side to side, or roll between the palms, to mix thoroughly.
3. Look for precipitation. If no cloudiness occurs, albumin is absent. Record your results on the Laboratory Report.

Test Strip Method

Shake the sample of urine and dip the test portion of an *Albustix* test strip into the urine. Touch the tip of the strip against the edge of the urine container to remove excess urine. Immediately compare the test area with the color chart on the bottle. Note that the color scale runs from yellow (negative) to turquoise (++++). Record your results on the Laboratory Report.

Detection of Mucin

Inflammation of the mucous membranes of the urinary tract and vagina may result in large amounts of mucin being present in urine. Since it can be confused with albumin, its presence should be confirmed. Mucin and mucoid are glycoproteins that will reduce Benedict's reagent in the presence of acid or alkali. Glacial acetic acid will be used here for the detection of mucin.

Materials:

test tubes and test tube holder
Bunsen burner
acetic acid (glacial)
sodium hydroxide (10%)
pipettes (5 ml size) or graduates
filter paper
funnel
ring stand
small graduate

1. If urine was positive for albumin, remove the albumin by boiling 5 ml of urine for a few minutes and filtering while hot.
2. After the urine has cooled, add 6 ml of **distilled water** to 2 ml of the urine to prevent precipitation of urates.
3. Add a few drops of **glacial acetic acid.** If mucin is present, the urine will become turbid.
4. To further prove that the precipitate is mucin, add a few drops of **sodium hydroxide.** The precipitate will disappear if it is mucin.

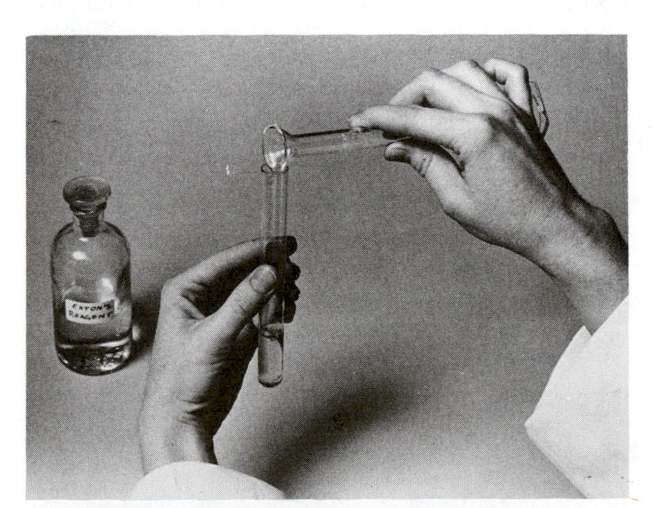

Figure 64.3 Albumin test. If albumin is present in the urine, the addition of Exton's reagent to it will cause a precipitate to form.

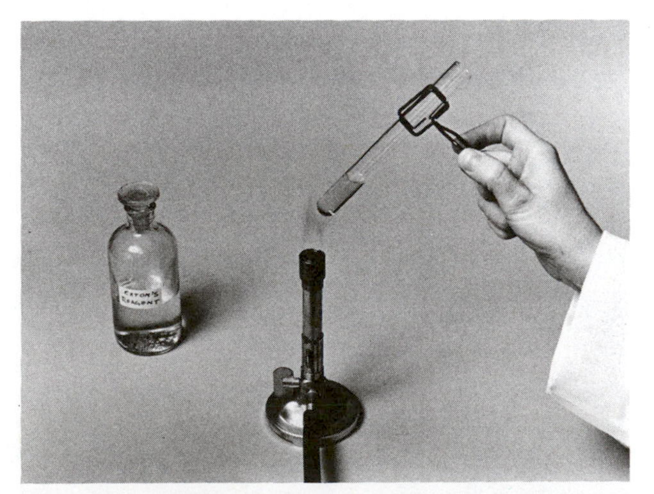

Figure 64.4 Albumin confirmation. If precipitate formed from the addition of Exton's reagent persists after heating, the presence of albumin is confirmed.

Detection of Glucose

Normally, urine will have no more than 0.03 gm of glucose per 100 ml of urine. When glucose exists in urine in amounts greater than this, *glycosuria* exists. This usually indicates that *diabetes mellitus* is present.

The renal threshold of glucose is around 160 mg per 100 ml. Glycosuria indicates that blood levels of glucose exceed this amount and the kidneys are unable to accomplish 100% reabsorption of this carbohydrate.

The detection of glucose in urine is usually performed with **Benedict's reagent.** In the presence of glucose a precipitate ranging from yellowish green to red will form. The color of the precipitate will indicate the approximate percentage of glucose present.

Materials:

test tubes
electric hot plate
beaker (250 ml size)
Benedict's qualitative reagent
pipettes (5 ml size), or graduates

1. Label one tube for each sample to be tested. One tube should be for a known positive sample, if available.
2. Pour 5 ml of Benedict's reagent into each tube.
3. Add 8 drops (0.5 ml) of urine to each tube of Benedict's reagent. Use separate clean pipettes for each urine sample.
4. Place the tubes in a beaker of warm water and bring to boil for 5 minutes.

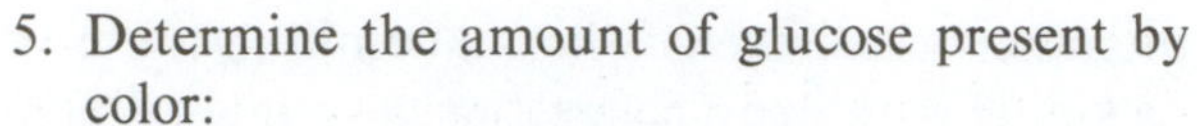

5. Determine the amount of glucose present by color:
   Negative = clear blue to cloudy green
   + = yellowish green (0.5 to 1.0 gm %)
   ++ = greenish yellow (1.0 to 1.5 gm %)
   +++ = yellow (1.5 to 2.5 gm %)
   ++++ = orange (2.5 to 4 gm %)
   red (4 gm % and over)

Test Strip Method

Shake the sample of urine and dip the test portion of a *Clinistix* test strip into the urine. **Ten seconds** after wetting, compare the color of the test area with the color chart on the label of the bottle. Note that there are three degrees of positivity. The light intensity generally indicates 0.25% or less glucose. The dark intensity indicates 0.5% or more glucose. The medium intensity has no quantitative significance. Record your results on the Laboratory Report.

Detection of Ketones

Normal catabolism of fats produces carbon dioxide and water as final end products. When there is inadequate carbohydrate in the diet, or when there is a defect in carbohydrate metabolism, the body begins to utilize an increasing amount of fatty acids. When this increased fat metabolism reaches a certain point, fatty acid utilization becomes incomplete, and intermediary products of fat metabolism occur in the blood and urine. These intermediary

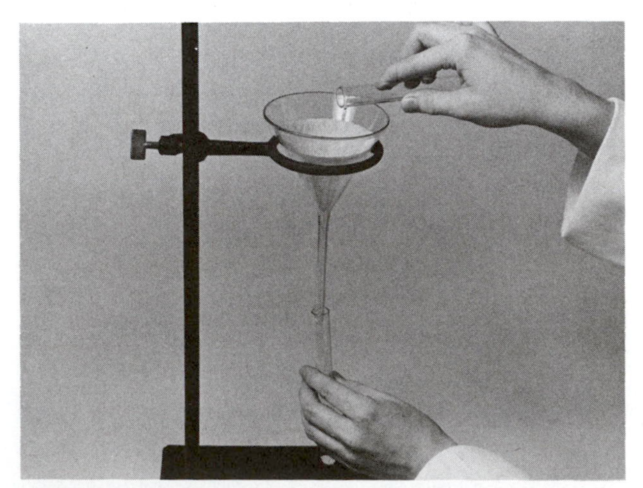

Figure 64.5 Mucin test. Before testing for mucin it is necessary to remove albumin from urine by filtering boiled urine. Acetic acid causes mucin to precipitate.

Figure 64.6 Glucose test. To detect glucose in urine it is necesary to boil 0.5 ml of urine in 5 ml of Benedict's for 5 minutes. A colored precipitate is positive.

substances are the ketones acetoacetic acid, acetone, and beta hydroxybutyric acid. The presence of these substances in urine is called *ketonuria.*

Diabetes mellitus is the most important disorder in which ketonuria occurs. Progressive diabetic ketosis is the cause of diabetic acidosis, which can eventually lead to coma or death. It is for this reason that the detection of ketonuria in diabetics is of great significance.

The method we will use for detection of ketones is *Rothera's test.*

Materials:

Rothera's reagent
test tubes
ammonium hydroxide (concentrated)

1. Add about 1 gm of **Rothera's reagent** to 5 ml of urine in a test tube.
2. Layer over the urine 1 to 2 ml of concentrated **ammonium hydroxide** by allowing it to flow gently down the side of the inclined test tube.
3. If a **pink-purple ring** develops at the interface, ketones are present. No ring, or a brown ring, is negative.

Test Strip Method

As with the previous rapid methods, dip the test portion of a *Ketostix* test strip into the urine sample, and tap dry on the edge of the urine container. **Fifteen seconds** after wetting, compare the color of the test strip with the color chart on the label of the bottle. Record your results on the Laboratory Report.

Detection of Hemoglobin

When red blood cells disintegrate (hemolyze) in the body, hemoglobin is released into the surrounding fluid. If the hemolysis occurs in the blood vessels,

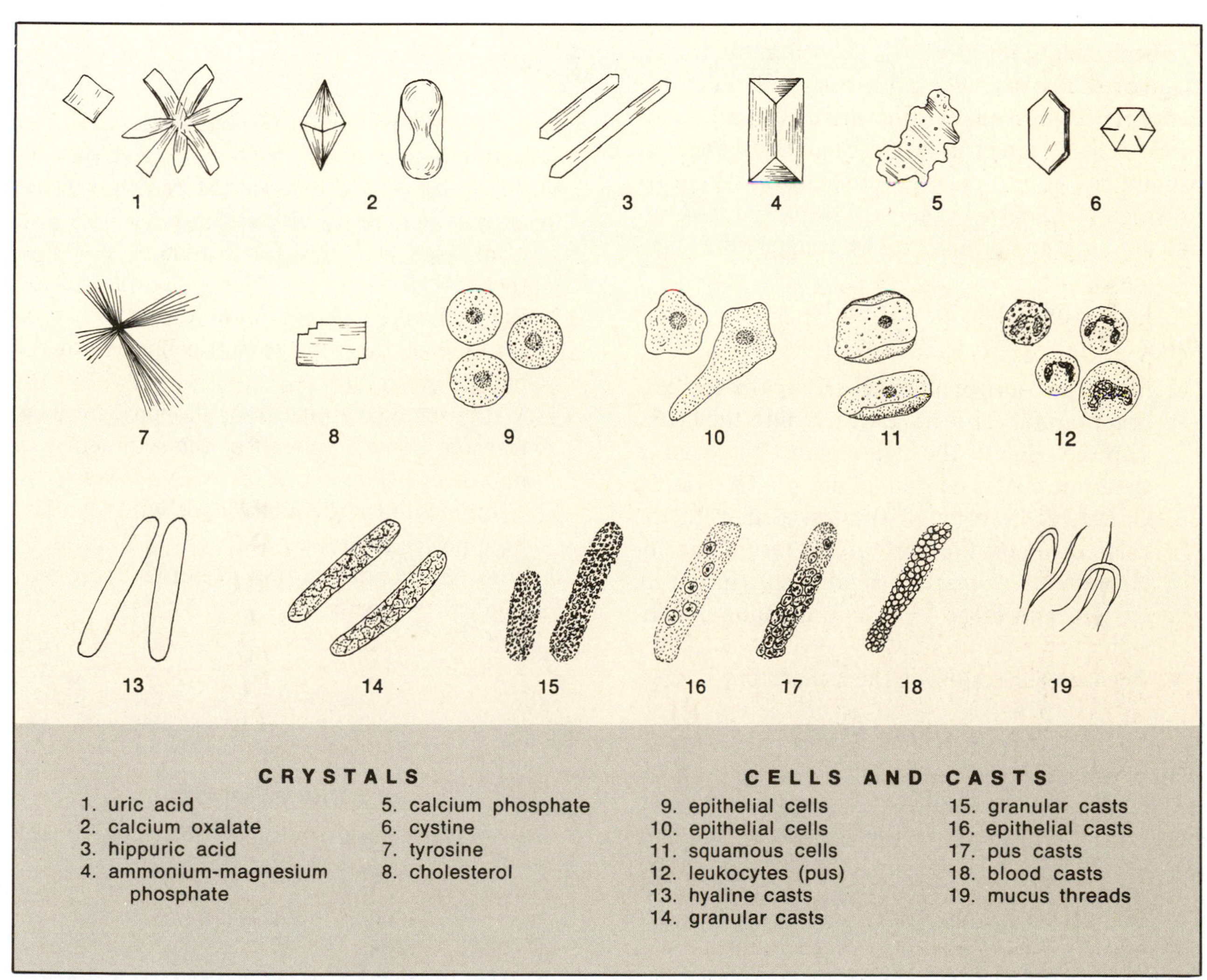

Figure 64.7 Microscopic elements in urine.

the hemoglobin becomes a constituent of plasma. Some of it will be excreted by the kidneys into urine. If the red blood cells enter the urinary tract due to disease or trauma, the cells will hemolyze in the urine. The presence of hemoglobin in urine is called *hemoglobinuria.*

Hemoglobinuria may be evidence of hemolytic anemia, transfusion reactions, yellow fever, smallpox, malaria, hepatitis, mushroom poisoning, renal infarction, burns, etc. The simplest way to test for hemoglobin is to use Hemastix test strips as follows:

1. Shake the sample of urine and dip the test portion of a *Hemastix* test strip into the urine.
2. Tap the edge of the strip against the edge of the urine container and let dry for **30 seconds.**
3. Compare the color of the test strip with the color chart on the bottle.
4. Record your results on the Laboratory Report.

Detection of Bilirubin

We have seen that bilirubin is formed in the second stage of hemoglobin degradation from hematin. It is normally present in urine in very small quantities. When present in large amounts, however, it usually indicates a disorder of the liver caused by infection by hepatoxic agents. The presence of significant amounts of bilirubin is designated as *bilirubinuria.*

The simplest way to test for bilirubin is to use *Bili-Labstix.* The procedure is as follows:

1. Shake the sample of urine and dip the test portion of a *Bili-Labstix* test strip into the urine.
2. Tap the edge of the strip against the edge of the urine container and let dry for **20 seconds.**
3. Compare the color of the test strip with the color chart on the bottle. The results are interpreted as negative, small (+), moderate (++), and large (+++) amounts of bilirubin.
4. Record your results on the Laboratory Report.

Microscopic Study

This is the most important phase of urine analysis, yet the scope of its implications goes considerably beyond the limitations of this course. A complete microscopic examination will include not only an analysis of the sediment, but also a bacteriological determination.

Normal urine will contain an occasional leukocyte, some epithelial cells, mucus, bacteria, and crystals of various kinds. The experienced technologist has to determine when these substances exist in excess amounts and be able to identify the various types of casts, cells, and crystals that predominate.

Figure 64.7 illustrates only a few of the elements that might be encountered in urine. No attempt shall be made here to identify all particulate matter in urine.

Materials:

microscope slides and cover glasses
capillary pipettes
centrifuge
centrifuge tubes (conical tipped)
pipettes (5 ml size) or graduates
wire loop

1. Pour 5 ml of urine into a centrifuge tube after shaking the urine sample to resuspend the sediment. Be sure to balance the centrifuge with an even number of loaded tubes.
2. Centrifuge the tubes for **5 minutes** at a slow speed (1500 rpm).
3. Pour off all urine and allow the sediment on the side of the tube to settle down into the bottom of the tube.
4. With a capillary pipette or flamed wire loop, transfer a small amount of the sediment to a microscope slide and cover with a cover glass.
5. Examine under the microscope with low- and high-power objectives. Reduce the lighting by adjusting the diaphragm. Refer to figure 64.7 to identify structures.

Laboratory Report

Complete the Laboratory Report for this exercise.

The Endocrine Glands 65

In various dissections of previous exercises, endocrine glands have been encountered and discussed briefly. In this exercise all the glands will be studied in a unified manner to review what has been stated previously, and to explore more in depth the histology of each gland. By studying stained microscope slides of the various glands, and comparing them with photomicrographs, you should have no difficulty identifying the specific cells in most glands that produce the various hormones. This laboratory experience should help to dispel some of the abstractness one usually encounters when attempting to assimilate a mass of endocrinological facts.

The Thyroid Gland

The thyroid gland consists of two lobes joined by a connecting isthmus. A posterior view of this gland is seen in figure 65.1. Note in figure 65.2 that, microscopically, it consists of large numbers of spherical sacs called **follicles.** These follicles are filled with a colloidal suspension of a glycoprotein, **thyroglobulin.** The principal hormones of this gland are the *thyroid hormone* and *calcitonin.*

The Thyroid Hormone

This hormone consists of thyroxine, triiodothyronine, and a small quantity of closely related iodinated hormones. They are formed within the thyroglobulin molecule, emerge from the molecule, and are absorbed by blood vessels in the gland. Once in the blood the hormones combine with blood proteins and are carried to all tissues of the body for utilization. Excess hormones are stored in thyroglobulin.

Action The combined activity of thyroxine and triiodothyronine increases the metabolic activity of most tissues of the body. Bone growth in children is accelerated; carbohydrate metabolism is enhanced; cardiac output and heart rate are increased; respiratory rate is increased; and mental activity is increased.

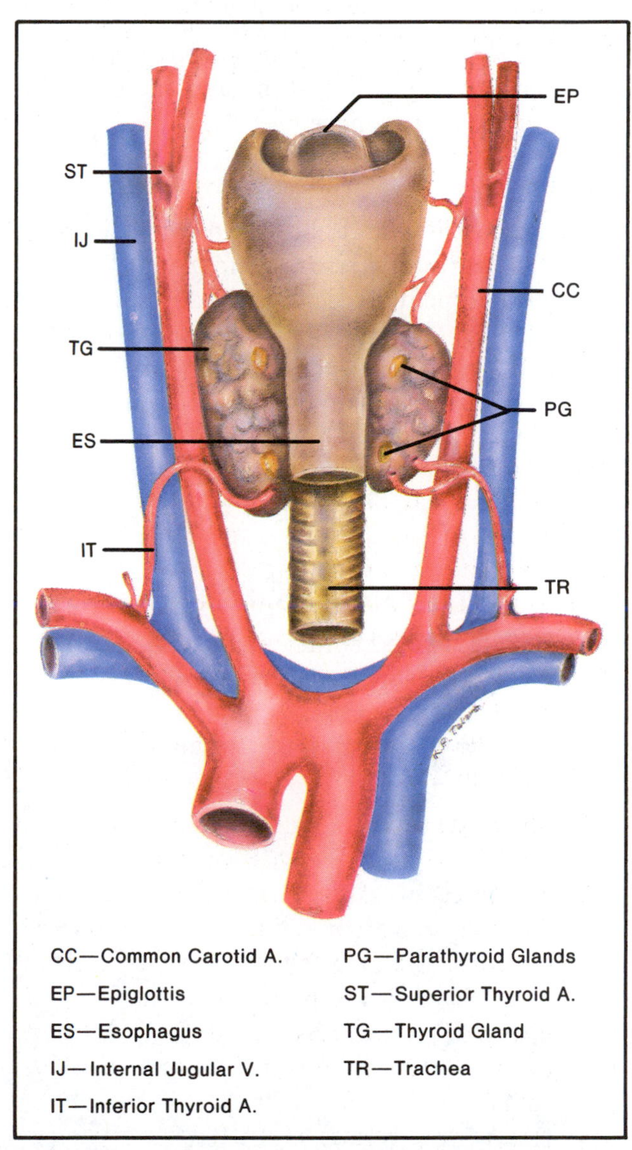

Figure 65.1 The thyroid and parathyroid glands.

Deficiency Symptoms In children hypothyroidism may result in *cretinism,* a condition characterized by physical and mental retardation. In adults, severe long-term deficiency causes *myxedema,* which is characterized by obesity, slow pulse, lack of energy, and mental depression.

Oversecretion Symptoms Hyperthyroidism is characterized by weight loss, rapid pulse, intolerance to heat, nervousness, and inability to sleep. *Exophthalmos,* or eyeball protrusion, is often present.

Goiters Enlarged thyroid glands ("goiters") may be present in both hypo- and hyperthyroidism. Hypothyroid goiters may be due to iodine deficiency (*endemic goiter*) or some other unknown cause (*idiopathic nontoxic goiter*). Hyperthyroid goiters may, or may not, be due to malignancy of the gland.

Regulation Thyrotropin, which is produced by the anterior pituitary gland, stimulates the thyroid to produce the thyroid hormone.

Calcitonin

This hormone is produced by the parafollicular cells in the interstitium of the thyroid gland. When injected experimentally, calcitonin causes a decrease in blood calcium levels. This is due to increased osteoblastic and decreased osteoclastic activity.

It probably plays an important role in bone remodeling during growth in children, but in adults it appears to have very little, if any, effect on blood calcium levels over the long term.

The Parathyroid Glands

There are four small parathyroid glands embedded in the posterior surface of the thyroid gland. See figure 65.1. Occasionally, parathyroid tissue is seen outside of the thyroid gland in the neck region.

Two kinds of cells are seen in this gland: **chief** and **oxyphil cells.** Both cells are shown in figure 65.2. Note that the chief cells are more numerous, smaller, and arranged in cords. They produce the only hormone of this gland, which has been designated as the *parathyroid hormone.* The function of the larger oxyphil cells is unknown.

Action The parathyroid hormone raises blood calcium levels and lowers the phosphorous levels in the blood. This is accomplished by activating osteoclasts in the bone and impeding the reabsorption of phosphorus in the renal tubules. To prevent calcium loss in the kidneys, the hormone also promotes the reabsorption of calcium in the renal tubules. Removal of all parathyroid tissue results in *tetany* and death.

Regulation Production of this hormone is regulated by the level of calcium in the blood. Excess blood calcium levels inhibit hormone production; lowered blood calcium levels cause the gland to produce more of the hormone.

The Pancreas

Figure 65.3 illustrates the histology of this gland. The endocrine-secreting portion of the pancreas is performed by clusters of cells called **pancreatic**

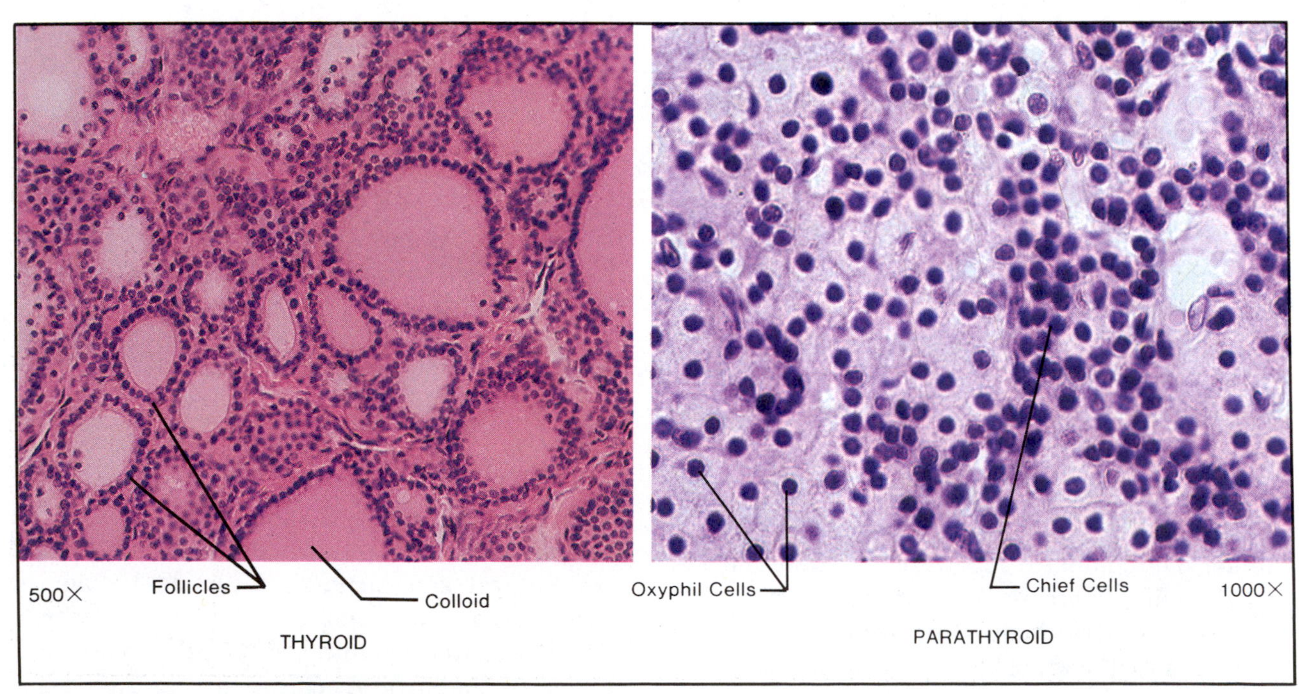

Figure 65.2 Thyroid and parathyroid glands.

islets (islets of Langerhans). These cells are located between the saclike glands *(acini)* that produce the pancreatic digestive enzymes. Insulin, glucagon, and somatostatin are produced by different types of cells in the islets.

Insulin

Insulin is an anabolic hormone that is produced by the **beta cells** of the pancreatic islets. These cells can be differentiated from other islet cells by the presence of many granules in the cytoplasm.

Insulin promotes the storage of glucose by converting it to glycogen. It also promotes the storage of fatty acids and amino acids. Insulin deficiency results in *diabetes mellitus* in which *hyperglycemia* (high blood sugar) is present.

Regulation Although a variety of stimulatory and inhibitory factors affect insulin production, feedback control by blood glucose levels on the beta cells is the major controlling factor. As glucose levels become elevated, insulin production is stepped up; when the glucose level is normal or low, the rate of insulin production is low.

Glucagon

Glucagon is a catabolic hormone produced by the **alpha cells** of the pancreatic islets. Its action is just the opposite of insulin in that it mobilizes glucose, fatty acids, and amino acids in tissues. Deficiency of this hormone will result in *hypoglycemia.* Glucagon also stimulates the production of the growth hormone, insulin, and pancreatic somatostatin.

Somatostatin

A third type of islet cells, called the **delta cells,** produce somatostatin, a growth-inhibiting hormone. With Mallory's stain the cytoplasm of these cells stains blue.

This hormone is growth inhibiting in that it inhibits the release of the growth hormone by the anterior pituitary gland. In addition, it inhibits the production of insulin and glucagon. Tumors involving the delta cells result in hyperglycemia and other diabetes-like symptoms. Removal of the tumors causes the symptoms to disappear.

The Adrenal Glands

Each adrenal gland has an outer **cortex** and inner **medulla.** Surrounding the entire gland is a **capsule** of fibrous connective tissue.

The embryological origins of the cortex and medulla explain their differences in function. The cells of the medulla originate from the neural crest of the embryo. This embryonic tissue also gives rise to ganglionic cells of the sympathetic nervous system; thus, we can expect a close kinship in function of the adrenal medulla and the sympathetic nervous system.

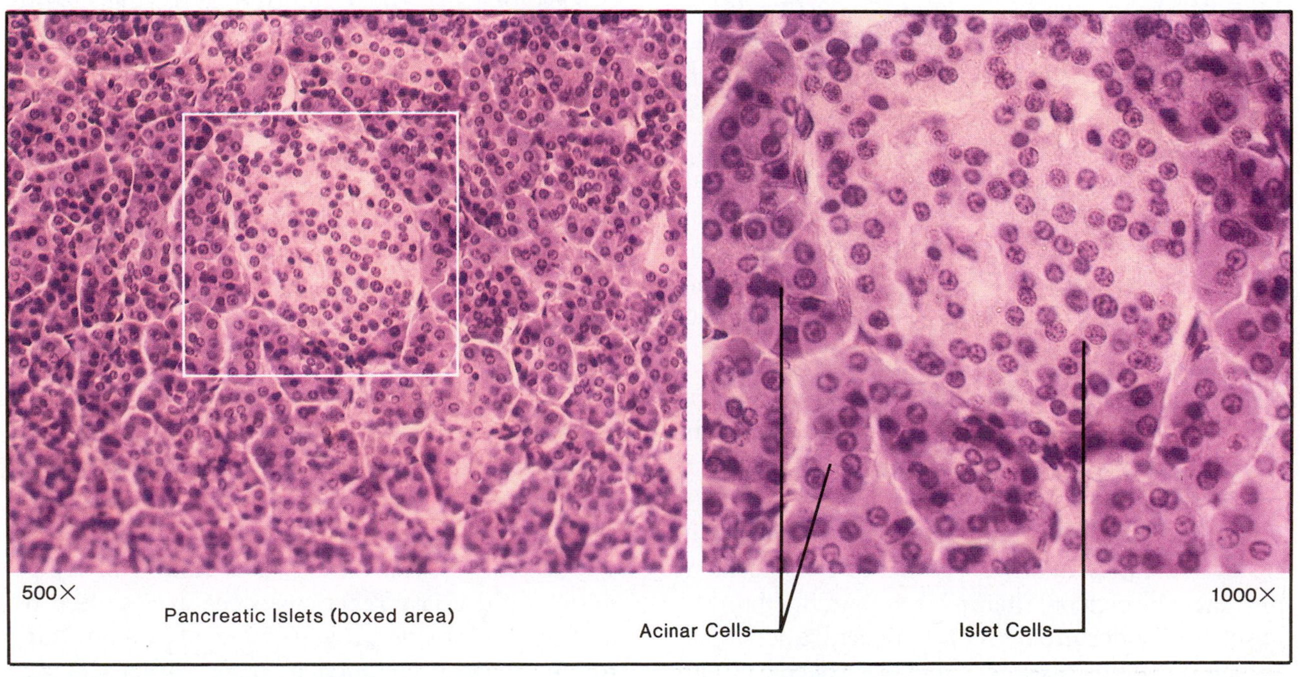

Figure 65.3 Pancreas histology.

Cells of the cortex, on the other hand, arise from embryonic tissue associated with the gonads. Hormones having some relationship to the reproductive organs might, thus, be expected from the cortex.

The Adrenal Cortex

Figures 65.4 and 65.5 reveal the three layers of the cortex. Note that immediately under the capsule lies a layer called the **zona glomerulosa.** The innermost layer of cells, which interfaces with the medulla, is the **zona reticularis.** The middle layer, which is the thickest, is the **zona fasciculata.** In all the layers are seen vascular channels, called **sinusoids,** that are larger in diameter than capillaries, but resemble capillaries in that their walls are one cell thick. It is in these channels that the hormones collect prior to passing directly into the general circulation.

The adrenal cortex secretes many different hormones, all of which belong to a group of substances called *steroids.* Collectively, they are referred to as **corticosteroids.** All of them are synthesized from cholesterol and have similar molecular structures. While destruction of the adrenal medulla is more or less inconsequential to survival, obliteration of the adrenal cortex results in death.

The corticosteroids fall into three groups: the mineralocorticoids, the glucocorticoids, and the androgenic hormones. The *mineralocorticoids* are hormones that control the excretion of Na^+ and K^+, which affects the electrolyte balance. The *glucocorticoids* affect primarily the metabolism of glucose and protein. The *androgenic hormones* produce some of the same effects on the body as the male sex hormone testosterone. Of the 30 or more corticoids that are produced by the adrenal cortex, the most important two are aldosterone and cortisol. Descriptions of these two hormones follow.

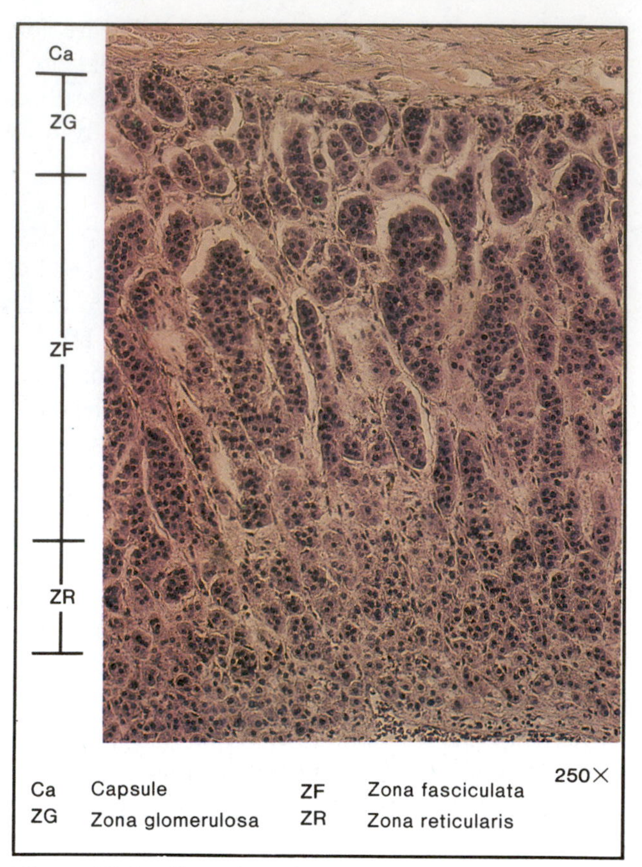

Figure 65.4 The adrenal cortex.

Aldosterone This mineralocorticoid is produced by cells of the zona glomerulosa. The most important function of this hormone is to increase the rate of renal tubular absorption of sodium. Any condition which impairs the production of aldosterone will result in death within two weeks if salt replacement or mineralocorticoid therapy is not provided. In the absence of aldosterone the K^+ concentration in extracellular fluids rises, Na^+ and Cl^- concentrations decrease, and the total volume of body fluids becomes greatly reduced. These conditions cause reduced cardiac output, shock, and death.

Factors that affect aldosterone production are ACTH, K^+ ion concentration, Na^+ ion concentration, and the renin-angiotensin system.

Cortisol *(Hydrocortisone, Compound F)* Approximately 95% of the glucocorticoid activity results from the action of cortisol. This hormone is produced primarily by cells in the zona fasciculata; some of it is also produced by the zona reticularis.

When an excess of cortisol exists, the following physiological activities take place in the body: (1) the rate of glucogenesis is stepped up; (2) glucose utilization by cells is decreased; (3) protein catabolism in all cells except the liver is increased; (4) amino acid transport to muscle cells is decreased; and (5) fatty acids are mobilized from adipose tissue.

Cushing's syndrome is a condition in which many of the above reactions occur. It is caused by an *excess of glucocorticoids* induced by a pituitary tumor. Protein depletion in these patients causes them to have poorly developed muscles, weak bones

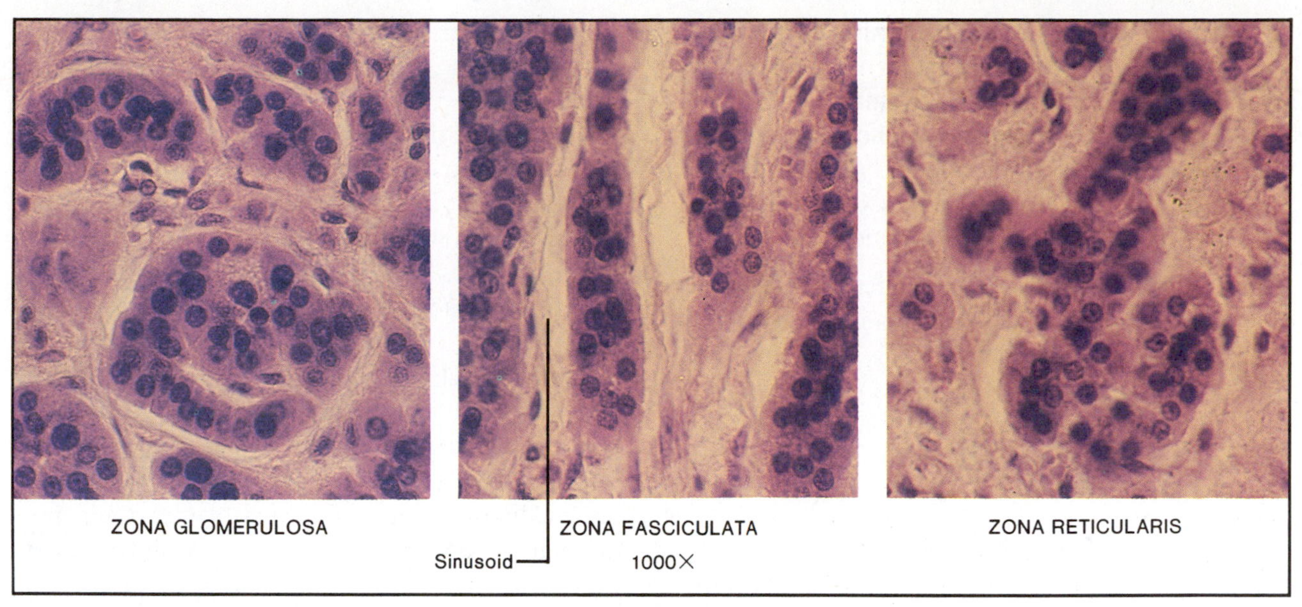

Figure 65.5 Histology of the three layers of the adrenal cortex.

due to bone dissolution, slow healing of wounds, hyperglycemia, and hair that is thin and scraggly.

Addison's disease is a condition in which total *glucocorticoid insufficiency* exists. It can be caused by the destruction of the adrenal cortex by cancer, tuberculosis, or some other infectious agent.

Although several substances such as vasopressin, serotonin, and angiotensin II stimulate the adrenal cortex, it is primarily the action of ACTH that induces the cells of the zona fasciculata and zona reticularis to produce cortisol. ACTH production by the anterior pituitary is initiated by stress through the hypothalamus. Stress induces the hypothalamus to produce CRF, which passes to the anterior pituitary, causing the latter to produce ACTH.

The Adrenal Medulla

This portion of the adrenal gland produces the two catecholamines, **norepinephrine** and **epinephrine.** Although different cells in the medulla produce each of these hormones, cell differences are difficult to detect microscopically.

Approximately 80% of the medullary secretion is epinephrine; 20% is norepinephrine. Norepinephrine is also produced by the autonomic nervous system.

Action The effects of norepinephrine and epinephrine on different tissues depend on the types of receptors that exist in the tissues. There are two classes of adrenergic receptors: alpha and beta. Two types of alpha receptors are designated as α_1 and α_2; the beta receptors are β_1 and β_2.

Both norepinephrine and epinephrine increase the force and rate of contraction of the isolated heart. These reactions are mediated by β_1 receptors. Norepinephrine causes vasoconstriction in all organs through α_1 receptors. Epinephrine causes vasoconstriction everywhere except in the muscles and liver, where beta receptors bring about vasodilation. Increased mental alertness is also induced by both catecholamines, which may be induced by the increased blood pressure.

Blood glucose levels are increased by both hormones. This is accomplished by glycogenolysis (glycogen $\longrightarrow$ glucose) in the liver. β_2 receptors account for this reaction.

Norepinephrine and epinephrine are equally potent in mobilizing free fatty acids through beta receptors. The metabolic rate is also increased by these hormones; how this takes place is not precisely understood at this time.

Control Production of both hormones is initiated when the adrenal medulla is stimulated by sympathetic nerve fibers.

The Thymus Gland

This gland consists of two long lobes and lies in the upper chest region above the heart. It is most highly developed before birth and during the growing years. After puberty it begins an involutionary process, which continues throughout life.

Histologically, this gland consists of a cortex and medulla. Figure 65.6 reveals its microscopic structure. The **cortex** consists of lymphoidlike tissue filled with a large number of lymphocytes. The **medulla** contains a smaller number of lymphocytes and structures called **thymic corpuscles** (Hassall's bodies) that are of unknown function.

Function It appears that the principal role of the thymus is to process lymphocytes into T-lymphocytes, as described on page 268. It is speculated that this gland produces one or more hormones that function in T-lymphocyte formation. Thymosin and several other extracts from this gland have been under study.

The Pineal Gland

The pineal gland is situated on the roof of the third ventricle under the posterior end of the corpus callosum. It is supported by a stalk which contains postganglionic sympathetic nerve fibers that do not seem to extend into the gland. The gland consists of neuroglial and parenchymal cells that suggest a secretory function. See figure 65.6.

In young animals and infants the gland is large and more glandlike; cells tend to be arranged in alveoli. Just before puberty the gland begins to regress and small concretions of calcium carbonate form that are called **pineal sand.** There is some evidence, though not conclusive, that the gland contains some gonadotropin peptides. Most attention, however, is on the production of melatonin by the gland.

Melatonin is an indole which is synthesized from serotonin. Its production appears to be regulated by daylight (circadian rhythm). During daylight its production is suppressed. At night the gland becomes active and produces considerable quantities of melatonin. Regulation occurs through the eyes. Light striking the retina sends messages to the pineal gland through a retina-hypothalamic pathway. Norepinephrine reaches the cells from the postganglionic sympathetic nerve endings in the pineal stalk. Beta adrenergic receptors in the cells are mediators in the inhibition of melatonin production. Although considerable speculation exists that melatonin inhibits the estrus cycle in humans, as it probably does in lower animals, there is no definitive proof at this time that this is the case.

The Testes

A section through the testis (figure 65.7) reveals that it consists of coils of **seminiferous tubules** where spermatozoa are produced, and **interstitial cells** where the male sex hormone, **testosterone,** is secreted. Note that the interstitial cells lie in the spaces between the seminiferous tubules.

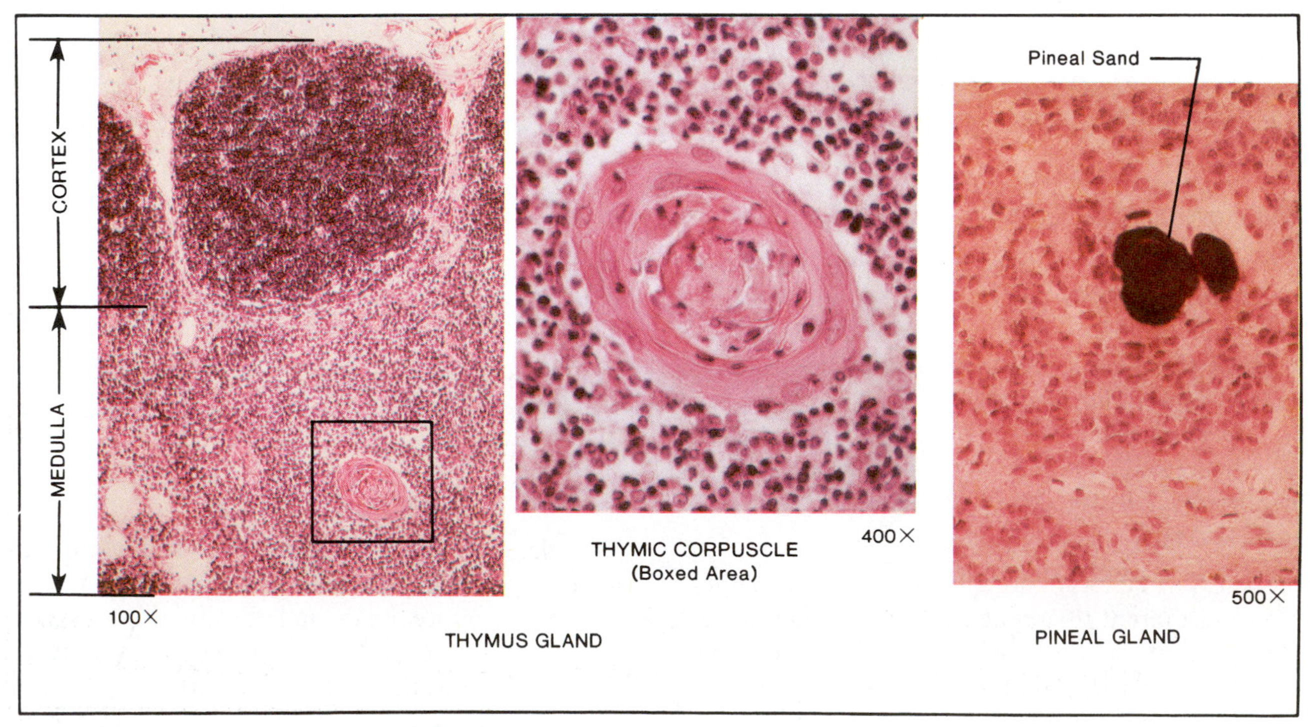

Figure 65.6 The thymus and pineal glands.

Testosterone is produced in considerable quantities during embryological development, but from early childhood to puberty very little of the hormone is produced. With the onset of puberty at the age of 10 or 11 years the testes begin to produce large quantities of testosterone as the development of sexual maturity takes place.

During embryological development, testosterone is responsible for the development of the male sex organs. If the testes are removed from a fetus at an early stage, the fetus will develop a clitoris and vagina instead of a penis and scrotum, even though the fetus is a male. The hormone also controls the development of the prostate gland, seminal vesicles, and male ducts while suppressing the formation of female genitals.

Embryological development of the testes occurs within the body cavity. During the last two months of development the testes descend into the scrotum through the inguinal canals. The descent of the testes is controlled by testosterone.

During puberty the testes become larger and produce a great deal of testosterone, which causes considerable enlargement of the penis and scrotum. At the same time, the male secondary sexual characteristics develop. These characteristics are (1) the appearance of hair on the face, axillae, chest, and pubic region; (2) the enlargement of the larynx accompanied by a voice change; (3) increased skin thickness; (4) increased muscular development; (5) increased bone thickness and roughness; and (6) baldness.

During embryological development the production of testosterone is regulated by **chorionic gonadotropin,** a hormone produced by the placenta. At the onset of puberty the testes are stimulated to produce testosterone by the **luteinizing hormone (LH),** which is produced by the anterior lobe of the pituitary gland. This hormone stimulates the interstitial cells to produce testosterone. Maturation of spermatozoa in the testes is controlled by the **follicle-stimulating hormone (FSH),** which is also produced by the anterior pituitary gland. Testosterone assists FSH in this process.

The Ovaries

During fetal development small groups of cells move inward from the germinal epithelium of the ovary and develop into primordial follicles. As seen in figure 65.8, the **germinal epithelium** is the layer of cells near the surface of the ovary, and the **primordial follicles** are the small round bodies that contain **ova.** The ovaries of a young female at the onset of puberty are believed to contain between 100,000 and 400,000 of these immature follicles. During all the reproductive years of a woman only about 400 of these follicles will reach maturity and expel their ova.

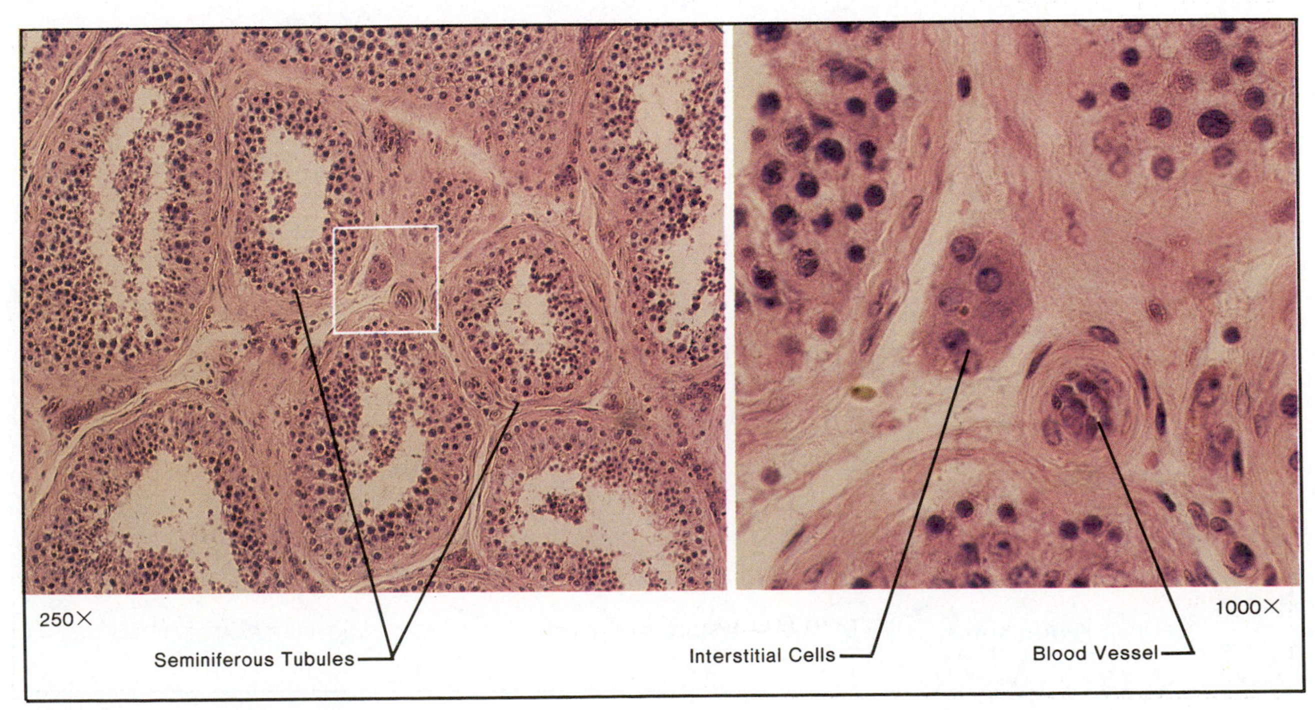

Figure 65.7 Testicular tissue.

During puberty selected primordial follicles enlarge to form **Graafian follicles,** and every 28 days a maturing Graafian follicle expels an ovum in the ovulation process. The development of these follicles and the subsequent corpus luteum results in the production of **estrogens** and **progesterone,** the principal female sex hormones.

Estrogens

The principal function of estrogens is to promote the growth of specific cells in the body and to control the development of female secondary sex characteristics. In addition to being produced in Graafian follicles, estrogens are also produced by the corpus luteum, placenta, adrenal cortex, and testes. The production of estrogens at puberty is initiated by the secretion of FSH by the anterior pituitary gland.

Of the six or seven estrogens that have been isolated from the plasma of women, **β-estradiol, estrone,** and **estriol** are the most abundant ones produced. The most potent of all the estrogens is **β-estradiol.**

During puberty when the estrogens are produced in large quantities the following changes occur: (1) the female sex organs become fully developed; (2) the vaginal epithelium changes from cuboidal to stratified epithelium; (3) the uterine lining (endometrium) becomes more glandular in preparation for implantation; (4) the breasts form; (5) osteoblastic activity increases with more rapid bone growth; (6) calcification of the epiphyses in long bones is hastened; (7) pelvic bones enlarge and change shape, increasing the size of the pelvic outlet; (8) fat deposition under the skin, in the hips, buttocks, and thighs is increased; and (9) skin vascularization is increased.

Progesterone

Once ovulation takes place in the ovary the Graafian follicle is replaced by a **corpus luteum.** Progesterone is the principal hormone produced by this body. The ovary in figure 65.13 shows the development of Graafian follicles and the corpus luteum.

The most important function of this hormone is to promote secretory changes in the endometrium in preparation for implantation of the fertilized ovum. In addition, the hormone causes mucosal changes in the uterine tubes and promotes the proliferation of alveolar cells in the breasts in preparation for milk production.

When the plasma level of progesterone falls due to regression of the corpus luteum, deterioration of the endometrium takes place and menstruation occurs. If pregnancy occurs the corpus luteum enlarges and produces additional quantities of progesterone. The conversion of a Graafian follicle into a corpus luteum is completely dependent on LH production.

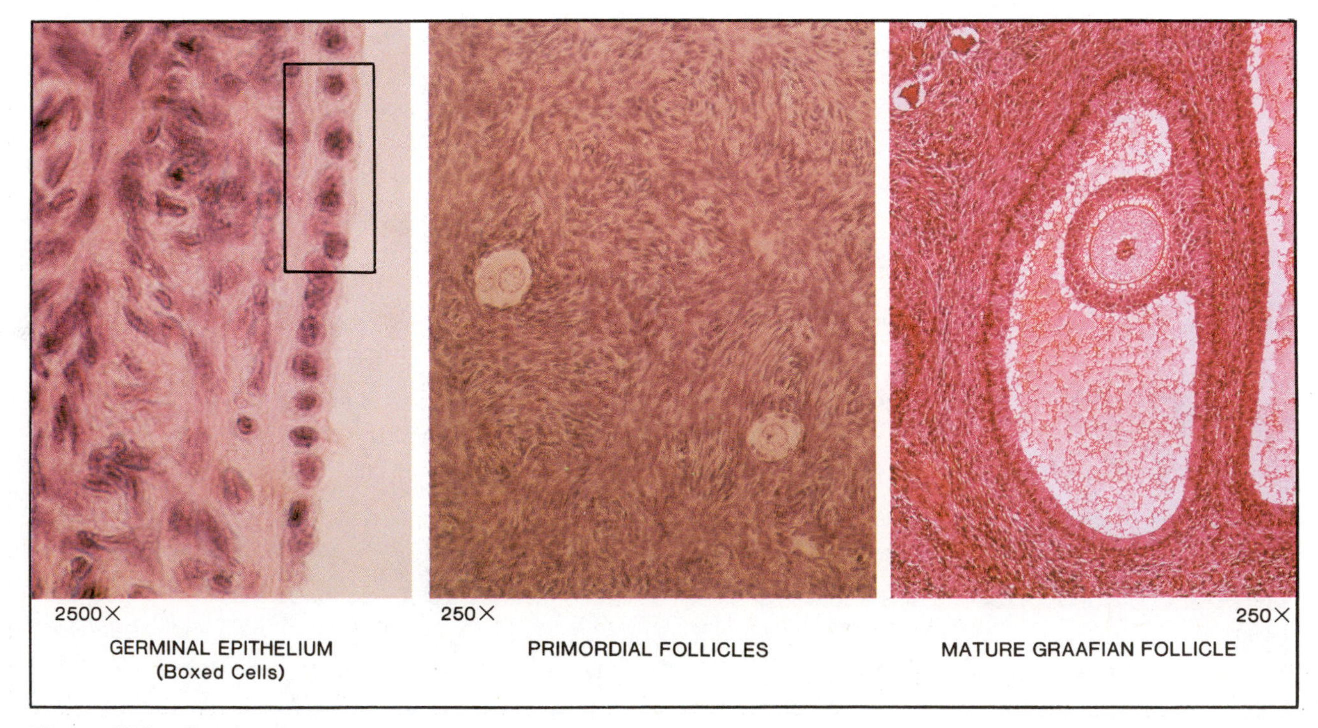

Figure 65.8 Ovarian tissue.

The Pituitary Gland

The pituitary gland, or hypophysis, consists of two lobes: anterior and posterior. The anterior lobe, or **adenohypophysis,** develops from the roof of the oral cavity in the embryo. This part of the hypophysis is made up of glandular cells. The posterior lobe, or **neurohypophysis,** develops as an outgrowth of the floor of the brain during embryological development. Unlike the adenohypophysis, the cells of this lobe are nonsecretory and resemble neuroglial tissue. The entire gland is invested by an extension of the dura mater.

Figure 65.9 reveals portions of the two lobes. Note that the posterior lobe in this region consists of two parts: the **pars intermedia** and the **pars nervosa.**

The Adenohypophysis

Two basic types of glandular cells are seen in this portion of the hypophysis: chromophobes and chromophils. Both types of cells can be seen in figure 65.10. **Chromophobes** are cells that lack an affinity for routine dyes used in staining tissues. **Chromophils** stain readily and are of two types: acidophils and basophils. **Acidophils** take on the pink stain of eosin; **basophils,** on the other hand, stain readily with the basic stains, such as methylene blue and crystal violet.

Six hormones are produced by these various cells in the anterior lobe. Except for somatotropin, all these hormones are targeted specifically for glands. Production of the hormones is regulated by the hypothalamus. Between the hypothalamus and adenohypophysis is a vascular connection, the *hypophyseal portal system,* that transports *releasing factors* from the hypothalamus to the secretory cells. The releasing factors are secreted by cells in the hypothalamus.

Somatotropin (*SH*) This hormone, also called the **growth hormone, (GH),** is produced by **somatotrophs,** a type of acidophil. In males and nonpregnant females most of the acidophils are of this type.

Somatotropin increases the growth rate of all cells in the body by enhancing amino acid uptake and protein synthesis. Excess production of the hormone during the growing years causes **gigantism** due to the stimulation of growth in the epiphyses of the long bones. An adult with excess SH production develops **acromegaly,** which is characterized by enlargement of the small bones of the hands and feet and the mandible and forehead. A deficiency of the hormone can cause **dwarfism.**

Prolactin (***Luteotropic Hormone, LTH***) This hormone is produced by another type of acidophil, called a **mammotroph.** With suitable staining methods it is possible to differentiate these cells from the somatotrophs.

Prolactin promotes the production of milk in the breasts after childbirth. The release of this hor-

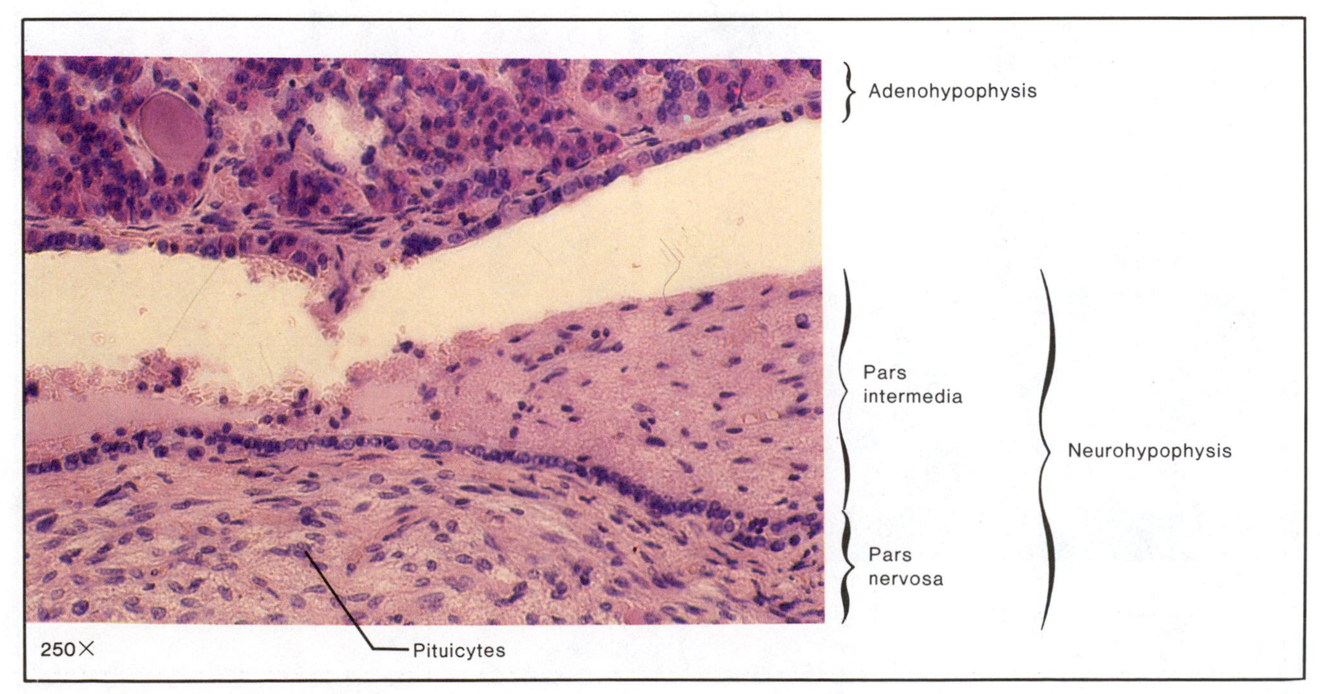

Figure 65.9 The pituitary gland.

mone prior to childbirth is inhibited by *PIF* (*prolactin inhibiting factor*), which is produced in the hypothalamus. The presence of large amounts of estrogens and progesterone prior to childbirth causes the hypothalamus to produce this inhibiting factor.

Thyrotropin (*TSH*) This hormone is produced by large irregular basophilic cells, called **thyrotrophs.** TSH regulates the rate of iodine uptake and the synthesis of the thyroid hormone by the thyroid gland. Excess amounts of TSH will cause hyperthyroidism. Hypothyroidism results from a deficiency of the hormone. TSH production is regulated by the hypothalamic *thyrotropin releasing factor* (*TRF*). TRF is secreted by nerve endings in the hypothalamus and passes through the hypophyseal portal system. Thyrotropin production is also regulated by a thyroid hormone feedback system.

Adrenocorticotropin (*ACTH*) This hormone is probably produced by a type of chromophobe. The suspected cell is a large chromophobe with large cytoplasmic granules, peripherally located. ACTH controls glucocorticoid production of the adrenal cortex. Its production is partially regulated by a glucocorticoid feedback mechanism.

Follicle-Stimulating Hormone (*FSH*) This hormone is produced by basophilic **gonadotrophs.** FSH promotes the development of Graafian follicles in the ovaries and the maturation of spermatozoa in the testes. FSH production by the adenohypophysis is regulated by estrogen and testosterone feedback mechanisms.

Luteinizing Hormone (*LH*) This hormone is produced by basophilic **gonadotrophs** that are somewhat larger than the FSH gonadotrophs. While the FSH cells tend to be located near the periphery of the gland, LH producing cells are more generally distributed throughout the gland. Maturation of the Graafian follicles and ovulation during the menstrual cycle are dependent on this hormone. Although FSH contributes much to early follicle development, ovulation is entirely dependent on LH. The production of estrogens and progesterone is dependent on LH also.

In addition, this hormone stimulates the interstitial cells to produce testosterone; thus, it has been referred to as the **interstitial cell-stimulating hormone (ICSH).** LH production by the adenohypophysis is regulated by estrogen and progesterone feedback mechanisms.

The Neurohypophysis

The neuroglial-like cells in this portion of the hypophysis are called **pituicytes.** These cells are small and have numerous processes. Unlike

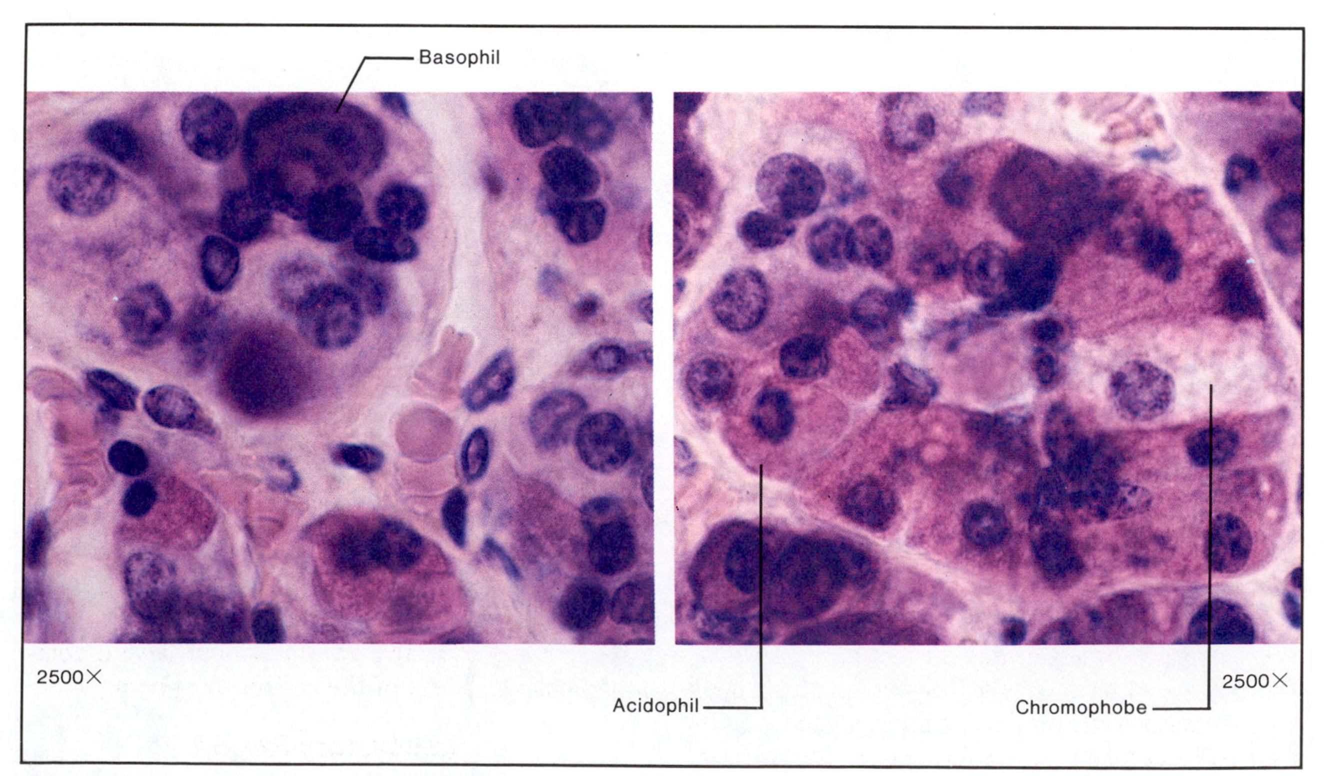

Figure 65.10 Cells of the adenohypophysis.

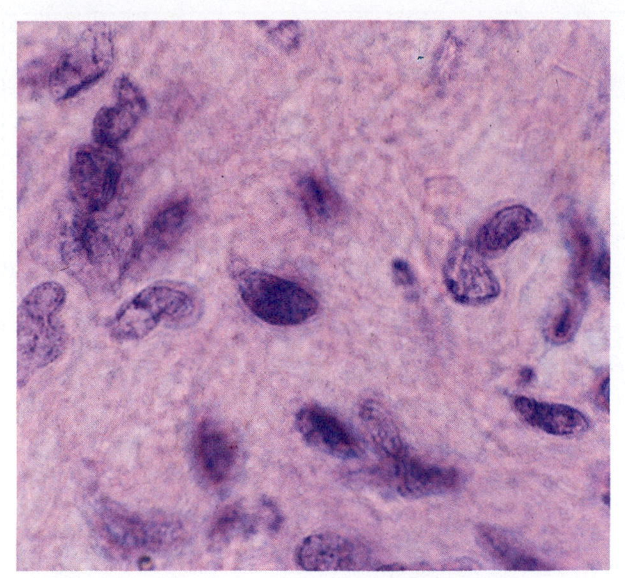

Figure 65.11 Pituicytes of the neurohypophysis (2500×).

neuroglia, pituicytes contain fat and pigment granules in their cytoplasm. Figure 65.11 is representative of these cells.

Pituicytes are nonsecretory. Their function is to provide support for nerve fibers from tracts that originate in the hypothalamus. Two hormones, the **antidiuretic hormone** and **oxytocin,** are liberated by the ends of these nerve tracts. The hormones are actually synthesized in the cell bodies of **neurosecretory neurons** of the hypothalamus and pass down the fibers into the neurohypophysis where they are released.

Antidiuretic Hormone (*ADH*) This hormone controls the permeability of collecting tubules in the nephrons to water reabsorption. When ADH is present in even minute amounts, water is easily reabsorbed back into the blood through the walls of the collecting tubules. When ADH is lacking, rapid water loss through the kidneys takes place. *Diabetes insipidus* is a disease in which the neurohypophysis fails to provide enough ADH. Without this hormone the osmotic balance of body fluids cannot remain stable very long.

Another function of ADH is to increase arterial blood pressure when a severe blood loss has reduced blood volume. A 25% loss of blood will increase ADH production by 25 to 50 times normal. ADH increases blood pressure by constricting arterioles. Because ADH has this potent pressor capability it is also called **vasopressin.** The mechanism for increased ADH production during blood loss lies in baroreceptors that are located in the carotid, aortic, and pulmonary regions.

Oxytocin Any substance that will cause uterine contractions during pregnancy is designated as being "oxytocic." During childbirth, pressure of the unborn child on the uterine cervix causes a neurogenic reflex to the neurohypophysis, stimulating it to produce the hormone oxytocin. This hormone, being oxytocic, increases the strength of uterine contractions to facilitate childbirth. Oxytocin also activates the myoepithelial cells of the mammary gland, resulting in the flow of milk.

Laboratory Assignment

Microscopic Studies

Do a systematic study of the various glands using the slides that are available. Make drawings, if required.

Materials:

prepared slides of the following glands:

thyroid (H14.11, H14.12, or H14.13)
parathyroid (H14.16)
thymus (H14.21 or H14.22)
adrenal (H14.26)
ovary (H9.310 or H9.3181)
testis (H9.415)
pituitary (H14.31)
pineal (H14.36)

While examining these slides, refer to the various photomicrographs in this exercise to identify the various kinds of cells that are mentioned in the text. Use low power for scanning and oil immersion wherever necessary.

Labeling

Figures 65.12 and 65.13 summarize many of the endocrinological facts learned in this exercise. Figure 65.12 pertains primarily to the various tissues that produce the hormones. Figure 65.13 depicts most of the hormones produced by the pituitary gland that regulate growth and glands of the body. It also includes many of the hormones produced by the glands that are regulated.

It will probably be necessary for you to refer back to various portions of the text in this exercise to seek out the various hormones. Once both of these diagrams are completed you should have a comprehensive picture of the endocrine system.

Laboratory Report

Complete the Laboratory Report for this exercise.

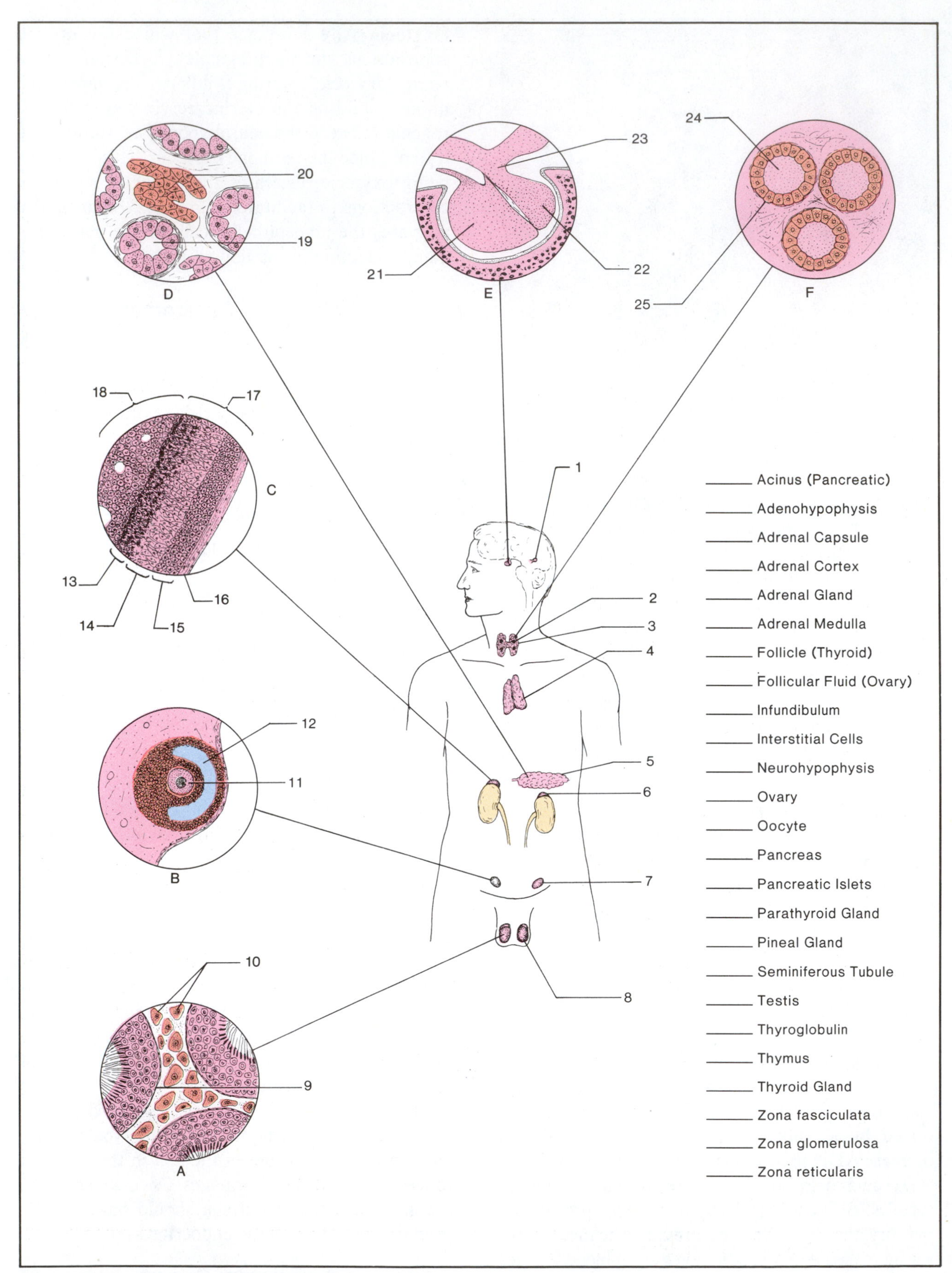

Figure 65.12 Secretory components of the endocrine system.

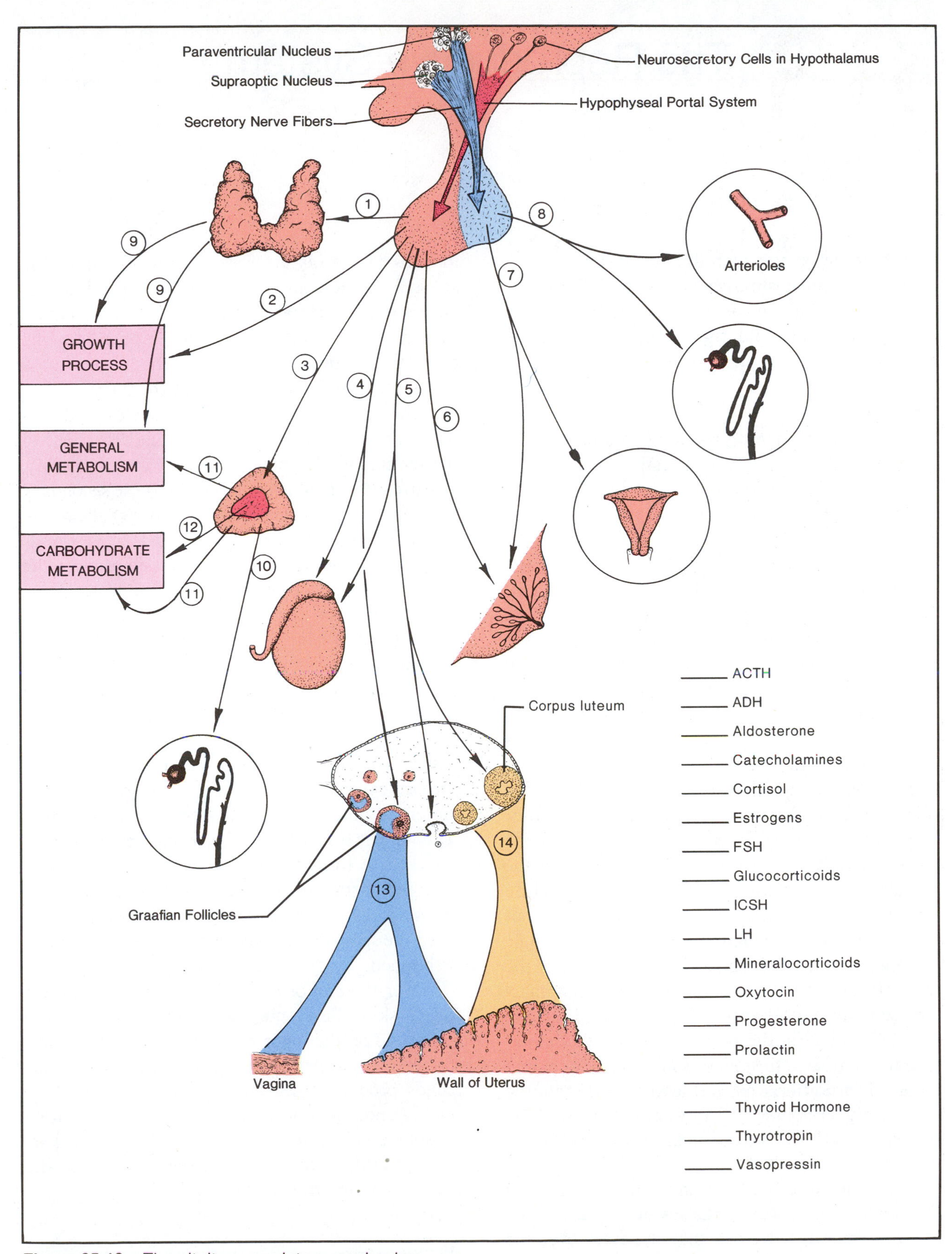

Figure 65.13 The pituitary regulatory mechanism.

66 The Reproductive System

This exercise on the reproductive system will include gross and histological studies of the reproductive organs, as well as dissection of the cat. The cytological nature and significance of spermatogenesis and oogenesis will also be studied.

Materials:

- models of human reproductive organs
- prepared slides of human testis
- ocular micrometer
- cat and dissection instruments

The Male Organs

Figure 66.1 is a sagittal section of the male reproductive system. The primary sex organs are the paired oval **testes** (*testicles*), which lie enclosed in a sac, the **scrotum.** The external nature of the testes provides a slightly lower temperature (94°–95° F), which favors the development of mature spermatozoa.

Lying over the superior and posterior surfaces of each testis is an elongated, flattened body, the **epididymis,** where immature sperm are stored after they leave the testis. The cutaway section of the testis in figure 66.1 reveals the relationship of the epididymis to the testis. Each testis is divided up into several chambers by partitions, or **septa,** that contain coiled-up **seminiferous tubules.** A single seminiferous tubule is shown held out of one of the compartments. All the seminiferous tubules anastomose to form a network of tubules called the **rete testis.** Cilia within the rete testis move the immature spermatozoa through this maze of tubules into 10 or 15 **vasa efferentia** that lead directly into the epididymis. The outer wall, or **tunica albuginea,** that surrounds each testis consists of fibrous connective tissue.

In the act of ejaculation the spermatozoa leave the epididymis by way of the **vas deferens.** Tracing this duct upward, we see that it passes over the pubic bone and bladder into the pelvic cavity. The terminus of the vas deferens is enlarged to form the **ampulla of the vas deferens.** A pair of glands, the **seminal vesicles,** and the ampulla empty into the **common ejaculatory duct.** This latter duct passes through the prostate gland and empties into the **prostatic urethra.** From here the sperm pass through the **penile urethra** out of the body.

The prostate gland, seminal vesicles, and bulbourethral glands contribute alkaline secretions to the seminal fluid, which stimulates sperm motility. The **prostate gland** is the largest of these secondary sex glands. The **bulbourethral** (Cowper's) **glands** are the smallest of the three. These pea-sized glands have ducts about one inch long that empty into the urethra at the base of the penis. The secretion of these glands is a clear mucoid fluid that lubricates the end of the penis and prepares the urethra for seminal fluid.

The **penis** consists of three cylinders of erectile tissue: two corpora cavernosa and one corpus spongiosum. The **corpus spongiosum** is a cylinder of tissue that surrounds the penile urethra. The two **corpora cavernosa** are located in the dorsal part of the organ and are separated on the midline by a **septum.** The distal end of the corpus spongiosum is enlarged to form a cone-shaped **glans penis.** The enlarged portion of the urethra within the glans is the **navicular fossa.** Erection of the penis occurs when the spongelike tissue of the three corpora fills with blood.

Over the end of the glans penis lies a circular fold of skin, the **prepuce.** Around the neck of the glans, and on the inner surface of the prepuce, are scattered small *preputial glands.* These sebaceous glands produce a secretion of peculiar odor that readily undergoes decomposition to form a whitish substance called *smegma. Circumcision* is a surgical procedure that involves the removal of the prepuce to facilitate sanitation.

Assignment:

Label figure 66.1.

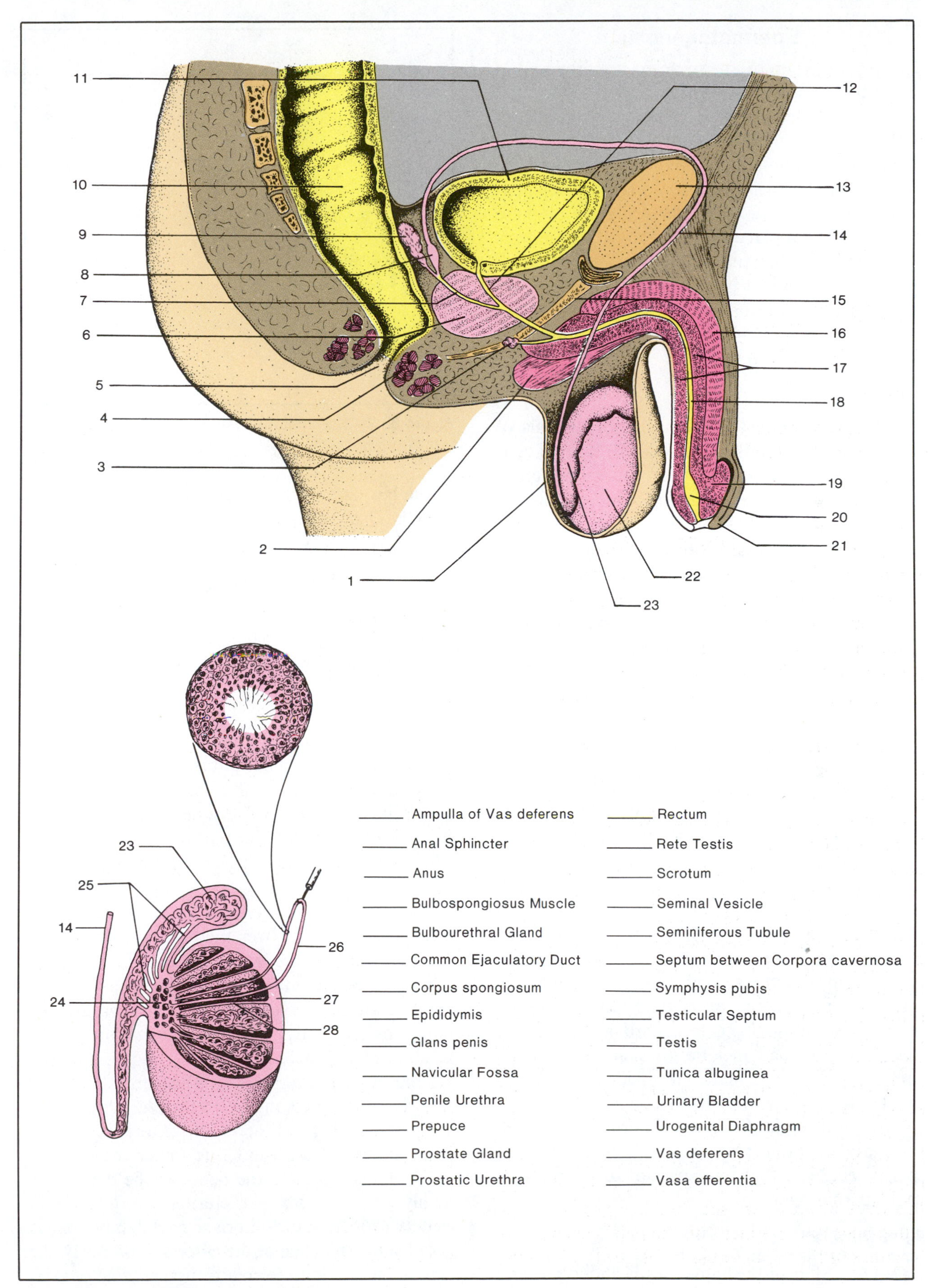

Figure 66.1 Male reproductive organs.

Spermatogenesis

The process whereby spermatozoa are produced in the testes is known as *spermatogenesis.* This function of the testes begins at puberty and continues without interruption throughout life. Figure 66.2 illustrates a section of a seminiferous tubule and a diagram of the various stages in the development of mature spermatozoa. The formation of these germ cells takes place as the result of both mitosis and meiosis. In this study of spermatogenesis, slides of human testes will be studied under the microscope. Before examining the slides, however, familiarize yourself with the characteristic differences between meiosis and mitosis.

Mitosis

All spermatozoa originate from **spermatogonia** of the **primary germinal epithelium.** This layer of cells is located at the periphery of the seminiferous tubule. These cells contain the same number of chromosomes as other body cells (i.e., 23 pairs) and are said to be *diploid.* Mitotic division of these cells occurs constantly, producing other spermatogonia. As the spermatogonia move toward the center of the tubule they enlarge to form **primary spermatocytes.** In this stage the homologous chromosomes unite to form chromosomal units called *tetrads.* This union of homologous chromosomes is called *synapsis.*

Meiosis

Each primary spermatocyte divides further to produce two secondary spermatocytes. This division, however, occurs by meiosis instead of mitosis. *Meiosis,* or reduction division, results in the distribution of a *haploid* number of chromosomes to each **secondary spermatocyte.** The result is that each secondary spermatocyte has only twenty-three chromosomes instead of forty-six. The individual chromosomes of the secondary spermatocytes are formed by the splitting of the tetrads along the line of previous conjugation to form chromosomes called *dyads.* The second meiotic division occurs when each secondary spermatocyte divides to produce two haploid **spermatids.** In this division the dyads split to form single chromosomes, or *monads.* Each spermatid metamorphoses directly into a mature sperm cell. Note that four spermatozoa form from each spermatogonium.

Assignment:

Examine a slide of the human testis under low- and high-power magnifications. Refer to figure HA–36 of the Histology Atlas to identify the various types of cells. Draw a representative section of a tubule, labeling all cell types.

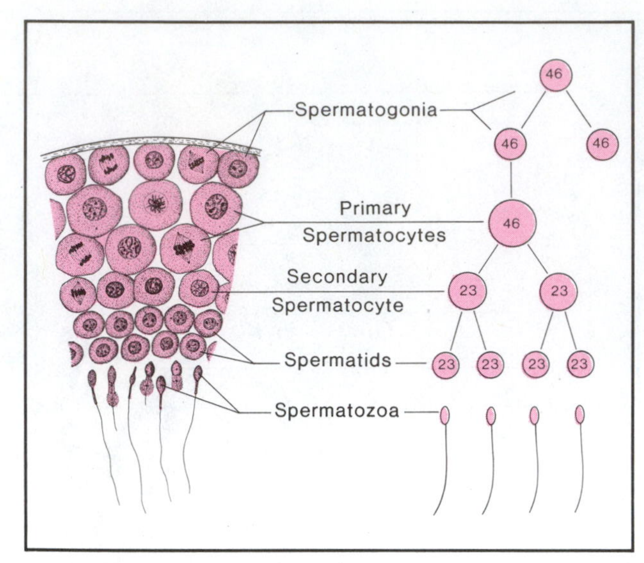

Figure 66.2 Spermatogenesis.

The Female Organs

Figures 66.3, 66.4, and 66.5 illustrate the female reproductive organs. Models and wall charts will be helpful in identifying all the structures.

External Genitalia

The **vulva** of the external female reproductive organs includes the mons pubis, labia majora, labia minora, and hymen. All of these structures are shown in figure 66.3. The **mons pubis** (*mons veneris*) is the most anterior portion and consists of a firm cushionlike elevation over the symphysis pubis. It is covered with hair.

Two folds of skin on each side of the **vaginal orifice** (label 11) lie over the opening. The larger exterior folds are the **labia majora.** Their exterior surfaces are covered with hair; their inner surfaces are smooth and moist. The labia majora are homologous (of similar embryological origin) to the scrotum of the male. Medial to the labia majora are the smaller **labia minora.** These folds meet anteriorly on the median line to form a fold of skin, the **prepuce of the clitoris.** The **clitoris** is a small protuberance of erectile tissue under the prepuce that is homologous to the penis on the male. It is highly sensitive to sexual excitation. The fold of skin that extends from the clitoris to each labium minus is called the **frenulum of the clitoris.** Posteriorly, the labia minora join to form a transverse fold of skin, the **posterior commissure,** or **fourchette.** Between

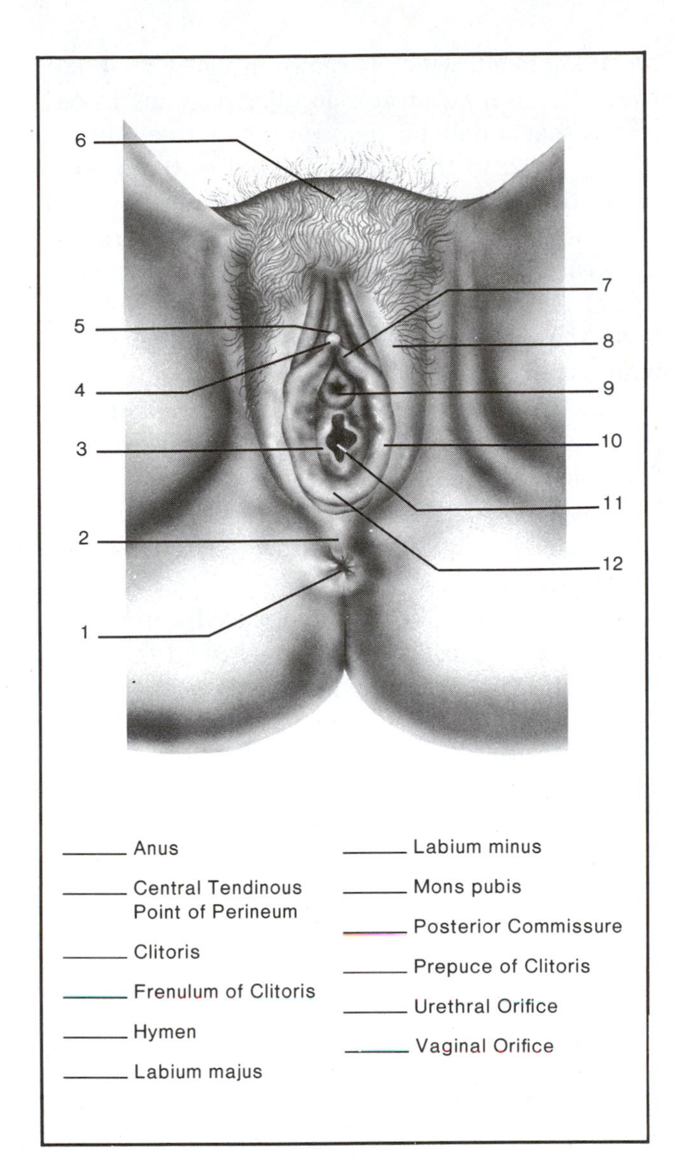

Figure 66.3 Female genitalia.

the anus and the posterior commissure is the **central tendinous point of the perineum.** The **perineum** corresponds to the outlet of the pelvis.

The *vestibule* is the area between the labia minora, extending from the clitoris to the fourchette. Situated within the vestibule are the vaginal orifice, hymen, urethral orifice, and openings of the vestibular glands. The **urethral orifice** lies about 2–3 centimeters posterior to the clitoris. Many small **paraurethral glands** surround this opening. They are homologous to the prostate gland of the male. On either side of the vaginal orifice are openings from the two **greater vestibular** (*Bartholin's*) **glands.** These glands are homologous to the bulbourethral glands. They provide mucous secretion for vaginal lubrication. The **hymen** is a thin fold of mucous membrane that separates the vagina from the vestibule. It may be completely absent or cover the vaginal orifice partially or completely. Its condition or absence is not a determinant of virginity.

Assignment:

Label figure 66.3.

Internal Organs

Figure 66.4 is a posterior view of the female reproductive organs which shows the relationship of the vagina, uterus, ovaries, and uterine tubes. It also reveals many of the principal supporting ligaments.

The Uterus The uterus is a pear-shaped, thick-walled, hollow organ which consists of a fundus, corpus, and cervix. The **fundus** is the dome-shaped portion at the large end. The body, or **corpus,** extends from the fundus to the cervix. The **cervix,** or neck, of the uterus is the narrowest portion; it is about one inch long. Within the cervix is an **endocervical canal** that has an outer opening, the **external os,** that opens into the **vagina;** an inner opening, the **internal os,** opens into the cavity of the corpus.

The thick muscular portion of the uterine wall is called the **myometrium.** The inner lining, or **endometrium,** consists of mucosal tissue. The anterior and posterior surfaces of the corpus, as well as the fundus, are covered with peritoneum.

The Uterine (*Fallopian*) **Tubes** Extending laterally from each side of the uterus is a uterine tube. The distal end of each tube is enlarged to form a funnel-shaped fimbriated **infundibulum** that surrounds an **ovary.** Although the infundibulum doesn't usually contact the ovary, one or more of the fingerlike **fimbriae** on the edge of the infundibulum usually do contact it. The infundibula and fimbriae receive oocytes produced by the ovaries.

The uterine tubes are lined with ciliated columnar cells which assist in transporting egg cells from the ovary into the uterus. The tubes also have a muscular wall which produces peristaltic movements that help to propel the egg cells. Fertilization usually occurs somewhere along the uterine tube. Note that each uterine tube narrows down to form an **isthmus** near the uterus.

The Urinary Bladder Figure 66.5 reveals the relationship of the urinary bladder to the reproductive organs. Note that it lies between the uterus and the **symphysis pubis.** The base of the bladder is in direct contact with the anterior vaginal wall. The **urethra** is positioned between the vagina and the symphysis pubis. The neck of the bladder lies on the superior surface of the **urogenital diaphragm** (label 25, figure 66.5).

The superior surface of the bladder is covered with peritoneum which is continuous with the peritoneum on the anterior face of the uterus. The small cavity lined with peritoneum between the bladder and uterus is known as the **anterior cul-de-sac.** The space between the uterus and rectum is the **recto-uterine pouch** (pouch of Douglas).

Ligaments The uterus, ovaries, and uterine tubes are held in place and supported by various ligaments formed from the peritoneum and connective and muscular tissue. Figure 66.4 reveals the majority of them. A few are shown in figure 66.5.

The largest supporting structure is the **broad ligament** (label 3, figure 66.4). It is an extension of the peritoneal layers that cover the fundus and corpus of the uterus. It extends up over the fundus and corpus of the uterus. It extends up over the uterine tubes to form a mesentery, the **mesosalpinx,** on each side of the uterus. This portion of the broad ligament, which is shown in figure 66.4 between the ovary and uterine tube, contains blood vessels that supply nutrients to the uterine tube.

Attached to the cervical region of the uterus are two **sacrouterine ligaments** that anchor the uterus to the sacral wall of the pelvic cavity. A **round ligament** (label 16, figure 66.5) extends from each side of the uterus to the body wall. This ligament is not shown in figure 66.4 because it is anterior in position.

Each ovary is held in place by ovarian and suspensory ligaments. The **ovarian ligament** extends from the medial surface of the ovary to the uterus. The **suspensory ligament** is a peritoneal fold on the other side of the ovary that attaches the ovary to the uterine tube.

Each uterine tube is held in place by an **infundibulopelvic ligament** that is attached to the posterior surface of the infundibulum.

Assignment:

Label figures 66.4 and 66.5.

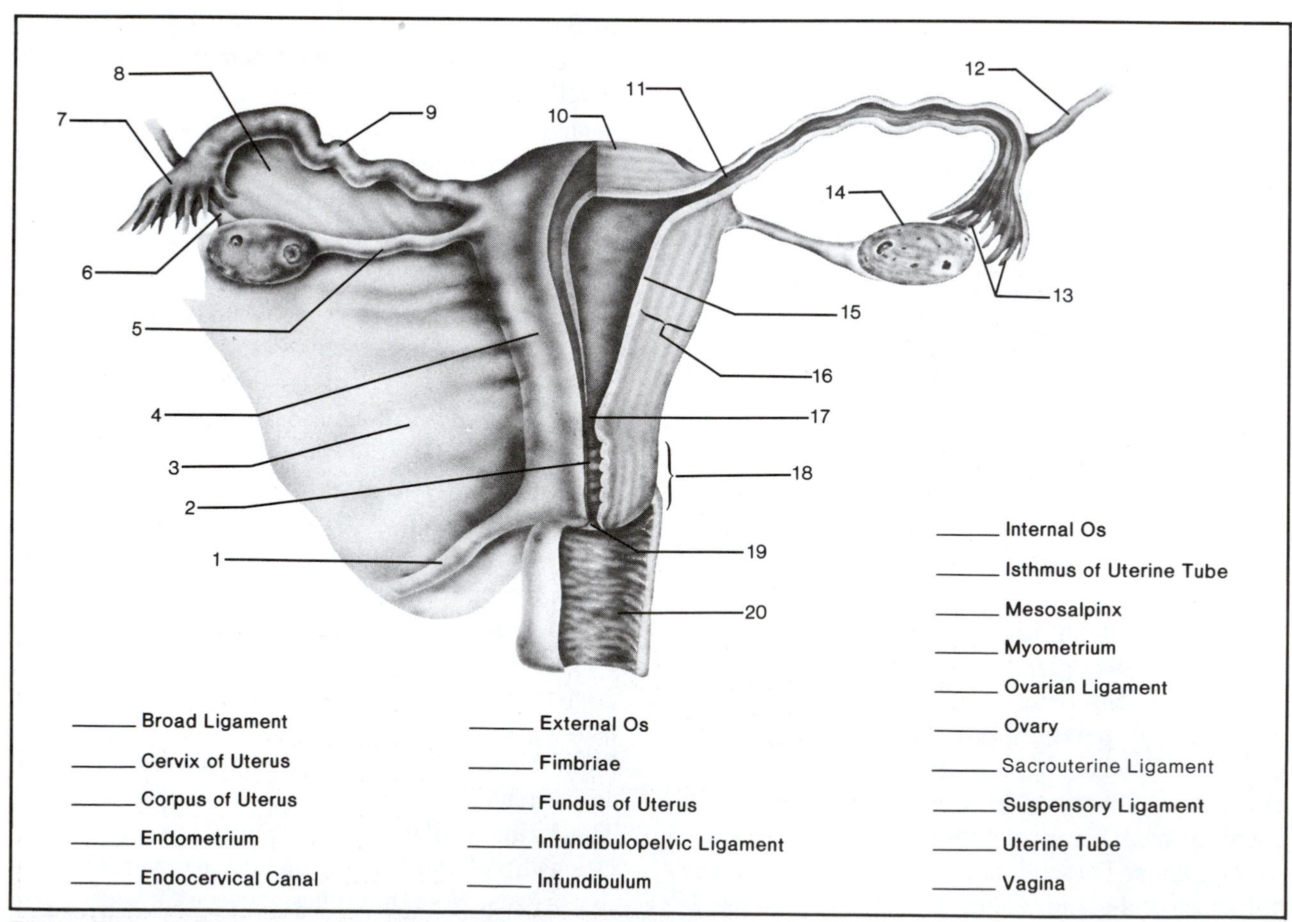

Figure 66.4 Posterior view of female reproductive organs.

Oogenesis

The development of egg cells in the ovary is called *oogenesis*. It was pointed out on page 327 that as many as 400,000 immature follicles develop in the ovaries from the germinal epithelium. At the onset of puberty all the potential ova in these follicles contain 46 chromosomes and are designated as **primary oocytes.** Only about 400 of these oocytes will mature during the reproductive lifetime of a woman.

Note in figure 66.6 that the primary oocyte, which forms from an oogonium of the germinal epithelium during embryological development, has 46 chromosomes. In the prophase stage of a dividing oocyte double-stranded chromosomes *(dyads)* unite to form 23 homologous pairs of chromosomes during synapsis. Each pair is made up of four chromatids; thus, it is called a *tetrad.* When the primary oocyte divides in the first meiotic division, each dyad member of each homologous pair enters a different cell. The first meiotic division produces a **secondary oocyte** with all the yolk of the primary oocyte and a **polar body** with no yolk. The secondary oocyte and polar body each contain 23 dyads.

The second meiotic division produces a **mature ovum** with 23 chromosomes that are *monads;* that is, each chromosome consists of a single chromatid. Another polar body is also formed in this division. The first polar body divides also to produce two new polar bodies with monads. The end result is that one primary oocyte produces one mature ovum and three polar bodies. The polar bodies never serve any direct function in the fertilization process.

When ovulation takes place the "ovum" is a secondary oocyte. Usually, it is necessary for the sperm to penetrate the cell to trigger the second meiotic division. When the 23 chromosomes in the sperm unite with the chromosomes of the mature ovum, a **zygote** of 46 chromosomes is formed.

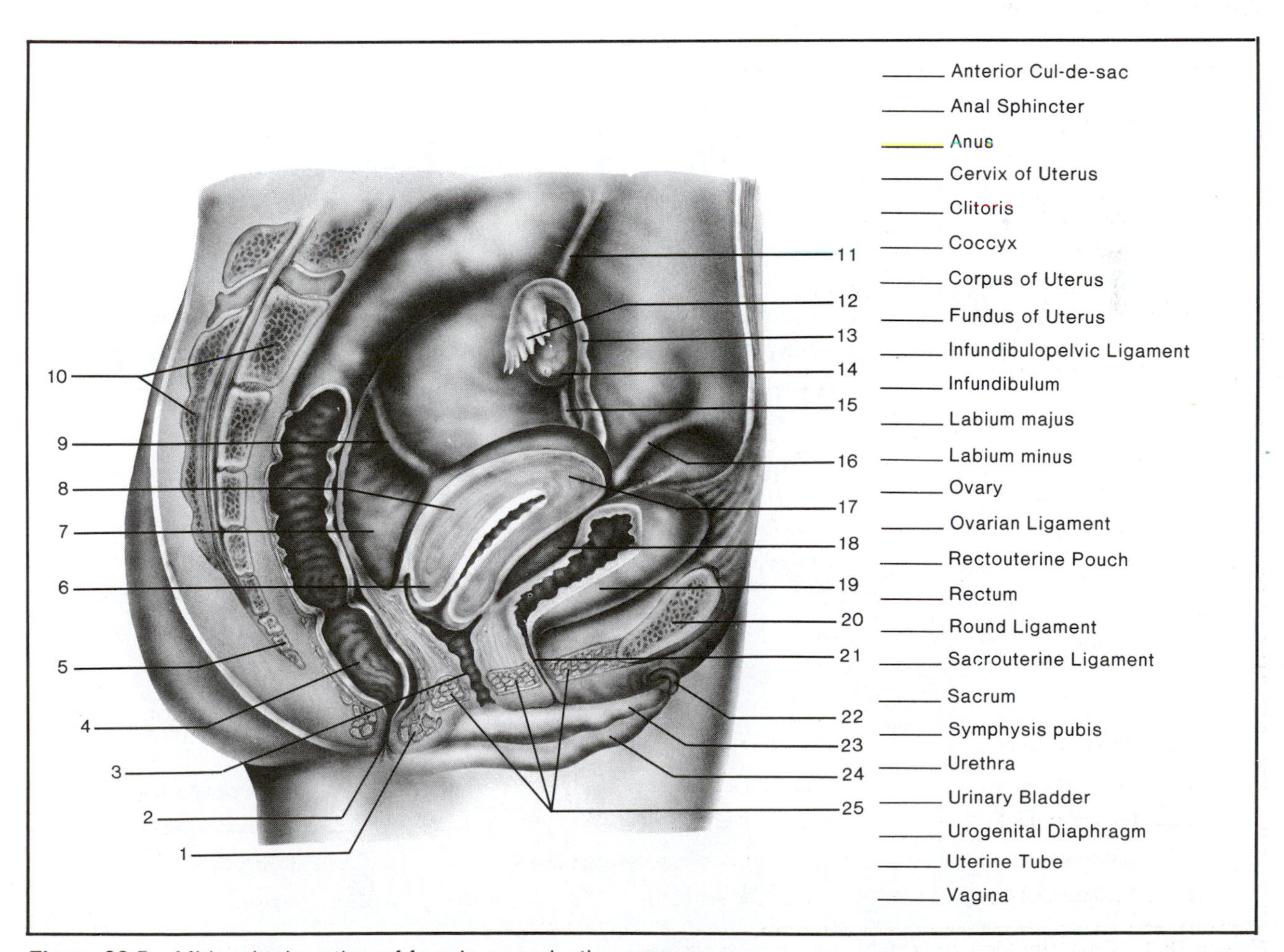

Figure 66.5 Midsagittal section of female reproductive organs.

Assignment:

Study figures 66.6 and 65.8 thoroughly to familiarize yourself with the terminology and kinds of divisions. Several questions on the Laboratory Report pertain to this phenomenon.

Histological Study

Assuming that spermatogenesis slides have been previously studied (page 336), we will focus here, primarily, on a histological study of the various reproductive organs of both sexes. Appropriate pages of the Histology Atlas and some of the illustrations in Exercise 65 will be used for reference.

Materials:

prepared slides:
- ovary (H9.310 or H9.3181)
- uterus (H9.331), vagina (H9.341)
- uterine tube (H9.321)
- vas deferens (H9.432)
- penis (H9.462), prostate gland (H9.443)
- seminal vesicle (H9.442)

Ovary Study various slides of the ovary to identify **primordial follicles,** the **germinal epithelium,** and **Graafian follicles.** Since the ovaries are not included in the Histology Atlas, use figure 65.8, page 328, for reference.

Note that the developing ova in primordial follicles are surrounded by an incomplete layer of low cuboidal or flattened epithelium; the ova in developing follicles, on the other hand, are larger and are surrounded by two or more layers of cuboidal follicular cells.

Uterus When studying a slide of the uterus, take into consideration the phase of the menstrual cycle that is represented on the slide. The photomicrographs in figure HA–31 were made of Turtox slide H9.331, which represents about the third week in the cycle. Both the endometrium and myometrium will look much different in the earlier and later phases.

Differentiate the **endometrium** from the **myometrium.** Identify the **uterine glands** and endometrial **epithelium.**

Uterine Tube Scan a slide of the uterine tube with the low-power objective to find the lumen of the tube. Refer to illustrations C and D, figure HA–31. Note the extremely irregular outline of the mucosa, which consists of a mixture of ciliated and nonciliated columnar epithelium. How many layers of smooth muscle tissue do you find in the muscularis?

Vagina Examine a slide of the mucosa of the vagina and compare it to illustration C, figure HA–32. How does the vaginal mucosa differ from the skin? What function do the lymphocytes perform in the lamina propria?

Uterine Cervix Study a longitudinal section through the cervical area of the uterus which shows where the vaginal lining meets the cervical region.

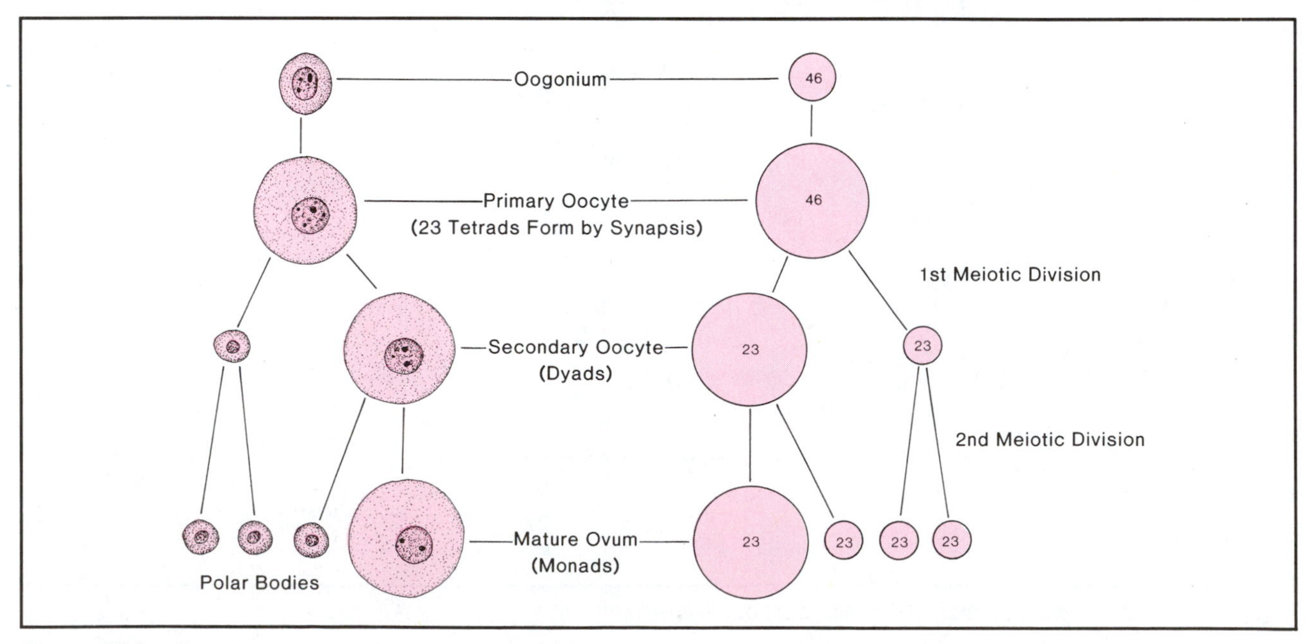

Figure 66.6 Oogenesis.

Consult illustration A, figure HA–32. Note the large number of spaces, or **cervical crypts,** that lie deep in the cervical tissue. Read the legend in figure HA–32 which pertains to the significance of these crypts.

Vas Deferens Examine a cross-sectional slide of the vas deferens, using minimum magnification to study the entire structure. Compare your slide with figure HA–34. Note that between its inner **epithelium** and the outer **adventitia** are three layers of smooth muscle tissue: an inner longitudinal layer, an outer longitudinal layer, and a middle layer of circular fibers. Note what the legend states about the **stereocilia** that are seen on the epithelium.

Penis Examine a cross-sectional preparation of the penis. The photomicrographs in illustrations C and D, HA–33, were made from an H9.462 Turtox slide. Examine the epithelium under high-dry magnification. Does the epithelium on your slide look like illustration D? If not, why not? Read the comments in the legend that pertain to epithelial differences.

Seminal Vesicle A cross section of a portion of a seminal vesicle is seen in illustrations A and B, figure HA–35. Examine a slide first with low-power magnification and note the striking way in which the mucosa is folded upon itself to produce a maze of pockets. Examine the epithelium under high-dry or oil immersion. What is the nature and function of the seminal fluid?

Prostate Gland Examine a slide of the gland under low power first to identify the structures seen in illustration C, figure HA–35. Note the large number of **tubulo-alveolar glands** that secrete the prostatic fluid. Why does the prostate gland tend to become hardened and restrict micturition in older men?

Laboratory Report

Complete the Laboratory Report for this exercise.

LABORATORY REPORT 1, 2

Student: ______________

Desk No: ________ Section: ______________

Anatomical Terminology, Body Cavities, and Membranes

A. Illustration Labels

Record the label numbers for all the illustrations in the appropriate places in the answer columns.

B. Terminology

From the list of positions, sections, and membranes, select those that are applicable to the following statements. In some cases more than one answer may apply.

Relative Positions		*Sections*	*Membranes*
anterior—1	lateral—7	frontal—13	mesentery—17
caudad—2	medial—8	midsagittal—14	pericardium, parietal—18
cephalad—3	posterior—9	sagittal—15	pericardium, visceral—19
distal—4	proximal—10	transverse—16	peritoneum, parietal—20
dorsal—5	superior—11		peritoneum, visceral—21
inferior—6	ventral—12		pleura, parietal—22
			pleura, pulmonary—23

Sections:

1. Divides body into front and back portions.
2. Divides body into unequal right and left sides.
3. Three sections that are longitudinal.
4. Divides body into equal right and left sides.
5. Two sections that expose both lungs and heart in each section.

Positions:

6. The tongue is ______________ to the palate.
7. The shoulder of a dog is ______________ to its hip.
8. A hat is worn on the ______________ surface of the head.
9. The cheeks are ______________ to the tongue.
10. The fingertips are ______________ to all structures of the hand.
11. The human shoulder is ______________ to the hip.
12. The shoulder is the ______________ (*proximal, distal*) portion of the arm.

Membranes:

13. Serous membrane attached to lung surface.
14. Serous membrane attached to thoracic wall.
15. Double-layered membrane that holds abdominal organs in place.
16. Double-layered membrane that surrounds the heart.

Select the surfaces on which the following are located:

17. Ear
18. Adam's apple
19. Kneecap
20. Palm of hand

Answers

Terms	Fig. 1.1
1. ________	________
2. ________	________
3. ________	________
4. ________	________
5. ________	________
6. ________	________
7. ________	________
8. ________	________
9. ________	________
10. ________	________
11. ________	________
12. ________	________
13. ________	________
14. ________	________
15. ________	________
16. ________	________
17. ________	________
18. ________	________
19. ________	________
20. ________	________

Fig. 2.1	Fig. 1.3
________	________
________	________
________	________
________	________
________	________
________	________
________	________
________	________
________	________
________	________
________	________
________	
________	
________	
________	
________	

C. Localized Areas

Identify the specific areas of the body described by the following statements.

antebrachium—1	calf—6	gluteal—11	iliac—16
antecubital—2	costal—7	groin—12	lumbar—17
axilla—3	cubital—8	ham—13	plantar—18
brachium—4	epigastric—9	hypochondriac—14	pectoral—19
buttocks—5	flank—10	hypogastric—15	popliteal—20
			umbilical—21

1. The elbow area.
2. The "rump" area.
3. The underarm area.
4. The sole of the foot.
5. The upper arm.
6. The forearm.
7. Upper chest region.
8. Depression on back of leg behind knee.
9. Side of abdomen between lower edge of rib cage and upper edge of hipbone.
10. Back portion of abdominal body wall that extends from lower edge of rib cage to hipbone.
11. Area over the ribs on the dorsum.
12. Posterior portion of lower leg.
13. Anterior surface of the elbow.
14. Abdominal area that surrounds the navel.
15. Abdominal area which is lateral to pubic region.
16. Area of abdominal wall that covers the stomach.
17. Abdominal area which is lateral to epigastric area.
18. Abdominal area which is lateral to umbilical area.
19. Area of arm on opposite side of elbow.
20. Abdominal areas between the transpyloric and transtubercular planes.

D. Body Cavities

From the following list of cavities select those applicable to the following statements.

abdominal—1	dorsal—4	spinal—8
abdominopelvic—2	pelvic—5	thoracic—9
cranial—3	pericardial—6	ventral—10
	pleural—7	

1. Cavity that consists of cranial and spinal cavities.
2. Most inferior portion of the abdominopelvic cavity.
3. Cavity separated into two major divisions by the diaphragm.
4. Cavity inferior to the diaphragm that consists of two parts.

Select the cavity in which the following organs are located:

5. Kidneys	9. Rectum	13. Spleen
6. Duodenum	10. Pancreas	14. Spinal cord
7. Brain	11. Lungs and heart	15. Urinary bladder
8. Heart	12. Liver	16. Stomach

Answers

Areas	Fig. 1.2
1. ________	________
2. ________	________
3. ________	________
4. ________	________
5. ________	________
6. ________	________
7. ________	________
8. ________	________
9. ________	________
10. ________	________
11. ________	________
12. ________	________
13. ________	________
14. ________	________
15. ________	________
16. ________	________
17. ________	________
18. ________	________
19. ________	________
20. ________	**Fig. 2.2**
	________
Cavities	________
1. ________	________
2. ________	________
3. ________	________
4. ________	________
5. ________	________
6. ________	________
7. ________	________
8. ________	**Fig. 2.3**
9. ________	
10. ________	________
11. ________	________
12. ________	________
13. ________	________
14. ________	________
15. ________	________
16. ________	________

LABORATORY REPORT **3**

Student: ____________________

Desk No: __________ Section: __________

Organ Systems: Rat Dissection

A. System Functions

Select the system (or systems) that perform the following functions in the body. Record the numbers in the answer column. More than one answer may apply.

circulatory—1	lymphatic—5	respiratory—9
digestive—2	muscular—6	reticuloendothelial—10
endocrine—3	nervous—7	skeletal—11
integumentary—4	reproductive—8	urinary—12

1. Removes carbon dioxide from the blood.
2. Provides a shield of protection for vital organs such as the brain and heart.
3. Movement of ribs in breathing.
4. Carries heat from the muscles to the surface of the body for dissipation.
5. Provides rigid support for the attachment of muscles.
6. Transportation of food from the intestines to all parts of the body.
7. Disposes of urea, uric acid, and other cellular wastes.
8. Movement of arms and legs.
9. Helps to maintain normal body temperature by dissipating excess heat.
10. Destroys microorganisms once they break through the skin.
11. Phagocytic cells of this system remove bacteria from lymph.
12. Carries messages from receptors to centers of interpretation.
13. Controls the rate of growth.
14. Returns tissue fluid to the blood.
15. Produces various kinds of blood cells.
16. Enables the body to adjust to changing internal and external environmental conditions.
17. Ensures continuity of the species.
18. Eliminates excess water from the body.
19. Carries hormones from glands to all tissues.
20. Acts as a shield against invasion by bacteria.
21. Carries fats from the intestines to the blood.
22. Coordination of all body movements.
23. Breaks food particles down into small molecules.
24. Controls the development of the ovaries and testes.
25. Production of spermatozoa and ova.

Answers
System Functions
1. ________
2. ________
3. ________
4. ________
5. ________
6. ________
7. ________
8. ________
9. ________
10. ________
11. ________
12. ________
13. ________
14. ________
15. ________
16. ________
17. ________
18. ________
19. ________
20. ________
21. ________
22. ________
23. ________
24. ________
25. ________

B. Organ Placement

Select the system at the right that includes the following organs.

Organs	Systems
1. Bones	circulatory—1
2. Brain	digestive—2
3. Bronchi	endocrine—3
4. Dermis	integumentary—4
5. Esophagus	lymphatic—5
6. Hair	muscular—6
7. Heart	nervous—7
8. Kidneys	reproductive—8
9. Lungs	respiratory—9
10. Pancreas	skeletal—10
11. Spleen	urinary—11
12. Stomach	
13. Teeth	
14. Trachea	
15. Toenails	
16. Tonsils	
17. Thyroid	
18. Ureters	
19. Uterus	
20. Veins	

Answers

Organ Placement

1. ________
2. ________
3. ________
4. ________
5. ________
6. ________
7. ________
8. ________
9. ________
10. ________
11. ________
12. ________
13. ________
14. ________
15. ________
16. ________
17. ________
18. ________
19. ________
20. ________

LABORATORY REPORT **4**

Student: ____________________

Desk No: ________ Section: ____________

Microscopy

A. Completion Questions

Record the answers to the following questions in the column at the right.

1. List three fluids that may be used for cleaning lenses.
2. How can one greatly increase the bulb life on a microscope lamp if the voltage is variable?
3. What characteristic of a microscope enables one to switch from one objective to another without altering the focus?
4. What effect (*increase* or *decrease*) does closing the diaphragm have on the following?
   a. Image brightness
   b. Image contrast
   c. Resolution
5. In general, at what position should the condenser be kept?
6. Express the maximum resolution of the compound microscope in terms of micrometers (μm).
7. If you are getting 225× magnification with a 45× high-dry objective, what would be the power of the eyepiece?
8. What is the magnification of objects observed through a 100× oil immersion objective with a 7.5× eyepiece?
9. Immersion oil must have the same refractive index as ________ to be of any value.
10. Substage filters should be of a ______________ color to get the maximum resolution of the optical system.

B. True–False

Record these statements as True or False in the answer column.

1. Eyepieces are of such simple construction that almost anyone can safely disassemble them for cleaning.
2. Lenses can be safely cleaned with almost any kind of tissue or cloth.
3. When swinging the oil immersion objective into position after using high-dry, one should always increase the distance between the lens and slide to prevent damaging the oil immersion lens.
4. Instead of starting first with the oil immersion lens, it is best to use one of the lower magnifications first, and then swing the oil immersion into position.
5. The 45× and 100× objectives have shorter working distances than the 10× objective.

Answers

Completion

1a. ____________________
b. ____________________
c. ____________________
2. ____________________
3. ____________________
4a. ____________________
b. ____________________
c. ____________________
5. ____________________
6. ____________________
7. ____________________
8. ____________________
9. ____________________
10. ____________________

True–False

1. ____________________
2. ____________________
3. ____________________
4. ____________________
5. ____________________

C. Multiple Choice

Select the best answer for the following statements.

1. The resolution of a microscope is increased by
   1. using blue light.
   2. stopping down the diaphragm.
   3. lowering the condenser.
   4. raising the condenser to its highest point.
   5. Both 1 and 4 are correct.
2. The magnification of an object seen through the 10× objective with a 10× ocular is
   1. ten times.
   2. twenty times.
   3. 1000 times.
   4. None of these are correct.
3. The most commonly used ocular is
   1. 5×.
   2. 10×.
   3. 15×.
   4. 20×.
4. Microscope lenses may be cleaned with
   1. lens tissue.
   2. a soft linen handkerchief.
   3. an air syringe.
   4. Both 1 and 3 are correct.
   5. 1, 2, and 3 are correct.
5. When changing from low power to high power, it is generally necessary to
   1. lower the condenser.
   2. open the diaphragm.
   3. close the diaphragm.
   4. Both 1 and 2 are correct.
   5. Both 1 and 3 are correct.

Answers
Multiple Choice
1. ________
2. ________
3. ________
4. ________
5. ________

LABORATORY REPORT 5, 6

Student: ______________________

Desk No: __________ Section: ______________

Cytology: Basic Cell Structure and Mitosis

A. *Figure 5.1*

Record the labels for figure 5.1 in the answer column.

B. *Cell Drawings*

Sketch in the space below one or two epithelial cells as seen under high-dry magnification. Label the **nucleus, cytoplasm,** and **cell membrane.** If mitosis drawings are to be made put them on a separate sheet of paper.

C. *Cell Structure*

From the list of structures below, select those that are described by the following statements. More than one structure may apply to some of the statements.

centrosome—1	Golgi apparatus—7	microtubules—12	polyribosomes—17
chromatin granules—2	lysosome—8	nuclear envelope—13	RER—18
cilia—3	microvilli—9	nucleolus—14	ribosomes—19
condensing vacuoles—4	mitochondria—10	nucleus—15	secretory granules—20
cytoplasm—5	microfilaments—11	plasma membrane—16	SER—21
flagellum—6			

1. Prominent body in nucleus that produces ribosomes.
2. Trilaminar unit membrane structure.
3. Outer membrane of the cell.
4. A nonmembranous organelle consisting of two bundles of microtubules.
5. Bodies within the nucleus that represent chromosomal material.
6. ER that lacks ribosomes.
7. Double-layered unit membrane structure.
8. Small bodies of RNA and protein attached to surfaces of the RER.
9. Extensive system of tubules, vesicles, and sacs within the cytoplasm.
10. All protoplasmic material between the plasma membrane and nucleus.
11. Sacs in the cytoplasm that contain digestive enzymes.
12. A unit membrane organelle in the cytoplasm that contains cristae.
13. Short, hairlike appendages on the surface of some cells.
14. ER that is covered by ribosomes.
15. Chains and rosettes of small bodies scattered throughout the cytoplasm.
16. Unit membrane structure that is continuous with the plasma and nuclear membranes.
17. Extranuclear bodies that possess DNA and are able to replicate themselves.
18. A layered concave structure in the cytoplasm that is similar to the basic structure of SER.
19. Surface protuberances that form a delicate brush border.
20. Bodies produced by Golgi and exocytosed by the cell.

Answers

Cell Structure	Fig. 5.1
1. ______	______
2. ______	______
3. ______	______
4. ______	______
5. ______	______
6. ______	______
7. ______	______
8. ______	______
9. ______	______
10. ______	______
11. ______	______
12. ______	______
13. ______	______
14. ______	______
15. ______	______
16. ______	
17. ______	
18. ______	
19. ______	
20. ______	

D. *Organelle Functions*

Select the organelles that perform the following functions. More than one structure may apply to some statements.

centrosome—1	microvilli—5	plasma membrane—9
flagellum—2	mitochondrion—6	polyribosomes—10
Golgi apparatus—3	nucleolus—7	RER—11
lysosome—4	nucleus—8	SER—12

1. Forms asters during mitosis.
2. Produces ribonucleoprotein for ribosomes.
3. The microcirculatory system of the cell.
4. Contains genetic code of the cell.
5. Place where oxidative respiration occurs.
6. Synthesize protein for endogenous use.
7. Brings about rapid hydrolysis (digestion) of cell after death.
8. Place where ATP is synthesized and stored.
9. Regulates the flow of materials into and out of cell.
10. Synthesizes protein for extracellular distribution.
11. Disposal organelles in cytoplasm that digest protein.
12. Increases absorptive area of the cell.
13. Synthesis, packaging, and transportation of substances.
14. Plays a role in flagellar development.
15. Accomplishes some motile function for some cells.

E. *True-False*

Evaluate the validity of the following statements concerning cell structure and function.

1. Secretory cells have well-developed Golgi.
2. Mitochondria are able to replicate themselves.
3. Practically all molecules that pass through the plasma membrane are actively assisted by the membrane.
4. Ribosomes are found in the cytoplasm, mitochondria, RER, and SER.
5. The inner and outer layers of the plasma membrane are essentially lipoidal in nature with scattered molecules of protein and glycoprotein.
6. A "brush border" consists of microvilli.
7. Spermatozoa move with cilia.
8. Ribosomes are primarily involved in oxidative respiration.
9. Cilia form from basal bodies derived from centrioles.
10. Asters develop from mitochondria.

F. *Mitosis (Ex. 6)*

Select the phase in which the following events occur.

interphase—1
prophase—2
metaphase—3
anaphase—4
telophase—5
daughter cells—6

1. Nuclear membrane disappears.
2. Cleavage furrow forms.
3. Spindle fibers begin to form.
4. New nuclear membrane forms.
5. Asters form from centrioles.
6. Centriole replication occurs.
7. Replication of DNA occurs.
8. Sister chromosomes separate and pass to opposite poles.
9. Formed after telophase is completed.
10. Chromosomes become oriented on equatorial plane.

Answers

Functions

1. ________
2. ________
3. ________
4. ________
5. ________
6. ________
7. ________
8. ________
9. ________
10. ________
11. ________
12. ________
13. ________
14. ________
15. ________

True-False

1. ________
2. ________
3. ________
4. ________
5. ________
6. ________
7. ________
8. ________
9. ________
10. ________

Mitosis

1. ________
2. ________
3. ________
4. ________
5. ________
6. ________
7. ________
8. ________
9. ________
10. ________

LABORATORY REPORT **7**

Student: ______________________

Desk No: ________ Section: ____________

Osmosis and Cell Membrane Integrity

A. Molecular Movement

1. **Brownian Movement**
   a. Were you able to see any movement of carbon particles under the microscope? __________

   b. What forces propel the movement of these particles? ______________________

   ____

   ____

   ____

2. **Diffusion** Plot the distance of diffusion of molecules for the two crystals on the following graphs.

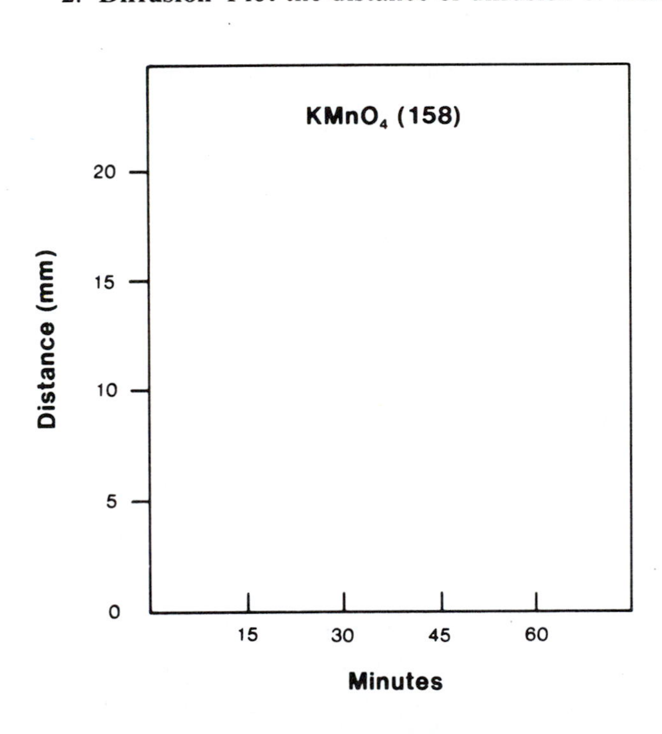

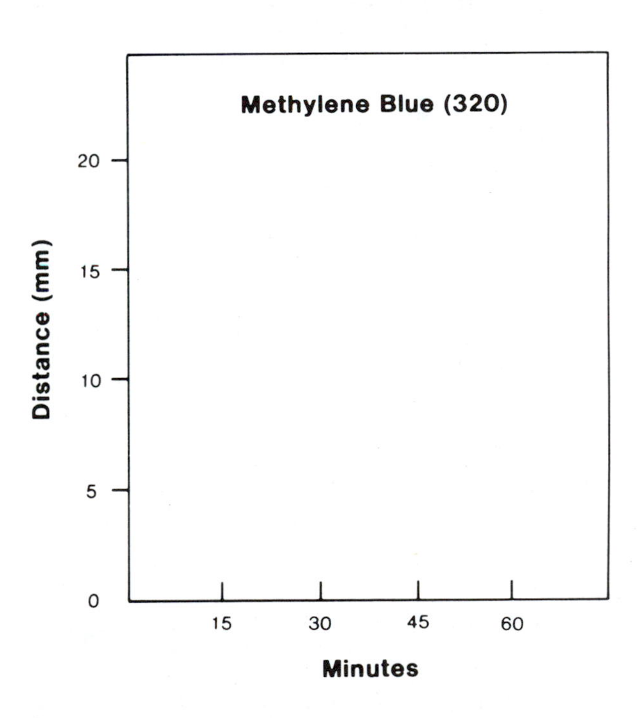

3. **Conclusion** What is the relationship of molecular weight to the rate of diffusion? __________

   ____

   ____

   ____

   ____

B. Osmotic Effects

As you examine each tube macroscopically and microscopically, record your results in the following table:

Solution	Macroscopic Appearance (lysis or no lysis)	Microscopic Appearance (crenation, lysis, or isotonic)
.15M Sodium Chloride		
.30M Sodium Chloride		
.28M Glucose $C_6H_{12}O_6$		
.30M Glycerine $C_3H_5(OH)_3$		
.30M Urea $CO(NH_2)_2$		

1. Are any of the above solutions isotonic for red blood cells? ____________________

   If so, which ones? ____________________

2. With the aid of the atomic weight table in Appendix A, calculate the percentages of solutes in those solutions that proved to be isotonic. ____________________

   ____

   ____

   ____

LABORATORY REPORT 8, 9

Student: ______________________

Desk No: ________ Section: ______________

Epithelial and Connective Tissues

A. Microscopic Study

If any drawings of tissues are to be made, make them on separate sheets of paper and label all identifiable structures.

B. Figure 9.5

Record the labels for this illustration in the answer column.

C. Epithelial Tissue Characteristics

From the list of tissues select those that are described by the following statements. More than one tissue may apply.

1. Elongated cells, single layer.
2. Cells with blocklike cross section.
3. Tissue with goblet cells.
4. Some cells are binucleate.
5. Have basal lamina.
6. Lack intercellular matrix.
7. Flattened cells, single layer.
8. Flattened cells, several layers.
9. Much intercellular matrix.
10. Reinforcing fibers between cells.
11. Cells near surface are dome-shaped.
12. Produce mucus on free surface.
13. Elongated cells, with some cells not extending up to free surface.
14. Cells have microvilli on free surface.

columnar, pseudostratified—1
columnar, simple—2
columnar, stratified—3
cuboidal—4
squamous, simple—5
squamous, stratified—6
transitional—7
all of above—8
none of above—9

D. Epithelial Tissue Locations

Identify the kind of epithelial tissues found in the following structures.

columnar, ciliated—1
columnar, plain—2
columnar, stratified—3
cuboidal—4
pseudostratified, ciliated—5
pseudostratified, plain—6
squamous, simple—7
squamous, stratified—8
transitional—9
none of above—10

1. Epidermis of skin
2. Lining of kidney calyces
3. Peritoneum
4. Auditory tube lining
5. Cornea of the eye
6. Capillary walls
7. Capsule of lens of the eye
8. Lining of the pharynx
9. Lining of the mouth
10. Pleural membranes
11. Intestinal lining
12. Lining of trachea
13. Lining of blood vessels
14. Lining of urinary bladder
15. Lining of stomach
16. Lining of male urethra
17. The conjunctiva
18. Lining of vagina
19. Anal lining
20. Parotid gland

Answers

Characteristics	Fig. 9.5
1. ______	______
2. ______	______
3. ______	______
4. ______	______
5. ______	______
6. ______	______
7. ______	______
8. ______	______
9. ______	______
10. ______	______
11. ______	______
12. ______	______
13. ______	______
14. ______	______
	______

Locations	
1. ______	11. ______
2. ______	12. ______
3. ______	13. ______
4. ______	14. ______
5. ______	15. ______
6. ______	16. ______
7. ______	17. ______
8. ______	18. ______
9. ______	19. ______
10. ______	20. ______

E. Connective Tissue Characteristics

From the list of tissues select those that are described by the following statements. More than one tissue may apply.

1. Contains all three types of fibers.
2. Reinforced mostly with collagenous fibers.
3. Cells are contained in lacunae.
4. Cells have large fat vacuoles.
5. Contain mast cells.
6. Nonyielding supporting tissue.
7. Devoid of blood supply.
8. Contain macrophages.
9. Most flexible of supporting tissues.
10. Converted to bone during growing years.
11. Matrix impregnated with calcium and phosphorous salts.
12. Toughest ordinary tissue (for stretching).
13. Cells have signet ringlike appearance.
14. Reinforced mostly with yellow fibers.
15. Has smooth glassy appearance.

adipose—1
areolar—2
bone—3
cartilage, elastic—4
cartilage, fibro—5
cartilage, hyaline—6
dense irregular—7
dense regular—8
reticular—9
none of above—10

F. Connective Tissue Locations

From the above list of tissues select the connective tissue that is found in the following:

1. Deep fasciae
2. Tendon sheaths
3. Intervertebral disks
4. Superficial fasciae
5. Epiglottis
6. Basal lamina
7. Lymph nodes
8. Aponeuroses
9. Nerve sheaths
10. Symphysis pubis
11. Auditory tube
12. Skeletal components
13. Periosteum (outer layer)
14. Sharpey's fibers
15. Ligaments and tendons
16. In hemopoietic tissue
17. In tracheal rings
18. In pinna of outer ear

G. Terminology

Select the terms on the right that are described by the following statements.

1. Protein in white fibers.
2. Cell that secretes mucus.
3. Material between fibroblasts.
4. Cartilage-producing cells.
5. Inner layer of periosteum.
6. Cell that produces fibers.
7. Synonym for Haversian system.
8. Protein in reticular fibers.
9. Bony plates in spongy bone.
10. Anchors periosteum to bone tissue.
11. Layers of bony matrix around central (Haversian) canal.
12. Layer of reticular fibers at base of epithelial tissues.
13. Cell found in loose fibrous tissue.
14. Minute canals radiating out from lacunae in bone tissue.
15. Canal in bone that contains capillaries.
16. Horizontal canals that connect adjacent central canals.

basement lamina—1
canaliculi—2
central canal—3
chondrocytes—4
collagen—5
fibroblast—6
goblet cell—7
lamellae—8
macrophage—9
mast cell—10
matrix—11
osteocyte—12
osteogenic layer—13
osteon—14
perforating canal—15
Sharpey's fibers—16
trabeculae—17

Answers

Characteristics	Terms
1. ______	1. ______
2. ______	2. ______
3. ______	3. ______
4. ______	4. ______
5. ______	5. ______
6. ______	6. ______
7. ______	7. ______
8. ______	8. ______
9. ______	9. ______
10. ______	10. ______
11. ______	11. ______
12. ______	12. ______
13. ______	13. ______
14. ______	14. ______
15. ______	15. ______
	16. ______

Locations
1. ______
2. ______
3. ______
4. ______
5. ______
6. ______
7. ______
8. ______
9. ______
10. ______
11. ______
12. ______
13. ______
14. ______
15. ______
16. ______
17. ______
18. ______

LABORATORY REPORT **10**

Student: ______________________________

Desk No: ________ Section: ______________

The Integument

Answers

Fig. 10.1

____
____
____
____
____
____
____
____
____
____
____
____
____
____
____
____
____
____
____
____
____
____

Questions

1. ___________
2. ___________
3. ___________
4. ___________
5. ___________
6. ___________
7. ___________
8. ___________

A. Figure 10.1

Record the labels for this illustration in the answer column.

B. Microscopic Study

If the space provided below is adequate for your histology drawings, use it. Use a separate sheet of paper if the space is inadequate.

C. Questions

Select the answer that completes the following statements. Only one answer applies for each question.

1. Granules of the stratum granulosum consist of
   (1) melanin. (2) keratin. (3) eleidin.
2. Pacinian corpuscles are sensitive to
   (1) temperature. (2) pressure. (3) touch.
3. Meissner's corpuscles are sensitive to
   (1) temperature. (2) pressure. (3) touch.
4. The epidermis consists of the following number of distinct layers:
   (1) three. (2) four. (3) five. (4) six.
5. The papillary layer is a part of the
   (1) dermis. (2) epidermis. (3) stratum spinosum.
6. Melanin granules are produced by the
   (1) stratum granulosum. (2) melanocytes.
   (3) corium. (4) stratum corneum.
7. New cells of the epidermis originate in the
   (1) dermis. (2) stratum germinativum.
   (3) stratum corneum. (4) corium.
8. The outermost layer of the epidermis is the
   (1) stratum corneum. (2) stratum germinativum.
   (3) stratum lucidum. (4) stratum spinosum.

9. The secretions of the apocrine glands usually empty
   (1) into a hair follicle. (2) directly out through skin surface.
   (3) Both 1 and 2 are correct. (4) None of these are correct.
10. Sebum is secreted by
    (1) sweat glands. (2) eccrine glands.
    (3) apocrine glands. (4) None of these are correct.
11. Keratin's function is primarily to
    (1) destroy bacteria. (2) provide waterproofing. (3) cool the body.
12. Meissner's corpuscles are located in the
    (1) papillary layer. (2) reticular layer. (3) hypodermis.
13. Arrector pili assist in
    (1) maintaining skin tonus. (2) limiting excessive sweating.
    (3) forcing out sebum. (4) None of these are correct.
14. The hair follicle is
    (1) a shaft of hair. (2) the root of a hair.
    (3) a tube in the skin. (4) None of these are correct.
15. Pacinian corpuscles are located in the
    (1) papillary layer. (2) reticular layer. (3) hypodermis.
16. Sweat is produced by
    (1) sebaceous glands. (2) eccrine glands.
    (3) ceruminous glands. (4) None of these are correct.
17. The coiled-up portion of an apocrine gland is located in the
    (1) hair follicle. (2) dermis. (3) epidermis. (4) hypodermis.
18. Apocrine glands are located
    (1) in the axillae. (2) on the scrotum.
    (3) in the ear canal. (4) All of these are correct.

Answers
Questions
9. ________
10. ________
11. ________
12. ________
13. ________
14. ________
15. ________
16. ________
17. ________
18. ________

LABORATORY REPORT **11**

Student: ____________________

Desk No: ________ Section: ____________

The Skeletal Plan

A. *Illustrations*

Record the labels for figures 11.1, 11.2, and 11.3 in the answer columns.

B. *Long Bone Structure*

Identify the terms described by the following statements.

1. Shaft portion of bone.
2. Hollow chamber in bone shaft.
3. Type of marrow in medullary canal.
4. Enlarged end of a bone.
5. Type of bone in diaphysis.
6. Fibrous covering of bone shaft.
7. Linear growth area of long bone.
8. Smooth gristle covering bone end.
9. Type of bone marrow in bone ends.
10. Lining of medullary canal.
11. Type of bone tissue in bone ends.
12. Lines the central canals.

articular cartilage—1
cancellous bone—2
compact bone—3
diaphysis—4
endosteum—5
epiphyseal disk—6
epiphysis—7
medullary canal—8
periosteum—9
red marrow—10
yellow marrow—11
none of these—12

C. *Bone Identification*

Select the structures on the right that match the statements on the left.

1. Shoulder blade
2. Collarbone
3. Breastbone
4. Shinbone
5. Kneecap
6. Upper arm bone
7. Bones of spine
8. Thighbone
9. Lateral bone of forearm
10. Joint between ossa coxae
11. Horseshoe-shaped bone
12. Bones of shoulder girdle
13. One half of pelvic girdle
14. Medial bone of forearm
15. Thin bone paralleling tibia (calf bone)

clavicle—1
femur—2
fibula—3
humerus—4
hyoid—5
os coxa—6
patella—7
radius—8
scapula—9
sternum—10
symphysis pubis—11
tibia—12
ulna—13
vertebrae—14

Answers

Structure	Bones
1. ______	1. ______
2. ______	2. ______
3. ______	3. ______
4. ______	4. ______
5. ______	5. ______
6. ______	6. ______
7. ______	7. ______
8. ______	8. ______
9. ______	9. ______
10. ______	10. ______
11. ______	11. ______
12. ______	12. ______
Fig. 11.2	13. ______
______	14. ______
______	15. ______
______	**Fig. 11.1**
______	______
______	______
______	______
______	______
______	______
______	______
______	______
______	______
______	______
______	______
______	
______	
______	
______	
______	

D. Medical

Select the condition that is described by the following statements. Since not all conditions are described in this manual it will be necessary for you to consult your lecture text or medical dictionary for some of the terminology.

closed reduction—1
Colles' fracture—2
comminuted fracture—3
compacted fracture—4
compound fracture—5
fissured fracture—6
greenstick fracture—7
open reduction—8
osteomalacia—9
osteomyelitis—10
osteoporosis—11
pathological fracture—12
Pott's fracture—13
rickets—14
simple fracture—15
none of these—16

1. Fracture due to weak bone structure, not trauma.
2. Fracture in which the skin is not broken.
3. Fracture caused by torsional forces.
4. Procedure used to set broken bones without using surgery.
5. Fracture caused by severe vertical forces.
6. Skeletal softness in adults.
7. Fracture in which a piece of bone is broken out of shaft.
8. Fracture characterized by two or more fragments.
9. Term applied to bone setting with the aid of surgery.
10. Skeletal softness in children due to vitamin D deficiency.
11. Infection of bone marrow.
12. Bone fracture extends only partially through a bone; incomplete fracture.
13. Fracture which occurs at an angle to the longitudinal axis of a bone.
14. Bone condition in which increased porosity occurs due to widening of central canals.
15. Outward displacement of foot due to fracture of lower part of fibula and malleolus.
16. Incomplete bone fracture in which fracture is apparent only on convex surface.
17. Displacement of hand backward and outward due to fracture of lower end of radius.

E. Terminology

The following processes, openings, depressions, and cavities exist on various bones of the skeleton. Identify the phrase at the right that best describes the term. More than one phrase may apply to each term.

1. Condyle
2. Fissure
3. Foramen
4. Fossa
5. Meatus
6. Sinus
7. Spine
8. Trochanter
9. Tubercle
10. Tuberosity

a narrow slit—1
a cavity in a bone—2
process to which muscle is attached—3
a small process for muscle attachment—4
a large process for muscle attachment—5
a rounded knucklelike process that articulates with another bone—6
a depression in a bone—7
a long tubelike passageway—8
a hole through which nerves and blood vessels pass—9
a sharp or long slender process—10

Answers

Medical

1. ________
2. ________
3. ________
4. ________
5. ________
6. ________
7. ________
8. ________
9. ________
10. ________
11. ________
12. ________
13. ________
14. ________
15. ________
16. ________
17. ________

Fig. 11.3

____
____
____
____
____
____
____
____
____
____

Terms

1. ________
2. ________
3. ________
4. ________
5. ________
6. ________
7. ________
8. ________
9. ________
10. ________

LABORATORY REPORT **12**

Student: ______________________

Desk No: ________ Section: ____________

The Skull

A. Illustrations

Record the labels for all the illustrations of this exercise in the answer columns.

B. Bone Names

Select the bones that are described by the following statements. More than one answer may apply.

1. Forehead bone
2. Upper jaw
3. Cheekbone
4. Lower jaw
5. Sides of skull
6. On walls of nasal fossa
7. Unpaired bones (usually)
8. Bones of hard palate (2 pair)
9. Back and bottom of skull
10. Contain sinuses (4 bones)
11. External bridge of nose
12. On median line in nasal cavity

ethmoid—1
frontal—2
lacrimal—3
mandible—4
maxilla—5
nasal—6
nasal conchae—7
occipital—8
palatine—9
parietal—10
sphenoid—11
temporal—12
vomer—13
zygomatic—14

C. Terminology

From the above list of bones select those which have the following structures.

1. Alveolar process
2. Angle
3. Antrum of Highmore
4. Chiasmatic groove
5. Coronoid process
6. Cribriform plate
7. Crista galli
8. External acoustic meatus
9. Incisive fossa
10. Inferior temporal line
11. Internal acoustic meatus
12. Mandibular condyle
13. Mandibular fossa
14. Mastoid process
15. Mylohyoid line
16. Occipital condyles
17. Ramus
18. Sella turcica
19. Styloid process
20. Superior nasal concha
21. Superior temporal line
22. Symphysis
23. Tear duct
24. Zygomatic arch

D. Sutures

Identify the sutures described by the following statements.

1. Between parietal and occipital.
2. Between two maxillae on median line.
3. Between the two parietals on the median line.
4. Between frontal and parietal.
5. Between parietal and temporal.
6. Between maxillary and palatine.

coronal—1
lambdoidal—2
median palatine—3
sagittal—4
squamosal—5
transverse palatine—6

Answers

Bones	Fig. 12.1	Fig. 12.2
1. ____	____	____
2. ____	____	____
3. ____	____	____
4. ____	____	____
5. ____	____	____
6. ____	____	____
7. ____	____	____
8. ____	____	____
9. ____	____	____
10. ____	____	____
11. ____	____	____
12. ____	____	____
Terms	____	____
	____	____
1. ____	____	____
2. ____	____	____
3. ____	____	____
4. ____	____	____
5. ____	____	____
6. ____	____	____
7. ____	____	
8. ____	____	**Sutures**
9. ____		1. ____
10. ____	**Fig. 12.5**	2. ____
11. ____	____	3. ____
12. ____	____	4. ____
13. ____	____	5. ____
14. ____	____	6. ____
15. ____	____	
16. ____	____	
17. ____	____	
18. ____	____	
19. ____	____	
20. ____	____	
21. ____	____	
22. ____		
23. ____		
24. ____		

E. *Foramina Location*

By referring to the illustrations select the bones on which the following foramina or canals are located.

1. Anterior palatine foramen	frontal—1
2. Carotid canal	mandible—2
3. Foramen magnum	maxilla—3
4. Foramen ovale	occipital—4
5. Hypoglossal canal	palatine—5
6. Infraorbital foramen	parietal—6
7. Internal acoustic meatus	sphenoid—7
8. Mandibular foramen	temporal—8
9. Mental foramen	zygomatic—9
10. Optic canal	
11. Supraorbital foramen	
12. Zygomaticofacial foramen	

F. *Foramen Function*

Select the cranial nerves and other structures that pass through the following foramina, canals, and fissures. Note that the number following each cranial nerve corresponds to the number designation of the particular cranial nerve. (Example: the abducens nerve is also known as the sixth cranial nerve.) Also, note that the fifth cranial nerve (trigeminal) has three divisions.

1. Carotid canal
2. Cribriform plate foramina
3. Foramen ovale
4. Foramen magnum
5. Hypoglossal canal
6. Internal acoustic meatus
7. Optic canal

Cranial Nerves
- olfactory—1
- optic—2
- oculomotor—3
- trochlear—4
- trigeminal branches
  - ophthalmic—5.1
  - maxillary—5.2
  - mandibular—5.3
- abducens—6
- facial—7
- vestibulocochlear—8
- spinal accessory—11
- hypoglossal—12

brain stem—13
internal carotid a.—14

G. *Completion*

Provide the name of the structures described by the following statements.

1. Lighten the skull.
2. Allow compression of skull bones during birth.
3. Provides attachment site for falx cerebri.
4. Allow sound waves to enter the skull.
5. Provides anchorage for sternocleidomastoideus muscle.
6. Point of attachment for some tongue muscles.
7. Points of articulation between skull and first vertebra.
8. Point of articulation between mandible and skull.

Answers

Location: 1. ____ 2. ____ 3. ____ 4. ____ 5. ____ 6. ____ 7. ____ 8. ____ 9. ____ 10. ____ 11. ____ 12. ____

Function: 1. ____ 2. ____ 3. ____ 4. ____ 5. ____ 6. ____ 7. ____

Fig 12.6

Fig. 12.3

Fig. 12.4

Fig. 12.7

Completion: 1. ____ 2. ____ 3. ____ 4. ____ 5. ____ 6. ____ 7. ____ 8. ____

Fig. 12.8

LABORATORY REPORT 13

Student: ____________________

Desk No: ________ Section: ____________

The Vertebral Column and Thorax

A. *Illustrations*

Record the labels for the four figures in the answer columns.

B. *Bone Names*

Select the proper anatomical terminology for the following bones.

1. Tailbone
2. Breastbone
3. False rib
4. Floating rib
5. True rib
6. Second cervical vertebra
7. First cervical vertebra
8. Inferior portion of breastbone
9. Midportion of breastbone
10. Superior portion of breastbone

atlas—1
axis—2
coccyx—3
gladiolus—4
manubrium—5
sacrum—6
sternum—7
vertebral ribs—8
vertebrochondral ribs—9
vertebrosternal ribs—10
xiphoid—11

C. *Structures*

Select the structures or foramina that perform the following functions.

body—1
intervertebral disks—2
intervertebral foramina—3
odontoid process—4
rib cage—5
spinal curves—6
spinous process—7
sternum—8
transverse process—9
transverse foramina—10

1. Provides anterior attachment for vertebrosternal ribs.
2. Openings through which spinal nerves exit.
3. Portion of vertebra that supports body weight.
4. Protects vital thoracic organs.
5. Impart springiness to vertebral column (two things).
6. Part of thoracic vertebra which provides attachment for rib.
7. Acts as a pivot for the atlas to move around.
8. Openings through which vertebral artery and vein pass.

Answers

Bone Names	Fig. 13.1
1. ______	______
2. ______	______
3. ______	______
4. ______	______
5. ______	______
6. ______	______
7. ______	______
8. ______	______
9. ______	______
10. ______	______
Structures	______
1. ______	______
2. ______	______
3. ______	______
4. ______	______
5. ______	______
6. ______	______
7. ______	______
8. ______	______
Fig. 13.2	______
______	______
______	______
______	______
______	______
______	______
______	______
______	______
______	______
______	______

D. *Vertebral Differences*

Select the vertebrae from the column on the right that is unique for the following characteristics.

1. Transverse foramina
2. Large massive body
3. Odontoid process
4. Downward sloping spinous process
5. Rib facets on transverse process

atlas—1
axis—2
cervical vertebrae—3
thoracic vertebrae—4
lumbar vertebrae—5

E. *Numbers*

Indicate the number of components that constitute the following.

1. Cervical vertebrae
2. Thoracic vertebrae
3. Lumbar vertebrae
4. Sacrum (fused vertebrae)
5. Coccyx (rudimentary vertebrae)
6. Pairs of ribs
7. Spinal curvatures
8. Pairs of vertebrochondral ribs
9. Pairs of vertebrosternal ribs
10. Pairs of vertebral ribs

F. *Medical*

Select the medical terms at the right that are described by the following statements. Consult your lecture text or medical dictionary.

1. Exaggerated lumbar curvature
2. Lateral curvature of the spine
3. "Slipped disk"
4. Exaggerated thoracic curvature

herniation—1
kyphosis—2
lordosis—3
scoliosis—4

Answers

Vertebral Differences	Fig. 13.3
1. ________	________
2. ________	________
3. ________	________
4. ________	________
5. ________	________
Numbers	________
1. ________	________
2. ________	________
3. ________	________
4. ________	________
5. ________	________
6. ________	________
7. ________	________
8. ________	________
9. ________	________
10. ________	________
Medical	________
1. ________	**Fig. 13.4**
2. ________	________
3. ________	________
4. ________	________

Student: ______________________

LABORATORY REPORT 14, 15

Desk No: ________ Section: ______________

The Appendicular Skeleton and Articulations

A. *Labels*

Record the labels for all the illustrations of Exercises 14 and 15 in the appropriate columns.

B. *Bone Names*

Select the names of the bones described in the following statements.

1. Lateral forearm bone.
2. Medial forearm bone.
3. Bones of palm of hand.
4. Upper arm bone.
5. Fingers.
6. Wrist bones.
7. Components of shoulder girdle.

carpals—1
clavicle—2
humerus—3
metacarpals—4
phalanges—5
radius—6
scapula—7
tarsals—8
ulna—9

C. *Landmarks* *(Upper Extremities)*

Identify the depressions and processes of the arm and shoulder described by the following statements.

acromion process—1
capitulum—2
coracoid process—3
coronoid fossa—4
coronoid process—5
deltoid tuberosity—6
glenoid cavity—7
greater tubercle—8
head—9
lateral epicondyle—10
lesser tubercle—11
medial epicondyle—12
olecranon fossa—13
olecranon process—14
radial notch—15
radial tuberosity—16
semilunar notch—17
styloid process—18
trochlea—19

1. Depression on scapula that articulates with humerus.
2. Depression on ulna that is in contact with head of radius.
3. Epicondyle of humerus adjacent to trochlea.
4. Distal medial condyle of humerus.
5. Distal lateral condyle of humerus.
6. Fossa on distal posterior surface of humerus.
7. Process on scapula that articulates with clavicle.
8. Process on anterior surface of ulna just below semilunar notch.
9. Process on scapula that lies anterior and superior to glenoid cavity.
10. Process on humerus to which deltoid muscle is attached.
11. Condyle of humerus that articulates with ulna.
12. Condyle of humerus that articulates with radius.
13. Part of humerus that articulates with scapula.
14. Epicondyle adjacent to capitulum.
15. Large process near head of humerus.
16. Fossa on distal anterior surface of humerus.
17. Small distal medial process of ulna.
18. Depression on proximal end of ulna that articulates with trochlea of humerus.
19. Proximal process of ulna, sometimes called the "funny bone."
20. Small process just above surgical neck of humerus.

Answers

Bone Names	Fig. 14.1
1. ______	______
2. ______	______
3. ______	______
4. ______	______
5. ______	______
6. ______	______
7. ______	______
	______
Landmarks	______
1. ______	______
2. ______	______
3. ______	______
4. ______	______
5. ______	**Fig. 14.2**
6. ______	
7. ______	______
8. ______	______
9. ______	______
10. ______	______
11. ______	______
12. ______	______
13. ______	______
14. ______	______
15. ______	______
16. ______	______
17. ______	______
18. ______	______
19. ______	______
20. ______	______
	______
	______
	______
	______
	______
	______
	______
	______
	______
	______
	______
	______
	______
	______
	______
	______

D. *Bones and Joints (Lower Extremities)*

Identify the bones and joints of the lower extremities that are described by the following statements.

1. Upper bone of os coxa.	calcaneous—1
2. Anterior bone of os coxa.	femur—2
3. Lower posterior bone of os coxa.	fibula—3
4. Heel bone.	ilium—4
5. Bones of toes.	ischium—5
6. Bones of fingers.	metatarsals—6
7. Bones surrounding the obturator foramen.	os coxa—7
8. One half of pelvic girdle.	patella—8
9. Pelvic girdle, sacrum, and coccyx.	pelvis—9
10. Longest, heaviest bone of leg.	phalanges—10
11. Strongest bone of lower leg.	pubis—11
12. Largest tarsal bone.	talus—12
13. Tarsal bone that articulates with tibia.	tibia—13
14. Five elongated bones that form instep of foot.	
15. Thin lateral bone of lower leg.	
16. Bone adjacent to the sacroiliac joint.	

E. *Numbers*

Supply the correct numbers for the following statements.

1. Number of phalanges in thumb.
2. Number of phalanges on each of the other four fingers.
3. Number of phalanges in the large toe.
4. Number of phalanges in each of the other toes.
5. Number designation for little finger metacarpal.
6. Number designation for thumb metacarpal.
7. Number designation for the small toe metatarsal.
8. Total number of tarsal bones in each foot.
9. Total number of metatarsals in each foot.
10. Total number of carpals in each wrist.

F. *Comparisons*

Compare the male and female pelves side-by-side to determine the validity (True or False) of these statements.

1. The greater pelvis (concavity of the expanded iliac bones above the pelvic brim) of the male is narrower than in the female.
2. The male sacrum and coccyx are more curved.
3. The muscular impressions on the female pelvis are less distinct (bone is smoother).
4. The sacrum of the female pelvis is shorter and wider.
5. The obturator foramina of the female pelvis are rounder and larger than in the male.
6. The iliac crests of the female pelvis protrude outward more than in the male.
7. The aperture of the female pelvis is larger and more oval than the male pelvis.
8. The acetabula of the female face more anteriorly.
9. The aperture of the female pelvis is heart-shaped.
10. The obturator foramina of the female pelvis tend toward being triangular and smaller than in the male.

Answers

Bones and Joints	Fig. 14.3
1. ______	______
2. ______	______
3. ______	______
4. ______	______
5. ______	______
6. ______	______
7. ______	______
8. ______	______
9. ______	______
10. ______	______
11. ______	______
12. ______	______
13. ______	______
14. ______	______
15. ______	______
16. ______	______

Numbers	Fig. 14.4
1. ______	______
2. ______	______
3. ______	______
4. ______	______
5. ______	______
6. ______	______
7. ______	______
8. ______	______
9. ______	______
10. ______	______
	______
Comparisons	______
1. ______	______
2. ______	______
3. ______	______
4. ______	______
5. ______	______
6. ______	______
7. ______	______
8. ______	______
9. ______	______
10. ______	______
	______
	______
	______
	______
	______
	______

G. *Landmarks (Lower Extremities)*

Identify the following processes, fossae, and other structures of bones of the lower extremities.

acetabulum—1
greater trochanter—2
head of femur—3
head of fibula—4
iliac crest—5
intertrochanteric crest—6
intertrochanteric line—7
ischial spine—8
lateral condyle—9
lateral malleolus—10
lesser trochanter—11
medial condyle—12
medial malleolus—13
obturator foramen—14
sacroiliac—15
symphysis pubis—16
tibial tuberosity—17
tuberosity of ischium—18

1. Joint between sacrum and ilium of os coxa.
2. Joint between pubic bones of ossa coxae.
3. Large prominent process of ischium.
4. Large process lateral to neck of femur.
5. Process just inferior to neck of femur on medial surface.
6. Proximal lateral process of tibia that articulates with femur.
7. Proximal medial process of tibia that articulates with femur.
8. Distal medial process of femur.
9. Distal lateral process of femur.
10. Fossa on os coxa that articulates with femur.
11. Uppermost process on posterior edge of ischium.
12. Uppermost ridge of os coxa.
13. Large opening in os coxa surrounded by pubis and ischium.
14. Part of femur that fits into depression of os coxa.
15. Oblique ridge between trochanters on posterior surface of femur.
16. Oblique ridge between trochanters on anterior surface of femur.
17. Distal lateral process of fibula that forms outer ankle bone.
18. Distal medial process of tibia that forms inner prominence of ankle.
19. Proximal anterior process of tibia which protrudes below the kneecap.
20. End of fibula that articulates with upper end of tibia.

H. *Joint Characteristics*

Select the joints from the list on the right that have the following characteristics. More than one answer often applies.

1. Immovable.
2. Slightly movable.
3. Freely movable in one plane only.
4. Freely movable in two planes.
5. Movement in all directions.
6. Rotational movement only.
7. Fused joint of hyaline cartilage.
8. Contains fibrocartilaginous pad.
9. Condyle end in elliptical cavity.
10. Held together with interosseous ligament.
11. Bone ends are flat or slightly convex.
12. Each bone end is concave in one direction; convex in another.
13. Bone ends joined by fibrous connective tissue.

ball and socket—1
condyloid—2
gliding—3
hinge—4
pivot—5
saddle—6
sutures—7
symphysis—8
synchondrosis—9
syndesmosis—10

Answers

Landmarks	Fig. 15.1
1. ______	______
2. ______	______
3. ______	______
4. ______	______
5. ______	______
6. ______	______
7. ______	______
8. ______	______
9. ______	______
10. ______	______
11. ______	**Fig. 15.2**
12. ______	
13. ______	______
14. ______	______
15. ______	______
16. ______	______
17. ______	______
18. ______	______
19. ______	______
20. ______	______
	______
Characteristics	**Fig. 15.3**
1. ______	
2. ______	______
3. ______	______
4. ______	______
5. ______	______
6. ______	______
7. ______	______
8. ______	______
9. ______	______
10. ______	______
11. ______	______
12. ______	______
13. ______	______

I. Joint Classification

By examining and manipulating the bones of a skeleton, classify the following joints. Two terms apply to each joint.

1. Elbow joint.
2. Knee joint.
3. Phalanges to metacarpals.
4. Phalanges to metatarsals.
5. Hip joint.
6. Shoulder joint.
7. Intervertebral joint (between vertebral bodies).
8. Metaphysis of long bone of child.
9. Interlocking joint between cranial bones.
10. Symphysis pubis.
11. Articulation between tibia and fibula at distal ends.
12. Atlas to skull.
13. Toe joints (between phalanges).
14. Finger joint (between phalanges).
15. Wrist (carpals to carpals).
16. Ankle (talus to tibia).
17. Ankle (tarsals to tarsals).
18. Axis-atlas rotation.

Basic Types

slightly movable—1
freely movable—2
immovable—3

Sub-types

ball and socket—4
condyloid—5
gliding—6
hinge—7
pivot—8
saddle—9
suture—10
symphysis—11
synchondrosis—12
syndesmosis—13

J. Medical

Consult your lecture text, medical dictionary, or other source for the following.

ankylosis—1
bursitis—2
chronic fibrositis—3
dislocation—4
gouty arthritis—5
osteoarthritis—6
rheumatism—7
rheumatoid arthritis—8
sprain—9
strain—10
subluxation—11
tenosynovitis—12

1. Arthritis due to excess levels of uric acid in blood.
2. Arthritis resulting from degenerative changes in joint (wear and tear).
3. Arthritis of confusing etiology, characterized by deformity and immobility of joints; often treated with cortisone.
4. Inflammation of fibrous connective tissue causing tenderness and stiffness.
5. Inflammation of tendon and its sheath.
6. General term applied to soreness and stiffness in muscles and joints.
7. Inflammation of synovial bursa.
8. Partial or incomplete dislocation.
9. Displacement of bones from normal orientation in a joint.
10. Fixation or fusion of a joint; usually preceded by infection.
11. Injury to joint in which ligaments are stretched without swelling or discoloration.
12. Injury to joint by displacement, resulting in swelling and discoloration.

Answers

Fig. 15.4	Classification
________	1. ________
________	2. ________
________	3. ________
________	4. ________
________	5. ________
________	6. ________
________	7. ________
________	8. ________
________	9. ________
________	10. ________
________	11. ________
________	12. ________
________	13. ________
________	14. ________
________	15. ________
________	16. ________
________	17. ________
________	18. ________
________	
________	

Medical

1. ________
2. ________
3. ________
4. ________
5. ________
6. ________
7. ________
8. ________
9. ________
10. ________
11. ________
12. ________

LABORATORY REPORT **16**

Student: ____________________

Desk No: ________ Section: ____________

Electronic Instrumentation

A. Transducers and Electrodes

1. What determines whether one uses an electrode or a transducer for monitoring a particular biological phenomenon?

____

____

2. Differentiate between:

Input Transducer ____

____

Output Transducer ____

____

3. Indicate what kind of pickup device (*electrode* or *transducer*) you would use for the following. If a transducer is to be used, indicate what specific type of transducer is necessary.

Sound ____

Light ____

Muscle Pull ____

Temperature ____

Skin Conductivity ____

Brain Waves ____

B. Unigraph Functions

Indicate after each of the following the specific control that you would use to produce the desired effect.

1. Turn on the unit ____
2. Change the speed of the paper ____
3. Increase the intensity of the tracing ____
4. Record on the chart the exact time that a new event occurs ____
5. Increase the sensitivity ____
6. Balance the transducer ____
7. Start and stop the movement of the paper ____

C. *Physiography® Functions*

Answer the following questions that pertain to the use of the Physiograph® Mark IIIS.

1. How do you increase or decrease the flow of ink to the pens? ______________________

   ____

2. Into what unit does one insert the cable from electrodes or transducers? ______________

   ____

3. At what sensitivity setting (*low* or *high*) does one usually start an experiment? __________
4. At what speeds can the paper move on a Physiograph®? ______________________
5. What button is pressed to start the chart moving? ______________________
6. How is ink removed from the pens on the instrument at the end of the laboratory period? _____

   ____

D. *Electronic Stimulator Functions*

Indicate how you would set the controls on the Grass and Narco stimulators to get the following outputs:

Grass SD9		*Narco SM–1*	
4.5 Volts			
Round Control Knob	________	Voltage Range Control	________
Decade Switch	________	Variable Control Knob	________
80 Impulses Per Second:			
Round Control Knob	________	Round Control Knob (Hz)	________
Decade Switch	________	Rate Switch	________
Impulse Duration (Width) of 1.5 Milliseconds:			
Round Control Knob	________	Round Control Knob	________
Decade Switch	________		

E. *The Intelitool System*

1. List several advantages that computerized hardware has over chart recorders.

   ____

   ____

2. What advantage does a Duograph or polygraph have over Intelitool computerized hardware?

   ____

3. Under what circumstances can "Review/Analyze" *not* be selected from the Main Menu?

   ____

4. What computer key is struck to get the program to take over before starting an experiment?

   ____

LABORATORY REPORT 17, 18

Student: ______________________

Desk No: ________ Section: ______________

Muscle Structure (Ex. 17)

A. *Figure 17.1*

Record the labels for figure 17.1 in the answer column.

B. *Muscle Terminology*

Match the terms of the right-hand column to the following statements.

1. Bundle of muscle fibers.
2. Sheath of fibrous connective tissue which surrounds each fasciculus.
3. Movable end of a muscle.
4. Immovable end of muscle.
5. Thin layer of connective tissue surrounding each muscle fiber.
6. Layer of connective tissue between the skin and deep fascia.
7. Flat sheet of connective tissue that attaches a muscle to the skeleton or another muscle.
8. Layer of connective tissue that surrounds a group of fasciculi.
9. Outer connective tissue that covers an entire muscle.
10. Band of fibrous connective tissue that connects muscle to bone or to another muscle.

aponeurosis—1
deep fascia—2
endomysium—3
epimysium—4
fasciculus—5
insertion—6
ligament—7
origin—8
perimysium—9
superficial fascia—10
tendon—11

Body Movements (Ex. 18)

A. *Figure 18.1*

Record the letters for the various types of movement in this illustration in the answer column.

B. *Muscle Types*

Determine the types of muscles described by the following statements.

1. Prime movers.
2. Opposing muscles.
3. Muscles that assist prime movers.
4. Muscles that hold structures steady to allow agonists to act smoothly.

agonists—1
antagonists—2
fixation muscles—3
synergists—4

Answers

Terms

1. ________
2. ________
3. ________
4. ________
5. ________
6. ________
7. ________
8. ________
9. ________
10. ________

Types

1. ________
2. ________
3. ________
4. ________

Fig. 17.1

____
____
____
____
____
____
____
____
____
____
____

C. Movements

Identify the types of movements described by the following statements.

1. Movement of limb away from median line of body.
2. Movement of limb toward median line of body.
3. Movement around central axis.
4. Turning of sole of foot inward.
5. Turning of sole of foot outward.
6. Flexion of toes.
7. Backward movement of head.
8. Extension of foot at the ankle.
9. Angle between two bones is increased.
10. Angle between two bones is decreased.
11. Movement of hand from palm-down to palm-up position.
12. Conelike rotational movement.
13. Movement of hand from palm-up to palm-down position.

abduction—1
adduction—2
circumduction—3
dorsiflexion—4
eversion—5
extension—6
flexion—7
hyperextension—8
inversion—9
plantar flexion—10
pronation—11
rotation—12
supination—13

Answers

Movements	Fig. 18.1
1. ______	______
2. ______	______
3. ______	______
4. ______	______
5. ______	______
6. ______	______
7. ______	______
8. ______	______
9. ______	______
10. ______	______
11. ______	______
12. ______	
13. ______	

LABORATORY REPORT **19**

Student: ____________________

Desk No: ________ Section: ____________

Nerve and Muscle Tissues

A. *Cell Drawings*

On a separate sheet of paper, sketch representative cells of the various types of neurons and muscle cells from slides that are available. Staple the drawings to this report when it is handed in for grading.

B. *Figure 19.1*

Record the label numbers for this illustration in the answer column.

C. *Nerve Cell Types*

From the list of cells below, select those that are described by the following statements. More than one cell may apply in some cases.

1. Afferent neurons.
2. Efferent neurons.
3. Specialized cells of retina.
4. Pseudounipolar neurons.
5. Multipolar neurons.
6. Ganglion cells.
7. Golgi Type II cells.
8. Neurons lacking in axons.
9. Predominant neurons in cerebellar cortex.
10. Linkage cells between Purkinje cells.
11. Linkage cells between pyramidal cells.
12. Linkage cells between sensory and motor neurons.
13. Predominant type of neuron in cerebral cortex.
14. Small cells covering the perikaryon of a sensory neuron.

amacrine cells—1
basket cells—2
interneurons—3
motor neuron—4
neurolemmocytes—5
Purkinje cells—6
pyramidal cells—7
satellite cells—8
sensory neuron—9
stellate cells—10

D. *True-False*

Indicate whether the following statements concerning nerve cells are True or False.

1. Neuroglial cells do not carry nerve impulses.
2. Perikarya of afferent neurons are located in the spinal ganglia.
3. Most neurons have many dendrites with only a single axon.
4. Perikarya of motor neurons are located in the white matter of the spinal cord.
5. Myelinated fibers carry nerve impulses at a slower speed than unmyelinated ones.
6. Unmyelinated fibers lack neurolemmocytes.
7. Part of the axon of a sensory neuron functions like a dendrite.
8. Some neurons lack axons.
9. Both the myelin sheath and neurolemma are formed by Schwann cells.
10. Motor neurons carry nerve impulses away from the CNS.

Answers

Nerve Cell Types	Fig. 19.1
1. ______	______
2. ______	______
3. ______	______
4. ______	______
5. ______	______
6. ______	______
7. ______	______
8. ______	______
9. ______	______
10. ______	______
11. ______	______
12. ______	______
13. ______	______
14. ______	______
True-False	______
	______
1. ______	______
2. ______	______
3. ______	
4. ______	
5. ______	
6. ______	
7. ______	
8. ______	
9. ______	
10. ______	

E. *Neuron Structure*

From the list of neuron structures select those that are described by the following statements. More than one structure may apply in some cases.

1. Nerve cell body.
2. Side branch of an axon.
3. Long process on motor neuron.
4. Short process on motor neuron.
5. Lacking on unmyelinated nerve fibers.
6. Process that receives nerve impulses.
7. Slender filaments in perikaryon.
8. Formed by neurolemmocytes.
9. Synonym for neurolemmal sheath.
10. Interruptions in neurolemma.
11. Outer membrane covering myelin sheath.
12. Bodies on the ER that synthesize protein.
13. Promotes regeneration of damaged nerve fiber.
14. Enables nerve fiber to carry impulses by saltatory conduction.
15. Bodies in perikaryon that stain well with basic aniline dyes.
16. Process that carries nerve impulses away from perikaryon.

axis cylinder—1
axon—2
collateral—3
dendrite—4
myelin sheath—5
neurolemma—6
neurolemmocytes—7
neurofibrils—8
Nissl bodies—9
nodes of Ranvier—10
perikaryon—11

F. *Muscle Cell Characteristics*

Identify the types of muscle fibers that possess these characteristics:

1. Attached to bones.
2. Spindle-shaped cells.
3. Nonsyncytial.
4. Syncytial structure.
5. Voluntarily controlled (generally).
6. Lack cross-striae.
7. Long multinucleated cells.
8. In walls of the venae cavae where they enter the right atrium.
9. In walls of other veins and arteries.
10. Involuntarily controlled (generally).
11. Bifurcated cells.
12. Nuclei near surface of cell.

skeletal—1
smooth—2
cardiac—3

G. *Muscle Cell Structures*

Identify the muscle cell structures that are described by the following statements.

endomysium—1
fascicle—2
intercalated disks—3
perimysium—4
sarcolemma—5
sarcoplasm—6

1. Cell membrane of skeletal muscle fibers.
2. Cell membranes between individual cardiac muscle fibers.
3. Connective tissue between individual skeletal muscle fibers.
4. Cytoplasmic material of skeletal muscle fibers.
5. Bundle of skeletal muscle fibers.

Answers

Structure

1. ________
2. ________
3. ________
4. ________
5. ________
6. ________
7. ________
8. ________
9. ________
10. ________
11. ________
12. ________
13. ________
14. ________
15. ________
16. ________

Characteristics

1. ________
2. ________
3. ________
4. ________
5. ________
6. ________
7. ________
8. ________
9. ________
10. ________
11. ________
12. ________

Structures

1. ________
2. ________
3. ________
4. ________
5. ________

Student: ______________________

LABORATORY REPORT **20, 21**

Desk No: ________ Section: ____________

The Neuromuscular Junction (Ex. 20)

A. Figure 20.1

Record the labels for this illustration in the answer column.

B. Summary

Supply the correct terms to complete the following statement concerning the neuromuscular junction.

The unmyelinated ending of a motor neuron consists of several knoblike structures called ___1___. These knobs lie in depressions of the sarcolemma that are called ___2___ gutters. Between these knobs and the sarcolemma is a ___3___ cleft that is filled with a gelatinous substance. When a nerve impulse reaches the end of the neuron, a neurotransmitter, called ___4___, bridges the gap to initiate an action potential in the muscle fiber.

Stimulation of the muscle fiber by the neurotransmitter is made possible by the presence of receptors in the ___5___ membrane, which are activated by the neurotransmitter. Although the synaptic vesicles contain large quantities of the neurotransmitter, the actual source of the neurotransmitter that bridges the gap is the ___6___ of the knoblike structures. Only after prolonged stimulation of the neuron do the ___7___ release the neurotransmitter.

Depolarization of the neurolemma results in ___8___ ions flowing into the neuron and ___9___ ions flowing outward. Depolarization in the muscle fiber progresses from the sarcolemma inward to the ___10___ system, and the release of ___11___ ions by the ___12___.

Within 1/500 of a second after its passage across the gap, the neurotransmitter is inactivated by the enzyme, ___13___. Inactivation of the neurotransmitter is achieved by it being split into its two components: ___14___ and ___15___. Resynthesis of the neurotransmitter from these two components is accomplished by the enzyme, ___16___, which is produced by the ___17___ of the neuron, and stored in the cytoplasm of knoblike structures.

Answers

Summary

1. ____________
2. ____________
3. ____________
4. ____________
5. ____________
6. ____________
7. ____________
8. ____________
9. ____________
10. ____________
11. ____________
12. ____________
13. ____________
14. ____________
15. ____________
16. ____________
17. ____________

Figure 20.1

________	________
________	________
________	________
________	________
________	________
________	________

The Physicochemical Nature of Muscle Contraction (Ex. 21)

A. Length Changes

Record muscle fiber lengths before and after adding activants.

Original length ____________ mm

Length after adding ATP, K^+, and Mg^{++} ____________ mm

B. *Drawings*

Sketch the appearance of the muscle fibers as seen under high-dry or oil immersion before and during contraction.

Before Contraction	During Contraction

C. *Structure*

Select the muscle fiber components that are described by the following statements.

1. Thick myofilament with cross-bridges.
2. Principal constituent of A band.
3. Two types of myofilaments that make up a myofibril.
4. Three components of actin.
5. Type of myofilament that makes up I band.
6. Protein of which cross-bridges are made.
7. Component of a myofibril that extends between two Z lines.

actin—1
F-actin protein—2
heavy meromyosin—3
light meromyosin—4
myosin—5
sarcomere—6
tropomyosin—7
troponin—8

D. *Physiology*

Supply the terms that are missing in the following explanation of muscle contraction.

actin—1	F-actin protein—5	sarcoplasmic reticulum—9
ATP—2	magnesium—6	troponin—10
calcium—3	myosin—7	tropomyosin—11
cross-bridges—4	potassium—8	

The interaction of actin and myosin during muscle contraction depends on the presence of ample amounts of ATP, calcium, and ___1___ ions. The calcium ions are released by the ___2___ during depolarization. When these ions are released, ___3___ myofilaments are pulled toward each other by the ratcheting action of ___4___ that are present on the ___5___ myofilaments. The function of the calcium ions is to combine with ___6___, causing the ___7___ molecule to be drawn deeper which, in turn, exposes active sites on the ___8___ filament. When active sites are exposed, the ___9___ pull the ___10___ filaments together, causing contraction. The energy for this ratcheting effect is derived from ___11___.

Answers

Structure

1. ________
2. ________
3. ________
4. ________
5. ________
6. ________
7. ________

Physiology

1. ________
2. ________
3. ________
4. ________
5. ________
6. ________
7. ________
8. ________
9. ________
10. ________
11. ________

Student: ______________________

LABORATORY REPORT 22, 24

Desk No: ________ Section: ______________

Muscle Contraction Experiments: Using Computerized Hardware (Ex. 22)

A. *Visible Twitches (Unrecorded)*

Record here the **threshold stimuli** that cause visible twitches without using the recorder. Note that the muscle is stimulated directly and through the nerve.

A visible twitch with probe on **muscle** .... ____________ volts

____________ msec

Extension of foot with probe on **muscle** .... ____________ volts

____________ msec

A visible twitch with probe on **nerve** .... ____________ volts

____________ msec

Extension of foot with probe on **nerve** .... ____________ volts

____________ msec

Comparisons

Determine how much greater the stimulus requirement is for direct muscle stimulation by dividing one voltage into the other as follows:

$$\frac{\text{Voltage by Direct Muscle Stimulation}}{\text{Voltage by Nerve Stimulation}} =$$

for visible twitch .... ____________

for foot extension .... ____________

B. *Recorded Twitches*

Attach a segment of the chart in the space provided below. Use Scotch tape, glue, or staples.

C. *Evaluation of Recorded Twitches*

From the chart record on the previous page, determine the following:

Latent Period (A) .... ____________ msec

Contraction Phase (B) .... ____________ msec

Relaxation Phase (C) .... ____________ msec

Entire Duration (A + B + C) .... ____________ msec

D. Summation Results

Attach a segment of the chart which illustrates multiple motor unit summation. Be sure to indicate the voltages that were used for each stimulation.

Multiple Motor Unit Summation **Wave Summation**

E. Definitions

Define the following terms:

Motor Unit ______________________________

Threshold Stimulus ______________________________

Recruitment ______________________________

____

Tetany ______________________________

Electromyography (Ex. 24)

A. Recordings

Attach samples of myograms in the space below.

B. Applications

List several applications in which electromyography can be very useful: ______________________________

____

____

LABORATORY REPORT 25

Student: ______________________

Desk No: __________ Section: ______________

Head and Neck Muscles

A. *Illustrations*

Record the labels for the illustrations of this exercise in the answer column.

B. *Facial Expressions*

Consult figure 25.1 and the text to identify the muscles described in the following statements.

1. Smiling
2. Horror
3. Irony
4. Sadness
5. Contempt or disdain
6. Pouting lips
7. Blinking or squinting
8. Horizontal wrinkling of forehead

frontalis—1
orbicularis oculi—2
orbicularis oris—3
platysma—4
quadratus labii inferioris—5
quadratus labii superioris—6
triangularis—7
zygomaticus—8

Origins

9. On frontal bone.
10. On zygomatic bone.

Insertions

11. On lips.
12. On eyebrows.
13. On eyelid.

C. *Mastication*

Consult figure 25.1 and the text to identify the muscles that perform the following movements in mastication of food.

1. Lowers the mandible.
2. Raises the mandible.
3. Retracts the mandible.
4. Protrudes the mandible.
5. Holds food in place.

buccinator—1
external pterygoid—2
internal pterygoid—3
masseter—4
temporalis—5

Origins

6. On zygomatic arch.
7. On sphenoid only.
8. On frontal, parietal, and temporal bones.
9. On maxilla, mandible, and pteromandibular raphé.
10. On maxilla, sphenoid, and palatine bones.

Insertions

11. On orbicularis oris.
12. On coronoid process of mandible.
13. On medial surface of body of mandible.
14. On lateral surface of ramus and angle of mandible.
15. On neck of mandibular condyle and articular disk.

Answers

Facial	Fig. 25.1
1. ________	________
2. ________	________
3. ________	________
4. ________	________
5. ________	________
6. ________	________
7. ________	________
8. ________	________
9. ________	________
10. ________	________
11. ________	________
12. ________	________
13. ________	________
	________
Mastication	________
	________

Mastication	Fig. 25.2
1. ________	
2. ________	
3. ________	
4. ________	________
5. ________	________
6. ________	________
7. ________	________
8. ________	________
9. ________	________
10. ________	________
11. ________	________
12. ________	________
13. ________	________
14. ________	________
15. ________	________
	________
	________

D. Neck Muscles

Consult figure 25.2 and the text to identify the muscles of the neck that are described below.

digastricus—1
longissimus capitis—2
mylohyoideus—3
omohyoideus—4
platysma—5
semispinalis capitis—6
splenius capitis—7
sternocleidomastoideus—8
sternohyoideus—9
sternothyroideus—10
trapezius—11

1. Broad sheetlike muscle on side of neck just under the skin.
2. A broad triangular muscle of the back that functions only partially as a neck muscle.
3. Provides support for the floor of the mouth.
4. A V-shaped muscle attached to hyoid and mandibular bones.
5. Posterior neck muscles that insert on the mastoid process.
6. Anterior neck muscle that is attached to the thyroid cartilage and part of the sternum.
7. Anterior neck muscle that extends from the hyoid bone to the clavicle and manubrium.
8. Long muscle of the neck that extends from the mastoid process to the clavicle and sternum.
9. Muscles that pull the head backward (hyperextension).
10. Muscles that pull the head forward.

Answers
Neck Muscles
1. ________
2. ________
3. ________
4. ________
5. ________
6. ________
7. ________
8. ________
9. ________
10. ________

LABORATORY REPORT **26**

Student: ____________________

Desk No: ________ Section: ________

Trunk and Shoulder Muscles

A. *Illustrations*

Record the labels for the illustrations of this exercise in the answer columns.

B. *Anterior Trunk Muscles*

By referring to figure 26.1 and the text, identify the muscles described by the following statements.

deltoideus—1
intercostalis externi—2
intercostalis interni—3
pectoralis major—4
pectoralis minor—5
serratus anterior—6
subscapularis—7

Origins

1. On upper eight or nine ribs.
2. On scapular spine and clavicle.
3. On third, fourth, and fifth ribs.
4. On clavicle, sternum, and costal cartilages.

Insertions

5. On coracoid process.
6. On deltoid tuberosity of humerus.
7. On anterior surface of scapula.
8. On upper anterior end of humerus.

Actions

9. Pulls scapula forward, downward, and inward.
10. Rotates arm medially.
11. Raises ribs (inhaling).
12. Lowers ribs (exhaling).
13. Adducts arm; also, rotates it medially.
14. Abducts arm.

C. *Posterior Trunk Muscles*

By referring to figure 26.2 and the text, identify the muscles described by the following statements.

infraspinatus—1
latissimus dorsi—2
levator scapulae—3
rhomboideus major—4
rhomboideus minor—5
sacrospinalis—6
supraspinatus—7
teres major—8
teres minor—9
trapezius—10

Origins

1. Near inferior angle of scapula.
2. On lateral (axillary) margin of scapula.
3. On fossa of scapula above spine.
4. On infraspinous fossa of scapula.
5. On thoracic and lumbar vertebrae, sacrum, lower ribs, and iliac crest.
6. On lower part of ligamentum nuchae and first thoracic vertebra.
7. On transverse processes of first four cervical vertebrae.

Answers

Anterior	Fig. 26.1
1. ______	______
2. ______	______
3. ______	______
4. ______	______
5. ______	______
6. ______	______
7. ______	______
8. ______	**Fig. 26.2**
9. ______	
10. ______	______
11. ______	______
12. ______	______
13. ______	______
14. ______	______
	______
Posterior	______
1. ______	______
2. ______	______
3. ______	______
4. ______	______
5. ______	______
6. ______	______
7. ______	

infraspinatus—1
latissimus dorsi—2
levator scapulae—3
rhomboideus major—4
rhomboideus minor—5
sacrospinalis—6
supraspinatus—7
teres major—8
teres minor—9
trapezius—10

Insertions

8. On lower third vertebral border of scapula.
9. On vertebral border of scapula near origin of scapular spine.
10. On crest of lesser tubercle of humerus.
11. On intertubercular groove of humerus.
12. On middle facet of greater tubercle of humerus.
13. On ribs and vertebrae.

Actions

14. Abducts arm.
15. Raises scapula and draws it medially.
16. Extends spine to maintain erectness.
17. Adducts and rotates arm medially.
18. Adducts and rotates arm laterally.

Answers

Posterior

8. ________
9. ________
10. ________
11. ________
12. ________
13. ________
14. ________
15. ________
16. ________
17. ________
18. ________

LABORATORY REPORT 27

Student: ______________________

Desk No: ________ Section: ____________

Arm Muscles

A. *Illustrations*

Record the labels for figures 27.1, 27.2, and 27.3 in the answer column.

B. *Arm Movements*

Consult figure 27.1 to select the muscles that apply to the following statements. Verify your answers by referring to the text material.

coracobrachialis—1
biceps brachii—2
brachialis—3
brachioradialis—4
triceps brachii—5

1. Extends the forearm.
2. Adducts the arm.
3. Flexes the forearm.
4. Carries the arm forward in flexion.
5. Flexes forearm and rotates the radius outward to supinate the hand.

Origins

6. On lower half of humerus.
7. On coracoid process of scapula.
8. Above the lateral epicondyle of the humerus.
9. Three heads: one on scapula, another on posterior surface of humerus, and the third just below the radial groove of humerus.
10. Two heads: one on coracoid process; other on supraglenoid tubercle of humerus.

Insertions

11. On the olecranon process.
12. On the radial tuberosity.
13. Near the middle, medial surface of humerus.
14. On front surface of coronoid process of ulna.
15. On the lateral surface of the radius just above the styloid process.

C. *Hand Movements*

Consult figures 27.2 and 27.3 to select the muscles that apply to the following statements. Verify as above.

abductor pollicis—1
extensor carpi radialis brevis—2
extensor carpi radialis longus—3
extensor carpi ulnaris—4
extensor digitorum communis—5
extensor pollicis longus—6
flexor carpi radialis—7
flexor carpi ulnaris—8
flexor digitorum profundus—9
flexor digitorum superficialis—10
flexor pollicis longus—11
pronator quadratus—12
pronator teres—13
supinator—14

1. Supinates the hand.
2. Pronates the hand.
3. Flexes and abducts hand.
4. Extends wrist and hand.
5. Flexes the thumb.
6. Extends all fingers except thumb.
7. Abducts the thumb.
8. Flexes all distal phalanges except the thumb.
9. Flexes all fingers except thumb.
10. Extends thumb.

Answers

Arm Movements

1. ________
2. ________
3. ________
4. ________
5. ________
6. ________
7. ________
8. ________
9. ________
10. ________
11. ________
12. ________
13. ________
14. ________
15. ________

Hand Movements

1. ________
2. ________
3. ________
4. ________
5. ________
6. ________
7. ________
8. ________
9. ________
10. ________

Fig. 27.1

____
____
____
____
____

Fig. 27.2

____
____
____
____
____
____
____

Fig. 27.3

____
____
____
____
____
____
____
____

C. Hand Movements (continued)

abductor pollicis—1
extensor carpi radialis brevis—2
extensor carpi radialis longus—3
extensor carpi ulnaris—4
extensor digitorum communis—5
extensor pollicis longus—6
flexor carpi radialis—7
flexor carpi ulnaris—8
flexor digitorum profundus—9
flexor digitorum superficialis—10
flexor pollicis longus—11
pronator quadratus—12
pronator teres—13
supinator—14

Origins

11. On interosseous membrane between radius and ulna.
12. On radius, ulna, and interosseous membrane.
13. On lateral epicondyle of humerus and part of ulna.
14. On humerus, ulna, and radius.
15. On medial epicondyle of humerus.

Insertions

16. On distal phalanges of second, third, fourth, and fifth fingers.
17. On distal phalanx of thumb.
18. On middle phalanges of second, third, fourth, and fifth fingers.
19. On radial tuberosity and oblique line of radius.
20. On upper lateral surface of radius.
21. On first metacarpal and trapezium.
22. On second metacarpal.
23. On proximal portion of second and third metacarpals.
24. On middle metacarpal.
25. On fifth metacarpal.

Answers

Hand Movements

11. ________
12. ________
13. ________
14. ________
15. ________
16. ________
17. ________
18. ________
19. ________
20. ________
21. ________
22. ________
23. ________
24. ________
25. ________

LABORATORY REPORT 28, 29

Student: ____________________

Desk No: ________ Section: ____________

Abdominal and Pelvic Muscles (Ex. 28)

A. *Illustrations*

Record the labels for figures 28.1 and 28.2 in the answer column.

B. *Abdomen*

Consult figure 28.1 to identify the muscles that apply to the following statements. Verify your selections from the text.

external oblique—1
internal oblique—2
rectus abdominis—3
transversus abdominis—4

1. Flexion of spine in lumbar region.
2. Maintain intraabdominal pressure.
3. Antagonists of diaphragm.

Origins

4. On pubic bone.
5. On external surface of lower eight ribs.
6. On lateral half of inguinal ligament and anterior two-thirds of iliac crest.
7. On inguinal ligament, iliac crest, and costal cartilages of lower six ribs.

Insertions

8. On linea alba and crest of pubis.
9. On cartilages of fifth, sixth, and seventh ribs.
10. On the linea alba where fibers of the aponeurosis interlace.
11. On costal cartilages of lower three ribs, linea alba, and crest of pubis.

C. *Pelvic Region*

Consult figure 28.2 to identify the muscles that apply to the following statements. Verify your selections from the text.

iliacus—1
psoas major—2
quadratus lumborum—3

1. Flexes femur on trunk.
2. Flexes lumbar region of vertebral column.
3. Extension of spine at lumbar vertebrae.

Origins

4. On lumbar vertebrae.
5. On iliac crest, iliolumbar ligament, and transverse processes of lower four lumbar vertebrae.
6. On iliac fossa.

Insertions

7. On inferior margin of last rib and transverse processes of upper four lumbar vertebrae.
8. On lesser trochanter of femur.

Answers

Abdomen

1. ________
2. ________
3. ________
4. ________
5. ________
6. ________
7. ________
8. ________
9. ________
10. ________
11. ________

Pelvic

1. ________
2. ________
3. ________
4. ________
5. ________
6. ________
7. ________
8. ________

Fig. 28.1

____
____
____
____
____
____
____

Fig. 28.2

____
____
____
____
____

Leg Muscles (Ex. 29)

A. *Illustrations*

Record the labels for this exercise in the answer column.

B. *Thigh Movements*

Consult figure 29.1 to select the muscles that apply to the following statements. More than one muscle may apply to a statement.

adductor brevis—1
adductor longus—2
adductor magnus—3
gluteus maximus—4
gluteus medius—5
gluteus minimus—6
piriformis—7

1. Gluteus muscle that extends femur.
2. Gluteus muscles that abduct femur.
3. Three muscles that pull femur toward median line.
4. Gluteus muscle that rotates femur outward.
5. Three muscles attached to linea aspera that flex femur.
6. Small muscle that abducts, extends, and rotates femur.

Origins

7. On anterior surface of sacrum.
8. On the pubis.
9. On external surface of ilium.
10. On ilium, sacrum, and coccyx.
11. On inferior surface of ischium and portion of pubis.

Insertions

12. On linea aspera of femur.
13. On anterior border of greater trochanter.
14. On upper border of greater trochanter of femur.
15. On lateral part of greater trochanter of femur.
16. On iliotibial tract and posterior part of femur.

C. *Thigh Muscles*

Identify the muscles of the thigh that are described by the following statements.

biceps femoris—1
gracilis—2
rectus femoris—3
sartorius—4
semimembranosus—5
semitendinosus—6
tensor fasciae latae—7
vastus intermedius—8
vastus lateralis—9
vastus medialis—10

1. The largest quadriceps muscle that is located on the side of the thigh.
2. That portion of the quadriceps that lies beneath the rectus femoris.
3. Portion of the quadriceps that is on the medial surface of the thigh.
4. Hamstring muscle that is on the lateral surface of the thigh.
5. The longest muscle of the thigh.
6. The most medial component of the hamstring muscles.
7. The smallest hamstring muscle.
8. A superficial muscle on the medial surface of the thigh that inserts with the sartorius on the tibia.
9. Four muscles of a group that extend the leg.
10. Flexes the thigh upon the pelvis.
11. Two muscles other than parts of quadriceps that rotate the thigh medially (inward).
12. Rotates the thigh laterally (outward).
13. Adducts the thigh.

Answers

Thigh Movements

1. ______
2. ______
3. ______
4. ______
5. ______
6. ______
7. ______
8. ______
9. ______
10. ______
11. ______
12. ______
13. ______
14. ______
15. ______
16. ______

Thigh and Lower Leg

1. ______
2. ______
3. ______
4. ______
5. ______
6. ______
7. ______
8. ______
9. ______
10. ______
11. ______
12. ______
13. ______

Fig. 29.1

Fig. 29.2

Fig. 29.3

Fig. 29.4

Fig. 29.5 Anterior

Fig. 29.5 Posterior

D. *Lower Leg and Foot Muscles*

Identify the muscles of the lower leg and foot that are described by the following statements. In some instances several answers are applicable.

extensor digitorum longus—1
extensor hallucis longus—2
flexor digitorum longus—3
flexor hallucis longus—4
gastrocnemius—5
peroneus brevis—6
peroneus longus—7
peroneus tertius—8
soleus—9
tibialis anterior—10
tibialis posterior—11

1. Located on the front of the lower leg; causes dorsiflexion and inversion of the foot.
2. Deep portion of triceps surae.
3. Superficial portion of triceps surae.
4. Three muscles that originate on the fibula that cause various movements of the foot.
5. Two muscles that join to form the Achilles tendon.
6. Part of the triceps surae that originates on the femur.
7. Inserts on the proximal superior portion of the fifth metatarsal bone of the foot.
8. Inserts on the proximal inferior portion of the fifth metatarsal bone of the foot.
9. Flexes the great toe.
10. On the posterior surfaces of the tibia and fibula; causes plantar flexion and inversion of foot.
11. Originates on the tibia and fascia of the tibialis posterior; causes flexion of 2nd, 3rd, 4th, and 5th toes.
12. Flexes the calf on the thigh.
13. Originates on the tibia, fibula, and interosseous membrane; extends toes (dorsiflexion) and inverts foot.
14. Two muscles that insert on the calcaneous.
15. Peroneus muscle that causes dorsiflexion and eversion of foot.
16. Two peroneus muscles that cause plantar flexion.
17. Muscle on tibia that provides support for foot arches.

E. *General Questions*

Record the answers to the following questions in the answer column.

1. List the four muscles that are collectively referred to as the quadriceps femoris.
2. List two muscles associated with the iliotibial tract.
3. List three muscles that constitute the hamstrings.
4. What two muscles are associated with the Achilles tendon?
5. What two muscles constitute the triceps surae?

F. *Figure 18.1*

Refer back to figure 18.1 and identify the muscles that cause each type of movement. Record answers in column.

Answers

Lower Leg and Foot

1. ________ 10. ________
2. ________ 11. ________
3. ________ 12. ________
4. ________ 13. ________
5. ________ 14. ________
6. ________ 15. ________
7. ________ 16. ________
8. ________ 17. ________
9. ________

General Questions

1a. ________
b. ________
c. ________
d. ________
2a. ________
b. ________
3a. ________
b. ________
c. ________
4a. ________
b. ________
5a. ________
b. ________

Figure 18.1

A. ________
B. ________
C. ________
D. ________
E. ________
F. ________
G. ________
H. ________
I. ________
J. ________
K. ________
L. ________
M. ________

LABORATORY REPORT **30**

Student: ______________________

Desk No: __________ Section: ______________

The Spinal Cord, Spinal Nerves, and Reflex Arcs

A. *Illustrations*

Record the labels for figures 30.1, 30.2, and 30.4 in the answer columns.

B. *Completion Questions*

Supply the information that is necessary to complete the following statements.

1. The dural sac of the spinal cord is continuous with the ______ mater that surrounds the brain.
2. The caudal end of the spinal cord is called the ____________ .
3. The caudal end of the spinal cord terminates at the lower border of the ____________ lumbar vertebra.
4. The cluster of nerves that extend downward from the end of the spinal cord is called the ____________ .
5. The coccygeal nerve emerges from the ____________ plexus.
6. The femoral nerve is formed by the union of the following three lumbar nerves: ____________ , ____________ , and ____________ .
7. The largest nerve emerging from the sacral plexus is the ____________ nerve.
8. The iliohypogastric and ilioinguinal nerves are branches of the first ____________ nerve.
9. The innermost meninx surrounding the spinal cord is the ____________ mater.
10. The meninx that lies just within the dura mater is the ____________ .
11. The spinal ganglion is an enlargement of the ____________ root.
12. The central canal of the spinal cord is lined with ____________ ependymal cells.
13. The cavity within the spinal cord that is continuous with the ventricles of the brain is called the ____________ canal.
14. The inner portion of the spinal cord, forming the pattern of a butterfly, consists of ____________ matter.
15. The space between the dura mater and vertebral bone is called the ____________ space.
16. ____________ (*vertebral* or *collateral*) ganglia are usually located close to the organ that is innervated.
17. A somatic reflex arc with one or more interneurons is said to be ____________ .
18. Cell bodies of afferent neurons of visceral reflexes are located in ____________ ganglia.
19. If a somatic reflex lacks an interneuron it is said to be ____________ .
20. List the two divisions of the autonomic nervous system.

Answers

Completion

1. ____________
2. ____________
3. ____________
4. ____________
5. ____________
6. ____________
7. ____________
8. ____________
9. ____________
10. ____________
11. ____________
12. ____________
13. ____________
14. ____________
15. ____________
16. ____________
17. ____________
18. ____________
19. ____________
20. ____________

Figure 30.1

________	________
________	________
________	________
________	________
________	________
________	________
________	________
________	________
________	________
________	________
________	________
________	________
________	________
________	

C. True–False

Indicate the validity of the following statements with a T or F.

1. The spinal cord during fetal life fills the entire length of the spinal cavity.
2. The lumbar plexus is formed by the union of L_1, L_2, L_3, and most of L_4 nerves.
3. The brachial plexus is formed by the union of the lower four cervical nerves and the first thoracic nerve.
4. The cervical plexus is formed by the union of the first three pairs of cervical nerves.
5. The filum terminale externa is a downward extension of the filum terminale interna.
6. The innermost fiber of the cauda equina (on the median line) is the filum terminale interna.
7. Cerebrospinal fluid fills the subarachnoid space.
8. The white matter of the spinal cord consists of myelinated fibers of nerve cells.
9. The posterior cutaneous nerve emerges from the lumbar plexus.
10. The sacral plexus is formed by the union of the femoral and sciatic nerves.
11. The effector in a somatic reflex is always skeletal muscle.
12. Most organs of the body are innervated by fibers of the sympathetic and parasympathetic divisions of the autonomic nervous system.
13. Collateral ganglia of the autonomic nervous system are united to form a chain along the vertebral column.
14. All visceral reflex arcs have two afferent neurons.
15. The central canal of the spinal cord is continuous with the ventricles of the brain.
16. The dura mater of the spinal cord extends peripherally to form a covering for spinal nerves.
17. The central canal of the spinal cord is lined with ciliated epithelium.
18. All somatic reflex arcs have at least one interneuron.
19. All visceral reflex arcs have two efferent neurons.
20. Spinal ganglia are located in the anterior roots of spinal nerves.

D. Numbers

Indicate the number of pairs of the following that are normally present.

1. Cervical nerves
2. Thoracic nerves
3. Lumbar nerves
4. Sacral nerves
5. Coccygeal nerves

Answers

True–False	Fig. 30.2
1. ______	______
2. ______	______
3. ______	______
4. ______	______
5. ______	______
6. ______	______
7. ______	______
8. ______	______
9. ______	______
10. ______	______
11. ______	______
12. ______	______
13. ______	______
14. ______	______
15. ______	______
16. ______	______
17. ______	**Fig. 30.4**
18. ______	
19. ______	______
20. ______	______
Numbers	______
	______
1. ______	______
2. ______	______
3. ______	
4. ______	
5. ______	

LABORATORY REPORT **31**

Student: ______________________

Desk No: __________ Section: ______________

Somatic Reflexes

A. *Functional Nature*

Unsuspended Frog: Does the spinal frog attempt to control the position of its hind legs? ___1___ . Do leg muscles of the spinal frog seem *flaccid* or *firm?* ___2___ . Does tonus exist in these muscles? ___3___ . Does the animal jump when prodded? ___4___ . Can the animal swim? ___5___ . Does the animal float or sink? ___6___ .

Suspended Frog: Does the frog attempt to remove the acid? ___7___ . Do you think that the frog experiences a burning sensation? ___8___ . Explain: ______________________________

____

Does the addition of acid to the abdomen or other leg cause a response in the animal? ___9___ . If you try to restrain the animal's attempt to remove the acid, does it attempt some other maneuver? ___10___ .

Make a statement as to the apparent functional nature of reflexes. ______

____

____

____

____

Answers

Functional Nature

1. ______
2. ______
3. ______
4. ______
5. ______
6. ______
7. ______
8. ______
9. ______
10. ______

Inhibition

1. ______
2. ______
3. ______
4. ______
5. ______

B. *Reaction Time*

Record the reaction times (seconds) for each concentration in the table below. Plot these times in the graph.

HCl (Percentage)	Time (Seconds)
.05	
.1	
.2	
.3	
.4	
.5	
1.0	

Conclusion: ______________________________

____

C. *Reflex Radiation*

Describe the response of the spinal frog to increased electrical stimulation of the foot. ____________

____

____

____

____

D. *Reflex Inhibition*

Was it possible to inhibit right foot withdrawal from the acid by electrically stimulating the left foot?

___1___. If so, how much voltage was required? ___2___. At what pH? ___3___.

At what acid concentration was reflex action inhibited? ___4___ At what voltage? ___5___

E. *Synaptic Fatigue*

Where do you believe fatigue occurred in this experiment? ____________

____

F. *Diagnostic Reflex Tests*

Record the responses for the first five reflexes with a 0 for *no response,* a + for *diminished response,* and + + for *good response.* For Hoffmann's reflex place a check in the proper space. List the nerves in the last column that are affected in abnormal responses.

REFLEX	RESPONSES		Nerves Affected When
	Right Side	Left Side	Response Is Abnormal
1. Biceps			
2. Triceps			
3. Brachioradialis			
4. Patellar			
5. Achilles			
6. Hoffmann's	Absent (Normal)	Present (Abnormal)	
7. Plantar Flexion	Present (Normal)	Babinski's Sign (Abnormal)	

LABORATORY REPORT **32**

Student: ______________________

Desk No: ________ Section: ____________

Brain Anatomy: External

A. *Illustrations*

Record the labels for all the illustrations of this exercise in the answer columns.

B. *Brain Dissections*

Answer the following questions that pertain to your observations of the sheep brain dissections.

1. Does the arachnoid mater appear to be attached to the dura mater? ____________ to the brain? ____________
2. Describe the appearance of the dura mater.

   ____

   ____

   ____
3. Within what structure is the sagittal sinus enclosed?

   ____
4. Differentiate:

   Gyrus: ______________________________________

   Sulcus: ______________________________________
5. Of what value are the sulci and gyri of the cerebrum?

   ____

   ____
6. List five structures that you were able to identify on the dorsal surface of the midbrain: ______________________

   ____

   ____
7. Is the cerebellum of the sheep brain divided on the median line as is the case of the human brain? ______________________

   ____
8. What significance is there to the size differences of the olfactory bulbs on the sheep and human brains?

   ____

   ____

   ____
9. Which cranial nerve is the largest in diameter?

   ____

Answers

Fig. 32.1	Fig. 32.5
________	________
________	________
________	________
________	________
________	________
________	________
________	________
________	________
________	________
________	________
________	________
	________
Fig. 32.2	________
________	________
________	________
________	________
________	________
________	________
________	________
________	________
________	**Fig. 32.6**
________	________
________	________
________	________
________	________
________	________
________	________
________	________
	________

C. *Meninges*

Select the meninx or meningeal space described by the following statements.

1. Fibrous outer meninx.
2. The middle meninx.
3. Meninx that surrounds the sagittal sinus.
4. Meninx that forms the falx cerebri.
5. Meninx attached to the brain surface.
6. Space that contains cerebrospinal fluid.

arachnoid mater—1
dura mater—2
epidural space—3
pia mater—4
subarachnoid space—5
subdural space—6

D. *Fissures and Sulci*

Select the fissure or sulcus that is described by the following statements.

1. Between the frontal and parietal lobes.
2. On the median line between the two cerebral hemispheres.
3. Between the occipital and parietal lobes.
4. Between the temporal and parietal lobes.

anterior central sulcus—1
central sulcus—2
lateral cerebral fissure—3
longitudinal cerebral fissure—4
parieto-occipital fissure—5

E. *Brain Functions*

Select the part of the brain that is responsible for the following activities.

1. Maintenance of posture.
2. Cardiac control.
3. Respiratory control.
4. Vasomotor control.
5. Consciousness.
6. Voluntary muscular movements.
7. Brightness and sound discrimination (animals only).
8. Contains fibers that connect the two cerebral hemispheres.
9. Contains nuclei of fifth, sixth, seventh, and eighth cranial nerves.
10. Coordination of complex muscular movements.
11. Function in man is unknown.

cerebellum—1
cerebrum—2
corpora quadrigemina—3
corpus callosum—4
medulla oblongata—5
midbrain—6
pineal body—7
pons Varolii—8

F. *Cerebral Functional Localization*

Select those areas of the cerebrum that are described by the following statements.

Location

1. On posterior central gyrus.
2. On parietal lobe.
3. On superior temporal gyrus.
4. On temporal lobe.
5. On frontal lobe.
6. On occipital lobe.
7. On anterior central gyrus.
8. On angular gyrus.
9. In area anterior to anterior central gyrus.

association area—1
auditory area—2
common integrative area—3
motor speech area—4
olfactory area—5
somatomotor area—6
somatosensory area—7
premotor area—8
visual area—9

Answers

Meninges

1. ________
2. ________
3. ________
4. ________
5. ________
6. ________

Fig. 32.7

____
____
____
____
____
____
____
____
____
____
____
____
____
____
____
____
____
____
____
____
____
____
____

Fissures and Sulci

1. ________
2. ________
3. ________
4. ________

Brain Functions

1. ________
2. ________
3. ________
4. ________
5. ________
6. ________
7. ________
8. ________
9. ________
10. ________
11. ________

Localization

1. ________
2. ________
3. ________
4. ________
5. ________
6. ________
7. ________
8. ________
9. ________

F. *Cerebral Functional Localization* (continued)

Function

10. Speech production.
11. Speech understanding.
12. Cutaneous sensibility.
13. Sight.
14. Sense of smell.
15. Visual interpretation.
16. Voluntary muscular movement.
17. Integration of sensory association areas.
18. Influences motor area function.

association area—1
auditory area—2
common integrative area—3
motor speech area—4
olfactory area—5
somatomotor area—6
somatosensory area—7
premotor area—8
visual area—9

G. *Cranial Nerves*

Record the number of the cranial nerve that applies to the following statements. More than one nerve may apply to some statements.

1. Sensory nerves.
2. Mixed nerves.
3. Emerges from midbrain.
4. Emerges from pons Varolii.
5. Emerges from medulla.

Structures Innervated

6. Cochlea of ear.
7. Heart.
8. Salivary glands.
9. Abdominal viscera.
10. Retina of eye.
11. Receptors in nasal membranes.
12. Taste buds at back of tongue.
13. Lateral rectus muscle of eye.
14. Taste buds of anterior two-thirds of tongue.
15. Semicircular canals of ear.
16. Thoracic viscera.
17. Three extrinsic eye muscles (superior rectus, medial rectus, inferior oblique) and levator palpebrae.
18. Superior oblique muscle of eye.
19. Pharynx, upper larynx, uvula, and palate.

1. Olfactory
2. Optic
3. Oculomotor
4. Trochlear
5. Trigeminal
6. Abducens
7. Facial
8. Vestibulocochlear
9. Glossopharyngeal
10. Vagus
11. Accessory
12. Hypoglossal

H. *Trigeminal Nerve*

Select the branch of the trigeminal nerve that innervates the following structures.

1. All lower teeth.
2. All upper teeth.
3. Tongue.
4. Lacrimal gland.
5. Outer surface of nose.
6. Buccal gum tissues of mandible.
7. Lower teeth, tongue, muscles of mastication, and gum surfaces.

incisal—1
infraorbital—2
inferior alveolar—3
lingual—4
long buccal—5
mandibular—6
maxillary—7
ophthalmic—8

Answers

Localization

10. ________
11. ________
12. ________
13. ________
14. ________
15. ________
16. ________
17. ________
18. ________

Cranial Nerves

1. ________
2. ________
3. ________
4. ________
5. ________
6. ________
7. ________
8. ________
9. ________
10. ________
11. ________
12. ________
13. ________
14. ________
15. ________
16. ________
17. ________
18. ________
19. ________

Trigeminal Nerve

1. ________
2. ________
3. ________
4. ________
5. ________
6. ________
7. ________

I. General Questions

Select the correct answers that complete the following statements.

1. Shallow furrows on the surface of the cerebrum are called
 (1) sulci. (2) fissures. (3) gyri.
2. Deep furrows on the surface of the cerebrum are called
 (1) sulci. (2) fissures. (3) gyri.
3. The infundibulum supports the
 (1) mammillary body. (2) hypophysis. (3) hypothalamus.
4. The mammillary bodies are a part of the
 (1) medulla. (2) midbrain. (3) hypothalamus.
5. The corpora quadrigemina are located on the
 (1) cerebrum. (2) medulla. (3) midbrain.
6. The major ganglion of the trigeminal nerve is the
 (1) gasserian. (2) sphenopalatine. (3) ciliary.
7. Convolutions on the surface of the cerebrum are called
 (1) sulci. (2) fissures. (3) gyri.
8. The hypophysis is located on the inferior surface of the
 (1) medulla. (2) midbrain. (3) hypothalamus.
9. The spinal bulb is the
 (1) medulla oblongata. (2) pons Varolii. (3) midbrain.
10. The pineal gland is located on the
 (1) medulla. (2) pons Varolii. (3) midbrain.

Answers
General Questions
1. ________
2. ________
3. ________
4. ________
5. ________
6. ________
7. ________
8. ________
9. ________
10. ________

LABORATORY REPORT 33

Student: ____________________

Desk No: ________ Section: ____________

Brain Anatomy: Internal

A. *Illustrations*

Record the labels for all the illustrations of this exercise in the answer columns.

B. *Location*

Identify that part of the brain in which the following structures are located.

cerebellum—1
cerebrum—2
diencephalon—3
medulla oblongata—4
midbrain—5
pons Varolii—6

1. Aqueduct of Sylvius
2. Caudate nuclei
3. Cerebral peduncles
4. Corpus callosum
5. Fornix
6. Globus pallidus
7. Hypothalamus
8. Lateral ventricles
9. Mammillary bodies
10. Putamen
11. Rhinencephalon
12. Thalamus
13. Third ventricle

C. *Functions*

Identify the structures that perform the following functions.

aqueduct of Sylvius—1
arachnoid granulations—2
caudate nuclei—3
cerebral peduncles—4
choroid plexus—5
corpus callosum—6
fornix—7
foramen of Monro—8
hypothalamus—9
infundibulum—10
intermediate mass—11
lateral aperture—12
lentiform nucleus—13
median aperture—14
rhinencephalon—15
thalamus—16

1. Relay station for all messages to cerebrum.
2. Temperature regulation.
3. Entire olfactory mechanism of cerebrum.
4. Assists in muscular coordination.
5. Fiber tracts of olfactory mechanism.
6. Connects the two halves of the thalamus.
7. Exerts steadying effect on voluntary movements.
8. Allows cerebrospinal fluid to pass from third to fourth ventricle.
9. Consists of ascending and descending tracts.
10. Supporting stalk of hypophysis.
11. Allows cerebrospinal fluid to return to blood.
12. A commissure which unites the cerebral hemispheres.
13. Secretes cerebrospinal fluid into ventricle.
14. Coordinates autonomic nervous system.

D. *General Questions*

Select the best answer that completes the following statements.

1. The brain stem consists of the
   (1) cerebrum, pons, midbrain, and medulla.
   (2) cerebellum, medulla, and pons.
   (3) pons, medulla, and midbrain.
2. The lateral ventricles are separated by the
   (1) thalamus. (2) fornix. (3) septum pellucidum.

Answers

Location

1. ______
2. ______
3. ______
4. ______
5. ______
6. ______
7. ______
8. ______
9. ______
10. ______
11. ______
12. ______
13. ______

Functions

1. ______
2. ______
3. ______
4. ______
5. ______
6. ______
7. ______
8. ______
9. ______
10. ______
11. ______
12. ______
13. ______
14. ______

General

1. ______
2. ______

Fig. 33.2

Fig. 33.3

3. Pain is perceived in the
   (1) somatomotor area. (2) thalamus. (3) pons.
4. The hypothalamus regulates
   (1) the hypophysis, appetite, and wakefulness.
   (2) body temperature, hypophysis, and vision.
   (3) reproductive functions, body temperature, and voluntary movements.
5. Feelings of pleasantness and unpleasantness appear to be associated with the
   (1) somatomotor area. (2) hypothalamus. (3) thalamus.
   (4) medulla oblongata.
6. The reticular formation consists of
   (1) gray matter. (2) white matter.
   (3) an interlacement of white and gray matter.
7. Alert consciousness is partially regulated by
   (1) caudate nuclei. (2) lentiform nuclei.
   (3) nuclei of the reticular formation.
8. The intermediate mass passes through the
   (1) midbrain. (2) third ventricle. (3) hypothalamus.
9. The reticular formation is seen in the
   (1) spinal cord. (2) cerebrum. (3) brain stem.
   (4) spinal cord, brain stem, and diencephalon.
10. Centers for vomiting, coughing, swallowing, and sneezing are located in the
    (1) pons. (2) medulla. (3) midbrain. (4) thalamus.

E. *Cerebrospinal Fluid*

You should be able to trace the path of the cerebrospinal fluid from its point of origin to where it is reabsorbed into the blood. If you can complete the following paragraph without referring to the text, you know the sequence fairly well. If you can state the sequence from memory, better yet.

Ventricles
lateral—1
third—2
fourth—3
Passageways
acoustic meatus—4
aqueduct of Sylvius—5
foramen magnum—6
foramen of Monro—7
lateral aperture—8
median aperture—9

Structures
arachnoid granulations—10
cerebellum—11
cerebrum—12
choroid plexus—13
cisterna cerebellomedullaris—14
cisterna superior—15
dura mater—16
sagittal sinus—17
septum pellucidum—18
subarachnoid space—19

Cerebrospinal fluid is secreted into each ventricle by a ___1___. From the ___2___ ventricles, which are located in the cerebral hemispheres, the fluid passes to the ___3___ ventricle through an opening called the ___4___. A canal, called the ___5___, allows the fluid to pass from the latter ventricle to the ___6___ ventricle. From this last ventricle the cerebrospinal fluid passes into a subarachnoid space called the ___7___ through three foramina: one ___8___ and two ___9___. From this cavity the fluid passes over the cerebellum into another subarachnoid space called the ___10___. It also passes ___11___ (*up, down*) the posterior side of the spinal cord and ___12___ (*up, down*) the anterior side of the spinal cord. From the cisterna superior the cerebrospinal fluid passes to the subarachnoid space around the ___13___. This fluid is reabsorbed back into the blood through delicate structures called the ___14___. The blood vessel that receives the cerebrospinal fluid is the ___15___.

Answers

General	Fig. 33.4
3. ______	______
4. ______	______
5. ______	______
6. ______	______
7. ______	______
8. ______	______
9. ______	______
10. ______	______
	______
Cerebrospinal Fluid	______
1. ______	______
2. ______	______
3. ______	______
4. ______	______
5. ______	______
6. ______	
7. ______	
8. ______	
9. ______	
10. ______	
11. ______	
12. ______	
13. ______	
14. ______	
15. ______	

LABORATORY REPORT 34, 35

Student: ____________________

Desk No: ________ Section: ____________

Anatomy of the Eye (Ex. 34)

A. *Illustrations*

Record the labels from figures 34.1, 34.2, and 34.3 in the answer columns.

B. *Structures*

Identify the structures described by the following statements.

1. Small nonphotosensitive area on retina.
2. Small pit in retina of eye.
3. Outer layer of wall of eye.
4. Fluid between lens and retina.
5. Fluid between lens and cornea.
6. Inner light-sensitive layer.
7. Round yellow spot on retina.
8. Delicate membrane that lines eyelids.
9. Middle vascular layer of wall of eyeball.
10. Drainage tubes for tears in eyelids.
11. Clear transparent portion of front of eyeball.
12. Tube that drains tears from lacrimal sac.
13. Chamber between iris and cornea.
14. Conical body in medial corner of the eye.
15. Chamber between the iris and lens.
16. Circular color band between lens and cornea.
17. Circular band of smooth muscle tissue surrounding lens.
18. Cartilaginous loop through which superior oblique muscle acts.
19. Connective tissue between lens perimeter and surrounding muscle.
20. A semicircular fold of conjunctiva in the medial canthus of the eye.

anterior chamber—1
aqueous humor—2
blind spot—3
caruncula—4
choroid coat—5
ciliary body—6
cornea—7
conjunctiva—8
fovea centralis—9
iris—10
lacrimal ducts—11
lacrimal sac—12
macula lutea—13
medial canthus—14
nasolacrimal duct—15
plica semilunaris—16
posterior chamber—17
pupil—18
retina—19
scleroid coat—20
suspensory ligament—21
trochlea—22
vitreous body—23

C. *Extrinsic Muscles*

Select the muscles of the eye that are described by the following statements.

1. Inserted on top of eyeball.
2. Inserted on side of eyeball.
3. Inserted on medial surface.
4. Inserted on bottom of eyeball.
5. Inserted on eyelid.
6. Raises the eyelid.
7. Rotates eyeball inward.
8. Rotates eyeball downward.
9. Rotates eyeball outward.
10. Rotates eyeball upward.

inferior oblique—1
inferior rectus—2
lateral rectus—3
medial rectus—4
superior levator palpebrae—5
superior oblique—6
superior rectus—7

Answers

Structures	Fig. 34.1
1. ______	______
2. ______	______
3. ______	______
4. ______	______
5. ______	______
6. ______	______
7. ______	______
8. ______	______
9. ______	______
10. ______	______
11. ______	______
12. ______	**Fig. 34.2**
13. ______	
14. ______	______
15. ______	______
16. ______	______
17. ______	______
18. ______	______
19. ______	______
20. ______	______
	______
Muscles	______
1. ______	______
2. ______	______
3. ______	______
4. ______	______
5. ______	______
6. ______	______
7. ______	______
8. ______	______
9. ______	______
10. ______	______
	______
	______

D. Functions

Select the part of the eye that performs the following functions. More than one answer may apply in some cases.

aqueous humor—1
blind spot—2
choroid coat—3
ciliary body—4
conjunctiva—5
iris—6
lacrimal ducts—7
lacrimal puncta—8
lacrimal sac—9
macula lutea—10
nasolacrimal duct—11
pupil—12
retina—13
scleral venous sinus—14
scleroid coat—15
suspensory ligament—16
trabeculae—17
trochlea—18
vitreous body—19

1. Place where nerve fibers of retina leave eyeball.
2. Furnishes blood supply to retina and sclera.
3. Large vessel in wall of the eye that collects aqueous humor from trabeculae.
4. Exerts force on the lens, changing its contour.
5. Controls the amount of light that enters the eye.
6. Maintains firmness and roundness of eyeball.
7. Part of retina where critical vision occurs.
8. Provides most of the strength to the wall of the eyeball.

Answers

Functions	Fig. 34.3
1. ______	______
	______
2. ______	______
	______
3. ______	______
	______
4. ______	______
	______
5. ______	______
	______
6. ______	
7. ______	
8. ______	

E. Beef Eye Dissection

Answer the following questions that pertain to the dissection of the beef eye.

1. What is the shape of the pupil? ______
2. Why do you suppose it is so difficult to penetrate the sclera with a sharp scalpel?
   ______
3. What is the function of the black pigment in the eye?
   ______
   ______
4. Compare the consistency of the two fluids in the eye.
   aqueous humor: ______
   vitreous humor: ______
5. When you hold the lens up and look through it what is unusual about the image?
   ______
6. Does the lens magnify printed matter when placed directly on it? ______
7. Compare the consistencies of the following portions of the lens.
   center: ______
   edge: ______
8. What is the reflective portion of the choroid coat called? ______
9. Is there a macula lutea on the retina of the beef eye? ______

F. Ophthalmoscopy

Record your ophthalmoscope measurements here, and answer the questions.

1. **Diopter Measurements**

   The diopter value of a lens is the reciprocal of its focal length (f) in meters, or $D=\frac{1}{f}$.

   A lens of one diopter (1D) has a focal length of 1 meter, $(D=\frac{1}{f}=\frac{1}{1}=1)$; a 2D lens has a focal length of .5 meter, or $\frac{1}{f}=\frac{1}{.5}=2$; etc.

   Record here your measurements for: 5D______; 10D______; 20D______; and 40D______.

   Calculated diopter distances: 5D______; 10D______; 20D______; and 40D______.

   Do your calculations match the measured distances? ____________________

2. If you were able to examine your laboratory partner's eye with "0" in the diopter window of the ophthalmoscope, what would this indicate about the curvature of the lens of your eye? ________

   about your laboratory partner's eye? ____________________

Visual Tests (Ex. 35)

A. The Purkinje Tree

Describe in a few sentences the image you observed. ____________________

B. Tabulations

Record the results that were obtained in the following five eye tests:

Test	Right Eye	Left Eye
Blind Spot (inches)		
Near Point (inches)		
Scheiner's Experiment (inches)		
Visual Acuity (X/20)		
Astigmatism (present or absent)		

C. Color Blindness Test

Have your laboratory partner record your responses to each test plate. The mark *x* indicates inability to read correctly.

<table>
<tr><th rowspan="3">Plate Number</th><th rowspan="3">Subject's Response</th><th rowspan="3">Normal Response</th><th colspan="4">Response if Red-Green Deficiency</th><th rowspan="3">Response if Totally Color-blind</th></tr>
<tr><th colspan="2">Protan</th><th colspan="2">Deutan</th></tr>
<tr><th>Strong</th><th>Mild</th><th>Strong</th><th>Mild</th></tr>
<tr><td>1</td><td></td><td>12</td><td colspan="4">12</td><td>12</td></tr>
<tr><td>2</td><td></td><td>8</td><td colspan="4">3</td><td>x</td></tr>
<tr><td>3</td><td></td><td>5</td><td colspan="4">2</td><td>x</td></tr>
<tr><td>4</td><td></td><td>29</td><td colspan="4">70</td><td>x</td></tr>
<tr><td>5</td><td></td><td>74</td><td colspan="4">21</td><td>x</td></tr>
<tr><td>6</td><td></td><td>7</td><td colspan="4">x</td><td>x</td></tr>
<tr><td>7</td><td></td><td>45</td><td colspan="4">x</td><td>x</td></tr>
<tr><td>8</td><td></td><td>2</td><td colspan="4">x</td><td>x</td></tr>
<tr><td>9</td><td></td><td>x</td><td colspan="4">2</td><td>x</td></tr>
<tr><td>10</td><td></td><td>16</td><td colspan="4">x</td><td>x</td></tr>
<tr><td>11</td><td></td><td>traceable</td><td colspan="4">x</td><td>x</td></tr>
<tr><td>12</td><td></td><td>35</td><td>5</td><td>(3)5</td><td>3</td><td>3(5)</td><td></td></tr>
<tr><td>13</td><td></td><td>96</td><td>6</td><td>(9)6</td><td>9</td><td>9(6)</td><td></td></tr>
<tr><td>14</td><td></td><td>can trace two lines</td><td>purple</td><td>purple (red)</td><td>red</td><td>red purple</td><td>x</td></tr>
</table>

x—Indicates inability of subject to respond in any way to test.

Conclusion: ______________________________

D. Pupillary Reflexes

Answer the following questions related to your observations in the two experiments on pupillary reflexes.

1. **Light Intensity and Pupil Size**
   a. Did the pupil of the unexposed eye become smaller when the right eye was exposed to light?

   ____

   b. Trace the pathways of the nerve impulses in effecting the response. ______________________________

   ____

   ____

   c. Can you suggest a possible benefit that might result from this phenomenon? ______________________________

   ____

2. **Accommodation to Distance**
   a. What pupillary size change occurred in this experiment? ______________________________

   ____

   b. Can you suggest a benefit that might result from this happening? ______________________________

   ____

LABORATORY REPORT **36**

Student: ____________________

Desk No: ________ Section: ____________

The Ear: Its Role in Hearing

A. *Figure 36.2*

Record the labels for this figure in the answer column.

B. *Structure*

Identify the structures described by the following statements.

basilar membrane—1	malleus—6	scala vestibuli—12
endolymph—2	perilymph—7	stapes—13
hair cells—3	reticular lamina—8	tectorial membrane—14
helicotrema—4	rods of Corti—9	tympanic membrane—15
incus—5	scala media—10	vestibular membrane—16
	scala tympani—11	

1. Ossicle that fits in oval window.
2. Ossicle activated by tympanic membrane.
3. Chamber of cochlea into which round window opens.
4. Chamber of cochlea into which oval window opens.
5. Membrane set in vibration by sound waves in the air.
6. Membrane that contains 20,000 fibers of varying lengths.
7. Membrane in which hair tips of hair cells are embedded.
8. Two fluids found in cochlea.
9. Fluid within scala tympani.
10. Fluid within cochlear duct.
11. Fluid within scala vestibuli.
12. Parts of organ of Corti.
13. Transfers vibrations from basilar membrane to reticular lamina.
14. Receptors that initiate action potentials in cochlear nerve.
15. Opening between scala vestibuli and scala tympani.

C. *Physiology of Hearing (True–False)*

Indicate the validity of each of the following statements.

1. The loudness of sound is directly proportional to the square of the amplitude.
2. The pitch of a sound is determined by the amplitude of the sound wave.
3. Quality and pitch are two different characteristics of sound.
4. Some hearing takes place through the bones of the skull rather than through the auditory ossicles.
5. While there is no cure for nerve deafness, conduction deafness can often be corrected.
6. Endolymph in the scala media flows into the scala vestibuli through the helicotrema.
7. Nerve deafness can be detected by placing a tuning fork on the mastoid process.
8. The frequency range used in speech is from 30 to 20,000 cps.
9. The bending of hairs of hair cells causes an alternating electrical charge known as the endocochlear potential.
10. Before using an audiometer, it is necessary for the operator to calibrate it.

Answers

Structure	Fig. 36.2
1. ______	______
2. ______	______
3. ______	______
4. ______	______
5. ______	______
6. ______	______
7. ______	______
8. ______	______
9. ______	______
10. ______	______
11. ______	______
12. ______	______
13. ______	______
14. ______	______
15. ______	______
	______
Physiology	______
1. ______	______
2. ______	______
3. ______	______
4. ______	______
5. ______	______
6. ______	______
7. ______	______
8. ______	______
9. ______	______
10. ______	

D. *Screening Test*

Record the distances at which the watch tick could be heard with your ears, as well as in three other persons.

Subject	Inches From Ear			
	Right Ear		Left Ear	
	Approaching	Receding	Approaching	Receding

E. *Tuning Fork Methods*

Record the results of the two tuning fork methods.

Rinne Test

Right Ear ______________________________

Left Ear ______________________________

Weber Test

Right Ear ______________________________

Left Ear ______________________________

F. *Audiometry*

Record the threshold levels in the right ear with a red "O" and in the left ear with a blue "X." Connect the points with appropriately colored lines.

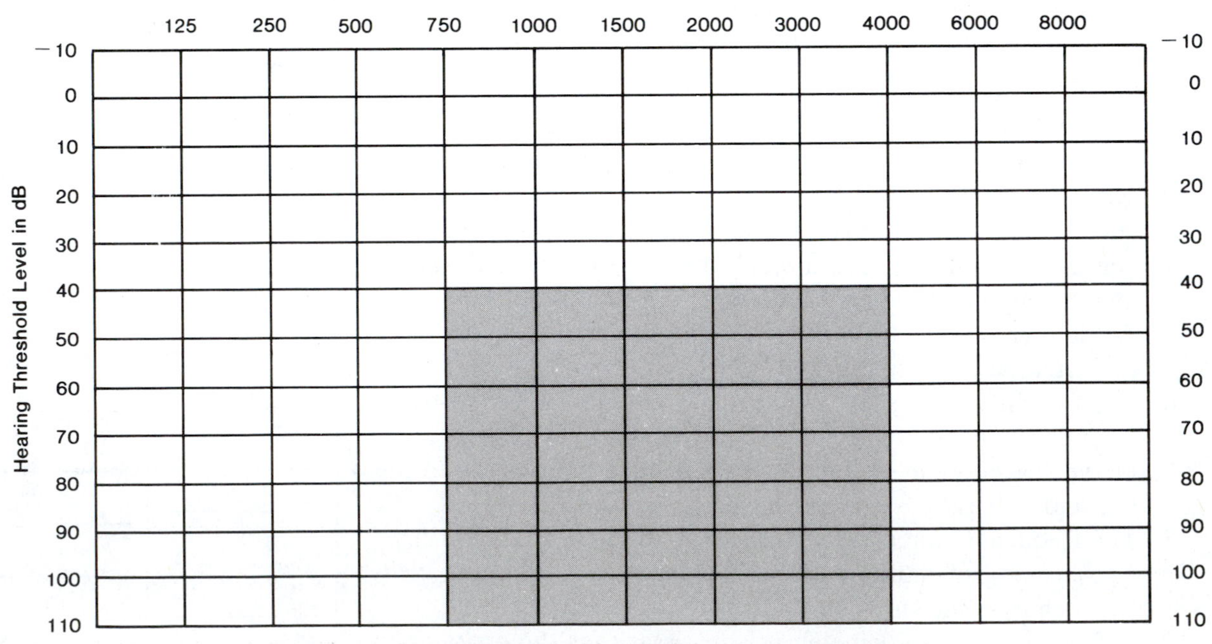

(Shaded area indicates critical area of speech interpretation)

LABORATORY REPORT **37**

Student: ______________________

Desk No: ________ Section: ____________

The Ear: Its Role in Equilibrium

A. *Anatomy*

Answer the following questions that pertain to the anatomy of the vestibular apparatus.

1. What are the sensory receptors for static equilibrium called?
2. Where are the static equilibrium sensory receptors located? (two places)
3. What are the small calcium carbonate crystals of the macula called?
4. Where are the sensory receptors of dynamic equilibrium located?
5. What nerve branch of the eighth cranial nerve supplies the vestibular apparatus?
6. What part of the vestibular apparatus connects directly to the cochlear duct?
7. What is the name of the fluid within the vestibular apparatus?
8. What structures in joints prevent a sense of malequilibrium when one tilts the head to one side?
9. What fluid lies between the semicircular ducts and the surrounding bone of the semicircular canals?
10. Within how many planes do the semicircular ducts lie?

Answers

Anatomy

1. ______________
2. ______________
   ____
3. ______________
4. ______________
5. ______________
6. ______________
7. ______________
8. ______________
9. ______________
10. ______________

B. *Nystagmus*

After having observed nystagmus, answer these questions:

1. What direction do the eyes move in relation to the direction of rotation? ______________
2. What seems to be the function of slow movement of the eye? ______________
3. What seems to be the function of fast movement of the eye? ______________
4. What part of the vestibular apparatus triggers the reflexes that control the eye muscles? ______________

C. Proprioceptive Influences

Record your observations here.

1. At-Rest Reactions

a. Was the subject able to place the heel of his foot on the toes of the other foot with eyes closed?

____

b. Was subject able to touch nose with eyes closed? ______________________________

c. Explain what mechanisms in the body make this possible: ______________________________

____

d. Describe the subject's ability to touch the pencil eraser with eyes closed after practicing with eyes open. ______________________________

____

2. Effects of Rotation

a. Was the subject, with eyes open, able to point directly to the pencil eraser with finger after rotation? ______________________________

____

b. When blindfolded subject tried to locate pencil eraser after rotation, what was the result?

____

____

LABORATORY REPORT 38–42

Student: ______________________

Desk No: ________ Section: ______________

Hematological Tests

TEST RESULTS

Except for blood typing, record all test results in table II on page 408. Calculations for blood cell counts should be performed as stated below.

A. Differential White Blood Cell Count

As you move the slide in the pattern indicated in figure 38.3, record all the different types of cells in table I. Refer to figure 38.1 for cell identification. Use this method of tabulation: ~~1111~~ ~~1111~~ 1 1. Identify and tabulate 100 leukocytes. Divide the total of each kind of cell by 100 to determine percentages.

Table I Leukocyte tabulation.

Neutrophils	Lymphocytes	Monocytes	Eosinophils	Basophils
Totals				
Percent				

B. Calculations for Blood Cell Counts

1. **Total White Blood Cell Count**
   Total WBCs counted in 4 W areas × 50 = **WBCs per cu mm**

2. **Red Blood Cell Count**
   Total RBCs counted in 5 R areas × 10,000 = **RBCs per cu mm**

C. Blood Typing

Record your blood type here: __________

1. If you needed a blood transfusion, what types of blood could you be given?

   ____

2. If your blood were to be given to someone in need, what type should that person have?

   ____

QUESTIONS

Although most of the answers to the following questions can be derived from this manual, it will be necessary to consult your lecture text or medical dictionary for some of the answers.

A. *Blood Components*

Identify the blood components described by the following statements. More than one answer may apply.

antibodies—1
basophils—2
eosinophils—3
erythrocytes—4
fibrin—5
fibrinogen—6
lymphocytes—7
lysozyme—8
monocytes—9
neutrophils—10
platelet factor—11
platelets—12
prothrombin—13
thromboplastin—14

1. Cells that transport O_2 and CO_2.
2. Phagocytic cell concerned, primarily, with local infections.
3. Phagocytic cell concerned, primarily, with generalized infections.
4. Protein substances in blood that inactivate foreign protein.
5. Noncellular formed elements essential for blood clotting.
6. Substances necessary for blood clotting.
7. Gelatinouslike material formed in blood clotting.
8. Substance released by injured cells that initiates blood clotting.
9. Enzyme in blood plasma that destroys some kinds of bacteria.
10. Leukocyte distinguished by having distinctive red stained granules.
11. Leukocyte that probably produces heparin.
12. Most numerous type of leukocyte.
13. Contains hemoglobin.
14. Most numerous type of blood cell.

B. *Blood Diseases*

Identify the pathological conditions characterized by the following statements. More than one condition may apply to some statements.

anemia—1
eosinophilia—2
erythroblastosis fetalis—3
hemophilia—4
leukemia—5
leukocytosis—6
leukopenia—7
lymphocytosis—8
neutropenia—9
neutrophilia—10
pernicious anemia—11
polycythemia—12
sickle-cell anemia—13

1. Too few red blood cells.
2. Too many red blood cells.
3. Too few leukocytes.
4. Too many neutrophils.
5. Too many lymphocytes.
6. Too little hemoglobin.
7. Defective RBCs due to stomach enzyme deficiency.
8. Rh factor incompatibility present at birth.
9. Hereditary blood disease of blacks; RBCs of abnormal shape.
10. Heritable bleeder's disease.
11. Cancerouslike condition in which there are too many leukocytes.

Answers

Components

1. ________
2. ________
3. ________
4. ________
5. ________
6. ________
7. ________
8. ________
9. ________
10. ________
11. ________
12. ________
13. ________
14. ________

Diseases

1. ________
2. ________
3. ________
4. ________
5. ________
6. ________
7. ________
8. ________
9. ________
10. ________
11. ________

C. Terminology

Differentiate between the following pairs of related terms:

Plasma ______________________________

Lymph ______________________________

Coagulation ______________________________

Agglutination ______________________________

Antibody ______________________________

Antigen ______________________________

D. General Questions

Indicate in the answer column whether the following statements are true or false.

1. The most common types of blood are types O and A.
2. A universal donor has O-Rh negative blood.
3. Diapedesis is the ability of leukocytes to move between the cells of the capillary walls.
4. Anemia is usually due to an iodine deficiency.
5. Hemoglobin combines with both oxygen and carbon dioxide.
6. The maximum life span of an erythrocyte is about 30 days.
7. White blood cells live much longer than red blood cells.
8. The vitamin which is essential to blood clotting is vitamin D.
9. An infant born with erythroblastosis would best be treated with a transfusion of Rh negative instead of Rh positive blood.
10. Worm infestations may cause eosinophilia.
11. Viral infections may cause an increase in the number of lymphocytes or monocytes.
12. Blood clotting can be enhanced with dicumarol.
13. Megakaryocytes of bone marrow produce blood platelets.
14. Bloods that are matched for ABO and Rh factor can be mixed with complete assurance of no incompatibility.
15. The rarest type of blood is AB negative.
16. Heparin is essential to fibrin formation.
17. Blood typing may be used to prove that an individual could not be the father in a paternity suit.
18. Blood typing may be used to prove that an individual is the father in a paternity suit.
19. Approximately 87% of the population is Rh negative.
20. A deficiency of calcium will result in the failure of blood to clot.
21. Fibrin is formed when prothrombin reacts with fibrinogen.

Answers

1. ________
2. ________
3. ________
4. ________
5. ________
6. ________
7. ________
8. ________
9. ________
10. ________
11. ________
12. ________
13. ________
14. ________
15. ________
16. ________
17. ________
18. ________
19. ________
20. ________
21. ________

E. *Summarization of Results*

Record blood cell counts and other test results in the following table:

Table II

Test	Normal Values	Test Results	Evaluation (over, under, normal)
Differential WBC Count	Neutrophils: 50%-70%		
	Lymphocytes: 20%-30%		
	Monocytes: 2%-6%		
	Eosinophils: 1%-5%		
	Basophils: 0.5%-1%		
Total WBC Count	5000-9000 per cu mm		
RBC Count	Males: 4.8-6.0 million/cu mm		
	Females: 4.1-5.1 million/cu mm		
Hemoglobin Percentage	Males: 13.4-16.4 gms/100 ml		
	Females: 12.2-15.2 gms/100 ml		
Hematocrit (VPRC)	Males: 40%-54% (Av. 47%)		
	Females: 37%-47% (Av. 42%)		
Coagulation Time	2 to 6 minutes		

F. *Materials*

Identify the various blood tests in which the following supplies are used.

1. Hemacytometer
2. Hemoglobinometer
3. Centrifuge
4. Capillary tubes
5. Wright's stain
6. Microscope slides
7. Hemolysis applicators
8. Seal-Ease
9. Diluting fluid
10. Typing antisera

hematocrit (VPRC)—1
clotting time—2
RBC and WBC counts—3
differential WBC count—4
hemoglobin determination—5
none of these—6

Materials

1. ________
2. ________
3. ________
4. ________
5. ________
6. ________
7. ________
8. ________
9. ________
10. ________

LABORATORY REPORT **43**

Student: ____________________

Desk No: __________ Section: __________

Anatomy of the Heart

A. *Illustrations*

Record the labels for figures 43.1 and 43.2 in the answer columns.

B. *Structures*

Identify the structures of the heart described by the following statements.

aorta—1
aortic semilunar valve—2
bicuspid valve—3
chordae tendineae—4
endocardium—5
epicardium—6
inferior vena cava—7
left atrium—8
left pulmonary artery—9
left ventricle—10
ligamentum arteriosum—11
mitral valve—12
myocardium—13
papillary muscles—14
parietal pericardium—15
pulmonary artery—16
pulmonic semilunar valve—17
right atrium—18
right pulmonary artery—19
right ventricle—20
superior vena cava—21
tricuspid valve—22
ventricular septum—23
visceral pericardium—24

1. Lining of the heart.
2. Partition between right and left ventricles.
3. Fibroserous saclike structure surrounding the heart.
4. Two chambers of the heart that contain deoxygenated blood.
5. Large artery that carries blood from the right ventricle.
6. Blood vessel that returns blood to heart from head and arms.
7. Remnant of a functional prenatal vessel between the pulmonary artery and the aorta.
8. Structure formed from ductus arteriosus.
9. Valve at the base of the pulmonary artery.
10. Two arteries that are branches of the pulmonary artery.
11. Synonym for bicuspid valve.
12. Structures on the cardiac wall to which the chordae tendineae are attached.
13. Muscular portion of cardiac wall.
14. Thin covering on surface of heart (2 names).
15. Large vein that empties blood into top of right atrium.
16. Two chambers of the heart that contain oxygenated blood.
17. Chamber of the heart that receives blood from the lungs.
18. Large artery that carries blood out from left ventricle.
19. Large vein that empties blood into the lower part of the right atrium.
20. Atrioventricular valves.
21. Valve at the base of the aorta.
22. Blood vessel in which openings to the coronary arteries are located.
23. Atrium into which the coronary sinus empties.
24. Valvular restraints that prevent the atrioventricular valve cusps from being forced back into the atria.

Answers

Structures	Fig. 43.1
1. ______	______
2. ______	______
3. ______	______
4. ______	______
5. ______	______
6. ______	______
7. ______	______
8. ______	______
9. ______	______
10. ______	______
11. ______	______
12. ______	______
13. ______	______
14. ______	______
15. ______	______
16. ______	______
17. ______	______
18. ______	______
19. ______	______
20. ______	______
21. ______	______
22. ______	______
23. ______	______
24. ______	______
	______
	______
	______
	______
	______
	______
	______
	______

C. Coronary Circulation

Identify the arteries and veins of the coronary circulatory system that are described by the following statements.

anterior interventricular artery—1
circumflex artery—2
coronary sinus—3
great cardiac vein—4
left coronary artery—5
middle cardiac vein—6
posterior descending right coronary artery—7
posterior interventricular vein—8
right coronary artery—9
small cardiac vein—10

1. Two principal branches of the left coronary artery.
2. Coronary vein that lies in the anterior interventricular sulcus.
3. Coronary artery that lies in the right atrioventricular sulcus.
4. Vein that receives blood from the great cardiac vein.
5. Vein that lies in the posterior interventricular sulcus.
6. Two major coronary arteries that take their origins in the wall of the aorta.
7. Large coronary vessel that empties into the right atrium.
8. Coronary vein that parallels the right coronary artery in the right atrioventricular sulcus.
9. Branch of the right coronary artery on the posterior surface of the heart.
10. Vein that lies alongside of the posterior descending right coronary artery.

D. Sheep Heart Dissection

After completing the sheep heart dissection answer the following questions. Some of these questions were encountered during the dissection; others pertain to structures not shown in figures 43.1 and 43.2.

1. Identify the following structures:

   Pectinate muscle: ______________________________

   ____

   Moderator band: ______________________________

   ____

   Ligamentum arteriosum: ______________________________

   ____

2. How many papillary muscles did you find in the

   right ventricle? ____________ left ventricle? ____________

3. How many pouches are present in each of the following?

   pulmonary semilunar valve: ____________

   aortic semilunar valve: ____________

4. Where does blood enter the myocardium? ______________________________

   ____

5. Where does blood leave the myocardium and return to the circulatory system? ______________________________

Answers

Coronary Circulation	Fig. 43.2 Anterior
1. ________	________
2. ________	________
3. ________	________
4. ________	________
5. ________	________
6. ________	________
7. ________	________
8. ________	________
9. ________	________
10. ________	________
Fig. 43.2 Posterior	________
________	________
________	________
________	________
________	________
________	________
________	________
________	________
________	
________	
________	
________	
________	
________	
________	
________	
________	

Student: ______________________

LABORATORY REPORT 44, 45

Desk No: ________ Section: ______________

Cardiovascular Sounds (Ex. 44)

A. Phonocardiogram

Attach below a segment of a phonocardiogram that you made on the recorder. Identify on the chart those segments that represent first and second sounds.

B. Questions

1. What causes the first sound? ______________________________

2. What causes the second sound? ______________________________

3. What causes the third sound? ______________________________

4. What causes a murmur? ______________________________

5. Why does the aortic valve close sooner than the pulmonic valve during inspiration? __________

6. What effect would complete right bundle branch block have on the splitting of the second sound?

7. Explain why pulmonary stenosis is associated with greater splitting of the second heart sound.

Electrocardiogram Monitoring: Using Chart Recorders (Ex. 45)

A. Tracings

Attach three tracings of your experiment in the spaces provided below. It will be necessary to trim excess paper from the tracings to fit them into the allotted spaces. Attach with Scotch tape.

Tracings	Evaluation
  Calibration	
  Subject At Rest	*Heart Rate: ________ per min QR Potential: ________ millivolts **Duration of cycle: ________ msec
  After Exercise	*Heart Rate: ________ per min QR Potential: ________ millivolts **Duration of cycle: ________ msec

*Space between two margin lines is 3 seconds.
**Time from beginning of P wave to end of T wave.

B. Questions

1. During what part of ECG wave pattern does atrial depolarization and contraction occur? ______

   ____

2. During what part of ECG wave pattern does ventricular depolarization and contraction occur?

   ____

3. Would a heart murmur necessarily show up on an ECG? ________________

   Explain. ____

   ____

47

Student: ____________________

Desk No: ________ Section: ____________

Pulse Monitoring

A. *Tracings*

Attach samples of tracings made in this experiment in the spaces provided below. Use the right-hand column for relevant comments.

TRACINGS	EVALUATION
Transducer at Heart Level	Pulse Rate: ________
Arm Slowly Raised Above Heart Level	Pulse Rate: ________
Arm Lowered to Floor	Pulse Rate: ________
Holding Breath for 15 Seconds	Pulse Rate: ________

TRACINGS	EVALUATION
Brachial Artery Occluded for 15–20 Seconds	
Chart Speed Increased to 25 mm per Second	
	Pulse Rate: ______
Subject Startled	

B. Pulse Tabulations

Record your pulse (while at rest) on the chalkboard. Once the pulse rates of all students are recorded, record the highest, lowest, and median pulses:

Highest ______ Lowest ______ Median ______

Significance ______________________________

____

C. Questions

1. What conceivably might be an explanation for changes occurring in the dicrotic notch with aging?

   ____

   ____

2. What might be an explanation for an alteration in the pulse amplitude during smoking? ______

   ____

3. The finger pulse is a manifestation of blood volume, velocity, direction, and pressure. Which one of these factors is directly responsible for the dicrotic notch and why? ______

   ____

LABORATORY REPORT 48, 49

Student: ____________________

Desk No: __________ Section: __________

Blood Pressure Monitoring (Ex. 48)

A. *Stethoscope Method*

Record here the blood pressures of the test subject.

	BEFORE EXERCISE	AFTER EXERCISE
Systolic		
Diastolic		
*Pulse Pressure		
**Mean Arterial Pressure		

*Pulse pressure = systolic pressure minus diastolic pressure.
**Mean arterial pressure ≃ diastolic pressure + ⅓ pulse pressure.

B. *Recording Method*

In the spaces provided below and on next page tape or staple chart recordings for the various portions of the experiment.

Calibration of Recorder

Subject in Chair

B. Recording Method (continued)

Subject Standing Erect	Subject Reclining

After Exercise

C. Questions

1. Give the systolic and diastolic pressures for normal blood pressure: ____________
2. What systolic pressure indicates hypertension? ____________
3. Why is a low diastolic pressure considered harmful? ____________
4. Why are diuretics of value in treating hypertension? ____________

   ____
5. Why is reduced salt intake of value in regulating blood pressure? ____________

   ____

   ____

Peripheral Circulation Control (Ex. 49)

A. Results

1. What observable change occurred in the frog's foot when histamine was applied to the foot?

   ____

   ____
2. What was the effect of histamine on the arterioles to produce this effect (*vasodilation* or *vasoconstriction*)? ____________
3. What observable change occurred in the frog's foot when epinephrine was applied to the foot?

   ____

   ____

LABORATORY REPORT **50**

Student: ______________________

Desk No: ________ Section: ________

The Arteries and Veins

A. *Illustrations*

Record the labels for figures 50.2 and 50.3 in the answer columns. Note that the labels for figure 50.2 are on both sides of this sheet.

B. *Arteries*

Refer to the left-hand illustration in figure 50.2 to identify the arteries described by the following statements.

anterior tibial—1
aorta—2
aortic arch—3
axillary—4
brachial—5
celiac—6
common iliac—7
deep femoral—8
external iliac—9
femoral—10
inferior mesenteric—11
innominate—12
internal iliac—13
left common carotid—14
left subclavian—15
popliteal—16
posterior tibial—17
radial—18
renal—19
right common carotid—20
subclavian—21
superior mesenteric—22
ulnar—23

1. Artery of the shoulder.
2. Artery of upper arm.
3. Artery of the armpit.
4. Gives rise to right common carotid and right subclavian.
5. Lateral artery of the forearm.
6. Medial artery of the forearm.
7. Three branches of the aortic arch.
8. Gives rise to femoral artery.
9. Major artery of the thigh.
10. Artery of knee region.
11. Artery of calf region.
12. Large branch of common iliac.
13. Small branch of common iliac.
14. Supplies the stomach, spleen, and liver.
15. Major artery of chest and abdomen.
16. Supplies the kidney.
17. Also known as the hypogastric artery.
18. Supplies blood to most of small intestines and part of colon.
19. Artery in interior portion of lower leg.
20. Branch of femoral that parallels medial surface of femur.
21. Curved vessel that receives blood from left ventricle.
22. Supplies the large intestine and rectum.

Answers

Arteries	Fig. 50.2 Arteries
1. ________	________
2. ________	________
3. ________	________
4. ________	________
5. ________	________
6. ________	________
7. ________	________
8. ________	________
9. ________	________
10. ________	________
11. ________	________
12. ________	________
13. ________	________
14. ________	________
15. ________	________
16. ________	________
17. ________	________
18. ________	________
19. ________	________
20. ________	________
21. ________	________
22. ________	________
	________

C. Veins

Refer to the right-hand illustration in figure 50.2 and to figure 50.3 to identify the veins described by the following statements.

accessory cephalic—1
axillary—2
basilic—3
brachial—4
cephalic—5
common iliac—6
coronary—7
dorsal venous arch—8
external iliac—9
external jugular—10
femoral—11
great saphenous—12
hepatic—13
inferior vena cava—14
innominate—15
inferior mesenteric—16
internal iliac—17
internal jugular—18
median cubital—19
popliteal—20
portal—21
posterior tibial—22
pyloric—23
subclavian—24
superior mesenteric—25
superior vena cava—26

1. Largest vein in neck.
2. Small vein in neck.
3. Vein of armpit.
4. Empty into innominate veins.
5. Collects blood from veins of head and arms.
6. Short vein between basilic and cephalic veins.
7. Collects blood from veins of chest, abdomen, and legs.
8. On posterior surface of humerus.
9. Large veins that unite to form the inferior vena cava.
10. On medial surface of upper arm.
11. On lateral portion of forearm.
12. On lateral portion of upper arm.
13. Vein from liver to inferior vena cava.
14. Receives blood from descending colon and rectum.
15. Empties into popliteal.
16. Collects blood from two innominate veins.
17. Collects blood from top of foot.
18. Superficial vein on medial surface of leg.
19. Vein of knee region.
20. Collects blood from intestines, stomach, and colon.
21. Vein that receives blood from posterior tibial.
22. Receives blood from ascending colon and part of ileum.
23. Empties into great saphenous.
24. Small vein which empties into common iliac at juncture of external iliac.
25. Vein that great saphenous empties into.
26. Two veins that drain blood from stomach into portal vein.

Answers

Veins	Fig. 50.2 Veins
1. ______	______
2. ______	______
3. ______	______
4. ______	______
5. ______	______
6. ______	______
7. ______	______
8. ______	______
9. ______	______
10. ______	______
11. ______	______
12. ______	______
13. ______	______
14. ______	______
15. ______	______
16. ______	______
17. ______	______
18. ______	______
19. ______	______
20. ______	______
21. ______	______
22. ______	______
23. ______	______
24. ______	______
25. ______	
26. ______	**Fig. 50.3**
	______
	______
	______
	______
	______
	______
	______
	______
	______
	______
	______
	______
	______
	______
	______

LABORATORY REPORT 51, 52

Student: ______________________

Desk No: ________ Section: ____________

Fetal Circulation (Ex. 51)

A. *Figure 51.1*

Record the labels for this illustration in the answer column.

B. *Questions*

Record the answers for the following questions in the answer column.

1. What blood vessel in the umbilical cord supplies the fetus with nutrients?
2. What blood vessel in the umbilical cord returns blood to the placenta from the fetus?
3. What blood vessel shunts blood from the pulmonary artery to the aorta?
4. What vessel in the liver carries blood from the umbilical vein to the inferior vena cava?
5. What structure in the liver is formed from the ductus venosus?
6. What structure forms from the ductus arteriosus?
7. What structure forms from the umbilical vein?
8. What structure enables blood to flow from the right atrium to the left atrium before birth?
9. Give two occurrences at childbirth that prevent excessive blood loss by the infant through cut umbilical cord.
10. What structures near the bladder form from the hypogastric arteries?

The Lymphatic System and the Immune Response (Ex. 52)

A. *Figure 52.1*

Record the labels for this illustration in the answer column.

B. *Components*

By referring to figures 52.1 and 52.2 identify the following structures in the lymphatic system.

1. Large vessel in thorax and abdomen that collects lymph from lower extremities.
2. Short vessel which collects lymph from right arm and right side of head.
3. Microscopic lymphatic vessels situated among cells of tissues.
4. Saclike structure that receives chyle from intestine.
5. Blood vessel that receives fluid from thoracic duct.
6. Filters of the lymphatic system.
7. Vessels of arms and legs that convey lymph to collecting ducts.
8. Depression in lymph node through which blood vessels enter node.
9. Outer covering of lymph node.
10. Structural units of lymph node that are packed with lymphocytes.
11. Cells involved in organ transplant rejections.
12. Cells that produce antibodies.

capsule—1
cisterna chyli—2
germinal centers—3
hilum—4
left subclavian vein—5
lymphatics—6
lymph capillaries—7
lymphoblasts—8
lymph nodes—9
plasma cells—10
right lymphatic duct—11
thoracic duct—12
none of these—13

Answers

Questions

1. ______
2. ______
3. ______
4. ______
5. ______
6. ______
7. ______
8. ______
9a. ______
b. ______
10. ______

Components	Fig. 51.1
1. ______	______
2. ______	______
3. ______	______
4. ______	______
5. ______	______
6. ______	______
7. ______	______
8. ______	______
9. ______	______
10. ______	______
11. ______	______
12. ______	______
	______
	______
	______
	______

C. Questions

1. List three forces that move lymph through the lymphatic vessels:
   a. ______
   b. ______ c. ______.
2. How does lymph in the lymphatics of the legs differ in composition from the lymph in the thoracic duct?
   ______
   ______
3. Where do stem cells for all lymphocytes originate? ______
   ______
4. Give the cell type that accounts for each type of immunity:
   Humoral immunity: ______
   Cellular immunity: ______
5. What tissues program stem cells for the following?
   Humoral immunity: ______
   Cellular immunity: ______
6. What type of tissue is common to the thymus gland and bursal tissues?
   ______

Answers
Fig. 52.1

LABORATORY REPORT **53**

Student: ______________________

Desk No: ________ Section: ______________

The Respiratory Organs

A. *Illustrations*

Record the labels for the illustrations of this exercise in the answer column.

B. *Sheep Pluck Dissection*

After completing the sheep pluck dissection, answer the following questions.

1. Are the cartilaginous rings of the trachea continuous all the way around the organ? ______________
2. Describe the texture of the surface of the lung. ______________
3. How many lobes exist on the right lung? ______________
   . . . on the left lung? ______________
4. Why does lung tissue collapse so readily when you quit blowing into it with a straw? ______________
5. What does the "pulmonary membrane" consist of? ______________

C. *Histological Study*

On a separate sheet of plain paper make drawings, as required by your instructor, of nasal, tracheal, and lung tissues.

Answers

Fig. 53.1	Fig. 53.2

Fig. 53.3

D. *Organ Identification*

Identify the respiratory structures according to the following statements.

alveoli—1
bronchioles—2
bronchi—3
cricoid cartilage—4
epiglottis—5
hard palate—6
larynx—7
lingual tonsils—8
nasal cavity—9
nasal conchae—10
nasopharynx—11
oral cavity proper—12
oral vestibule—13
oropharynx—14
palatine tonsils—15
pharyngeal tonsils—16
pleural cavity—17
soft palate—18
thyroid cartilage—19
trachea—20

1. Partition between nasal and oral cavities.
2. Cartilage of Adam's apple.
3. Small air sacs of lung tissue.
4. Cavity above soft palate.
5. Cavity above hard palate.
6. Cavity that contains the tongue.
7. Cavity near palatine tonsils.
8. Voice box.
9. Small tubes leading into alveoli.
10. Another name for adenoids.
11. Tube between larynx and bronchi.
12. Fleshy lobes in nasal cavity.
13. Cavity between lips and teeth.
14. Tonsils located on sides of pharynx.
15. Tonsils attached to base of tongue.
16. Flexible flaplike cartilage over larynx.
17. Most inferior cartilaginous ring of larynx.
18. Tubes formed by bifurcation of trachea.
19. Potential cavity between lung and thoracic wall.

E. *Organ Function*

Select the structures in the above list that perform the following functions.

1. Initiates swallowing.
2. Prevents food from entering the larynx when swallowing.
3. Provides supporting walls for vocal folds.
4. Assists in destruction of harmful bacteria in the oral region.
5. Warms the air as it passes through the nasal cavity.
6. Prevents food from entering nasal cavity during chewing and swallowing.
7. Provides surface for gas exchange in the lungs.
8. Essential for speech.

Answers

Organ Identification

1. ________
2. ________
3. ________
4. ________
5. ________
6. ________
7. ________
8. ________
9. ________
10. ________
11. ________
12. ________
13. ________
14. ________
15. ________
16. ________
17. ________
18. ________
19. ________

Organ Function

1. ________
2. ________
3. ________
4. ________
5. ________
6. ________
7. ________
8. ________

LABORATORY REPORT **54**

Student: ______________________

Desk No: ________ Section: ______________

Hyperventilation and Rebreathing

A. *Questions for Nonrecording Portion of Experiment*

1. Why does breathing into a bag after hyperventilating restore normal breathing more rapidly than breathing into the air? ______________________

2. How long were you able to hold your breath with only one deep inspiration? ______________

3. How long were you able to hold your breath after hyperventilating?

4. After exercising, how long were you able to hold your breath? ______________

Generalizations on above: ______________________

B. *Tracings*

In the spaces provided below tape or staple chart recordings for the various portions of this experiment.

Normal Respiration (Slow Speed)	Speed: ____ cm/sec

B. *Tracings (continued)*

Normal Respiration (Fast Speed)	Speed: ______ cm/sec

Rebreathing and Recovery

Hyperventilation

C. *Generalizations*

What generalizations can you make from the above tracings?

____

____

____

____

____

____

____

Student: ____________________

LABORATORY REPORT 55, 56

Desk No: ________ Section: ____________

The Diving Reflex (Ex. 55)

A. Tracings

Tape or staple the chart recordings for the various portions of this experiment in the spaces provided below:

Tidal Breathing before Immersion

Holding Breath after Tidal Inspiration

Immersion in Water at 70° Fahrenheit

Immersion in Water at 60° Fahrenheit

***A. Tracings** (continued)*

Hyperventilation Followed by Holding Breath (No Immersion)

Hyperventilation Followed by Immersion   Total Time Immersion: ______

B. Generalizations

The Valsalva Maneuver (Ex. 56)

A. Tracing

In the space below tape or staple a chart recording made during a Valsalva maneuver.

Student: ______________________

LABORATORY REPORT 57, 58

Desk No: ________ Section: ______________

Spirometry: Lung Capacities (Ex. 57)

A. *Tabulation*

Record in the following table the results of your spirometer readings and calculations.

Lung Capacities	Normal (ml)	Your Capacities
Tidal Volume (TV)	500	
Minute Respiratory Volume (MRV) (MRV = TV × Resp. Rate)	6,000	
Expiratory Reserve Volume (ERV)	1,100	
Vital Capacity (VC) (VC = TV + ERV + IRV)	See Appendix A (Tables II and III)	
Inspiratory Capacity (IC) (IC = VC − ERV)	3,000	
Inspiratory Reserve Volume (IRV) (IRV = IC − TV)	2,500	

B. *Evaluation*

List below any of your lung capacities that are significantly low (22% or more). ______________

____

____

____

Spirometry: The FEV_T Test (Ex. 58)

A. *Expirogram*

Trim the tracing of your expirogram to as small a piece as you can without removing volume values and tape it in space shown. Do not tape sides and bottom.

Tape Here

B. *Calculations*

Record your computations for determining the various percentages here.

1. **Vital Capacity** From the expirogram determine the total volume of air expired (Vital Capacity) .... ________
2. **Corrected Vital Capacity** By consulting table IV, appendix A, determine the corrected vital capacity (Step 4 on page 285).

   VC × Conversion Factor = ... ________

   Math:
3. **One-Second Timed Vital Capacity (FEV_1)** After determining the FEV_1 (Steps 1, 2, and 3 on page 285) record your results here .... ________
4. **Corrected FEV_1** Correct the FEV_1 for the temperature of the spirometer (Step 5 on page 285).

   FEV_1 × Conversion Factor = ... ________
5. **FEV_1 Percentage** Divide the corrected FEV_1 by the corrected vital capacity to determine the percentage expired in the first second (Step 6 on page 285).

   $\frac{\text{Corrected } FEV_1}{\text{Corrected Vital Capacity}} =$ .... ________

   Math:
6. **Percent of Predicted VC** Determine what percentage your vital capacity is of the predicted vital capacity (Step 7 on page 285).

   $\frac{\text{Your Corrected VC}}{\text{Predicted VC}} =$ .... ________

   Math:

C. *Questions*

1. Why isn't a measure of one's vital capacity as significant as the FEV_T? ________

   ____

   ____

   ____
2. What respiratory diseases are most readily detected with this test? ________

   ____

   ____

LABORATORY REPORT **60**

Student: ______________________

Desk No: ________ Section: ______________

Anatomy of the Digestive System

A. Illustrations

Record the labels for the illustrations of this exercise in the answer columns.

B. Microscopic Studies

Record here the drawings of the microscopic examinations that are required for this exercise. If there is insufficient space for all required drawings, utilize a separate sheet of paper for all of them.

Answers

Fig. 60.1	Fig. 60.2

Fig. 60.3

C. Alimentary Canal

Identify the parts of the digestive system described by the following statements.

anus—1
cecum—2
colon—3
colon, sigmoid—4
duct, common bile—5
duct, cystic—6
duct, hepatic—7
duct, pancreatic—8
duodenum—9
esophagus—10
ileum—11
pharynx—12
rectum—13
stomach, fundus—14
stomach, pyloric portion—15
valve, cardiac—16
valve, ileocecal—17
valve, pyloric—18

1. Place where swallowing (peristalsis) begins.
2. Entrance opening of stomach.
3. Exit opening of stomach.
4. Tube between mouth and stomach.
5. Distal coiled portion of small intestine.
6. First twelve inches of small intestine.
7. Proximal pouch or compartment of large intestine.
8. Section of large intestine between descending colon and rectum.
9. Most active portion of stomach.
10. Duct that drains the liver.
11. Valve between small and large intestines.
12. Duct that drains the gallbladder.
13. Part of small intestine where most digestion occurs.
14. Part of small intestine where most absorption occurs.
15. Duct that joins the common bile duct before entering the intestine.
16. Structure on which appendix is located.
17. Duct that conveys bile to intestine.
18. Exit of alimentary canal.
19. Part of tract where most water absorption (conservation) occurs.
20. Last six inches of alimentary canal.

Answers

Alimentary Canal	Fig. 60.4
1. ______	______
2. ______	______
3. ______	______
4. ______	______
5. ______	______
6. ______	______
7. ______	______
8. ______	______
9. ______	______
10. ______	______
11. ______	______
12. ______	______
13. ______	
14. ______	
15. ______	
16. ______	
17. ______	
18. ______	
19. ______	
20. ______	

D. *Oral Cavity*

Identify the structures of the mouth that are described by the following statements.

arch, glossopalatine—1
arch, pharyngopalatine—2
buccae—3
duct, parotid—4
duct, submandibular—5
ducts, lesser sublingual—6
frenulum, labial—7
frenulum, lingual—8
gingiva—9
glands, parotid—10
glands, sublingual—11
glands, submandibular—12
mucosa—13
papillae, filiform—14
papillae, foliate—15
papillae, fungiform—16
papillae, vallate—17
tonsil, lingual—18
tonsil, palatine—19
tonsil, pharyngeal—20
uvula—21

1. Lining of the mouth.
2. Name for the cheeks.
3. Duct that drains the parotid gland.
4. Fold of skin between the lip and gums.
5. Duct that drains the submandibular gland.
6. Fold of skin between the tongue and floor of mouth.
7. Portion of mucosa around the teeth.
8. Tonsils seen at back of mouth.
9. Tonsils located at root of tongue.
10. Fingerlike projection at end of soft palate.
11. Vertical ridges on side of tongue.
12. Rounded papillae on dorsum of tongue.
13. Small tactile papillae on surface of tongue.
14. Large papillae at back of tongue.
15. Ducts that drain sublingual gland.
16. Place where taste buds are located.
17. Salivary glands located inside and below the mandible.
18. Membrane located in front of palatine tonsil.
19. Membrane located in back of palatine tonsil.
20. Salivary glands located under the tongue.
21. Salivary glands located in cheeks.

E. *The Teeth*

Select the correct answer that completes each of the following statements.

1. All deciduous teeth usually erupt by the time a child is
   (1) one year. (2) two years. (3) four years old.
2. A complete set of primary teeth consists of
   (1) ten teeth. (2) twenty teeth. (3) thirty-two teeth.
3. The permanent teeth with the longest roots are the
   (1) incisors. (2) cuspids. (3) molars.
4. The smallest permanent molars are the
   (1) first molars. (2) second molars. (3) third molars.
5. A complete set of permanent teeth consists of
   (1) twenty-four teeth. (2) twenty-eight teeth.
   (3) thirty-two teeth.
6. The primary dentition lacks
   (1) molars. (2) incisors. (3) bicuspids.

Answers

Oral Cavity	Fig. 60.5
1. ______	______
2. ______	______
3. ______	______
4. ______	______
5. ______	______
6. ______	______
7. ______	______
8. ______	______
9. ______	______
10. ______	**Fig. 60.7**
11. ______	
12. ______	______
13. ______	______
14. ______	______
15. ______	______
16. ______	______
17. ______	______
18. ______	______
19. ______	______
20. ______	______
21. ______	______
	______
Teeth	______
1. ______	______
2. ______	______
3. ______	______
4. ______	______
5. ______	______
6. ______	

7. A child's first permanent tooth usually erupts during the
   (1) second year. (2) third year.
   (3) fifth year. (4) sixth year.
8. Trifurcated roots exist on
   (1) upper cuspids. (2) upper molars. (3) lower molars.
9. Bifurcated roots exist on
   (1) upper first bicuspids and lower molars.
   (2) cuspids and lower bicuspids. (3) all molars.
10. A tooth is considered "dead" or devitalized if
   (1) the enamel is destroyed. (2) the pulp is destroyed.
   (3) the periodontal membrane is infected.
11. Dentin is produced by cells called
   (1) odontoblasts. (2) ameloblasts. (3) cementocytes.
12. Cementum is secreted by cells called
   (1) odontoblasts. (2) ameloblasts. (3) cementocytes.
13. The wisdom tooth is
   (1) a supernumerary tooth. (2) a succedaneous tooth.
   (3) the third molar.
14. Caries (cavities) are caused primarily by
   (1) using the wrong toothpaste. (2) acid production by bacteria. (3) fluorides in water.
15. Most teeth that have to be extracted have become useless because of
   (1) caries. (2) gingivitis. (3) periodontitis (pyorrhea).
16. Enamel covers the
   (1) entire tooth. (2) clinical crown only.
   (3) anatomical crown only.
17. The tooth receives nourishment through
   (1) the apical foramen. (2) the periodontal membrane.
   (3) both the apical foramen and the periodontal membrane.
18. The tooth is held in the alveolus by
   (1) bone. (2) dentin. (3) periodontal membrane.

Answers
Teeth
7. ________
8. ________
9. ________
10. ________
11. ________
12. ________
13. ________
14. ________
15. ________
16. ________
17. ________
18. ________

Student: ______________________

LABORATORY REPORT 61, 62

Desk No: __________ Section: ______________

The Chemistry of Hydrolysis (Ex. 61)

A. Carbohydrate Digestion

1. **IKI Test** Record here the presence (+) or absence (−) of starch as revealed by IKI test on spot plates.

1A	2A	3A	4A	5A

a. Did the pancreatic extract hydrolyze starch? ______________________

b. Which tube is your evidence for this conclusion? ______________________

c. How do tubes 1A and 5A compare in color? ______________________

d. What do you conclude from question *c?* ______________________

e. What is the value of tube 4A? ______________________

____

f. What is the value of tube 2A? ______________________

____

2. **Barfoed's and Benedict's Tests** After performing Barfoed's and Benedict's tests on tubes 1A and 5A, what are your conclusions about the degree of digestion of starch by amylase? __________

____

____

B. Protein Digestion

Record the color and optical densities (OD) of each tube in the following chart.

	1P	2P	3P	4P	5P
Color					
O D					

1. Did pancreatic juice hydrolyze the BAPNA? ______________________

2. Which tube is your evidence for this conclusion? ______________________

3. What other tube shows protein hydrolysis? ______________________

4. What can you conclude from the previous observation in question 3? ______________________

____

5. Why is there no hydrolysis in tube 4P? ______________________

6. Why is there no hydrolysis in tube 3P? ______________________

7. Why didn't tube 2P show hydrolysis? ______________________

____

C. Fat Digestion

Determine the pH of each tube by comparing with a bromthymol blue color standard. Record these values in the table below.

1L	2L	3L	4L	5L	6L

1. Did the pancreatic extract hydrolyze fat in the presence of bile? ______________________

2. Did the pancreatic extract hydrolyze fat in the absence of bile? ______________________

3. How do tubes 1L and 6L compare as to degree of hydrolysis? ______________________

____

4. What can you conclude from the previous observation in question 3? ______________________

____

5. What is the value of tube 2L? ______________________

____

D. Summarization

Make an itemized statement of all the facts you learned in this experiment.

Enzyme Review

Consult your text and lecture notes to answer the following questions concerning digestive enzymes.

A. *Digestive Juices*

Select the enzymes that are present in the following digestive juices. Use the enzymes listed in the right column.

1. Saliva
2. Gastric juice
3. Pancreatic fluid
4. Intestinal juice (succus entericus)

B. *Substrates*

Select the enzymes that act on the following substrates.

1. Casein
2. Fats
3. Lactose
4. Maltose
5. Peptides
6. Proteins
7. Proteoses and peptones
8. Starch
9. Sucrose

Carbohydrases
- amylase, pancreatic—1
- amylase, salivary—2
- lactase—3
- maltase—4
- sucrase—5

Proteases
- carboxypeptidase—6
- chymotrypsin—7
- erepsin (peptidase)—8
- pepsin—9
- rennin—10
- trypsin—11

Lipases
- lipase, pancreatic—12

C. *End Products*

Select the enzymes from the above list that produce the following end products in digestion.

1. Amino acids
2. Fatty acids
3. Fructose
4. Galactose
5. Glucose
6. Glycerol
7. Maltose
8. Polypeptides

Answers

Juices

1. ________
2. ________
3. ________
4. ________

Substrates

1. ________
2. ________
3. ________
4. ________
5. ________
6. ________
7. ________
8. ________
9. ________

End Products

1. ________
2. ________
3. ________
4. ________
5. ________
6. ________
7. ________
8. ________

Factors Affecting Hydrolysis (Ex. 62)

A. *Tabulations*

If the class has been subdivided into separate groups to perform different parts of this experiment, these tabulations should be recorded on the chalkboard so that all students can copy the results onto their own Laboratory Report sheets.

1. **Temperature** After IKI solution has been added to all ten tubes, record the color and degree of amylase activity in each tube in the following tables. Refer to the carbohydrate differentiation separation outline in Appendix C pertaining to the IKI test.

Table I Amylase action at 20° C.

Tube No.	1	2	3	4	5	6	7	8	9	10
Color										
Starch Hydrolysis										

Table II Amylase action at 37° C.

Tube No.	1	2	3	4	5	6	7	8	9	10
Color										
Starch Hydrolysis										

2. **Hydrogen Ion Concentration** After IKI solution has been added to all nine tubes, record, as above, the color and degree of amylase action for each tube in the following table.

Table III pH and amylase action at 37° C.

Time	2 Minutes			4 Minutes			6 Minutes		
Tube No.	1	2	3	4	5	6	7	8	9
pH	5	7	9	5	7	9	5	7	9
Color									
Starch Hydrolysis									

B. *Conclusions*

Answer the following questions that are related to the results of this experiment.

1. How long did it take for complete starch hydrolysis at 20° C? ______ at 37° C? ______
2. Why would you expect the above results to turn out as they have? ______

______

______

3. What effect does boiling have on amylase? ______

______

4. At which pH was starch hydrolysis most rapid? ______

How does this pH compare with the pH of normal saliva? ______

______

5. What is the pH of gastric juice? ______

What do you suppose happens to the action of amylase in the stomach? ______

______

LABORATORY REPORT **63**

Student: ______________________

Desk No: ________ Section: ____________

Anatomy of the Urinary System

A. Illustrations

Record the labels for the three illustrations of this exercise in the answer columns. Note that figure 63.2 is on the reverse side of this sheet.

B. Microscopy

If the space provided below is adequate for your histology drawings, use it. Use a separate sheet of paper if several sketches are to be made.

C. Anatomy

Identify the structures of the urinary system that are described by the following statements.

calyces—1	medulla—6	renal pelvis—11
collecting tubule—2	nephron—7	renal pyramids—12
cortex—3	renal capsule—8	ureters—13
glomerular capsule—4	renal column—9	urethra—14
glomerulus—5	renal papilla—10	

1. Tubes that drain the kidneys.
2. Tube that drains the urinary bladder.
3. Portion of kidney that contains renal corpuscles.
4. Cone-shaped areas of medulla.
5. Portion of kidney that consists primarily of collecting tubules.
6. Basic functioning unit of the kidney.
7. Two portions of renal corpuscle.
8. Distal tip of renal pyramid.
9. Short tubes that receive urine from renal papillae.
10. Funnellike structure that collects urine from calyces of each kidney.
11. Structure that receives urine from several nephrons.
12. Cup-shaped membranous structure that surrounds glomerulus.
13. Thin fibrous outer covering of kidney.
14. Cortical tissue between renal pyramids.
15. Tuft of capillaries that produces dilute urine.

Answers

Anatomy	Fig. 63.1
1. ______	______
2. ______	______
3. ______	______
4. ______	______
5. ______	______
6. ______	______
7. ______	______
8. ______	______
9. ______	______
10. ______	______
11. ______	**Fig. 63.3**
12. ______	______
13. ______	______
14. ______	______
15. ______	______
	______
	______
	______
	______
	______
	______
	______
	______
	______
	______
	______

D. Physiology

Select the best answer that completes the following statements concerning the physiology of urine production. For definitions of medical terms consult a medical dictionary or your lecture text.

1. Blood enters the glomerulus through the
   (1) efferent arteriole. (2) arcuate artery.
   (3) afferent arteriole. (4) None of these are correct.
2. Surgical removal of a kidney is called
   (1) nephrectomy. (2) nephrotomy. (3) nephrolithotomy.
3. Water reabsorption from the glomerular filtrate into the peritubular blood is facilitated by
   (1) antidiuretic hormone. (2) renin.
   (3) aldosterone. (4) Both 1 and 3 are correct.
4. The following substances are reabsorbed through the walls of the nephron into the peritubular blood:
   (1) glucose and water. (2) urea and water.
   (3) glucose, amino acids, salts, and water.
   (4) glucose, amino acids, urea, salts, and water.
5. The amount of urine produced is affected by
   (1) blood pressure.
   (2) environmental temperature.
   (3) amount of solute in glomerular filtrate.
   (4) 1, 2, 3 and additional factors are correct.
6. The amount of urine normally produced in 24 hours is about
   (1) 100 ml. (2) 500 ml. (3) 1.5 liters. (4) 4.5 liters.
7. The reabsorption of sodium ions from the glomerular filtrate into the peritubular blood draws the following back into the blood:
   (1) potassium ions. (2) chloride ions.
   (3) water. (4) chloride ions and water.
8. Cells of the walls of collecting tubules secrete the following substances into urine:
   (1) amino acids. (2) uric acid.
   (3) ammonia. (4) Both 2 and 3 are correct.
9. Renal diabetes is due to
   (1) a lack of ADH. (2) a lack of insulin.
   (3) faulty reabsorption of glucose in the nephron.
   (4) Both 1 and 2 are correct.
10. The presence of glucose in the urine and a low level of insulin in the blood would be diagnosed as
    (1) diabetes insipidus. (2) diabetes mellitus.
    (3) renal diabetes. (4) None of these are correct.
11. Approximately eighty percent of water in the glomerular filtrate is reabsorbed in the
    (1) proximal convoluted tubule. (2) Henle's loop.
    (3) distal convoluted tubule. (4) None of these are correct.
12. The desire to micturate normally occurs when the following amount of urine is present in the bladder:
    (1) 100 ml. (2) 200 ml. (3) 300 ml. (4) 400 ml.
13. Removal of calculus (kidney stones) is called
    (1) nephrectomy. (2) nephrotomy.
    (3) nephrolithotomy. (4) Both 2 and 3 are correct.

Answers

Physiology

1. ________
2. ________
3. ________
4. ________
5. ________
6. ________
7. ________
8. ________
9. ________
10. ________
11. ________
12. ________
13. ________

Fig. 63.2

____
____
____
____
____
____
____
____
____
____
____
____
____
____
____
____
____
____
____
____
____

LABORATORY REPORT **64**

Student: ____________________

Desk No: ________ Section: ____________

Urine: Composition and Tests

A. *Test Results*

Record on the chart below the results of any urine test performed. If the tests are done as a demonstration, the results will be tabulated in columns A and B. If students perform the tests on their own urine, the last column will be used for their results.

TEST	Normal Values	Abnormal Values	A Positive Test Control	B Unknown Sample	C Student's Urine
COLOR	Colorless Pale straw Straw Amber	Milky Reddish amber Brownish yellow Green Smoky brown			
CLOUDINESS	Clear	+ slight ++ moderate +++ cloudy ++++ very cloudy			
SP. GRAVITY	1.001–1.060	Above 1.060			
pH	4.8–7.5	Below 4.8 Above 7.5			
ALBUMIN (Protein)	None	+ barely visible ++ granular +++ flocculent ++++ large flocculent			
MUCIN	None	Visible amounts			
GLUCOSE	None	+ yellow green ++ greenish yellow +++ yellow ++++ orange			
KETONES	None	+ pink-purple ring			
HEMOGLOBIN	None	See color chart			
BILIRUBIN					

B. *Microscopy*

Record here by sketch and word any structures seen on microscopic examination.

C. *Interpretation*

Indicate the probable significance of each of the following in urine. More than one condition may apply in some instances.

bladder infection—1
cirrhosis of liver—2
diabetes insipidus—3
diabetes mellitus—4
exercise (extreme)—5
kidney tumor—6
glomerulonephritis—7
gonorrhea—8
hepatitis—9
normally present—10

1. Albumin (constantly)
2. Albumin (periodic)
3. Bacteria
4. Bile pigments
5. Blood
6. Creatinine
7. Glucose
8. Mucin
9. Porphyrin
10. Pus cells (neutrophils)
11. Sodium chloride
12. Specific gravity (high)
13. Specific gravity (low)
14. Urea
15. Uric acid

D. *Terminology*

Select the terms described by the following statements.

albuminuria—1
anuria—2
catheter—3
cystitis—4
diuretic—5
enuresis—6
glycosuria—7
nephritis—8
pyelitis—9
uremia—10

1. Absence of urine production.
2. High urea level in blood.
3. Sugar in urine.
4. Albumin in urine.
5. Inflammation of nephrons.
6. Inflammation of pelvis of kidney.
7. Bladder infection.
8. Device for draining bladder.
9. Involuntary bed-wetting during sleep.
10. Chemical that stimulates urine production.

Answers

Interpretation

1. ________
2. ________
3. ________
4. ________
5. ________
6. ________
7. ________
8. ________
9. ________
10. ________
11. ________
12. ________
13. ________
14. ________
15. ________

Terminology

1. ________
2. ________
3. ________
4. ________
5. ________
6. ________
7. ________
8. ________
9. ________
10. ________

LABORATORY REPORT **65**

Student: ______________________

Desk No: __________ Section: ______________

The Endocrine Glands

A. *Illustrations*

Record the labels for figures 65.12 and 65.13 in the answer columns.

B. *Microscopy*

Make the drawings of the various microscopic studies on separate drawing paper. Include these drawings with the Laboratory Report.

C. *Sources*

Select the glandular tissues that produce the following hormones.

1. ACTH
2. ADH
3. Aldosterone
4. Calcitonin
5. Chorionic gonadotropin
6. Cortisol
7. Epinephrine
8. Estrogens
9. FSH
10. Glucagon
11. ICSH
12. Insulin
13. LH
14. Melatonin
15. Norepinephrine
16. Oxytocin
17. Parathyroid hormone
18. PIF
19. Prolactin
20. Progesterone
21. Somatostatin
22. Testosterone
23. Thymosin
24. Thyrotropin
25. Thyroxine
26. Triiodothyronine
27. Vasopressin

adrenal cortex
- zona fasciculata—1
- zona glomerulosa—2
- zona reticularis—3

adrenal medulla—4
hypothalamus—5

ovary
- corpus luteum—6
- Graafian follicle—7

pancreas
- alpha cells—8
- beta cells—9
- delta cells—10
- acini cells—11

parathyroid gland—12
pineal gland—13

pituitary gland
- adenohypophysis—14
- neurohypophysis—15

placenta—16
testis—17
thymus gland—18

thyroid gland
- follicular cells—19
- parafollicular cells—20

D. *Physiology*

Select the hormones that produce the following physiological effects. Note that a separate list of hormones is provided for each group.

GROUP I

1. Increases metabolism.
2. Increases blood pressure.
3. Increases strength of heartbeat.
4. Promotes sodium absorption in nephron.
5. Promotes gluconeogenesis.
6. Causes vasoconstriction in all organs.
7. Causes vasoconstriction in all organs except muscles and liver.
8. Causes vasodilation in skeletal muscles.
9. Inhibits the estrus cycle in lower animals.

aldosterone—1
epinephrine—2
glucocorticoids—3
melatonin—4
norepinephrine—5
none of these—6

Answers

Sources	Fig. 65.12
1. ______	______
2. ______	______
3. ______	______
4. ______	______
5. ______	______
6. ______	______
7. ______	______
8. ______	______
9. ______	______
10. ______	______
11. ______	______
12. ______	______
13. ______	______
14. ______	______
15. ______	______
16. ______	______
17. ______	______
18. ______	______
19. ______	______
20. ______	______
21. ______	______
22. ______	______
23. ______	______
24. ______	______
25. ______	
26. ______	
27. ______	
Physiology Group I	
1. ______	
2. ______	
3. ______	
4. ______	
5. ______	
6. ______	
7. ______	
8. ______	
9. ______	

D. *Physiology (continued)*

Select the hormones that produce the following physiological effects.

GROUP II

1. Inhibits insulin production.
2. Inhibits production of SH.
3. Stimulates osteoclasts.
4. Mobilizes fatty acids.
5. Inhibits glucagon production.
6. Raises blood calcium level.
7. Increases cardiac output and respiratory rate.
8. Impedes phosphorus reabsorption in renal tubules.
9. Promotes storage of glucose, fatty acids, and amino acids.
10. Mobilizes glucose, fatty acids, and amino acids.
11. Stimulates metabolism of all cells in body.
12. Lowers blood phosphorus levels.
13. Decreases blood calcium when injected intravenously.

calcitonin—1
glucagon—2
insulin—3
parathyroid hormone—4
somatostatin—5
thyroid hormone—6
none of these—7

GROUP III

1. Promotes breast enlargement.
2. Promotes rapid bone growth.
3. Stimulates testicular descent.
4. Causes milk ejection from breasts.
5. Promotes maturation of spermatozoa.
6. Causes constriction of arterioles.
7. Stimulates production of thyroxine.
8. Stimulates glucocorticoid production.
9. Inhibits prolactin production.
10. Causes thickening and vascularization of endometrium.
11. Causes interstitial cells to produce testosterone.
12. Causes myometrium contraction during childbirth.
13. Promotes milk production after childbirth.
14. Promotes stratification of vaginal epithelium.
15. Suppresses development of female genitalia in male.
16. Prepares the endometrium for egg cell implantation.
17. Promotes growth of all cells in the body.

ACTH—1
ADH—2
chorionic gonadotropin—3
estrogens—4
FSH—5
LH—6
oxytocin—7
PIF—8
progesterone—9
somatotropin—10
testosterone—11
thyrotropin—12
none of these—13

E. *Hormonal Imbalance*

Select the hormones that are responsible for the following disorders. After each hormone number indicate with a (+) or (−) whether the condition is due to *excess* (+) or *deficiency* (−). More than one hormone may apply in some cases.

1. Acromegaly
2. Addison's disease
3. Cretinism
4. Cushing's syndrome
5. Diabetes insipidus
6. Diabetes mellitus
7. Dwarfism
8. Exophthalmos
9. Gigantism
10. Hyperglycemia
11. Hypothyroidism
12. Myxedema
13. Tetany

ADH—1
glucagon—2
glucocorticoids—3
insulin—4
mineralocorticoids—5
parathyroid hormone—6
somatotropin—7
thyroid hormone—8
none of these—9

Answers

Physiology Group II	Fig. 65.13
1. ______	______
2. ______	______
3. ______	______
4. ______	______
5. ______	______
6. ______	______
7. ______	______
8. ______	______
9. ______	______
10. ______	______
11. ______	______
12. ______	______
13. ______	______
Group III	______
	______
1. ______	______
2. ______	______
3. ______	______
4. ______	
5. ______	**Imbalance**
6. ______	1. ______
7. ______	2. ______
8. ______	3. ______
9. ______	4. ______
10. ______	5. ______
11. ______	6. ______
12. ______	7. ______
13. ______	8. ______
14. ______	9. ______
15. ______	10. ______
16. ______	11. ______
17. ______	12. ______
	13. ______

LABORATORY REPORT **66**

Student: ______________________

Desk No: ________ Section: ______________

The Reproductive System

A. Labels

Record the labels for the illustrations of this exercise in the answer columns.

B. Microscopy

Draw the various stages of spermatogenesis on a separate piece of paper to be included with this Laboratory Report.

C. Male Organs

Identify the structures described by the following statements.

common ejaculatory duct—1
corpora cavernosa—2
corpus spongiosum—3
bulbourethral gland—4
epididymis—5
glans penis—6
penis—7
prepuce—8
prostate gland—9
prostatic urethra—10
seminal vesicle—11
seminiferous tubule—12
testes—13
vas deferens—14

1. Copulatory organ of male.
2. Source of spermatozoa.
3. Erectile tissue of penis (3).
4. Fold of skin over end of penis.
5. Structure that stores spermatozoa.
6. Contribute fluids to semen (3).
7. Duct that passes from urinary bladder through the prostate gland.
8. Coiled-up structures in testes where spermatozoa originate.
9. Tube that carries sperm from epididymis to ejaculatory duct.
10. Distal end of penis.

D. Female Organs

Identify the structures described by the following statements.

cervix—1
clitoris—2
external os—3
fimbriae—4
greater vestibular glands—5
hymen—6
infundibulum—7
internal os—8
labia majora—9
labia minora—10
mons pubis—11
myometrium—12
paraurethral glands—13
ovaries—14
uterine tubes—15
vagina—16

1. Copulatory organ of female.
2. Source of ova in female.
3. Muscular wall of uterus.
4. Neck of uterus.
5. Provides vaginal lubrication during coitus.
6. Fingerlike projections around edge of infundibulum.
7. Protuberance of erectile tissue sensitive to sexual excitation.
8. Entrance to uterus at vaginal end.
9. Homologous to bulbourethral glands of male.
10. Homologous to penis of male.
11. Homologous to scrotum of male.
12. Homologous to prostate gland of male.
13. Ducts that convey ovum to uterus.
14. Membranous fold of tissue surrounding entrance to vagina.
15. Open funnellike end of uterine tube.

Answers

Male Organs	Fig. 66.1
1. ______	______
2. ______	______
3. ______	______
4. ______	______
5. ______	______
6. ______	______
7. ______	______
8. ______	______
9. ______	______
10. ______	______
	______
Female Organs	______
1. ______	______
2. ______	______
3. ______	______
4. ______	______
5. ______	______
6. ______	______
7. ______	______
8. ______	______
9. ______	______
10. ______	______
11. ______	______
12. ______	______
13. ______	______
14. ______	______
15. ______	______
	______

E. Ligaments

Identify the following supporting structures that hold the ovaries, uterus, and uterine tubes in place.

broad ligament—1	round ligament—5
infundibulopelvic ligament—2	sacrouterine ligament—6
mesosalpinx—3	suspensory ligament—7
ovarian ligament—4	none of these—8

1. A ligament that extends from the uterus to the ovary.
2. Ligament between the cervix and sacral part of the pelvic wall.
3. A ligament that extends from the infundibulum to the wall of the pelvic cavity.
4. A ligament that extends from the corpus of the uterus to the body wall.
5. A large flat ligament of peritoneal tissue that extends from the uterine tube to the cervical part of the uterus.
6. A mesentery between the ovary and uterine tubes that contains blood vessels leading to the uterine tube.
7. A fold of peritoneum that attaches the ovary to the uterine tube.

F. Germ Cells

Identify the types of cells of oogenesis and spermatogenesis described by the following statements. More than one answer may apply.

Spermatogenesis	*Oogenesis*
primary spermatocyte—1	oogonia—6
secondary spermatocyte—2	polar bodies—7
spermatids—3	primary oocyte—8
spermatogonia—4	secondary oocyte—9
spermatozoa—5	

1. Cells that are haploid.
2. Cells that contain tetrads.
3. Cells that contain monads.
4. Cells that contain dyads.
5. Cells that are diploid.
6. Cells at periphery of ovary that produce all ova.
7. Cells that undergo mitosis.
8. Cells in which first meiotic division occurs.
9. Cells at periphery of seminiferous tubule that give rise to all spermatozoa.
10. Cells that develop directly into mature spermatozoa.
11. Cells in which second meiotic division occurs.

G. Physiology of Reproduction and Development

Select the best answer that completes the following statements.

1. The ovum gets from the ovary to the uterus by
   (1) amoeboid movement. (2) ciliary action.
   (3) peristalsis and ciliary action of uterine tube.
2. The vaginal lining is normally
   (1) alkaline. (2) acid. (3) neutral.
3. The seminal vesicle secretion is
   (1) alkaline. (2) acid. (3) neutral.
4. The birth canal consists of the
   (1) vagina. (2) uterus. (3) the vagina and uterus.
5. Spermatozoan viability is enhanced by a temperature that is
   (1) 98.6° F. (2) above 98.6° F. (3) below 98.6° F.
6. The prostatic secretion is
   (1) alkaline. (2) acid. (3) neutral.

Answers

Ligaments	Fig. 66.3
1. ______	______
2. ______	______
3. ______	______
4. ______	______
5. ______	______
6. ______	______
7. ______	______
	______
Germ Cells	______
	______
1. ______	______
2. ______	______
3. ______	
4. ______	**Fig. 66.4**
5. ______	
6. ______	______
7. ______	______
8. ______	______
9. ______	______
10. ______	______
11. ______	______
	______
Physiology	______
1. ______	______
2. ______	______
3. ______	______
4. ______	______
5. ______	______
6. ______	______
	______
	______
	______
	______
	______

7. The fetal stage of the human is
   (1) the first 8 weeks. (2) from the ninth week until birth.
   (3) the entire prenatal term.
8. Circumcision involves excisement of the
   (1) prepuce. (2) perineum. (3) glans penis.
9. Fertilization of the human ovum usually occurs in the
   (1) uterus. (2) vagina. (3) uterine tube.
10. Implantation of the fertilized ovum occurs usually
    (1) within 2 days after fertilization.
    (2) within 6 to 8 days after fertilization.
    (3) after 10 days.
11. The innermost fetal membrane surrounding the embryo is the
    (1) amnion. (2) chorion. (3) decidua.
12. The life span of an ovum without fertilization is
    (1) 6 hours. (2) 24–28 hours. (3) 7 days.
13. Uterine muscle consists of
    (1) smooth muscle. (2) striated muscle.
    (3) both smooth and striated muscle.
14. The outermost fetal membrane surrounding the embryo is the
    (1) amnion. (2) chorion. (3) decidua.
15. The lining of the uterus is the
    (1) myometrium. (2) endometrium. (3) epimetrium.
16. Milk production by the breasts usually occurs
    (1) immediately after birth. (2) on the second day.
    (3) on the third or fourth day.
17. The afterbirth refers to the
    (1) placenta. (2) damaged uterus.
    (3) the placenta, umbilical vessels, and fetal membranes.
18. Parturition is the
    (1) process of giving birth. (2) period of pregnancy.
    (3) first few days of postnatal life.
19. The following substances pass from the mother to the fetus through the placenta:
    (1) nutrients, gases, and blood cells.
    (2) nutrients and all blood cells except red blood cells.
    (3) nutrients, gases, hormones, and antibodies.
20. Abortion is correctly known as
    (1) criminal emptying of the uterus.
    (2) interruption of pregnancy during fetal life.
    (3) interruption of pregnancy during embryonic life.

H. Menstrual Cycle

Select the correct answer to the following statements.

1. The cessation of menstrual flow in a woman in her late forties is known as
   (1) menarche. (2) menopause. (3) amenorrhea.
2. The first menstrual flow is known as
   (1) menarche. (2) climacteric. (3) menopause.
3. A distinct rise in the basal body temperature usually occurs
   (1) at ovulation. (2) during the proliferative phase.
   (3) during the quiescent period.

Answers

Physiology	Fig. 66.5
7. ______	______
8. ______	______
9. ______	______
10. ______	______
11. ______	______
12. ______	______
13. ______	______
14. ______	______
15. ______	______
16. ______	______
17. ______	______
18. ______	______
19. ______	______
20. ______	______
	______
Menstrual Cycle	______
	______
1. ______	______
2. ______	______
3. ______	______
	______
	______
	______
	______
	______
	______

4. Menstrual flow is the result of
   (1) a deficiency of progesterone and estrogen.
   (2) a deficiency of estrogen.
   (3) an excess of progesterone and estrogen.
5. Ovulation usually (not always) occurs the following number of days before the next menstrual period:
   (1) three. (2) seven. (3) fourteen. (4) eighteen.
6. Pain during ovulation, as experienced by some women, is known as
   (1) dysmenorrhea. (2) oligomenorrhea.
   (3) mittelschmerz.
7. The proliferative phase in the menstrual cycle begins at about the
   (1) second day. (2) fifth day. (3) seventh day of menstrual cycle.
8. Repair of the endometrium after menstruation is due to
   (1) estrogen. (2) progesterone. (3) luteinizing hormone.
9. Progesterone secretion ceases entirely within the following number of days after the onset of menses:
   (1) ten. (2) fourteen. (3) twenty-six. (4) twenty-eight.
10. Excessive blood flow during menstruation is known as
    (1) oligomenorrhea. (2) dysmenorrhea. (3) menorrhagia.
11. Excessive discomfort and pain during menstruation is known as
    (1) dysmenorrhea. (2) oligomenorrhea. (3) menorrhagia.
12. Occasional or irregular menses is known as
    (1) amenorrhea. (2) oligomenorrhea. (3) dysmenorrhea.
13. Absence of menstruation is called
    (1) dysmenorrhea. (2) oligomenorrhea. (3) amenorrhea.

I. Medical

Select the type of surgery or condition that is described by the following statements.

anteflexion—1	oophorectomy—8
cryptorchidism—2	oophorhysterectomy—9
endometritis—3	oophoroma—10
episiotomy—4	retroflexion—11
gonorrhea—5	salpingitis—12
hysterectomy—6	salpingectomy—13
mastectomy—7	syphilis—14
	vasectomy—15

1. Surgical removal of breast.
2. Failure of testes to descend into scrotum.
3. Surgical removal of uterus.
4. Spirochaetal sexually transmitted disease.
5. Sterilization procedure in males.
6. Type of incision made in perineum at childbirth to prevent excessive damage to anal sphincter.
7. Surgical removal of one or both ovaries.
8. Ovarian malignancy.
9. Sexually transmitted disease that affects the mucous membranes rather than the blood.
10. Surgical removal or sectioning of uterine tubes.
11. Most common sexually transmitted disease.
12. Inflammation of uterine wall.
13. Malpositioned uterus (2 types).
14. Sexually transmitted disease caused by a coccoidal (spherical) organism.
15. Surgical removal of ovaries and uterus.

Answers

Menstrual Cycle

4. ________
5. ________
6. ________
7. ________
8. ________
9. ________
10. ________
11. ________
12. ________
13. ________

Medical

1. ________
2. ________
3. ________
4. ________
5. ________
6. ________
7. ________
8. ________
9. ________
10. ________
11. ________
12. ________
13. ________
14. ________
15. ________

Appendix A
Tables

Table I International atomic weights.

Element	Symbol	Atomic Number	Atomic Weight
Aluminum	Al	13	26.97
Antimony	Sb	51	121.76
Arsenic	As	33	74.91
Barium	Ba	56	137.36
Beryllium	Be	4	9.013
Bismuth	Bi	83	209.00
Boron	B	5	10.82
Bromine	Br	35	79.916
Cadmium	Cd	48	112.41
Calcium	Ca	20	40.08
Carbon	C	6	12.010
Chlorine	Cl	17	35.457
Chromium	Cr	24	52.01
Cobalt	Co	27	58.94
Copper	Cu	29	63.54
Fluorine	F	9	19.00
Gold	Au	79	197.2
Hydrogen	H	1	1.0080
Iodine	I	53	126.92
Iron	Fe	26	55.85
Lead	Pb	82	207.21
Magnesium	Mg	12	24.32
Manganese	Mn	25	54.93
Mercury	Hg	80	200.61
Nickel	Ni	28	58.69
Nitrogen	N	7	14.008
Oxygen	O	8	16.0000
Palladium	Pd	46	106.7
Phosphorus	P	15	30.98
Platinum	Pt	78	195.23
Potassium	K	19	39.096
Radium	Ra	88	226.05
Selenium	Se	34	78.96
Silicon	Si	14	28.06
Silver	Ag	47	107.880
Sodium	Na	11	22.997
Strontium	Sr	38	87.63
Sulfur	S	16	32.066
Tin	Sn	50	118.70
Titanium	Ti	22	47.90
Tungsten	W	74	183.92
Uranium	U	92	238.07
Vanadium	V	23	50.95
Zinc	Zn	30	65.38
Zirconium	Zr	40	91.22

Table II Predicted vital capacities for females.

AGE	HEIGHT IN CENTIMETERS AND INCHES																		
CM	152	154	156	158	160	162	164	166	168	170	172	174	176	178	180	182	184	186	188
IN	59.8	60.6	61.4	62.2	63.0	63.7	64.6	65.4	66.1	66.9	67.7	68.5	69.3	70.1	70.9	71.7	72.4	73.2	74.0
16	3,070	3,110	3,150	3,190	3,230	3,270	3,310	3,350	3,390	3,430	3,470	3,510	3,550	3,590	3,630	3,670	3,715	3,755	3,800
17	3,055	3,095	3,135	3,175	3,215	3,255	3,295	3,335	3,375	3,415	3,455	3,495	3,535	3,575	3,615	3,655	3,695	3,740	3,780
18	3,040	3,080	3,120	3,160	3,200	3,240	3,280	3,320	3,360	3,400	3,440	3,480	3,520	3,560	3,600	3,640	3,680	3,720	3,760
20	3,010	3,050	3,090	3,130	3,170	3,210	3,250	3,290	3,330	3,370	3,410	3,450	3,490	3,525	3,565	3,605	3,645	3,695	3,720
22	2,980	3,020	3,060	3,095	3,135	3,175	3,215	3,255	3,290	3,330	3,370	3,410	3,450	3,490	3,530	3,570	3,610	3,650	3,685
24	2,950	2,985	3,025	3,065	3,100	3,140	3,180	3,220	3,260	3,300	3,335	3,375	3,415	3,455	3,490	3,530	3,570	3,610	3,650
26	2,920	2,960	3,000	3,035	3,070	3,110	3,150	3,190	3,230	3,265	3,300	3,340	3,380	3,420	3,455	3,495	3,530	3,570	3,610
28	2,890	2,930	2,965	3,000	3,040	3,070	3,115	3,155	3,190	3,230	3,270	3,305	3,345	3,380	3,420	3,460	3,495	3,535	3,570
30	2,860	2,895	2,935	2,970	3,010	3,045	3,085	3,120	3,160	3,195	3,235	3,270	3,310	3,345	3,385	3,420	3,460	3,495	3,535
32	2,825	2,865	2,900	2,940	2,975	3,015	3,050	3,090	3,125	3,160	3,200	3,235	3,275	3,310	3,350	3,385	3,425	3,460	3,495
34	2,795	2,835	2,870	2,910	2,945	2,980	3,020	3,055	3,090	3,130	3,165	3,200	3,240	3,275	3,310	3,350	3,385	3,425	3,460
36	2,765	2,805	2,840	2,875	2,910	2,950	2,985	3,020	3,060	3,095	3,130	3,165	3,205	3,240	3,275	3,310	3,350	3,385	3,420
38	2,735	2,770	2,810	2,845	2,880	2,915	2,950	2,990	3,025	3,060	3,095	3,130	3,170	3,205	3,240	3,275	3,310	3,350	3,385
40	2,705	2,740	2,775	2,810	2,850	2,885	2,920	2,955	2,990	3,025	3,060	3,095	3,135	3,170	3,205	3,240	3,275	3,310	3,345
42	2,675	2,710	2,745	2,780	2,815	2,850	2,885	2,920	2,955	2,990	3,025	3,060	3,100	3,135	3,170	3,205	3,240	3,275	3,310
44	2,645	2,680	2,715	2,750	2,785	2,820	2,855	2,890	2,925	2,960	2,995	3,030	3,060	3,095	3,130	3,165	3,200	3,235	3,270
46	2,615	2,650	2,685	2,715	2,750	2,785	2,820	2,855	2,890	2,925	2,960	2,995	3,030	3,060	3,095	3,130	3,165	3,200	3,235
48	2,585	2,620	2,650	2,685	2,715	2,750	2,785	2,820	2,855	2,890	2,925	2,960	2,995	3,030	3,060	3,095	3,130	3,160	3,195
50	2,555	2,590	2,625	2,655	2,690	2,720	2,755	2,785	2,820	2,855	2,890	2,925	2,955	2,990	3,025	3,060	3,090	3,125	3,155
52	2,525	2,555	2,590	2,625	2,655	2,690	2,720	2,755	2,790	2,820	2,855	2,890	2,925	2,955	2,990	3,020	3,055	3,090	3,125
54	2,495	2,530	2,560	2,590	2,625	2,655	2,690	2,720	2,755	2,790	2,820	2,855	2,885	2,920	2,950	2,985	3,020	3,050	3,085
56	2,460	2,495	2,525	2,560	2,590	2,625	2,655	2,690	2,720	2,755	2,790	2,820	2,855	2,885	2,920	2,950	2,980	3,015	3,045
58	2,430	2,460	2,495	2,525	2,560	2,590	2,625	2,655	2,690	2,720	2,750	2,785	2,815	2,850	2,880	2,920	2,945	2,975	3,010
60	2,400	2,430	2,460	2,495	2,525	2,560	2,590	2,625	2,655	2,685	2,720	2,750	2,780	2,810	2,845	2,875	2,915	2,940	2,970
62	2,370	2,405	2,435	2,465	2,495	2,525	2,560	2,590	2,620	2,655	2,685	2,715	2,745	2,775	2,810	2,840	2,870	2,900	2,935
64	2,340	2,370	2,400	2,430	2,465	2,495	2,525	3,555	2,585	2,620	2,650	2,680	2,710	2,740	2,770	2,805	2,835	2,865	2,895
66	2,310	2,340	2,370	2,400	2,430	2,460	2,495	2,525	2,555	2,585	2,615	2,645	2,675	2,705	2,735	2,765	2,800	2,825	2,860
68	2,280	2,310	2,340	2,370	2,400	2,430	2,460	2,490	2,520	2,550	2,580	2,610	2,640	2,670	2,700	2,730	2,760	2,795	2,820
70	2,250	2,280	2,310	2,340	2,370	2,400	2,425	2,455	2,485	2,515	2,545	2,575	2,605	2,635	2,665	2,695	2,725	2,755	2,780
72	2,220	2,250	2,280	2,310	2,335	2,365	2,395	2,425	2,455	2,480	2,510	2,540	2,570	2,600	2,630	2,660	2,685	2,715	2,745
74	2,190	2,220	2,245	2,275	2,305	2,335	2,360	2,390	2,420	2,450	2,475	2,505	2,535	2,565	2,590	2,620	2,650	2,680	2,710

From: Archives of Environmental Health
February 1966, Vol. 12, pp. 146–189
E. A. Gaensler, MD and G. W. Wright, MD

Table III Predicted vital capacities for males.

AGE	HEIGHT IN CENTIMETERS AND INCHES																		
	CM	152	154	156	158	160	162	164	166	168	170	172	174	176	178	180	182	184	186	188
	IN	59.8	60.6	61.4	62.2	63.0	63.7	64.6	65.4	66.1	66.9	67.7	68.5	69.3	70.1	70.9	71.7	72.4	73.2	74.0
16		3,920	3,975	4,025	4,075	4,130	4,180	4,230	4,285	4,335	4,385	4,440	4,490	4,540	4,590	4,645	4,695	4,745	4,800	4,850
18		3,890	3,940	3,995	4,045	4,095	4,145	4,200	4,250	4,300	4,350	4,405	4,455	4,505	4,555	4,610	4,660	4,710	4,760	4,815
20		3,860	3,910	3,960	4,015	4,065	4,115	4,165	4,215	4,265	4,320	4,370	4,420	4,470	4,520	4,570	4,625	4,675	4,725	4,775
22		3,830	3,880	3,930	3,980	4,030	4,080	4,135	4,185	4,235	4,285	4,335	4,385	4,435	4,485	4,535	4,585	4,635	4,685	4,735
24		3,785	3,835	3,885	3,935	3,985	4,035	4,085	4,135	4,185	4,235	4,285	4,330	4,380	4,430	4,480	4,530	4,580	4,630	4,680
26		3,755	3,805	3,855	3,905	3,955	4,000	4,050	4,100	4,150	4,200	4,250	4,300	4,350	4,395	4,445	4,495	4,545	4,595	4,645
28		3,725	3,775	3,820	3,870	3,920	3,970	4,020	4,070	4,115	4,165	4,215	4,265	4,310	4,360	4,410	4,460	4,510	4,555	4,605
30		3,695	3,740	3,790	3,840	3,890	3,935	3,985	4,035	4,080	4,130	4,180	4,230	4,275	4,325	4,375	4,425	4,470	4,520	4,570
32		3,665	3,710	3,760	3,810	3,855	3,905	3,950	4,000	4,050	4,095	4,145	4,195	4,240	4,290	4,340	4,385	4,435	4,485	4,530
34		3,620	3,665	3,715	3,760	3,810	3,855	3,905	3,950	4,000	4,045	4,095	4,140	4,190	4,225	4,285	4,330	4,380	4,425	4,475
36		3,585	3,635	3,680	3,730	3,775	3,825	3,870	3,920	3,965	4,010	4,060	4,105	4,155	4,200	4,250	4,295	4,340	4,390	4,435
38		3,555	3,605	3,650	3,695	3,745	3,790	3,840	3,885	3,930	3,980	4,025	4,070	4,120	4,165	4,210	4,260	4,305	4,350	4,400
40		3,525	3,575	3,620	3,665	3,710	3,760	3,805	3,850	3,900	3,945	3,990	4,035	4,085	4,130	4,175	4,220	4,270	4,315	4,360
42		3,495	3,540	3,590	3,635	3,680	3,725	3,770	3,820	3,865	3,910	3,955	4,000	4,050	4,095	4,140	4,185	4,230	4,280	4,325
44		3,450	3,495	3,540	3,585	3,630	3,675	3,725	3,770	3,815	3,860	3,905	3,950	3,995	4,040	4,085	4,130	4,175	4,220	4,270
46		3,420	3,465	3,510	3,555	3,600	3,645	3,690	3,735	3,780	3,825	3,870	3,915	3,960	4,005	4,050	4,095	4,140	4,185	4,230
48		3,390	3,435	3,480	3,525	3,570	3,615	3,655	3,700	3,745	3,790	3,835	3,880	3,925	3,970	4,015	4,060	4,105	4,150	4,190
50		3,345	3,390	3,430	3,475	3,520	3,565	3,610	3,650	3,695	3,740	3,785	3,830	3,870	3,915	3,960	4,005	4,050	4,090	4,135
52		3,315	3,353	3,400	3,445	3,490	3,530	3,575	3,620	3,660	3,705	3,750	3,795	3,835	3,880	3,925	3,970	4,010	4,055	4,100
54		3,285	3,325	3,370	3,415	3,455	3,500	3,540	3,585	3,630	3,670	3,715	3,760	3,800	3,845	3,890	3,930	3,975	4,020	4,060
56		3,255	3,295	3,340	3,380	3,425	3,465	3,510	3,550	3,595	3,640	3,680	3,725	3,765	3,810	3,850	3,895	3,940	3,980	4,025
58		3,210	3,250	3,290	3,335	3,375	3,420	3,460	3,500	3,545	3,585	3,630	3,670	3,715	3,755	3,800	3,840	3,880	3,925	3,965
60		3,175	3,220	3,260	3,300	3,345	3,385	3,430	3,470	3,500	3,555	3,595	3,635	3,680	3,720	3,760	3,805	3,845	3,885	3,930
62		3,150	3,190	3,230	3,270	3,310	3,350	3,390	3,440	3,480	3,520	3,560	3,600	3,640	3,680	3,730	3,770	3,810	3,850	3,890
64		3,120	3,160	3,200	3,240	3,280	3,320	3,360	3,400	3,440	3,490	3,530	3,570	3,610	3,650	3,690	3,730	3,770	3,810	3,850
66		3,070	3,110	3,150	3,190	3,230	3,270	3,310	3,350	3,390	3,430	3,470	3,510	3,550	3,600	3,640	3,680	3,720	3,760	3,800
68		3,040	3,080	3,120	3,160	3,200	3,240	3,280	3,320	3,360	3,400	3,440	3,480	3,520	3,560	3,600	3,640	3,680	3,720	3,760
70		3,010	3,050	3,090	3,130	3,170	3,210	3,250	3,290	3,330	3,370	3,410	3,450	3,480	3,520	3,560	3,600	3,640	3,680	3,720
72		2,980	3,020	3,060	3,100	3,140	3,180	3,210	3,250	3,290	3,330	3,370	3,410	3,450	3,490	3,530	3,570	3,610	3,650	3,680
74		2,930	2,970	3,010	3,050	3,090	3,130	3,170	3,200	3,240	3,280	3,320	3,360	3,400	3,440	3,470	3,510	3,550	3,590	3,630

From: Archives of Environmental Health
February 1966, Vol. 12, pp. 146–189
E. A. Gaensler, MD and G. W. Wright, MD

Table IV Conversion factors for temperature differentials (spirometry).

°C	°F	Conversion Factor
20	68.0	1.102
21	69.8	1.096
22	71.6	1.091
23	73.4	1.085
24	75.2	1.080
25	77.0	1.075
26	78.8	1.068
27	80.6	1.063
28	82.4	1.057
29	84.2	1.051
30	86.0	1.045
31	87.8	1.039
32	89.6	1.032
33	91.4	1.026
34	93.2	1.020
35	95.0	1.014
36	96.8	1.007
37	98.6	1.000

Appendix Solutions and Reagents

Physiological Solutions

Working with freshly dissected or excised vertebrate tissues requires that they be perfused or immersed in an environment that approximates as nearly as possible the ionic, osmotic, and pH qualities of their own tissue fluids. Doing so prevents dysfunction and allows for maintenance and observation of the tissue over longer intervals. A partial list of physiological solutions follows.

Physiological Saline: A solution of sodium chloride in water that can allow for moisture maintenance and tonicity of most tissues for short periods. Consists of 0.9% for mammals and 0.7% for amphibians.

Frog Ringer's: An all-purpose solution for amphibian tissues (nerve, muscle, skin, etc.).

Turtle Ringer's: Replaces the tissue fluids of most reptiles.

Mammalian Ringer's: An all-purpose solution for general applications in mammalian tissue studies, both short and long term.

Locke's Solution: Devised for use primarily with isolated smooth muscle and cardiac muscle of mammals.

Tyrode's Solution: Designed for use with mammalian smooth muscle preparations.

Krebs-Henseleit Solution: May be used for work with mammalian nerves and for metabolic measurements of other mammalian tissues.

In this appendix, the individual recipes for some of the aforementioned solutions will be given in alphabetical order; however, where large quantities of solutions are required during the semester, it will be more convenient to make up stock solutions from which the various solutions can be quickly made. The longer shelf life of stock solutions enables one to be able to make up fresh physiological solutions daily as needed.

Stock Solutions

Make up five flasks in which the amounts listed are *grams per liter of distilled water.* The molarity of each solution is also given.

Table I Physiological solutions from stock solutions.

Stock	Frog Ringer's	Turtle Ringer's	Mammalian Ringer's	Tyrode's	Locke's	Krebs-Henseleit
A $NaCl$	103 ml	117 ml	155 ml	137 ml	155 ml	111 ml
B KCl	30 ml	40 ml	65 ml	27 ml	56 ml	47 ml
C $CaCl_2$	20 ml	20 ml	22 ml	18 ml	22 ml	25 ml
$NaHCO_3$ (dry)	.2 gm	.2 gm	.2 gm	.1 gm	.2 gm	.2 gm
D NaH_2PO_4	---	---	8.3 ml	2.5 ml	---	---
E $MgSO_4$	---	---	---	---	---	12.5 ml
*Glucose (dry)	1.0 gm	1.0 gm	1.0 gm	1.0 gm	1.0 gm	1.0 gm

*Addition of glucose is optional. Keep refrigerated. Lasts only 1 week in refrigerator.

Sol'n	Compound	Amount	Molarity
A	NaCl	58.54 gm	1M
B	KCl	7.45 gm	0.1M
C	$CaCl_2$	11.1 gm	0.1M
D	NaH_2PO_4	12.0 gm	0.1M
E	$MgSO_4$	12.0 gm	0.1M

In addition to the above solutions, $NaHCO_3$ in dry form is used. Its instability in solution precludes using a stock solution of this ingredient. Glucose, in dry form, is also added for certain applications. *Refrigeration of all glucose solutions is necessary to inhibit bacterial action.*

Dilutions

To prepare a liter of any of the physiological solutions, fill a volumetric flask half full of distilled water and add the amounts given in table I. Each compound is added in sequence and mixed thoroughly before adding succeeding compounds. After all ingredients are added, the flask is filled to the one liter mark.

Caution If excess dissolved CO_2 is present in the water, it will tend to precipitate out the calcium in the form of $CaCO_3$. This may be prevented by (1) pre-aerating the water with an oil-free oxygen or air line for 5–10 minutes or (2) quickly adding solution C ($CaCl_2$) with rapid agitation at the very last. If a precipitate does occur, the solution is not usable.

Other Solutions and Reagents

BAPNA (Ex. 61)

BAPNA is benzoyl DL arginine p-nitroaniline HCl. To prepare this substrate for Ex. 61, dissolve 43 mg of BAPNA in 1 ml of dimethyl sulfoxide (caution) and make up to a volume of 100 ml with TRIS buffer solution.

Barfoed's Reagent

Dissolve 12 gm copper acetate in 450 ml of boiling distilled water. Do not filter if precipitate forms. To this hot solution, quickly add 13 ml of 8.5% lactic acid. Most of precipitate will dissolve. Cool the mixture and dilute to 500 ml. Remove any final precipitate by filtration.

Benedict's Solution (Qualitative)

Sodium citrate	173.0 gm
Sodium carbonate, anhydrous	100.0 gm
Copper sulfate, pure crystalline $CuSO_4 \cdot 5H_2O$	17.3 gm

Dissolve the sodium citrate and sodium carbonate in 700 ml distilled water with aid of heat and filter.

Then, dissolve the copper sulfate in 100 ml of distilled water with aid of heat and pour this solution slowly into the first solution, stirring constantly. Make up to 1 liter volume with distilled water.

Buffer Solution, pH 5 (Ex. 62)

Make up 1 liter each of the following solutions:

M/10 potassium acid phthalate
20.418 gm $KHC_8H_4O_4$ to distilled water to make one liter.
M/10 sodium hydroxide
4.0 gm sodium hydroxide to distilled water to make one liter.

Use 50 ml of the first solution and 23.9 ml of the second solution to make 73.9 ml of pH 5 buffer solution.

Buffer Solution, pH 7 (Ex. 62)

Make up 1 liter each of the following solutions:

M/15 potassium acid phosphate
9.08 gm KH_2PO_4 to distilled water to make one liter.
M/15 disodium phosphate
9.47 gm Na_2HPO_4 (anhydrous) to distilled water to make one liter.

Use 38.9 ml of the first solution and 61.1 ml of the second solution to make 100 ml of pH 7 buffer solution.

Buffer Solution, pH 9 (Ex. 62)

Make up 1 liter each of the following solutions:

.2M boric acid
5.96 gm H_3BO_3 to distilled water to make one liter.
.2M potassium chloride
14.89 gm KCl to distilled water to make one liter.
.2M sodium hydroxide
8 gm NaOH to distilled water to make one liter.

Use 50 ml each of the first two solutions, 21.5 ml of the third solution and dilute to 200 ml of distilled water.

Exton's Reagent (for albumin test)

Sodium sulfate (anhydrous)	88 gm
Sulfosalicylic acid	50 gm
Distilled water to make one liter.	

Dissolve the sodium sulfate in 80 ml of water with heat. Cool, add the sulfosalicylic acid, and make up to volume with water.

Frog Ringer's Solution

This solution is used with most amphibian muscle and nerve preparations. Frog skin and toad bladder preparations can also be handled with this saline solution.

NaCl	6.02 gm
KCl	0.22 gm
$CaCl_2$	0.22 gm
$NaHCO_3$	2.00 gm
Glucose	1.00 gm
Distilled water to make 1 liter.	

Iodine (IKI) Solution

Potassium iodide	20 gm
Iodine crystals	4 gm
Distilled water	1 liter

Dissolve the potassium iodide in 1 liter of distilled water and add the iodine crystals, stirring to dissolve. Store in dark bottles.

Locke's Solution

NaCl	9.06 gm
KCl	0.42 gm
$CaCl_2$	0.24 gm
$NaHCO_3$	2.00 gm
Glucose	1.00 gm
Distilled water to make 1 liter.	

Mammalian Ringer's Solution

NaCl	9.00 gm
KCl	0.48 gm
$CaCl_2$	0.24 gm
$NaHCO_3$	0.20 gm
NaH_2PO_4	0.10 gm
Glucose	1.00 gm
Distilled water to make one liter.	

Methylene Blue (Loeffler's)

Solution A: Dissolve 0.3 gm of methylene blue (90% dye content) in 30.0 ml ethyl alcohol (95%).

Solution B: Dissolve 0.01 gm potassium hydroxide in 100.0 ml distilled water. Mix solutions A and B.

Red Blood Cell (RBC) Diluting Fluid (Hayem's)

Mercuric chloride	1.0 gm
Sodium sulfate (anhydrous)	4.4 gm
Sodium chloride	2.0 gm
Distilled water	400.0 ml

Rothera's Reagent

The addition of 1 gm of this reagent to 5 ml of urine with ammonium hydroxide is used to detect ketones.

Sodium nitroprusside	7.5 gm
Ammonium sulfate	700.0 gm

Mix and pulverize in mortar with pestle.

Starch Solution (1 %)

Add 10 grams of cornstarch to 1 liter of distilled water. Bring to boil, cool, and filter. Keep refrigerated.

TRIS, pH 8.2 (Ex. 61)

The molecular weight of 2-amino-2(hydroxymethyl)-1,3-propanediol, or TRIS, is 121.14. Its formula is $(HOCH_2)_3CNH_2$. To make up this solution for Ex. 61, make up a .05M solution first by dissolving 6.066 gms in a liter of distilled water. To 500 ml of the 0.5M TRIS buffer, add 130 ml of .05N HCl.

White Blood Cell (WBC) Diluting Fluid

Hydrochloric acid	5 ml
Distilled water	495 ml

Add 2 small crystals of thymol as a preservative.

C Appendix Tests and Methods

Carbohydrate Differentiation

Several experiments in this book require that sugars and starches be differentiated. An understanding of the application of these tests can be derived from the separation outline at the bottom of this page.

Barfoed's Test To 5 ml of reagent in test tube, add 1 ml of unknown. Place in boiling water bath for 5 minutes. Interpretation is based on separation outline.

Benedict's Test This is a semiquantitative test for amounts of reducing sugar present. To 5 ml of reagent in test tube, add 8 drops of unknown. Place in boiling water bath for 5 minutes. A green, yellow, or orange-red precipitate determines the amount of reducing sugar present. See Ex. 61.

Electronic Equipment Adjustments

Lengthy instructions pertaining to recorder and transducer manipulations that are used in several experiments are outlined here for reference.

Balancing Transducer on Unigraph
(Ex. 22)

The Unigraph is designed to function in a sensitivity range of .1 MV/CM to 2 MV/CM. When an experiment is begun, the sensitivity control is set at 2 MV/CM, its least sensitive position. As the experiment progresses it may become necessary to shift to a more sensitive setting, such as 1 MV/CM or .5 MV/CM. If the Wheatstone bridge in the transducer is not balanced, the shifting from one sensitivity position to another will cause the stylus to change position. This produces an unsatisfactory record. Thus, it is essential that one go through the following steps to balance the transducer.

1. Connect the transducer jack to the transducer socket in the end of the Unigraph. The small upper socket is the one to use.
2. Before plugging in the Unigraph, make sure that the power switch is off, the chart control switch is on STBY, the gain selector knob is on 2 MV/CM, the sensitivity knob is turned com-

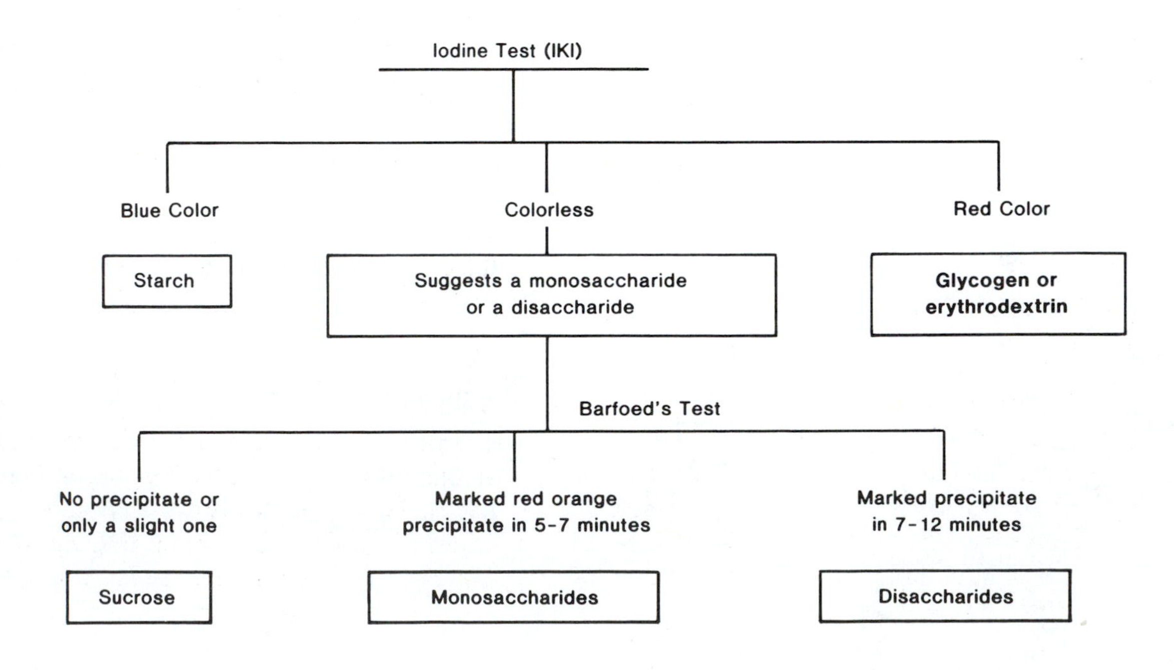

pletely counterclockwise, the mode selector control is on TRANS, and the Hi Filter and Mean switches are positioned toward Normal. Now plug in the Unigraph.

3. Turn on the Unigraph power switch and set the heat control knob at the two o'clock position.
4. Place the speed control lever at the slow position and the c.c. lever at Chart On.
5. As the chart moves along, bring the stylus to the center of the paper by turning the centering knob.
6. Turn the sensitivity control completely clockwise to maximum sensitivity. If the bridge is unbalanced, the stylus will move away from the center. To return the stylus to the center, unlock the TRANS-BAL control by pushing the small lever on the control and turning the knob in either direction to return the stylus to the center of the chart.
7. Set the gain control to 1. If the bridge is unbalanced, the stylus will not be centered. Center with the TRANS-BAL control.
8. Set the gain control to .5 and re-center the stylus with the TRANS-BAL control.
   Repeat this same procedure for .2 and .1 settings of the gain control. The bridge is now completely balanced at its highest gain and sensitivity. Lock the TRANS-BAL control with the lock lever.
   Note: Although we have gone through this lengthy process to achieve a balanced transducer, we may find that we will still get baseline shifts when going from one setting to another. However, they will be minor compared to one that has not been through this procedure. Any minor deviations should be corrected with the centering control.
9. Return the gain control to 2 and the chart control switch to STBY.

Calibration of Unigraph with Strain Gauge (Ex. 22)

When the transducer has been balanced, it is necessary to calibrate the tracing on the Unigraph to a known force on the transducer. Calibration establishes a linear relationship between the magnitude of deflection and the degree of strain (extent of bending) of the transducer leaf before the muscle is attached to the transducer. Calibrations must be made for each gain at maximum sensitivity that one expects to use in the experiment (your muscle preparations will probably require gain settings of 2, 1, or .5). After calibration, it is possible to directly determine the force of each contraction by simply reading the maximum height of the tracing.

Materials:

small paper clip weighing around 0.5 gram

1. Weigh a convenient object, such as a small paper clip, to the nearest 0.1 mg.
2. With the chart control switch on STBY, suspend the paper clip on to the two smallest leaves of the transducer.
3. Put the c.c. switch to Chart On and note the extent of deflection on the chart. Record about 1 centimeter on the chart and place the c.c. switch on STBY. Label this deflection in mm/mg.
4. Set the gain control at 1, c.c. to Chart On, and record for another centimeter. Place the c.c. switch on STBY. Label the deflection.
5. Repeat the above procedure for the other two gain settings (.5, .2, and .1). Note how the stylus vibrates as you increase the sensitivity of the Unigraph. This is normal.
6. Return the c.c. switch to STBY, gain to 2, and remove the paper clip. The Unigraph is now calibrated and the muscle can now be attached to it.

Calibration for EMG on Unigraph (Ex. 24)

Calibration of the Unigraph for EMG measurements is necessary to establish that one centimeter deflection of the stylus equals 0.1 millivolt. When calibration is accomplished, the recording is made with the EEG mode since there is no EMG mode. An EMG can be made very well in this mode at lower sensitivities than would be used when making an EEG. Proceed as follows to calibrate:

1. Turn the mode selector control to CC/Cal. (Capacitor Coupled Calibrate).
2. Turn the small main switch to ON.
3. Turn the stylus heat control to the two o'clock position.
4. Check the speed selector lever to see that it is in the slow position. The free end of the lever should be pointed toward the styluses.
5. Move the chart-stylus switch to Chart On and observe the width of the tracing that appears

on the paper. Adjust the stylus heat control to produce the desired width of tracing.

6. Set the gain knob to .1 MV/CM.
7. Push down the .1 MV button to determine the amount of deflection. Hold the button down for 2 seconds before releasing. If it is released too quickly the tracing will not be perfect. Adjust the deflection so that it deflects exactly 1 centimeter in each direction by turning the sensitivity knob. The unit is now calibrated so that you get 1 cm deflection for 0.1 millivolt.
8. Reset the gain knob to .5 MV/CM. This reduces the sensitivity of the instrument.
9. Return the mode selector control to EEG.
10. Return the chart-stylus switch to STBY. Place the speed selector in the fast position. The instrument is now ready for use.

Balancing the Myograph to Physiograph® Transducer Coupler (Ex. 22)

Balancing (matching) of the transducer signal with the amplifier is necessary to get the recording pen to stay on a preset baseline when the RECORD button is in either the OFF or ON position. Proceed as follows:

1. Before starting, make sure that the RECORD switch is OFF and that the myograph jack is inserted into the transducer coupler.
2. Set the sensitivity control (outer knob) to its lowest numbered setting (highest sensitivity).
3. Set the paper speed at 0.5 cm/sec and lower the pen lifter.
4. Adjust the pen position so that the pen is writing *exactly* on the center line.
5. Place the RECORD switch in the ON position. The pen will probably be moving a large distance either up or down.
6. With the BALANCE control, adjust the pen so that it is again writing *exactly* on the center line.
7. Check your match, or balance, by placing the RECORD button in the OFF position. **Your system is balanced if the pen remains on the center line.**
8. If the pen does not remain on the center line, relocate the pen to the center line again with

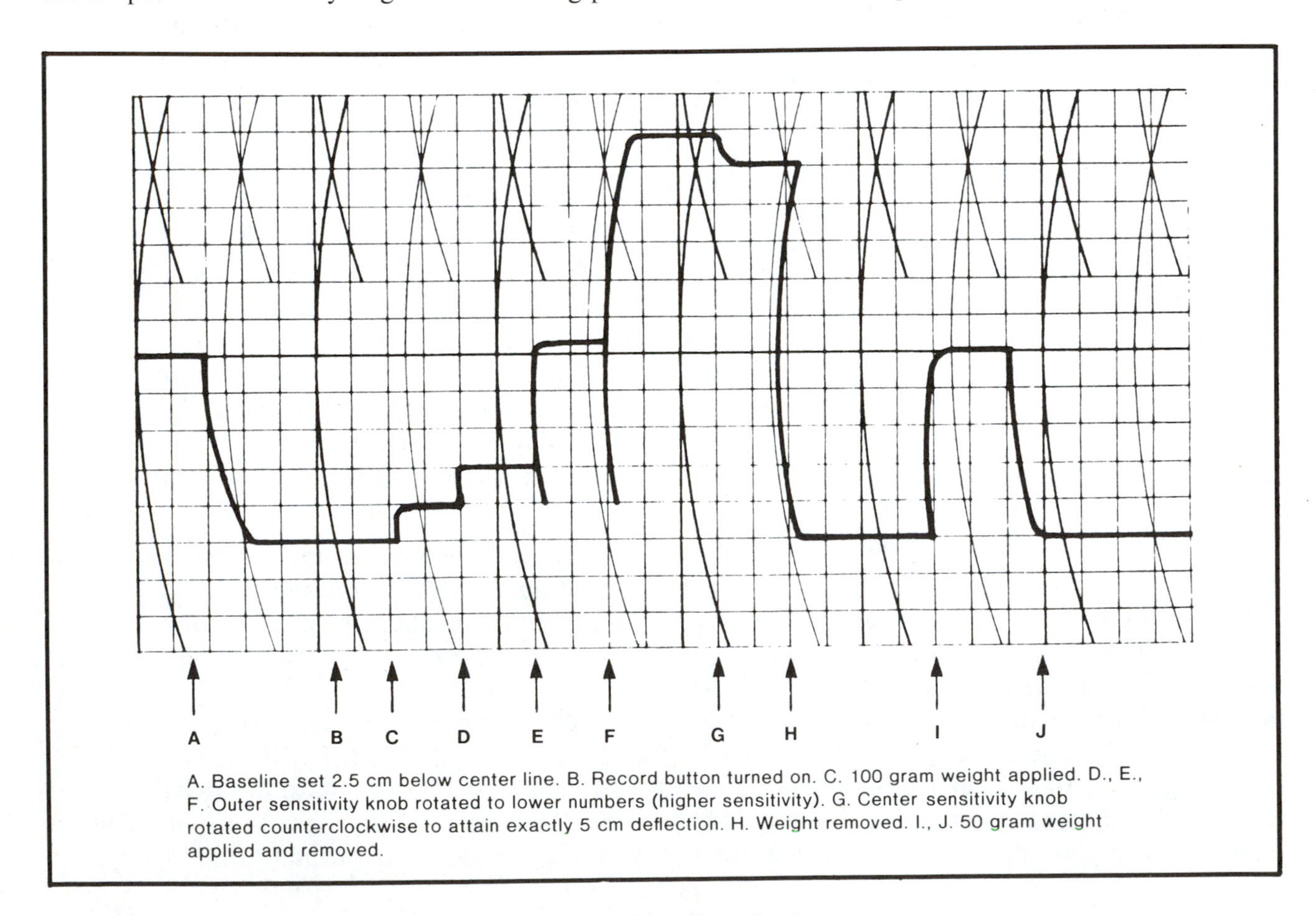

A. Baseline set 2.5 cm below center line. B. Record button turned on. C. 100 gram weight applied. D., E., F. Outer sensitivity knob rotated to lower numbers (higher sensitivity). G. Center sensitivity knob rotated counterclockwise to attain exactly 5 cm deflection. H. Weight removed. I., J. 50 gram weight applied and removed.

Figure C-1 Sample record of calibration at 100 grams per 5 centimeters.

the position control, place the RECORD button in the ON position, and repeat the balancing procedure. You have a balanced system *only* when the pen remains on a preset baseline with the RECORD button in both OFF and ON positions.

Calibration of Myograph Transducer on Physiograph® (Ex. 22)

Calibration of the transducer is necessary so that we know exactly how much tension, in grams, is being exerted by the muscle during contraction. The channel amplifier must be balanced first. Our calibration here will be to get 5 cm pen deflection with 100 grams. *These instructions apply to couplers that lack built-in calibration buttons.* Figure C–1 on the previous page illustrates the various steps.

Materials:

100 gram weight
50 gram weight

1. Start the paper drive at 0.1 cm/sec and lower the pen lifter so that we are recording on the desired channel.
2. Check to see that the channel amplifier is balanced.
3. With the pen position control, set the baseline exactly 2.5 cm (5 blocks) below the channel center line.
4. Rotate the outer knob of the sensitivity control fully counterclockwise to the lowest sensitivity level (i.e., 1000 setting). Be sure that the inner knob of the sensitivity control is in its fully clockwise "clicked" position.
5. Place the RECORD button in the ON position.
6. Suspend a 100 gram weight to the actuator of the myograph. Note that the addition of the weight has caused the baseline to move upward approximately one block (0.5 cm). See step C, figure C–1.
7. Rotate the *outer* knob of the sensitivity control clockwise until the pen exceeds 5 cm (10 blocks) of deflection from the original baseline. See step F, figure C–1.
8. Rotate the *inner* knob of the sensitivity control counterclockwise to bring the pen back downward until it is *exactly* 5 cm (10 blocks) of deflection from the original baseline.
9. Remove the weight. The pen should return to the original baseline. If it does not return exactly to the baseline, reset the baseline with the position control, reapply the weight, and rotate the inner knob until the pen is writing exactly 5 cm above the original baseline.
10. Attach a 50 gram weight to the myograph to see if you get 5 blocks of pen deflection, as shown in the last deflection in figure C–1. The unit is now properly calibrated.

Calibration of Myograph Transducer with Calibration Button (Ex. 22)

When the coupler has a calibration button, the use of a 100 gram weight is not necessary. To calibrate, all that is necessary is to press the 100 gram button, and follow the steps above to produce the 5 cm pen deflection.

D Appendix
Physiograph® Sample Records

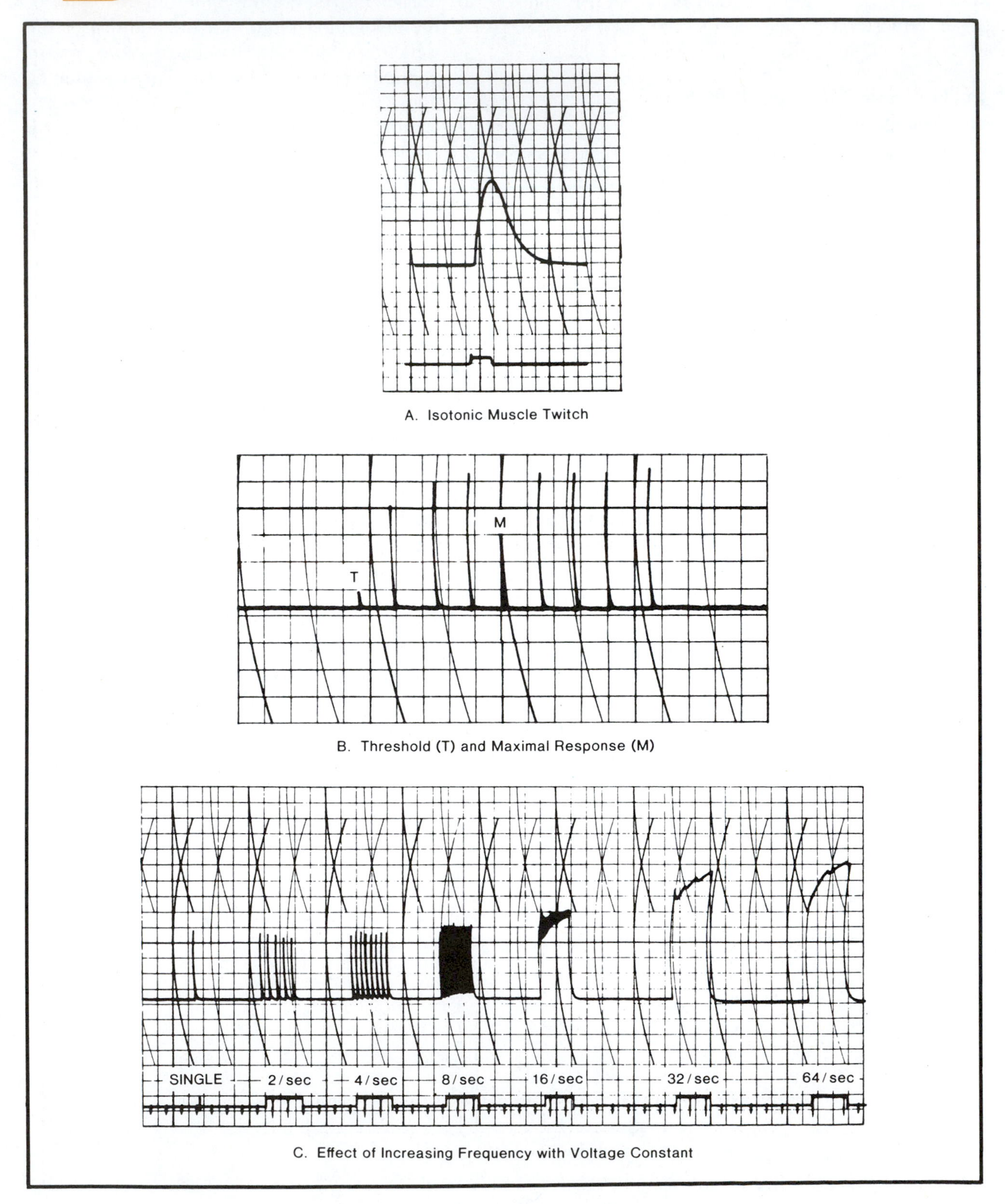

A. Isotonic Muscle Twitch

B. Threshold (T) and Maximal Response (M)

C. Effect of Increasing Frequency with Voltage Constant

Figure SR-1 Physiograph® sample records of frog muscle contraction (Ex. 22).

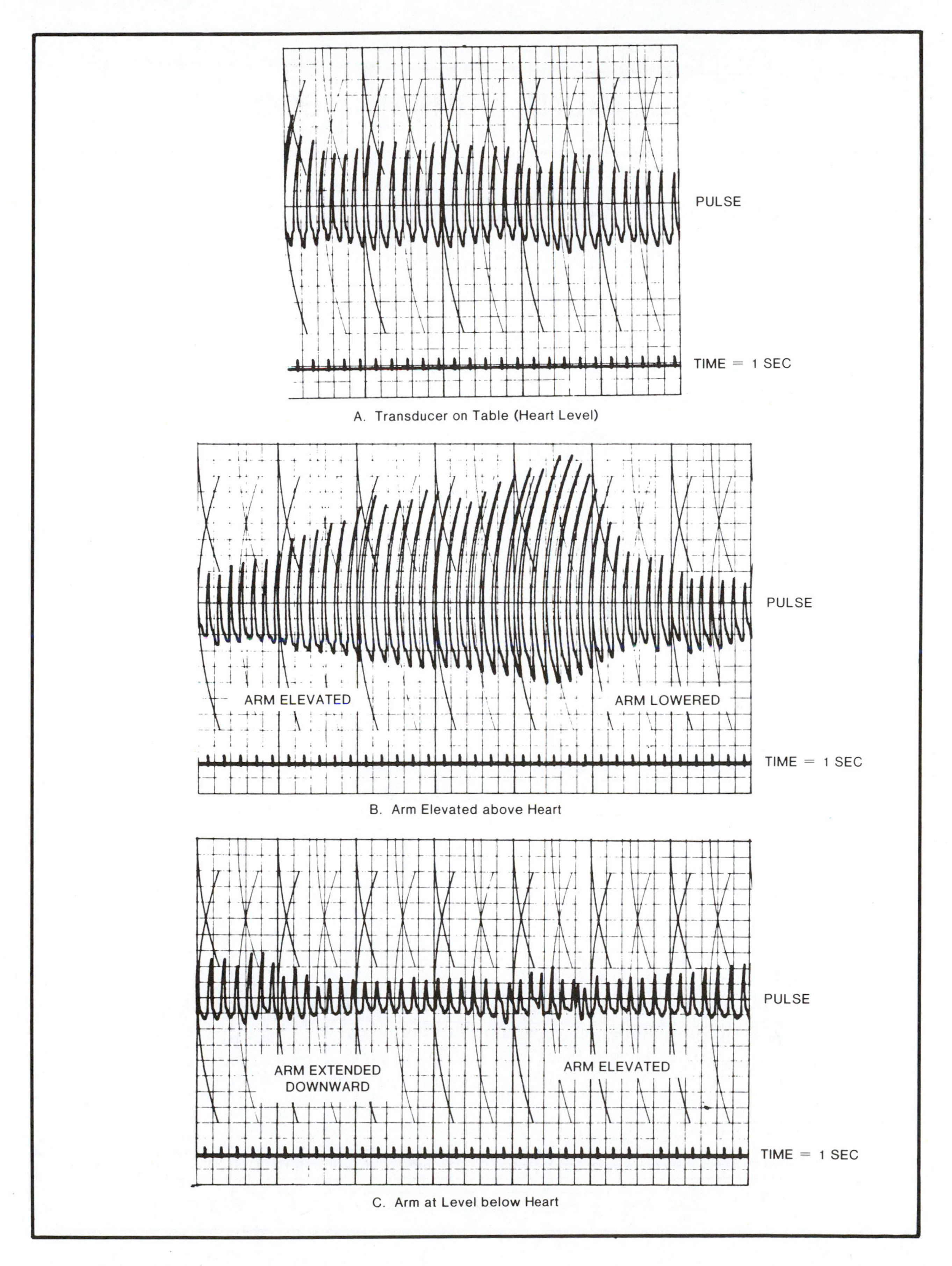

Figure SR-2 Effects of hand position on pulse (Ex. 47).

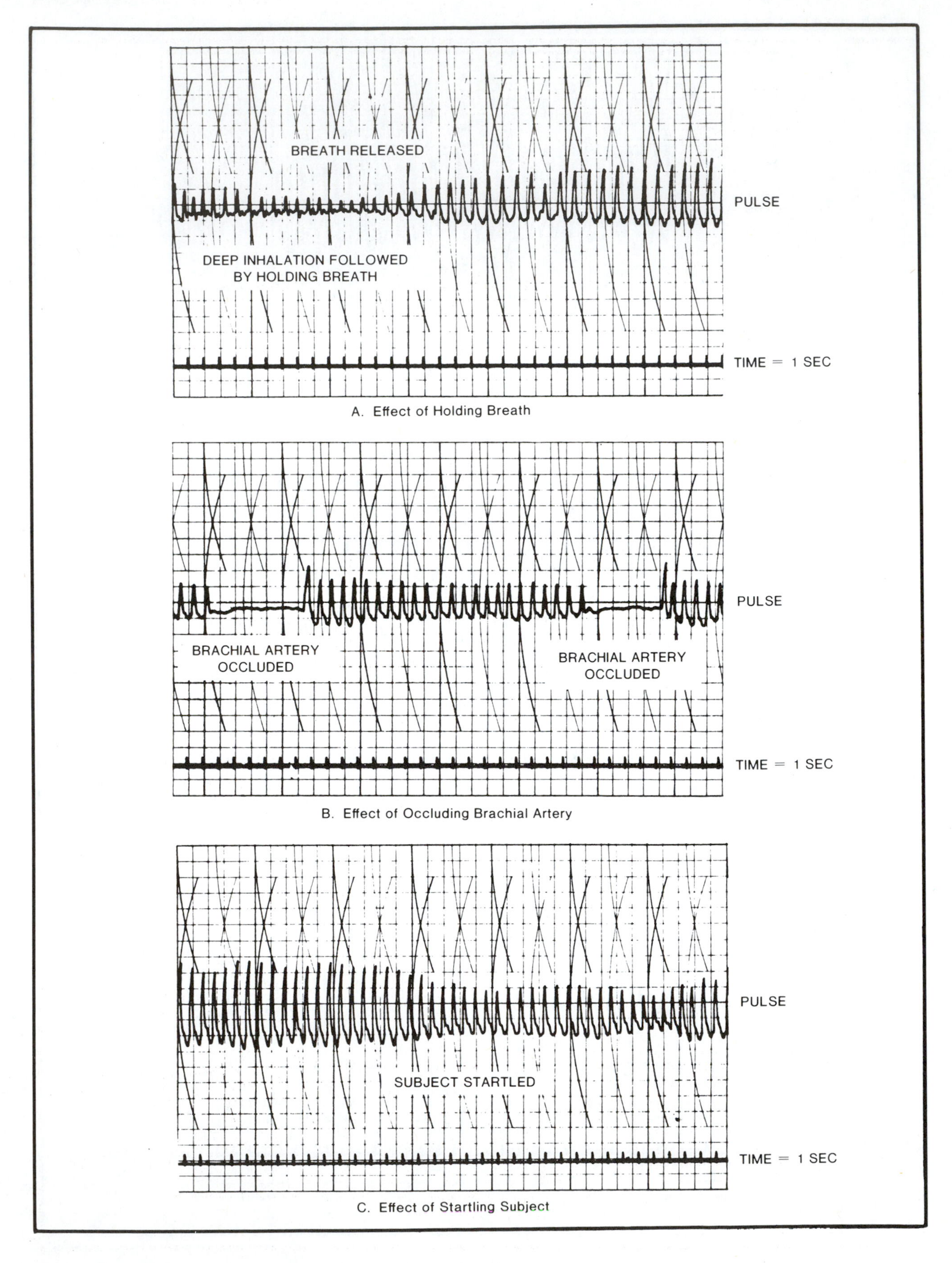

A. Effect of Holding Breath

B. Effect of Occluding Brachial Artery

C. Effect of Startling Subject

Figure SR-3 Pulse variations due to various factors (Ex. 47).

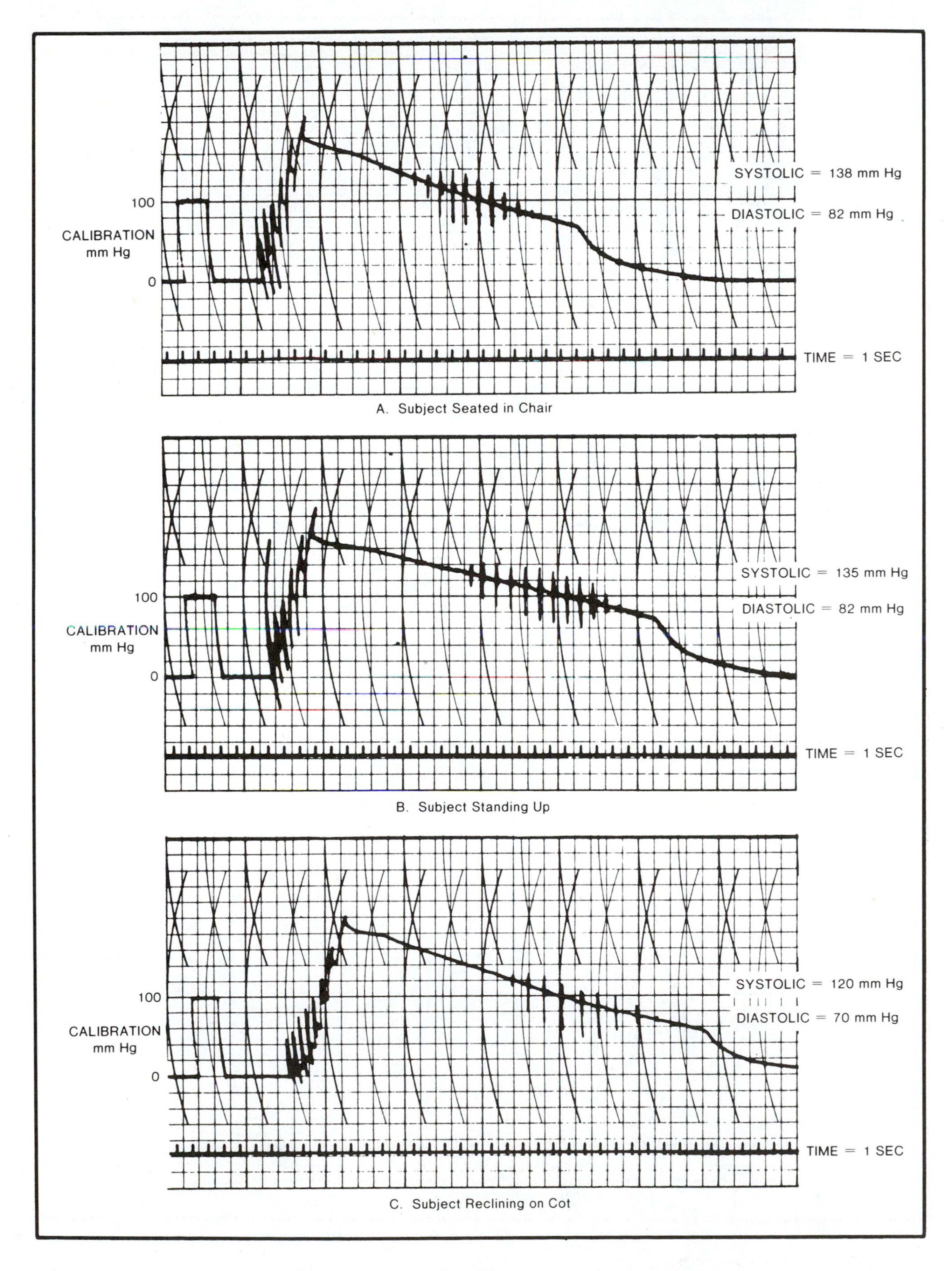

Figure SR-4 Effects of posture on blood pressure (Ex. 48).

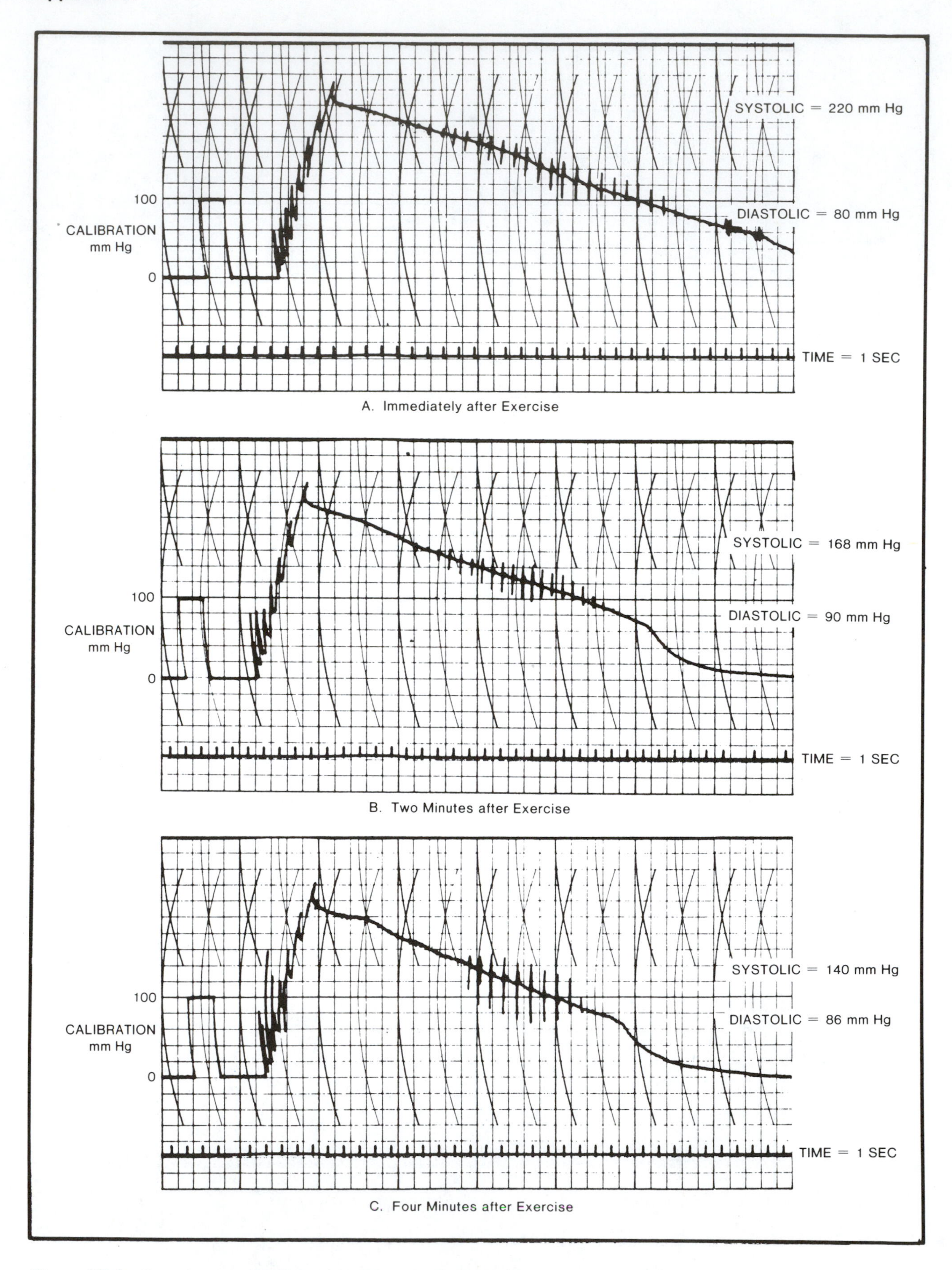

Figure SR-5 Sample records of blood pressure monitoring after exercise (Ex. 48).

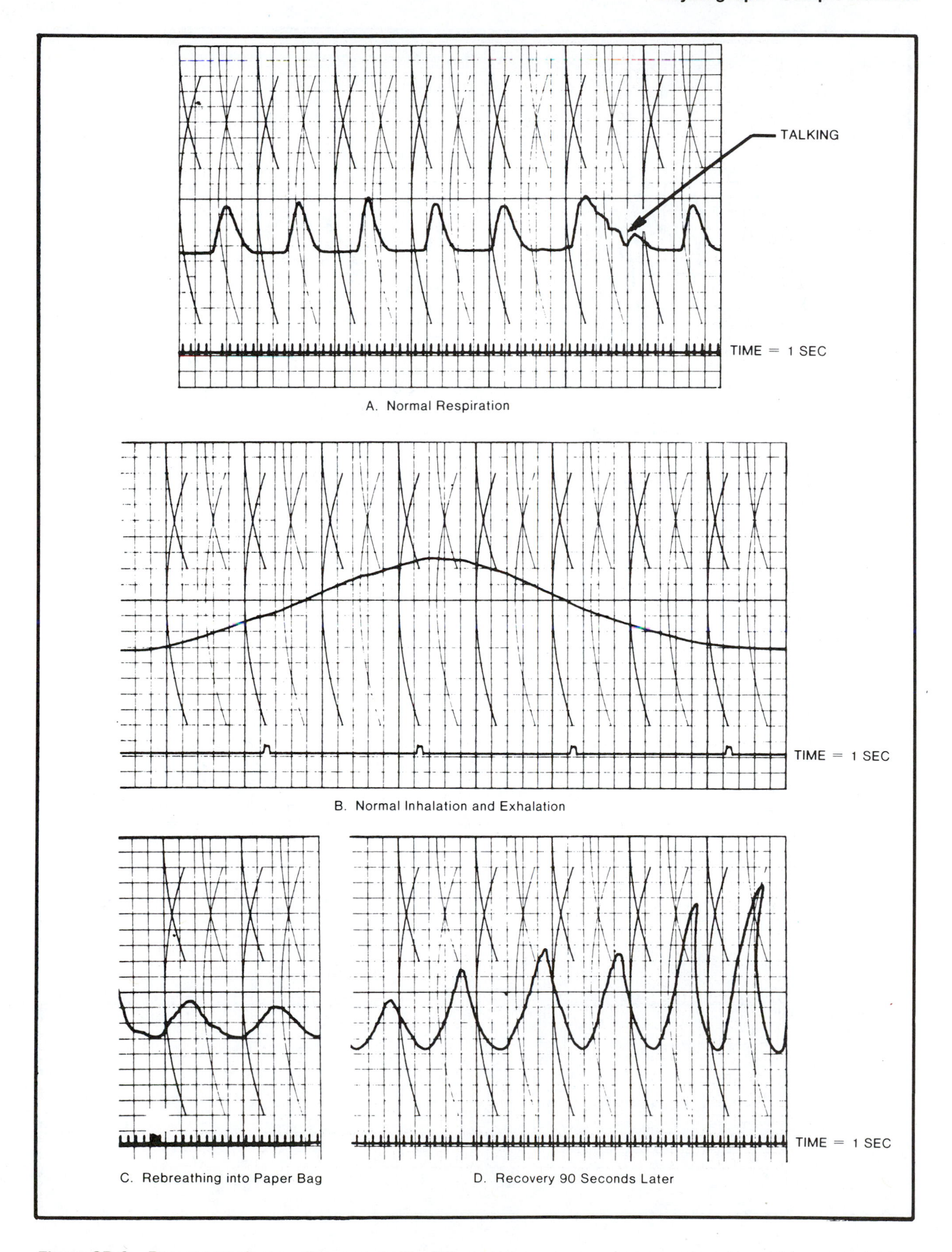

Figure SR-6 Pneumograph recording records (Ex. 54).

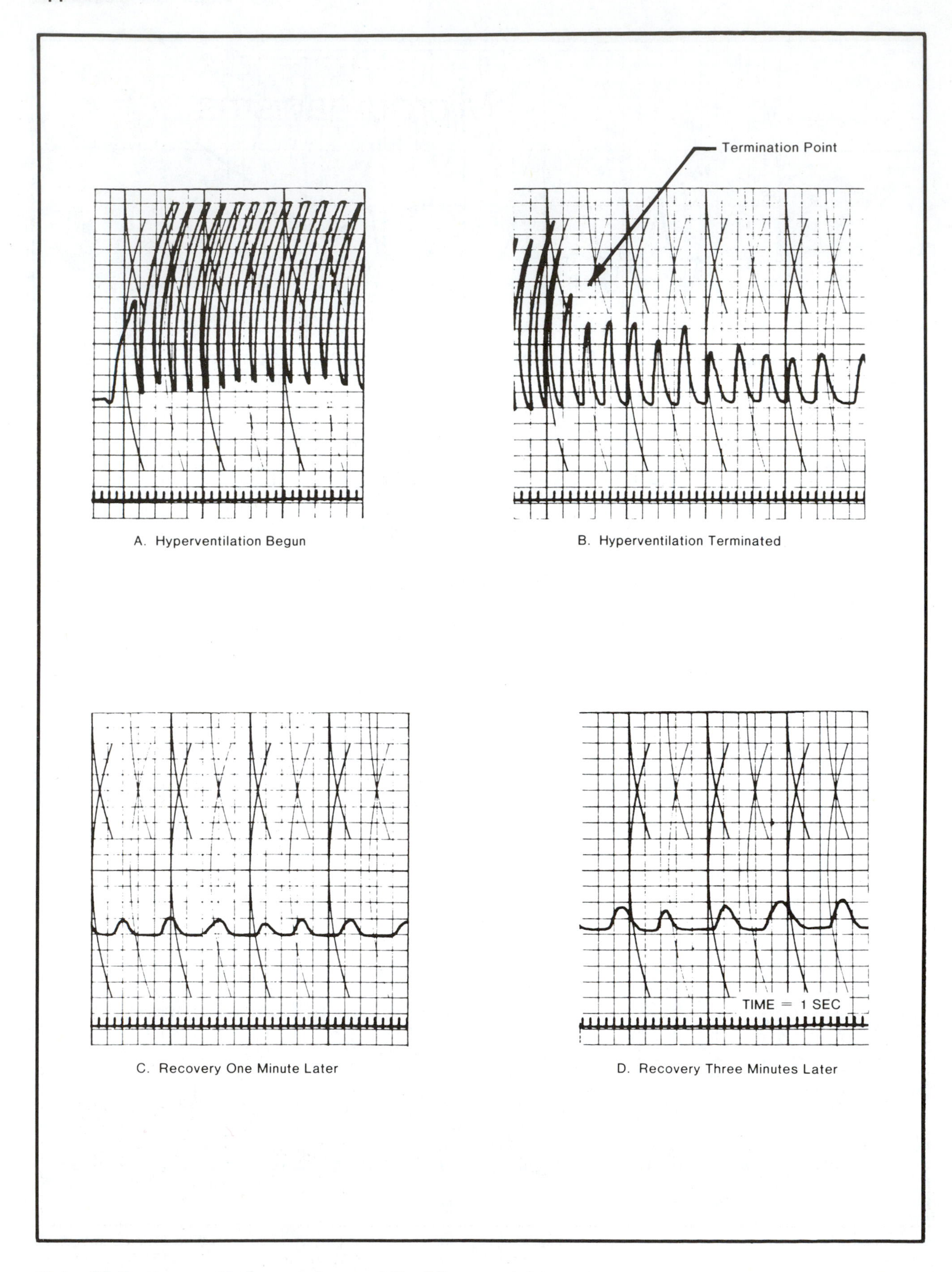

Figure SR-7 Hyperventilation sample record (Ex. 54).

Appendix Microorganisms

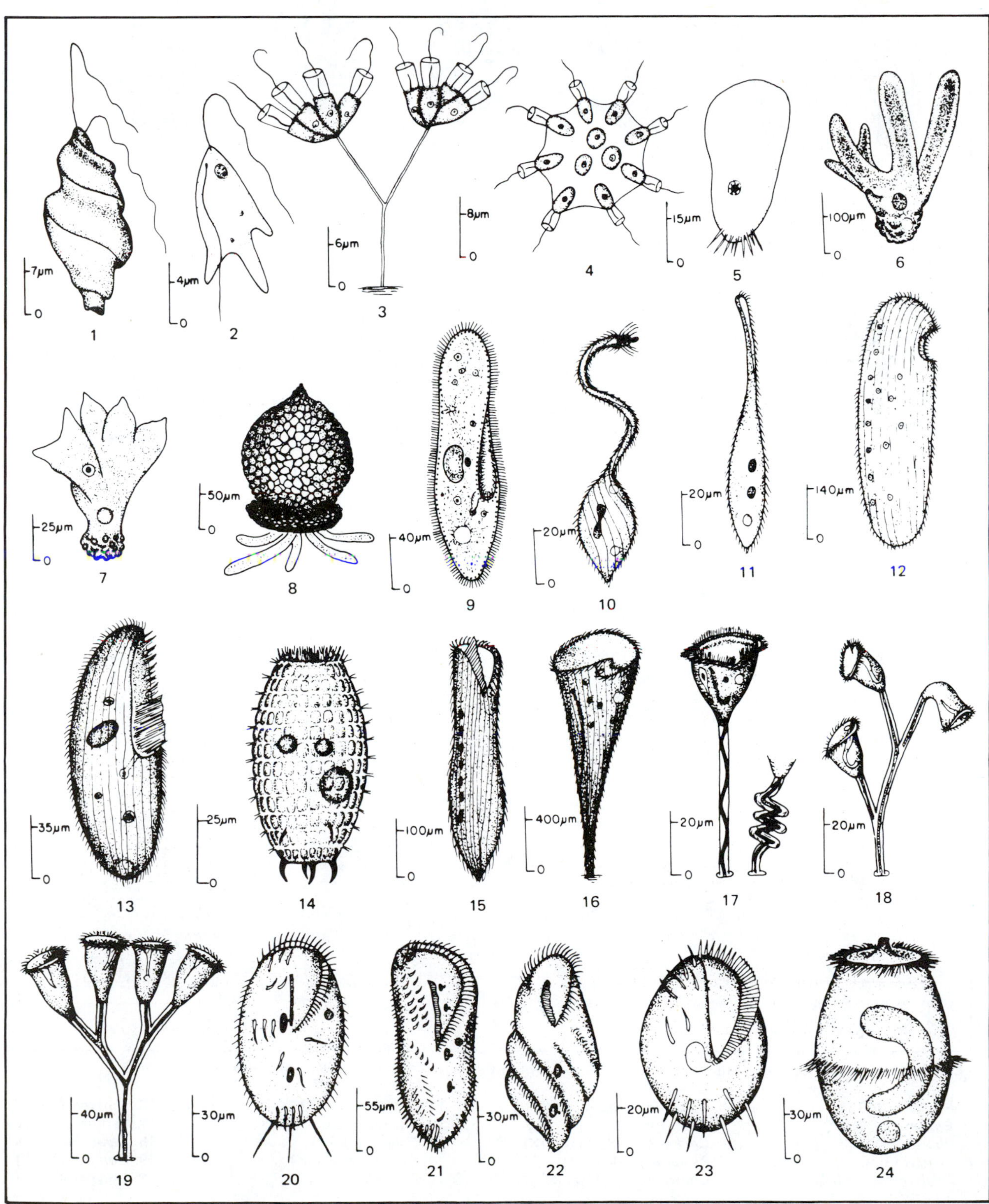

1. *Heteronema*
2. *Cercomonas*
3. *Codosiga*
4. *Protospongia*
5. *Trichamoeba*
6. *Amoeba*
7. *Mayorella*
8. *Diffugia*
9. *Paramecium*
10. *Lacrymaria*
11. *Lionotus*
12. *Loxodes*
13. *Blepharisma*
14. *Coleps*
15. *Condylostoma*
16. *Stentor*
17. *Vorticella*
18. *Carchesium*
19. *Zoothamnium*
20. *Stylonychia*
21. *Onychodromos*
22. *Hypotrichidium*
23. *Euplotes*
24. *Didinium*

Figure MO-1 Protozoans.

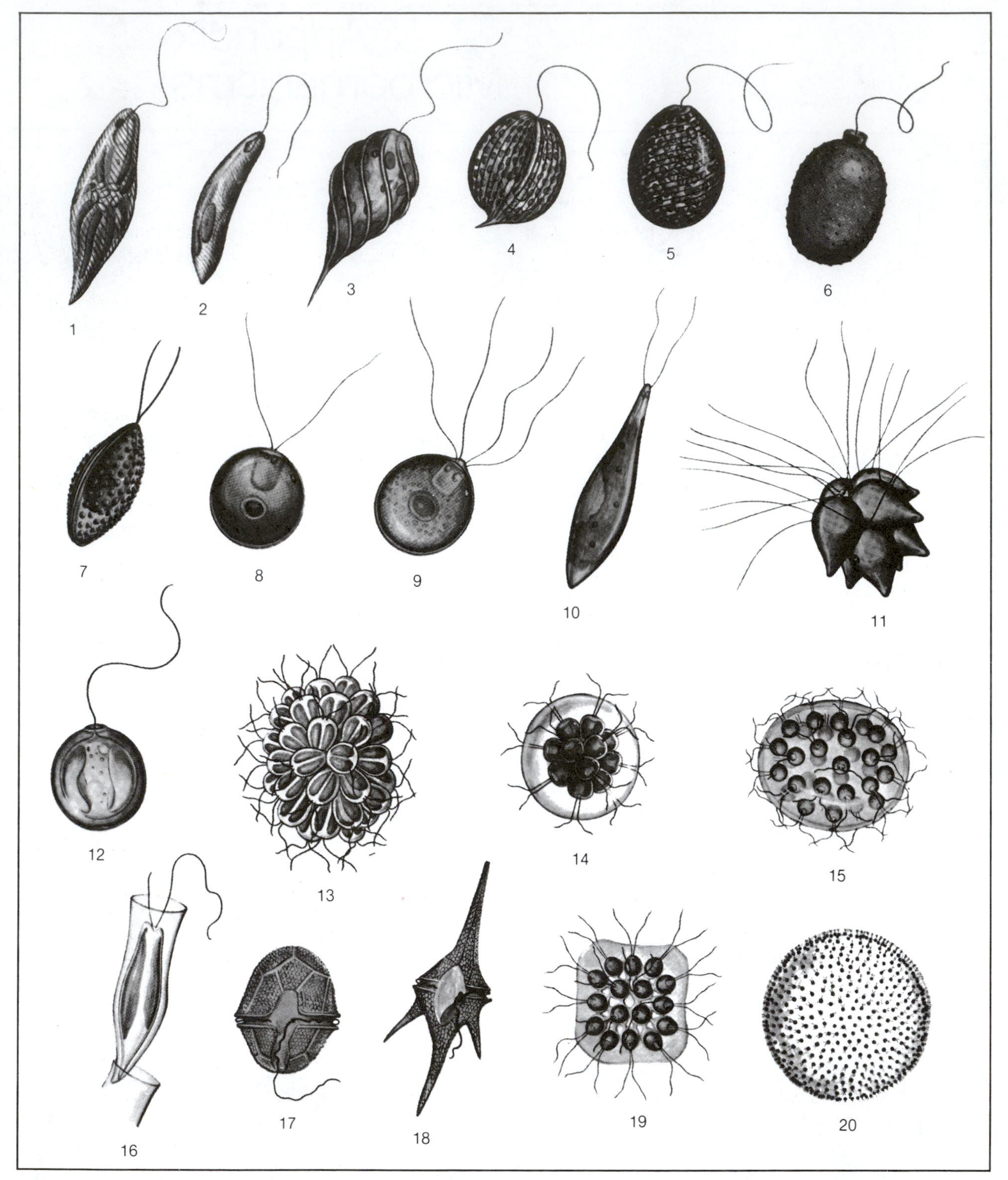

Courtesy of the U.S. Environmental Protection Agency, Office of Research & Development, Cincinnati, Ohio 45268.

1. *Euglena* (700X)
2. *Euglena* (700X)
3. *Phacus* (1000X)
4. *Phacus* (350X)
5. *Lepocinclis* (350X)
6. *Trachelomonas* (1000X)
7. *Phacotus* (1500X)
8. *Chlamydomonas* (1000X)
9. *Carteria* (1500X)
10. *Chlorogonium* (1000X)
11. *Pyrobotrys* (1000X)
12. *Chrysococcus* (3000X)
13. *Synura* (350X)
14. *Pandorina* (350X)
15. *Eudorina* (175X)
16. *Dinobyron* (1000X
17. *Peridinium* (350X)
18. *Ceratium* (175X)
19. *Gonium* (350X)
20. *Volvox* (100X)

Figure MO-2 Flagellated algae.

Reading References

Anatomy and Physiology

Anthony, Catherine P., and Thibodeau, Gary A., 1983. *Textbook of anatomy and physiology.* 11th ed. St. Louis: C. V. Mosby Co.

De Coursey, Russell M., 1980. *The human organism.* 5th ed. New York: McGraw-Hill.

Hole, John W., Jr., 1987. *Human anatomy and physiology.* 4th ed. Dubuque, IA: Wm. C. Brown Publishers.

Langley, L. L., et al., 1980. *Dynamic anatomy and physiology.* 5th ed. New York: McGraw-Hill.

Tortora, Gerard J., and Anagnostakos, Nicholas P., 1986. *Principles of anatomy and physiology.* 5th ed. New York: Harper-Row.

Anatomy

Clemente, Carmine D., ed., 1984. *Gray's anatomy of the human body.* 30th ed. Philadelphia: Lea and Febiger.

Crouch, James, 1985. *Functional human anatomy.* 4th ed. Philadelphia: Lea and Febiger.

Greenblatt, Gordon M., 1981. *Cat musculature: a photographic atlas.* 2d ed. Chicago: Univ. of Chicago Press.

House, E. L., and Pansky, B., 1979. *A functional approach to neuroanatomy.* 3d ed. New York: McGraw-Hill.

Netter, Frank H., 1957. *The CIBA collection of medical illustrations.* 7 vols. Summit, New Jersey: CIBA Pharmaceutical Products, Inc.

Schlossberg, Leon, and Zuidema, George D., 1986. *The Johns Hopkins atlas of human functional anatomy.* 3d ed. Baltimore: Johns Hopkins.

Physiology

Ganong, Arthur C., 1985. *Review of medical physiology.* 12th ed. Los Altos, CA: Appleton and Lange Medical Publishers.

Guyton, Arthur C., 1981. *Textbook of medical physiology.* 6th ed. Philadelphia: W. B. Saunders Co.

Guyton, Arthur C., 1984. *Physiology of the human body.* 6th ed. Philadelphia: W. B. Saunders Co.

Malasanos, Lois, et al., 1985. *Health assessment.* 3d ed. St. Louis: C. V. Mosby Co.

Ruch, Theodore C., and Patton, Harry D., eds., 1982. *Physiology and biophysics.* 4 vols. Philadelphia: W. B. Saunders Co.

Histology

Arey, Leslie B., 1974. *Human histology, a textbook in outline form.* 4th ed. Philadelphia: W. B. Saunders Co.

Bloom, William, and Fawcett, Don W., 1975. *A textbook of histology.* 10th ed. Philadelphia: W. B. Saunders Co.

Cormack, David H., 1984. *Introduction to histology.* Philadelphia: Lippincott Co.

DiFiore, Mariano S., 1981. *Atlas of human histology.* 5th ed. Philadelphia: Lea and Febiger.

Ham, Arthur W., and Cormack, David H., 1979. *Histology.* 8th ed. Philadelphia: Lippincott Co.

Rhodin, Johannes G., 1975. *An atlas of histology.* New York: Oxford University Press.

Chemistry

Campbell, P. N., and Kilby, B. A., eds., 1975. *Basic biochemistry for medical students.* Orlando, FL.: Academic Press.

Gilman, Alfred G., et al., eds., 1980. *The pharmacological basis of therapeutics.* 6th ed. New York: Macmillan Pubs.

Richterich, R., 1981. *Clinical chemistry: theory, practice, and interpretation.* New York: Wiley Interscience.

Tietz, Norbert W., ed., 1975. *Fundamentals of clinical chemistry.* 2d ed. Philadelphia: W. B. Saunders Co.

Varley, H., 1967. *Practical clinical biochemistry.* 4th ed. London: Wm. Heineman Medical Books, Ltd.

Wynter, C. I., 1975. *Chemical analyses for medical technologists.* Springfield, IL: C. C. Thomas Publishers.

Instrumentation

Cobbold, Richard S., 1974. *Transducers for biochemical measurements: principles and applications.* New York: Wiley Interscience.

Dewhurst, D. J., 1975. *An introduction to biomedical instrumentation.* 2d ed. Elmsford, NY: Pergamon Press.

DuBorg, Joseph L., 1978. *Introduction to biomedical electronics.* New York: McGraw-Hill.

Geddes, L. A., 1972. *Electrodes and the measurement of bioelectric events.* New York: Wiley and Sons.

Geddes, L. A., and Baker, L. E., 1975. *Principles of applied medical instrumentation.* New York: Wiley Interscience.

Miller, H., and Harrison, D. C., 1974. *Biomedical electrode technology.* Orlando, FL: Academic Press.

Strong, Peter, 1971. *Biophysical measurements.* Beaverton, OR: Tektronix.

Tischler, Morris, 1980. *Experiments in general and biomedical instrumentation.* New York: McGraw-Hill.

Muscle Physiology

Anderson, K. Lange, et al., 1970. *Fundamentals of exercise testing.* Albany, NY: World Health Organization.

Basmajian, John V., and DeLuca, Carlo J., 1985. *Muscles alive.* 5th ed. Baltimore: Williams and Wilkins.

Cohen, C., 1975. *The protein switch of muscle contraction.* Scientific American Offprint #1329. San Francisco: W. H. Freeman & Co.

Gross, John A., and Flicek, Barbara D., 1987. *Electromyographic technologists handbook.* St. Louis: Green.

Hoyle, G., 1970. *How is muscle turned on and off?* Scientific American Offprint #1175. San Francisco: W. H. Freeman & Co.

Jabre, Joe F., and Hackett, Earl R., 1983. *EMG manual.* Springfield, IL: C. C. Thomas.

Murray, J. M., and Weber, A., 1974. *The cooperative action of muscle proteins.* Scientific Offprint #1290. San Francisco: W. H. Freeman & Co.

Murray, Jim, and Karpovich, Peter, 1983. *Weight training in athletics.* Englewood Cliffs, NJ: Prentice-Hall.

Shepard, Roy J., 1984. *Biochemistry of physical activity.* Springfield, IL: C. C. Thomas.

Shepard, Roy J., 1985. *Physiology and biochemistry of exercise.* New York: Praeger.

Eye and Ear

Bellows, John G., 1972. *Contemporary ophthalmology.* Melbourne, FL: R. E. Krieger Co.

Bellows, John G., ed., 1975. *Cataract and abnormalities of the lens.* Orlando, FL: Grune.

Browning, G. G., 1986. *Clinical otology and audiology.* Stoneham, MA: Butterworth.

Christman, Ernst H., 1972. *A primer on refraction.* Springfield, IL: C. C. Thomas.

Corboy, J. M., 1979. *The retinoscopy book: a manual for beginners.* Thorofare, NJ: Slack, Inc.

Glorig, Aram, ed., 1977. *Audiometry: principles and practices.* Melbourne, FL: R. E. Kreiger Co.

Keeney, A. H., 1976. *Ocular examination: basics and technique.* St. Louis: C. V. Mosby Co.

Lloyd, L. L., and Kaplan, Harriet, 1978. *Audiometric interpretation: a manual of basic audiometry.* Baltimore: University Park Press.

Masland, Richard H., December, 1986, Scientific American. *The functional architecture of the retina.* New York: Scientific American, Inc.

Moses, Robert A., 1980. *Adler's physiology of the eye, clinical application.* 7th ed. St. Louis: C. V. Mosby Co.

Schnapf, Julie L., and Baylor, Denis A., April, 1987, Scientific American. *How photoreceptor cells respond to light.* New York: Scientific American, Inc.

Miscellaneous

Adamovich, David R., 1984. *The heart: fundamentals of electrocardiography, exercise physiology, and exercise stress testing.* Hempstead, NY: Sports Medicine Books.

Baker, Brian H., 1982. *Performing the electrocardiogram.* Springfield, IL: C. C. Thomas.

Basmajian, John V., 1983. *Biofeedback: principles and practice for clinicians.* 2d ed. Baltimore: Williams & Wilkins.

Blowers, Margaret G., and Sims, Roberta S., 1983. *How to read an ECG.* 3d ed. Oradell, NJ: Medical Economics.

Diggs, L. W., et al. *The morphology of human blood cells.* 4th ed. Chicago: Abbott Laboratories.

Hughes, John R., 1982. *EEG in clinical practice.* Stoneham, MA: Butterworth.

Kalashnikov, V., 1986. *Beat the box: the insider's guide to outwitting the lie detector.* Staten Island, NY: Gordon Press.

Keeler, Eloise, 1984. *Lie detector manual.* Marshfield, MA: TelShare Pub. Co.

Matte, James, 1980. *The art and science of the polygraph technique.* Springfield, IL: C. C. Thomas.

Moscovits, Toni, 1978. *Physiology of diving.* Lorain, OH: Dayton Labs.

Reid, J. E., and Inbau, F. E., 1977. *Truth and deception, the polygraph technique.* Baltimore: Williams & Wilkins Co.

Short, Charles, 1987. *Principles and practice of veterinary anesthesia.* Baltimore: Williams & Wilkins.

Spehlmann, R., 1981. *EEG primer.* New York: Elsevier.

Wintrobe, Maxwell M., et al., 1981. *Clinical hematology.* 8th ed. Philadelphia: Lea & Febiger.

Index